Important Concepts Guide

1.1 REAL NUMBERS AND THEIR GRAPHS

Natural numbers: $\{1, 2, 3, 4, 5, . . .\}$

Whole numbers: $\{0, 1, 2, 3, 4, 5, . . .\}$

Integers: $\{. . . , -3, -2, -1, 0, 1, 2, 3, . . .\}$

Rational numbers: $\left\{\frac{a}{b} \mid a \text{ and } b \text{ are integers and } b \neq 0\right\}$

Irrational numbers: Numbers that cannot be expressed as $\frac{a}{b}$ where a and b are integers and $b \neq 0$

Real numbers: All numbers that are either a rational number or an irrational number

Prime numbers: $\{2, 3, 5, 7, 11, 13, 17, . . .\}$

Composite numbers: $\{4, 6, 8, 9, 10, 12, 14, 15, . . .\}$

Even integers: $\{. . . , -6, -4, -2, 0, 2, 4, 6, . . .\}$

Odd integers: $\{. . . , -5, -3, -1, 1, 3, 5, . . .\}$

1.2 FRACTIONS

If there are no divisions by 0, then

$$\frac{ax}{bx} = \frac{a}{b} \qquad \frac{a}{b} \cdot \frac{c}{d} = \frac{ac}{bd}$$

$$\frac{a}{b} \div \frac{c}{d} = \frac{ad}{bc} \qquad \frac{a}{d} + \frac{b}{d} = \frac{a+b}{d}$$

$$\frac{a}{d} - \frac{b}{d} = \frac{a-b}{d} \qquad -\frac{a}{b} = \frac{-a}{b} = \frac{a}{-b}$$

1.3 EXPONENTS AND ORDER OF OPERATIONS

If n is a natural number, then

$$x^n = \overbrace{x \cdot x \cdot x \cdot \cdots \cdot x}^{n \text{ factors of } x}$$

To simplify expressions, do all calculations within each pair of grouping symbols, working from the innermost pair to the outermost pair.

1. Find the values of any exponential expressions.
2. Do all multiplications and divisions from left to right.
3. Do all additions and subtractions from left to right.

In a fraction, simplify the numerator and denominator separately and then simplify the fraction, if possible.

Figure	Perimeter	Area
Square	$P = 4s$	$A = s^2$
Rectangle	$P = 2l + 2w$	$A = lw$
Triangle	$P = a + b + c$	$A = \frac{1}{2}bh$
Trapezoid	$P = a + b + c + d$	$A = \frac{1}{2}h(b + d)$
Circle	$C = \pi D = 2\pi r$	$A = \pi r^2$

Figure	Volume
Rectangular solid	$V = lwh$
Cylinder	$V = Bh^*$
Pyramid	$V = \frac{1}{3}Bh^*$
Cone	$V = \frac{1}{3}Bh^*$
Sphere	$V = \frac{4}{3}\pi r^3$

*B is the area of the base.

1.4 ADDING AND SUBTRACTING REAL NUMBERS

1. *To add two positive numbers*, add their absolute values and the answer is positive.
2. *To add two negative numbers*, add their absolute values and the answer is negative.

To add a positive and a negative number, subtract the smaller absolute value from the larger.

1. If the positive number has the larger absolute value, the answer is positive.
2. If the negative number has the larger absolute value, the answer is negative.

$$a - b = a + (-b)$$

1.5 MULTIPLYING AND DIVIDING REAL NUMBERS

To multiply two real numbers, multiply their absolute values.

1. *If the numbers are positive*, the product is positive.
2. *If the numbers are negative*, the product is positive.
3. *If one number is positive and the other is negative*, the product is negative.
4. Any number multiplied by 0 is 0: $a \cdot 0 = 0 \cdot a = 0$
5. Any number multiplied by 1 is the number itself: $a \cdot 1 = 1 \cdot a = a$

To divide two real numbers, find the quotient of their absolute values.

1. If the numbers are positive, the quotient is positive.
2. If the numbers are negative, the quotient is positive.
3. If one number is positive and the other is negative, the quotient is negative.
4. $\frac{a}{0}$ is undefined; $\frac{0}{0}$ is indeterminate.
5. $\frac{a}{1} = a$, and if $a \neq 0$, $\frac{0}{a} = 0$ and $\frac{a}{a} = 1$.

(continues on back endsheet)

Beginning
Algebra

10th Edition

Beginning Algebra

A Guided Approach

Rosemary M. Karr
Collin College

Marilyn B. Massey
Collin College

R. David Gustafson
Rock Valley College

CENGAGE
Learning®

Australia • Brazil • Japan • Korea • Mexico • Singapore • Spain • United Kingdom • United States

Beginning Algebra: A Guided Approach,
Tenth Edition
**Rosemary M. Karr, Marilyn B. Massey,
R. David Gustafson**

Product Manager: Marc Bove

Content Developer: Stefanie Beeck

Content Coordinator: Lauren Crosby

Product Assistant: Jennifer Cordoba

Media Developer: Bryon Spencer

Brand Manager: Gordon Lee

Market Development Manager: Danae April

Content Project Manager: Jennifer Risden

Art Director: Vernon Boes

Manufacturing Planner: Becky Cross

Rights Acquisitions Specialist: Tom McDonough

Production Service: Lachina Publishing Services

Photo Researcher: Q2A/Bill Smith Group

Text Researcher: Pablo D'Stair

Copy Editor: Lachina Publishing Services

Illustrator: Lori Heckelman; Lachina Publishing
    Services

Text Designer: Terri Wright

Cover Designer: Terri Wright

Cover Image: Kevin Twomey

Compositor: Lachina Publishing Services

Design images: Fig leaf: © Michael Breuer/
    Masterfile; Green background: © John Fox/
    Getty Images

For product information and technology assistance, contact us at
Cengage Learning Customer & Sales Support, 1-800-354-9706.
For permission to use material from this text or product,
submit all requests online at **www.cengage.com/permissions.**
Further permissions questions can be e-mailed to
permissionrequest@cengage.com.

Library of Congress Control Number: 2013932540

ISBN-13: 978-1-4354-6247-2

ISBN-10: 1-4354-6247-5

Cengage Learning
200 First Stamford Place, 4th Floor
Stamford, CT 06902
USA

Cengage Learning is a leading provider of customized learning solutions with office locations around the globe, including Singapore, the United Kingdom, Australia, Mexico, Brazil, and Japan. Locate your local office at **www.cengage.com/global.**

Cengage Learning products are represented in Canada by Nelson Education, Ltd.

To learn more about Cengage Learning Solutions, visit **www.cengage.com.**

Purchase any of our products at your local college store or at our preferred online store **www.cengagebrain.com.**

Printed in the United States of America
1 2 3 4 5 6 7 17 16 15 14 13

About the Authors

ROSEMARY M. KARR graduated from Eastern Kentucky University (EKU) with a Bachelor's degree in Mathematics, attained her Master of Arts degree at EKU in Mathematics Education, and earned her Ph.D. from the University of North Texas. After two years of teaching high school mathematics, she joined the faculty at Eastern Kentucky University, where she earned tenure as Assistant Professor of Mathematics. A professor at Collin College in Plano, Texas, since 1990, Professor Karr has written more than 10 solutions manuals, presented numerous papers, and been an active member in several educational associations including President of the National Association for Developmental Education. She has been honored several times by Collin College, and has received such national recognitions as U.S. Professor of the Year (2007) and induction as Fellow by the Council of Learning Assistance and Developmental Education Associations (2012).

MARILYN B. MASSEY teaches mathematics at Collin College in McKinney, Texas, where she joined the faculty in 1991. She has been President of the Texas Association for Developmental Education, served as academic chair of the Developmental Mathematics Department, and received an Excellence in Teaching Award from the National Conference for College Teaching and Learning. Professor Massey has presented at numerous state and national conferences. She earned her Bachelor's degree from the University of North Texas and Master of Arts degree in Mathematics Education from the University of Texas at Dallas.

R. DAVID GUSTAFSON is Professor Emeritus of Mathematics at Rock Valley College in Illinois and also has taught extensively at Rockford College and Beloit College. He is coauthor of several best-selling mathematics textbooks, including Gustafson/Frisk/Hughes, *College Algebra*; Gustafson/Karr/Massey, *Beginning Algebra, Intermediate Algebra, Beginning and Intermediate Algebra: A Combined Approach*; and the Tussy/Gustafson and Tussy/Gustafson/Koenig developmental mathematics series. His numerous professional honors include Rock Valley Teacher of the Year and Rockford's Outstanding Educator of the Year. He has been very active in AMATYC as a Midwest Vice-president and has been President of IMACC, AMATYC's Illinois affiliate. He earned a Master of Arts degree in Mathematics from Rockford College in Illinois, as well as a Master of Science degree from Northern Illinois University.

To my husband and family, for their
unwavering support of my work
—R.M.K.

To my life partner, Ron; son, Christopher;
and parents, Dale and Martha
—M.B.M.

To Craig, Jeremy, Paula, Gary, Bob, Jennifer,
John-Paul, Gary, and Charlie
—R.D.G.

Contents

Preface

This tenth edition of *Beginning Algebra* is an exciting and innovative revision. The new edition reflects a thorough update, has new pedagogical features that make the text easier to read, and has an entirely new and fresh interior design. This series is known for its integrated approach to student applications and the development of student study skills, for the clarity of its writing, and for making algebra relevant and engaging. The revisions to this already successful text will further promote student achievement. Coauthors Rosemary Karr and Marilyn Massey who joined the team with Dave Gustafson in the last edition have now assumed primary responsibility, bringing their extensive experience in developmental education to bear on this revision.

This new edition has expanded on the learning plan that helps students transition to the next level, teaching them the problem-solving strategies that will serve them well in their everyday lives. Most textbooks share the goals of clear writing, well-developed examples, and ample exercises, whereas the Karr/Massey/Gustafson series develops student success beyond the demands of traditional required coursework. The tenth edition's learning tools have been developed with your students in mind.

Through their collective teaching experience, the authors have developed an acute awareness of students' approaches to homework and have determined that exercise sets should serve as *more* than just a group of problems to work. The authors' philosophy is to guide the student through new material in a gentle progression of thought development that slowly reduces the student's dependence on external factors and relates new concepts to previously learned material. They have written the textbook to guide students through the material while providing a decreasing level of support throughout each section. Initially, the authors provide a map to the content through learning objectives that serve as advanced organizers for what students can expect to learn in that section. The vocabulary encourages students to speak the language of mathematics. *Getting Ready* exercises at the beginning of each section prepare students for the upcoming concepts by reviewing relevant previous skills. The instructor may guide the students through the examples, but the students will independently attempt the self checks. Students will begin to use their "mathematical voice" to explain problems to one another or work collectively to find a solution to a problem not previously encountered, a primary goal of the *Now Try This* feature. The guidance shifts from instructor to fellow students to individual through carefully designed exercise sets.

■ Changes for the Tenth Edition

The major changes to this edition are the inclusion of *Self Checks* for *all* examples, study skills strategies for each chapter, modification of the *Warm-Up* exercises, revision of many exercises, and the addition of two new appendices. These appendices—"Permutations, Combinations, and Probability" and "Measurement Conversions"—reflect changes in the curriculum for some beginning algebra courses as a result of a national call for new "transition" courses.

- The *Warm-Ups* have been significantly revised to reflect their original purpose, now including skills needed for the development of section concepts.

- Interactive study skills strategies called *Reach for Success* have been incorporated throughout the textbook for a more holistic learning experience. The authors feel that most students already know effective study strategies and could list many of the successful techniques. It is the authors' intention to encourage the students' thoughtful consideration and implementation of these skills. The interactive approach to these skills engages the student, encouraging an active participation to develop and reinforce them. The incorporation of student advising and the development of study skills are critical for academic success. That is why this feature is titled *Reach for Success*. The authors hope to help the students do just that. These worksheets (and others) are available online on the text-specific website for flexibility in the order of assignment.

- *Everyday Connections* have been revised to more accurately reflect a topic of the selected chapter.

- *Self Checks* have been included for *every* example in the textbook. *Self Checks* are especially needed for applications where students need to develop problem-solving strategies.

- Additional *Teaching Tips* have been included at the request of reviewers.

Calculators

The use of calculators is assumed throughout the text, although the calculator icon makes the exercises easily identified for possible omission if the instructor chooses to do so. The authors believe that students should learn calculator skills in the mathematics classroom. They will then be prepared to use calculators in science and business classes and for nonacademic purposes. The directions within each exercise set indicate which exercises require the use of a calculator.

Since most beginning algebra students now have graphing calculators, keystrokes are given for both scientific and graphing calculators. Removable cards for the *Basic Calculator Keystroke Guide for the TI-83/84 Family of Calculators* and the *Casio FX-9750G2* are bound into the book as a resource for those students learning how to use a graphing calculator.

A Guided Tour

Chapter Openers showcase the variety of career paths available in the world of mathematics. They include brief overviews of each career, job outlook statistics from the U.S. Department of Labor, and potential job growth and annual earnings.

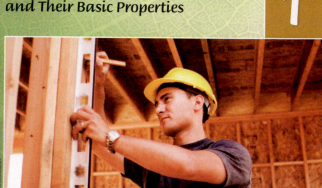

Real Numbers and Their Basic Properties

1

Careers and Mathematics

CARPENTERS

Carpenters are involved in many kinds of construction activities. They cut, fit, and assemble wood and other materials for the construction of buildings, highways, bridges, docks, industrial plants, boats, and many other structures. About 32% of all carpenters—the largest construction trade—are self-employed. Future carpenters should take classes in English, algebra, geometry, physics, mechanical drawing, and blueprint reading.

Job Outlook:
Job opportunities will be the best for those with the most training and skills. Between 3 and 4 years of both on-the-job training and classroom instruction usually are needed to become a skilled carpenter. Overall, the employment of carpenters is expected to increase by 10% through 2016, about as fast as the average for all occupations.

Hourly Earnings:
$17.57–$30.45

For More Information:
http://www.bls.gov/oco/ocos202.htm

For a Sample Application:
See Problem 154 in Section 1.2.

REACH FOR SUCCESS
1.1 Real Numbers and Their Graphs
1.2 Fractions
1.3 Exponents and Order of Operations
1.4 Adding and Subtracting Real Numbers
1.5 Multiplying and Dividing Real Numbers
1.6 Algebraic Expressions
1.7 Properties of Real Numbers
■   *Projects*
     REACH FOR SUCCESS EXTENSION
     CHAPTER REVIEW
     CHAPTER TEST

In this chapter
In Chapter 1, we will discuss the various types of numbers that we will use throughout this course. Then we will review the basic arithmetic of fractions, explain how to add, subtract, multiply, and divide real numbers, introduce algebraic expressions, and summarize the properties of real numbers.

1

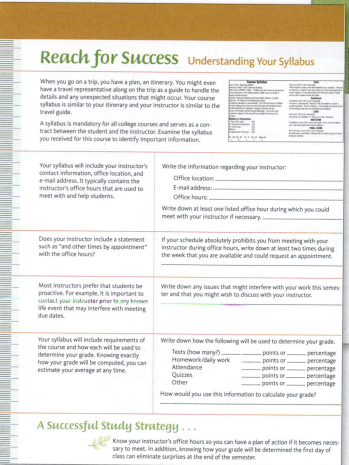

Reach for Success   Understanding Your Syllabus

When you go on a trip, you have a plan, an itinerary. You might even have a travel representative along on the trip as a guide to handle the details and any unexpected situations that might occur. Your course syllabus is similar to your itinerary and your instructor is similar to your travel guide.

A syllabus is mandatory for all college courses and serves as a contract between the student and the instructor. Examine the syllabus you received for this course to identify important information.

Your syllabus will include your instructor's contact information, office location, and e-mail address. It typically contains the instructor's office hours that are used to meet with and help students.

Write the information regarding your instructor:
Office location: _____
E-mail address: _____
Office hours: _____

Write down at least one listed office hour during which you could meet with your instructor if necessary. _____

Does your instructor include a statement such as "and other times by appointment" with the office hours?

If your schedule absolutely prohibits you from meeting with your instructor during office hours, write down at least two times during the week that you are available and could request an appointment.

Most instructors prefer that students be proactive. For example, it is important to contact your instructor prior to any known life event that may interfere with meeting due dates.

Write down any issues that might interfere with your work this semester and that you might wish to discuss with your instructor.

Your syllabus will include requirements of the course and how each will be used to determine your grade. Knowing exactly how your grade will be computed, you can estimate your average at any time.

Write down how the following will be used to determine your grade.
Tests (how many?) _____ points or _____ percentage
Homework/daily work _____ points or _____ percentage
Attendance _____ points or _____ percentage
Quizzes _____ points or _____ percentage
Other _____ points or _____ percentage

How would you use this information to calculate your grade?

A Successful Study Strategy . . .
Know your instructor's office hours so you can have a plan of action if it becomes necessary to meet. In addition, knowing how your grade will be determined the first day of class can eliminate surprises at the end of the semester.

Reach for Success is a new feature where study skills have been written for inclusion in every chapter. This feature is designed as an opening activity to help the student prepare for a successful semester.

Appearing at the beginning of each section, **Learning Objectives** are mapped to the appropriate content, as well as to relevant exercises in the **Guided Practice** section. These allow students to identify specific mathematical processes that may need additional reinforcement, while the instructor can use these to identify appropriate exercises to use as homework.

In order to work mathematics, one must be able to speak the language. Not only are **Vocabulary** words identified at the beginning of each section, these words are also boldfaced within the section. Exercises include vocabulary questions, and a glossary has been included to facilitate the students' reference to these words. An optional Spanish glossary is available upon request.

Getting Ready questions appear at the beginning of each section, linking past concepts to the upcoming material.

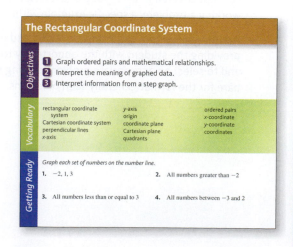

The Rectangular Coordinate System

Objectives
1. Graph ordered pairs and mathematical relationships.
2. Interpret the meaning of graphed data.
3. Interpret information from a step graph.

Vocabulary
rectangular coordinate system
Cartesian coordinate system
perpendicular lines
x-axis
y-axis
origin
coordinate plane
Cartesian plane
quadrants
ordered pairs
x-coordinate
y-coordinate
coordinates

Getting Ready
Graph each set of numbers on the number line.
1. −2, 1, 3
2. All numbers greater than −2
3. All numbers less than or equal to 3
4. All numbers between −3 and 2

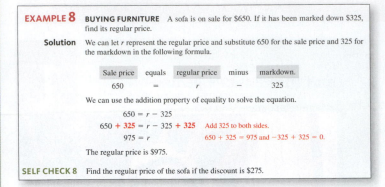

EXAMPLE 8   **BUYING FURNITURE**   A sofa is on sale for $650. If it has been marked down $325, find its regular price.

Solution   We can let r represent the regular price and substitute 650 for the sale price and 325 for the markdown in the following formula.

Sale price	equals	regular price	minus	markdown.
650	=	r	−	325

We can use the addition property of equality to solve the equation.

$$650 = r - 325$$
$$650 + 325 = r - 325 + 325 \quad \text{Add 325 to both sides.}$$
$$975 = r \qquad \qquad 650 + 325 = 975 \text{ and } -325 + 325 = 0.$$

The regular price is $975.

SELF CHECK 8   Find the regular price of the sofa if the discount is $275.

Examples are worked out in each chapter, highlighting the concept being discussed. Author notes are included in many of the examples, prompting students to consider the thought process used when approaching a problem and working toward a solution.

All examples end with a *Self Check* problem, so that students can check reading comprehension. Answers are found at the end of the section.

Each section ends with *Now Try This,* which is intended to increase conceptual understanding through active classroom participation and involvement. These problems can be worked independently or in small groups and transition to the Exercise Sets, as well as to material in future sections. They reinforce topics, digging a little deeper than the examples. To encourage students to think more critically, the answers will *only* be included in the *Annotated Instructor's Edition* of the text.

NOW TRY THIS

Simplify each expression. Write your answer with positive exponents only. Assume no variables are 0.

1. $-2(x^2 y^5)^0$

2. $-3x^{-2}$

3. $9^2 - 9^0$

4. Explain why the instructions above include the statement "Assume no variables are 0."

- **Warm-Ups** get students into the homework mind-set, asking quick memory-testing questions.

- **Review** provides exercises to remind students of previous skills.

- **Vocabulary and Concepts** exercises emphasize the main concepts taught in this section.

- **Guided Practice** exercises are keyed to the objectives to increase student success by directing students to the concept covered in that group of exercises. Should a student encounter difficulties working a problem, a specific example within the objective is also cross-referenced.

- **Additional Practice** problems are mixed and not linked to objectives or examples, providing the student with the opportunity to distinguish among problem types and to select an appropriate problem-solving strategy. This will help students prepare for the format generally seen on exams.

4.2   Exercises

WARM-UPS   *Identify the base in each expression.*

1. $3x^{-2}$

2. $(3x)^{-2}$

3. $(5x)^0$

4. $5x^0$

Simplify by dividing out common factors. Assume no variable is 0.

5. $\dfrac{3 \cdot a \cdot a}{3 \cdot a \cdot a \cdot a}$

6. $\dfrac{-2 \cdot x \cdot x \cdot x}{-2 \cdot x \cdot x}$

7. $\dfrac{a \cdot a \cdot b}{a \cdot b \cdot b}$

8. $\dfrac{x \cdot x \cdot y \cdot y}{x \cdot x \cdot x \cdot y \cdot y}$

REVIEW

9. If $a = -2$ and $b = 3$, evaluate $\dfrac{3a^2 + 4b + 8}{a + 2b^2}$.

10. Evaluate: $|-3 + 5 \cdot 2|$

Solve each equation.

11. $5\left(x - \dfrac{1}{2}\right) = \dfrac{7}{2}$

12. $\dfrac{5(2 - x)}{6} = \dfrac{x + 6}{2}$

13. Solve $P = L + \dfrac{s}{f}i$ for s.

14. Solve $P = L + \dfrac{s}{f}i$ for i.

VOCABULARY AND CONCEPTS   *Fill in the blanks.*

15. If x is any nonzero real number, then $x^0 = $ __. If x is any nonzero real number, then $x^{-n} = $ __.

16. Since $\dfrac{6^4}{6^4} = 6^{4-4} = 6^0$ and $\dfrac{6^4}{6^4} = 1$, we define 6^0 to be __.

- *Applications* ask students to apply their new skills in real-life situations, ***Writing About Math*** builds students' mathematical communication skills, and ***Something to Think About*** encourages students to take what they have learned in a section and use those concepts to work through a situation in a new way.

- Many exercises are available online through Enhanced WebAssign®. These homework problems are algorithmic, ensuring that your students will learn mathematical processes, not just how to work with specific numbers.

- Chapter-ending ***Projects*** encourage in-depth exploration of key concepts.

- A ***Chapter Review*** grid presents the material cleanly and simply, giving students an efficient means of reviewing material.

- ***Chapter Tests*** allow students to pinpoint their strengths and challenges with the material. Answers are included at the back of the student edition.

- ***Cumulative Review Exercises*** follow the end of chapter material for every even-numbered chapter, and keep students' skills current before moving on to the next topic.

■ **Additional Features**

Everyday connections
Paralympic Medals

The International Olympic Committee sets the specifications for all medals awarded in the Olympics and Paralympics. They must be at least 60 millimeters in diameter and at least 3 millimeters thick. Gold medals must be 92.5% pure silver and plated with at least 6 grams of gold. The gold medal for the 2010 Vancouver Winter Paralympics was 100 millimeters in diameter, 6 millimeters thick, and because each medal was unique, each weighed between 500 and 576 grams.

Photo by Adrian Pang, available under a Creative Commons Attribution license.

When the medal was awarded in early 2010, 6 grams of gold and 481 grams of silver were worth $478.75. In late 2011, when the price of silver had doubled and gold increased by 150%, the same medal was worth $850.40.

1. What was the price of one gram of gold and one gram of silver in 2010?

2. What was the price of one gram of gold and one gram of silver in 2011?

Source: http://www.olympic.org/Documents/
Reference_documents_Factsheets/
Winter_Games_Medals_FACTSHEET_EN.pdf

Everyday Connections boxes reveal the real-world power of mathematics. Each Everyday Connections box invites students to see how the material covered in the chapter is relevant to their lives.

Perspective boxes highlight interesting facts from mathematics history or important mathematicians, past and present. These brief but interesting biographies connect students to discoveries of the past and their importance to the present.

Perspective

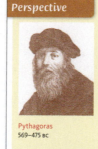

Pythagoras
569–475 BC

THE FIRST IRRATIONAL NUMBER

The Greek mathematician and philosopher Pythagoras believed that every aspect of the natural world could be represented by ratios of whole numbers (i.e., rational numbers). However, one of his students accidentally disproved this claim by examining a surprisingly simple example. The student examined a right triangle whose legs were each 1 unit long and posed the following question. "How long is the third side of the triangle?"

Using the well-known theorem of Pythagoras, the length of the third side can be determined by using the formula $c^2 = 1^2 + 1^2$.

In other words, $c^2 = 2$. Using basic properties of arithmetic, it turns out that the numerical value of c cannot be expressed as a rational number. So we have an example of an aspect of the natural world that corresponds to an irrational number, namely $c = \sqrt{2}$.

Accent on technology
▸ Finding Present Value

To find out how much money P (called the *present value*) must be invested at an annual rate i (expressed as a decimal) to have $\$A$ in n years, we use the formula $P = A(1 + i)^{-n}$. To find out how much we must invest at 6% to have $\$50,000$ in 10 years, we substitute 50,000 for A, 0.06 (6%) for i, and 10 for n to get

$$P = A(1 + i)^{-n}$$
$$P = 50{,}000(1 + 0.06)^{-10}$$

To evaluate P with a calculator, we enter these numbers and press these keys:

(1 + .06) y^x 10 +/− × 50000   Using a calculator with a y^x and a +/− key.

50000 (1 + .06) ∧ (−) 10 **ENTER**   Using a TI84 graphing calculator.

Either way, we see that we must invest $27,919.74 to have $50,000 in 10 years.

For instructions regarding the use of a Casio graphing calculator, please refer to the Casio Keystroke Guide in the back of the book.

Accent on Technology boxes teach students the calculator skills to prepare them for using these tools in science and business classes, as well as for nonacademic purposes. Calculator examples are given in these boxes, and keystrokes are given for both scientific and graphing calculators. For instructors who do not use calculators in the classroom, the material on calculators is easily omitted without interrupting the flow of ideas.

Comment notations alert students to common errors as well as provide helpful and pertinent information about the concepts they are learning.

For the instructor, **Teaching Tips** are provided in the margins of the *Annotated Instructor's Edition* as interesting historical information, alternative approaches for teaching the material, and class activities.

> **COMMENT** The product rule for exponents applies only to exponential expressions with the same base. An expression such as $x^2 y^3$ cannot be simplified, because x^2 and y^3 have different bases.

> **Teaching Tip**
>
> If your class is full and the chairs are arranged in rows and columns, introduce graphing this way:
> 1. Number the rows and columns.
> 2. Ask row 3, column 1 to stand. This is the point (3, 1).
> 3. Ask (1, 3) to stand. That this is a different point emphasizes the ordered pair idea.
> 4. Repeat with more examples.

∽ Supplements

FOR THE STUDENT	FOR THE INSTRUCTOR
	Annotated Instructor Edition (ISBN: 978-1-4354-6248-9) The **Annotated Instructor Edition** features answers to all problems in the book.
Student Solutions Manual (ISBN: 978-1-285-18366-4) *Author: Michael Welden, Mt. San Jacinto College* The **Student Solutions Manual** provides worked-out solutions to the odd-numbered problems in the book.	***Complete Solutions Manual*** (ISBN: 978-1-285-18376-3) *Author: Michael Welden, Mt. San Jacinto College* The **Complete Solutions Manual** provides worked-out solutions to all of the problems in the text.
Student Workbook (ISBN: 978-1-285-18377-0) *Author: Maria H. Andersen, former math faculty at Muskegon Community College and now working in the learning software industry* The **Student Workbook** contains all of the assessments, activities, and worksheets from the **Instructor's Resource Binder** for classroom discussions, in-class activities, and group work.	***Instructor's Resource Binder*** (ISBN: 978-0-538-73675-6) *Author: Maria H. Andersen, former math faculty at Muskegon Community College and now working in the learning software industry* The **Instructor's Resource Binder** contains uniquely designed Teaching Guides, which include instruction tips, examples, activities worksheets, overheads, and assessments, with answers provided.
	Instructor Companion Website Everything you need for your course in one place! This collection of book-specific lecture and class tools is available online via **www.cengage.com/login**. Formerly found on the PowerLecture, access and download PowerPoint® presentations, images, and more.
	Solution Builder This online instructor database offers complete worked solutions to all exercises in the text, allowing you to create customized, secure solutions printouts (in PDF format) matched exactly to the problems you assign in class. For more information, visit **www.cengage.com/solutionbuilder**.
ENHANCED Web**Assign** ***Enhanced WebAssign***® (Printed Access Card ISBN: 978-1-285-85770-1 Online Access Code ISBN: 978-1-285-85773-2) **Enhanced WebAssign** (assigned by the instructor) provides instant feedback on homework assignments to students. This online homework system is easy to use and includes a multimedia eBook, video examples, and problem-specific tutorials.	**ENHANCED** Web**Assign** ***Enhanced WebAssign***® (Printed Access Card ISBN: 978-1-285-85770-1 Online Access Code ISBN: 978-1-285-85773-2) Instant feedback and ease of use are just two reasons why **WebAssign** is the most widely used homework system in higher education. **WebAssign's** homework delivery system allows you to assign, collect, grade, and record homework assignments via the web. And now this proven system has been enhanced to include a multimedia eBook, video examples, and problem-specific tutorials. **Enhanced WebAssign** is more than a homework system—it is a complete learning system for math students.
	Text-Specific Videos *Author: Rena Petrello* These videos are available at no charge to qualified adopters of the text and feature 10–20 minute problem-solving lessons that cover each section of every chapter.

To the Student

Congratulations! You now own a state-of-the-art textbook that has been written especially for you. We have tried to write a book that you can read and understand. The text includes carefully written narrative and an extensive number of worked examples with **Self Checks**. **Now Try This** exercises can be worked with your classmates, and **Guided Practice** exercises tell you exactly which example to use as a resource for each question. These are just a few of the many features included in this text with your success in mind. **Study Skill strategies** have been added at the start and end of each chapter. We urge you to take the time to complete these activities to increase your understanding of what it takes to be a successful student.

To get the most out of this course, you must read and study the textbook properly. We recommend that you work the examples on paper first, and then work the Self Checks. Only after you thoroughly understand the concepts taught in the examples should you attempt to work the exercises. A **Student Solutions Manual** is available, which contains the worked-out solutions to the odd-numbered exercises.

Since the material presented in **Beginning Algebra**, **Tenth Edition**, will be of value to you in later years, we suggest that you keep this text. It will be a good reference in the future and will keep at your fingertips the material that you have learned here.

■ Hints on Studying Algebra

The phrase "practice makes perfect" is not quite true. It is "*perfect* practice that makes perfect." For this reason, it is important that you learn how to study algebra to get the most out of this course.

Although we all learn differently, here are some hints on studying algebra that most students find useful.

Planning a Strategy for Success To get where you want to be, you need a goal and a plan. Your goal should be to pass this course with a grade of A or B. To earn one of these grades, you must have a plan to achieve it. A good plan involves several points:

- Getting ready for class,
- Attending class,
- Doing homework,
- Making use of the extensive extra help available, if your instructor has set up a course in WebAssign, and
- Having a strategy for taking tests.

Getting Ready for Class To get the most out of every class session, you will need to prepare for class. One of the best things you can do is to preview the material in the text that your instructor will be discussing in class. Perhaps you will not understand all of what you read, but you will be better able to understand your instructor when he or she discusses the material in class.

Do your work every day. If you fall behind, you will become frustrated and discouraged. Make a commitment to always prepare for class, and then do it.

Attending Class The classroom experience is your opportunity to learn from your instructor and interact with your classmates. Make the most of it by attending every class. Sit near the front of the room where you can see and hear easily. Remember that it is your responsibility to follow the discussion, even though it takes concentration and hard work.

Pay attention to your instructor, and jot down the important things that he or she says. However, do not spend so much time taking notes that you fail to concentrate on what your instructor is explaining. Listening and understanding the big picture is much more important than just copying solutions to problems.

Do not be afraid to ask questions when your instructor asks for them. Asking questions will make you an active participant in the class. This will help you pay attention and keep you alert and involved.

Doing Homework It requires practice to excel at tennis, master a musical instrument, or learn a foreign language. In the same way, it requires practice to learn mathematics. Since practice in mathematics is homework, homework is your opportunity to practice your skills and experiment with ideas. Consider creating note cards for important concepts.

It is important for you to pick a definite time to study and do homework. Set a formal schedule and stick to it. Try to study in a place that is comfortable and quiet. If you can, do some homework shortly after class, or at least before you forget what was discussed in class. This quick follow-up will help you remember the skills and concepts your instructor taught that day.

Each formal study session should include three parts:

1. Begin every study session with a review period. Look over previous chapters and see if you can do a few problems from previous sections. Keeping skills current will greatly reduce the amount of time you will need to prepare for tests.
2. After reviewing, read the assigned material. Resist the temptation to dive into the exercises without reading and understanding the examples. Instead, work the examples and Self Checks with pencil and paper. Only after you completely understand the underlying principles behind them should you try to work the exercises.

   Once you begin to work the exercises, check your answers with the printed answers in the back of your text. If one of your answers differs from the printed answer, see if the two can be reconciled. Sometimes answers have more than one form. If you decide that your answer is incorrect, compare your work to the example in the text that most closely resembles the exercise, and try to find your mistake. If you cannot find an error, consult the *Student Solutions Manual.* If nothing works, mark the problem and ask about it in your next class meeting.
3. After completing the written assignment, preview the next section. This preview will be helpful when you hear the material discussed during the next class period.

You probably already know the general rule of thumb for college homework: two to three hours of practice for every hour you spend in class. If mathematics is difficult for you, plan on spending even more time on homework.

To make doing homework more enjoyable, study with one or more friends. The interaction will clarify ideas and help you remember them. If you choose to study alone, a good study technique is to explain the material to yourself out loud or use a white board where you can stand back and look at the big picture.

Accessing Additional Help Access any help that is available from your instructor. Often, your instructor can clear up difficulties in a short period of time. Find out whether your college has a free tutoring program. Peer tutors often can be a great help or consider setting up your own study group.

Taking Tests Students often become nervous before taking a test because they are worried that they may not do well.

To build confidence in your ability to take tests, rework many of the problems in the exercise sets, work the exercises in the Chapter Reviews, and take the Chapter Tests. Check all answers with the answers printed at the back of your text.

Guess what the instructor will ask, build your own tests, and work them. Once you know your instructor, you will be surprised at how good you can get at selecting test questions. With this preparation, you will have some idea of what will be on the test, and you will have more confidence in your ability to do well.

When you take a test, work slowly and deliberately. Write down any formulas at the top of the page. Scan the test and work the problems you find easiest first. Tackle the hardest problems last.

We wish you well.

Acknowledgements

We are grateful to the following people who reviewed this and previous editions. They all had valuable suggestions that have been incorporated into the texts.

Kent Aeschliman, *Oakland Community College*
Carol Anderson, *Rock Valley College*
A. Elena Bogardus, *Camden County College*
Cynthia Broughton, *Arizona Western College*
David Byrd, *Enterprise State Junior College*
Pablo Chalmeta, *New River Community College*
Jerry Chen, *Suffolk County Community College*
Michael F. Cullinan, *Glendale Community College*
Amy Cuneo, *Santiago Canyon College*
Lou D'Alotto, *York College-CUNY*
Thomas DeAgostino, *Jackson Community College*
Kristin Dillard, *San Bernardino Valley College*
Kirsten Dooley, *Midlands Technical College*
Karen Driskell, *Calhoun Community College*
Joan Evans, *Texas Southern University*
Hamidullah Farhat, *Hampton University*
Harold Farmer, *Wallace Community College-Hanceville*
Mark Fitch, *University of Alaska, Anchorage*
Mark Foster, *Santa Monica College*
Tom Fox, *Cleveland State Community College*
Jeremiah Gilbert, *San Bernardino Valley College*
Joseph Guiciardi, *Community College of Allegheny County/Boyce Campus*
Harvey Hanna, *Ferris State University*
Jonathan P. Hexter, *Piedmont Virginia Community College*
Kathy Holster, *South Plains College*
Dorothy K. Holtgrefe, *Seminole Community College*
Mark Hopkins, *Oakland Community College*
Kelly Jackson, *Camden County College*
Mike Judy, *Fullerton College*
Lynette King, *Gadsden State Community College*
Chad T. Lower, *Pennsylvania College of Technology*
Janet Mazzarella, *Southwestern College*
Donald J. McCarthy, *Glendale Community College*
Robert McCoy, *University of Alaska, Anchorage*
Andrew P. McKintosh, *Glendale Community College*
Christian R. Miller, *Glendale Community College*
Feridoon Moinian, *Cameron University*
Brent Monte, *Irvine Valley College*
Daniel F. Mussa, *Southern Illinois University*
Joanne Peeples, *El Paso Community College*
Mary Ann Petruska, *Pensacola Junior College*
Linda Pulsinelli, *Western Kentucky University*
Kimberly Ricketts, *Northwest-Shoals Community College*
Janet Ritchie, *SUNY-Old Westbury*
Joanne Roth, *Oakland Community College*

Richard Rupp, *Del Mar College*
Rebecca Sellers, *Jefferson State Community College*
Kathy Spradlin, *Liberty University*
John Squires, *Cleveland State Community College*
April D. Strom, *Glendale Community College*
Robert Vilardi, *Troy University*
Victoria Wacek, *Missouri Western State College*
Judy Wells, *University of Southern Indiana*
Hattie White, *St. Phillip's College*
George J. Witt, *Glendale Community College*
Margaret Yoder, *Eastern Kentucky University*

We are grateful to the staff at Cengage Learning, especially our editor Marc Bove. We also thank Vernon Boes, Jennifer Risden, Stefanie Beeck, and Bryon Spencer.

We are indebted to Lachina Publishing Services, our production service; to Rhoda Oden, who read the entire manuscript and worked every problem; and to Michael Welden, who prepared the *Student Solutions Manual*.

—Rosemary M. Karr
  Marilyn B. Massey
  R. David Gustafson

Index of Applications

Examples that are applications are shown with boldface numbers.
Exercises that are applications are shown with regular face numbers.

Real Numbers and Their Basic Properties

© iStockphoto.com/Andrew Rich

Careers and Mathematics

CARPENTERS

Carpenters are involved in many kinds of construction activities. They cut, fit, and assemble wood and other materials for the construction of buildings, highways, bridges, docks, industrial plants, boats, and many other structures. About 32% of all carpenters—the largest construction trade—are self-employed. Future carpenters should take classes in English, algebra, geometry, physics, mechanical drawing, and blueprint reading.

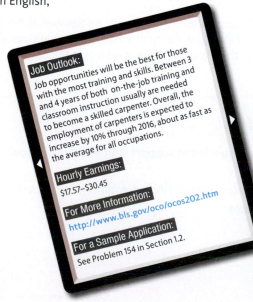

Job Outlook:
Job opportunities will be the best for those with the most training and skills. Between 3 and 4 years of both on-the-job training and classroom instruction usually are needed to become a skilled carpenter. Overall, the employment of carpenters is expected to increase by 10% through 2016, about as fast as the average for all occupations.

Hourly Earnings:
$17.57–$30.45

For More Information:
http://www.bls.gov/oco/ocos202.htm

For a Sample Application:
See Problem 154 in Section 1.2.

In this chapter

In Chapter 1, we will discuss the various types of numbers that we will use throughout this course. Then we will review the basic arithmetic of fractions, explain how to add, subtract, multiply, and divide real numbers, introduce algebraic expressions, and summarize the properties of real numbers.

Reach for Success   Analyzing Your Time

We can all agree that we only have 24 hours in a day, right? The expression "My, how time flies!" is certainly relevant. It might surprise you where the time goes when you begin to monitor the hours.

How can you schedule your time to allow for a successful semester? Begin by completing the pie chart to analyze a typical day.

©Brad Collett/Shutterstock.com

HOW MANY HOURS A DAY DO YOU SPEND:

For every hour spent on each activity shade that many pie segments. (It might be helpful to use a different color to shade for each question.) There are no right or wrong answers. You just need an honest assessment of your time.

_____ working at outside employment (include commute time)?

_____ sleeping?

_____ preparing meals, eating, and exercising?

_____ getting ready for work/class?

_____ in class (include lab hours, commuting, or tutoring time)?

_____ with family and friends?

_____ on the Internet, phone, playing video games, texting, watching TV, going to movies, or other entertainment?

24-Hour Pie Chart

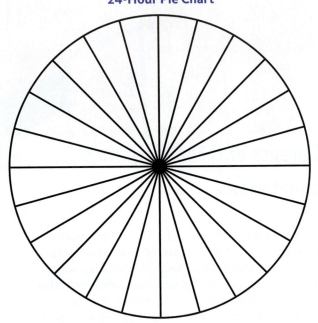

Now we have a mathematics problem! How many hours are left for studying? Remember, the rule of thumb is for every **1** hour in class, you will need at least **2** hours of studying. For mathematics, you could need even more. The key is to spend as much time as you need to understand the material. There is no magic formula!

A Successful Study Strategy . . .

 Adjust your schedule as needed to support your educational goals.

At the end of the chapter you will find an additional exercise to guide you in planning for a successful semester.

Section 1.1

Real Numbers and Their Graphs

Objectives

1. List the numbers in a set of real numbers that are natural, whole, integers, rational, irrational, composite, prime, even, or odd.
2. Insert a symbol $<$, $>$, or $=$ to define the relationship between two rational numbers.
3. Graph a real number or a subset of real numbers on the number line.
4. Find the absolute value of a real number.

Vocabulary

set	irrational numbers	inequality symbols
natural numbers	real numbers	variables
positive integers	prime numbers	number line
whole numbers	composite numbers	origin
ellipses	even integers	coordinate
negative numbers	odd integers	negatives
integers	sum	opposites
subsets	difference	intervals
set-builder notation	product	absolute value
rational numbers	quotient	

Getting Ready

1. Give an example of a number that is used for counting.
2. Give an example of a number that is used when dividing a pizza.
3. Give an example of a number that is used for measuring temperatures that are below zero.
4. What other types of numbers can you think of?

We will begin by discussing various sets of numbers.

1. List the numbers in a set of real numbers that are natural, whole, integers, rational, irrational, composite, prime, even, or odd.

A **set** is a collection of objects. For example, the set

$\{1, 2, 3, 4, 5\}$ Read as "the set with elements 1, 2, 3, 4, and 5."

contains the numbers 1, 2, 3, 4, and 5. The *members*, or *elements*, of a set are listed within braces $\{\ \}$.

Two basic sets of numbers are the **natural numbers** (often called the **positive integers**) and the **whole numbers**.

THE SET OF NATURAL NUMBERS (POSITIVE INTEGERS)	$\{1, 2, 3, 4, 5, 6, 7, 8, 9\ 10, \ldots\}$

THE SET OF WHOLE NUMBERS	$\{0, 1, 2, 3, 4, 5, 6, 7, 8, 9, 10, \ldots\}$

The three dots in the previous definitions, called **ellipses**, indicate that each list of numbers continues on forever.

We can use whole numbers to describe many real-life situations. For example, some cars might get 30 miles per gallon (mpg) of gas, and some students might pay $1,750 in tuition.

Numbers that show a loss or a downward direction are called **negative numbers**, and they are denoted with a $-$ sign. For example, a debt of $1,500 can be denoted as $1,500, and a temperature of 20° below zero can be denoted as $-20°$.

The negatives of the natural numbers and the whole numbers together form the set of **integers**.

THE SET OF INTEGERS	$\{\ldots, -5, -4, -3, -2, -1, 0, 1, 2, 3, 4, 5, \ldots\}$

Because the set of natural numbers and the set of whole numbers are included within the set of integers, these sets are called **subsets** of the set of integers.

Integers cannot describe every real-life situation. For example, a student might study $3\frac{1}{2}$ hours, or a TV set might cost $217.37. To describe these situations, we need fractions, more formally called *rational numbers*.

We cannot list the set of rational numbers as we have listed the previous sets in this section. Instead, we will use **set-builder notation**. This notation uses a variable (or variables) to represent the elements in a set and a rule to determine the possible values of the variable.

THE SET OF RATIONAL NUMBERS	**Rational numbers** are fractions that have an integer numerator and a nonzero integer denominator. Using set-builder notation, the rational numbers are

$$\left\{ \frac{a}{b} \;\middle|\; a \text{ is an integer and } b \text{ is a nonzero integer.} \right\}$$

COMMENT Because division by 0 is undefined, expressions such as $\frac{6}{0}$ and $\frac{8}{0}$ do not represent any number.

The previous notation is read as "the set of all numbers $\frac{a}{b}$ such that a is an integer and b is a nonzero integer."

Some examples of rational numbers are

$$\frac{3}{2}, \frac{17}{12}, 5, -\frac{43}{8}, 0.25, \text{ and } -0.66666\ldots$$

The decimals 0.25 and $-0.66666\ldots$ are rational numbers, because 0.25 can be written as the fraction $\frac{1}{4}$, and $-0.66666\ldots$ can be written as the fraction $-\frac{2}{3}$.

Since every integer can be written as a fraction with a denominator of 1, every integer is also a rational number. Since every integer is a rational number, the set of integers is a subset of the rational numbers.

Since π and $\sqrt{2}$ cannot be written as fractions with an integer numerator and a nonzero integer denominator, they are not rational numbers. They are called **irrational numbers**. We can find their decimal approximations with a calculator. For example,

$\pi \approx 3.141592654$ Using a scientific calculator, press π. Using a graphing calculator, press **2nd** ^ (π) **ENTER**. Read $\approx$ as "is approximately equal to."

$\sqrt{2} \approx 1.414213562$ Using a scientific calculator, press 2 $\sqrt{\ }$. Using a graphing calculator, press **2nd** x^2 ($\sqrt{\ }$) 2 **ENTER**.

If we combine the rational and the irrational numbers, we have the set of **real numbers**.

THE SET OF REAL NUMBERS	$\{x \mid x \text{ is either a rational number or an irrational number.}\}$

COMMENT The symbol $\mathbb{R}$ is often used to represent the set of real numbers.

The previous notation is read as "the set of all numbers x such that x is either a rational number or an irrational number."

Figure 1-1 illustrates how the various sets of numbers are interrelated.

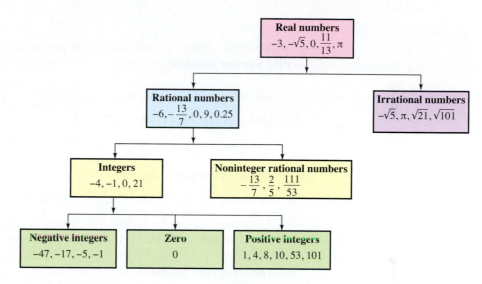

Figure 1-1

EXAMPLE 1 Which numbers in the set $\left\{-3, 0, \frac{1}{2}, 1.25, \sqrt{3}, 5\right\}$ are **a.** natural numbers **b.** whole numbers **c.** negative integers **d.** rational numbers **e.** irrational numbers **f.** real numbers?

Solution **a.** The only natural number is 5.

b. The whole numbers are 0 and 5.

c. The only negative integer is -3.

d. The rational numbers are $-3, 0, \frac{1}{2}, 1.25$, and 5. $\left(1.25 \text{ is rational, because } 1.25 \text{ can be written in the form } \frac{125}{100}.\right)$

e. The only irrational number is $\sqrt{3}$.

f. All of the numbers are real numbers.

SELF CHECK 1 Which numbers in the set $\left\{-2, 0, 1.5, \sqrt{5}, 7\right\}$ are **a.** natural numbers **b.** rational numbers?

Pythagoras
569–475 BC

THE FIRST IRRATIONAL NUMBER

The Greek mathematician and philosopher Pythagoras believed that every aspect of the natural world could be represented by ratios of whole numbers (i.e., rational numbers). However, one of his students accidentally disproved this claim by examining a surprisingly simple example. The student examined a right triangle whose legs were each 1 unit long and posed the following question. "How long is the third side of the triangle?"

Using the well-known theorem of Pythagoras, the length of the third side can be determined by using the formula $c^2 = 1^2 + 1^2$.

In other words, $c^2 = 2$. Using basic properties of arithmetic, it turns out that the numerical value of c cannot be expressed as a rational number. So we have an example of an aspect of the natural world that corresponds to an irrational number, namely $c = \sqrt{2}$.

A natural number greater than 1 that can be divided evenly only by 1 and itself is called a **prime number**.

The set of prime numbers:

$\{2, 3, 5, 7, 11, 13, 17, 19, 23, 29, \ldots\}$

A nonprime natural number greater than 1 is called a **composite number**.

The set of composite numbers:

$\{4, 6, 8, 9, 10, 12, 14, 15, 16, 18, 20, 21, 22, \ldots\}$

An integer that can be divided evenly by 2 is called an **even integer**. An integer that cannot be divided evenly by 2 is called an **odd integer**.

The set of even integers:

$\{\ldots, -10, -8, -6, -4, -2, 0, 2, 4, 6, 8, 10, \ldots\}$

The set of odd integers:

$\{\ldots, -9, -7, -5, -3, -1, 1, 3, 5, 7, 9, \ldots\}$

EXAMPLE 2   Which numbers in the set $\{-3, -2, 0, 1, 2, 3, 4, 5, 9\}$ are
a. prime numbers   **b.** composite numbers   **c.** even integers   **d.** odd integers?

Solution   **a.** The prime numbers are 2, 3, and 5.

b. The composite numbers are 4 and 9.

c. The even integers are $-2, 0, 2$, and 4.

d. The odd integers are $-3, 1, 3, 5$, and 9.

 **SELF CHECK 2**   Which numbers in the set $\{-5, 0, 1, 2, 4, 5\}$ are
a. prime numbers   **b.** even integers?

2 Insert a symbol $<$, $>$, or $=$ to define the relationship between two rational numbers.

To show that two expressions represent the same number, we use an $=$ sign. Since $4 + 5$ and 9 represent the same number, we can write

$4 + 5 = 9$ Read as "the sum of 4 and 5 is equal to 9." The answer to any addition problem is called a *sum*.

Likewise, we can write

$5 - 3 = 2$ Read as "the difference between 5 and 3 equals 2," or "5 minus 3 equals 2." The answer to any subtraction problem is called a *difference*.

$4 \cdot 5 = 20$ Read as "the product of 4 and 5 equals 20," or "4 times 5 equals 20." The answer to any multiplication problem is called a *product*.

and

$30 \div 6 = 5$ Read as "the quotient obtained when 30 is divided by 6 is 5," or "30 divided by 6 equals 5." The answer to any division problem is called a *quotient*.

We can use **inequality symbols** to show that expressions are not equal.

Symbol	Read as	Symbol	Read as
$\approx$	"is approximately equal to"	$\neq$	"is not equal to"
$<$	"is less than"	$>$	"is greater than"
$\leq$	"is less than or equal to"	$\geq$	"is greater than or equal to"

EXAMPLE 3 **Inequality symbols**

a. $\pi \approx 3.14$ Read as "pi is approximately equal to 3.14."

b. $6 \neq 9$ Read as "6 is not equal to 9."

c. $8 < 10$ Read as "8 is less than 10."

d. $12 > 1$ Read as "12 is greater than 1."

e. $5 \leq 5$ Read as "5 is less than or equal to 5." (Since $5 = 5$, this is a true statement.)

f. $9 \geq 7$ Read as "9 is greater than or equal to 7." (Since $9 > 7$, this is a true statement.)

 **SELF CHECK 3** Determine whether each statement is true or false.
a. $12 \neq 12$ **b.** $7 \geq 7$ **c.** $125 < 137$

Inequality statements can be written so that the inequality symbol points in the opposite direction. For example,

$5 < 7$ and $7 > 5$

both indicate that 5 is less than 7. Likewise,

$12 \geq 3$ and $3 \leq 12$

both indicate that 12 is greater than or equal to 3.

COMMENT In algebra, we usually do not use the times sign ($\times$) to indicate multiplication. It might be mistaken for the variable x.

In algebra, we use letters, called **variables**, to represent real numbers. For example,

- If x represents 4, then $x = 4$.
- If y represents any number greater than 3, then $y > 3$.
- If z represents any number less than or equal to -4, then $z \leq -4$.

3 Graph a real number or a subset of real numbers on the number line.

COMMENT The number 0 is neither positive nor negative.

We can use the **number line** shown in Figure 1-2 to represent sets of numbers. The number line continues forever to the left and to the right. Numbers to the left of 0 (the **origin**) are negative, and numbers to the right of 0 are positive.

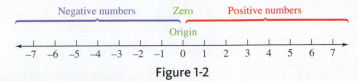

Figure 1-2

The number that corresponds to a point on the number line is called the **coordinate** of that point. For example, the coordinate of the origin is 0.

Many points on the number line do not have integer coordinates. For example, the point midway between 0 and 1 has the coordinate $\frac{1}{2}$, and the point midway between -3 and -2 has the coordinate $-\frac{5}{2}$ (see Figure 1-3).

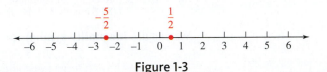

Figure 1-3

Numbers represented by points that lie on opposite sides of the origin and at equal distances from the origin are called **negatives** (or **opposites**) of each other. For example, 5 and -5 are negatives (or opposites). We need parentheses to express the opposite of a negative number. For example, $-(-5)$ represents the opposite of -5, which we know to be 5. Thus,

$$-(-5) = 5$$

This suggests the following rule.

DOUBLE NEGATIVE RULE

If x represents a real number, then

$$-(-x) = x$$

If one point lies to the *right* of a second point on a number line, its coordinate is the *greater*. Since the point with coordinate 1 lies to the right of the point with coordinate -2 (see Figure 1-4(a)), it follows that $1 > -2$.

If one point lies to the *left* of another, its coordinate is the *smaller* (see Figure 1-4(b)). The point with coordinate -6 lies to the left of the point with coordinate -3 so it follows that $-6 < -3$.

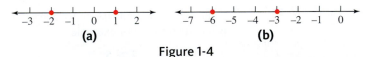

(a)   (b)

Figure 1-4

Figure 1-5 shows the graph of the natural numbers from 2 to 8. The points on the line are called graphs of their corresponding coordinates.

Figure 1-5

EXAMPLE 4 Graph the set of integers between -3 and 3.

Solution The integers between -3 and 3 are $-2, -1, 0, 1,$ and 2. The graph is shown in Figure 1-6.

Figure 1-6

SELF CHECK 4 Graph the set of integers between -4 and 0.

Graphs of many sets of real numbers are **intervals** on the number line. For example, two graphs of all real numbers x such that $x > -2$ are shown in Figure 1-7. The parenthesis and the open circle at -2 show that this point is not included in the graph. The arrow pointing to the right shows that all numbers to the right of -2 are included.

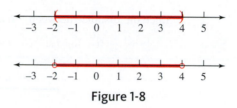

Figure 1-7

Figure 1-8 shows two graphs of the set of real numbers x between -2 and 4. This is the graph of all real numbers x such that $x > -2$ and $x < 4$. The parentheses or open circles at -2 and 4 show that these points are not included in the graph. However, all the numbers between -2 and 4 are included.

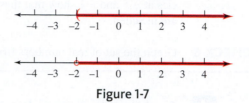

Figure 1-8

EXAMPLE 5   Graph all real numbers x such that $x < -3$ or $x > 1$.

Solution   The graph of all real numbers less than -3 includes all points on the number line that are to the left of -3. The graph of all real numbers greater than 1 includes all points that are to the right of 1. The two graphs are shown in Figure 1-9.

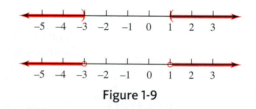

Figure 1-9

 SELF CHECK 5   Graph all real numbers x such that $x < -1$ or $x > 0$. Use parentheses.

Leonardo Fibonacci
(late 12th and early 13th centuries)

Fibonacci, an Italian mathematician, is also known as Leonardo da Pisa. In his work *Liber abaci*, he advocated the adoption of Arabic numerals, the numerals that we use today. He is best known for a sequence of numbers that bears his name. Can you find the pattern in this sequence?

1, 1, 2, 3, 5, 8, 13, . . .

Perspective

Algebra is an extension of arithmetic. In algebra, the operations of addition, subtraction, multiplication, and division are performed on numbers and letters, with the understanding that the letters represent numbers.

The origins of algebra are found in a papyrus written before 1600 BC by an Egyptian priest named Ahmes. This papyrus contains 84 algebra problems and their solutions.

Further development of algebra occurred in the ninth century in the Middle East. In AD 830, an Arabian mathematician named al-Khowarazmi wrote a book called *Ihm al-jabr wa'l muqabalah*. This title was shortened to *al-Jabr*. We now know the subject as *algebra*. The French mathematician François Vieta (1540–1603) later simplified algebra by developing the symbolic notation that we use today.

The Ahmes Papyrus

EXAMPLE 6   Graph the set of all real numbers from -5 to -1.

Solution   The set of all real numbers from -5 to -1 includes -5 and -1 and all the numbers in between. In the graphs shown in Figure 1-10, the brackets or the solid circles at -5 and -1 show that these points are included.

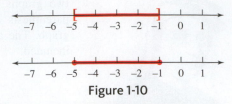

Figure 1-10

 SELF CHECK 6   Graph the set of real numbers from -2 to 1.

4   Find the absolute value of a real number.

On a number line, the distance between a number x and 0 is called the **absolute value** of x. For example, the distance between 5 and 0 is 5 units (see Figure 1-11). Thus, the absolute value of 5 is 5:

$$|5| = 5 \quad \text{Read as "The absolute value of 5 is 5."}$$

Since the distance between -6 and 0 is 6,

$$|-6| = 6 \quad \text{Read as "The absolute value of } -6 \text{ is 6."}$$

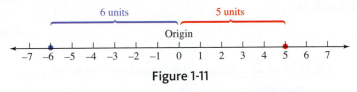

Figure 1-11

Because the absolute value of a real number represents that number's distance from 0 on the number line, the absolute value of every real number x is either positive or 0. In symbols, we say

$$|x| \geq 0 \quad \text{for every real number } x$$

EXAMPLE 7   Evaluate: **a.** $|6|$ **b.** $|-3|$ **c.** $|0|$ **d.** $-|2 + 3|$

Solution   **a.** $|6| = 6$, because 6 is six units from 0.

b. $|-3| = 3$, because -3 is three units from 0.

c. $|0| = 0$, because 0 is zero units from 0.

d. $-|2 + 3| = -|5| = -5$, because the opposite of the absolute value of 5 is -5.

 SELF CHECK 7   Evaluate: **a.** $|8|$ **b.** $|-8|$ **c.** $-|-8|$

SELF CHECK ANSWERS

1. a. 7   **b.** $-2, 0, 1.5, 7$ **2. a.** 2, 5   **b.** 0, 2, 4   **3. a.** false   **b.** true   **c.** true

4. ← ┤ ● ● ● ┤ →
          -4 -3 -2 -1 0

5. ← ─[─(┤ →
          -2 -1 0 1

6. ← ┤ [┤ ┤] ┤ →
          -3 -2 -1 0 1 2

7. a. 8   **b.** 8   **c.** -8

NOW TRY THIS

Given the set $\{\sqrt{10}, 4.2, \sqrt{16}, 0, |-1|, 9\}$, *list*

1. the integer(s)

2. the irrational number(s)

3. the rational number(s)

4. the prime number(s)

5. the composite number(s)

1.1 Exercises

WARM-UPS *Describe each set of numbers.*

1. Natural numbers
2. Whole numbers
3. Integers
4. Rational numbers
5. Real numbers
6. Prime numbers
7. Composite numbers
8. Even integers
9. Odd integers
10. Irrational numbers

Find each value.

11. $-|-7|$
12. $|-12|$

VOCABULARY AND CONCEPTS *Fill in the blanks.*

13. A ___ is a collection of objects.

14. The numbers 1, 2, 3, 4, 5, . . . form the set of _____ numbers. This set is also called the set of _____.

15. The set of _____ numbers is the set {0, 1, 2, 3, 4, 5, . . .}.

16. The dots following the sets in Exercises 14 and 15 are called _____.

17. The set of _____ is the set {. . . , −3, −2, −1, 0, 1, 2, 3, . . .}.

18. Numbers that show a loss or a downward direction are called _____.

19. Since every whole number is also an integer, the set of whole numbers is called a _____ of the set of integers.

20. The set $\{x \mid x \text{ is a whole number.}\}$ is read as "_____ _____."

21. Fractions that have an integer numerator and a nonzero integer denominator are called _____ numbers.

22. $\sqrt{2}$ is an example of an _____ number.

23. The set that includes the rational and irrational numbers is called the set of ___ numbers.

24. If a natural number is greater than 1 and can be divided exactly only by 1 and itself, it is called a _____ number.

25. A composite number is a _____ number that is greater than 1 and is not _____.

26. An integer that can be evenly divided by 2 is called an ____ integer.

27. An integer that cannot be evenly divided by 2 is called an ____ integer.

28. The symbol ≠ means _____.

29. The symbol _ means "is less than."

30. The symbol ≥ means _____.

31. In algebra, we use letters, called _____, to represent real numbers.

32. The figure $\xleftarrow{\quad -3 \ -2 \ -1 \ \ 0 \ \ 1 \ \ 2 \ \ 3 \quad}\rightarrow$ is called a _____ line. The point with a coordinate of 0 is called the _____.

33. The negative, or opposite, of −7 is _.

34. The graphs of inequalities are _____ on the number line.

35. A _____ or ____ circle shows that a point is not included in a graph.

36. A _____ or _____ circle shows that a point is included in a graph.

37. The _____ between 0 and 6 on a number line is called the absolute value of _.

38. The result of an addition is called the ___. The result of a subtraction is called a _____. The result of a multiplication is called a _____. The result of a division is called a _____.

GUIDED PRACTICE *Which numbers in the set* $\{-3, -\frac{1}{2}, -1, 0, 1, 2, \frac{5}{3}, \sqrt{7}, 3.25, 6, 9\}$ *are in each category?* **SEE EXAMPLES 1–2. (OBJECTIVE 1)**

39. natural numbers
40. whole numbers
41. positive integers
42. negative integers
43. integers
44. rational numbers
45. real numbers
46. irrational numbers
47. odd integers
48. even integers
49. composite numbers
50. prime numbers

Place one of the symbols =, <, or > in each box to make a true statement. **SEE EXAMPLE 3. (OBJECTIVE 2)**

51. 7 ☐ 10
52. 3 ☐ 2 + 1
53. 9 ☐ 2 + 5
54. −5 ☐ −4
55. −6 ☐ −8
56. 2 + 3 ☐ 17
57. 5 + 7 ☐ 10
58. 3 + 3 ☐ 9 − 3

Graph each pair of numbers on a number line. In each pair, indicate which number is the greater and which number lies farther to the right. **(OBJECTIVE 3)**

59. 2, 4
60. 5, 9
61. 11, 6
62. 15, 10

63. −5, −2

64. 4, 10

65. 8, 0

66. −7, −1

Graph each set of numbers on a number line. Use brackets or parentheses where applicable. SEE EXAMPLES 4–6. (OBJECTIVE 3)

67. The natural numbers between 2 and 8

68. The prime numbers between 5 and 15

69. The real numbers between 3 and 8

70. The odd integers between −5 and 5 that are exactly divisible by 3

71. The real numbers greater than or equal to 8

72. The real numbers greater than or equal to 3 or less than or equal to −3

73. The odd numbers from 10 to 20

74. The even integers greater than or equal to 10 and less than or equal to 20.

Find each absolute value. SEE EXAMPLE 7. (OBJECTIVE 4)

75. $|36|$

76. $|-17|$

77. $|0|$

78. $|120|$

79. $-|-23|$

80. $|18 - 12|$

81. $|12 - 4|$

82. $|100 - 100|$

ADDITIONAL PRACTICE *Simplify each expression. Then classify the result as a natural number, an even integer, an odd integer, a prime number, a composite number, and/or a whole number.*

83. $6 + 3$

84. $7 - 2$

85. $15 - 15$

86. $13 - 6$

87. $3 \cdot 8$

88. $6 \cdot 12$

89. $24 \div 8$

90. $7 \div 7$

Place one of the symbols =, <, or > in each box to make a true statement.

91. $5 + 6 \quad\boxed{}\quad 13 - 1$

92. $19 - 3 \quad\boxed{}\quad 8 + 6$

93. $4 \cdot 3 \quad\boxed{}\quad 3 \cdot 4$

94. $7 \cdot 9 \quad\boxed{}\quad 9 \cdot 6$

95. $0 \div 6 \quad\boxed{}\quad 1$

96. $2 + 7 \quad\boxed{}\quad 7 + 2$

97. $45 \div 9 \quad\boxed{}\quad 36 \div 12$

98. $5 \cdot 12 \quad\boxed{}\quad 300 \div 5$

99. $3 + 2 + 5 \quad\boxed{}\quad 5 + 2 + 3$

100. $8 + 5 + 2 \quad\boxed{}\quad 5 + 2 + 8$

Write each sentence as a mathematical expression.

101. Nine is greater than four.

102. Five is less than thirty-two.

103. Eight is less than or equal to eight.

104. Twenty-five is not equal to twenty-three.

105. The sum of adding three and four is equal to seven.

106. Thirty-seven is greater than the product of multiplying three and four.

107. $\sqrt{2}$ is approximately equal to 1.41.

108. x is greater than or equal to 5.

Write each inequality as an equivalent inequality in which the inequality symbol points in the opposite direction.

109. $3 \le 7$

110. $5 > 2$

111. $6 > 0$

112. $34 \le 40$

113. $3 + 8 > 8$

114. $8 - 3 < 8$

115. $6 - 2 < 10 - 4$

116. $8 \cdot 2 \ge 8 \cdot 1$

117. $2 \cdot 3 < 3 \cdot 4$

118. $8 \div 2 \ge 9 \div 3$

119. $\dfrac{12}{4} < \dfrac{24}{6}$

120. $\dfrac{2}{3} \le \dfrac{3}{4}$

Graph each set of numbers on a number line. Use brackets or parentheses where applicable.

121. The even integers that are also prime numbers

122. The numbers that are whole numbers but not natural numbers

123. The natural numbers between 15 and 25 that are multiples of 6

124. The real numbers greater than −2 and less than 3

125. The real numbers greater than or equal to −5 and less than 4

126. The real numbers between −7 and 7, including −7 and 7

Find each absolute value.

127. $|21 - 19|$

128. $|25 - 21|$

WRITING ABOUT MATH

129. Explain why there is no greatest natural number.

130. Explain why 2 is the only even prime number.

131. Explain how to determine the absolute value of a number.

132. Explain why zero is an even integer.

SOMETHING TO THINK ABOUT *Consider the following sets: the integers, natural numbers, even and odd integers, positive and negative numbers, prime and composite numbers, and rational numbers.*

133. Find a number that fits in as many of these categories as possible.

134. Find a number that fits in as few of these categories as possible.

Section 1.2

Fractions

Objectives

1. Simplify a fraction.
2. Multiply and divide two fractions.
3. Add and subtract two or more fractions.
4. Add and subtract two or more mixed numbers.
5. Add, subtract, multiply, and divide two or more decimals.
6. Round a decimal to a specified number of places.
7. Use the appropriate operation for an application.

Vocabulary

numerator	proper fraction	mixed number
denominator	improper fraction	terminating decimal
lowest terms	reciprocal	repeating decimal
simplest form	equivalent fractions	divisor
factors of a product	least (or lowest) common	dividend
prime-factored form	denominator	percent

Getting Ready

1. Add:
$$\begin{array}{r} 132 \\ 45 \\ 73 \end{array}$$

2. Subtract:
$$\begin{array}{r} 321 \\ 173 \end{array}$$

3. Multiply:
$$\begin{array}{r} 437 \\ 38 \end{array}$$

4. Divide: $37\overline{)3{,}885}$

In this section, we will review arithmetic fractions. This will help us prepare for algebraic fractions, which we will encounter later in the book.

1   Simplify a fraction.

In the fractions

$$\frac{1}{2}, \frac{3}{5}, \frac{2}{17}, \quad \text{and} \quad \frac{37}{7}$$

the number above the bar is called the **numerator**, and the number below the bar is called the **denominator**.

We often use fractions to indicate parts of a whole. In Figure 1-12(a) on the next page, a rectangle has been divided into 5 equal parts, and 3 of the parts are shaded. The fraction $\frac{3}{5}$ indicates how much of the figure is shaded. In Figure 1-12(b), $\frac{5}{7}$ of the rectangle is shaded. In either example, the denominator of the fraction shows the total number of equal parts into which the whole is divided, and the numerator shows how many of these equal parts are being considered.

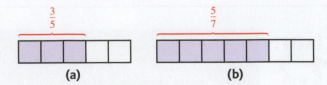

Figure 1-12

We can also use fractions to indicate division. For example, the fraction $\frac{8}{2}$ indicates that 8 is to be divided by 2:

$$\frac{8}{2} = 8 \div 2 = 4$$

COMMENT Note that $\frac{8}{2} = 4$, because $4 \cdot 2 = 8$, and that $\frac{0}{7} = 0$, because $0 \cdot 7 = 0$. However, $\frac{6}{0}$ is undefined, because no number multiplied by 0 gives 6. Remember that the denominator of a fraction cannot be 0.

A fraction is said to be in **lowest terms** (or **simplest form**) when no integer other than 1 will divide both its numerator and its denominator exactly. The fraction $\frac{6}{11}$ is in lowest terms because only 1 divides both 6 and 11 exactly. The fraction $\frac{6}{8}$ is not in lowest terms, because 2 divides both 6 and 8 exactly.

We can simplify a fraction that is not in lowest terms by dividing its numerator and its denominator by the same number. For example, to simplify $\frac{6}{8}$, we divide the numerator and the denominator by 2.

$$\frac{6}{8} = \frac{6 \div 2}{8 \div 2} = \frac{3}{4}$$

From Figure 1-13, we see that $\frac{6}{8}$ and $\frac{3}{4}$ are equal fractions, because each one represents the same part of the rectangle.

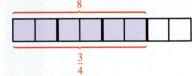

Figure 1-13

When a composite number has been written as the product of other natural numbers, we say that it has been factored. For example, 15 can be written as the product of 5 and 3.

$$15 = 5 \cdot 3$$

The numbers 5 and 3 are called **factors** of 15. When a composite number is written as the product of prime numbers, we say that it is written in **prime-factored form**.

EXAMPLE 1   Write 210 in prime-factored form.

Solution   We can write 210 as the product of 21 and 10 and proceed as follows:

$$210 = \mathbf{21 \cdot 10}$$
$$210 = \mathbf{3 \cdot 7 \cdot 2 \cdot 5} \qquad \text{Factor 21 as } 3 \cdot 7 \text{ and factor 10 as } 2 \cdot 5.$$

Since 210 is now written as the product of prime numbers, its prime-factored form is $210 = 2 \cdot 3 \cdot 5 \cdot 7$.

 SELF CHECK 1   Write 70 in prime-factored form.

To simplify a fraction, we factor its numerator and denominator and divide out all factors that are common to the numerator and denominator. For example,

$$\frac{6}{8} = \frac{3 \cdot 2}{4 \cdot 2} = \frac{3 \cdot \overset{1}{\cancel{2}}}{4 \cdot \underset{1}{\cancel{2}}} = \frac{3}{4} \quad \text{and} \quad \frac{15}{18} = \frac{5 \cdot 3}{6 \cdot 3} = \frac{5 \cdot \overset{1}{\cancel{3}}}{6 \cdot \underset{1}{\cancel{3}}} = \frac{5}{6}$$

COMMENT Remember that a fraction is in lowest terms only when its numerator and denominator have no common factors.

EXAMPLE 2 Simplify, if possible: **a.** $\dfrac{6}{30}$ **b.** $\dfrac{33}{40}$

Solution **a.** To simplify $\dfrac{6}{30}$, we factor the numerator and denominator and divide out the common factor of 6.

$$\frac{6}{30} = \frac{6 \cdot 1}{6 \cdot 5} = \frac{\overset{1}{\cancel{6}} \cdot 1}{\underset{1}{\cancel{6}} \cdot 5} = \frac{1}{5}$$

b. To simplify $\dfrac{33}{40}$, we factor the numerator and denominator and divide out any common factors.

$$\frac{33}{40} = \frac{3 \cdot 11}{2 \cdot 2 \cdot 2 \cdot 5}$$

Since the numerator and denominator have no common factors, $\dfrac{33}{40}$ is in lowest terms.

SELF CHECK 2 Simplify, if possible: $\dfrac{14}{35}$

The preceding examples illustrate the *fundamental property of fractions*.

THE FUNDAMENTAL PROPERTY OF FRACTIONS	If a, b, and x are real numbers, $$\frac{a \cdot x}{b \cdot x} = \frac{a}{b} \quad (b \neq 0 \text{ and } x \neq 0)$$

2 Multiply and divide two fractions.

To multiply fractions, we use the following rule.

MULTIPLYING FRACTIONS	To multiply fractions, we multiply their numerators and multiply their denominators. In symbols, if $a, b, c,$ and d are real numbers, $$\frac{a}{b} \cdot \frac{c}{d} = \frac{a \cdot c}{b \cdot d} \quad (b \neq 0 \text{ and } d \neq 0)$$

For example,

$$\frac{4}{7} \cdot \frac{2}{3} = \frac{4 \cdot 2}{7 \cdot 3} \qquad \frac{4}{5} \cdot \frac{13}{9} = \frac{4 \cdot 13}{5 \cdot 9}$$

$$= \frac{8}{21} \qquad\qquad\quad = \frac{52}{45}$$

To justify the rule for multiplying fractions, we consider the square in Figure 1-14. Because the length of each side of the square is 1 unit and the area is the product of the lengths of two sides, the area is 1 square unit.

If this square is divided into 3 equal parts vertically and 7 equal parts horizontally, it is divided into 21 equal parts, and each represents $\dfrac{1}{21}$ of the total area. The area

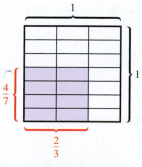

Figure 1-14

of the shaded rectangle in the square is $\frac{8}{21}$, because it contains 8 of the 21 parts. The width, w, of the shaded rectangle is $\frac{4}{7}$; its length, l, is $\frac{2}{3}$; and its area, A, is the product of l and w:

$$A = l \cdot w$$

$$\frac{8}{21} = \frac{2}{3} \cdot \frac{4}{7}$$

This suggests that we can find the product of

$$\frac{4}{7} \quad \text{and} \quad \frac{2}{3}$$

by multiplying their numerators and multiplying their denominators.

Fractions whose numerators are less than their denominators, such as $\frac{8}{21}$, are called **proper fractions**. Fractions whose numerators are greater than or equal to their denominators, such as $\frac{52}{45}$, are called **improper fractions**.

EXAMPLE 3   Perform each multiplication.

a. $\dfrac{3}{7} \cdot \dfrac{13}{5} = \dfrac{3 \cdot 13}{7 \cdot 5}$ Multiply the numerators and multiply the denominators. There are no common factors.

$= \dfrac{39}{35}$ Multiply in the numerator and multiply in the denominator.

b. $5 \cdot \dfrac{3}{15} = \dfrac{5}{1} \cdot \dfrac{3}{15}$ Write 5 as the improper fraction $\frac{5}{1}$.

$= \dfrac{5 \cdot 3}{1 \cdot 15}$ Multiply the numerators and multiply the denominators.

$= \dfrac{5 \cdot 3}{1 \cdot 5 \cdot 3}$ To simplify the fraction, factor the denominator.

$= \dfrac{\overset{1}{\cancel{5}} \cdot \overset{1}{\cancel{3}}}{1 \cdot \underset{1}{\cancel{5}} \cdot \underset{1}{\cancel{3}}}$ Divide out the common factors of 3 and 5.

$= 1$ $\frac{1 \cdot 1}{1 \cdot 1 \cdot 1} = 1$

 **SELF CHECK 3**   Multiply: $\dfrac{5}{9} \cdot \dfrac{7}{10}$

EXAMPLE 4   **TRAVEL**   Out of 36 students in a history class, three-fourths have signed up for a trip to Europe. If there are 28 places available on the flight, will there be room for one more student?

Solution   We first find three-fourths of 36.

$$\frac{3}{4} \cdot 36 = \frac{3}{4} \cdot \frac{36}{1}$$ Write 36 as $\frac{36}{1}$.

$$= \frac{3 \cdot 36}{4 \cdot 1}$$ Multiply the numerators and multiply the denominators.

$$= \frac{3 \cdot 4 \cdot 9}{4 \cdot 1}$$ To simplify, factor the numerator.

$$= \frac{3 \cdot \overset{1}{\cancel{4}} \cdot 9}{\underset{1}{\cancel{4}} \cdot 1}$$ Divide out the common factor of 4.

$$= \frac{27}{1}$$

$$= 27$$

Twenty-seven students plan to go on the trip. Since there is room for 28 passengers, there is room for one more.

🍃 **SELF CHECK 4** If seven-ninths of the 36 students had signed up, would there be room for one more?

One number is called the **reciprocal** of another if their product is 1. For example, $\frac{3}{5}$ is the reciprocal of $\frac{5}{3}$, because

$$\frac{3}{5} \cdot \frac{5}{3} = \frac{15}{15} = 1$$

DIVIDING FRACTIONS To divide two fractions, we multiply the first fraction by the reciprocal of the second fraction. In symbols, if a, b, c, and d are real numbers,

$$\frac{a}{b} \div \frac{c}{d} = \frac{a}{b} \cdot \frac{d}{c} = \frac{a \cdot d}{b \cdot c} \quad (b \neq 0, c \neq 0, \text{ and } d \neq 0)$$

EXAMPLE 5 Perform each division.

a. $\dfrac{3}{5} \div \dfrac{6}{5} = \dfrac{3}{5} \cdot \dfrac{5}{6}$ Multiply $\frac{3}{5}$ by the reciprocal of $\frac{6}{5}$.

$$= \frac{3 \cdot 5}{5 \cdot 6}$$ Multiply the numerators and multiply the denominators.

$$= \frac{3 \cdot 5}{5 \cdot 2 \cdot 3}$$ Factor the denominator.

$$= \frac{\overset{1}{\cancel{3}} \cdot \overset{1}{\cancel{5}}}{\underset{1}{\cancel{5}} \cdot 2 \cdot \underset{1}{\cancel{3}}}$$ Divide out the common factors of 3 and 5.

$$= \frac{1}{2}$$

b. $\dfrac{15}{7} \div 10 = \dfrac{15}{7} \div \dfrac{10}{1}$ Write 10 as the improper fraction $\frac{10}{1}$.

$$= \frac{15}{7} \cdot \frac{1}{10}$$ Multiply $\frac{15}{7}$ by the reciprocal of $\frac{10}{1}$.

$$= \frac{15 \cdot 1}{7 \cdot 10}$$ Multiply the numerators and multiply the denominators.

$$= \frac{3 \cdot \overset{1}{\cancel{5}}}{7 \cdot 2 \cdot \underset{1}{\cancel{5}}}$$ Factor the numerator and the denominator, and divide out the common factor of 5.

$$= \frac{3}{14}$$

🍃 **SELF CHECK 5** Perform the division: $\dfrac{13}{6} \div \dfrac{26}{8}$

3 Add and subtract two or more fractions.

To add fractions with like denominators, we use the following rule.

ADDING FRACTIONS WITH THE SAME DENOMINATOR	To add fractions with the same denominator, we add the numerators and keep the common denominator. In symbols, if a, b, and d are real numbers, $$\frac{a}{d} + \frac{b}{d} = \frac{a + b}{d} \quad (d \neq 0)$$

For example,

$$\frac{3}{7} + \frac{2}{7} = \frac{3 + 2}{7}$$ Add the numerators and keep the common denominator.

$$= \frac{5}{7}$$

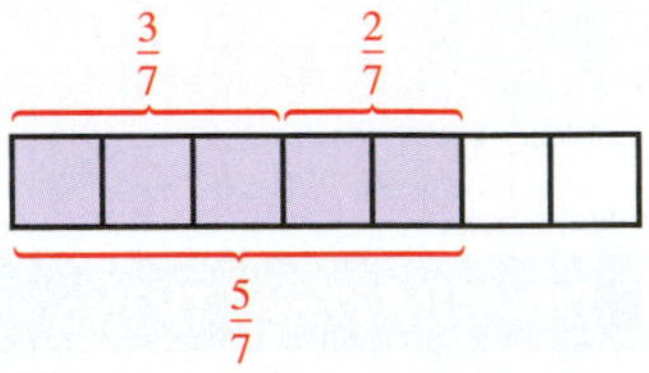

Figure 1-15

Figure 1-15 illustrates why $\frac{3}{7} + \frac{2}{7} = \frac{5}{7}$.

To add fractions with unlike denominators, we write the fractions so that they have the same denominator. For example, we can multiply both the numerator and denominator of $\frac{1}{3}$ by 5 to obtain an **equivalent fraction** with a denominator of 15:

$$\frac{1}{3} = \frac{1 \cdot 5}{3 \cdot 5} = \frac{5}{15}$$

To write $\frac{1}{5}$ as an equivalent fraction with a denominator of 15, we multiply the numerator and the denominator by 3:

$$\frac{1}{5} = \frac{1 \cdot 3}{5 \cdot 3} = \frac{3}{15}$$

Since 15 is the smallest number that can be used as a common denominator for $\frac{1}{3}$ and $\frac{1}{5}$, it is called the **least** (or **lowest**) **common denominator** (the **LCD**).

To add the fractions $\frac{1}{3}$ and $\frac{1}{5}$, we write each fraction as an equivalent fraction having a denominator of 15, and then we add the results:

$$\frac{1}{3} + \frac{1}{5} = \frac{1 \cdot 5}{3 \cdot 5} + \frac{1 \cdot 3}{5 \cdot 3}$$

$$= \frac{5}{15} + \frac{3}{15}$$

$$= \frac{5 + 3}{15}$$

$$= \frac{8}{15}$$

In the next example, we will add the fractions $\frac{3}{10}$ and $\frac{5}{28}$.

EXAMPLE 6 Add: $\dfrac{3}{10} + \dfrac{5}{28}$

Solution To find the LCD, we find the prime factorization of each denominator and use each prime factor the greatest number of times it appears in either factorization.

$$\left. \begin{array}{l} 10 = 2 \cdot 5 \\ 28 = 2 \cdot 2 \cdot 7 \end{array} \right\} \quad \text{LCD} = 2 \cdot 2 \cdot 5 \cdot 7 = 140$$

Since 140 is the smallest number that 10 and 28 divide exactly, we write both fractions as fractions with denominators of 140.

$$\frac{3}{10} + \frac{5}{28} = \frac{3 \cdot \mathbf{14}}{10 \cdot \mathbf{14}} + \frac{5 \cdot \mathbf{5}}{28 \cdot \mathbf{5}}$$ Write each fraction as a fraction with a denominator of 140.

$$= \frac{42}{140} + \frac{25}{140}$$ Do the multiplications.

$$= \frac{42 + 25}{140}$$ Add the numerators and keep the denominator.

$$= \frac{67}{140}$$

Since 67 is a prime number, it has no common factor with 140. Thus, $\frac{67}{140}$ is in lowest terms.

 **SELF CHECK 6**   Add: $\frac{3}{8} + \frac{5}{12}$

To subtract fractions with like denominators, we use the following rule.

SUBTRACTING FRACTIONS WITH THE SAME DENOMINATOR

To subtract fractions with the same denominator, we subtract their numerators and keep their common denominator. In symbols, if a, b, and d are real numbers,

$$\frac{a}{d} - \frac{b}{d} = \frac{a - b}{d} \quad (d \neq 0)$$

For example,

$$\frac{7}{9} - \frac{2}{9} = \frac{7 - 2}{9} = \frac{5}{9}$$

To subtract fractions with unlike denominators, we write them as equivalent fractions with a common denominator. For example, to subtract $\frac{2}{5}$ from $\frac{3}{4}$, we write $\frac{3}{4} - \frac{2}{5}$, find the LCD of 4 and 5, which is 20, and proceed as follows:

$$\frac{3}{4} - \frac{2}{5} = \frac{3 \cdot \mathbf{5}}{4 \cdot \mathbf{5}} - \frac{2 \cdot \mathbf{4}}{5 \cdot \mathbf{4}}$$ Write each fraction as a fraction with a denominator of 20.

$$= \frac{15}{20} - \frac{8}{20}$$ Do the multiplications.

$$= \frac{15 - 8}{20}$$ Add the numerators and keep the denominator.

$$= \frac{7}{20}$$

EXAMPLE 7   Subtract 5 from $\frac{23}{3}$.

Solution

$$\frac{23}{3} - 5 = \frac{23}{3} - \frac{5}{1}$$ Write 5 as the improper fraction $\frac{5}{1}$.

$$= \frac{23}{3} - \frac{5 \cdot \mathbf{3}}{1 \cdot \mathbf{3}}$$ Write $\frac{5}{1}$ as a fraction with a denominator of 3.

$$= \frac{23}{3} - \frac{15}{3}$$ Do the multiplications.

$$= \frac{23 - 15}{3}$$ Subtract the numerators and keep the denominator.

$$= \frac{8}{3}$$

SELF CHECK 7   Subtract: $\dfrac{5}{6} - \dfrac{3}{4}$

4   **Add and subtract two or more mixed numbers.**

The **mixed number** $3\frac{1}{2}$ represents the sum of 3 and $\frac{1}{2}$. We can write $3\frac{1}{2}$ as an improper fraction as follows:

$$3\frac{1}{2} = 3 + \frac{1}{2}$$

$$= \frac{6}{2} + \frac{1}{2}   \qquad 3 = \frac{6}{2}$$

$$= \frac{6 + 1}{2}   \qquad \text{Add the numerators and keep the denominator.}$$

$$= \frac{7}{2}$$

To write the fraction $\frac{19}{5}$ as a mixed number, we divide 19 by 5 to get 3, with a remainder of 4.

$$\frac{19}{5} = 3 + \frac{4}{5} = 3\frac{4}{5}$$

EXAMPLE 8   Add: $2\frac{1}{4} + 1\frac{1}{3}$

Solution   We first change each mixed number to an improper fraction.

$$2\frac{1}{4} = 2 + \frac{1}{4}   \qquad\qquad 1\frac{1}{3} = 1 + \frac{1}{3}$$

$$= \frac{8}{4} + \frac{1}{4}   \qquad\qquad\qquad = \frac{3}{3} + \frac{1}{3}$$

$$= \frac{9}{4}   \qquad\qquad\qquad\qquad = \frac{4}{3}$$

Then we add the fractions.

$$2\frac{1}{4} + 1\frac{1}{3} = \frac{9}{4} + \frac{4}{3}$$

$$= \frac{9 \cdot 3}{4 \cdot 3} + \frac{4 \cdot 4}{3 \cdot 4}   \qquad \text{Write each fraction with the LCD of 12.}$$

$$= \frac{27}{12} + \frac{16}{12}$$

$$= \frac{43}{12}$$

Finally, we change $\frac{43}{12}$ to a mixed number.

$$\frac{43}{12} = 3 + \frac{7}{12} = 3\frac{7}{12}$$

SELF CHECK 8   Add: $5\frac{1}{7} + 4\frac{2}{3}$

EXAMPLE 9   **FENCING LAND**   How much fencing will be needed to enclose the area within the triangular lot shown in Figure 1-16?

Solution   We can find the sum of the lengths by adding the whole-number parts and the fractional parts of the dimensions separately.

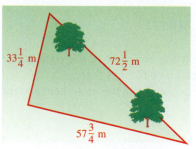

$33\frac{1}{4} + 57\frac{3}{4} + 72\frac{1}{2} = 33 + 57 + 72 + \frac{1}{4} + \frac{3}{4} + \frac{1}{2}$

$= 162 + \frac{1}{4} + \frac{3}{4} + \frac{2}{4}$ Write $\frac{1}{2}$ as $\frac{2}{4}$ to obtain a common denominator.

$= 162 + \frac{6}{4}$ Add the fractions by adding the numerators and keeping the common denominator.

$= 162 + \frac{3}{2}$ $\frac{6}{4} = \frac{2 \cdot 3}{2 \cdot 2} = \frac{\overset{1}{\cancel{2}} \cdot 3}{\underset{1}{\cancel{2}} \cdot 2} = \frac{3}{2}$

$= 162 + 1\frac{1}{2}$ Write $\frac{3}{2}$ as a mixed number.

$= 163\frac{1}{2}$

Figure 1-16

To enclose the area, $163\frac{1}{2}$ meters of fencing will be needed.

COMMENT   Remember to include the proper units in your answer. The Mars Climate Orbiter crashed due to lack of unit communication between the Jet Propulsion Lab and Lockheed/Martin engineers, resulting in a loss of $125 million.

 SELF CHECK 9   Find the length of fencing needed to enclose a rectangular plot that is $85\frac{1}{2}$ feet wide and $140\frac{2}{3}$ feet deep.

5   **Add, subtract, multiply, and divide two or more decimals.**

Rational numbers can always be changed to decimal form. For example, to write $\frac{1}{4}$ and $\frac{5}{22}$ as decimals, we use long division.

$$
\begin{array}{r}
0.25 \\
4\overline{)1.00} \\
\underline{8} \\
20 \\
\underline{20}
\end{array}
\qquad
\begin{array}{r}
0.22727\ldots \\
22\overline{)5.00000} \\
\underline{4\,4} \\
60 \\
\underline{44} \\
160 \\
\underline{154} \\
60 \\
\underline{44} \\
160
\end{array}
$$

The decimal 0.25 is called a **terminating decimal**. The decimal 0.2272727. . . (often written as $0.2\overline{27}$) is called a **repeating decimal**, because it repeats the block of digits 27. Every rational number can be changed into either a terminating or a repeating decimal.

Terminating decimals	Repeating decimals
$\dfrac{1}{2} = 0.5$	$\dfrac{1}{3} = 0.33333\ldots$ or $0.\overline{3}$
$\dfrac{3}{4} = 0.75$	$\dfrac{1}{6} = 0.16666\ldots$ or $0.1\overline{6}$
$\dfrac{5}{8} = 0.625$	$\dfrac{5}{22} = 0.2272727\ldots$ or $0.2\overline{27}$

The decimal 0.5 has one *decimal place*, because it has one digit to the right of the decimal point. The decimal 0.75 has two decimal places, and 0.625 has three.

To *add* or *subtract* decimals, we align their decimal points and then add or subtract.

EXAMPLE 10   Add 25.568 and 2.74 using a vertical format.

Solution   We align the decimal points and add the numbers, column by column,

$$
\begin{array}{r}
25.568 \\
+\ \ \underline{2.74} \\
28.308
\end{array}
$$

 SELF CHECK 10   Subtract 2.74 from 25.568 using a vertical format.

To perform the previous operations with a calculator, we enter these numbers and press these keys:

25.568 $+$ 2.74 $=$ and   25.568 $-$ 2.74 $=$ Using a scientific calculator

25.568 $+$ 2.74 **ENTER**   and   25.568 $-$ 2.74 **ENTER**   Using a graphing calculator

To *multiply* decimals, we multiply the numbers and place the decimal point so that the number of decimal places in the answer is equal to the sum of the decimal places in the factors.

EXAMPLE 11   Multiply: 9.25 by 3.453

Solution   We multiply the numbers and place the decimal point so that the number of decimal places in the answer is equal to the sum of the decimal places in the factors.

$$
\begin{array}{r}
3.453 \quad \text{Here there are three decimal places.} \\
\times\ \ \underline{9.25} \quad \text{Here there are two decimal places.} \\
17265 \\
6906 \\
\underline{31\ 077\ \ \ } \\
31.94025 \quad \text{The product has } 3 + 2 = 5 \text{ decimal places.}
\end{array}
$$

 SELF CHECK 11   Multiply: 2.45 by 9.25

To perform the multiplication of Example 11 with a calculator, we enter these numbers and press these keys:

3.453 $\times$ 9.25 $=$ Using a scientific calculator

3.453 $\times$ 9.25 **ENTER**   Using a graphing calculator

To *divide* decimals, we move the decimal point in the **divisor** to the right to make the divisor a whole number. We then move the decimal point in the **dividend** the same number of places to the right.

EXAMPLE 12   Divide: 30.258 by 1.23

Solution   We will write the division using a long division format in which the divisor is 1.23 and the dividend is 30.258.

$$1.23\overline{)30.258}$$

Move the decimal point in both the divisor and the dividend two places to the right.

We align the decimal point in the quotient with the repositioned decimal point in the dividend and use long division.

$$
\begin{array}{r}
24.6 \\
123\overline{)3025.8} \\
\underline{246} \\
565 \\
\underline{492} \\
73\ 8 \\
\underline{73\ 8}
\end{array}
$$

 **SELF CHECK 12**   Divide 579.36 by 1.2.

To perform the previous division with a calculator, we enter these numbers and press these keys:

30.258 ÷ 1.23 =   Using a scientific calculator

30.258 ÷ 1.23 **ENTER**   Using a graphing calculator

6 Round a decimal to a specified number of places.

We often round long decimals to a specific number of decimal places. For example, the decimal 25.36124 rounded to one place (or to the nearest tenth) is 25.4. Rounded to two places (or to the nearest one-hundredth), the decimal is 25.36.

Throughout this text, we use the following procedures to round decimals.

ROUNDING DECIMALS

1. Determine to how many decimal places you want to round.
2. Look at the first digit to the right of that decimal place.
3. If that digit is 4 or less, drop it and all digits that follow. If it is 5 or greater, add 1 to the digit in the position to which you want to round, and drop all digits that follow.

EXAMPLE 13   Round 2.4863 to two decimal places.

Solution   Since we are to round to two digits, we look at the digit to the right of the 8, which is 6. Since 6 is greater than 5, we add 1 to the 8 and drop all of the digits that follow. The rounded number is 2.49.

 **SELF CHECK 13**   Round 6.5731 to three decimal places.

Everyday connections
2010 Gubernatorial Elections

In the 2010 gubernatorial elections, six of the most closely contested races were in Connecticut, Florida, Illinois, Minnesota, Ohio, and Oregon.*

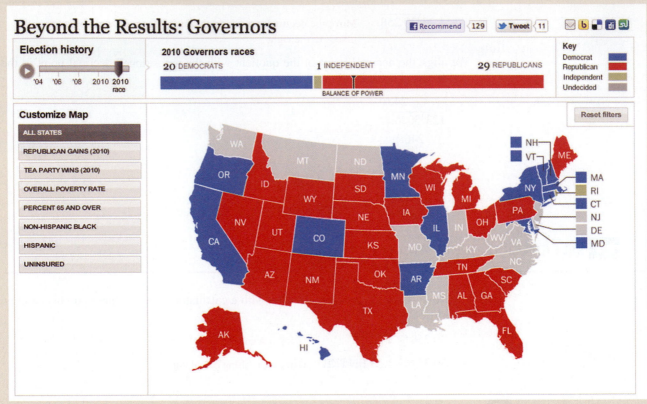

From The Washington Post (11/10/2010). Copyright © 2010 the Washington Post. Used by permission and protected by the Copyright Laws of the United States. www.washingtonpost.com

State	Democrats	Republicans
Connecticut	564,885	557,123
Florida	2,522,857	2,589,915
Illinois	1,721,812	1,702,399
Minnesota	919,231	910,480
Ohio	1,752,507	1,849,609
Oregon	680,840	665,930

1. Find the number of votes cast for Democrats and the number of votes cast for Republicans in the six states.

2. Find the average number of votes cast for Democrats and the average number of votes cast for Republicans, rounded to the nearest vote.

3. Find the difference in the number of Democratic and Republican votes cast in Connecticut, Illinois, and Minnesota.

4. Use the differences found in Question 3 to determine the percent of the total number of votes cast in each of the three states. This percentage is referred to as the margin of victory.

5. Which of the three states in Question 3 had the most closely contested race?

*All of these states had third-, and some fourth-, party candidates whose votes are not considered here.

7   Use the appropriate operation for an application.

A **percent** is the numerator of a fraction with a denominator of 100. For example, $6\frac{1}{4}$ percent, written $6\frac{1}{4}\%$, is the fraction $\frac{6.25}{100}$, or the decimal 0.0625. In problems involving percent, the word *of* usually indicates multiplication. For example, $6\frac{1}{4}\%$ of 8,500 is the product 0.0625(8,500).

EXAMPLE 14   **AUTO LOANS**   Juan signs a one-year note to borrow $8,500 to buy a car. If the rate of interest is $6\frac{1}{4}\%$, how much interest will he pay?

Solution   For the privilege of using the bank's money for one year, Juan must pay $6\frac{1}{4}\%$ of $8,500. We calculate the interest, i, as follows:

$$i = 6\frac{1}{4}\% \text{ of } 8,500$$
$$= 0.0625 \cdot 8,500 \qquad \text{In this case, the word } of \text{ means } times.$$
$$= 531.25$$

Juan will pay $531.25 interest.

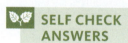 **SELF CHECK 14**   If the rate is 9%, how much interest will he pay?

 SELF CHECK ANSWERS   **1.** $2 \cdot 5 \cdot 7$ **2.** $\frac{2}{5}$ **3.** $\frac{7}{18}$ **4.** no   **5.** $\frac{2}{3}$ **6.** $\frac{19}{24}$ **7.** $\frac{1}{12}$ **8.** $9\frac{17}{21}$ **9.** $452\frac{1}{3}$ ft   **10.** 22.828   **11.** 22.6625   **12.** 482.8   **13.** 6.573   **14.** $765

NOW TRY THIS

Perform each operation.

1. $\frac{7}{3} + \frac{7}{9} - \frac{5}{6} - \frac{41}{18}$

2. $25 - 13.583$

3. Robert's answer to a problem asking to find the length of a piece of lumber is $\frac{5}{2}$ feet. Is this the best form for the answer given the context of the problem? If not, write the answer in the most appropriate form.

4. $\frac{5}{x-3} - \frac{1}{x-3}$ $(x \neq 3)$

1.2   Exercises

WARM-UPS *Find the largest common factor of each pair of numbers.*

1. 3, 6

2. 5, 10

3. 12, 18

4. 15, 27

Perform each operation.

5. $\frac{3}{4} \cdot \frac{1}{2}$

6. $\frac{5}{6} \cdot \frac{5}{7}$

7. $\frac{3}{4} \div \frac{4}{3}$

8. $\frac{3}{5} \div \frac{5}{2}$

9. $\frac{4}{9} + \frac{7}{9}$

10. $\frac{10}{11} - \frac{2}{11}$

11. $\frac{2}{3} - \frac{1}{2}$

12. $\frac{3}{4} + \frac{1}{2}$

13. $5.1 + 0.62$

14. $3.45 - 2.21$

15. $0.2 \cdot 2.5$

16. $0.4 \cdot 16$

Round each decimal to two decimal places.

17. 5.165329

18. 5.164493

REVIEW *Determine whether the following statements are true or false.*

19. 6 is an integer.

20. $\frac{1}{2}$ is a natural number.

21. 21 is a prime number.

22. No prime number is an even number.

23. $-5 > -2$

24. $-3 < -2$

25. $9 \le |-9|$

26. $|-11| \ge 10$

Place an appropriate symbol in each box to make the statement true.

27. $3 + 7$ ▢ 10

28. $\frac{3}{7}$ ▢ $\frac{2}{7} = \frac{1}{7}$

29. $|-2|$ ▢ 2

30. $4 + 8$ ▢ 11

VOCABULARY AND CONCEPTS *Fill in the blanks.*

31. The number above the bar in a fraction is called the _____.

32. The number below the bar in a fraction is called the _____.

33. The fraction $\frac{17}{0}$ is said to be _____.

34. To _____ a fraction, we divide its numerator and denominator by the same number.

35. To write a number in prime-factored form, we write it as the product of _____ numbers.

36. If the numerator of a fraction is less than the denominator, the fraction is called a _____ fraction.

37. If the numerator of a fraction is greater than the denominator, the fraction is called an _____ fraction.

38. A fraction is written in _____ or simplest form when its numerator and denominator have no common factors.

39. If the product of two numbers is _, the numbers are called reciprocals.

40. $\frac{ax}{bx} = $ _____

41. To multiply two fractions, _____ the numerators and multiply the denominators.

42. To divide two fractions, multiply the first fraction by the _____ of the second fraction.

43. To add fractions with the same denominator, add the _____ and keep the common _____.

44. To subtract fractions with the same _____, subtract the numerators and keep the common denominator.

45. To add fractions with unlike denominators, first find the _____ and write each fraction as an _____ fraction.

46. $75\frac{2}{3}$ means 75 ___ $\frac{2}{3}$. The number $75\frac{2}{3}$ is called a _____ number.

47. 0.75 is an example of a _____ decimal and it has _ decimal places.

48. $5.3\overline{27}$ is an example of a _____ decimal.

49. In the figure $2\overline{)6}^{\,3}$, 2 represents the _____, 6 represents the _____, and 3 represents the _____.

50. A _____ is the numerator of a fraction whose denominator is 100.

GUIDED PRACTICE *Write each number in prime-factored form.* SEE EXAMPLE 1. (OBJECTIVE 1)

51. 24

52. 105

53. 48

54. 315

Write each fraction in lowest terms. If the fraction is already in lowest terms, so indicate. SEE EXAMPLE 2. (OBJECTIVE 1)

55. $\frac{6}{12}$

56. $\frac{3}{9}$

57. $\frac{15}{20}$

58. $\frac{33}{55}$

59. $\frac{27}{18}$

60. $\frac{35}{14}$

61. $\frac{72}{64}$

62. $\frac{26}{21}$

Perform each multiplication. Simplify each result when possible. SEE EXAMPLE 3. (OBJECTIVE 2)

63. $\frac{1}{3} \cdot \frac{2}{5}$

64. $\frac{3}{4} \cdot \frac{5}{7}$

65. $\frac{4}{3} \cdot \frac{6}{5}$

66. $\frac{7}{8} \cdot \frac{6}{15}$

67. $12 \cdot \frac{5}{6}$

68. $10 \cdot \frac{5}{12}$

69. $\frac{10}{21} \cdot 14$

70. $\frac{5}{24} \cdot 16$

Perform each division. Simplify each result when possible. SEE EXAMPLE 5. (OBJECTIVE 2)

71. $\frac{2}{5} \div \frac{3}{2}$

72. $\frac{4}{5} \div \frac{3}{7}$

73. $\frac{3}{4} \div \frac{6}{5}$

74. $\frac{3}{8} \div \frac{15}{28}$

75. $9 \div \frac{3}{8}$

76. $23 \div \frac{46}{5}$

77. $\frac{54}{20} \div 3$

78. $\frac{39}{27} \div 13$

Perform each operation. Simplify each result when possible.
SEE EXAMPLES 6–7. (OBJECTIVE 3)

79. $\dfrac{3}{5} + \dfrac{3}{5}$ **80.** $\dfrac{4}{7} - \dfrac{2}{7}$

81. $\dfrac{5}{17} - \dfrac{3}{17}$ **82.** $\dfrac{2}{11} + \dfrac{9}{11}$

83. $\dfrac{1}{42} + \dfrac{1}{6}$ **84.** $\dfrac{17}{25} - \dfrac{2}{5}$

85. $\dfrac{7}{10} - \dfrac{1}{14}$ **86.** $\dfrac{8}{25} + \dfrac{1}{10}$

Perform each operation. Simplify each result when possible.
SEE EXAMPLE 8. (OBJECTIVE 4)

87. $4\dfrac{3}{5} + \dfrac{3}{5}$ **88.** $2\dfrac{1}{8} + \dfrac{3}{8}$

89. $3\dfrac{1}{3} - 1\dfrac{2}{3}$ **90.** $6\dfrac{1}{5} - 4\dfrac{2}{5}$

91. $3\dfrac{3}{4} - 2\dfrac{1}{2}$ **92.** $15\dfrac{5}{6} + 11\dfrac{5}{8}$

93. $8\dfrac{2}{9} - 7\dfrac{2}{3}$ **94.** $3\dfrac{4}{5} - 3\dfrac{1}{10}$

Change each fraction to decimal form and determine whether the decimal is a terminating or repeating decimal. **(OBJECTIVE 5)**

95. $\dfrac{3}{5}$ **96.** $\dfrac{5}{9}$

97. $\dfrac{9}{22}$ **98.** $\dfrac{8}{5}$

Perform each operation. **SEE EXAMPLES 10–12. (OBJECTIVE 5)**

99. $43.54 + 315.7$ **100.** $345.213 - 27.35$

101. $67.235 - 22.45$ **102.** $21.36 + 4.573$

103. $7.2 \cdot 15.6$ **104.** $4.21 \cdot 2.73$

105. $0.23 \overline{)1.0465}$ **106.** $4.7 \overline{)10.857}$

Round each of the following to two decimal places and then to three decimal places. **SEE EXAMPLE 13. (OBJECTIVE 6)**

107. 496.2583 **108.** 13.0547

109. $6,025.3982$ **110.** 1.6048

ADDITIONAL PRACTICE *Perform each operation.*

111. $\dfrac{5}{12} \cdot \dfrac{18}{5}$ **112.** $\dfrac{5}{4} \cdot \dfrac{12}{10}$

113. $\dfrac{17}{34} \cdot \dfrac{3}{6}$ **114.** $\dfrac{21}{14} \cdot \dfrac{3}{6}$

115. $\dfrac{2}{13} \div \dfrac{8}{13}$ **116.** $\dfrac{4}{7} \div \dfrac{20}{21}$

117. $\dfrac{21}{35} \div \dfrac{3}{14}$ **118.** $\dfrac{23}{25} \div \dfrac{46}{5}$

119. $\dfrac{3}{5} + \dfrac{2}{3}$ **120.** $\dfrac{4}{3} + \dfrac{7}{2}$

121. $\dfrac{9}{4} - \dfrac{5}{6}$ **122.** $\dfrac{2}{15} + \dfrac{7}{9}$

123. $3 - \dfrac{3}{4}$ **124.** $5 + \dfrac{21}{5}$

125. $\dfrac{17}{3} + 4$ **126.** $\dfrac{13}{9} - 1$

Use a calculator to perform each operation and round each answer to two decimal places.

127. $474.81 + 23.4532$

128. $843.45213 - 712.765$

129. $25.25 \cdot 132.179$

130. $234.874 \cdot 242.46473$

131. $0.456 \overline{)4.5694323}$

132. $43.225 \overline{)32.465748}$

133. $55.77443 - 0.568245$

134. $0.62317 + 1.3316$

APPLICATIONS SEE EXAMPLES 4, 9, AND 14. (OBJECTIVE 7)

135. Spring plowing A farmer has plowed $12\dfrac{1}{3}$ acres of a $43\dfrac{1}{2}$-acre field. How much more needs to be plowed?

136. Fencing a garden The four sides of a garden measure $7\dfrac{2}{3}$ feet, $15\dfrac{1}{4}$ feet, $19\dfrac{1}{2}$ feet, and $10\dfrac{3}{4}$ feet. Find the length of the fence needed to enclose the garden.

137. Making clothes A designer needs $4\dfrac{1}{3}$ yards of material for each dress he makes. How much material will he need to make 15 dresses?

138. Track and field Each lap around a stadium track is $\dfrac{1}{4}$ mile. How many laps would a runner have to complete to run 26 miles?

139. Disaster relief After hurricane damage estimated at \$187.75 million, a county sought relief from three agencies. Local agencies gave \$46.8 million and state agencies gave \$72.5 million. How much must the federal government contribute to make up the difference?

140. Minority population 26.5% of the 12,419,000 citizens of Illinois are nonwhite. How many are nonwhite?

The following circle graph shows the various sources of retirement income for a typical retired person. Use this information in Exercises 141–142.

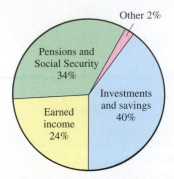

141. **Retirement income**   If a retiree has $36,000 of income, how much is expected to come from pensions and Social Security?

142. **Retirement income**   If a retiree has $52,000 of income, how much is expected to come from earned income?

143. **Quality control**   In the manufacture of active-matrix color LCD computer displays, many units must be rejected as defective. If 23% of a production run of 17,500 units is defective, how many units are acceptable?

144. **Freeze-drying**   Almost all of the water must be removed when food is preserved by freeze-drying. Find the weight of the water removed from 750 pounds of a food that is 36% water.

145. **Planning for growth**   This year, sales at Positronics Corporation totaled $18.7 million. If the projection of 12% annual growth is true, what will be next year's sales?

146. **Speed skating**   In tryouts for the Olympics, a speed skater had times of 44.47, 43.24, 42.77, and 42.05 seconds. Find the average time. Give the result to the nearest hundredth. (*Hint:* Add the numbers and divide by 4.)

147. **Cost of gasoline**   Samuel drove his car 16,275.3 miles last year, averaging 25.5 miles per gallon of gasoline. If the average cost of gasoline was $3.45 per gallon, find the fuel cost to drive the car.

148. **Paying taxes**   A woman earns $48,712.32 in taxable income. She must pay 15% tax on the first $23,000 and 28% on the rest. In addition, she must pay a Social Security tax of 15.4% on the total amount. How much tax will she need to pay?

149. **Sealing asphalt**   A rectangular parking lot is 253.5 feet long and 178.5 feet wide. A 55-gallon drum of asphalt sealer covers 4,000 square feet and costs $97.50. Find the cost to seal the parking lot. (Sealer can be purchased only in full drums.)

150. **Inventory costs**   Each TV a retailer buys costs $3.25 per day for warehouse storage. What does it cost to store 37 TVs for three weeks?

151. **Manufacturing profits**   A manufacturer of computer memory boards has a profit of $37.50 on each standard-capacity memory board, and $57.35 on each high-capacity board. The sales department has orders for 2,530 standard boards and 1,670 high-capacity boards. Which order will produce the greater profit?

152. **Dairy production**   A Holstein cow will produce 7,600 pounds of milk each year, with a $3\frac{1}{2}$% butterfat content. Each year, a Guernsey cow will produce about 6,500 pounds of milk that is 5% butterfat. Which cow produces more butterfat?

153. **Feeding dairy cows**   Each year, a typical dairy cow will eat 12,000 pounds of food that is 57% silage. To feed 30 cows, how much silage will a farmer use in a year?

154. **Comparing bids**   Two carpenters bid on a home remodeling project. The first bids $9,350 for the entire job. The second will work for $27.50 per hour, plus $4,500 for materials. He estimates that the job will take 150 hours. Which carpenter has the lower bid?

155. **Choosing a furnace**   A high-efficiency home heating system can be installed for $4,170, with an average monthly heating bill of $57.50. A regular furnace can be installed for $1,730, but monthly heating bills average $107.75. After three years, which system has cost more altogether?

156. **Choosing a furnace**   Refer to Exercise 155. Decide which furnace system will have cost more after five years.

WRITING ABOUT MATH

157. Describe how you would find the common denominator of two fractions.

158. Explain how to convert an improper fraction into a mixed number.

159. Explain how to convert a mixed number into an improper fraction.

160. Explain how you would decide which of two decimal fractions is the larger.

SOMETHING TO THINK ABOUT

161. In what situations would it be better to leave an answer in the form of an improper fraction?

162. When would it be better to change an improper-fraction answer into a mixed number?

163. Can the product of two proper fractions be larger than either of the fractions?

164. How does the product of one proper and one improper fraction compare with the two factors?

Section 1.3

Exponents and Order of Operations

Objectives

1 Identify the base and the exponent to simplify an exponential expression.
2 Evaluate a numeric expression following the order of operations.
3 Use the correct geometric formula for an application.

Vocabulary

base	perimeter	radius
exponent	area	volume
exponential expression	circumference	linear units
power of x	diameter	square units
grouping symbol	center	cubic units

Getting Ready

Perform each operation.

1. $2 \cdot 2$ **2.** $3 \cdot 3$ **3.** $3 \cdot 3 \cdot 3$ **4.** $2 \cdot 2 \cdot 2$

5. $\dfrac{1}{2} \cdot \dfrac{1}{2}$ **6.** $\dfrac{1}{3} \cdot \dfrac{1}{3} \cdot \dfrac{1}{3}$ **7.** $\dfrac{2}{5} \cdot \dfrac{2}{5} \cdot \dfrac{2}{5}$ **8.** $\dfrac{3}{10} \cdot \dfrac{3}{10} \cdot \dfrac{3}{10}$

In algebra we encounter many expressions that contain exponents, a shortcut method of showing repeated multiplication. In this section, we will introduce exponential notation and discuss the rules for the order of operations.

1 **Identify the base and the exponent to simplify an exponential expression.**

To show how many times a number is to be used as a factor in a product, we use exponents. In the expression 2^3, 2 is called the **base** and 3 is called the **exponent**.

$$\text{Base} \rightarrow 2^3 \leftarrow \text{Exponent}$$

The exponent of 3 indicates that the base of 2 is to be used as a factor three times:

COMMENT Note that $2^3 = 8$. This is not the same as $2 \cdot 3 = 6$.

$$\overset{\text{3 factors of 2}}{\overbrace{2^3 = 2 \cdot 2 \cdot 2}} = 8$$

In the expression x^5 (called an **exponential expression** or a **power of x**), x is the base and 5 is the exponent. The exponent of 5 indicates that a base of x is to be used as a factor five times.

$$\overset{\text{5 factors of } x}{\overbrace{x^5 = x \cdot x \cdot x \cdot x \cdot x}}$$

In expressions such as 7, x, or y, the exponent is understood to be 1:

$$7 = 7^1 \qquad x = x^1 \qquad y = y^1$$

In general, we have the following definition.

NATURAL-NUMBER EXPONENTS	If n is a natural number, then

$$\overset{n \text{ factors of } x}{\overbrace{x^n = x \cdot x \cdot x \cdots \cdots x}}$$

EXAMPLE 1   Write each expression without exponents.

a. $4^2 = 4 \cdot 4 = 16$ Read 4^2 as "4 squared" or as "4 to the second power."

b. $5^3 = 5 \cdot 5 \cdot 5 = 125$ Read 5^3 as "5 cubed" or as "5 to the third power."

c. $6^4 = 6 \cdot 6 \cdot 6 \cdot 6 = 1{,}296$ Read 6^4 as "6 to the fourth power."

d. $\left(\dfrac{2}{3}\right)^5 = \dfrac{2}{3} \cdot \dfrac{2}{3} \cdot \dfrac{2}{3} \cdot \dfrac{2}{3} \cdot \dfrac{2}{3} = \dfrac{32}{243}$ Read $\left(\frac{2}{3}\right)^5$ as "$\frac{2}{3}$ to the fifth power."

 SELF CHECK 1   Write each expression without exponents: **a.** 7^2 **b.** $\left(\frac{3}{4}\right)^3$

We can find powers using a calculator. For example, to find 2.35^4, we enter these numbers and press these keys:

2.35 $\boxed{y^x}$ 4 $\boxed{=}$ Using a scientific calculator

2.35 $\boxed{\wedge}$ 4 **ENTER**   Using a graphing calculator

Either way, the display will read 30.49800625. Some scientific calculators have an $\boxed{x^y}$ key rather than a $\boxed{y^x}$ key.

In the next example, the base of an exponential expression is a variable.

EXAMPLE 2   Write each expression without exponents.

a. $y^6 = y \cdot y \cdot y \cdot y \cdot y \cdot y$ Read y^6 as "y to the sixth power."

b. $x^3 = x \cdot x \cdot x$ Read x^3 as "x cubed" or as "x to the third power."

c. $z^2 = z \cdot z$ Read z^2 as "z squared" or as "z to the second power."

d. $a^1 = a$ Read a^1 as "a to the first power."

e. $2(3x)^2 = 2(3x)(3x)$ Read $2(3x)^2$ as "2 times $(3x)$ to the second power."

 SELF CHECK 2   Write each expression without exponents.   **a.** a^3 **b.** b^4

2   **Evaluate a numeric expression following the order of operations.**

Suppose you are asked to contact a friend if you see a Rolex watch for sale while traveling in Switzerland. After locating the watch, you send the following message to your friend.

You receive this response.

The first statement says to buy the watch at any price. The second says not to buy it, because it is too expensive. The placement of the exclamation point makes these statements read differently, resulting in different interpretations.

When reading a mathematical statement, the same kind of confusion is possible. To illustrate, we consider the expression $2 + 3 \cdot 4$, which contains the operations of addition and multiplication. We can calculate this expression in two different ways. We can perform the multiplication first and then perform the addition. Or we can perform the addition first and then perform the multiplication. However, we will get different results.

Multiply first		*Add first*	
$2 + 3 \cdot 4 = 2 + 12$	Multiply 3 and 4.	$2 + 3 \cdot 4 = 5 \cdot 4$	Add 2 and 3.
$= 14$	Add 2 and 12.	$= 20$	Multiply 5 and 4.

Different results

To eliminate the possibility of getting different answers, we will agree to perform multiplications before additions. The correct calculation of $2 + 3 \cdot 4$ is

$2 + 3 \cdot 4 = 2 + 12$ Do the multiplication first.
$ = 14$

To indicate that additions are to be done before multiplications, we use **grouping symbols** such as parentheses ( ), brackets [ ], or braces { }. The operational symbols $\sqrt{}$, $|\,|$, and fraction bars are also grouping symbols. In the expression $(2 + 3)4$, the parentheses indicate that the addition is to be done first:

$(2 + 3)4 = 5 \cdot 4$ Do the addition within the parentheses first.
$ = 20$

To guarantee that calculations will have one correct result, we will always perform calculations in the following order.

RULES FOR THE ORDER OF OPERATIONS	Use the following steps to perform all calculations within each pair of grouping symbols, working from the innermost pair to the outermost pair.

1. Find the values of any exponential expressions.
2. Perform all multiplications and divisions, working from left to right.
3. Perform all additions and subtractions, working from left to right.
4. Because a fraction bar is a grouping symbol, simplify the numerator and the denominator in a fraction separately. Then simplify the fraction, whenever possible.

COMMENT   Note that $4(2)^3 \neq (4 \cdot 2)^3$:

$$4(2)^3 = 4 \cdot 2 \cdot 2 \cdot 2 = 4(8) = 32 \quad \text{and} \quad (4 \cdot 2)^3 = 8^3 = 8 \cdot 8 \cdot 8 = 512$$

Likewise, $4x^3 \neq (4x)^3$ because

$$4x^3 = 4xxx \quad \text{and} \quad (4x)^3 = (4x)(4x)(4x) = 64xxx$$

EXAMPLE 3   Evaluate: $5^3 + 2(8 - 3 \cdot 2)$

Solution   We perform the work within the parentheses first and then simplify.

$5^3 + 2(8 - 3 \cdot 2) = 5^3 + 2(8 - 6)$	Do the multiplication within the parentheses.
$= 5^3 + 2(2)$	Do the subtraction within the parentheses.
$= 125 + 2(2)$	Find the value of the exponential expression.

$$= 125 + 4 \qquad \text{\color{red}Do the multiplication.}$$
$$= 129 \qquad \text{\color{red}Do the addition.}$$

 SELF CHECK 3   Evaluate: $5 + 4 \cdot 3^2$

EXAMPLE 4   Evaluate: $\dfrac{3(3 + 2) + 5}{17 - 3(4)}$

Solution   We simplify the numerator and denominator separately and then simplify the fraction.

$$\frac{3(\mathbf{3 + 2}) + 5}{17 - 3(4)} = \frac{3(\mathbf{5}) + 5}{17 - 3(4)} \qquad \text{\color{red}Do the addition within the parentheses.}$$

$$= \frac{15 + 5}{17 - 12} \qquad \text{\color{red}Do the multiplications.}$$

$$= \frac{20}{5} \qquad \text{\color{red}Do the addition and the subtraction.}$$

$$= 4 \qquad \text{\color{red}Do the division.}$$

 SELF CHECK 4   Evaluate: $\dfrac{4 + 2(5 - 3)}{2 + 3(2)}$

EXAMPLE 5   Evaluate: $\dfrac{3(4^2) - 2(3)}{2(4 + 3)}$

Solution $\dfrac{3(4^2) - 2(3)}{2(4 + 3)} = \dfrac{3(16) - 2(3)}{2(7)} \qquad$ {\color{red}Find the value of 4^2 in the numerator and do the addition in the denominator.}

$$= \frac{48 - 6}{14} \qquad \text{\color{red}Do the multiplications.}$$

$$= \frac{42}{14} \qquad \text{\color{red}Do the subtraction.}$$

$$= 3 \qquad \text{\color{red}Do the division.}$$

 SELF CHECK 5   Evaluate: $\dfrac{2^2 + 6(5)}{2(2 + 5) + 3}$

3   ## Use the correct geometric formula for an application.

To find perimeters and areas of geometric figures, we often must substitute numbers for variables in a formula. The **perimeter** of a geometric figure is the distance around it, and the **area** of a geometric figure is the amount of surface that it encloses. The perimeter of a circle is called its **circumference**.

EXAMPLE 6   **CIRCLES**   Use the information in Figure 1-17 to find:
    a. the circumference   **b.** the area of the circle.

Solution   **a.** The formula for the circumference of a circle is

$$C = \pi D$$

where C is the circumference, π can be approximated by $\frac{22}{7}$, and D is the **diameter**—a line segment that passes through the center of the circle and joins two points on the

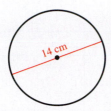

Figure 1-17

circle. We can approximate the circumference by substituting $\frac{22}{7}$ for π and 14 for D in the formula and simplifying.

$$C = \pi D$$

$$C \approx \frac{22}{7} \cdot 14 \qquad \text{Read} \approx \text{as "is approximately equal to."}$$

$$C \approx \frac{22 \cdot \overset{2}{\cancel{14}}}{\underset{1}{\cancel{7}} \cdot 1} \qquad \text{Multiply the fractions and simplify.}$$

$$C \approx 44$$

The circumference is approximately 44 centimeters. To use a calculator, we enter these numbers and press these keys:

$\boxed{\pi}\ \boxed{\times}\ 14\ \boxed{=}$ Using a scientific calculator

$\boxed{\text{2nd}}\ \boxed{\wedge}\ \boxed{(\pi)}\ \boxed{\times}\ 14\ \boxed{\textbf{ENTER}}$ Using a graphing calculator

Either way, the display will read 43.98229715. The result is not 44, because a calculator uses a better approximation for π than $\frac{22}{7}$.

COMMENT   A segment drawn from the **center** of a circle to a point on the circle is called a **radius**. Since the diameter D of a circle is twice as long as its radius r, we have $D = 2r$. If we substitute $2r$ for D in the formula $C = \pi D$, we obtain an alternate formula for the circumference of a circle: $C = 2\pi r$.

b. The formula for the area of a circle is

$$A = \pi r^2$$

where A is the area, $\pi \approx \frac{22}{7}$, and r is the radius of the circle. We can approximate the area by substituting $\frac{22}{7}$ for π and 7 for r in the formula and simplifying.

$$A = \pi r^2$$

$$A \approx \frac{22}{7} \cdot 7^2$$

$$A \approx \frac{22}{7} \cdot \frac{49}{1} \qquad \text{Evaluate the exponential expression.}$$

$$A \approx \frac{22 \cdot \overset{7}{\cancel{49}}}{\underset{1}{\cancel{7}} \cdot 1} \qquad \text{Multiply the fractions and simplify.}$$

$$A \approx 154$$

The area is approximately 154 square centimeters.

To use a calculator, we enter these numbers and press these keys:

$\boxed{\pi}\ \boxed{\times}\ 7\ \boxed{x^2}\ \boxed{=}$ Using a scientific calculator

$\boxed{\text{2nd}}\ \boxed{\wedge}\ \boxed{(\pi)}\ \boxed{\times}\ 7\ \boxed{x^2}\ \boxed{\textbf{ENTER}}$ Using a graphing calculator

The display will read 153.93804.

 **SELF CHECK 6**   Given a circle with a diameter of 28 meters, find an estimate to the nearest whole number of

a. the circumference   **b.** the area.

(Use $\frac{22}{7}$ to estimate π.) Check your results with a calculator.

Table 1-1 shows the formulas for the perimeter and area of several geometric figures.

Euclid

325–265 BC

Although Euclid is best known for his study of geometry, many of his writings deal with number theory. In about 300 BC, the Greek mathematician Euclid proved that the number of prime numbers is unlimited—that there are infinitely many prime numbers. This is an important branch of mathematics called number theory.

TABLE 1-1

Figure	Name	Perimeter	Area
	Square	$P = 4s$	$A = s^2$
	Rectangle	$P = 2l + 2w$	$A = lw$
	Triangle	$P = a + b + c$	$A = \frac{1}{2}bh$
	Trapezoid	$P = a + b + c + d$	$A = \frac{1}{2}h(b + d)$
	Circle	$C = \pi D = 2\pi r$ $(D = 2r)$	$A = \pi r^2$

The **volume** of a three-dimensional geometric solid is the amount of space it encloses. Table 1-2 shows the formulas for the volume of several solids.

TABLE 1-2

Figure	Name	Volume
	Rectangular solid	$V = lwh$
	Cylinder	$V = Bh$, where B is the area of the base
	Pyramid	$V = \frac{1}{3}Bh$, where B is the area of the base
	Cone	$V = \frac{1}{3}Bh$, where B is the area of the base
	Sphere	$V = \frac{4}{3}\pi r^3$

When working with geometric figures, measurements are often given in **linear units** such as feet (ft), centimeters (cm), or meters (m). If the dimensions of a two-dimensional geometric figure are given in feet, we can calculate its perimeter by finding the sum of the lengths of its sides. This sum will be in feet.

If we calculate the area of a two-dimensional figure, the result will be in **square units**. For example, if we calculate the area of the figure whose sides are measured in centimeters, the result will be in square centimeters (cm²).

If we calculate the volume of a three-dimensional figure, the result will be in **cubic units**. For example, the volume of a three-dimensional geometric figure whose sides are measured in meters will be in cubic meters (m³).

EXAMPLE 7   **WINTER DRIVING**   Find the number of cubic feet of road salt in the conical pile shown in Figure 1-18. Round the answer to two decimal places.

18.75 ft

14.3 ft

Figure 1-18

Solution   We can find the area of the circular base by substituting $\frac{22}{7}$ for π and 14.3 for the radius.

$$A = \pi r^2$$

$$\approx \frac{22}{7}(14.3)^2$$

$$\approx 642.6828571 \qquad \text{Use a calculator.}$$

We then substitute 642.6828571 for B and 18.75 for h in the formula for the volume of a cone.

$$V = \frac{1}{3}Bh$$

$$\approx \frac{1}{3}(642.6828571)(18.75)$$

$$\approx 4{,}016.767857 \qquad \text{Use a calculator.}$$

To two decimal places, there are 4,016.77 cubic feet of salt in the pile.

   **SELF CHECK 7**   To the nearest hundredth, find the number of cubic feet of water that can be contained in a spherical tank that has a radius of 9 feet. (Use $\pi \approx \frac{22}{7}$.)

   **SELF CHECK ANSWERS**   **1. a.** 49 **b.** $\frac{27}{64}$ **2. a.** $a \cdot a \cdot a$ **b.** $b \cdot b \cdot b \cdot b \cdot b$ **3.** 41 **4.** 1 **5.** 2 **6. a.** 88 m **b.** 616 m²
7. 3,054.86 ft²

NOW TRY THIS

Simplify each expression.

1. $28 - 7(4 - 1)$

2. $\dfrac{5 - |4 - 1|}{2}$

3. Insert the appropriate operations and one set of parentheses (if necessary) so that the expression yields the given value.

  a. $16 \quad 3 \quad 5 = 2$

  b. $4 \quad 2 \quad 6 = 12$

1.3 Exercises

WARM-UPS *Perform each operation.*

1. $2 \cdot 2 \cdot 2 \cdot 2 \cdot 2$

2. $3 \cdot 3 \cdot 3 \cdot 3$

3. $4 \cdot 4 \cdot 4$

4. $5 \cdot 5 \cdot 5$

5. $\dfrac{2}{3} \cdot \dfrac{2}{3} \cdot \dfrac{2}{3}$

6. $\dfrac{4}{5} \cdot \dfrac{4}{5}$

Identify the base in each expression.

7. y^3

8. $(2x)^4$

9. $7(4x)^2$

10. $3y^2$

REVIEW

11. On the number line, graph the prime numbers between 10 and 20.

12. Write the inequality $7 \leq 12$ as an inequality using the symbol $\geq$.

13. Classify the number 17 as a prime number or a composite number.

14. Evaluate: $\dfrac{3}{5} - \dfrac{1}{2}$

VOCABULARY AND CONCEPTS *Fill in the blanks.*

15. An _____ indicates how many times a base is to be used as a factor in a product.

16. In the exponential expression (power of x) x^7, x is called the ____ and 7 is called an _____.

17. Parentheses, brackets, and braces are called _____ symbols.

18. A line segment that passes through the center of a circle and joins two points on the circle is called a _____. A line segment drawn from the center of a circle to a point on the circle is called a _____.

19. The distance around a rectangle is called the _____, and the distance around a circle is called the _____.

20. The region enclosed by a two-dimensional geometric figure is called the ____ and is designated by _____ units, and the region enclosed by a three-dimensional geometric figure is called the _____ and is designated by ____ units.

Write the appropriate formula to find each quantity and state the correct units.

21. The perimeter of a square _____; ____

22. The area of a square _____; _____

23. The perimeter of a rectangle _____; ____

24. The area of a rectangle _____; _____

25. The perimeter of a triangle _____; ____

26. The area of a triangle _____; _____

27. The perimeter of a trapezoid _____; ____

28. The area of a trapezoid _____; _____

29. The circumference of a circle _____; ____

30. The area of a circle _____; _____

31. The volume of a rectangular solid _____; _____

32. The volume of a cylinder _____; _____

33. The volume of a pyramid _____; _____

34. The volume of a cone _____; _____

35. The volume of a sphere _____; _____

36. In Exercises 32–34, B is the ____ of the base.

GUIDED PRACTICE *Write each expression without using exponents and find the value of each expression.* SEE EXAMPLE 1. (OBJECTIVE 1)

37. 6^2

38. 9^2

39. $\left(-\dfrac{1}{5}\right)^4$

40. $\left(\dfrac{1}{2}\right)^6$

Write each expression without using exponents. SEE EXAMPLE 2. (OBJECTIVE 1)

41. x^3

42. y^4

43. $8z^4$

44. $5t^2$

45. $(4x)^3$

46. $(3z)^4$

47. $3(6y)^2$

48. $2(4t)^3$

Find the value of each expression. SEE EXAMPLES 3–5. (OBJECTIVE 2)

49. $4(3^2)$　　　　　　**50.** $4(2^3)$

51. $(2 \cdot 5)^4$　　　　　**52.** $(2 \cdot 2)^3$

53. $5(4)^2$　　　　　　**54.** $4(5)^2$

55. $(3 \cdot 2)^3$　　　　　**56.** $(2 \cdot 3)^2$

57. $3 \cdot 5 - 4$　　　　　**58.** $3 + 6 \cdot 4$

59. $3(5 - 4)$　　　　　**60.** $3(5 + 8)$

61. $2 + 3 \cdot 5 - 4$　　　**62.** $10 + 2 \cdot 4 + 3$

63. $48 \div (4 + 2)$　　　**64.** $16 \div (5 + 3)$

65. $3^2 + 2(1 + 4) - 2$　**66.** $4 \cdot 3 + 2(5 - 2) - 2^3$

67. $\dfrac{3}{5} \cdot \dfrac{10}{3} + \dfrac{1}{2} \cdot 12$　　**68.** $\dfrac{15}{4}\left(1 + \dfrac{3}{5}\right)$

69. $\left[\dfrac{1}{3} - \left(\dfrac{1}{2}\right)^2\right]^2$　　**70.** $\left[\left(\dfrac{2}{3}\right)^2 - \dfrac{1}{3}\right]^2$

71. $\dfrac{(3 + 5)^2 + 2}{2(8 - 5)}$　　**72.** $\dfrac{25 - (2 \cdot 3 - 1)}{2 \cdot 9 - 8}$

73. $\dfrac{(5 - 3)^2 + 2}{4^2 - (8 + 2)}$　　**74.** $\dfrac{(4^2 - 2) + 7}{5(2 + 4) - 3^2}$

75. $\dfrac{3 \cdot 7 - 5(3 \cdot 4 - 11)}{4(3 + 2) - 3^2 + 5}$　　**76.** $\dfrac{2 \cdot 5^2 - 2^2 + 3}{2(5 - 2)^2 - 11}$

Find the perimeter of each figure. (OBJECTIVE 3)

77.

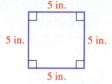

5 in.
5 in.　5 in.
5 in.

78.

10 cm
3 cm　3 cm
10 cm

79.

3 m　5 m
7 m

80.

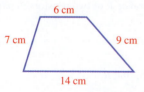

6 cm
7 cm　9 cm
14 cm

Find the area of each figure. (OBJECTIVE 3)

81.

6 m
6 m

82.

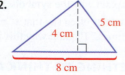

5 cm
4 cm
8 cm

83.

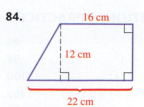

5 ft
11 ft

84.

16 cm
12 cm
22 cm

Find the circumference of each circle. Use $\pi \approx \frac{22}{7}$. SEE EXAMPLE 6. (OBJECTIVE 3)

85.

14 m

86.

21 cm

Find the area of each circle. Use $\pi \approx \frac{22}{7}$. SEE EXAMPLE 6. (OBJECTIVE 3)

87.

42 ft

88.

7 m

Find the volume of each solid. Use $\pi \approx \frac{22}{7}$ *where applicable.* SEE EXAMPLE 7. (OBJECTIVE 3)

89.

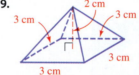

2 cm
3 cm　3 cm
3 cm
3 cm

90.

6 ft
2 ft
3 ft

91.

6 m

92.

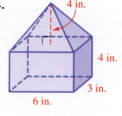

14 in.
12 in.

93.

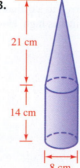

21 cm
14 cm
8 cm

94.

4 in.
4 in.
3 in.
6 in.

ADDITIONAL PRACTICE *Simplify each expression.*

95. 6^2

96. 7^3

97. $2 + 4^2$

98. $4^2 - 2^2$

99. $(2 + 4)^2$

100. $(7 - 3)^3$

101. $(7 + 9) \div (2 \cdot 4)$

102. $(7 + 9) \div 2 \cdot 4$

103. $(5 + 7) \div 3 \cdot 4$

104. $(5 + 7) \div (3 \cdot 4)$

105. $24 \div 4 \cdot 3 + 3$

106. $36 \div 9 \cdot 4 - 2$

107. $6^2 - (8 - 5)^2$

108. $3^3 + (3 - 1)^3$

109. $(2 \cdot 3 - 4)^3$

110. $(3 \cdot 5 - 2 \cdot 6)^2$

111. $\dfrac{2[4 + 2(3 - 1)]}{3[3(2 \cdot 3 - 4)]}$

112. $\dfrac{3[9 - 2(7 - 3)]}{(8 - 5)(9 - 7)}$

Use a calculator to find each power.

113. 7.9^3

114. 0.45^4

115. 25.3^2

116. 7.567^3

Insert parentheses in the expression $3 \cdot 8 + 5 \cdot 3$ *to make its value equal to the given number.*

117. 39

118. 117

119. 87

120. 69

APPLICATIONS *Use a calculator. For* π, *use the* π *key. Round to two decimal places.* SEE EXAMPLE 7. (OBJECTIVE 3)

121. Volume of a tank Find the number of cubic feet of water in the spherical tank at the top of the water tower.

21.35 ft

122. Storing solvents A hazardous solvent fills a rectangular tank with dimensions of 12 inches by 9.5 inches by 7.3 inches. For disposal, it must be transferred to a cylindrical canister 7.5 inches in diameter and 18 inches high. How much solvent will be left over?

123. Buying fencing How many meters of fencing are needed to enclose the square pasture shown in the illustration?

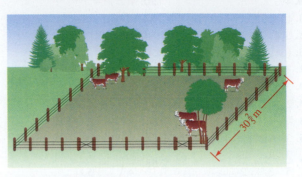

$30\frac{2}{3}$ m

124. Installing carpet What will it cost to carpet the area shown in the illustration with carpet that costs $29.79 per square yard? (One square yard is 9 square feet.)

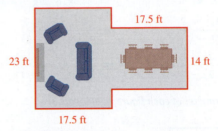

17.5 ft

23 ft

14 ft

17.5 ft

125. Volume of a classroom Thirty students are in a classroom with dimensions of 40 feet by 40 feet by 9 feet. How many cubic feet of air are there for each student?

126. Wallpapering One roll of wallpaper covers about 33 square feet. At $27.50 per roll, how much would it cost to paper two walls 8.5 feet high and 17.3 feet long? (*Hint:* Wallpaper can be purchased only in full rolls.)

127. Focal length The focal length f of a double-convex thin lens is given by the formula

$$f = \frac{rs}{(r + s)(n - 1)}$$

If $r = 8$, $s = 12$, and $n = 1.6$, find f.

128. Resistance The total resistance R of two resistors in parallel is given by the formula

$$R = \frac{rs}{r + s}$$

If $r = 170$ and $s = 255$, find R.

WRITING ABOUT MATH

129. Explain why the symbols $3x$ and x^3 have different meanings.

130. Students often say that x^n means "x multiplied by itself n times." Explain why this is not correct.

SOMETHING TO THINK ABOUT

131. If x were greater than 1, would raising x to higher and higher powers produce bigger numbers or smaller numbers?

132. What would happen in Exercise 131 if x were a positive number that was less than 1?

Section 1.4

Adding and Subtracting Real Numbers

Objectives

1. Add two or more real numbers with like signs.
2. Add two or more real numbers with unlike signs.
3. Subtract two real numbers.
4. Use signed numbers and one or more operations to model an application.
5. Use a calculator to add or subtract two real numbers.

Vocabulary

like signs   unlike signs

Getting Ready

Perform each operation.

1. $14.32 + 3.2$
2. $5.54 - 2.6$
3. $4.2 - (3 - 0.8)$
4. $(5.42 - 4.22) - 0.2$
5. $(437 - 198) - 143$
6. $437 - (198 - 143)$

In this section, we will discuss how to add and subtract real numbers. Recall that the result of an addition is called a *sum* and the result of a subtraction is called a *difference*. To develop the rules for adding real numbers, we will use the number line.

1   Add two or more real numbers with like signs.

Since the positive direction on the number line is to the right, positive numbers can be represented by arrows pointing to the right. Negative numbers can be represented by arrows pointing to the left.

To add $+2$ and $+3$, we can represent $+2$ with an arrow the length of 2, pointing to the right. We can represent $+3$ with an arrow of length 3, also pointing to the right. To add the numbers, we place the arrows end to end, as in Figure 1-19. Since the endpoint of the second arrow is the point with coordinate $+5$, we have

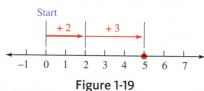

Figure 1-19

$$(+2) + (+3) = +5$$

As a check, we can think of this problem in terms of money. If you had \$2 and earned \$3 more, you would have a total of \$5.

The addition

$$(-2) + (-3)$$

COMMENT

$0 + a = a + 0 = a$

can be represented by the arrows shown in Figure 1-20. Since the endpoint of the final arrow is the point with coordinate -5, we have

$$(-2) + (-3) = -5$$

Figure 1-20

As a check, we can think of this problem in terms of money. If you lost \$2 and then lost \$3 more, you would have lost a total of \$5.

Because two real numbers with **like signs** can be represented by arrows pointing in the same direction, we have the following rule.

ADDING REAL NUMBERS WITH LIKE SIGNS	**1.** *To add two positive numbers*, add their absolute values and the answer is positive. **2.** *To add two negative numbers*, add their absolute values and the answer is negative.

EXAMPLE 1   Add:

a. $(+4) + (+6) = +(4 + 6)$
$\qquad\qquad\quad = 10$

b. $(-4) + (-6) = -(4 + 6)$
$\qquad\qquad\quad = -10$

c. $+5 + (+10) = +(5 + 10)$
$\qquad\qquad\quad = 15$

d. $-\dfrac{1}{2} + \left(-\dfrac{3}{2}\right) = -\left(\dfrac{1}{2} + \dfrac{3}{2}\right)$
$\qquad\qquad\qquad = -\dfrac{4}{2}$
$\qquad\qquad\qquad = -2$

 SELF CHECK 1   Add: **a.** $(+0.5) + (+1.2)$ **b.** $(-3.7) + (-2.3)$

2   Add two or more real numbers with unlike signs.

Real numbers with **unlike signs** can be represented by arrows on a number line pointing in opposite directions. For example, the addition

$$(-6) + (+2)$$

COMMENT   We do not need to write a $+$ sign in front of a positive number.

$$+4 = 4 \text{ and } +5 = 5$$

However, we must always write a $-$ sign in front of a negative number.

can be represented by the arrows shown in Figure 1-21. Since the endpoint of the final arrow is the point with coordinate -4, we have

$$(-6) + (+2) = -4$$

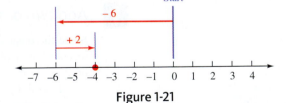

Figure 1-21

As a check, we can think of this problem in terms of money. If you lost \$6 and then earned \$2, you would still have a loss of \$4.

The addition

$$(+7) + (-4)$$

COMMENT   On a number line, a negative number moves to the left, and a positive number moves to the right.

can be represented by the arrows shown in Figure 1-22. Since the endpoint of the final arrow is the point with coordinate $+3$, we have

$$(+7) + (-4) = +3$$

Figure 1-22

As a check, you can think of this problem in terms of money. If you had $7 and then lost $4, you would still have a gain of $3.

Because two real numbers with unlike signs can be represented by arrows pointing in opposite directions, we have the following rule.

ADDING REAL NUMBERS WITH UNLIKE SIGNS

To add a positive and a negative number, subtract the smaller absolute value from the larger.

1. If the positive number has the larger absolute value, the answer is positive.
2. If the negative number has the larger absolute value, the answer is negative.
3. If a number and its opposite are added, the answer is zero.

EXAMPLE 2   Add:

a. $(+6) + (-5) = +(6 - 5)$
$$= 1$$

b. $(-2) + (+3) = +(3 - 2)$
$$= 1$$

c. $+6 + (-9) = -(9 - 6)$
$$= -3$$

d. $-\dfrac{2}{3} + \left(+\dfrac{1}{2}\right) = -\left(\dfrac{2}{3} - \dfrac{1}{2}\right)$
$$= -\left(\dfrac{4}{6} - \dfrac{3}{6}\right)$$
$$= -\dfrac{1}{6}$$

 SELF CHECK 2   Add: **a.** $(+3.5) + (-2.6)$ **b.** $(-7.2) + (+4.7)$

When adding three or more real numbers, we use the rules for the order of operations.

EXAMPLE 3   Add:

a. $[(\mathbf{+3}) + (\mathbf{-7})] + (-4) = [\mathbf{-4}] + (-4)$ *Do the work within the brackets first.*
$$= -8$$

b. $-3 + [(\mathbf{-2}) + (\mathbf{-8})] = -3 + [\mathbf{-10}]$ *Do the work within the brackets first.*
$$= -13$$

c. $2.75 + [\mathbf{8.57} + (\mathbf{-4.8})] = 2.75 + \mathbf{3.77}$ *Do the work within the brackets first.*
$$= 6.52$$

 SELF CHECK 3   Add: $-2 + [(+5.2) + (-12.7)]$

Sometimes numbers are added vertically, as shown in the next example.

EXAMPLE 4   Add:

a. $+5$
$\underline{+2}$
$+7$

b. $+5$
$\underline{-2}$
$+3$

c. -5
$\underline{+2}$
-3

d. -5
$\underline{-2}$
-7

 SELF CHECK 4   Add: **a.** $+3.2$
$\underline{-5.4}$

b. -13.5
$\underline{-4.3}$

3   Subtract two real numbers.

In arithmetic, subtraction is a take-away process. For example,

$$7 - 4 = 3$$

can be thought of as taking 4 objects away from 7 objects, leaving 3 objects.

For algebra, a better approach treats the subtraction problem

$$7 - 4$$

as the equivalent addition problem:

$$7 + (-4)$$

In either case, the answer is 3.

$$7 - 4 = 3 \text{ and } 7 + (-4) = 3$$

Thus, to subtract 4 from 7, we can add the negative (or opposite) of 4 to 7. In general, to subtract one real number from another, we add the negative (or opposite) of the number that is being subtracted.

SUBTRACTING REAL NUMBERS	If a and b are two real numbers, then $$a - b = a + (-b)$$

EXAMPLE 5   Evaluate: **a.** $12 - 4$ **b.** $-13 - 5$ **c.** $-14 - (-6)$

Solution   **a.** $12 - 4 = 12 + (-4)$ To subtract 4, add the opposite of 4.
                         $= 8$

b. $-13 - 5 = -13 + (-5)$ To subtract 5, add the opposite of 5.
                     $= -18$

c. $-14 - (-6) = -14 + [-(-6)]$ To subtract -6, add the opposite of -6.
                         $= -14 + 6$ The opposite of -6 is 6.
                         $= -8$

SELF CHECK 5   Evaluate: **a.** $-12.7 - 8.9$ **b.** $15.7 - (-11.3)$

To use a vertical format for subtracting real numbers, we add the opposite of the number that is to be subtracted by changing the sign of the lower number and proceeding as in addition.

EXAMPLE 6   Perform each subtraction by doing an equivalent addition.

a. The subtraction $\begin{array}{r} 5 \\ -\underline{-4} \end{array}$ becomes the addition $\begin{array}{r} 5 \\ +\underline{+4} \\ 9 \end{array}$

b. The subtraction $\begin{array}{r} -8 \\ -\underline{+3} \end{array}$ becomes the addition $\begin{array}{r} -8 \\ +\underline{-3} \\ -11 \end{array}$

SELF CHECK 6   Perform the subtraction: $\begin{array}{r} 5.8 \\ -\underline{-4.6} \end{array}$

When adding or subtracting three or more real numbers, we use the order of operations.

EXAMPLE 7 Simplify: **a.** $3 - [4 + (-6)]$ **b.** $[-5 + (-3)] - [-2 - (+5)]$

Solution **a.** $3 - [\mathbf{4 + (-6)}] = 3 - (\mathbf{-2})$ Do the addition within the brackets first.

$= 3 + [-(-2)]$ To subtract -2, add the opposite of -2.

$= 3 + 2$ $-(-2) = 2$

$= 5$

b. $[-5 + (-3)] - [-2 - (+5)]$

$= [-5 + (-3)] - [-2 + (-5)]$ To subtract -5, add the opposite of 5.

$= -8 - (-7)$ Do the work within the brackets.

$= -8 + [-(-7)]$ To subtract -7, add the opposite of -7.

$= -8 + 7$ $-(-7) = 7$

$= -1$

SELF CHECK 7 Simplify: $[7.2 - (-3)] - [3.2 + (-1.7)]$

EXAMPLE 8 Evaluate: **a.** $\dfrac{-3 - (-5)}{7 + (-5)}$ **b.** $\dfrac{6 + (-5)}{-3 - (-5)} - \dfrac{-3 - 4}{7 + (-5)}$

Solution **a.** $\dfrac{-3 - (-5)}{7 + (-5)} = \dfrac{-3 + [-(-5)]}{7 + (-5)}$ To subtract -5, add the opposite of -5.

$= \dfrac{-3 + 5}{2}$ $-(-5) = 5; 7 + (-5) = 2$

$= \dfrac{2}{2}$

$= 1$

b. $\dfrac{6 + (-5)}{-3 - (-5)} - \dfrac{-3 - 4}{7 + (-5)} = \dfrac{1}{-3 + 5} - \dfrac{-3 + (-4)}{2}$ $6 + (-5) = 1;$
$-(-5) = +5;$
$-3 - 4 = -3 + (-4);$
$7 + (-5) = 2$

$= \dfrac{1}{2} - \dfrac{-7}{2}$ $-3 + (-4) = -7; -3 + 5 = 2$

$= \dfrac{1 - (-7)}{2}$ Subtract the numerators and keep the denominator.

$= \dfrac{1 + [-(-7)]}{2}$ To subtract -7, add the opposite of -7.

$= \dfrac{1 + 7}{2}$ $-(-7) = 7$

$= \dfrac{8}{2}$

$= 4$

SELF CHECK 8 Evaluate: $\dfrac{7 - (-3)}{-5 - (-3) + 3}$

4 Use signed numbers and one or more operations to model an application.

Words such as *found, gain, credit, up, increase, forward, rises, in the future,* and *to the right* indicate a positive direction. Words such as *lost, loss, debit, down, decrease, backward, falls, in the past,* and *to the left* indicate a negative direction.

EXAMPLE 9   **ACCOUNT BALANCES**   The treasurer of a math club opens a checking account by depositing $350 in the bank. The bank debits the account $9 for check printing, and the treasurer writes a check for $22. Find the balance after these transactions.

Solution   The deposit can be represented by $+350$. The debit of $9 can be represented by -9, and the check written for $22 can be represented by -22. The balance in the account after these transactions is the sum of 350, -9, and -22.

$$350 + (-9) + (-22) = 341 + (-22) \qquad \text{Work from left to right.}$$
$$= 319$$

The balance is $319.

 SELF CHECK 9   Find the balance if another deposit of $17 is made.

EXAMPLE 10   **TEMPERATURE CHANGES**   At noon, the temperature was 7° above zero. At midnight, the temperature was 4° below zero. Find the difference between these two temperatures.

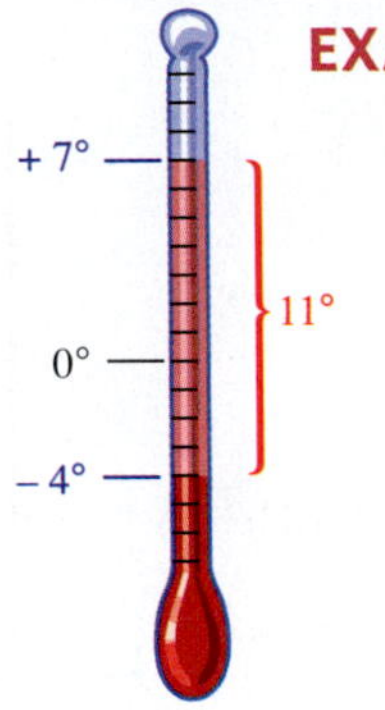

+7°

11°

0°

−4°

Solution   A temperature of 7° above zero can be represented as $+7$. A temperature of 4° below zero can be represented as -4. To find the difference between these temperatures, we can set up a subtraction problem and simplify.

$$7 - (-4) = 7 + [-(-4)] \qquad \text{To subtract } -4, \text{ add the opposite of } -4.$$
$$= 7 + 4 \qquad\qquad -(-4) = 4$$
$$= 11$$

Figure 1-23

The difference between the temperatures is 11°. Figure 1-23 shows this difference.

 **SELF CHECK 10**   Find the difference between temperatures of 32° and $-10°$.

---

5 Use a calculator to add or subtract two real numbers.

A calculator can add positive and negative numbers.

COMMENT   A common error is to use the subtraction key $\boxed{-}$ on a calculator rather than the negative key $\boxed{(-)}$.

- You do not have to do anything special to enter positive numbers. When you press 5, for example, a positive 5 is entered.
- To enter -5 into a calculator with a $\boxed{+/-}$ key, called the *plus-minus* or *change-of-sign* key, you must enter 5 and then press the $\boxed{+/-}$ key. To enter -5 into a calculator with a $\boxed{(-)}$ key, you must press the $\boxed{(-)}$ key and then press 5.

EXAMPLE 11   To evaluate $-345.678 + (-527.339)$, we enter these numbers and press these keys:

345.678 $\boxed{+/-}$ $\boxed{+}$ 527.339 $\boxed{+/-}$ $\boxed{=}$ Using a calculator with a $\boxed{+/-}$ key

$\boxed{(-)}$ 345.678 $\boxed{+}$ $\boxed{(-)}$ 527.339 **ENTER**   Using a graphing calculator

The display will read -873.017.

 **SELF CHECK 11**   Evaluate: $-783.291 - (-28.3264)$

SELF CHECK ANSWERS   **1. a.** 1.7 **b.** −6   **2. a.** 0.9 **b.** −2.5   **3.** −9.5   **4. a.** −2.2   **b.** −17.8   **5. a.** −21.6   **b.** 27
6. 10.4   **7.** 8.7   **8.** 10   **9.** $336   **10.** 42°   **11.** −754.9646

NOW TRY THIS

1. Evaluate each expression.

   a. $-2 - |5 - 8|$

   b. $\dfrac{|6 - (-4)|}{|-1 - 9|}$

2. Determine the signs necessary to obtain the given value.

   a. ▢ $3 + (\,▢\,)5 = -2$

   b. ▢ $6 + (\,▢\,)8 = -14$

   c. ▢ $56 + (\,▢\,)24 = -32$

1.4 Exercises

WARM-UPS *Find each value.*

1. $2 + 3$

2. $2 + (-5)$

3. $-4 + 7$

4. $-5 + (-6)$

5. $6 - 2$

6. $-8 - 4$

7. $-5 - (-7)$

8. $12 - (-4)$

REVIEW *Simplify each expression.*

9. $5 + 3(7 - 2)$

10. $(5 + 3)(7 - 2)$

11. $5 + 3(7) - 2$

12. $(5 + 3)7 - 2$

VOCABULARY AND CONCEPTS *Fill in the blanks.*

13. Positive and negative numbers can be represented by _____ on the number line.

14. The numbers +5 and +8 and the numbers −5 and −8 are said to have ___ signs.

15. The numbers +7 and −9 are said to have _____ signs.

16. To find the sum of two real numbers with like signs, ___ their absolute values and ____ their common sign.

17. To find the sum of two real numbers with unlike signs, _____ their absolute values and use the sign of the number with the _____ absolute value.

18. $a - b =$ _____

19. To subtract a number, we ___ its _____.

20. $a +$ ____ $= (-a) +$ _ $= 0$.

GUIDED PRACTICE *Find each sum. SEE EXAMPLE 1. (OBJECTIVE 1)*

21. $5 + 9$

22. $(-6) + (-4)$

23. $(-7) + (-2)$

24. $(+4) + 11$

25. $\dfrac{1}{5} + \left(+\dfrac{1}{7}\right)$

26. $\left(-\dfrac{3}{4}\right) + \left(-\dfrac{1}{4}\right)$

27. $44.902 + 33.098$

28. $-421.377 + (-122.043)$

Find each sum. SEE EXAMPLE 2. (OBJECTIVE 2)

29. $7 + (-3)$

30. $8 + (-5)$

31. $(-0.4) + 0.9$

32. $(-1.2) + (-5.3)$

33. $\dfrac{2}{3} + \left(-\dfrac{1}{4}\right)$

34. $-\dfrac{1}{2} + \dfrac{1}{3}$

35. $73.82 + (-108.4)$

36. $-721.964 + (38.291)$

Evaluate each expression. SEE EXAMPLE 3. (OBJECTIVES 1 AND 2)

37. $5 + [4 + (-2)]$

38. $-2 + [(-5) + 3]$

39. $-2 + (-4 + 5)$

40. $5 + [-4 + (-6)]$

41. $(-7 + 5) + 2$

42. $-12 + (-2 + 10)$

43. $-9 + [-6 + (-4)]$

44. $-27 + [-12 + (-13)]$

Add vertically. SEE EXAMPLE 4. (OBJECTIVES 1 AND 2)

45. $\begin{array}{r} 5 \\ +\underline{-4} \end{array}$

46. $\begin{array}{r} -18 \\ +\underline{-11} \end{array}$

47. $\begin{array}{r} -1.3 \\ +\underline{\;3.5} \end{array}$

48. $\begin{array}{r} 1.3 \\ +\underline{-2.5} \end{array}$

Find each difference. SEE EXAMPLE 5. (OBJECTIVE 3)

49. $8 - 4$

50. $-8 - 4$

51. $8 - (-4)$

52. $-8 - (-4)$

53. $0 - (-5)$

54. $0 - 75$

55. $\dfrac{5}{3} - \dfrac{7}{6}$

56. $-\dfrac{5}{9} - \dfrac{5}{3}$

Subtract vertically. SEE EXAMPLE 6. (OBJECTIVE 3)

57. $\begin{array}{r} 8 \\ \underline{-4} \end{array}$

58. $\begin{array}{r} 8 \\ \underline{-\,-3} \end{array}$

59. $\begin{array}{r} -10 \\ \underline{-\,-3} \end{array}$

60. $\begin{array}{r} -13 \\ \underline{-\;\;5} \end{array}$

Simplify each expression. SEE EXAMPLE 7. (OBJECTIVE 3)

61. $5 - [(-2) - 4]$

62. $-3 - [5 - (-4)]$

63. $4 - [(-3) - 5]$

64. $(3 - 5) - [5 - (-3)]$

Simplify each expression. SEE EXAMPLE 8. (OBJECTIVE 3)

65. $\dfrac{5 - (-4)}{3 - (-6)}$

66. $\dfrac{2 + (-3)}{-3 - (-4)}$

67. $\dfrac{-6 - (-3)}{5 + (-8)}$

68. $\dfrac{2 + (-3)}{-3 - (-5)} + \dfrac{-4 + 1}{8 + (-6)}$

Use a calculator to evaluate each quantity. Round the answers to two decimal places. SEE EXAMPLE 11. (OBJECTIVE 5)

69. $4.26 - 6.34 + 0.56$

70. $6.34 - 0.56 - 4.26$

71. $(2.34)^2 - (3.47)^2 - (0.72)^2$

72. $(0.72)^2 - (2.34)^2 + (3.47)^3$

ADDITIONAL PRACTICE *Simplify each expression.*

73. $\left(\dfrac{5}{2} - 3\right) - \left(\dfrac{3}{2} - 5\right)$

74. $\left(\dfrac{7}{3} - \dfrac{5}{6}\right) - \left[\dfrac{5}{6} - \left(-\dfrac{7}{3}\right)\right]$

75. $(5.2 - 2.5) - (5.25 - 5)$

76. $(3.7 - 8.25) - (3.75 + 2.5)$

77. $4 + (-12)$

78. $11 + (-15)$

79. $[-4 + (-3)] + [2 + (-2)]$

80. $[3 + (-1)] + [-2 + (-3)]$

81. $-4 + (-3 + 2) + (-3)$

82. $5 + [2 + (-5)] + (-2)$

83. $-|8 + (-4)| + 7$

84. $\left|\dfrac{3}{5} + \left(-\dfrac{4}{5}\right)\right|$

85. $-5.2 + |-2.5 + (-4)|$

86. $6.8 + |8.6 + (-1.1)|$

87. $-3\dfrac{1}{2} - 5\dfrac{1}{4}$

88. $2\dfrac{1}{2} - \left(-3\dfrac{1}{2}\right)$

89. $-6.7 - (-2.5)$

90. $25.3 - 17.5$

91. $\dfrac{-4 - 2}{-[2 + (-3)]}$

92. $\dfrac{-3 + (-2)}{2 - (-1)} - \dfrac{1 - 7}{-4 - (-7)}$

93. $\left(\dfrac{3}{4} - \dfrac{4}{5}\right) - \left(\dfrac{2}{3} + \dfrac{1}{4}\right)$

94. $\left(3\dfrac{1}{2} - 2\dfrac{1}{2}\right) - \left[5\dfrac{1}{3} - \left(-5\dfrac{2}{3}\right)\right]$

APPLICATIONS *Use the appropriate signed numbers and operations for each.* SEE EXAMPLES 9–10. (OBJECTIVE 4)

95. College tuition A student owes \$735 in tuition. If she is awarded a scholarship that will pay \$500 of the bill, what will she still owe?

96. Dieting Scott weighed 212 pounds but lost 24 pounds during a three-month diet. What does Scott weigh now?

97. Temperatures The temperature rose 13 degrees in 1 hour and then dropped 4 degrees in the next hour. What signed number represents the net change in temperature?

98. Mountain climbing A team of mountaineers climbed 2,347 feet one day and then came down 597 feet to make camp. What signed number represents their net change in altitude?

99. Temperatures The temperature fell from zero to 14° below one night. By 5:00 P.M. the next day, the temperature had risen 10 degrees. What was the temperature at 5:00 P.M.?

100. History In 1897, Joseph Thompson discovered the electron. Fifty-four years later, the first fission reactor was built. Nineteen years before the reactor was erected, James Chadwick discovered the neutron. In what year was the neutron discovered?

101. History The Greek mathematician Euclid was alive in 300 BC. The English mathematician Sir Isaac Newton was alive in AD 1700. How many years apart did they live?

102. Banking A student deposited \$415 in a new checking account, wrote a check for \$176, and deposited another \$212. Find the balance in his account.

103. Military history An army retreated 2,300 meters. After regrouping, it moved forward 1,750 meters. The next day it gained another 1,875 meters. What was the army's net gain?

104. Football A football player gained and lost the yardage shown in the illustration on six consecutive plays. How many total yards were gained or lost on the six plays?

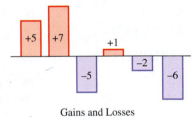

Gains and Losses

105. Aviation A pilot flying at 32,000 feet is instructed to descend to 28,000 feet. How many feet must he descend?

106. Stock market Tuesday's high and low prices for Transitronics stock were 37.125 and 31.625. Find the range of prices for this stock.

107. Temperatures Find the difference between a temperature of 32° above zero and a temperature of 27° above zero.

108. Temperatures Find the difference between a temperature of 3° below zero and a temperature of 21° below zero.

109. Stock market At the opening bell on Monday, the Dow Jones Industrial Average was 12,153. At the close, the Dow was down 23 points, but news of a half-point drop in interest rates on Tuesday sent the market up 57 points. What was the Dow average after the market closed on Tuesday?

110. Stock market On a Monday morning, the Dow Jones Industrial Average opened at 11,917. For the week, the Dow rose 29 points on Monday and 12 points on Wednesday. However, it fell 53 points on Tuesday and 27 points on both Thursday and Friday. Where did the Dow close on Friday?

111. Buying stock A woman owned 500 shares of Microsoft stock, bought another 500 shares on a price dip, and then sold 300 shares when the price rose. How many shares does she now own?

112. Small business Maria earned $2,532 in a part-time business. However, $633 of the earnings went for taxes. Find Maria's net earnings.

Use a calculator to help answer each question.

113. Balancing the books On January 1, Sally had $437.45 in the bank. During the month, she had deposits of $25.17, $37.93, and $45.26, and she had withdrawals of $17.13, $83.44, and $22.58. How much was in her account at the end of the month?

114. Small business The owner of a small business has a gross income of $97,345.32. However, he paid $37,675.66 in expenses plus $7,537.45 in taxes, $3,723.41 in health-care premiums, and $5,767.99 in pension payments. What was his profit?

115. Closing real estate transactions A woman sold her house for $115,000. Her fees at closing were $78 for preparing a deed, $446 for title work, $216 for revenue stamps, and a sales commission of $7,612.32. In addition, there was a deduction of $23,445.11 to pay off her old mortgage. As part of the deal, the buyer agreed to pay half of the title work. How much money did the woman receive after closing?

116. Winning the lottery Mike won $500,000 in a state lottery. He will get $\frac{1}{20}$ of the sum each year for the next 20 years. After

he receives his first installment, he plans to pay off a car loan of $7,645.12 and give his son $10,000 for college. By paying off the car loan, he will receive a rebate of 2% of the loan. If he must pay income tax of 28% on his first installment, how much will he have left to spend?

WRITING ABOUT MATH

117. Explain why the sum of two negative numbers is always negative, and the sum of two positive numbers is always positive.

118. Explain why the sum of a negative number and a positive number could be either negative or positive.

SOMETHING TO THINK ABOUT

119. Think of two numbers. First, add the absolute values of the two numbers, and write your answer. Second, add the two numbers, take the absolute value of that sum, and write that answer. Do the two answers agree? Can you find two numbers that produce different answers? When do you get answers that agree, and when don't you?

120. "Think of a very small number," requests the teacher. "One one-millionth," answers Charles. "Negative one million," responds Mia. Explain why either answer might be considered correct.

Section 1.5

Multiplying and Dividing Real Numbers

Objectives

1. Multiply two or more real numbers.
2. Divide two real numbers.
3. Use signed numbers and an operation to model an application.
4. Use a calculator to multiply or divide two real numbers.

Getting Ready

Find each product or quotient.

1. $8 \cdot 7$　　2. $9 \cdot 6$　　3. $8 \cdot 9$　　4. $7 \cdot 9$

5. $\frac{81}{9}$　　6. $\frac{48}{8}$　　7. $\frac{64}{8}$　　8. $\frac{56}{7}$

In this section, we will develop the rules for multiplying and dividing real numbers. We will see that the rules for multiplication and division are very similar.

1 Multiply two or more real numbers.

Because the times sign, ×, looks like the letter *x*, it is seldom used in algebra. Instead, we will use a dot, parentheses, or no symbol at all to denote multiplication. Each of the following expressions indicates the *product* obtained when two real numbers *x* and *y* are multiplied.

$$x \cdot y \quad (x)(y) \quad x(y) \quad (x)y \quad xy$$

To develop rules for multiplying real numbers, we rely on the definition of multiplication. The expression $5 \cdot 4$ indicates that 4 is to be used as a term in a sum five times.

$$5(4) = 4 + 4 + 4 + 4 + 4 = 20 \qquad \text{Read } 5(4) \text{ as "5 times 4."}$$

Likewise, the expression $5(-4)$ indicates that -4 is to be used as a term in a sum five times.

$$5(-4) = (-4) + (-4) + (-4) + (-4) + (-4) = -20 \qquad \text{Read } 5(-4) \text{ as "5 times negative 4."}$$

If multiplying by a positive number indicates repeated addition, it is reasonable that multiplication by a negative number indicates repeated subtraction. The expression $(-5)4$, for example, means that 4 is to be used as a term in a repeated subtraction five times.

$$
\begin{aligned}
(-5)4 &= -(4) - (4) - (4) - (4) - (4) \\
&= (-4) + (-4) + (-4) + (-4) + (-4) \\
&= -20
\end{aligned}
$$

Likewise, the expression $(-5)(-4)$ indicates that -4 is to be used as a term in a repeated subtraction five times.

$$
\begin{aligned}
(-5)(-4) &= -(-4) - (-4) - (-4) - (-4) - (-4) \\
&= -(-4) + [-(-4)] + [-(-4)] + [-(-4)] + [-(-4)] \\
&= 4 + 4 + 4 + 4 + 4 \\
&= 20
\end{aligned}
$$

The expression $0(-2)$ indicates that -2 is to be used zero times as a term in a repeated addition. Thus,

$$0(-2) = 0$$

Finally, the expression $(-3)(1) = -3$ suggests that the product of any number and 1 is the number itself.

The previous results suggest the following rules.

RULES FOR MULTIPLYING SIGNED NUMBERS	To multiply two real numbers, multiply their absolute values.

1. *If the numbers are positive*, the product is positive.

2. *If the numbers are negative*, the product is positive.

3. *If one number is positive and the other is negative*, the product is negative.

4. Any number multiplied by 0 is 0: $a \cdot 0 = 0 \cdot a = 0$

5. Any number multiplied by 1 is the number itself: $a \cdot 1 = 1 \cdot a = a$

EXAMPLE 1 Find each product: **a.** $4(-7)$ **b.** $(-5)(-4)$ **c.** $(-7)(6)$ **d.** $8(6)$ **e.** $(-3)^2$ **f.** $(-3)^3$ **g.** $(-3)(5)(-4)$ **h.** $(-4)(-2)(-3)$

Solution

a. $\begin{aligned}4(-7) &= (-4 \cdot 7) \\ &= -28\end{aligned}$

b. $\begin{aligned}(-5)(-4) &= +(5 \cdot 4) \\ &= +20\end{aligned}$

c. $\begin{aligned}(-7)(6) &= -(7 \cdot 6) \\ &= -42\end{aligned}$

d. $\begin{aligned}8(6) &= +(8 \cdot 6) \\ &= +48\end{aligned}$

e. $\begin{aligned}(-3)^2 &= (-3)(-3) \\ &= +9\end{aligned}$

f. $\begin{aligned}(-3)^3 &= (-3)(-3)(-3) \\ &= 9(-3) \\ &= -27\end{aligned}$

g. $(-3)(5)(-4) = (-15)(-4)$
$= +60$

h. $(-4)(-2)(-3) = 8(-3)$
$= -24$

 **SELF CHECK 1**   Find each product: **a.** $-7(5)$ **b.** $-12(-7)$ **c.** $(-5)^2$
d. $-2(-4)(-9)$

EXAMPLE 2   Evaluate: **a.** $2 + (-3)(4)$ **b.** $-3(2 - 4)$ **c.** $(-2)^2 - 3^2$
d. $-(-2)^2 + 4$

Solution   **a.** $2 + (-3)(4) = 2 + (-12)$
$= -10$

b. $-3(2 - 4) = -3[2 + (-4)]$
$= -3(-2)$
$= 6$

c. $(-2)^2 - 3^2 = 4 - 9$
$= -5$

d. $-(-2)^2 + 4 = -4 + 4$
$= 0$

 SELF CHECK 2   Evaluate: **a.** $-4 - (-3)(5)$ **b.** $(-3.2)^2 - 2(-5)^3$

EXAMPLE 3   Find each product: **a.** $\left(-\dfrac{2}{3}\right)\left(-\dfrac{6}{5}\right)$ **b.** $\left(\dfrac{3}{10}\right)\left(-\dfrac{5}{9}\right)$

Solution   **a.** $\left(-\dfrac{2}{3}\right)\left(-\dfrac{6}{5}\right) = +\left(\dfrac{2}{3} \cdot \dfrac{6}{5}\right)$

$= \dfrac{2 \cdot \overset{2}{\cancel{6}}}{\underset{1}{\cancel{3}} \cdot 5}$

$= \dfrac{4}{5}$

b. $\left(\dfrac{3}{10}\right)\left(-\dfrac{5}{9}\right) = -\left(\dfrac{3}{10} \cdot \dfrac{5}{9}\right)$

$= -\dfrac{\overset{1}{\cancel{3}} \cdot \overset{1}{\cancel{5}}}{\underset{2}{\cancel{10}} \cdot \underset{3}{\cancel{9}}}$

$= -\dfrac{1}{6}$

 SELF CHECK 3   Evaluate: **a.** $\dfrac{3}{5}\left(-\dfrac{10}{9}\right)$ **b.** $-\left(\dfrac{15}{8}\right)\left(-\dfrac{16}{5}\right)$

---

2   Divide two real numbers.

Recall that the result in a division is called a *quotient*. We know that 8 divided by 4 has a quotient of 2, and 18 divided by 6 has a quotient of 3.

$\dfrac{8}{4} = 2$, because $2 \cdot 4 = 8$ $\dfrac{18}{6} = 3$, because $3 \cdot 6 = 18$

These examples suggest that the following rule

$\dfrac{a}{b} = c$ if and only if $c \cdot b = a$

is true for the division of any real number a by any nonzero real number b as illustrated in the following.

$\dfrac{+10}{+2} = +5$, because $(+5)(+2) = +10$.

$\dfrac{-10}{-2} = +5$, because $(+5)(-2) = -10$.

$$\frac{+10}{-2} = -5, \text{ because } (-5)(-2) = +10.$$

$$\frac{-10}{+2} = -5, \text{ because } (-5)(+2) = -10.$$

Furthermore,

$$\frac{-10}{0} \text{ is undefined, because no number multiplied by 0 gives } -10.$$

However,

$$\frac{0}{-10} = 0, \text{ because } 0(-10) = 0.$$

These examples suggest the rules for dividing real numbers.

RULES FOR DIVIDING SIGNED NUMBERS

To divide two real numbers, find the quotient of their absolute values.

1. *If the numbers are positive*, the quotient is positive.
2. *If the numbers are negative*, the quotient is positive.
3. *If one number is positive and the other is negative*, the quotient is negative.
4. $\dfrac{a}{0}$ is undefined.   5. $\dfrac{a}{1} = a$.
6. If $a \neq 0$, then $\dfrac{0}{a} = 0$.   7. If $a \neq 0$, then $\dfrac{a}{a} = 1$.

EXAMPLE 4   Find each quotient: **a.** $\dfrac{36}{18}$ **b.** $\dfrac{-44}{11}$ **c.** $\dfrac{27}{-9}$ **d.** $\dfrac{-64}{-8}$

Solution   **a.** $\dfrac{36}{18} = +\dfrac{36}{18} = 2$ The quotient of two numbers with like signs is positive.

b. $\dfrac{-44}{11} = -\dfrac{44}{11} = -4$ The quotient of two numbers with unlike signs is negative.

COMMENT
$$\frac{-a}{b} = \frac{a}{-b} = -\frac{a}{b}$$

c. $\dfrac{27}{-9} = -\dfrac{27}{9} = -3$ The quotient of two numbers with unlike signs is negative.

d. $\dfrac{-64}{-8} = +\dfrac{64}{8} = 8$ The quotient of two numbers with like signs is positive.

 SELF CHECK 4   Find each quotient: **a.** $\dfrac{-72.6}{12.1}$ **b.** $\dfrac{-24.51}{-4.3}$

EXAMPLE 5   Evaluate: **a.** $\dfrac{16(-4)}{-(-64)}$ **b.** $\dfrac{(-4)^3(16)}{-64}$

Solution   **a.** $\dfrac{16(-4)}{-(-64)} = \dfrac{-64}{+64}$ **b.** $\dfrac{(-4)^3(16)}{-64} = \dfrac{(-64)(16)}{(-64)}$

$$= -1 \qquad\qquad\qquad\qquad = 16$$

 SELF CHECK 5   Evaluate: $\dfrac{-64 + 16}{-(-4)^2}$

When multiplying or dividing three or more real numbers, we use the order of operations.

EXAMPLE 6   Evaluate: **a.** $\dfrac{(-50)(10)(-5)}{-50 - 5(-5)}$ **b.** $\dfrac{3(-50)(10) + 2(10)(-5)}{2(-50 + 10)}$

Solution   **a.** $\dfrac{(-50)(10)(-5)}{-50 - 5(-5)} = \dfrac{(-500)(-5)}{-50 + 25}$ Multiply.

$\qquad\qquad = \dfrac{2,500}{-25}$ Multiply and add.

$\qquad\qquad = -100$ Divide.

b. $\dfrac{3(-50)(10) + 2(10)(-5)}{2(-50 + 10)} = \dfrac{-150(10) + (20)(-5)}{2(-40)}$ Multiply and add.

$\qquad\qquad = \dfrac{-1,500 - 100}{-80}$ Multiply.

$\qquad\qquad = \dfrac{-1,600}{-80}$ Subtract.

$\qquad\qquad = 20$ Divide.

SELF CHECK 6   Evaluate: $\dfrac{2(-50)(10) - 3(-5) - 5}{3[10 - (-5)]}$

---

3   Use signed numbers and an operation to model an application.

EXAMPLE 7   **STOCK REPORTS**   In its annual report, a corporation reports its performance on a per-share basis. When a company with 35 million shares loses $2.3 million, find the per-share loss.

Solution   A loss of $2.3 million can be represented by $-2,300,000$. Because there are 35 million shares, the per-share loss can be represented by the quotient $\dfrac{-2,300,000}{35,000,000}$.

$$\dfrac{-2,300,000}{35,000,000} \approx -0.065714286 \qquad \text{Use a calculator.}$$

The company lost about 6.6¢ per share.

SELF CHECK 7   If the company earns $1.5 million in the following year, find its per-share gain for that year.

---

4   Use a calculator to multiply or divide two real numbers.

A calculator can be used to multiply and divide positive and negative numbers. To evaluate $(-345.678)(-527.339)$, we enter these numbers and press these keys:

345.678 $\boxed{+/-}$ $\boxed{\times}$ 527.339 $\boxed{+/-}$ $\boxed{=}$ Using a calculator with a $\boxed{+/-}$ key

$\boxed{(-)}$ 345.678 $\boxed{\times}$ $\boxed{(-)}$ 527.339 $\boxed{\textbf{ENTER}}$ Using a graphing calculator

The display will read $\boxed{182289.4908}$.

To evaluate $\dfrac{-345.678}{-527.339}$, we enter these numbers and press these keys:

345.678 $\boxed{+/-}$ $\boxed{\div}$ 527.339 $\boxed{+/-}$ $\boxed{=}$ Using a calculator with a $\boxed{+/-}$ key

$\boxed{(-)}$ 345.678 $\boxed{\div}$ $\boxed{(-)}$ 527.339 $\boxed{\textbf{ENTER}}$ Using a graphing calculator

The display will read $\boxed{0.655513816}$.

NOW TRY THIS

Perform each operation. If the result is undefined, so indicate.

1. $-2 - 3(1 - 6)$

2. $-3^2 - 4(3)(-1)$

3. $\dfrac{5^2 - 2(6)(-1)}{45 - 5 \cdot 9}$

Determine the value of x that will make each fraction undefined.

4. $\dfrac{12}{x}$ **5.** $\dfrac{7}{x + 1}$

1.5 Exercises 🌿

WARM-UPS *Find each product or quotient.*

1. $1(3)$

2. $2(5)$

3. $2(3)(4)$

4. $5(3)(2)$

5. $\dfrac{12}{6}$

6. $\dfrac{10}{2}$

7. $\dfrac{3(6)}{2}$

8. $\dfrac{2 \cdot 3}{6}$

9. $12 \div 4(3)$

10. $16 \div 2(4)$

REVIEW

11. A concrete block weighs $37\frac{1}{2}$ pounds. How much will 30 of these blocks weigh?

12. If one brick weighs 1.3 pounds, how much will 500 bricks weigh?

13. Evaluate: $3^3 - 8(3)^2$

14. Place $<$, $=$, or $>$ in the box to make a true statement:
$$-2(-3 + 4) \quad \boxed{} \quad -3[3 - (-4)]$$

VOCABULARY AND CONCEPTS *Fill in the blanks.*

15. The product of two positive numbers is _____.

16. The product of a _____ number and a negative number is negative.

17. The product of two negative numbers is _____.

18. The quotient of a _____ number and a positive number is negative.

19. The quotient of two negative numbers is _____.

20. Any number multiplied by _ is 0.

21. $a \cdot 1 = $ _

22. The quotient $\dfrac{a}{0}$ is _____.

23. If $a \neq 0, \dfrac{0}{a} = $ _.   **24.** If $a \neq 0, \dfrac{a}{a} = $ _.

GUIDED PRACTICE *Perform each operation.* SEE EXAMPLE 1.
(OBJECTIVE 1)

25. $(4)(9)$

26. $(-5)(-6)$

27. $(-8)(-7)$

28. $(9)(-6)$

29. $(-10)(+9)$

30. $(-3)(11)$

31. $(-32)(-14)$

32. $(-27)(14)$

33. $(-2)(3)(4)$

34. $(5)(0)(-3)$

35. $(-5)^2$

36. $(-2)^3$

37. $(-4)^3$

38. $(-6)^2$

39. $(-3)(5)(-6)$

40. $(-1)(-3)(-6)$

Perform each operation. SEE EXAMPLE 2. (OBJECTIVE 1)

41. $2 + (-1)(-3)$

42. $-3 - (-1)(2)$

43. $(-1 + 2)(-3)$

44. $3[-2 - (-4)]$

45. $[-1 - (-3)][-1 + (-3)]$

46. $[2 + (-3)][-1 - (-3)]$

47. $2(-1)^2 - 3(-2)^2$

48. $(-1)^2(3) + (-3)(2)$

Perform each operation. SEE EXAMPLE 3. (OBJECTIVE 1)

49. $\left(\dfrac{2}{3}\right)(-36)$

50. $\left(-\dfrac{3}{4}\right)(12)$

51. $\left(-\dfrac{20}{3}\right)\left(-\dfrac{3}{5}\right)$

52. $\left(-\dfrac{2}{5}\right)\left(\dfrac{15}{2}\right)$

Perform each operation. SEE EXAMPLE 4. (OBJECTIVE 2)

53. $\dfrac{80}{-20}$

54. $\dfrac{-66}{33}$

55. $\dfrac{-110}{-55}$

56. $\dfrac{200}{40}$

57. $\dfrac{-120}{30}$

58. $\dfrac{-250}{-25}$

59. $\dfrac{320}{-16}$

60. $\dfrac{180}{-36}$

Perform each operation. SEE EXAMPLE 5. (OBJECTIVE 2)

61. $\dfrac{-3(6)}{-(-2)}$

62. $\dfrac{4(-3)^2}{-2}$

63. $\dfrac{(-2)^3(10)}{-(-5)}$

64. $\dfrac{-18}{-2(3)}$

Perform each operation. If the result is undefined, so indicate.
SEE EXAMPLE 6. (OBJECTIVE 2)

65. $\dfrac{18 - 20}{-2}$

66. $\dfrac{16 - 2}{2 - 9}$

67. $\dfrac{-3(-2)(-4)}{-4 - 2(-5)}$

68. $\dfrac{2(15)^2 - 2}{-2^3 + 1}$

69. $\dfrac{6 - 3(2)^2}{-1(7 - 4)}$

70. $\dfrac{2(-25)(10) + 4(5)(-5)}{5(125 - 25)}$

71. $\dfrac{-4(5)(2) + 2(-10)(3)}{-2(-4) - 8}$

72. $\dfrac{-5(-2) + 4}{-4(2) + 8}$

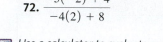 *Use a calculator to evaluate each expression. Refer to the calculator tear out card for entering fractions.* (OBJECTIVE 4)

73. $\dfrac{(-6) + 4(-3)}{4 - 6}$

74. $\dfrac{4 - 2(4)(-3) + (-3)}{4 - (-6) - 3}$

75. $\dfrac{4(-6)^2(-3) + 4^2(-6)}{2(-6) - 2(-3)}$

76. $\dfrac{[4^2 - 2(-6)](-3)^2}{-4(-3)}$

ADDITIONAL PRACTICE *Simplify each expression.*

77. $-4\left(\dfrac{-3}{4}\right)$

78. $(5)\left(-\dfrac{2}{5}\right)$

79. $(-1)(2^3)$

80. $[2(-3)]^2$

81. $(-2)(-2)(-2)(-3)(-4)$

82. $(-5)(4)(3)(-2)(-1)$

83. $(2)(-5)(-6)(-7)$

84. $(-3)(-5)(-5)(-2)$

85. $(-7)^2$

86. $(-2)^3$

87. $-(-3)^2$

88. $-(-1)(-3)^2$

89. $(-1)^2[2 - (-3)]$

90. $2^2[-1 - (-3)]$

91. $-3(-1) - (-3)(2)$

92. $-1(2)(-3) + 6$

93. $(-1)^3(-2)^2 + (-3)^2$

94. $(-2)^3[3 - (-5)]$

95. $\dfrac{4 + (-12)}{(-2)^2 - 4}$

96. $\dfrac{-2(3)(4)}{3 - 1}$

97. $\dfrac{-2(5)(4)}{-3 + 1}$

98. $\dfrac{-3 + 2 - (-10)}{4(-3) + 2(6)}$

99. $\dfrac{1}{2} - \dfrac{2}{3} - \dfrac{3}{4}$

100. $-\dfrac{2}{3} + \dfrac{1}{2} + \dfrac{3}{4}$

101. $\dfrac{1}{2} - \dfrac{2}{3}$

102. $-\dfrac{2}{3} - \dfrac{3}{4}$

103. $\left(\dfrac{1}{2} - \dfrac{2}{3}\right)\left(\dfrac{1}{2} + \dfrac{2}{3}\right)$

104. $\left(\dfrac{1}{2} + \dfrac{3}{4}\right)\left(\dfrac{1}{2} - \dfrac{3}{4}\right)$

105. $\left(\dfrac{1}{4} - \dfrac{2}{3}\right)\left(\dfrac{3}{4} - \dfrac{1}{3}\right)$

106. $\left(\dfrac{2}{5} - \dfrac{1}{4}\right)\left(\dfrac{1}{5} - \dfrac{3}{4}\right)$

APPLICATIONS *Use signed numbers and one or more operations to answer each question.* SEE EXAMPLE 7. (OBJECTIVE 3)

107. Loss of revenue A manufacturer's website normally produces sales of $425 per hour, but was offline for 12 hours due to a systems virus. How much revenue was lost?

108. Mowing lawns Justin worked all day mowing lawns and was paid $8 per hour. If he had $94 at the end of an 8-hour day, how much did he have before he started working?

109. Temperatures Suppose that the temperature is dropping at the rate of 3 degrees each hour. If the temperature has dropped 18 degrees, what signed number expresses how many hours the temperature has been falling?

110. Dieting A man lost 37.5 pounds. If he lost 2.5 pounds each week, how long has he been dieting?

111. Inventories A spreadsheet is used to record inventory losses at a warehouse. The items, their cost, and the number missing are listed in the table.

   a. Find the value of the lost MP3 players.
   b. Find the value of the lost cell phones.
   c. Find the value of the lost GPS systems.
   d. Find the total losses.

	A	B	C	D
	Item	**Cost**	**Number of units**	**$ Losses**
1	MP3 player	75	−32	
2	Cell phone	57	−17	
3	GPS system	87	−12	

112. Toy inventories A spreadsheet is used to record inventory losses at a warehouse. The item, the number of units, and the dollar losses are listed in the table.

   a. Find the cost of a truck.
   b. Find the cost of a drum.
   c. Find the cost of a ball.

	A	B	C	D
	Item	**Cost**	**Number of units**	**$ Losses**
1	Truck		−12	−$60
2	Drum		−7	−$49
3	Ball		−13	−$39

 Use a calculator to help answer each question.

113. Stock market Over a 7-day period, the Dow Jones Industrial Average had gains of 26, 35, and 17 points. In that period, there were also losses of 25, 31, 12, and 24 points. What is the average daily performance over the 7-day period?

114. Astronomy Light travels at the rate of 186,000 miles per second. How long will it take light to travel from the Sun to Venus? (*Hint:* The distance from the Sun to Venus is 67,000,000 miles.)

115. Saving for school   A student has saved $15,000 to attend graduate school. If she estimates that her expenses will be $613.50 a month while in school, does she have enough to complete an 18-month master's degree program?

116. Earnings per share   Over a five-year period, a corporation reported profits of $19 million, $15 million, and $12 million. It also reported losses of $11 million and $39 million. What is the average gain (or loss) each year?

WRITING ABOUT MATH

117. Explain how you would decide whether the product of several numbers is positive or negative.

118. Describe two situations in which negative numbers are useful.

SOMETHING TO THINK ABOUT

119. If the quotient of two numbers is undefined, what would their product be?

120. If the product of five numbers is negative, how many of the factors could be negative?

121. If x^5 is a negative number, can you determine whether x is also negative?

122. If x^6 is a positive number, can you determine whether x is also positive?

Section 1.6

Algebraic Expressions

Objectives

1. Translate an English phrase into an algebraic expression.
2. Evaluate an algebraic expression when given values for its variables.
3. Identify the number of terms in an algebraic expression and identify the numerical coefficient of each term.

Vocabulary

algebraic expression   constant   term   numerical coefficient

Getting Ready

Identify each of the following as a sum, difference, product, or quotient.

1. $x + 3$ **2.** $57x$

3. $\dfrac{x}{9}$ **4.** $19 - y$

5. $\dfrac{x - 7}{3}$ **6.** $x - \dfrac{7}{3}$

7. $5(x + 2)$ **8.** $5x + 10$

Algebraic expressions are a fundamental concept in the study of algebra. They convey mathematical operations and are the building blocks of many equations, the main topic of the next chapter.

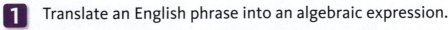

1 Translate an English phrase into an algebraic expression.

Variables and numbers can be combined with the operations of arithmetic to produce algebraic expressions. For example, if x and y are variables, the **algebraic expression** $x + y$ represents the sum of x and y, and the algebraic expression $x - y$ represents their difference.

There are many other ways to express addition or subtraction with algebraic expressions, as shown in Tables 1-3 and 1-4.

TABLE 1-3	
The phrase	**translates into the algebraic expression**
the *sum* of t and 12	$t + 12$
5 *plus* s	$5 + s$
7 *added to* a	$a + 7$
10 *more than* q	$q + 10$
12 *greater than* m	$m + 12$
l *increased by* m	$l + m$
exceeds p by 50	$p + 50$

TABLE 1-4	
The phrase	**translates into the algebraic expression**
the *difference* of 50 and r	$50 - r$
1,000 *minus* q	$1,000 - q$
15 *less than* w	$w - 15$
t *decreased by* q	$t - q$
12 *reduced by* m	$12 - m$
l *subtracted from* 250	$250 - l$
2,000 *less* p	$2,000 - p$

EXAMPLE 1   Let x represent a certain number. Write an expression that represents
a. the number that is 5 more than x **b.** the number 12 decreased by x.

Solution   **a.** The number "5 more than x" is the number found by adding 5 to x. It is represented by $x + 5$.

     b. The number "12 decreased by x" is the number found by subtracting x from 12. It is represented by $12 - x$.

 SELF CHECK 1   Let y represent a certain number. Write an expression that represents y increased by 25.

EXAMPLE 2   **INCOME TAXES**   Bob worked x hours preparing his income tax return. He worked 3 hours less than that on his son's return. Write an expression that represents
a. the number of hours he spent preparing his son's return
b. the total number of hours he worked.

Solution   **a.** Because he worked x hours on his own return and 3 hours less on his son's return, he worked $(x - 3)$ hours on his son's return.

     b. Because he worked x hours on his own return and $(x - 3)$ hours on his son's return, the total time he spent on taxes was $[x + (x - 3)]$ hours.

 SELF CHECK 2   Javier deposited $\$d$ in a bank account. Later, he withdrew $\$500$. Write an expression that represents the number of dollars in his account.

There are several ways to indicate the product of two numbers with algebraic expressions, as shown in Table 1-5.

TABLE 1-5	
The phrase	**translates into the algebraic expression**
the *product* of a and b	ab
25 *times* B	$25B$
twice x	$2x$
$\frac{1}{2}$ *of* z	$\frac{1}{2}z$
12 *multiplied by* m	$12m$

EXAMPLE 3   Let x represent a certain number. Denote a number that is
   a. twice as large as x **b.** 5 more than 3 times x **c.** 4 less than $\frac{1}{2}$ of x.

Solution   **a.** The number "twice as large as x" is found by multiplying x by 2. It is represented by $2x$.

   b. The number "5 more than 3 times x" is found by adding 5 to the product of 3 and x. It is represented by $3x + 5$.

   c. The number "4 less than $\frac{1}{2}$ of x" is found by subtracting 4 from the product of $\frac{1}{2}$ and x. It is represented by $\frac{1}{2}x - 4$.

 SELF CHECK 3   Find the product of 40 and t.

EXAMPLE 4   **STOCK VALUATIONS**   Jim owns x shares of Transitronics stock, valued at $29 a share; y shares of Positone stock, valued at $32 a share; and 300 shares of Baby Bell, valued at $42 a share.
   a. How many shares of stock does he own?
   b. What is the value of his stock?

Solution   **a.** Because there are x shares of Transitronics, y shares of Positone, and 300 shares of Baby Bell, his total number of shares is $x + y + 300$.

   b. The value of x shares of Transitronics is $\$29x$, the value of y shares of Positone is $\$32y$, and the value of 300 shares of Baby Bell is $\$42(300)$. The total value of the stock is $\$(29x + 32y + 12{,}600)$.

 SELF CHECK 4   If water softener salt costs $\$p$ per bag, find the cost of 25 bags.

There are also several ways to indicate the quotient of two numbers with algebraic expressions, as shown in Table 1-6.

TABLE 1-6	
The phrase	**translates into the algebraic expression**
the *quotient* of 470 and A	$\dfrac{470}{A}$
B *divided by* 9	$\dfrac{B}{9}$
the *ratio* of h *to* 5	$\dfrac{h}{5}$
x *split into* 5 equal parts	$\dfrac{x}{5}$

EXAMPLE 5   Let x and y represent two numbers. Write an algebraic expression that represents the sum obtained when 3 times the first number is added to the quotient obtained when the second number is divided by 6.

Solution   Three times the first number x is denoted as $3x$. The quotient obtained when the second number y is divided by 6 is the fraction $\frac{y}{6}$. Their sum is expressed as $3x + \frac{y}{6}$.

 SELF CHECK 5   If the cost c of a meal is split equally among 4 people, what is each person's share?

EXAMPLE 6 **CUTTING ROPES** A 5-foot section is cut from the end of a rope that is *l* feet long. If the remaining rope is divided into three equal pieces, find an expression for the length of each of the equal pieces.

Solution After a 5-foot section is cut from one end of *l* feet of rope, the rope that remains is $(l - 5)$ feet long. When that remaining rope is cut into 3 equal pieces, each piece will be $\frac{l - 5}{3}$ feet long. See Figure 1-24.

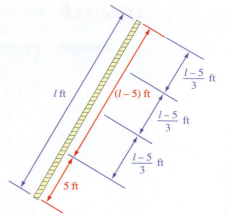

Figure 1-24

 SELF CHECK 6 If a 7-foot section is cut from a rope that is *l* feet long and the remaining rope is divided into two equal pieces, find an expression for the length of each piece.

2 Evaluate an algebraic expression when given values for its variables.

Since variables represent numbers, algebraic expressions also represent numbers. We can evaluate algebraic expressions when we know the values of the variables.

EXAMPLE 7 If $x = 8$ and $y = 10$, evaluate: **a.** $x + y$ **b.** $y - x$ **c.** $3xy$ **d.** $\frac{5x}{y - 5}$

Solution We substitute 8 for *x* and 10 for *y* in each expression and simplify.

a. $x + y = 8 + 10$
$= 18$

b. $y - x = 10 - 8$
$= 2$

c. $3xy = (3)(8)(10)$
$= (24)(10)$ Do the multiplications from left to right.
$= 240$

d. $\frac{5x}{y - 5} = \frac{5(8)}{10 - 5}$
$= \frac{40}{5}$ Simplify the numerator and the denominator separately.
$= 8$ Simplify the fraction.

COMMENT When substituting a number for a variable in an expression, it is a good idea to write the number within parentheses. This will avoid mistaking 5(8) for 58.

 SELF CHECK 7 If $a = 2$ and $b = 5$, evaluate: $\frac{6b + 18}{a + 2b}$

EXAMPLE 8   If $x = -4$, $y = 8$, and $z = -6$, evaluate: **a.** $\dfrac{7x^2 y}{2(y - z)}$ **b.** $\dfrac{3xz^2}{y(x + z)}$

Solution   We substitute -4 for x, 8 for y, and -6 for z in each expression and simplify.

a. $\dfrac{7x^2 y}{2(y - z)} = \dfrac{7(-4)^2 (8)}{2[8 - (-6)]}$

$= \dfrac{7(16)(8)}{2(14)}$ $(-4)^2 = 16;\ 8 - (-6) = 14$

$= \dfrac{7 \,(\overset{1}{2})\,(\overset{1}{2})\,(4)(8)}{2(2)(7)}$ Factor the numerator and denominator and divide
$\phantom{= \dfrac{7}{2(2)}}$ out all common factors.

$= 32$ $4 \cdot 8 = 32$

b. $\dfrac{3xz^2}{y(x + z)} = \dfrac{3(-4)(-6)^2}{8[-4 + (-6)]}$

$= \dfrac{3(-4)(36)}{8(-10)}$ $(-6)^2 = 36;\ -4 + (-6) = -10$

$= \dfrac{3(\overset{1}{\cancel{-2}})(2)(\overset{1}{\cancel{4}})(9)}{2(\overset{1}{\cancel{4}})(\overset{1}{\cancel{-2}})(5)}$ Factor the numerator and denominator and divide
 out all common factors.

$= \dfrac{27}{5}$ $3(9) = 27;\ 1(5) = 5$

 SELF CHECK 8   If $a = -3$, $b = -2$, and $c = -5$, evaluate: $\dfrac{b(a + c^2)}{abc}$

---

3   ## Identify the number of terms in an algebraic expression and identify the numerical coefficient of each term.

Numbers without variables, such as 7, 21, and 23, are called **constants**. Expressions such as 37, xyz, and $32t$, which are constants, variables, or products of constants and variables, are called algebraic **terms**.

- The expression $3x + 5y$ contains two terms. The first term is $3x$, and the second term is $5y$.
- The expression $xy + (-7)$ contains two terms. The first term is xy, and the second term is -7.
- The expression $3 + x + 2y$ contains three terms. The first term is 3, the second term is x, and the third term is $2y$.

Numbers and variables that are part of a product are called factors. For example,

- The product $7x$ has two factors, which are 7 and x.
- The product $-3xy$ has three factors, which are -3, x, and y.
- The product $\frac{1}{2}abc$ has four factors, which are $\frac{1}{2}$, a, b, and c.

The number factor of a product is called its **numerical coefficient**. The numerical coefficient (or just the *coefficient*) of $7x$ is 7. The coefficient of $-3xy$ is -3, and the coefficient of $\frac{1}{2}abc$ is $\frac{1}{2}$. The coefficient of terms such as x, ab, and rst is understood to be 1.

$$x = \mathbf{1}x, \qquad ab = \mathbf{1}ab, \qquad \text{and} \qquad rst = \mathbf{1}rst$$

EXAMPLE 9
a. The expression $5x + y$ has two terms. The coefficient of its first term is 5. The coefficient of its second term is 1.

b. The expression $-17wxyz$ has one term, which contains the five factors -17, w, x, y, and z. Its coefficient is -17.

c. The expression 37 has one term, the constant 37.

d. The expression $3x^2 - 2x$ has two terms. The coefficient of the first term is 3. Since $3x^2 - 2x$ can be written as $3x^2 + (-2x)$, the coefficient of the second term is -2.

 **SELF CHECK 9**   How many terms does the expression $3x^2 - 2x + 7$ have? Find the sum of the coefficients.

 **SELF CHECK ANSWERS**   **1.** $y + 25$ **2.** $d - 500$ **3.** $40t$ **4.** $\$25p$ **5.** $\frac{c}{4}$ **6.** $\frac{l-7}{2}$ ft **7.** 4 **8.** $\frac{22}{15}$ **9.** 3; 8

NOW TRY THIS

If $a = -2$, $b = -1$, and $c = 8$, evaluate each expression.

1. $3a^2$

2. $\dfrac{a - b}{c - a}$

3. $b^2 - 4ac$

4. Write $2(a - b)$ as an English phrase.

1.6 Exercises

WARM-UPS *Identify each as a sum, difference, product, or quotient.*

1. 5 more than p
2. twice y
3. $\frac{1}{2}$ of q
4. 6 less than y
5. the ratio of 7 to z
6. x divided by 12
7. 8 decreased by x
8. r greater than 9

REVIEW *Evaluate each expression.*

9. $0.14 \cdot 3,800$
10. $\dfrac{3}{5} \cdot 4,765$
11. $\dfrac{-4 + (7 - 9)}{(-9 - 7) + 4}$
12. $\dfrac{5}{4}\left(1 - \dfrac{3}{5}\right)$

VOCABULARY AND CONCEPTS *Fill in the blanks.*

13. The answer to an addition problem is called a ____.
14. The answer to a _____ problem is called a difference.
15. The answer to a _____ problem is called a product.
16. The answer to a division problem is called a _____.
17. An _____ expression is a combination of variables, numbers, and the operation symbols for addition, subtraction, multiplication, or division.

18. To _____ an algebraic expression, we substitute values for the variables and simplify.
19. A ____ is the product of constants and/or variables and the numerical part is called the _____.
20. Terms that have no variables are called _____.

GUIDED PRACTICE *Let x and y represent two real numbers. Write an algebraic expression to denote each quantity.* SEE EXAMPLE 1. (OBJECTIVE 1)

21. The sum of x and y
22. The sum of twice x and twice y
23. The number that is 3 less than x
24. The difference obtained when twice x is subtracted from y

Let x, y, and z represent three real numbers. Write an algebraic expression to denote each quantity. SEE EXAMPLE 3. (OBJECTIVE 1)

25. The product of twice x and y
26. The product of x and twice y
27. The product of 3, x, and y
28. The product of 3 and $2z$

Let x, y, and z represent three real numbers. Write an algebraic expression to denote each quantity. Assume that no denominators are 0. SEE EXAMPLE 5. (OBJECTIVE 1)

29. The quotient obtained when y is divided by x

30. The quotient obtained when the sum of x and y is divided by y

31. The quotient obtained when the product of 3 and z is divided by the product of 4 and x

32. The quotient obtained when the sum of x and y is divided by the sum of y and z

Evaluate each expression if x = −2, y = 5, and z = −3. SEE EXAMPLES 7–8. (OBJECTIVE 2)

33. $x + y$

34. $x - z$

35. $4xyz$

36. $2x^2z$

37. $\dfrac{x^2y}{z-1}$

38. $\dfrac{xy-2}{z}$

39. $\dfrac{4z^2y}{3(x-z)}$

40. $\dfrac{x+y+z}{4y^2x}$

41. $\dfrac{x(y+z)-25}{(x+z)^2-y^2}$

42. $\dfrac{(x+y)(y+z)}{x+z+y}$

43. $\dfrac{3(x+z^2)+4}{y(x-z)}$

44. $\dfrac{x(y^2-2z)-1}{z(y-x^2)}$

Give the number of terms in each algebraic expression and also give the numerical coefficient of the first term. SEE EXAMPLE 9. (OBJECTIVE 3)

45. $-7c$

46. $4c - 9d$

47. $-xy - 5z + 8$

48. cd

49. $-3xy + yz - zw + 5$

50. $-2xyz + cde - 14$

51. $9abc - 5ab - c$

52. $5uvw - 4uv + 8uw$

53. $5x - 4y + 3z + 2$

54. $7abc - 9ab + 2bc + a - 1$

ADDITIONAL PRACTICE *Let x, y, and z represent three real numbers. Write an algebraic expression to denote each quantity. Assume that no denominators are 0.*

55. The sum obtained when the quotient of x divided by y is added to z

56. z decreased by 3

57. z less the product of x and y

58. z less than the product of x and y

59. The quotient obtained when the product of x and y is divided by the sum of x and z

60. The sum of the product xy and the quotient obtained when y is divided by z

61. The number obtained when x decreased by 4 is divided by the product of 3 and y

62. The number obtained when $2z$ minus $5y$ is divided by the sum of x and $3y$

Let x, y, and z represent three real numbers. Write each algebraic expression as an English phrase. Assume that no denominators are 0.

63. $y + 4$

64. $x - 5$

65. $xy(x + y)$

66. $(x + y + z)(xyz)$

67. $\dfrac{x+2}{z}$

68. $5 + \dfrac{y}{z}$

69. $\dfrac{y}{z}$

70. xy

71. $2xy$

72. $\dfrac{x+y}{2}$

73. $\dfrac{5}{x+y}$

74. $\dfrac{3x}{y+z}$

Let x = 8, y = 4, and z = 2. Write each phrase as an algebraic expression, and evaluate it. Assume that no denominators are 0.

75. The sum of x and z

76. The product of x, y, and z

77. z less than y

78. The quotient obtained when y is divided by z

79. 3 less than the product of y and z

80. 7 less than the sum of x and y

81. The quotient obtained when the product of x and y is divided by z

82. The quotient obtained when 10 greater than x is divided by z

Consider the algebraic expression 29xyz + 23xy + 19x.

83. What are the factors of the third term?

84. What are the factors of the second term?

85. What factor is common to the first and third terms?

86. What factor is common to all three terms?

Consider the algebraic expression 3xyz + 5xy + 17xz.

87. What are the factors of the first term?

88. What are the factors of the second term?

89. What are the factors of the third term?

90. What factor is common to all three terms?

Consider the algebraic expression 5xy + yt + 8xyt.

91. Find the numerical coefficients of each term.

92. What factor is common to all three terms?

93. What factors are common to the first and third terms?

94. What factors are common to the second and third terms?

Consider the algebraic expression 3xy + y + 25xyz.

95. Use the numerical coefficient of each term to find their product.

96. Use the numerical coefficient of each term to find their sum.

97. What factors are common to the first and third terms?

98. What factor is common to all three terms?

APPLICATIONS *Write an algebraic expression to denote each quantity. Assume that no denominators are 0.* **SEE EXAMPLES 2, 4, AND 6. (OBJECTIVE 1)**

99. Course loads A man enrolls in college for *c* hours of credit, and his sister enrolls for 6 more hours than her brother. Write an expression that represents the number of hours the sister is taking.

100. Antique cars An antique Ford has 25,000 more miles on its odometer than a newer car. If the newer car has traveled *m* miles, find an expression that represents the mileage on the Ford.

101. Heights of trees
  a. If *h* represents the height (in feet) of the oak tree, write an expression that represents the height of the crab apple tree.
  b. If *c* represents the height (in feet) of the crab apple tree, write an expression that represents the height of the oak.

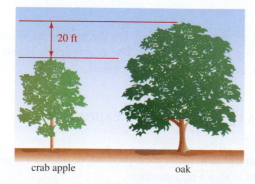

crab apple   oak

102. T-bills Write an expression that represents the value of *t* T-bills, each worth $9,987.

103. Real estate Write an expression that represents the value of *n* vacant lots if each lot is worth $35,000.

104. Cutting ropes A rope *x* feet long is cut into 5 equal pieces. Find an expression for the length of each piece.

105. Invisible tape If *x* inches of tape have been used off the roll shown below, how many inches of tape are left on the roll?

106. Plumbing A plumber cuts a pipe that is 12 feet long into *x* equal pieces. Find an expression for the length of each piece.

107. Comparing assets A girl had *d* dollars, and her brother had $5 more than three times that amount. How much did the brother have?

108. Comparing investments Wendy has *x* shares of stock. Her sister has 2 fewer shares than twice Wendy's shares. How many shares does her sister have?

109. Sorting records In electronic data processing, the process of sorting records into sequential order is a common task. One sorting technique, called a **selection sort**, requires *C* comparisons to sort *N* records, where *C* and *N* are related by the formula

$$C = \frac{N(N-1)}{2}$$

How many comparisons are necessary to sort 10,000 records?

110. Sorting records How many comparisons are necessary to sort 50,000 records? See Exercise 109.

WRITING ABOUT MATH

111. Distinguish between the meanings of these two phrases: "3 less than *x*" and "3 is less than *x*."

112. Distinguish between *factor* and *term*.

113. What is the purpose of using variables? Why aren't ordinary numbers enough?

114. In words, *xy* is "the product of *x* and *y*." However, $\frac{x}{y}$ is "the quotient obtained when *x* is divided by *y*." Explain why the extra words are needed.

SOMETHING TO THINK ABOUT

115. If the value of *x* were doubled, what would happen to the value of 37*x*?

116. If the values of both *x* and *y* were doubled, what would happen to the value of $5xy^2$?

Section 1.7

Properties of Real Numbers

Objectives

1. Apply the closure properties by evaluating an expression for given values for variables.
2. Apply the commutative and associative properties.
3. Apply the distributive property of multiplication over addition to rewrite an expression.
4. Recognize the identity elements and find the additive and multiplicative inverse of a nonzero real number.
5. Identify the property that justifies a given statement.

Vocabulary

closure properties
commutative properties
associative properties

distributive property
identity elements
additive inverse

reciprocal
multiplicative inverse

Getting Ready

Perform each operation.

1. $3 + (5 + 9)$
2. $(3 + 5) + 9$
3. $23.7 + 14.9$
4. $14.9 + 23.7$
5. $7(5 + 3)$
6. $7 \cdot 5 + 7 \cdot 3$
7. $125.3 + (-125.3)$
8. $125.3\left(\dfrac{1}{125.3}\right)$
9. $777 + 0$
10. $777 \cdot 1$

To understand algebra, we must know the properties that govern the operations of addition, subtraction, multiplication, and division of real numbers. These properties enable us to write expressions in equivalent forms, often making our work easier.

1 Apply the closure properties by evaluating an expression for given values for variables.

The **closure properties** guarantee that the sum, difference, product, or quotient (except for division by 0) of any two real numbers is also a real number.

CLOSURE PROPERTIES   If a and b are real numbers, then

$a + b$ is a real number. $a - b$ is a real number.

ab is a real number. $\dfrac{a}{b}$ is a real number $(b \neq 0)$.

EXAMPLE 1 Let $x = 8$ and $y = -4$. Find the real-number answers to show that each expression represents a real number.

a. $x + y$ **b.** $x - y$ **c.** xy **d.** $\frac{x}{y}$

Solution We substitute 8 for x and -4 for y in each expression and simplify.

a. $x + y = 8 + (-4)$
$= 4$

b. $x - y = 8 - (-4)$
$= 8 + 4$
$= 12$

c. $xy = 8(-4)$
$= -32$

d. $\dfrac{x}{y} = \dfrac{8}{-4}$
$= -2$

SELF CHECK 1 Let $a = -6$ and $b = 3$. Find the real-number answers to show that each expression represents a real number.
a. $a - b$ **b.** $\frac{a}{b}$

2 Apply the commutative and associative properties.

The **commutative properties** (from the word *commute*, which means to go back and forth) guarantee that addition or multiplication of two real numbers can be done in either order.

COMMUTATIVE PROPERTIES	If a and b are real numbers, then
	$a + b = b + a$ commutative property of addition
	$ab = ba$ commutative property of multiplication

EXAMPLE 2 Let $x = -3$ and $y = 7$. Show that **a.** $x + y = y + x$ **b.** $xy = yx$

Solution **a.** We can show that the sum $x + y$ is the same as the sum $y + x$ by substituting -3 for x and 7 for y in each expression and simplifying.

$$x + y = -3 + 7 = 4 \quad \text{and} \quad y + x = 7 + (-3) = 4$$

COMMENT Since $5 - 3 \neq 3 - 5$ and $5 \div 3 \neq 3 \div 5$, the commutative property cannot be applied to a subtraction or a division.

b. We can show that the product xy is the same as the product yx by substituting -3 for x and 7 for y in each expression and simplifying.

$$xy = -3(7) = -21 \quad \text{and} \quad yx = 7(-3) = -21$$

SELF CHECK 2 Let $a = 6$ and $b = -5$. Show that
a. $a + b = b + a$
b. $ab = ba$

The **associative properties** guarantee that three real numbers can be regrouped in an addition or multiplication.

ASSOCIATIVE PROPERTIES	If a, b, and c are real numbers, then
	$(a + b) + c = a + (b + c)$ associative property of addition
	$(ab)c = a(bc)$ associative property of multiplication

Because of the associative property of addition, we can group (or associate) the numbers in a sum in any way that we wish. For example,

$$(\mathbf{3 + 4}) + 5 = \mathbf{7} + 5 \qquad\qquad 3 + (\mathbf{4 + 5}) = 3 + \mathbf{9}$$
$$= 12 \qquad\qquad\qquad\qquad = 12$$

The answer is 12 regardless of how we group the three numbers.

The associative property of multiplication permits us to group (or associate) the numbers in a product in any way that we wish. For example,

$$(\mathbf{3 \cdot 4}) \cdot 7 = \mathbf{12} \cdot 7 \qquad\qquad 3 \cdot (\mathbf{4 \cdot 7}) = 3 \cdot \mathbf{28}$$
$$= 84 \qquad\qquad\qquad\qquad = 84$$

The answer is 84 regardless of how we group the three numbers.

COMMENT Since $(2 - 5) - 3 \neq 2 - (5 - 3)$ and $(2 \div 5) \div 3 \neq 2 \div (5 \div 3)$, the associative property cannot be applied to subtraction or division.

3 ## Apply the distributive property of multiplication over addition to rewrite an expression.

The **distributive property** shows how to multiply the sum of two numbers by a third number. Because of this property, we can often add first and then multiply, or multiply first and then add.

For example, $2(3 + 7)$ can be calculated in two different ways. We will add and then multiply, or we can multiply each number within the parentheses by 2 and then add.

$$2(\mathbf{3 + 7}) = 2(\mathbf{10}) \qquad\qquad 2(3 + 7) = \mathbf{2} \cdot 3 + \mathbf{2} \cdot 7$$
$$= 20 \qquad\qquad\qquad\qquad = 6 + 14$$
$$\qquad\qquad\qquad\qquad\qquad = 20$$

Either way, the result is 20.

In general, we have the following property.

DISTRIBUTIVE PROPERTY OF MULTIPLICATION OVER ADDITION	If a, b, and c are real numbers, then $$a(b + c) = ab + ac$$

Figure 1-25

Because multiplication is commutative, the distributive property also can be written in the form

$$(\boldsymbol{b + c})\boldsymbol{a} = \boldsymbol{ba} + \boldsymbol{ca}$$

We can interpret the distributive property geometrically. Since the area of the largest rectangle in Figure 1-25 is the product of its width a and its length $b + c$, its area is $a(b + c)$. The areas of the two smaller rectangles are ab and ac. Since the area of the largest rectangle is equal to the sum of the areas of the smaller rectangles, we have $a(b + c) = ab + ac$.

The previous discussion shows that multiplication distributes over addition. Multiplication also distributes over subtraction. For example, $2(3 - 7)$ can be calculated in two different ways. We will subtract and then multiply, or we can multiply each number within the parentheses by 2 and then subtract.

$$2(\mathbf{3 - 7}) = 2(\mathbf{-4}) \qquad\qquad 2(3 - 7) = \mathbf{2} \cdot 3 - \mathbf{2} \cdot 7$$
$$= -8 \qquad\qquad\qquad\qquad = 6 - 14$$
$$\qquad\qquad\qquad\qquad\qquad = -8$$

Either way, the result is -8. In general, we have

$$\boldsymbol{a(b - c) = ab - ac}$$

EXAMPLE 3   Evaluate each expression in two different ways:

  a. $3(5 + 9)$ **b.** $4(6 - 11)$ **c.** $-2(-7 + 3)$

Solution   **a.** $3(\mathbf{5 + 9}) = 3(\mathbf{14})$ $\mathbf{3}(5 + 9) = \mathbf{3 \cdot 5 + 3 \cdot 9}$
                    $= 42$ $= 15 + 27$
                                                   $= 42$

  b. $4(\mathbf{6 - 11}) = 4(\mathbf{-5})$ $4(6 - 11) = \mathbf{4 \cdot 6 - 4 \cdot 11}$
                    $= -20$ $= 24 - 44$
                                                   $= -20$

  c. $-2(\mathbf{-7 + 3}) = -2(\mathbf{-4})$ $-2(-7 + 3) = \mathbf{-2}(-7) + (\mathbf{-2})(3)$
                    $= 8$ $= 14 + (-6)$
                                                   $= 8$

 SELF CHECK 3   Evaluate $-5(-7 + 20)$ in two different ways.

The distributive property can be extended to three or more terms. For example, if a, b, c, and d are real numbers, then

$$a(b + c + d) = ab + ac + ad$$

EXAMPLE 4   Write $3.2(x + y + 2.7)$ without using parentheses.

Solution $\mathbf{3.2}(x + y + 2.7) = \mathbf{3.2}x + \mathbf{3.2}y + (\mathbf{3.2})(2.7)$ Distribute the multiplication by 3.2.
                    $= 3.2x + 3.2y + 8.64$

 SELF CHECK 4   Write $-6.3(a + 2b + 3.7)$ without using parentheses.

4   Recognize the identity elements and find the additive and multiplicative inverse of a nonzero real number.

The numbers 0 and 1 play special roles in mathematics. The number 0 is the only number that can be added to another number (say, a) and give an answer that is the same number a:

$$0 + a = a + 0 = a$$

The number 1 is the only number that can be multiplied by another number (say, a) and give an answer that is the same number a:

$$1 \cdot a = a \cdot 1 = a$$

Because adding 0 to a number or multiplying a number by 1 leaves that number the same (identical), the numbers 0 and 1 are called **identity elements**.

IDENTITY ELEMENTS   0 is the **identity element for addition (additive identity)**.

1 is the **identity element for multiplication (multiplicative identity)**.

If the sum of two numbers is 0, the numbers are called **negatives** (or **opposites** or **additive inverses**) of each other. Since $3 + (-3) = 0$, the numbers 3 and -3 are negatives (or opposites or additive inverses) of each other. In general, because

$$a + (-a) = 0$$

the numbers represented by a and $-a$ are negatives (or opposites or additive inverses) of each other.

If the product of two numbers is 1, the numbers are called **reciprocals**, or **multiplicative inverses**, of each other. Since $7\left(\frac{1}{7}\right) = 1$, the numbers 7 and $\frac{1}{7}$ are reciprocals. Since $(-0.25)(-4) = 1$, the numbers -0.25 and -4 are reciprocals. In general, because

$$a\left(\frac{1}{a}\right) = 1 \qquad \text{provided } a \neq 0$$

the numbers represented by a and $\frac{1}{a}$ are reciprocals (or multiplicative inverses) of each other.

ADDITIVE AND MULTIPLICATIVE INVERSES	Because $a + (-a) = 0$, the numbers a and $-a$ are called **negatives**, **opposites**, or **additive inverses**. Because $a\left(\frac{1}{a}\right) = 1$ $(a \neq 0)$, the numbers a and $\frac{1}{a}$ are called **reciprocals** or **multiplicative inverses**.

EXAMPLE 5 Find the additive and multiplicative inverses of $\frac{2}{3}$.

Solution The additive inverse of $\frac{2}{3}$ is $-\frac{2}{3}$ because $\frac{2}{3} + \left(-\frac{2}{3}\right)$.

The multiplicative inverse of $\frac{2}{3}$ is $\frac{3}{2}$ because $\frac{2}{3}\left(\frac{3}{2}\right) = 1$.

 SELF CHECK 5 Find the additive and multiplicative inverses of $-\frac{1}{5}$.

5 Identify the property that justifies a given statement.

EXAMPLE 6 The property in the right column justifies the statement in the left column.

  a. $3 + 4$ is a real number.   closure property of addition

  b. $\dfrac{8}{3}$ is a real number.   closure property of division

  c. $3 + 4 = 4 + 3$ commutative property of addition

  d. $-3 + (2 + 7) = (-3 + 2) + 7$ associative property of addition

  e. $(5)(-4) = (-4)(5)$ commutative property of multiplication

  f. $(ab)c = a(bc)$ associative property of multiplication

  g. $3(a + 2) = 3a + 3 \cdot 2$ distributive property

  h. $3 + 0 = 3$ additive identity property

  i. $3(1) = 3$ multiplicative identity property

  j. $2 + (-2) = 0$ additive inverse property

  k. $\left(\dfrac{2}{3}\right)\left(\dfrac{3}{2}\right) = 1$ multiplicative inverse property

 SELF CHECK 6 Which property justifies each statement?
  a. $a + 7 = 7 + a$
  b. $3(y + 2) = 3y + 3 \cdot 2$
  c. $3 \cdot (2 \cdot p) = (3 \cdot 2) \cdot p$

The properties of the real numbers are summarized as follows.

PROPERTIES OF REAL NUMBERS		
	For all real numbers a, b, and c,	
	Closure properties	$a + b$ is a real number. $a \cdot b$ is a real number. $a - b$ is a real number. $a \div b$ is a real number $(b \neq 0)$.

	Addition	*Multiplication*
Commutative properties	$a + b = b + a$	$a \cdot b = b \cdot a$
Associative properties	$(a + b) + c = a + (b + c)$	$(ab)c = a(bc)$
Identity properties	$a + 0 = a$	$a \cdot 1 = a$
Inverse properties	$a + (-a) = 0$	$a\left(\dfrac{1}{a}\right) = 1$ $(a \neq 0)$
Distributive property	$a(b + c) = ab + ac$	

SELF CHECK ANSWERS

1. a. -9 b. -2 2. a. $a + b = 1$ and $b + a = 1$ b. $ab = -30$ and $ba = -30$ 3. -65
4. $-6.3a - 12.6b - 23.31$ 5. $\frac{1}{5}$, -5 6. a. commutative property of addition b. distributive property c. associative property of multiplication

NOW TRY THIS

1. Give the additive and multiplicative inverses of 1.2.
2. Use the commutative property of multiplication to write the expression $x(y + w)$.
3. Simplify: $-4(2a - 3b + 5)$
4. Use the distributive property to complete the following multiplication:
   $9x - 12y - 3 = 3 (\,\blacksquare - \blacksquare - \blacksquare\,)$
5. Find the additive inverse of $x - y$. Try to find a second way to write it (there are three).

1.7 Exercises

WARM-UPS *Give an example of each property.*

1. The associative property of multiplication
2. The additive identity property
3. The distributive property
4. The inverse property for multiplication

Provide an example to illustrate each statement.

5. Subtraction is not commutative.
6. Division is not associative.

REVIEW

7. Write as a mathematical inequality: The sum of x and the square of y is greater than or equal to z.
8. Write as an English phrase: $3(x + z)$

Fill each box with an appropriate symbol.

9. For any number x, $|x|\ \blacksquare\ 0$.
10. $x - y = x + (\,\blacksquare\,)$

Fill in the blanks.

11. The product of two negative numbers is a _____ number.

12. The sum of two negative numbers is a _____ number.

VOCABULARY AND CONCEPTS   *Fill in the blanks.*

13. Closure property: If a and b are real numbers, $a + b$ is a ____ number.

14. Closure property: If a and b are real numbers, $\frac{a}{b}$ is a real number, provided that _____.

15. Commutative property of addition: $a + b = b + \_$

16. Commutative property of multiplication: $a \cdot b = \_ \cdot a$

17. Associative property of addition: $(a + b) + c = a + \_\_\_\_\_$

18. Associative property of multiplication: $(ab)c = \_ \cdot (bc)$

19. Distributive property: $a(b + c) = ab + \_\_$

20. $0 + a = \_$

21. $a \cdot 1 = \_$

22. 0 is the _____ element for _____.

23. 1 is the identity _____ for _____.

24. If $a + (-a) = 0$, then a and $-a$ are called _____ inverses.

25. If $a\left(\dfrac{1}{a}\right) = 1$, then $\_$ and $\_$ are called *reciprocals* or _____ inverses.

26. $a(b + c + d) = ab + \_\_\_\_\_$

GUIDED PRACTICE   *Let $x = 12$ and $y = -2$. Show that each expression represents a real number by finding the real-number answer.* SEE EXAMPLE 1. (OBJECTIVE 1)

27. $x + y$

28. $y - x$

29. xy

30. $\dfrac{x}{y}$

31. x^2

32. y^2

33. $\dfrac{x}{y^2}$

34. $\dfrac{2x}{3y}$

Let $x = 5$, $y = 7$, and $z = -1$. Show that the two expressions have the same value. SEE EXAMPLE 2. (OBJECTIVE 2)

35. $x + y$; $y + x$

36. xy; yx

37. $3x + 2y$; $2y + 3x$

38. $3xy$; $3yx$

39. $x(x + y)$; $(x + y)x$

40. $xy + y^2$; $y^2 + xy$

41. $x^2(yz^2)$; $(x^2y)z^2$

42. $x(y^2z^3)$; $(xy^2)z^3$

Use the distributive property to write each expression without parentheses. Simplify each result, if possible. SEE EXAMPLES 3–4. (OBJECTIVE 3)

43. $3(x + 5)$

44. $7(y + 2)$

45. $5(z - 4)$

46. $4(a - 3)$

47. $-2(3x + y)$

48. $-3(4a + b)$

49. $x(x + 3)$

50. $y(y + z)$

51. $-x(a + b)$

52. $-a(x + y)$

53. $-4(x^2 + x + 2)$

54. $-2(a^2 - a + 3)$

Give the additive and the multiplicative inverses of each number, if possible. SEE EXAMPLE 5. (OBJECTIVE 4)

55. 5

56. 3

57. $\dfrac{1}{3}$

58. $-\dfrac{1}{3}$

59. 0

60. -4

61. $-\dfrac{2}{3}$

62. 0.5

63. -0.2

64. 0.75

65. $\dfrac{5}{4}$

66. -1.25

Use the given property to rewrite the expression in a different form. SEE EXAMPLE 6. (OBJECTIVE 5)

67. $8(x + 2)$; distributive property

68. $a + b$; commutative property of addition

69. xy^3; commutative property of multiplication

70. $2 + (5 + 3)$; associative property of addition

71. $(x + y)z$; commutative property of addition

72. $7(x + 2)$; distributive property

73. $(xy)z$; associative property of multiplication

74. $1x$; multiplicative identity property

ADDITIONAL PRACTICE

Let $x = 2$, $y = -3$, and $z = 1$. Show that the two expressions have the same value.

75. $(x + y) + z$; $x + (y + z)$

76. $(xy)z$; $x(yz)$

77. $(xz)y$; $x(yz)$

78. $(x + y) + z$; $y + (x + z)$

Use the distributive property to write each expression without parentheses.

79. $-6(a + 4)$

80. $2x(a - x)$

81. $-3x(x - a)$

82. $-a(a + b)$

Which property of real numbers justifies each statement?

83. $3 + x = x + 3$

84. $(3 + x) + y = 3 + (x + y)$

85. $xy = yx$

86. $(4)(-7) = (-7)(4)$

87. $-5(x + 4) = -5x + (-5)(4)$

88. $x(y + z) = (y + z)x$

89. $(x + y) + z = z + (x + y)$

90. $3(x + y) = 3x + 3y$

91. $5 \cdot 1 = 5$

92. $x + 0 = x$

93. $3 + (-3) = 0$ **94.** $9 \cdot \dfrac{1}{9} = 1$

95. $0 + x = x$ **96.** $5 \cdot \dfrac{1}{5} = 1$

WRITING ABOUT MATH

97. Explain why division is not commutative.

98. Describe two ways of calculating the value of $3(12 + 7)$.

SOMETHING TO THINK ABOUT

99. Suppose there were no numbers other than the odd integers.
- Would the closure property for addition still be true?
- Would the closure property for multiplication still be true?
- Would there still be an identity for addition?
- Would there still be an identity for multiplication?

100. Suppose there were no numbers other than the even integers. Answer the four parts of Exercise 99 again.

Projects

PROJECT 1

The circumference of any circle and its diameter are related. When you divide the circumference by the diameter, the quotient is always the same number, **pi**, denoted by the Greek letter π.

- Carefully measure the circumference of several circles—a quarter, a dinner plate, a bicycle tire—whatever you can find that is round. Then calculate approximations of π by dividing each circle's circumference by its diameter.

- Use the π key on the calculator to obtain a more accurate value of π. How close were your approximations?

PROJECT 2

a. The fraction $\dfrac{22}{7}$ is often used as an approximation of π. To how many decimal places is this approximation accurate?

b. Experiment with your calculator and try to do better. Find another fraction (with no more than three digits in either its numerator or its denominator) that is closer to π. Who in your class has done best?

PROJECT 3

Write an essay answering this question.

When three professors attending a convention in Las Vegas registered at the hotel, they were told that the room rate was $120. Each professor paid his $40 share.

Later the desk clerk realized that the cost of the room should have been $115. To fix the mistake, she sent a bellhop to the room to refund the $5 overcharge. Realizing that $5 could not be evenly divided among the three professors, the bellhop refunded only $3 and kept the other $2.

Since each professor received a $1 refund, each paid $39 for the room, and the bellhop kept $2. This gives $39 + $39 + $39 + $2, or $119. What happened to the other $1?

Reach for Success
EXTENSION OF ANALYZING YOUR TIME

Now that you've analyzed a day, let's move on to take a look at a typical week. Fill in the chart below to account for every hour of every day. To simplify the process, you can use the following abbreviations:

W (work time including commute), **S** (sleeping), **P** (preparing for work; preparing meals, eating, exercising), **C** (class time including commute), **F** (time with family and friends), **ST** (study time), and **E** (entertainment)

Remember, there are no right or wrong answers. This information is to give you a complete picture of how you are spending your time in a typical week.

	SUNDAY	MONDAY	TUESDAY	WEDNESDAY	THURSDAY	FRIDAY	SATURDAY
6:00–7:00							
7:00–8:00							
8:00–9:00							
9:00–10:00							
10:00–11:00							
11:00–Noon							
Noon–1:00							
1:00–2:00							
2:00–3:00							
3:00–4:00							
4:00–5:00							
5:00–6:00							
6:00–7:00							
7:00–8:00							
8:00–9:00							
9:00–10:00							
10:00–11:00							
11:00–12:00							
12:00–1:00							
1:00–2:00							
2:00–3:00							
3:00–4:00							
4:00–5:00							
5:00–6:00							

After reviewing your weekly schedule, do you have enough time to meet the "rule of thumb" of studying at least 2 hours per week for every 1 hour you are in class? If yes, congratulations!

If not, is it possible for you to find the additional hours to help you be successful in this and all your other classes?

Are you able/willing to adjust your schedule to find this time? _____ Explain your answer.

1   Review

SECTION 1.1   Real Numbers and Their Graphs

DEFINITIONS AND CONCEPTS	EXAMPLES
Natural numbers: $\{1, 2, 3, 4, 5, \ldots\}$	Which numbers in the set $\left\{-5, 0, \frac{2}{3}, 1.5, \sqrt{9}, \pi, 6\right\}$ are **a.** natural numbers **b.** whole numbers **c.** integers **d.** rational numbers **e.** irrational numbers **f.** real numbers **g.** prime numbers **h.** composite numbers **i.** even integers **j.** odd integers?
Whole numbers: $\{0, 1, 2, 3, 4, 5, \ldots\}$	**a.** $\sqrt{9}, 6, \left(\sqrt{9} \text{ is a natural number since } \sqrt{9} = 3.\right)$ **b.** $0, \sqrt{9}, 6$
Integers: $\{\ldots, -3, -2, -1, 0, 1, 2, 3, \ldots\}$	**c.** $-5, 0, \sqrt{9}, 6$
Rational numbers: $\left\{\dfrac{a}{b} \;\middle|\; a \text{ is an integer and } b \text{ is a nonzero integer.}\right\}$	**d.** $-5, 0, \frac{2}{3}, 1.5, \sqrt{9}, 6$ **e.** π
Irrational numbers: $\{x \mid x \text{ is a number such as } \pi \text{ or } \sqrt{2} \text{ that cannot be written as a fraction with an integer numerator and a nonzero integer denominator.}\}$	**f.** $-5, 0, \frac{2}{3}, 1.5, \sqrt{9}, \pi, 6$ **g.** $\sqrt{9}$
Real numbers: $\{\text{Rational numbers or irrational numbers}\}$	**h.** 6 **i.** $0, 6$
Prime numbers: $\{2, 3, 5, 7, 11, 13, 17, \ldots\}$	**j.** $-5, \sqrt{9}$
Composite numbers: $\{4, 6, 8, 9, 10, 12, 14, 15, \ldots\}$	
Even integers: $\{\ldots, -6, -4, -2, 0, 2, 4, 6, \ldots\}$	
Odd integers: $\{\ldots, -5, -3, -1, 1, 3, 5, \ldots\}$	
Double negative rule: $\quad -(-x) = x$	$-(-3) = 3$
Sets of numbers can be graphed on the number line.	**1.** Graph the set of integers between -2 and 4. $-3 \;-2 \;-1 \;\;0 \;\;1 \;\;2 \;\;3 \;\;4$ **2.** Graph all real numbers x such that $x < -2$ or $x > 1$. $-3 \;-2 \;-1 \;\;0 \;\;1 \;\;2 \;\;3$
The **absolute value** of x, denoted as $\lvert x \rvert$, is the distance between x and 0 on the number line. $\lvert x \rvert \geq 0$	Evaluate: $-\lvert -8 \rvert$ $\quad -\lvert -8 \rvert = -(8) = -8$

REVIEW EXERCISES

Consider the set $\{0, 1, 2, 3, 4, 5\}$.

1. Which numbers are natural numbers?
2. Which numbers are prime numbers?
3. Which numbers are odd natural numbers?
4. Which numbers are composite numbers?

Consider the set $\left\{-6, -\frac{2}{3}, 0, \sqrt{2}, 2.6, \pi, 5\right\}$.

5. Which numbers are integers?
6. Which numbers are rational numbers?
7. Which numbers are prime numbers?

8. Which numbers are real numbers?
9. Which numbers are even integers?
10. Which numbers are odd integers?
11. Which numbers are irrational?
12. Which numbers are negative numbers?

Place one of the symbols $=$, $<$, *or* $>$ *in each box to make a true statement.*

13. $-3 \;\square\; 5 - 5$

14. $\dfrac{12}{4} \;\square\; 7$

15. $\dfrac{36}{4} \;\square\; -2$

16. $2 - 2 \;\square\; 8 - \dfrac{24}{3}$

Simplify each expression.

17. $-(-9)$ **18.** $-(12 - 4)$

Draw a number line and graph each set of numbers.

19. The composite numbers from 14 to 20

20. The whole numbers between 19 and 25

21. The real numbers less than or equal to -3 or greater than 2

22. The real numbers greater than -4 and less than 3

Find each absolute value.

23. $|29 - 24|$ **24.** $|-25|$

SECTION 1.2   Fractions

DEFINITIONS AND CONCEPTS	EXAMPLES
To simplify a fraction, factor the numerator and the denominator. Then divide out all common factors.	Simplify: $\dfrac{12}{32}$ $\dfrac{12}{32} = \dfrac{4 \cdot 3}{4 \cdot 8} = \dfrac{\overset{1}{\cancel{4}} \cdot 3}{\underset{1}{\cancel{4}} \cdot 8} = \dfrac{3}{8}$
To multiply two fractions, multiply their numerators and multiply their denominators.	$4 \cdot \dfrac{5}{6} = \dfrac{4}{1} \cdot \dfrac{5}{6}$ $= \dfrac{4 \cdot 5}{1 \cdot 6}$ $= \dfrac{\overset{1}{\cancel{2}} \cdot 2 \cdot 5}{1 \cdot \underset{1}{\cancel{2}} \cdot 3}$ $= \dfrac{10}{3}$
To divide two fractions, multiply the first by the reciprocal of the second.	$\dfrac{2}{3} \div \dfrac{5}{6} = \dfrac{2}{3} \cdot \dfrac{6}{5}$ $= \dfrac{2 \cdot 6}{3 \cdot 5}$ $= \dfrac{2 \cdot 2 \cdot \overset{1}{\cancel{3}}}{\underset{1}{\cancel{3}} \cdot 5}$ $= \dfrac{4}{5}$
To add (or subtract) two fractions with like denominators, add (or subtract) their numerators and keep their common denominator.	$\dfrac{9}{11} + \dfrac{2}{11} = \dfrac{9 + 2}{11}$ $= \dfrac{11}{11}$ $= 1$
To add (or subtract) two fractions with unlike denominators, find equivalent fractions with the same denominator (LCD), add (or subtract) their numerators, and keep the common denominator.	Subtract: $\dfrac{11}{12} - \dfrac{3}{4}$ Begin by finding the LCD. $\left. \begin{array}{l} 12 = 2 \cdot 2 \cdot 3 \\ 4 = 2 \cdot 2 \end{array} \right\}$ LCD $= 2 \cdot 2 \cdot 3 = 12$ Write $\frac{3}{4}$ as a fraction with a denominator of 12 and then do the subtraction. $\dfrac{11}{12} - \dfrac{3}{4} = \dfrac{11}{12} - \dfrac{3 \cdot 3}{4 \cdot 3}$ $= \dfrac{11}{12} - \dfrac{9}{12}$ $= \dfrac{11 - 9}{12}$

$$= \frac{2}{12}$$

$$= \frac{\overset{1}{\cancel{2}}}{\underset{1}{\cancel{2}} \cdot 6}$$

$$= \frac{1}{6}$$

Before working with mixed numbers, convert them to improper fractions.	Write $5\frac{7}{9}$ as an improper fraction. $$5\frac{7}{9} = 5 + \frac{7}{9} = \frac{45}{9} + \frac{7}{9} = \frac{52}{9}$$
A **percent** is the numerator of a fraction with a denominator of 100.	$5\frac{1}{2}\%$ can be written as $\frac{5.5}{100}$, or as the decimal 0.055.

REVIEW EXERCISES

Simplify each fraction.

25. $\dfrac{45}{27}$ **26.** $\dfrac{48}{18}$

Perform each operation and simplify the answer, if possible.

27. $\dfrac{31}{15} \cdot \dfrac{10}{62}$ **28.** $\dfrac{25}{36} \cdot \dfrac{12}{15} \cdot \dfrac{3}{5}$

29. $\dfrac{18}{21} \div \dfrac{6}{7}$ **30.** $\dfrac{14}{24} \div \dfrac{7}{12} \div \dfrac{2}{5}$

31. $\dfrac{7}{12} + \dfrac{9}{12}$ **32.** $\dfrac{13}{24} - \dfrac{5}{24}$

33. $\dfrac{1}{5} + \dfrac{1}{4}$ **34.** $\dfrac{5}{7} + \dfrac{4}{9}$

35. $\dfrac{2}{3} - \dfrac{1}{7}$ **36.** $\dfrac{4}{5} - \dfrac{2}{3}$

37. $3\dfrac{2}{3} + 5\dfrac{1}{4}$ **38.** $7\dfrac{5}{12} - 4\dfrac{1}{2}$

Perform each operation.

39. $48.29 + 31.90$ **40.** $36.85 - 15.86$

41. $4.32 \cdot 1.5$ **42.** $21.83 \div 5.9$

Perform each operation and round to two decimal places.

43. $2.7(4.92 - 3.18)$ **44.** $\dfrac{3.3 + 2.5}{0.22}$

45. $\dfrac{12.5}{14.7 - 11.2}$ **46.** $(3 - 0.7)(3.63 - 2)$

47. Farming   One day, a farmer plowed $17\frac{1}{2}$ acres and on the second day, $15\frac{3}{4}$ acres. How much is left to plow if the fields total 100 acres?

48. Study times   Four students recorded the time they spent working on a take-home exam: 5.2, 4.7, 9.5, and 8 hours. Find the average time spent.

49. Absenteeism   During the height of the flu season, 20% of the 425 university faculty members were sick. How many were ill?

50. Packaging   Four steel bands surround the shipping crate in the illustration. Find the total length of strapping needed.

4.2 ft

2.7 ft

1.2 ft

SECTION 1.3   Exponents and Order of Operations

DEFINITIONS AND CONCEPTS	EXAMPLES
If *n* is a natural number, then $$x^n = \underbrace{x \cdot x \cdot x \cdot x \cdots \cdot x}_{n \text{ factors of } x}$$	$x^5 = x \cdot x \cdot x \cdot x \cdot x$ $b^7 = b \cdot b \cdot b \cdot b \cdot b \cdot b \cdot b$
Order of operations Within each pair of grouping symbols (working from the innermost pair to the outermost pair), perform the following operations: **1.** Evaluate all exponential expressions. **2.** Perform multiplications and divisions, working from left to right. **3.** Perform additions and subtractions, working from left to right.	Evaluate: $6^2 - 5(12 - 2 \cdot 5)$ $6^2 - 5(12 - 2 \cdot 5) = 6^2 - 5(12 - 10)$ Do the multiplication within the parentheses. $= 6^2 - 5(2)$ Do the subtraction within the parentheses. $= 36 - 5(2)$ Find the value of the exponential expression. $= 36 - 10$ Do the multiplication. $= 26$ Do the subtraction.

4. Because the bar in a fraction is a grouping symbol, simplify the numerator and the denominator of a fraction separately. Then simplify the fraction, whenever possible.

To simplify $\dfrac{2^3 + 4 \cdot 2}{2 + 6}$, we first simplify the numerator and the denominator.

$$\dfrac{2^3 + 4 \cdot 2}{2 + 6} = \dfrac{8 + 8}{8} \quad \text{Find the power. Then find the product. Find the sum in the denominator.}$$

$$= \dfrac{16}{8} \quad \text{Find the sum in the numerator.}$$

$$= 2 \quad \text{Find the quotient.}$$

To find perimeters, areas, and volumes of geometric figures, substitute numbers for variables in the formulas. Be sure to include the proper units in the answer.

Find the perimeter of a rectangle whose length is 4 feet and whose width is 1 foot.

$$P = 2l + 2w \quad \text{This is the formula for the perimeter of a rectangle.}$$

$$= 2(4) + 2(1) \quad \text{Substitute 4 for } l \text{ and 1 for } w.$$

$$= 8 + 2$$

$$= 10$$

The perimeter is 10 feet.

REVIEW EXERCISES

Find the value of each expression.

51. 3^4

52. $\left(\dfrac{2}{3}\right)^2$

53. $(0.5)^2$

54. $5^2 + 2^3$

55. $3^2 + 4^2$

56. $(3 + 4)^2$

57. Geometry Find the area of a triangle with a base of $6\frac{1}{2}$ feet and a height of 7 feet.

58. 📟 **Petroleum storage** Find the volume of the cylindrical storage tank in the illustration. Round to the nearest tenth.

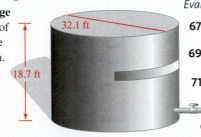

32.1 ft

18.7 ft

Simplify each expression.

59. $7 + 3^3$

60. $6 + 2 \cdot 4$

61. $5 + 6 \div 2$

62. $(8 + 6) \div 2$

63. $5^3 - \dfrac{81}{3}$

64. $(5 - 2)^2 + 5^2 + 2^2$

65. $\dfrac{4 \cdot 3 + 3^4}{31}$

66. $\dfrac{4}{3} \cdot \dfrac{9}{2} + \dfrac{1}{2} \cdot 18$

Evaluate each expression.

67. $8^2 - 6$

68. $(8 - 6)^2$

69. $\dfrac{10 + 2}{10 - 6}$

70. $\dfrac{6(8) - 12}{4 + 8}$

71. $2^2 + 2(3^2)$

72. $\dfrac{2^2 + 3}{2^3 - 1}$

SECTION 1.4   Adding and Subtracting Real Numbers

DEFINITIONS AND CONCEPTS	EXAMPLES
To add two positive numbers, add their absolute values and make the answer positive.	$(+1) + (+6) = +7$
To add two negative numbers, add their absolute values and make the answer negative.	$(-1) + (-6) = -7$
To add a positive and a negative number, subtract the smaller absolute value from the larger. **1.** If the positive number has the larger absolute value, the answer is positive. **2.** If the negative number has the larger absolute value, the answer is negative.	$(-1) + (+6) = +5$ $(+1) + (-6) = -5$
If a and b are two real numbers, then $a - b = a + (-b)$	$-8 - 2 = -8 + (-2) \quad \text{To subtract 2, add the opposite of 2.}$ $= -10$

REVIEW EXERCISES

Simplify each expression.

73. $(+15) + (+9)$ **74.** $(-17) + (-16)$

75. $(-2.7) + (-3.8)$ **76.** $\dfrac{1}{2} + \left(-\dfrac{1}{6}\right)$

77. $(+12) + (-24)$ **78.** $(-44) + (+60)$

79. $3.7 + (-2.5)$ **80.** $-5.6 + (+2.06)$

81. $15 - (-4)$ **82.** $-8 - (-15)$

83. $[-5 + (-5)] - (-5)$

84. $1 - [5 - (-3)]$

85. $-\dfrac{7}{10} - \left(-\dfrac{2}{5}\right)$ **86.** $\dfrac{2}{3} - \left(\dfrac{1}{3} - \dfrac{2}{3}\right)$

87. $\left| \dfrac{3}{7} - \left(-\dfrac{4}{7}\right) \right|$ **88.** $\dfrac{3}{7} - \left| -\dfrac{4}{7} \right|$

SECTION 1.5   Multiplying and Dividing Real Numbers

DEFINITIONS AND CONCEPTS	EXAMPLES
To multiply two real numbers, multiply their absolute values. 1. If the numbers are positive, the product is positive. 2. If the numbers are negative, the product is positive. 3. If one number is positive and the other is negative, the product is negative. 4. $a \cdot 0 = 0 \cdot a = 0$ 5. $a \cdot 1 = 1 \cdot a = a$	$3(7) = 21$ $-3(-7) = 21$ $-3(7) = -21 \qquad 3(-7) = -21$ $5(0) = 0$ $-7(1) = -7$
To divide two real numbers, find the quotient of their absolute values. 1. If the numbers are positive, the quotient is positive. 2. If the numbers are negative, the quotient is positive. 3. If one number is positive and the other is negative, the quotient is negative. 4. $\dfrac{a}{0}$ is undefined. 5. If $a \neq 0$, then $\dfrac{0}{a} = 0$.	$\dfrac{+6}{+2} = +3$ because $(+2)(+3) = +6$ $\dfrac{-6}{-2} = +3$ because $(-2)(+3) = -6$ $\dfrac{+6}{-2} = -3$ because $(-2)(-3) = +6$ $\dfrac{-6}{+2} = -3$ because $(+2)(-3) = -6$ $\dfrac{2}{0}$ is undefined because no number multiplied by 0 gives 2 $\dfrac{0}{2} = 0$ because $(2)(0) = 0$

REVIEW EXERCISES

Simplify each expression.

89. $(+5)(+8)$ **90.** $(-5)(-12)$

91. $\left(-\dfrac{3}{14}\right)\left(-\dfrac{7}{6}\right)$ **92.** $(3.75)(0.37)$

93. $5(-7)$ **94.** $(-15)(7)$

95. $\left(-\dfrac{1}{2}\right)\left(\dfrac{4}{3}\right)$ **96.** $(2.1)(-8.2)$

97. $\dfrac{+36}{+12}$ **98.** $\dfrac{-14}{-2}$

99. $\dfrac{(-2)(-7)}{4}$ **100.** $\dfrac{-22.5}{-3.75}$

101. $\dfrac{(-2)(-9)}{-3}$

102. $\dfrac{(-6)(12)}{-4}$

103. $\left(\dfrac{-10}{2}\right)^2 - (-1)^3$

104. $\dfrac{[-3 + (-4)]^2}{10 + (-3)}$

105. $\left(\dfrac{-3 + (-3)}{3}\right)\left(\dfrac{-15}{5}\right)$

106. $\dfrac{-2 - (-8)}{5 + (-1)}$

SECTION 1.6   Algebraic Expressions

DEFINITIONS AND CONCEPTS	EXAMPLES
Variables and numbers can be combined with operations of arithmetic to produce **algebraic expressions**.	$5x \qquad 3x^2 + 7x \qquad 5(3x - 8)$
We can **evaluate** algebraic expressions when we know the values of the variables.	Evaluate: $5x - 2$ when $x = 3$ $5x + 2 = 5(3) - 2$ Substitute 3 for x. $\qquad\quad = 15 - 2$ $\qquad\quad = 13$
Numbers written without variables are called **constants**.	Identify the constant in $6x^2 - 4x + 2$ The constant is 2.
Expressions that are constants, variables, or products of constants and variables are called **algebraic terms**.	Identify the algebraic terms: $6x^2 - 4x + 2$ The terms are $6x^2$, $-4x$, and 2.
Numbers and variables that are part of a product are called **factors**.	Identify the factors in $7x$. The factors are 7 and x.
The number factor of a product is called its **numerical coefficient**.	Identify the numerical coefficient of $7x$. The numerical coefficient is 7.

REVIEW EXERCISES

Let x, y, and z represent three real numbers. Write an algebraic expression that represents each quantity.

107. The product of x and z

108. The sum of x and twice y

109. Twice the sum of x and y

110. x decreased by the product of y and z

Write each algebraic expression as an English phrase.

111. $5xz$

112. $5 - yz$

113. $xy - 4$

114. $\dfrac{x + y + z}{2xyz}$

Let x = 2, y = −3, and z = −1 and evaluate each expression.

115. $x + z$

116. $x + y + z$

117. $5x + (y - z)$

118. $z^2 - y$

119. $x - (y - z)$

120. $(x - y) - z$

Let x = 2, y = −3, and z = −1 and evaluate each expression.

121. yz

122. xyz

123. $(x + y)(y + z)$

124. $\dfrac{3(x - y)}{x + (y - z)}$

125. $y^2z + x$

126. $yz^3 + (xy)^2$

127. $\dfrac{2y^2}{3x - 6}$

128. $\dfrac{|xy|}{3z}$

129. How many terms does the expression $3x + 4y + 9$ have?

130. What is the numerical coefficient of the term $7xy$?

131. What is the numerical coefficient of the term xy?

132. Find the sum of the numerical coefficients in $2x^3 + 4x^2 + 3x$.

SECTION 1.7   Properties of Real Numbers

DEFINITIONS AND CONCEPTS	EXAMPLES
The closure properties: $\qquad a + b$ is a real number. $\qquad a - b$ is a real number. $\qquad ab$ is a real number. $\qquad \dfrac{a}{b}$ is a real number $(b \neq 0)$.	$5 + (-2) = 3$ is a real number. $5 - 2 = 3$ is a real number. $5(-2) = -10$ is a real number. $\dfrac{10}{-5} = -2$ is a real number.

The commutative properties:	
$a + b = b + a$ of addition	The commutative property of addition justifies the statement $x + 3 = 3 + x$.
$ab = ba$ of multiplication	The commutative property of multiplication justifies the statement $x \cdot 3 = 3 \cdot x$.
The associative properties:	
$(a + b) + c = a + (b + c)$ of addition	The associative property of addition justifies the statement $(x + 3) + 4 = x + (3 + 4)$.
$(ab)c = a(bc)$ of multiplication	The associative property of multiplication justifies the statement $(x \cdot 3) \cdot 4 = x \cdot (3 \cdot 4)$.
The distributive property of multiplication over addition:	Use the distributive property to write the expression $7(2x - 8)$ without parentheses.
$a(b + c) = ab + ac$	$7(2x - 8) = 7(2x) - 7(8) = 14x - 56$
$a(b - c) = ab - ac$	
The identity elements:	
0 is the identity for addition (additive identity).	$0 + 5 = 5$
1 is the identity for multiplication (multiplicative identity).	$1 \cdot 5 = 5$
The additive and multiplicative inverse properties:	
$a + (-a) = 0$	The additive inverse of -2 is 2 because $-2 + 2 = 0$.
$a\left(\dfrac{1}{a}\right) = 1 \quad (a \ne 0)$	The multiplicative inverse of -2 is $-\dfrac{1}{2}$ because $-2\left(-\dfrac{1}{2}\right) = 1$.

REVIEW EXERCISES

Determine which property of real numbers justifies each statement. Assume that all variables represent real numbers.

133. $b + c$ is a real number
134. $3 \cdot (4 \cdot 5) = (4 \cdot 5) \cdot 3$
135. $3 + (4 + 5) = (3 + 4) + 5$
136. $5(x + 2) = 5 \cdot x + 5 \cdot 2$

137. $b + c = c + b$
138. $3 \cdot (4 \cdot 5) = (3 \cdot 4) \cdot 5$
139. $3 + (x + 1) = (x + 1) + 3$
140. $x \cdot 1 = x$
141. $8 + (-8) = 0$
142. $x + 0 = x$

1   Test

1. List the prime numbers between 30 and 50.

2. What is the only even prime number?

3. Graph the composite numbers less than 10 on a number line.

4. Graph the real numbers from 5 to 15 on a number line.

5. Evaluate: $-|-17|$

6. Evaluate: $-|9| + |-9|$

Place one of the symbols $=$, $<$, or $>$ in each box to make a true statement.

7. $3(4 - 2) \; \boxed{} \; -2(2 - 5)$ **8.** $1 + 4 \cdot 3 \; \boxed{} \; -2(-7)$

9. 25% of 136 $\boxed{}$ $\dfrac{1}{2}$ of 66   **10.** $-8.5 \; \boxed{} \; -|-8.5|$

Simplify each expression.

11. $\dfrac{26}{40}$ **12.** $\dfrac{9}{11} \cdot \dfrac{44}{45}$

13. $\dfrac{14}{21} \div \dfrac{28}{9}$ **14.** $\dfrac{24}{16} + 3$

15. $\dfrac{17 - 5}{36} - \dfrac{2(13 - 5)}{12}$ **16.** $\dfrac{|-7 - (-6)|}{-7 - |-6|}$

17. Find 13% of 256 and round the answer to one decimal place.

18. Find the area of a rectangle 18.9 feet wide and 21.25 feet long. Round the answer to two decimal places.

19. Find the area of the triangle in the illustration.

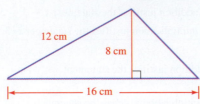

20. To the nearest cubic inch, find the volume of the solid in the illustration.

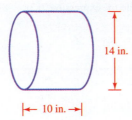

Let $x = -2, y = 3,$ and $z = 4$. Evaluate each expression.

21. $xy + z$

22. $x(y + z)$

23. $\dfrac{z + 4y}{2x}$

24. $|x^3 - z|$

25. $x^3 + y^2 + z$

26. $|x| - 3|y| - 4|z|$

27. Let x and y represent two real numbers. Write an algebraic expression to denote the quotient obtained when the product of the two numbers is divided by their sum.

28. Let x and y represent two real numbers. Write an algebraic expression to denote the difference obtained when the sum of x and y is subtracted from the product of 5 and y.

29. A man lives 12 miles from work and 7 miles from the grocery store. If he made x round trips to work and y round trips to the store, write an expression to represent how many miles he drove.

30. A baseball costs $\$a$ and a glove costs $\$b$. Write an expression to represent how much it will cost a community center to buy 12 baseballs and 8 gloves.

31. What is the numerical coefficient of the term $-5x^2y^3$?

32. How many terms are in the expression $3x^2y + 5xy^2 + x + 7$?

Write each expression without using parentheses.

33. $3(x + 2)$

34. $-p(r - t)$

35. What is the identity element for addition?

36. What is the multiplicative inverse of $\frac{1}{5}$?

Determine which property of the real numbers justifies each statement.

37. $(xy)z = z(xy)$

38. $3(x + y) = 3x + 3y$

39. $2 + x = x + 2$

40. $7 \cdot \dfrac{1}{7} = 1$

Equations and Inequalities

© iStockphoto.com/John Cowie

Careers and Mathematics

SECURITIES AND FINANCIAL SERVICES SALES AGENTS

Many investors use securities and financial sales agents when buying or selling stocks, bonds, shares in mutual funds, annuities, or other financial products. The overwhelming majority of people in this occupation are college graduates, with courses in business administration, economics, mathematics, and finance. After working for a few years, many agents get a Master's degree in Business Administration (MBA).

Job Outlook:
Employment of people in this field is expected to grow rapidly over the next decade, especially in banking. However, there will be keen competition for these jobs.

Annual Earnings:
$42,630–$126,290

For More Information:
http://www.bls.gov/oco/ocos122.htm

For a Sample Application:
See Problem 59 in Section 2.5.

In this chapter

In this chapter, we will learn how to solve basic linear equations and apply that knowledge to solving many types of problems. We also will consider special equations called formulas and conclude by solving linear inequalities.

Reach for Success   Reading This Mathematics Textbook

You might be asking yourself, "Who reads their math book anyway?" The better question might be, "Why should I read it?" or "How do I read it?" Why is easy to answer . . . to improve understanding and, thus, your grade! How is not as easy to answer. It takes practice.

Get acquainted with some of the features of this textbook.

Turn to the first page of Section 2.2 in this chapter. On which page did you find this? _____

Note that the objectives are provided at the beginning of each section. Why do you think these are listed?

_____

Note the vocabulary list at the beginning of each section. Where might you find the definitions?

1. _____   2. _____

What is the purpose of the self-check exercises that directly follow each section's worked-out examples?

_____

A mathematics textbook is structured differently than texts in other subjects. Usually, each chapter is divided into 5 to 7 sections, consisting of 6 to 10 pages each. Thus, you only need to study a few pages at a time. Can you see an advantage to this structure? _____

_____

_____

One advantage to fewer pages is that you can study a smaller "chunk" of mathematics at a time. Use your syllabus (or ask your instructor) to identify which objectives are covered in each section for your course.

Which objectives are you responsible for learning in the second section of this chapter? _____

_____

Did you know that the answers to the odd-numbered exercises are in the back of the book? _____
Why could this be important to you? _____

_____

What is the purpose of the glossary? _____

When would you use it?

_____

What is the purpose of the index? _____

When would you use it?

_____

A Successful Study Strategy . . .

 Read each textbook section prior to the classroom discussion. Even having an idea of the topic for the day will better prepare you for learning.

At the end of the chapter you will find an additional exercise to help guide you to planning for a successful semester.

Section 2.1

Solving Basic Linear Equations in One Variable

Objectives

1. Determine whether a statement is an expression or an equation.
2. Determine whether a number is a solution of an equation.
3. Solve a linear equation in one variable by applying the addition or subtraction property of equality.
4. Solve a linear equation in one variable by applying the multiplication or division property of equality.
5. Solve a linear equation in one variable involving markdown and markup.
6. Solve a percent problem involving a linear equation in one variable using the formula $rb = a$.
7. Solve an application involving percents.

Vocabulary

equation	linear equation	discount
expression	addition property of equality	markup
variable	equivalent equations	rate
solution	multiplication property	base
root	of equality	amount
solution set	markdown	

Getting Ready

Fill in the blanks.

1. $3 + \boxed{} = 0$
2. $(-7) + \boxed{} = 0$
3. $(-x) + \boxed{} = 0$
4. $\frac{1}{3} \cdot 3 = \boxed{}$
5. $x \cdot \boxed{} = 1 \quad x \neq 0$
6. $\frac{-6}{-6} = \boxed{}$
7. $\frac{4(2)}{\boxed{}} = 2$
8. $5 \cdot \frac{4}{5} = \boxed{}$
9. $\frac{-5(3)}{-5} = \boxed{}$

To answer questions such as "How many?," "How far?," "How fast?," and "How heavy?," we will often use mathematical statements called *equations*. In this chapter, we will discuss this important concept.

1 Determine whether a statement is an expression or an equation.

An **equation** is a statement indicating that two quantities are equal. Some examples of equations are

$$x + 5 = 21 \qquad 2x - 5 = 11 \qquad \text{and} \qquad 3x^2 - 4x + 5 = 0$$

The **expression** $3x + 2$ is not an equation, because we do not have two quantities that are being compared to one another. Some examples of expressions are

$$6x - 1 \qquad 3x^2 - x - 2 \qquad \text{and} \qquad -8(x + 1)$$

EXAMPLE 1   Determine whether the following are expressions or equations.

    a. $9x^2 - 5x = 4$ **b.** $3x + 2$ **c.** $6(2x - 1) + 5$

Solution   **a.** $9x^2 - 5x = 4$ is an equation because it indicates that $9x^2 - 5x$ is equal to 4.

    b. $3x + 2$ is an expression because we do not have $3x + 2$ set equal to another quantity.

    c. $6(2x - 1) + 5$ is an expression because we do not have $6(2x - 1) + 5$ set equal to another quantity.

 SELF CHECK 1   Is $8(x + 1) = 4$ an expression or an equation?

---

2 Determine whether a number is a solution of an equation.

In the equation $x + 5 = 21$, the expression $x + 5$ is called the *left side* and 21 is called the *right side*. The letter x is called the **variable** (or the **unknown**).

An equation can be true or false. The equation $16 + 5 = 21$ is true, but the equation $10 + 5 = 21$ is false. The equation $2x - 5 = 11$ might be true or false, depending on the value of x. For example, when $x = 8$, the equation is true, because when we substitute 8 for x we obtain 11.

$$2(\mathbf{8}) - 5 = 16 - 5$$
$$= 11$$

Any number that makes an equation true when substituted for its variable is said to *satisfy* the equation. A number that makes an equation true is called a **solution** or a **root** of the equation. Since 8 is the only number that satisfies the equation $2x - 5 = 11$, it is the only solution.

The **solution set** of an equation is the set of numbers that make the equation true. In the previous equation, the solution set is {8}.

EXAMPLE 2   Determine whether 6 is a solution of $3x - 5 = 2x$.

Solution   To see whether 6 is a solution, we can substitute 6 for x and simplify.

$$3\mathbf{x} - 5 = 2\mathbf{x}$$
$$3 \cdot \mathbf{6} - 5 \stackrel{?}{=} 2 \cdot \mathbf{6} \qquad \text{Substitute 6 for } x.$$
$$18 - 5 \stackrel{?}{=} 12 \qquad \text{Do the multiplication.}$$
$$13 = 12 \qquad \text{False.}$$

Since $13 = 12$ is a false statement, 6 is not a solution.

 SELF CHECK 2   Determine whether 1 is a solution of $2x + 3 = 5$.

---

3 Solve a linear equation in one variable by applying the addition or subtraction property of equality.

To solve an equation means to find its solutions. To develop an understanding of how to solve basic equations of the form $ax + b = c$, $a \neq 0$, called **linear equations**, we will refer to the scales shown in Figure 2-1. We can think of the scale shown in Figure 2-1(a) as representing the equation $x - 5 = 2$. The weight on the left side of the scale is $(x - 5)$ grams, and the weight on the right side is 2 grams. Because these weights are equal, the scale is in balance. To find the value of x, we need to isolate it by adding 5 grams to the left side of the scale. To keep the scale in balance, we must also add 5 grams to the right side. After adding 5 grams to both sides of the scale, we can see from Figure 2-1(b) that x grams will be balanced by 7 grams. We say that we have solved the equation and that the solution is 7, or we can say that the solution set is {7}.

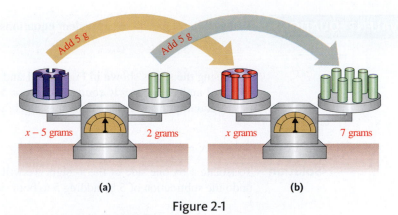

Figure 2-1

Figure 2-1 suggests the **addition property of equality**: *If the same quantity is added to equal quantities, the results will be equal quantities.*

We can think of the scale shown in Figure 2-2(a) as representing the equation $x + 4 = 9$. The weight on the left side of the scale is $(x + 4)$ grams, and the weight on the right side is 9 grams. Because these weights are equal, the scale is in balance. To find the value of x, we need to isolate it by removing 4 grams from the left side. To keep the scale in balance, we must also remove 4 grams from the right side. In Figure 2-2(b), we can see that x grams will be balanced by 5 grams. We have found that the solution is 5, or that the solution set is {5}.

François Vieta (Viete)
1540–1603
By using letters in place of unknown numbers, Vieta simplified algebra and brought its notation closer to the notation that we use today. The one symbol he didn't use was the equal sign.

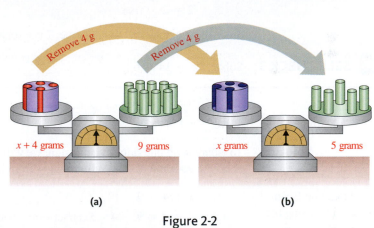

Figure 2-2

Figure 2-2 suggests the *subtraction property of equality: If the same quantity is subtracted from equal quantities, the results will be equal quantities.*

The previous discussion justifies the following properties.

ADDITION PROPERTY OF EQUALITY	Suppose that a, b, and c are real numbers. If $a = b$, then $a + c = b + c$.

SUBTRACTION PROPERTY OF EQUALITY	Suppose that a, b, and c are real numbers. If $a = b$, then $a - c = b - c$.

COMMENT The subtraction property of equality is a special case of the addition property. Instead of subtracting a number from both sides of an equation, we could add the opposite of the number to both sides.

When we use the properties described above, the resulting equation will have the same solution set as the original one. We say that the equations are *equivalent*.

EQUIVALENT EQUATIONS	Two equations are called **equivalent equations** when they have the same solution set.

Using the scales shown in Figures 2-1 and 2-2, we found that $x - 5 = 2$ is equivalent to $x = 7$ and $x + 4 = 9$ is equivalent to $x = 5$. In the next two examples, we use properties of equality to solve these equations algebraically.

EXAMPLE 3   Solve: $x - 5 = 2$

Solution   To isolate x on one side of the = sign, we will use the addition property of equality to undo the subtraction of 5 by adding 5 to both sides of the equation.

$$x - 5 = 2$$
$$x - 5 \mathbf{+ 5} = 2 \mathbf{+ 5} \quad \text{Add 5 to both sides of the equation.}$$
$$x + 0 = 7 \quad \text{Apply the additive inverse property.}$$
$$x = 7 \quad \text{Apply the additive identity property.}$$

We check by substituting 7 for x in the original equation and simplifying.

$$x - 5 = 2$$
$$7 - 5 \overset{?}{=} 2 \quad \text{Substitute 7 for } x.$$
$$2 = 2 \quad \text{True.}$$

Since the previous statement is true, 7 is a solution. The solution set of this equation is $\{7\}$.

 SELF CHECK 3   Solve: $b - 14 = 6$

EXAMPLE 4   Solve: $x + 4 = 9$

Solution   To isolate x on one side of the = sign, we will use the subtraction property of equality to undo the addition of 4 by subtracting 4 from both sides of the equation.

COMMENT   Note that Example 4 can be solved by using the addition property of equality. We would add -4 to both sides to undo the addition of 4.

$$x + 4 = 9$$
$$x + 4 \mathbf{- 4} = 9 \mathbf{- 4} \quad \text{Subtract 4 from both sides.}$$
$$x + 0 = 5 \quad \text{Apply the additive inverse property.}$$
$$x = 5 \quad \text{Apply the additive identity property.}$$

We can check by substituting 5 for x in the original equation and simplifying.

$$x + 4 = 9$$
$$5 + 4 \overset{?}{=} 9 \quad \text{Substitute 5 for } x.$$
$$9 = 9 \quad \text{True.}$$

Since the solution 5 checks, the solution set is $\{5\}$.

 SELF CHECK 4   Solve: $a + 175 = 122$

4   ### Solve a linear equation in one variable by applying the multiplication or division property of equality.

We can think of the scale shown in Figure 2-3(a) as representing the equation $\frac{x}{3} = 12$. The weight on the left side of the scale is $\frac{x}{3}$ grams, and the weight on the right side is 12 grams. Because these weights are equal, the scale is in balance. To find the value of x, we can triple (or multiply by 3) the weight on each side. When we do this, the scale will remain in balance. From the scale shown in Figure 2-3(b), we can see that x grams will be balanced by 36 grams. Thus, $x = 36$. Since 36 is the solution of the equation, the solution set is {36}.

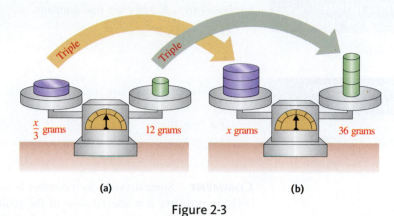

(a)   (b)

Figure 2-3

Perspective

To answer questions such as "How many?," "How far?," "How fast?," and "How heavy?," we often make use of equations. This concept has a long history, and the techniques that we will study in this chapter have been developed over many centuries.

The mathematical notation that we use today to solve equations is the result of thousands of years of development. The ancient Egyptians used a word for variables, best translated as *heap*. Others used the word *res*, which is Latin for *thing*. In the fifteenth century, the letters *p*: and *m*: were used for *plus* and *minus*. What we would now write as $2x + 3 = 5$ might have been written by those early mathematicians as

   2 *res p*:3 *aequalis* 5

Figure 2-3 suggests the **multiplication property of equality**: *If equal quantities are multiplied by the same quantity, the results will be equal quantities.*

We will now consider how to solve the equation $2x = 6$. Since $2x$ means $2 \cdot x$, the equation can be written as $2 \cdot x = 6$. We can think of the scale shown in Figure 2-4(a) as representing this equation. The weight on the left side of the scale is $2 \cdot x$ grams, and the weight on the right side is 6 grams. Because these weights are equal, the scale is in balance. To find the value of x, we remove half of the weight from each side. This is equivalent to dividing the weight on both sides by 2. When we do this, the scale will remain in balance. From the scale shown in Figure 2-4(b), we can see that x grams will be balanced by 3 grams. Thus, $x = 3$. Since 3 is a solution of the equation, the solution set is {3}.

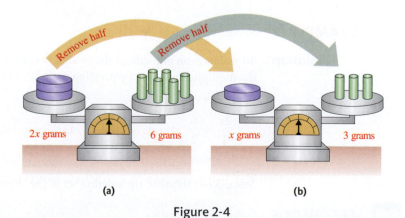

(a)   (b)

Figure 2-4

Figure 2-4 suggests the *division property of equality: If equal quantities are divided by the same quantity, the results will be equal quantities.*

The previous discussion justifies the following properties.

MULTIPLICATION PROPERTY OF EQUALITY	Suppose that a, b, and c are real numbers.
	If $a = b$, then $ca = cb$.

DIVISION PROPERTY OF EQUALITY	Suppose that a, b, and c are real numbers and $c \neq 0$.
	If $a = b$, then $\dfrac{a}{c} = \dfrac{b}{c}$.

COMMENT Since dividing by a number is the same as multiplying by its reciprocal, the division property is a special case of the multiplication property. However, because the reciprocal of 0 is undefined, we must exclude the possibility of division by 0.

When we use the multiplication and division properties, the resulting equations will be equivalent to the original ones.

To solve the previous equations algebraically, we proceed as in the next examples.

EXAMPLE 5 Solve: $\dfrac{x}{3} = 12$

Solution To isolate x on one side of the $=$ sign, we use the multiplication property of equality to undo the division by 3 by multiplying both sides of the equation by 3.

$$\frac{x}{3} = 12$$

$$\mathbf{3} \cdot \frac{x}{3} = \mathbf{3} \cdot 12 \qquad \text{\color{red}{Multiply both sides by 3.}}$$

$$x = 36 \qquad \text{\color{red}{$3 \cdot \frac{x}{3} = x$ and $3 \cdot 12 = 36$.}}$$

Since 36 is a solution, the solution set is $\{36\}$. Verify that the solution checks.

 SELF CHECK 5 Solve: $\dfrac{x}{5} = -7$

EXAMPLE 6 Solve: $2x = 6$

Solution To isolate x on one side of the $=$ sign, we use the division property of equality to undo the multiplication by 2 by dividing both sides by 2.

$$2x = 6$$

$$\frac{2x}{\mathbf{2}} = \frac{6}{\mathbf{2}} \qquad \text{\color{red}{Divide both sides by 2.}}$$

$$x = 3 \qquad \text{\color{red}{$\frac{2}{2} = 1$ and $\frac{6}{2} = 3$.}}$$

Since 3 is a solution, the solution set is $\{3\}$. Verify that the solution checks.

 SELF CHECK 6 Solve: $-5x = 15$

COMMENT Note that we could have solved the equation in Example 6 by using the multiplication property of equality. To isolate x, we could have multiplied both sides by $\frac{1}{2}$.

EXAMPLE 7   Solve: $3x = \dfrac{1}{5}$

Solution   To isolate x on the left side of the equation, we could undo the multiplication by 3 by dividing both sides by 3. However, it is easier to isolate x by multiplying both sides by the reciprocal of 3, which is $\dfrac{1}{3}$.

$$3x = \frac{1}{5}$$

$$\mathbf{\frac{1}{3}}(3x) = \mathbf{\frac{1}{3}}\left(\frac{1}{5}\right) \qquad \text{Multiply both sides by } \tfrac{1}{3}.$$

$$\left(\frac{1}{3} \cdot 3\right)x = \frac{1}{15} \qquad \text{Apply the associative property of multiplication.}$$

$$1x = \frac{1}{15} \qquad \text{Apply the multiplicative inverse property.}$$

$$x = \frac{1}{15} \qquad \text{Apply the multiplicative identity property.}$$

Since the solution is $\dfrac{1}{15}$, the solution set is $\left\{\dfrac{1}{15}\right\}$. Verify that the solution checks.

 SELF CHECK 7   Solve: $-5x = \dfrac{1}{3}$

5 Solve a linear equation in one variable involving markdown and markup.

When the price of merchandise is reduced, the amount of reduction is called the **markdown** or the **discount**. To find the sale price of an item, we subtract the markdown from the regular price.

EXAMPLE 8   **BUYING FURNITURE**   A sofa is on sale for $650. If it has been marked down $325, find its regular price.

Solution   We can let r represent the regular price and substitute 650 for the sale price and 325 for the markdown in the following formula.

Sale price	equals	regular price	minus	markdown.
650	=	r	−	325

We can use the addition property of equality to solve the equation.

$$650 = r - 325$$
$$650 \mathbf{\,+\,325} = r - 325 \mathbf{\,+\,325} \qquad \text{Add 325 to both sides.}$$
$$975 = r \qquad\qquad 650 + 325 = 975 \text{ and } -325 + 325 = 0.$$

The regular price is $975.

 SELF CHECK 8   Find the regular price of the sofa if the discount is $275.

To make a profit, a merchant must sell an item for more than he paid for it. The retail price of the item is the sum of its wholesale cost and the **markup**.

EXAMPLE 9   **BUYING CARS**   A car with a sticker price of $17,500 has a markup of $3,500. Find the invoice price (the wholesale price) to the dealer.

Solution   We can let w represent the wholesale price and substitute 17,500 for the retail price and 3,500 for the markup in the following formula.

Retail price	equals	wholesale cost	plus	markup.
17,500	=	w	+	3,500

We can use the subtraction property of equality to solve the equation.

$$17,500 = w + 3,500$$
$$17,500 - \mathbf{3,500} = w + 3,500 - \mathbf{3,500} \qquad \text{Subtract 3,500 from both sides.}$$
$$14,000 = w \qquad\qquad\qquad 17,500 - 3,500 = 14,000 \text{ and}$$
$$3,500 - 3,500 = 0.$$

The invoice price is $14,000.

SELF CHECK 9   Find the invoice price of the car if the markup is $6,700.

6   **Solve a percent problem involving a linear equation in one variable using the formula $rb = a$.**

A percent is the numerator of a fraction with a denominator of 100. For example, $6\frac{1}{4}$ percent (written as $6\frac{1}{4}\%$) is the fraction $\frac{6.25}{100}$, or the decimal 0.0625. In problems involving percent, the word *of* usually means multiplication. For example, $6\frac{1}{4}\%$ of 8,500 is the product of 0.0625 and 8,500.

$$6\frac{1}{4}\% \text{ of } 8,500 = 0.0625 \cdot 8,500$$
$$= 531.25$$

In the statement $6\frac{1}{4}\%$ of $8,500 = 531.25$, the percent $6\frac{1}{4}\%$ is called a **rate**, 8,500 is called the **base**, and their product, 531.25, is called the **amount**. Every percent problem is based on the equation rate $\cdot$ base = amount.

PERCENT FORMULA   If r is the rate, b is the base, and a is the amount, then

$$rb = a.$$

COMMENT   Note that the percent formula can be written in the equivalent form $a = rb$.

Percent problems involve questions such as the following.

- What is 30% of 1,000?   We must find the amount.
- 45% of what number is 405?   We must find the base.
- What percent of 400 is 60?   We must find the rate.

When we substitute the values of the rate, base, and amount into the percent formula, we will obtain an equation that we can solve.

EXAMPLE 10   What is 30% of 1,000?

Solution   In this example, the rate r is 30% and the base is 1,000. We must find the amount.

Rate	$\cdot$	base	=	amount.
30%	of	1,000	is	the amount.

We can substitute these values into the percent formula and solve for a.

$$rb = a$$
$$30\% \cdot 1{,}000 = a \qquad \text{Substitute 30\% for } r \text{ and 1,000 for } b.$$
$$0.30 \cdot 1{,}000 = a \qquad \text{Change 30\% to the decimal 0.30.}$$
$$300 = a \qquad \text{Multiply.}$$

Thus, 30% of 1,000 is 300.

 SELF CHECK 10   Find 45% of 800.

EXAMPLE 11   45% of what number is 405?

Solution   In this example, the rate r is 45% and the amount a is 405. We must find the base.

| Rate | · | base | = | amount. |

45%   of   what number   is   405?

We can substitute these values into the percent formula and solve for b.

$$rb = a$$
$$45\% \cdot b = 405 \qquad \text{Substitute 45\% for } r \text{ and 405 for } a.$$
$$0.45 \cdot b = 405 \qquad \text{Change 45\% to a decimal.}$$
$$\frac{0.45b}{0.45} = \frac{405}{0.45} \qquad \text{To undo the multiplication by 0.45, divide both sides by 0.45.}$$
$$b = 900 \qquad \tfrac{0.45}{0.45} = 1 \text{ and } \tfrac{405}{0.45} = 900.$$

Thus, 45% of 900 is 405.

 SELF CHECK 11   35% of what number is 280?

EXAMPLE 12   What percent of 400 is 60?

Solution   In this example, the base b is 400 and the amount a is 60. We must find the rate.

| Rate | · | base | = | amount. |

What percent   of   400   is   60?

We can substitute these values in the percent formula and solve for r.

$$rb = a$$
$$r \cdot 400 = 60 \qquad \text{Substitute 400 for } b \text{ and 60 for } a.$$
$$\frac{400r}{400} = \frac{60}{400} \qquad \text{To undo the multiplication by 400, divide both sides by 400.}$$
$$r = 0.15 \qquad \tfrac{400}{400} = 1 \text{ and } \tfrac{60}{400} = 0.15.$$
$$r = 15\% \qquad \text{To change the decimal into a percent, we multiply by 100 and insert a \% sign.}$$

Thus, 15% of 400 is 60.

 SELF CHECK 12   What percent of 600 is 150?

7 Solve an application involving percents.

The ability to solve linear equations enables us to solve many applications. This is what makes the algebra relevant to our lives.

EXAMPLE 13   **INVESTING**   At a stockholders meeting, members representing 4.5 million shares voted in favor of a proposal for a mandatory retirement age for the members of the board of directors. If these shares represented 75% of the number of shares outstanding, how many shares were outstanding?

Solution   Let b represent the number of outstanding shares. Then 75% of b is 4.5 million. We can substitute 75% for r and 4.5 million for a in the percent formula and solve for b.

$$rb = a$$
$$75\% \cdot b = 4{,}500{,}000 \qquad \text{4.5 million} = 4{,}500{,}000.$$
$$0.75b = 4{,}500{,}000 \qquad \text{Change 75\% to a decimal.}$$
$$\frac{0.75b}{0.75} = \frac{4{,}500{,}000}{0.75} \qquad \text{To undo the multiplication of 0.75, divide both sides by 0.75.}$$
$$b = 6{,}000{,}000 \qquad \text{Divide.}$$

There were 6 million shares outstanding.

 **SELF CHECK 13**   If the 4.5 million shares represented 60% of the number of shares outstanding, how many shares were outstanding?

EXAMPLE 14   **QUALITY CONTROL**   After examining 240 sweaters, a quality-control inspector found 5 with defective stitching, 8 with mismatched designs, and 2 with incorrect labels. What percent were defective?

Solution   Let r represent the percent that are defective. Then the base b is 240 and the amount a is the number of defective sweaters, which is $5 + 8 + 2 = 15$. We can find r by using the percent formula.

$$rb = a$$
$$r \cdot 240 = 15 \qquad \text{Substitute 240 for } b \text{ and 15 for } a.$$
$$\frac{240r}{240} = \frac{15}{240} \qquad \text{To undo the multiplication of 240, divide both sides by 240.}$$
$$r = 0.0625 \qquad \text{Divide.}$$
$$r = 6.25\% \qquad \text{To change 0.0625 to a percent, multiply by 100 and insert a \% sign.}$$

The defect rate is 6.25%.

 **SELF CHECK 14**   If the inspector examined 400 sweaters and found the same number of defects as in Example 14, what percent were defective?

SELF CHECK
ANSWERS
1. equation   2. yes   3. 20   4. −53   5. −35   6. −3   7. $-\frac{1}{15}$ 8. \$925   9. \$10,800   10. 360
11. 800   12. 25%   13. 7,500,000 or 7.5 million   14. 3.75%

NOW TRY THIS

Solve each equation.

1. $\dfrac{x}{12} = 0$

2. $\dfrac{2}{3}x = 24$

3. $-25 + x = 25$

2.1 Exercises

WARM-UPS *State the property of equality you would use to solve each equation. Do not solve.*

1. $x - 5 = 15$
2. $x - 3 = 13$
3. $w + 5 = 7$
4. $x + 32 = 36$
5. $-8x = -24$
6. $-7x = 14$
7. $\dfrac{x}{5} = 2$
8. $\dfrac{x}{2} = -10$

REVIEW *Perform the operations. Simplify the result when possible.*

9. $\dfrac{4}{5} + \dfrac{2}{3}$
10. $\dfrac{5}{6} \cdot \dfrac{12}{25}$
11. $\dfrac{5}{9} \div \dfrac{3}{5}$
12. $\dfrac{15}{7} - \dfrac{10}{3}$
13. $3 + 5 \cdot 6$
14. $3 \cdot 4^2$
15. $3 + 4^3(-5)$
16. $\dfrac{5(-4) - 3(-2)}{10 - (-4)}$

VOCABULARY AND CONCEPTS *Fill in the blanks.*

17. An _____ is a statement that two quantities are equal. An _____ is a mathematical statement without an = sign.
18. A _____ or ___ of an equation is a number that satisfies the equation.
19. If two equations have the same solutions, they are called _____ equations.
20. To solve a linear equation, we isolate the _____, or unknown, on one side of the equation.
21. If the same quantity is added to _____ quantities, the results will be equal quantities.
22. If the same quantity is subtracted from equal quantities, the results will be _____ quantities.
23. If equal quantities are multiplied or divided by the same nonzero quantity, the results are _____ quantities.
24. An equation in the form $x + b = c$ is called a _____ equation.
25. Sale price = _____ − markdown
26. Retail price = wholesale cost + _____
27. A *percent* is the numerator of a fraction whose denominator is ___.
28. Rate · ____ = amount

GUIDED PRACTICE *Determine whether each statement is an expression or an equation.* SEE EXAMPLE 1. (OBJECTIVE 1)

29. $x = -4$
30. $y = 3$
31. $6x + 7$
32. $2(x - 3) + 1$
33. $x^2 + 2x = 3$
34. $3(x + 1) = 9$
35. $3(x - 4)$
36. $5(2 + x)$

Determine whether the given number is a solution of the equation. SEE EXAMPLE 2. (OBJECTIVE 2)

37. $x + 3 = 6; 3$
38. $x + 5 = 8; 3$
39. $2y - 5 = y; 4$
40. $x - 7 = 2; 9$
41. $\dfrac{y}{7} = 4; 28$
42. $\dfrac{c}{-5} = -2; -10$
43. $\dfrac{x}{5} = x; 0$
44. $\dfrac{x}{7} = 7x; 0$
45. $3k + 5 = 5k - 1; 3$
46. $2s - 1 = s + 7; 6$
47. $\dfrac{5 + x}{10} - x = \dfrac{1}{2}; 0$
48. $\dfrac{x - 5}{6} = 12 - x; 11$

Use the addition property of equality to solve each equation. Check all solutions. SEE EXAMPLE 3. (OBJECTIVE 3)

49. $a - 6 = 9$
50. $y - 9 = 20$
51. $b - 5 = -19$
52. $m - 5 = -12$
53. $4 = c - 9$
54. $1 = y - 5$
55. $r - \dfrac{1}{5} = \dfrac{3}{10}$
56. $\dfrac{4}{3} = -\dfrac{2}{3} + x$

Use the subtraction property of equality to solve each equation. Check all solutions. SEE EXAMPLE 4. (OBJECTIVE 3)

57. $y + 8 = 11$
58. $x + 4 = 12$
59. $a + 9 = -12$
60. $c + 6 = -9$
61. $41 = 45 + q$
62. $0 = r + 10$
63. $k + \dfrac{2}{3} = \dfrac{1}{5}$
64. $b + \dfrac{4}{7} = \dfrac{15}{14}$

Use the multiplication property of equality to solve each equation. Check all solutions. SEE EXAMPLE 5. (OBJECTIVE 4)

65. $\dfrac{x}{6} = 3$
66. $\dfrac{y}{11} = 4$
67. $\dfrac{b}{3} = 5$
68. $\dfrac{a}{5} = -3$
69. $\dfrac{a}{3} = \dfrac{1}{9}$
70. $\dfrac{a}{13} = \dfrac{1}{26}$

71. $\dfrac{u}{5} = -\dfrac{3}{10}$ **72.** $\dfrac{t}{-7} = \dfrac{1}{2}$

Use the division property of equality to solve each equation. Check all solutions. SEE EXAMPLE 6. (OBJECTIVE 4)

73. $7x = 28$ **74.** $25x = 625$

75. $11x = -121$ **76.** $-8a = -32$

77. $-4x = 36$ **78.** $-16y = 64$

79. $4w = 108$ **80.** $-66 = -6w$

Use the multiplication or division property of equality to solve each equation. Check all solutions. SEE EXAMPLE 7. (OBJECTIVE 4)

81. $5x = \dfrac{5}{8}$ **82.** $6x = \dfrac{2}{3}$

83. $\dfrac{1}{7}w = 14$ **84.** $-19x = -57$

85. $-1.2w = -102$ **86.** $1.5a = -15$

87. $0.25x = 1{,}228$ **88.** $-0.2y = 51$

Solve each application involving markdown or markup. SEE EXAMPLES 8–9. (OBJECTIVE 5)

89. Buying boats   A boat is on sale for $7,995. Find its regular price if it has been marked down $1,350.

90. Buying houses   A house that was priced at $105,000 has been discounted $7,500. Find the new asking price.

91. Buying clothes   A sport jacket that sells for $175 has a markup of $85. Find the wholesale price.

92. Buying vacuum cleaners   A vacuum that sells for $97 has a markup of $37. Find the wholesale price.

Use the formula rb = a or a = rb to find each value. SEE EXAMPLES 10–12. (OBJECTIVE 6)

93. What number is 40% of 200?

94. What number is 45% of 340?

95. What number is 50% of 38?

96. What number is 25% of 300?

97. 35% of what number is 182?

98. 26% of what number is 78?

99. 48 is 15% of what number?

100. 13.3 is 3.5% of what number?

101. 28% of what number is 42?

102. 44% of what number is 143?

103. What percent of 357.5 is 71.5?

104. What percent of 254 is 13.208?

ADDITIONAL PRACTICE *Solve each equation. Be sure to check each answer.*

105. $p + 0.27 = 3.57$ **106.** $m - 5.36 = 1.39$

107. $\dfrac{x}{15} = -4$ **108.** $\dfrac{y}{16} = -5$

109. $-57 = b - 29$ **110.** $-93 = 67 + y$

111. $y - 2.63 = -8.21$ **112.** $s + 8.56 = 5.65$

113. $\dfrac{y}{-3} = -\dfrac{5}{6}$ **114.** $\dfrac{y}{-8} = -\dfrac{3}{16}$

115. $-18 + y = 18$ **116.** $-43 + a = -43$

117. $-3 = \dfrac{x}{11}$ **118.** $\dfrac{w}{-12} = 4$

119. $b + 7 = \dfrac{20}{3}$ **120.** $x + \dfrac{5}{7} = -\dfrac{2}{7}$

121. $3x = -\dfrac{1}{4}$ **122.** $-8x = -8$

123. $-\dfrac{3}{5} = x - \dfrac{2}{5}$ **124.** $d + \dfrac{2}{3} = \dfrac{3}{2}$

125. $\dfrac{1}{7}x = \dfrac{5}{7}$ **126.** $-17x = -51$

127. $-27w = 81$ **128.** $15 = \dfrac{r}{-5}$

129. $18x = -9$ **130.** $-12x = 3$

Find each value.

131. 0.48 is what percent of 8?

132. 3.6 is what percent of 28.8?

133. 34 is what percent of 17?

134. 39 is what percent of 13?

APPLICATIONS *Solve each application involving percents. SEE EXAMPLES 13–14. (OBJECTIVE 7)*

135. Selling microwave ovens   The 5% sales tax on a microwave oven amounts to $13.50. What is the microwave's selling price?

136. Hospitals   18% of hospital patients stay for less than 1 day. If 1,008 patients in January stayed for less than 1 day, what total number of patients did the hospital treat in January?

137. Sales taxes   Sales tax on a $12 compact disc is $0.72. At what rate is sales tax computed?

138. Home prices   The average price of homes in one neighborhood decreased 8% since last year, a drop of $7,800. What was the average price of a home last year?

Solve.

139. Banking   The amount A in an account is given by the formula

$$A = p + i$$

where p is the principal and i is the interest. How much interest was earned if an original deposit (the principal) of $4,750 has grown to be $5,010?

140. Selling real estate   The money m received from selling a house is given by the formula

$$m = s - c$$

where s is the selling price and c is the agent's commission. Find the selling price of a house if the seller received $217,000 and the agent received $13,020.

141. Customer satisfaction   One-third of the movie audience left the theater in disgust. If 78 angry patrons walked out, how many were there originally?

142. Off-campus housing   One-seventh of the senior class is living in off-campus housing. If 217 students live off campus, how large is the senior class?

143. Shopper dissatisfaction Refer to the survey results shown in the table. What percent of those surveyed were not pleased?

Shopper survey results	
First-time shoppers	1,731
Major purchase today	539
Shopped within previous month	1,823
Satisfied with service	4,140
Seniors	2,387
Total surveyed	9,200

144. Shopper satisfaction Refer to the survey results shown in the table above. What percent of those surveyed were satisfied with their service?

145. Union membership If 2,484 union members represent 90% of a factory's work force, how many workers are employed?

146. Charities Out of $237,000 donated to a certain charity, $5,925 is used to pay for fund-raising expenses. What percent of the donations is overhead?

147. Stock splits After a 3-for-2 stock split, each shareholder will own 1.5 times as many shares as before. If 555 shares are owned after the split, how many were owned before?

148. Stock splits After a 2-for-1 stock split, each shareholder owned twice as many shares as before. If 2,570 shares are owned after the split, how many were owned before?

149. Depreciation Find the original cost of a car that is worth $10,250 after depreciating $7,500.

150. Appreciation Find the original purchase price of a house that is worth $150,000 and has appreciated $57,000.

151. Taxes Find the tax paid on an item that was priced at $37.10 and cost $39.32.

152. Buying carpets How much did it cost to install $317 worth of carpet that cost $512?

153. Buying paint After reading this ad, a decorator bought 1 gallon of primer, 1 gallon of paint, and a brush. If the total cost was $30.44, find the cost of the brush.

154. Painting a room After reading the ad above, a woman bought 2 gallons of paint, 1 gallon of primer, and a brush. If the total cost was $46.94, find the cost of the brush.

155. Buying real estate The cost of a condominium is $57,595 less than the cost of a house. If the house costs $202,744, find the cost of the condominium.

156. Buying airplanes The cost of a twin-engine plane is $175,260 less than the cost of a two-seater jet. If the jet cost $321,435, find the cost of the twin-engine plane.

WRITING ABOUT MATH

157. Explain what it means for a number to satisfy an equation.

158. How can you tell whether a number is the solution to an equation?

SOMETHING TO THINK ABOUT

159. The Ahmes papyrus mentioned on page 9 contains this statement: *A circle nine units in diameter has the same area as a square eight units on a side.* From this statement, determine the ancient Egyptians' approximation of π.

160. Calculate the Egyptians' *percent of error*: What percent of the actual value of π is the difference of the estimate obtained in Exercise 159 and the actual value of π?

Section 2.2
Solving More Linear Equations in One Variable

Objectives

1. Solve a linear equation in one variable requiring more than one property of equality.
2. Solve an application requiring more than one property of equality.
3. Solve an application involving percent of increase or decrease.

Vocabulary

clearing fractions   percent of increase   percent of decrease

Getting Ready

Perform the operations.

1. $7 + 3 \cdot 5$

2. $3(5 + 7)$

3. $\dfrac{3 + 7}{2}$

4. $3 + \dfrac{7}{2}$

5. $\dfrac{3(5 - 8)}{9}$

6. $3 \cdot \dfrac{5 - 8}{9}$

7. $\dfrac{3 \cdot 5 - 8}{9}$

8. $3 \cdot \dfrac{5}{9} - 8$

We have solved linear equations in one variable by using one of the addition, subtraction, multiplication, and division properties of equality. To solve some equations, we need to use several of these properties in succession.

 Solve a linear equation in one variable requiring more than one property of equality.

In the following examples, we will combine the addition or subtraction property with the multiplication or division property to solve more complicated equations.

Everyday connections
Renting a Car

Rental car rates for various cars are given for three different companies. In each formula, the constant represents the base charge.

Economy car from Dan's Rentals

$C = 19.95x + 39.99$

Luxury car from Spencer's Cars

$C = 55x + 124.15$

SUV from Tyler's Auto Rentals

$C = 35.88x + 89.95$

1. Interpret the meaning of the variables x and C.

2. What does the coefficient of x represent in each equation?

We can solve an equation to compare the companies to one another.

3. Suppose you have \$900 available to spend on a rental car. Find the number of days you can afford to rent from each company. (*Hint*: Substitute 900 for C.)

Dan's Rentals

Spencer's Cars

Tyler's Auto Rentals

EXAMPLE 1   Solve: $-12x + 5 = 17$

Solution   The left side of the equation indicates that x is to be multiplied by -12 and then 5 is to be added to that product. To isolate x, we must undo these operations in the reverse order.

- To undo the addition of 5, we subtract 5 from both sides.
- To undo the multiplication by -12, we divide both sides by -12.

$$-12x + 5 = 17$$

$$-12x + 5 - 5 = 17 - 5$$ To undo the addition of 5, subtract 5 from both sides.

$$-12x = 12$$ $5 - 5 = 0$ and $17 - 5 = 12$.

$$\frac{-12x}{-12} = \frac{12}{-12}$$ To undo the multiplication by -12, divide both sides by -12.

$$x = -1$$ $\frac{-12}{-12} = 1$ and $\frac{12}{-12} = -1$.

Check: $$-12x + 5 = 17$$

$$-12(-1) + 5 \overset{?}{=} 17$$ Substitute -1 for x.

$$12 + 5 \overset{?}{=} 17$$ Multiply.

$$17 = 17$$ True.

Since $17 = 17$, the solution -1 checks and the solution set is $\{-1\}$.

SELF CHECK 1   Solve: $2x + 3 = 15$

EXAMPLE 2   Solve: $\dfrac{x}{3} - 7 = -3$

Solution   The left side of the equation indicates that x is to be divided by 3 and then 7 is to be subtracted from that quotient. To isolate x, we must undo these operations in the reverse order.

- To undo the subtraction of 7, we add 7 to both sides.
- To undo the division by 3, we multiply both sides by 3.

$$\frac{x}{3} - 7 = -3$$

$$\frac{x}{3} - 7 + 7 = -3 + 7$$ To undo the subtraction of 7, add 7 to both sides.

$$\frac{x}{3} = 4$$ $-7 + 7 = 0$ and $-3 + 7 = 4$.

$$3 \cdot \frac{x}{3} = 3 \cdot 4$$ To undo the division by 3, multiply both sides by 3.

$$x = 12$$ $3 \cdot \frac{x}{3} = x$ and $3 \cdot 4 = 12$.

Check: $$\frac{x}{3} - 7 = -3$$

$$\frac{12}{3} - 7 \overset{?}{=} -3$$ Substitute 12 for x.

$$4 - 7 \overset{?}{=} -3$$ Simplify.

$$-3 = -3$$

Since $-3 = -3$, the solution 12 checks and the solution set is $\{12\}$.

SELF CHECK 2   Solve: $\dfrac{x}{4} - 3 = 5$

EXAMPLE 3   Solve: $\dfrac{x-7}{3} = 9$

Solution   The left side of the equation indicates that 7 is to be subtracted from x and that the difference is to be divided by 3. To isolate x, we must undo these operations in the reverse order.

• To undo the division by 3, we multiply both sides by 3.
• To undo the subtraction of 7, we add 7 to both sides.

$$\dfrac{x-7}{3} = 9$$

$$\mathbf{3}\left(\dfrac{x-7}{3}\right) = \mathbf{3}(9) \qquad \text{To undo the division by 3, multiply both sides by 3.}$$

$$x - 7 = 27 \qquad 3 \cdot \tfrac{1}{3} = 1 \text{ and } 3(9) = 27.$$

$$x - 7 \mathbf{+ 7} = 27 \mathbf{+ 7} \qquad \text{To undo the subtraction of 7, add 7 to both sides.}$$

$$x = 34 \qquad -7 + 7 = 0 \text{ and } 27 + 7 = 34.$$

Since the solution is 34, the solution set is $\{34\}$. Verify that the solution checks.

 SELF CHECK 3   Solve: $\dfrac{a-3}{5} = -2$

EXAMPLE 4   Solve: $\dfrac{3x}{4} + \dfrac{2}{3} = -7$

Solution   The left side of the equation indicates that x is to be multiplied by 3, then $3x$ is to be divided by 4, and then $\frac{2}{3}$ is to be added to that result. To isolate x, we must undo these operations in the reverse order.

COMMENT   We undo the multiplication by 3 by multiplying both sides by $\frac{1}{3}$ to simplify the arithmetic. Dividing both sides by 3 would introduce a complex fraction.

• To undo the addition of $\frac{2}{3}$, we subtract $\frac{2}{3}$ from both sides.
• To undo the division by 4, we multiply both sides by 4.
• To undo the multiplication by 3, we multiply both sides by $\frac{1}{3}$.

$$\dfrac{3x}{4} + \dfrac{2}{3} = -7$$

$$\dfrac{3x}{4} + \dfrac{2}{3} \mathbf{- \dfrac{2}{3}} = -7 \mathbf{- \dfrac{2}{3}} \qquad \text{To undo the addition of } \tfrac{2}{3}, \text{ subtract } \tfrac{2}{3} \text{ from both sides.}$$

$$\dfrac{3x}{4} = -\dfrac{23}{3} \qquad \tfrac{2}{3} - \tfrac{2}{3} = 0 \text{ and } -7 - \tfrac{2}{3} = -\tfrac{23}{3}.$$

$$\mathbf{4}\left(\dfrac{3x}{4}\right) = \mathbf{4}\left(-\dfrac{23}{3}\right) \qquad \text{To undo the division by 4, multiply both sides by 4.}$$

$$3x = -\dfrac{92}{3} \qquad 4 \cdot \tfrac{3x}{4} = 3x \text{ and } 4\left(-\tfrac{23}{3}\right) = -\tfrac{92}{3}.$$

$$\dfrac{\mathbf{1}}{\mathbf{3}}(3x) = \dfrac{\mathbf{1}}{\mathbf{3}}\left(-\dfrac{92}{3}\right) \qquad \text{To undo the multiplication by 3, multiply both sides by } \tfrac{1}{3}.$$

$$x = -\dfrac{92}{9} \qquad \tfrac{1}{3} \cdot 3x = x \text{ and } \tfrac{1}{3}\left(-\tfrac{92}{3}\right) = -\tfrac{92}{9}.$$

Since the solution is $-\dfrac{92}{9}$, the solution set is $\left\{-\dfrac{92}{9}\right\}$. Verify that the solution checks.

SELF CHECK 4   Solve: $\dfrac{2x}{3} - \dfrac{4}{5} = 3$

An alternate method for solving Example 4 is to **clear fractions** first. In this method, we multiply both sides of the equation by the least common multiple of the denominators.

Solve: $\dfrac{3x}{4} + \dfrac{2}{3} = -7$

The least common multiple for 4 and 3 is 12. We will multiply both sides of the equation by 12.

$$12\left(\frac{3x}{4} + \frac{2}{3}\right) = 12(-7)$$ Multiply both sides by 12.

$$12\left(\frac{3x}{4}\right) + 12\left(\frac{2}{3}\right) = 12(-7)$$ Use the distributive property to remove the parentheses.

$$\overset{3}{12}\left(\frac{3x}{\underset{1}{4}}\right) + \overset{4}{12}\left(\frac{2}{\underset{1}{3}}\right) = -84$$ Simplify each fraction.

$$9x + 8 = -84$$ Simplify.

$$9x + 8 - 8 = -84 - 8$$ Subtract 8 from both sides.

$$9x = -92$$ Simplify.

$$\frac{9x}{9} = \frac{-92}{9}$$ Divide both sides by 9.

$$x = -\frac{92}{9}$$

This method is one you will use when solving rational equations later in this textbook.

2   Solve an application requiring more than one property of equality.

EXAMPLE 5   **ADVERTISING**   A store manager hires a student to distribute advertising circulars door to door. The student will be paid $24 a day plus 12¢ for every ad she distributes. How many ads must she distribute to earn $42 in one day?

Solution   We can let *a* represent the number of ads that the student must distribute. Her earnings can be expressed in two ways: as $24 plus the 12¢-apiece pay for distributing the ads, and as $42.

$24	plus	*a* ads at $0.12 each	is	$42	12¢ = $0.12
24	+	0.12*a*	=	42	

We can solve this equation as follows:

$$24 + 0.12a = 42$$

$$24 - 24 + 0.12a = 42 - 24$$ To undo the addition of 24, subtract 24 from both sides.

$$0.12a = 18$$ $24 - 24 = 0$ and $42 - 24 = 18$.

$$\frac{0.12a}{0.12} = \frac{18}{0.12}$$ To undo the multiplication by 0.12, divide both sides by 0.12.

$$a = 150$$ $\frac{0.12}{0.12} = 1$ and $\frac{18}{0.12} = 150$.

The student must distribute 150 ads. Check the result.

SELF CHECK 5   How many ads must the student deliver in one day to earn $48?

3 Solve an application involving percent of increase or decrease.

We have seen that the retail price of an item is the sum of the cost and the markup.

| Retail price | equals | cost | plus | markup. |

Often, the markup is expressed as a percent of the cost.

| Markup | equals | Percent of markup | times | cost. |

Suppose a store manager buys toasters for $21 and sells them at a 17% markup. To find the retail price, the manager begins with his cost and adds 17% of that cost.

Retail price	=	cost	+	markup.
	=	cost	+	Percent of markup · cost
	=	21	+	0.17 · 21

$$= 21 + 3.57$$

$$= 24.57$$

The retail price of a toaster is $24.57.

EXAMPLE 6   **ANTIQUE CARS**   In 1956, a Chevrolet BelAir automobile sold for $4,000. Today, it is worth about $28,600. Find the percent that its value has increased, called the **percent of increase**.

Solution   We let p represent the percent of increase, expressed as a decimal.

| Current price | equals | original price | plus | p(original price). |

$$28,600 = 4,000 + p(4,000)$$

$$28,600 - \textbf{4,000} = 4,000 - \textbf{4,000} + 4,000p \quad \text{To undo the addition of 4,000,}$$
subtract 4,000 from both sides.

$$24,600 = 4,000p \quad \begin{array}{l} 28,600 - 4,000 = 24,600 \text{ and} \\ 4,000 - 4,000 = 0. \end{array}$$

$$\frac{24,600}{\textbf{4,000}} = \frac{4,000p}{\textbf{4,000}} \quad \begin{array}{l} \text{To undo the multiplication by 4,000,} \\ \text{divide both sides by 4,000.} \end{array}$$

$$6.15 = p \quad \text{Simplify.}$$

To convert 6.15 to a percent, we multiply by 100 and insert a % sign. Since the percent of increase is 615%, the car has appreciated 615%.

SELF CHECK 6   Find the percent of increase if the car sells for $30,000.

We have seen that when the price of merchandise is reduced, the amount of reduction is the markdown (also called the *discount*).

| Sale price | equals | regular price | minus | markdown. |

Usually, the markdown is expressed as a percent of the regular price.

| Markdown | equals | percent of markdown | times | regular price. |

Suppose that a TV set that regularly sells for $570 has been marked down 25%. That means the customer will pay 25% less than the regular price. To find the sale price, we use the formula

Sale price	=	regular price	−	markdown		
	=	regular price	−	percent of markdown	·	regular price
	=	$570	−	25%	of	$570

$$= \$570 - (0.25)(\$570)$$ Write 25% as a decimal.

$$= \$570 - \$142.50$$ Multiply.

$$= \$427.50$$ Subtract.

The TV set is selling for $427.50.

EXAMPLE 7 **BUYING CAMERAS** A camera that was originally priced at $452 is on sale for $384.20. Find the percent of markdown.

Solution We let p represent the percent of markdown, expressed as a decimal, and substitute $384.20 for the sale price and $452 for the regular price.

Sale price	equals	regular price	minus	percent of markdown	times	regular price.
384.20	=	452	−	p	·	452

$$384.20 - \textbf{452} = 452 - \textbf{452} - p(452)$$ To undo the addition of 452, subtract 452 from both sides.

$$-67.80 = -p(452)$$ $384.20 - 452 = -67.80$; $452 - 452 = 0$

$$\frac{-67.80}{\textbf{−452}} = \frac{-p(452)}{\textbf{−452}}$$ To undo the multiplication by −452, divide both sides by −452.

$$0.15 = p$$ $\frac{-67.80}{-452} = 0.15$ and $\frac{-452}{-452} = 1$.

The camera is on sale at a 15% markdown.

SELF CHECK 7 If the camera is reduced another $22.60, find the percent of markdown.

COMMENT When a price increases from $100 to $125, the percent of increase is 25%. When the price *decreases* from $125 to $100, the **percent of decrease** is 20%. These different results occur because the percent of increase is a percent of the original (smaller) price, $100. The percent of decrease is a percent of the original (larger) price, $125.

SELF CHECK ANSWERS

1. 6 **2.** 32 **3.** −7 **4.** $\frac{57}{10}$ **5.** 200 **6.** 650% **7.** 20%

NOW TRY THIS

Solve each equation.

1. $\frac{2}{7}x + 3 = 3$

2. $10 - \frac{2}{3}x = -6$

3. $-0.2x - 4.3 = -10.7$

2.2   Exercises 🌿

WARM-UPS *What would you do first when solving each equation?*

1. $6x - 9 = -15$

2. $15 = \dfrac{x}{5} + 3$

3. $\dfrac{x}{7} - 3 = 0$

4. $\dfrac{x - 3}{7} = -7$

5. $\dfrac{x - 7}{3} = 5$

6. $\dfrac{3x - 5}{2} + 2 = 0$

Solve each equation.

7. $7z - 7 = 14$

8. $\dfrac{p - 1}{2} = 6$

REVIEW *Refer to the formulas given in Section 1.3.*

9. Find the perimeter of a rectangle with sides measuring 8.5 and 16.5 cm.

10. Find the area of a rectangle with sides measuring 2.3 in. and 3.7 in.

11. Find the area of a trapezoid with a height of 8.5 in. and bases measuring 6.7 in. and 12.2 in.

12. Find the volume of a rectangular solid with dimensions of 8.2 cm by 7.6 cm by 10.2 cm.

VOCABULARY AND CONCEPTS *Fill in the blanks.*

13. Retail price = ___ + markup

14. Markup = percent of markup · ___

15. Markdown = _____ of markdown · regular price

16. Another word for markdown is _____.

17. The percent that an object has increased in value is called the _____.

18. The percent that an object has deceased in value is called the _____.

GUIDED PRACTICE *Solve each equation. Check all solutions.*
SEE EXAMPLE 1. (OBJECTIVE 1)

19. $5x - 1 = 4$

20. $5x + 3 = 8$

21. $-6x + 2 = 14$

22. $4x - 4 = 8$

23. $6x + 2 = -4$

24. $4x - 4 = 4$

25. $3x - 8 = 1$

26. $7x - 19 = 2$

27. $4x - 7 = 5$

28. $-8x + 5 = 21$

29. $-3x - 6 = 12$

30. $5x + 9 = -16$

31. $-2x - 8 = -2$

32. $-3x + 17 = -4$

Solve each equation. Check all solutions. **SEE EXAMPLE 2. (OBJECTIVE 1)**

33. $\dfrac{z}{9} + 5 = -1$

34. $\dfrac{y}{5} - 3 = 3$

35. $\dfrac{x}{4} + 7 = 3$

36. $\dfrac{a}{5} - 3 = -4$

37. $\dfrac{x}{3} - 10 = -1$

38. $\dfrac{x}{7} + 3 = 5$

39. $\dfrac{p}{11} + 9 = 6$

40. $\dfrac{r}{12} + 2 = 4$

Solve each equation. Check all solutions. **SEE EXAMPLE 3. (OBJECTIVE 1)**

41. $\dfrac{b + 5}{3} = 11$

42. $\dfrac{a + 2}{13} = 3$

43. $\dfrac{x + 5}{2} = 4$

44. $\dfrac{r - 3}{8} = -2$

45. $\dfrac{3x - 12}{2} = 9$

46. $\dfrac{5x + 10}{7} = 0$

47. $\dfrac{4k - 1}{5} = 3$

48. $\dfrac{2k - 1}{3} = -5$

Solve each equation. Check all solutions. **SEE EXAMPLE 4. (OBJECTIVE 1)**

49. $\dfrac{k}{5} - \dfrac{1}{2} = \dfrac{3}{2}$

50. $\dfrac{y}{3} - \dfrac{6}{5} = -\dfrac{1}{5}$

51. $\dfrac{w}{16} + \dfrac{5}{4} = 1$

52. $\dfrac{m}{7} - \dfrac{1}{14} = \dfrac{1}{14}$

53. $\dfrac{3x}{2} - 6 = 9$

54. $\dfrac{5x}{7} + 3 = 8$

55. $\dfrac{9y}{2} + 3 = -15$

56. $\dfrac{5z}{3} + 3 = -2$

ADDITIONAL PRACTICE *Solve each equation. Check all solutions.*

57. $43p + 72 = 158$

58. $96q + 23 = -265$

59. $-47 - 21n = 58$

60. $-151 + 13m = -229$

61. $2y - \dfrac{5}{3} = \dfrac{4}{3}$

62. $9y + \dfrac{1}{2} = \dfrac{3}{2}$

63. $-0.4y - 12 = -20$

64. $-0.8y + 64 = -32$

65. $\dfrac{2x}{3} + \dfrac{1}{2} = 3$

66. $\dfrac{4x}{5} - \dfrac{1}{3} = 1$

67. $\dfrac{3x}{4} - \dfrac{2}{5} = 2$

68. $\dfrac{5x}{6} + \dfrac{3}{5} = 3$

69. $\dfrac{u - 4}{7} = 1$

70. $\dfrac{v - 7}{3} = -1$

71. $\dfrac{x - 5}{3} = -4$

72. $\dfrac{3 + y}{5} = -3$

73. $\dfrac{3z + 2}{17} = 0$

74. $\dfrac{8n - 7}{4} = -1$

75. $\dfrac{17k - 28}{21} + \dfrac{4}{3} = 0$

76. $\dfrac{5a - 2}{3} = \dfrac{1}{6}$

77. $-\dfrac{x}{3} - \dfrac{1}{2} = -\dfrac{5}{2}$

78. $\dfrac{15 - 5a}{3} = 3$

79. $\dfrac{10 - 3w}{9} = \dfrac{2}{3}$

80. $\dfrac{3p - 5}{5} + \dfrac{1}{2} = -\dfrac{19}{2}$

APPLICATIONS *Solve.* SEE EXAMPLE 5. (OBJECTIVE 2)

81. Apartment rentals A student moves into a bigger apartment that rents for $450 per month. That rent is $200 less than twice what she had been paying. Find her former rent.

82. Auto repairs A mechanic charged $20 an hour to repair the water pump on a car, plus $95 for parts. If the total bill was $155, how many hours did the repair take?

83. Boarding dogs A sportsman boarded his dog at a kennel for a $20 registration fee plus $14 a day. If the stay cost $104, how many days was the owner gone?

84. Water billing The city's water department charges $7 per month, plus 42¢ for every 100 gallons of water used. Last month, one homeowner used 1,900 gallons and received a bill for $17.98. Was the billing correct?

Solve. SEE EXAMPLES 6–7. (OBJECTIVE 3)

85. Clearance sales Sweaters already on sale for 20% off the regular price cost $36 when purchased with a promotional coupon that allows an additional 10% discount. Find the original price. (*Hint:* When you save 20%, you are paying 80%.)

86. Furniture sales A $1,250 sofa is marked down to $900. Find the percent of markdown.

87. Value of coupons The percent discount offered by this coupon depends on the amount purchased. Find the range of the percent discount.

Value coupon
Save $15
on purchases of $100 to $250.

88. Furniture pricing A bedroom set selling for $1,900 cost $1,000 wholesale. Find the percent markup.

Solve.

89. Integers Six less than 3 times a number is 9. Find the number.

90. Integers Seven less than 5 times a number is 23. Find the number.

91. Integers If a number is increased by 7 and that result is divided by 2, the number 5 is obtained. Find the original number.

92. Integers If twice a number is decreased by 5 and that result is multiplied by 4, the result is 36. Find the number.

93. Telephone charges A call to Tucson from a pay phone in Chicago costs 85¢ for the first minute and 27¢ for each additional minute or portion of a minute. If a student has an $8.68 balance on a phone card, how long can she talk?

94. Monthly sales A clerk's sales in February were $2,000 less than 3 times her sales in January. If her February sales were $7,000, by what amount did her sales increase?

95. Ticket sales A music group charges $1,500 for each performance, plus 20% of the total ticket sales. After a concert, the group received $2,980. How much money did the ticket sales raise?

96. Getting an A To receive a grade of A, the average of four 100-point exams must be 90 or better. If a student received scores of 88, 83, and 92 on the first three exams, what minimum score does he need on the fourth exam to earn an A?

97. Getting an A The grade in history class is based on the average of five 100-point exams. One student received scores of 85, 80, 95, and 78 on the first four exams. With an average of 90 needed, what chance does he have for an A?

98. Excess inventory From the portion of the following ad, determine the sale price of a shirt.

Clearance Sale
Save 40%

	Regularly	Sale
Sweaters	$45.95	$27.57
Shirts	$37.50	$

WRITING ABOUT MATH

99. In solving the equation $5x - 3 = 12$, explain why you would add 3 to both sides first, rather than dividing by 5 first.

100. To solve the equation $\dfrac{3x - 4}{7} = 2$, what operations would you perform, and in what order?

SOMETHING TO THINK ABOUT

101. Suppose you must solve the following equation but you can't read one number. If the solution of the equation is 1, what is the equation?

$$\frac{7x + \blacksquare}{22} = \frac{1}{2}$$

102. A store manager first increases his prices by 30% to get a new retail price and then advertises as shown at the right. What is the real percent discount to customers?

SALE
30% savings
off retail price!!

Section 2.3

Simplifying Expressions to Solve Linear Equations in One Variable

Objectives

1 Simplify an expression using the order of operations and combining like terms.

2 Solve a linear equation in one variable requiring simplifying one or both sides.

3 Solve a linear equation in one variable that is an identity or a contradiction.

Vocabulary

coefficient   conditional equation   empty set
like terms   identity
unlike terms   contradiction

Getting Ready

Use the distributive property to remove parentheses.

1. $(3 + 4)x$ **2.** $(7 + 2)x$

3. $(8 - 3)w$ **4.** $(10 - 4)y$

Simplify each expression by performing the operations within the parentheses.

5. $(3 + 4)x$ **6.** $(7 + 2)x$

7. $(8 - 3)w$ **8.** $(10 - 4)y$

When algebraic expressions with the same variables occur, we can combine them.

1 ## Simplify an expression using the order of operations and combining like terms.

Recall that a *term* is either a number or the product of numbers and variables. Some examples of terms are $7x$, $-3xy$, y^2, and 8. The number part of each term is called its **coefficient**.

- The coefficient of $7x$ is 7.
- The coefficient of $-3xy$ is -3.
- The coefficient of y^2 is the understood factor of 1.
- The coefficient of 8 is 8.

LIKE TERMS

Like terms, or *similar terms*, are terms with the same variables having the same exponents.

The terms $3x$ and $5x$ are like terms, as are $9x^2$ and $-3x^2$. The terms $4xy$ and $3x^2$ are **unlike terms**, because they have different variables. The terms $4x$ and $5x^2$ are unlike terms, because the variables have different exponents.

COMMENT Terms are separated by + and − signs.

The distributive property can be used to combine terms of algebraic expressions that contain sums or differences of like terms. For example, the terms in $3x + 5x$ and $9xy^2 − 11xy^2$ can be combined as follows:

expressions with like terms   expressions with like terms

$$3x + 5x = (3 + 5)x \qquad 9xy^2 − 11xy^2 = (9 − 11)xy^2$$
$$= 8x \qquad\qquad\qquad\qquad\qquad = −2xy^2$$

These examples suggest the following rule.

COMBINING LIKE TERMS   To combine like terms, add their coefficients and keep the same variables and exponents.

COMMENT If the terms of an expression are unlike terms, they cannot be combined. For example, since the terms in $9xy^2 − 11x^2y$ have variables with different exponents, they are unlike terms and cannot be combined.

EXAMPLE 1   Simplify: $3(x + 2) + 2(x − 8)$

Solution   To simplify the expression, we will use the distributive property to remove parentheses and then combine like terms.

$$3(x + 2) + 2(x − 8)$$
$$= 3x + 3 \cdot 2 + 2x + 2 \cdot (−8) \quad \text{Use the distributive property to remove parentheses.}$$
$$= 3x + 6 + 2x − 16 \quad\quad\quad\quad 3 \cdot 2 = 6 \text{ and } 2 \cdot 8 = 16.$$
$$= 3x + 2x + 6 − 16 \quad\quad\quad \text{Use the commutative property of addition:}$$
$$\quad\quad\quad\quad\quad\quad\quad\quad\quad\quad 6 + 2x = 2x + 6$$
$$= 5x − 10 \quad\quad\quad\quad\quad\quad\quad \text{Combine like terms.}$$

 SELF CHECK 1   Simplify: $−5(a + 3) + 2(a − 5)$

EXAMPLE 2   Simplify: $3(x − 3) − 5(x + 4)$

Solution   To simplify the expression, we will use the distributive property to remove parentheses and then combine like terms.

$$3(x − 3) − 5(x + 4)$$
$$= 3(x − 3) + (−5)(x + 4) \quad\quad a − b = a + (−b).$$
$$= 3x − 3 \cdot 3 + (−5)x + (−5)4 \quad \text{Use the distributive property to remove parentheses.}$$
$$= 3x − 9 + (−5x) + (−20) \quad\quad 3 \cdot 3 = 9 \text{ and } (−5)(4) = −20.$$
$$= −2x − 29 \quad\quad\quad\quad\quad\quad\quad \text{Combine like terms.}$$

 SELF CHECK 2   Simplify: $−3(b − 2) − 4(b − 4)$

COMMENT In algebra, you will simplify expressions and solve equations. Recognizing which one to do is a skill that we will apply throughout this course.

An expression can be simplified only by combining its like terms. Since an equation contains two expressions set equal to each other, it can be solved. Remember that

Expressions are to be simplified. Equations are to be solved.

2 Solve a linear equation in one variable requiring simplifying one or both sides.

To solve a linear equation in one variable, we must isolate the variable on one side. This is often a multistep process that may require combining like terms. As we solve equations, we will follow these steps, if necessary.

SOLVING EQUATIONS

1. Clear the equation of any fractions or decimals.
2. Use the distributive property to remove any grouping symbols.
3. Combine like terms on each side of the equation.
4. Undo the operations of addition and subtraction to collect the variables on one side and the constants on the other.
5. Combine like terms and undo the operations of multiplication and division to isolate the variable.
6. Check the solution in the original equation.

EXAMPLE 3   Solve: $3(x + 2) - 5x = 0$

Solution   To solve the equation, we will remove parentheses, combine like terms, and solve for x.

$$3(x + 2) - 5x = 0$$

$$3x + 3 \cdot 2 - 5x = 0 \qquad \text{Use the distributive property to remove parentheses.}$$

$$3x - 5x + 6 = 0 \qquad \text{Use the commutative property of addition and simplify.}$$

$$-2x + 6 = 0 \qquad \text{Combine like terms.}$$

$$-2x + 6 - 6 = 0 - 6 \qquad \text{Subtract 6 from both sides.}$$

$$-2x = -6 \qquad \text{Combine like terms.}$$

$$\frac{-2x}{-2} = \frac{-6}{-2} \qquad \text{Divide both sides by } -2.$$

$$x = 3 \qquad \text{Simplify.}$$

Check:
$$3(x + 2) - 5x = 0$$
$$3(3 + 2) - 5 \cdot 3 \overset{?}{=} 0 \qquad \text{Substitute 3 for } x.$$
$$3 \cdot 5 - 5 \cdot 3 \overset{?}{=} 0 \qquad \text{Perform the operation inside the parentheses.}$$
$$15 - 15 \overset{?}{=} 0 \qquad \text{Multiply.}$$
$$0 = 0 \qquad \text{True.}$$

Since the solution 3 checks, the solution set is $\{3\}$.

 SELF CHECK 3   Solve: $-2(y - 3) - 4y = 0$

In the next example, we will isolate the variable on the right side of the equation to keep the coefficient of x positive.

EXAMPLE 4   Solve: $3(x - 5) = 4(x + 9)$

Solution   To solve the equation, we will remove parentheses, collect all like terms involving x on one side, combine like terms, and solve for x.

$$3(x - 5) = 4(x + 9)$$

$$3x - 15 = 4x + 36 \qquad \text{Use the distributive property to remove parentheses.}$$

$$3x - 15 - 3x = 4x + 36 - 3x \qquad \text{Subtract } 3x \text{ from both sides.}$$

$$-15 = x + 36 \qquad \text{Combine like terms.}$$

$$-15 - 36 = x + 36 - 36$$ Subtract 36 from both sides.

$$-51 = x$$ Combine like terms.

Check: $$3(x - 5) = 4(x + 9)$$

$$3(-51 - 5) \stackrel{?}{=} 4(-51 + 9)$$ Substitute -51 for x.

$$3(-56) \stackrel{?}{=} 4(-42)$$

$$-168 = -168$$ True.

Since the solution -51 checks, the solution set is $\{-51\}$.

🌱 **SELF CHECK 4**   Solve: $4(z + 3) = -3(z - 4)$

Note: The solution to an equation is the same whether the variable is isolated on the left or the right side of the equation.

EXAMPLE 5   Solve: $\dfrac{3x + 11}{5} = x + 3$

Solution   We first multiply both sides by 5 to clear the equation of fractions. When we multiply the right side by 5, we must multiply the *entire* right side by 5.

$$\frac{3x + 11}{5} = x + 3$$

$$5\left(\frac{3x + 11}{5}\right) = 5(x + 3)$$ Multiply both sides by 5.

$$3x + 11 = 5x + 15$$ Use the distributive property to remove parentheses.

$$3x + 11 - 11 = 5x + 15 - 11$$ Subtract 11 from both sides.

$$3x = 5x + 4$$ Combine like terms.

$$3x - 5x = 5x + 4 - 5x$$ Subtract $5x$ from both sides.

$$-2x = 4$$ Combine like terms.

$$\frac{-2x}{-2} = \frac{4}{-2}$$ Divide both sides by -2.

$$x = -2$$ Simplify.

COMMENT   Remember that when you multiply one side of an equation by a nonzero number, you must multiply the other side by the same number to maintain the equality.

Check: $$\frac{3x + 11}{5} = x + 3$$

$$\frac{3(-2) + 11}{5} \stackrel{?}{=} (-2) + 3$$ Substitute -2 for x.

$$\frac{-6 + 11}{5} \stackrel{?}{=} 1$$ Simplify.

$$\frac{5}{5} \stackrel{?}{=} 1$$

$$1 = 1$$ True.

Since the solution -2 checks, the solution set is $\{-2\}$.

🌱 **SELF CHECK 5**   Solve: $\dfrac{2x - 5}{4} = x - 2$

EXAMPLE 6   Solve: $0.2x + 0.4(50 - x) = 19$

Solution   Since $0.2 = \frac{2}{10}$ and $0.4 = \frac{4}{10}$, this equation contains fractions. To clear the fractions, we will multiply both sides by 10.

$$0.2x + 0.4(50 - x) = 19$$

$\mathbf{10}[0.2x + 0.4(50 - x)] = \mathbf{10}(19)$	Multiply both sides by 10.
$\mathbf{10}[0.2x] + \mathbf{10}[0.4(50 - x)] = \mathbf{10}(19)$	Use the distributive property on the left side.
$2x + 4(50 - x) = 190$	Multiply.
$2x + 200 - 4x = 190$	Use the distributive property to remove parentheses.
$-2x + 200 = 190$	Combine like terms.
$-2x = -10$	Subtract 200 from both sides.
$x = 5$	Divide both sides by -2.

Since the solution is 5, the solution set is $\{5\}$. Verify that the solution checks.

 SELF CHECK 6   Solve: $0.3(20 - x) + 0.5x = 15$

 Solve a linear equation in one variable that is an identity or a contradiction.

The equations solved in Examples 3–6 are called **conditional equations**. For these equations, each has exactly one solution.

An equation that is true for all values of its variable is called an **identity**. For example, the equation $x + x = 2x$ is an identity because it is true for all values of x. The solution of an identity is the set of *all real numbers* and is denoted by the symbol $\mathbb{R}$.

An equation that is not true for any value of its variable is called a **contradiction**. For example, the equation $x = x + 1$ is a contradiction because there is no value of x that will make the statement true. Since there are no solutions to a contradiction, its set of solutions is empty. This is denoted by the symbol $\varnothing$ or $\{\ \}$ and is called the **empty set**. Table 2.1 summarizes the three possibilities.

TABLE 2-1

Type of equation	Examples		Solution sets
Conditional	$2x + 4 = 8$	$\dfrac{x}{2} - 4 = 12$	$\{2\}$ and $\{32\}$
Identity	$x + x = 2x$	$2(x + 3) = 2x + 6$	$\mathbb{R}$ and $\mathbb{R}$
Contradiction	$x - 1 = x$	$2(x + 3) = 2x + 5$	$\varnothing$ and $\varnothing$

EXAMPLE 7   Solve: $3(x + 8) + 5x = 2(12 + 4x)$

Solution   To solve this equation, we will remove parentheses, combine terms, and solve for x.

$\mathbf{3}(x + 8) + 5x = \mathbf{2}(12 + 4x)$	
$3x + 24 + 5x = 24 + 8x$	Use the distributive property to remove parentheses.
$8x + 24 = 24 + 8x$	Combine like terms.
$8x + 24 - \mathbf{8x} = 24 + 8x - \mathbf{8x}$	Subtract $8x$ from both sides.
$24 = 24$	Combine like terms.

Since the result $24 = 24$ is true for every number x, every number is a solution of the original equation. This equation is an identity. The solution set is the set of real numbers, $\mathbb{R}$.

 SELF CHECK 7   Solve: $-2(x - 3) - 18x = 2(3 - 10x)$

EXAMPLE 8   Solve: $3(x + 7) - x = 2(x + 10)$

Solution   To solve this equation, we will remove parentheses, combine terms, and solve for x.

$$3(x + 7) - x = 2(x + 10)$$
$$3x + 21 - x = 2x + 20 \qquad \text{Use the distributive property to remove parentheses.}$$
$$2x + 21 = 2x + 20 \qquad \text{Combine like terms.}$$
$$2x + 21 - 2x = 2x + 20 - 2x \qquad \text{Subtract } 2x \text{ from both sides.}$$
$$21 = 20 \qquad \text{Combine like terms.}$$

Since the result $21 = 20$ is false, the original equation is a contradiction. Since the original equation has no solution, the solution set is $\varnothing$.

 SELF CHECK 8   Solve: $5(x - 2) - 2x = 3(x + 7)$

 **SELF CHECK ANSWERS**

1. $-3a - 25$ **2.** $-7b + 22$ **3.** 1   **4.** 0   **5.** $\frac{3}{2}$ **6.** 45   **7.** identity, $\mathbb{R}$ **8.** contradiction, $\varnothing$

NOW TRY THIS

Identify each of the following as an expression or an equation. Simplify or solve as appropriate.

1. $4\left(x - \dfrac{7}{4}\right) + 3(x + 2)$

2. $4\left(x - \dfrac{7}{4}\right) = 3(x + 2)$

3. $6x - 2(3x - 9)$

2.3   Exercises

WARM-UPS   *Identify each statement as an expression or an equation.*

1. $5x - 2$

2. $2x + 3 = 5$

3. $3x + 7 = -1$

4. $9 + x$

5. $6 - 4x = 7$

6. $7x + 8$

Identify each equation as an identity or a contradiction.

7. $2x + 5 = 2x + 5$

8. $-3x - 9 = -3x + 9$

9. $6x - 2 = 6x + 4$

10. $4x + 9 = 4x + 9$

REVIEW   *Evaluate each expression when $x = -3, y = -5,$ and $z = 0$.*

11. $x^2 z(y^3 - z)$

12. $y - x^3$

13. $\dfrac{x - y^2}{2y - 1 + x}$

14. $\dfrac{3y + x^2}{x} + z$

Perform the operations.

15. $\dfrac{6}{7} - \dfrac{5}{8}$

16. $\dfrac{6}{7} \cdot \dfrac{5}{8}$

17. $\dfrac{6}{7} \div \dfrac{5}{8}$

18. $\dfrac{6}{7} + \dfrac{5}{8}$

VOCABULARY AND CONCEPTS   *Fill in the blanks.*

19. If terms have the same _____ with the same exponents, they are called ___ terms. Terms that have different variables or have a variable with different exponents are called _____ terms. The number part of a term is called its _____.

20. To combine like terms, ___ their coefficients and ____ the same variables and exponents.

21. If an equation is true for all values of its variable, it is called an _____. If an equation is true for no values of its variable, it is called a _____.

22. If an equation is true for some values of its variable, but not all, it is called a _____ equation.

GUIDED PRACTICE *Simplify each expression, when possible.*
SEE EXAMPLE 1. (OBJECTIVE 1)

23. $8x + 12x$ **24.** $12y - 15y$
25. $8x^2 - 5x^2$ **26.** $17x^2 + 3x^2$
27. $9x + 3y$ **28.** $5x + 5y$
29. $4(x + 3) - 2x$ **30.** $9(y - 3) + 2y$

Simplify each expression. SEE EXAMPLE 2. (OBJECTIVE 1)

31. $5(z - 3) + 2z$ **32.** $7(y + 4) - 10y$

33. $12(x + 11) - 11$ **34.** $-3(3 + z) + 2z$

35. $6(y - 2) - 3(y + 1)$ **36.** $9(z + 2) + 5(3 - z)$

37. $5x - 2(y - x) + 4y$ **38.** $3y - 6(y + z) + y$

Solve each equation. Check all solutions. SEE EXAMPLE 3. (OBJECTIVE 2)

39. $8(x + 5) + 6(7 - x) = 0$
40. $3(x + 15) + 4(11 - x) = 0$
41. $12x - 4(5 + x) = 4$ **42.** $5(x - 6) - 8x = 15$

Solve each equation. Check all solutions. SEE EXAMPLE 4. (OBJECTIVE 2)

43. $3x + 2 = 2x$ **44.** $7x + 5 = 6x$
45. $5x - 3 = 4x$ **46.** $4x + 3 = 5x$
47. $9y - 3 = 6y$ **48.** $8y + 4 = 4y$
49. $10y - 10 = 5y$ **50.** $9y - 8 = y$
51. $3(a + 2) = 4a$ **52.** $4(a - 5) = 3a$
53. $6(b + 4) = 8b$ **54.** $7(b - 3) = 10b$
55. $2 + 3(x - 5) = 4(x - 1)$
56. $2 - (4x + 7) = 3 + 2(x + 2)$
57. $3(a + 2) = 2(a - 7)$
58. $9(n - 1) = 6(n + 2) - n$

Solve each equation. Check all solutions. SEE EXAMPLE 5. (OBJECTIVE 2)

59. $\dfrac{3(t - 7)}{2} = t - 6$ **60.** $\dfrac{4(p + 8)}{3} = p - 4$

61. $\dfrac{2(t - 1)}{6} - 2 = \dfrac{t + 2}{6}$

62. $\dfrac{2(2r - 1)}{6} + 5 = \dfrac{3(r + 7)}{6}$

Solve each equation. Check all solutions. SEE EXAMPLE 6. (OBJECTIVE 2)

63. $3.1(x - 2) = 1.3x + 2.8$
64. $0.6x - 0.8 = 0.8(2x - 1) - 0.7$
65. $2.7(y + 1) = 0.3(3y + 33)$
66. $1.5(5 - y) = 3y + 12$

Solve each equation. If it is an identity or a contradiction, so indicate.
SEE EXAMPLES 7–8. (OBJECTIVE 3)

67. $7x + 5(3 - x) = 2(x + 5) + 5$
68. $21(b - 1) + 3 = 3(7b - 6)$
69. $2(s + 2) = 2(s + 1) + 3$
70. $4(2z + 3) = 2(4z - 6) + 11$
71. $\dfrac{5(x + 3)}{3} - x = \dfrac{2(x + 8)}{3}$

72. $5(x + 2) = 5x - 2$

73. $x + 7 = \dfrac{2x + 6}{2} + 4$

74. $2(y - 3) - \dfrac{y}{2} = \dfrac{3}{2}(y - 4)$

ADDITIONAL PRACTICE *Identify each statement as
an expression or an equation, and then either simplify or solve as
appropriate.*

75. $2(x - y) - (x + y) + y$ **76.** $3z + 2(y - z) + y$

77. $\dfrac{4(2x - 10)}{3} = 2(x - 4)$ **78.** $\dfrac{11(x - 12)}{2} = 9 - 2x$

79. $2\left(4x + \dfrac{9}{2}\right) - 3\left(x + \dfrac{2}{3}\right)$ **80.** $\dfrac{5(2 - m)}{3} = m + 6$

81. $\dfrac{8(5 - q)}{5} = -2q$ **82.** $\dfrac{20 - a}{2} = \dfrac{3}{2}(a + 4)$

83. $\dfrac{3x + 14}{2} = x - 2 + \dfrac{x + 18}{2}$

84. $7\left(3x - \dfrac{2}{7}\right) - 5\left(2x - \dfrac{3}{5}\right) + x$

85. $6 - 5r = 7r$ **86.** $y + 4 = -7y$

87. $22 - 3r = 8r$ **88.** $14 + 7s = s$

89. $8(x + 3) - 3x$ **90.** $2x + 2(x + 3)$

91. $19.1x - 4(x + 0.3) = -46.5$
92. $18.6x + 7.2 = 1.5(48 - 2x)$
93. $3.2(m + 1.3) - 2.5(m - 7.2)$
94. $6.7(t - 2.1) + 5.5(t + 1)$
95. $14.3(x + 2) + 13.7(x - 3) = 15.5$
96. $1.25(x - 1) = 0.5(3x - 1) - 1$
97. $10x + 3(2 - x) = 5(x + 2) - 4$
98. $19.1x - 4(x + 0.3)$

Solve each equation and round the result to the nearest tenth.

99. $\dfrac{3.7(2.3x - 2.7)}{1.5} = 5.2(x - 1.2)$

100. $\dfrac{-2.1(1.7x + 0.9)}{3.1} = -7.1(x - 1.3)$

WRITING ABOUT MATH

101. Explain why $3x^2y$ and $5x^2y$ are like terms.
102. Explain why $3x^2y$ and $3xy^2$ are unlike terms.
103. Discuss whether $7xxy^3$ and $5x^2yyy$ are like terms.
104. Discuss whether $\dfrac{3}{2}x$ and $\dfrac{3x}{2}$ are like terms.

SOMETHING TO THINK ABOUT

105. What number is equal to its own double?
106. What number is equal to one-half of itself?

Section 2.4

Formulas

Objectives

1. Solve a formula for an indicated variable using the properties of equality.
2. Evaluate a formula for specified values for the variables.
3. Solve an application using a given formula and specified values for the variables.

Vocabulary

literal equations   formulas

Getting Ready

Fill in the blanks.

1. $\dfrac{3x}{\boxed{}} = x$

2. $\dfrac{-5y}{\boxed{}} = y$

3. $\dfrac{rx}{\boxed{}} = x$

4. $\dfrac{-ay}{\boxed{}} = y$

5. $\boxed{} \cdot \dfrac{x}{7} = x$

6. $\boxed{} \cdot \dfrac{y}{12} = y$

7. $\boxed{} \cdot \dfrac{x}{d} = x$

8. $\boxed{} \cdot \dfrac{y}{s} = y$

Equations with several variables are called **literal equations**. Often these equations are **formulas** such as $A = lw$, the formula for finding the area of a rectangle.

Suppose that we want to find the lengths of several rectangles whose areas and widths are known. It would be tedious to substitute values for A and w into the formula and then repeatedly solve the formula for l. It would be much easier to solve the formula $A = lw$ for l first, then substitute values for A and w, and compute l directly.

1 Solve a formula for an indicated variable using the properties of equality.

To *solve a formula for a variable* means to isolate that variable on one side of the equation, with all other numbers and variables on the other side. We can isolate the variable by using the equation-solving techniques we have learned in the previous three sections.

EXAMPLE 1   Solve $A = lw$ for l.

Solution   To isolate l, we undo the multiplication by w by dividing both sides of the equation by w.

$$A = lw$$

$$\frac{A}{w} = \frac{lw}{w} \qquad \text{To undo the multiplication by } w, \text{ divide both sides by } w.$$

$$\frac{A}{w} = l \qquad \frac{w}{w} = 1.$$

 SELF CHECK 1   Solve $A = lw$ for w.

EXAMPLE 2 Recall that the formula $A = \frac{1}{2}bh$ gives the area of a triangle with base b and height h. Solve the formula for b.

Solution To isolate b, we will clear the fraction by multiplying both sides by 2. Then we will undo the multiplication by h by dividing both sides by h.

$$A = \frac{1}{2}bh$$

$$\mathbf{2} \cdot A = \mathbf{2} \cdot \left(\frac{1}{2}bh\right) \qquad \text{To clear the fraction, multiply both sides by 2.}$$

$$2A = bh \qquad\qquad 2 \cdot \frac{1}{2} = 1.$$

$$\frac{2A}{\mathbf{h}} = \frac{bh}{\mathbf{h}} \qquad\qquad \text{To undo the multiplication by } h, \text{ divide both sides by } h.$$

$$\frac{2A}{h} = b \qquad\qquad \frac{h}{h} = 1.$$

If the area A and the height h of a triangle are known, the base b is given by the formula $b = \frac{2A}{h}$.

🌿 **SELF CHECK 2** Solve $A = \frac{1}{2}bh$ for h.

EXAMPLE 3 The formula $C = \frac{5}{9}(F - 32)$ is used to convert Fahrenheit temperature readings into their Celsius equivalents. Solve the formula for F.

Solution To isolate F, we will undo the multiplication by $\frac{5}{9}$ by multiplying both sides by the reciprocal of $\frac{5}{9}$, which is $\frac{9}{5}$. Then we will use the distributive property to remove parentheses and finally undo the subtraction of 32 by adding 32 to both sides.

$$C = \frac{5}{9}(F - 32)$$

$$\frac{\mathbf{9}}{\mathbf{5}} \cdot C = \frac{\mathbf{9}}{\mathbf{5}} \cdot \left[\frac{5}{9}(F - 32)\right] \qquad \text{To eliminate } \frac{5}{9}, \text{ multiply both sides by } \frac{9}{5}.$$

$$\frac{9}{5}C = \left(\frac{9}{5} \cdot \frac{5}{9}\right)(F - 32) \qquad \text{Apply the associative property of multiplication.}$$

$$\frac{9}{5}C = 1(F - 32) \qquad \frac{9}{5} \cdot \frac{5}{9} = \frac{9 \cdot 5}{5 \cdot 9} = 1.$$

$$\frac{9}{5}C = F - 32 \qquad \text{Use the distributive property to remove parentheses.}$$

$$\frac{9}{5}C + \mathbf{32} = F - 32 + \mathbf{32} \qquad \text{To undo the subtraction of 32, add 32 to both sides.}$$

$$\frac{9}{5}C + 32 = F \qquad \text{Combine like terms.}$$

The formula $F = \frac{9}{5}C + 32$ is used to convert degrees Celsius to degrees Fahrenheit.

🌿 **SELF CHECK 3** Solve $x = \frac{2}{3}(y + 5)$ for y.

EXAMPLE 4 Recall that the area A of the trapezoid shown in Figure 2-5 is given by the formula

$$A = \frac{1}{2}h(B + b)$$

where B and b are its bases and h is its height. Solve the formula for b.

Solution There are two different ways to solve this formula.

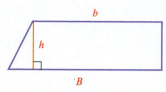

Figure 2-5

Method 1: $A = \frac{1}{2}(B + b)h$

$\mathbf{2 \cdot} A = \mathbf{2 \cdot} \left[\frac{1}{2}(B + b)h\right]$ Multiply both sides by 2.

$2A = \left(2 \cdot \frac{1}{2}\right)(B + b)h$ Apply the associative property of multiplication.

$2A = Bh + bh$ Simplify and use the distributive property to remove parentheses.

$2A \mathbf{- Bh} = Bh + bh \mathbf{- Bh}$ Subtract Bh from both sides.

$2A - Bh = bh$ Combine like terms.

$\dfrac{2A - Bh}{h} = \dfrac{bh}{h}$ Divide both sides by h.

$\dfrac{2A - Bh}{h} = b$ $\frac{h}{h} = 1$.

Method 2: $A = \frac{1}{2}(B + b)h$

$\mathbf{2 \cdot} A = \mathbf{2 \cdot} \left[\frac{1}{2}(B + b)h\right]$ Multiply both sides by 2.

$2A = (B + b)h$ Simplify.

$\dfrac{2A}{h} = \dfrac{(B + b)h}{h}$ Divide both sides by h.

$\dfrac{2A}{h} = B + b$ $\frac{h}{h} = 1$.

$\dfrac{2A}{h} \mathbf{- B} = B + b \mathbf{- B}$ Subtract B from both sides.

$\dfrac{2A}{h} - B = b$ Combine like terms.

Although they look different, the results of Methods 1 and 2 are equivalent.

 SELF CHECK 4 Solve $A = \frac{1}{2}h(B + b)$ for B.

2 Evaluate a formula for specified values for the variables.

EXAMPLE 5 Solve the formula $P = 2l + 2w$ for l. Evaluate that formula for l when $P = 56$ and $w = 11$.

Solution We first solve the formula $P = 2l + 2w$ for l.

$$P = 2l + 2w$$

$P \mathbf{- 2w} = 2l + 2w \mathbf{- 2w}$ Subtract $2w$ from both sides.

Albert Einstein
1879–1955
Einstein was a theoretical physicist best known for his theory of relativity. Although Einstein was born in Germany, he became a Swiss citizen and earned his doctorate at the University of Zurich in 1905. In 1910, he returned to Germany to teach. He fled Germany because of the Nazi government and became a United States citizen in 1940. He is famous for his formula $E = mc^2$.

$$P - 2w = 2l \qquad \text{Combine like terms.}$$

$$\frac{P - 2w}{2} = \frac{2l}{2} \qquad \text{Divide both sides by 2.}$$

$$\frac{P - 2w}{2} = l \qquad \tfrac{2}{2} = 1.$$

We will then substitute 56 for P and 11 for w and simplify.

$$l = \frac{\mathbf{P} - 2\mathbf{w}}{2}$$

$$l = \frac{\mathbf{56} - 2(\mathbf{11})}{2}$$

$$= \frac{56 - 22}{2}$$

$$= \frac{34}{2}$$

$$= 17$$

Thus, $l = 17$.

 SELF CHECK 5   Solve $P = 2l + 2w$ for w. Evaluate that fomula for w when $P = 46$ and $l = 16$.

---

3 **Solve an application using a given formula and specified values for the variables.**

EXAMPLE 6   Recall that the volume V of the right-circular cone shown in Figure 2-6 is given by the formula

$$V = \frac{1}{3}Bh$$

where B is the area of its circular base and h is its height. Solve the formula for h and find the height of a right-circular cone with a volume of 64 cubic centimeters and a base area of 16 square centimeters.

Solution   We first solve the formula for h.

$$V = \frac{1}{3}Bh$$

$$\mathbf{3 \cdot} V = \mathbf{3 \cdot} \left(\frac{1}{3} Bh \right) \qquad \text{Multiply both sides by 3.}$$

$$3V = Bh \qquad\qquad 3 \cdot \tfrac{1}{3} = 1.$$

$$\frac{3V}{\mathbf{B}} = \frac{Bh}{\mathbf{B}} \qquad\qquad \text{Divide both sides by } B.$$

$$\frac{3V}{B} = h \qquad\qquad \tfrac{B}{B} = 1.$$

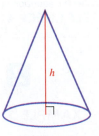

Figure 2-6

We then substitute 64 for V and 16 for B and simplify.

$$h = \frac{3V}{B}$$

$$h = \frac{3(64)}{16}$$

$$= 3(4)$$

$$= 12$$

The height of the cone is 12 centimeters.

SELF CHECK 6 Solve $V = \frac{1}{3}Bh$ for B, and find the area of the base when the volume is 42 cubic feet and the height is 6 feet.

SELF CHECK ANSWERS 1. $w = \frac{A}{l}$ 2. $h = \frac{2A}{b}$ 3. $y = \frac{3}{2}x - 5$ 4. $B = \frac{2A - hb}{h}$ or $B = \frac{2A}{h} - b$ 5. $w = \frac{P - 2l}{2}$, 7 6. $B = \frac{3V}{h}$, 21 ft^2

NOW TRY THIS

A student's test average for four tests can be modeled by the equation

$$A = \frac{T_1 + T_2 + T_3 + T_4}{4}$$

where T_1 is the grade for Test 1, T_2 is the grade for Test 2, and so on.

1. Solve the equation for T_4.
2. Julio has test grades of 82, 88, and 71. What grade would he need on Test 4 to have a test average of 80?
3. Leslie has test grades of 75, 80, and 89. What grade would Leslie need on Test 4 to have a test average of 90? Interpret your answer.

2.4 Exercises

WARM-UPS *Solve the equation* $ab + c = 0$.

1. for a
2. for c

Solve the equation $a = \frac{b}{c}$.

3. for b
4. for c

REVIEW *Simplify each expression, if possible.*

5. $7x - 4y - 4x$
6. $3ab^2 + 7a^2b$
7. $\frac{2}{3}(a + 3) - \frac{5}{3}(6 + a)$
8. $\frac{2}{11}(22x - y) + \frac{9}{11}y$

VOCABULARY AND CONCEPTS *Fill in the blanks.*

9. Equations that contain several variables are called _____ equations.
10. The equation $A = lw$ is an example of a _____.
11. To solve a formula for a variable means to _____ the variable on one side of the equation.
12. To solve the formula $d = rt$ for t, divide both sides of the formula by _.
13. To solve $A = p + i$ for p, _____ i from both sides.
14. To solve $t = \frac{d}{r}$ for d, _____ both sides by r.

GUIDED PRACTICE *Solve for the indicated variable.*
SEE EXAMPLE 1. (OBJECTIVE 1)

15. $E = IR$ for I

16. $i = prt$ for t

17. $V = lwh$ for w

18. $C = 2\pi r$ for r

19. $x = y + 12$ for y

20. $P = a + b + c$ for c

Solve for the indicated variable. SEE EXAMPLE 2. (OBJECTIVE 1)

21. $V = \dfrac{1}{3}Bh$ for h

22. $V = \dfrac{1}{3}Bh$ for B

23. $V = \dfrac{1}{3}\pi r^2 h$ for h

24. $I = \dfrac{E}{R}$ for R

Solve for the indicated variable. SEE EXAMPLE 3. (OBJECTIVE 1)

25. $y = \dfrac{1}{2}(x + 2)$ for x

26. $x = \dfrac{1}{5}(y - 7)$ for y

27. $A = \dfrac{5}{2}(B + 3)$ for B

28. $y = \dfrac{5}{2}(x - 10)$ for x

Solve for the indicated variable. SEE EXAMPLE 4. (OBJECTIVE 1)

29. $p = \dfrac{h}{2}(q + r)$ for q

30. $p = \dfrac{h}{2}(q + r)$ for r

31. $G = 2b(r - 1)$ for r

32. $F = f(1 - M)$ for M

Solve each formula for the indicated variable. Then evaluate the new formula for the values given. SEE EXAMPLE 5. (OBJECTIVE 2)

33. $d = rt$ Find t if $d = 455$ and $r = 65$.

34. $d = rt$ Find r if $d = 275$ and $t = 5$.

35. $P = a + b + c$ Find b if $P = 37, a = 15$, and $c = 6$.

36. $y = mx + b$ Find x if $y = 30, m = 3$, and $b = 0$.

ADDITIONAL PRACTICE *Solve each formula for the indicated variable.*

37. $3x + 2y = 5$ for y

38. $y = mx + b$ for b

39. $C = \pi d$ for d

40. $P = I^2R$ for R

41. $P = 2l + 2w$ for w

42. $V = lwh$ for l

43. $A = P + Prt$ for t

44. $A = \dfrac{1}{2}(B + b)h$ for h

45. $K = \dfrac{wv^2}{2g}$ for w

46. $V = \pi r^2 h$ for h

47. $K = \dfrac{wv^2}{2g}$ for g

48. $P = \dfrac{RT}{mV}$ for V

49. $F = \dfrac{GMm}{d^2}$ for M

50. $C = 1 - \dfrac{A}{a}$ for A

51. Given that $i = prt$, find p if $i = 90, t = 4$, and $r = 0.03$.

52. Given that $i = prt$, find r if $i = 120, p = 500$, and $t = 6$.

53. Given that $K = \dfrac{1}{2}h(a + b)$, find h if $K = 48, a = 7$, and $b = 5$.

54. Given that $\dfrac{x}{2} + y = z^2$, find x if $y = 3$ and $z = 3$.

APPLICATIONS *Solve.* SEE EXAMPLE 6 (OBJECTIVE 3)

55. Volume of a cone   The volume V of a cone is given by the formula $V = \dfrac{1}{3}\pi r^2 h$. Solve the formula for h, and then calculate the height h if V is 36π cubic inches and the radius r is 6 inches.

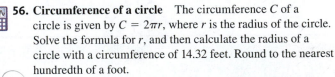**56. Circumference of a circle**   The circumference C of a circle is given by $C = 2\pi r$, where r is the radius of the circle. Solve the formula for r, and then calculate the radius of a circle with a circumference of 14.32 feet. Round to the nearest hundredth of a foot.

57. Ohm's law   The formula $E = IR$, called **Ohm's law,** is used in electronics. Solve for I, and then calculate the current I if the voltage E is 48 volts and the resistance R is 12 ohms. Current has units of *amperes*.

58. Growth of money   At a simple interest rate r, an amount of money P grows to an amount A in t years according to the formula $A = P(1 + rt)$. Solve the formula for P. After $t = 3$ years, a girl has an amount $A = \$4{,}357$ on deposit. What amount P did she start with? Assume an interest rate of 6%.

59. Power loss   The power P lost when an electric current I passes through a resistance R is given by the formula $P = I^2R$. Solve for R. If P is 2,700 watts and I is 14 amperes, calculate R to the nearest hundredth of an ohm.

60. Geometry   The perimeter P of a rectangle with length l and width w is given by the formula $P = 2l + 2w$. Solve this formula for w. If the perimeter of a certain rectangle is 58.37 meters and its length is 17.23 meters, find its width. Round to two decimal places.

61. **Force of gravity** The masses of the two objects in the illustration are m and M. The force of gravitation F between the masses is given by

$$F = \frac{GmM}{d^2}$$

where G is a constant and d is the distance between them. Solve for m.

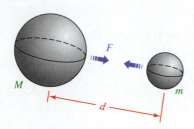

62. **Thermodynamics** In thermodynamics, the Gibbs free-energy equation is given by

$$G = U - TS + pV$$

Solve this equation for the pressure, p.

63. **Pulleys** The approximate length L of a belt joining two pulleys of radii r and R feet with centers D feet apart is given by the formula

$$L = 2D + 3.25(r + R)$$

Solve the formula for D. If a 25-foot belt joins pulleys with radii of 1 foot and 3 feet, how far apart are their centers?

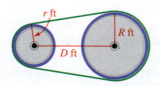

64. **Geometry** The measure a of an interior angle of a regular polygon with n sides is given by $a = 180°\left(1 - \frac{2}{n}\right)$. Solve the formula for n. How many sides does a regular polygon have if an interior angle is 108°? (*Hint:* Distribute first.)

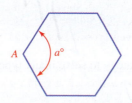

One common retirement plan for self-employed people is called a Simplified Employee Pension Plan. It allows for a maximum annual contribution of 15% of taxable income (earned income minus deductible expenses). However, since the Internal Revenue Service considers the SEP contribution to be a deductible expense, the taxable income must be reduced by the amount of the contribution. Therefore, to calculate the maximum contribution C, we take 15% of what's left after we subtract the contribution C from the taxable income T.

$$C = 0.15(T - C)$$

65. **Calculating SEP contributions** Find the maximum allowable contribution to a SEP plan by solving the equation $C = 0.15(T - C)$ for C.

66. **Calculating SEP contributions** Find the maximum allowable contribution to a SEP plan for a person who earns $75,000 and has deductible expenses of $27,540. See Exercise 65.

WRITING ABOUT MATH

67. The formula $P = 2l + 2w$ is also an equation, but an equation such as $2x + 3 = 5$ is not a formula. What equations do you think should be called formulas?

68. To solve the equation $s - A(s - 5) = r$ for the variable s, one student simply added $A(s - 5)$ to both sides to get $s = r + A(s - 5)$. Explain why this is not correct.

SOMETHING TO THINK ABOUT

69. The energy of an atomic bomb comes from the conversion of matter into energy, according to Einstein's formula $E = mc^2$. The constant c is the speed of light, about 300,000 meters per second. Find the energy in a mass m of 1 kilogram. Energy has units of **joules**.

70. When a car of mass m collides with a wall, the energy of the collision is given by the formula $E = \frac{1}{2}mv^2$. Compare the energy of two collisions: a car striking a wall at 30 mph, and at 60 mph.

Section 2.5

Introduction to Problem Solving

Objectives

1. Solve a number application using a linear equation in one variable.
2. Solve a geometry application using a linear equation in one variable.
3. Solve an investment application using a linear equation in one variable.

Vocabulary

angle
degree
right angle
straight angle

adjacent angles
complementary angles
supplementary angles
isosceles triangle

vertex angle of an isosceles triangle
base angles of an isosceles triangle

Getting Ready

1. If one part of a pipe is x feet long and the other part is $(x + 2)$ feet long, find an expression that represents the total length of the pipe.

2. If one part of a board is x feet long and the other part is three times as long, find an expression that represents the length of the board.

3. What is the formula for the perimeter of a rectangle?

4. Define a triangle.

In this section, we will use the equation-solving skills we have learned in the previous four sections to solve many types of applications. The key to successful problem solving is to understand the situation thoroughly and then devise a plan to solve it. To do so, we will use the following problem-solving strategy.

PROBLEM SOLVING

1. **Analyze the problem** and **identify a variable** by asking yourself "What am I asked to find?" Choose a variable to represent the quantity to be found and then express all other unknown quantities in the problem as expressions involving that variable.

2. **Form an equation** by expressing a quantity in two different ways. This may require reading the problem several times to understand the given facts. What information is given? Is there a formula that applies to this situation? Often a sketch, chart, or diagram will help you visualize the facts of the problem.

3. **Solve the equation** found in Step 2.

4. **State the conclusion.**

5. **Check the result** to be certain it satisfies the given conditions.

In this section, we will use this five-step strategy to solve many types of applications.

1 Solve a number application using a linear equation in one variable.

EXAMPLE 1   **PLUMBING**   A plumber wants to cut a 17-foot pipe into three parts. (See Figure 2-7.) If the longest part is to be 3 times as long as the shortest part, and the middle-sized part is to be 2 feet longer than the shortest part, how long should each part be?

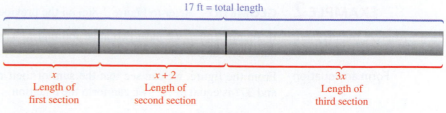

Figure 2-7

Analyze the problem We are asked to find the length of three pieces of pipe. The information is given in terms of the length of the shortest part. Therefore, we let x represent the length, in feet, of the shortest part and express the other lengths in terms of x. Then $3x$ represents the length of the longest part, and $x + 2$ represents the length of the middle-sized part.

Form an equation The sum of the lengths of these three parts is equal to the total length of the pipe.

The length of part 1	plus	the length of part 2	plus	the length of part 3	equals	the total length.
x	$+$	$x + 2$	$+$	$3x$	$=$	17

Solve the equation We can solve this equation as follows.

$$x + x + 2 + 3x = 17 \qquad \text{This is the equation to solve.}$$
$$5x + 2 = 17 \qquad \text{Combine like terms.}$$
$$5x = 15 \qquad \text{Subtract 2 from both sides.}$$
$$x = 3 \qquad \text{Divide both sides by 5.}$$

State the conclusion The shortest part is 3 feet long. Because the middle-sized part is 2 feet longer than the shortest, it is 5 feet long. Because the longest part is 3 times longer than the shortest, it is 9 feet long.

Check the result Because the sum of 3 feet, 5 feet, and 9 feet is 17 feet, the solution checks.

 **SELF CHECK 1** A plumber wants to cut a 17-foot pipe into three parts. If the longest part is to be 2 times as long as the shortest part, and the middle-sized part is to be 1 foot longer than the shortest part, how long should each part be?

COMMENT Remember to include any units (feet, inches, pounds, etc.) when stating the conclusion to an application.

2 Solve a geometry application using a linear equation in one variable.

The geometric figure shown in Figure 2-8(a) is an **angle**. Angles are measured in **degrees**. The angle shown in Figure 2-8(b) measures 45 degrees (denoted as 45°). If an angle measures 90°, as in Figure 2-8(c), it is a **right angle**. If an angle measures 180°, as in Figure 2-8(d), it is a **straight angle**. **Adjacent angles** are two angles that share a common side.

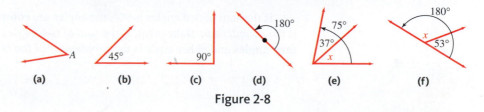

Figure 2-8

EXAMPLE 2   **GEOMETRY**   Refer to Figure 2-8(e) on the previous page and find the value of x.

Analyze the problem   In Figure 2-8(e), we have two adjacent angles. The unknown angle measure is designated as x degrees.

Form an equation   From the figure, we can see that the sum of their measures is 75°. Since the sum of x and 37° is equal to 75°, we can form the equation.

The angle that measures $x°$	plus	the angle that measures 37°	equals	the angle that measures 75°.
x	$+$	37	$=$	75

Solve the equation   We can solve this equation as follows.

$$x + 37 = 75 \qquad \text{This is the equation to solve.}$$
$$x + 37 - 37 = 75 - 37 \qquad \text{Subtract 37 from both sides.}$$
$$x = 38 \qquad 37 - 37 = 0 \text{ and } 75 - 37 = 38.$$

State the conclusion   The value of x is 38°.

Check the result   Since the sum of 38° and 37° is 75°, the solution checks.

🍃 **SELF CHECK 2**   Find the value of x in the figure to the right.

EXAMPLE 3   **GEOMETRY**   Refer to Figure 2-8(f) and find the value of x.

Analyze the problem   In Figure 2-8(f), we have two adjacent angles. The unknown angle measure is designated as x degrees.

Form an equation   From the figure, we can see that the sum of their measures is 180°. Since the sum of x and 53° is equal to 180°, we can form the equation.

The angle that measures $x°$	plus	the angle that measures 53°	equals	the angle that measures 180°.
x	$+$	53	$=$	180

Solve the equation   We can solve this equation as follows.

$$x + 53 = 180 \qquad \text{This is the equation to solve.}$$
$$x + 53 - 53 = 180 - 53 \qquad \text{Subtract 53 from both sides.}$$
$$x = 127 \qquad 53 - 53 = 0 \text{ and } 180 - 53 = 127.$$

State the conclusion   The value of x is 127°.

Check the result   Since the sum of 127° and 53° is 180°, the solution checks.

🍃 **SELF CHECK 3**   Find the value of x in the figure to the right.

If the sum of two angles is 90°, the angles are **complementary angles** and either angle is the *complement* of the other. If the sum of two angles is 180°, the angles are **supplementary angles** and either angle is the *supplement* of the other.

EXAMPLE 4   **COMPLEMENTARY ANGLES**   Find the complement of an angle measuring 30°.

Analyze the problem   To find the complement of a 30° angle, we must find an angle whose measure plus 30° equals 90°. We can let x represent the complement of 30°.

Form an equation   Since the sum of two complementary angles is 90°, we can form the equation.

The angle that measures $x°$	plus	the angle that measures 30°	equals	90°.
x	$+$	30	$=$	90

Solve the equation   We can solve this equation as follows.

$$x + 30 = 90 \qquad \text{This is the equation to solve.}$$
$$x + 30 - 30 = 90 - 30 \qquad \text{Subtract 30 from both sides.}$$
$$x = 60 \qquad 30 - 30 = 0 \text{ and } 90 - 30 = 60.$$

State the conclusion   The complement of a 30° angle is a 60° angle.

Check the result   Since the sum of 60° and 30° is 90°, the solution checks.

 SELF CHECK 4   Find the complement of an angle measuring 40°.

EXAMPLE 5   **SUPPLEMENTARY ANGLES**   Find the supplement of an angle measuring 50°.

Analyze the problem   To find the supplement of a 50° angle, we must find an angle whose measure plus 50° equals 180°. We can let x represent the supplement of 50°.

Form an equation   Since the sum of two supplementary angles is 180°, we can form the equation.

The angle that measures $x°$	plus	the angle that measures 50°	equals	180°.
x	$+$	50	$=$	180

Solve the equation   We can solve this equation as follows.

$$x + 50 = 180 \qquad \text{This is the equation to solve.}$$
$$x + 50 - 50 = 180 - 50 \qquad \text{Subtract 50 from both sides.}$$
$$x = 130 \qquad 50 - 50 = 0 \text{ and } 180 - 50 = 130.$$

State the conclusion   The supplement of a 50° angle is a 130° angle.

Check the result   Since the sum of 50° and 130° is 180°, the solution checks.

 SELF CHECK 5   Find the supplement of an angle measuring 80°.

EXAMPLE 6   **RECTANGLES**   The length of a rectangle is 4 meters longer than twice its width. If the perimeter of the rectangle is 26 meters, find its dimensions.

Analyze the problem   Because we are asked to find the dimensions of the rectangle, we will need to find both the width and the length. If we let w represent the width of the rectangle in meters, then $(4 + 2w)$ will represent its length.

Form an equation   To visualize the problem, we sketch the rectangle as shown in Figure 2-9 on the next page. Recall that the formula for finding the perimeter of a rectangle is $P = 2l + 2w$. Therefore, the perimeter of the rectangle in the figure is $2(4 + 2w) + 2w$. We also are told that the perimeter is 26.

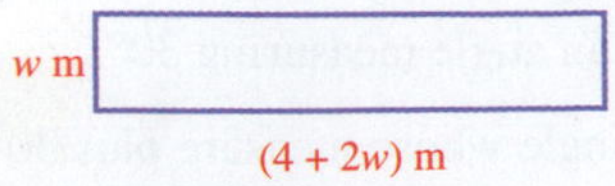

(4 + 2w) m

Figure 2-9

We can form the equation as follows.

2	times	the length	plus	2	times	the width	equals	the perimeter.
2	·	(4 + 2w)	+	2	·	w	=	26

Solve the equation   We can solve this equation as follows.

$$2(4 + 2w) + 2w = 26 \quad \text{This is the equation to solve.}$$
$$8 + 4w + 2w = 26 \quad \text{Use the distributive property to remove parentheses.}$$
$$6w + 8 = 26 \quad \text{Combine like terms.}$$
$$6w = 18 \quad \text{Subtract 8 from both sides.}$$
$$w = 3 \quad \text{Divide both sides by 6.}$$

State the conclusion   The width of the rectangle is 3 meters, and the length, $4 + 2w$, is 10 meters.

Check the result   If the rectangle has a width of 3 meters and a length of 10 meters, the length is 4 meters longer than twice the width ($4 + 2 \cdot 3 = 10$), and the perimeter is 26 meters. The solution checks.

 SELF CHECK 6   The length of a rectangle is 1 meter longer than twice its width. If the perimeter of the rectangle is 32 meters, find its dimensions.

EXAMPLE 7   **ISOSCELES TRIANGLES**   The vertex angle of an isosceles triangle is 56°. Find the measure of each base angle.

Analyze the problem   An **isosceles triangle** has two sides of equal length, which meet to form the **vertex angle**. See Figure 2-10. The angles opposite those sides, called **base angles**, have equal measures. If we let x represent the measure of one base angle, the measure of the other base angle is also x.

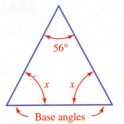

Figure 2-10

Form an equation   From geometry, we know that in any triangle the sum of the measures of its three angles is 180°. Therefore, we can form the equation.

The measure of one base angle	plus	the measure of the other base angle	plus	the measure of the vertex angle	equals	180°.
x	+	x	+	56	=	180

Solve the equation   We can solve this equation as follows.

$$x + x + 56 = 180 \quad \text{This is the equation to solve.}$$
$$2x + 56 = 180 \quad \text{Combine like terms.}$$
$$2x = 124 \quad \text{Subtract 56 from both sides.}$$
$$x = 62 \quad \text{Divide both sides by 2.}$$

State the conclusion   The measure of each base angle is 62°.

Check the result   The measure of each base angle is 62°, and the vertex angle measures 56°. Since $62° + 62° + 56° = 180°$, the sum of the measures of the three angles is 180°. The solution checks.

 SELF CHECK 7   The vertex angle of an isosceles triangle is 42°. Find the measure of each base angle.

3 Solve an investment application using a linear equation in one variable.

EXAMPLE 8   **INVESTMENTS**   A teacher invests part of $12,000 at 6% annual simple interest, and the rest at 9%. If the annual income from these investments was $945, how much did the teacher invest at each rate?

Analyze the problem   We are asked to find the amount of money the teacher has invested in two different accounts. If we let x represent the amount of money invested at 6% annual interest, the remainder, $(12,000 - x)$, represents the amount invested at 9% annual interest.

Form an equation   The interest i earned by an amount p invested at an annual rate r for t years is given by the formula $i = prt$. In this example, $t = 1$ year. Hence, if x dollars were invested at 6%, the interest earned would be $0.06x$ dollars. If x dollars were invested at 6%, the rest of the money, $(12,000 - x)$ dollars, would be invested at 9%. The interest earned on that money would be $0.09(12,000 - x)$ dollars. The total interest earned in dollars can be expressed in two ways: as 945 and as the sum $0.06x + 0.09(12,000 - x)$.

We can form an equation as follows.

The interest earned at 6%	plus	the interest earned at 9%	equals	the total interest.
$0.06x$	$+$	$0.09(12,000 - x)$	$=$	945

Solve the equation   We can solve this equation as follows.

$$0.06x + 0.09(12,000 - x) = 945$$ This is the equation to solve.

$$6x + 9(12,000 - x) = 94,500$$ Multiply both sides by 100 to clear the equation of decimals.

$$6x + 108,000 - 9x = 94,500$$ Use the distributive property to remove parentheses.

$$-3x + 108,000 = 94,500$$ Combine like terms.

$$-3x = -13,500$$ Subtract 108,000 from both sides.

$$x = 4,500$$ Divide both sides by -3.

State the conclusion   The teacher invested $4,500 at 6% and $12,000 - $4,500 or $7,500 at 9%.

Check the result   The first investment earned 6% of $4,500, or $270. The second investment earned 9% of $7,500, or $675. Because the total return was $270 + $675, or $945, the solutions check.

 **SELF CHECK 8**   A teacher invests part of $15,000 at 6% annual simple interest, and the rest at 9%. If the annual income from these investments was $945, how much did the teacher invest at each rate?

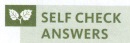 **SELF CHECK ANSWERS**   **1.** 4 ft, 8 ft, 5 ft   **2.** 48°   **3.** 62°   **4.** 50°   **5.** 100°   **6.** 5 meters by 11 meters   **7.** 69°   **8.** $13,500 at 6%, $1,500 at 9%

NOW TRY THIS

1. Mark invested $100,000 in two accounts. Part was in bonds that paid 7% annual interest and the rest in stocks that lost 5% of their value. How much did he originally invest in each account if his total earned interest for the year was $2,200? How much money does he have in each account now?

2. Sharlette invested $80,000 in two accounts. Part was in an account paying 6.25% and the rest in an account paying 4.48%. If her total interest was $4,115, how much was invested in each account?

2.5 Exercises

WARM-UPS

If the length of a board is 12 feet and one part of it is x feet,

1. Write an expression for the other part.
2. Write an expression for a length that is 5 feet shorter than twice the first part.

If an instructor invests $18,000, $x in an account paying 4%, and the rest in an account paying 7%,

3. Write an expression for the interest earned on the money in the 7% account.
4. Write an expression for the total interest earned.

REVIEW *Refer to the formulas in Section 1.3.*

5. Find the volume of a pyramid that has a height of 6 centimeters and a square base, 10 centimeters on each side.
6. Find the volume of a cone with a height of 6 centimeters and a circular base with radius 6 centimeters. Use $\pi \approx \frac{22}{7}$.

Simplify each expression.

7. $5(x - 2) + 3(x + 4)$
8. $6(y + 2) - 4(y - 3)$
9. $\frac{1}{2}(x + 1) - \frac{1}{2}(x + 4)$
10. $\frac{3}{2}\left(x + \frac{2}{3}\right) + \frac{1}{2}(x + 8)$.

11. The amount A on deposit in a bank account bearing simple interest is given by the formula

    $$A = P + Prt$$

    Find A when $P = \$1,200$, $r = 0.08$, and $t = 3$.

12. The distance s that a certain object falls from a height of 350 ft in t seconds is given by the formula

    $$s = 350 - 16t^2 + vt$$

    Find s when $t = 3$ and $v = -4$.

VOCABULARY AND CONCEPTS *Fill in the blanks.*

13. The perimeter of a rectangle is given by the formula $P = $ _____.
14. An _____ triangle is a triangle with two sides of equal length.
15. The sides of equal length of an isosceles triangle meet to form the _____ angle.
16. The angles opposite the sides of equal length of an isosceles triangle are called ____ angles.
17. Angles are measured in _____.
18. If an angle measures 90°, it is called a ____ angle.
19. If an angle measures 180°, it is called a _____ angle.
20. If the sum of the measures of two angles is 90°, the angles are called _____ angles.
21. If the sum of the measures of two angles is 180°, the angles are called _____ angles.
22. The sum of the measures of the angles of any triangle is ____.

APPLICATIONS *SEE EXAMPLE 1. (OBJECTIVE 1)*

23. **Carpentry** The 12-foot board in the illustration has been cut into two parts, one twice as long as the other. How long is each part?

24. **Plumbing** A 20-foot pipe has been cut into two parts, one 3 times as long as the other. How long is each part?

25. **Robotics** If the robotic arm shown in the illustration will extend a total distance of 30 feet, how long is each section?

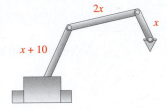

26. **Statue of Liberty** If the figure part of the Statue of Liberty is 3 feet shorter than the height of its pedestal base, find the height of the figure.

27. **Window designs** The perimeter of the triangular window shown in the illustration is 24 feet. How long is each section?

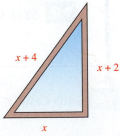

28. **Football** In 1967, Green Bay beat Kansas City by 25 points in the first Super Bowl. If a total of 45 points were scored, what was the final score of the game?

29. **Publishing** A book can be purchased in hardcover for $16.95 or in paperback for $5.95. How many of each type were printed if 14 times as many paperbacks were printed as hardcovers and a total of 210,000 books were printed?

30. **Concert tours** A rock group plans three concert tours over a period of 26 weeks. The tour in Britain will be twice as long

as the tour in France and the tour in Germany will be 2 weeks shorter than the tour in France. How many weeks will they be in France?

Find the value of x. SEE EXAMPLE 2. (OBJECTIVE 2)

31.

32.

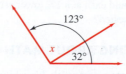

33.

34.

Find the value of x. SEE EXAMPLE 3. (OBJECTIVE 2)

35.

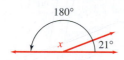

36.

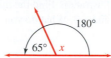

37.

38.

Find each value. SEE EXAMPLES 4–5. (OBJECTIVE 2)

39. Find the complement of an angle measuring 46°.

40. Find the supplement of an angle measuring 46°.

41. Find the supplement of the complement of an angle measuring 40°.

42. Find the complement of the supplement of an angle measuring 125°.

Solve. SEE EXAMPLE 6. (OBJECTIVE 2)

43. Circuit boards   The perimeter of the circuit board in the illustration is 90 centimeters. Find the dimensions of the board.

w cm

(*w* + 7) cm

44. Swimming pools   The width of a rectangular swimming pool is 11 meters less than the length, and the perimeter is 94 meters. Find its dimensions.

45. Framing pictures   The length of a rectangular picture is 6 inches less than twice the width. If the perimeter is 60 inches, find the dimensions of the frame.

46. Land areas   The perimeter of a square piece of land is twice the perimeter of an equilateral (equal-sided) triangular lot. If one side of the square is 60 meters, find the length of a side of the triangle.

Solve. SEE EXAMPLE 7. (OBJECTIVE 2)

47. Triangular bracing   The outside perimeter of the triangular brace shown in the illustration is 57 feet. If all three sides are of equal length, find the length of each side.

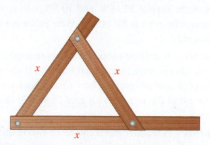

48. Trusses   The truss in the illustration is in the form of an isosceles triangle. Each of the two equal sides is 5 feet less than the third side. If the perimeter is 29 feet, find the length of each side.

49. Guy wires   The two guy wires in the illustration form an isosceles triangle. One of the two equal angles of the triangle is 4 times the third angle (the vertex angle). Find the measure of the vertex angle.

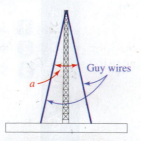

50. Equilateral triangles   Find the measure of each angle of an equilateral triangle. (*Hint*: The three angles of an equilateral triangle are equal.)

Solve. SEE EXAMPLE 8. (OBJECTIVE 3)

51. Investments   A student invested some money at an annual rate of 5%. If the annual income from the investment is $300, how much did he invest?

52. Investments   A student invested 90% of her savings in the stock market. If she invested $4,050, what are her total savings?

53. Investments   A broker invested $24,000 in two mutual funds, one earning 9% annual interest and the other earning

14%. After 1 year, his combined interest is $3,135. How much was invested at each rate?

54. **Investments**   A rollover IRA of $18,750 was invested in two mutual funds, one earning 12% interest and the other earning 10%. After 1 year, the combined interest income is $2,117. How much was invested at each rate?

55. **Investments**   One investment pays 8% and another pays 11%. If equal amounts are invested in each, the combined interest income for 1 year is $712.50. How much is invested at each rate?

56. **Investments**   When equal amounts are invested in each of three accounts paying 7%, 8%, and 10.5%, one year's combined interest income is $1,249.50. How much is invested in each account?

57. **Investments**   A college professor wants to supplement her retirement income with investment interest. If she invests $15,000 at 6% annual interest, how much more would she have to invest at 7% to achieve a goal of $1,250 in supplemental income?

58. **Investments**   A teacher has a choice of two investment plans: an insured fund that has paid an average of 11% interest per year, or a riskier investment that has averaged a 13% return. If the same amount invested at the higher rate would generate an extra $150 per year, how much does the teacher have to invest?

59. **Investments**   A financial counselor recommends investing twice as much in CDs (certificates of deposit) as in a

bond fund. A client follows his advice and invests $21,000 in CDs paying 1% more interest than the fund. The CDs would generate $840 more interest than the fund. Find the two rates. (*Hint*: 1% = 0.01.)

60. **Investments**   The amount of annual interest earned by $8,000 invested at a certain rate is $200 less than $12,000 would earn at a 1% lower rate. At what rate is the $8,000 invested?

WRITING ABOUT MATH

61. Write a paragraph describing the problem-solving process.

62. List as many types of angles as you can think of. Then define each type.

SOMETHING TO THINK ABOUT

63. If two lines intersect as in the illustration, angle 1 (denoted as ∠1) and ∠2, and ∠3 and ∠4, are called **vertical angles**. Let the measure of ∠1 be various numbers and compute the values of the other three angles. What do you discover?

64. If two lines meet and form a right angle, the lines are said to be **perpendicular**. See the illustration. Find the measures of ∠1, ∠2, and ∠3. What do you discover?

Section 2.6
Motion and Mixture Applications

Objectives

1. Solve a motion application using a linear equation in one variable.
2. Solve a liquid mixture application using a linear equation in one variable.
3. Solve a dry mixture application using a linear equation in one variable.

Getting Ready

1. At 30 mph, how far would a bus go in 2 hours?

2. At 55 mph, how far would a car travel in 7 hours?

3. If 8 gallons of a mixture of water and alcohol is 70% alcohol, how much alcohol does the mixture contain?

4. At $7 per pound, how many pounds of chocolate would be worth $63?

In this section, we consider uniform motion and mixture applications. In these problems, we will use the following three formulas:

$$r \cdot t = d \qquad \text{The rate multiplied by the time equals the distance.}$$
$$r \cdot b = a \qquad \text{The rate multiplied by the base equals the amount.}$$
$$v = p \cdot n \qquad \text{The value equals the price multiplied by the number.}$$

1   Solve a motion application using a linear equation in one variable.

EXAMPLE 1

TRAVELING   Chicago and Green Bay are about 200 miles apart. If a car leaves Chicago traveling toward Green Bay at 55 mph at the same time as a truck leaves Green Bay bound for Chicago at 45 mph following the same route, how long will it take them to meet?

Analyze the problem

We are asked to find the amount of time it takes for the two vehicles to meet, so we will let t represent the time in hours.

Form an equation

Motion applications are based on the relationship $d = rt$, where d is the distance traveled, r is the rate, and t is the time. We can organize the information of this problem in a chart or a diagram, as shown in Figure 2-11.

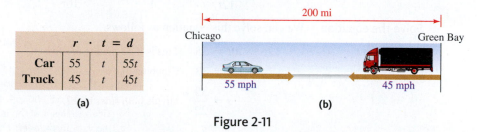

	r	$\cdot$	t	$=$	d
Car	55		t		$55t$
Truck	45		t		$45t$

(a)   (b)

Figure 2-11

We know that the two vehicles travel for the same amount of time, t hours. The faster car will travel $55t$ miles, and the slower truck will travel $45t$ miles. At the time they meet, the total distance traveled can be expressed in two ways: as the sum $55t + 45t$, and as 200 miles.

After referring to Figure 2-11, we can form the equation.

The distance the car travels	plus	the distance the truck travels	equals	the total distance traveled.
$55t$	$+$	$45t$	$=$	200

Solve the equation

We can solve this equation as follows.

$55t + 45t = 200$ This is the equation to solve.
$100t = 200$ Combine like terms.
$t = 2$ Divide both sides by 100.

State the conclusion

The vehicles will meet in 2 hours.

Check the result

In 2 hours, the car will travel $55 \cdot 2 = 110$ miles, while the truck will travel $45 \cdot 2 = 90$ miles. The total distance traveled will be $110 + 90 = 200$ miles. Since this is the total distance between Chicago and Green Bay, the solution checks.

 **SELF CHECK 1**

Dallas and Austin are about 200 miles apart. If a car leaves Dallas traveling toward Austin at 65 mph at the same time as a truck leaves Austin bound for Dallas at 60 mph following the same route, how long will it take them to meet?

EXAMPLE 2

SHIPPING   Two ships leave port, one heading east at 12 mph and one heading west at 10 mph. How long will it take before they are 33 miles apart?

Analyze the problem

We are asked to find the amount of time in hours, so we will let t represent the time.

Form an equation

The ships leave port at the same time and travel in opposite directions. We know that both travel for the same amount of time, t hours. The faster ship will travel $12t$ miles, and the slower ship will travel $10t$ miles. We can organize the information of this example in a chart or a diagram, as shown in Figure 2-12 on the next page. When the ships are 33 miles apart, the total distance traveled can be expressed in two ways: as the sum $(12t + 10t)$, and as 33 miles.

r	$\cdot$	t	$=$	d
Faster ship	12	t		$12t$
Slower ship	10	t		$10t$

(a)

(b)

Figure 2-12

After referring to Figure 2-12, we can form the equation.

The distance the faster ship travels	plus	the distance the slower ship travels	equals	the total distance traveled.
$12t$	$+$	$10t$	$=$	33

Solve the equation   We can solve this equation as follows.

$$12t + 10t = 33 \qquad \text{This is the equation to solve.}$$

$$22t = 33 \qquad \text{Combine like terms.}$$

$$t = \frac{33}{22} \qquad \text{Divide both sides by 22.}$$

$$t = \frac{3}{2} \qquad \text{Simplify the fractions: } \frac{33}{22} = \frac{3 \cdot \cancel{11}}{2 \cdot \cancel{11}} = \frac{3}{2}$$

State the conclusion   The ships will be 33 miles apart in $\frac{3}{2}$ hours (or $1\frac{1}{2}$ hours).

Check the result   In 1.5 hours, the faster ship travels $12 \cdot 1.5 = 18$ miles, while the slower ship travels $10 \cdot 1.5 = 15$ miles. Since the total distance traveled is $18 + 15 = 33$ miles, the solution checks.

SELF CHECK 2   Two ships leave port, one heading east at 15 mph and one heading west at 12 mph. How long will it take before they are 54 miles apart?

EXAMPLE 3   **TRAVELING**   A car leaves Beloit, heading east at 50 mph. One hour later, a second car leaves Beloit, heading east along the same route at 65 mph. How long will it take for the second car to overtake the first car?

Analyze the problem   The cars travel different amounts of time. In fact, the first car travels for one extra hour because it had a 1-hour head start. It is convenient to let the variable t represent the time traveled by the second car. Then $(t + 1)$ represents the number of hours the first car travels.

Form an equation   We know that car 1 travels at 50 mph and car 2 travels at 65 mph. Using the formula $r \cdot t = d$, when car 2 overtakes car 1, car 2 will have traveled $65t$ miles and car 1 will have traveled $50(t + 1)$ hours.

We can organize the information in a chart or a diagram, as shown in Figure 2-13.

r	$\cdot$	t	$=$	d
Car 1	50	$(t + 1)$		$50(t + 1)$
Car 2	65	t		$65t$

(a)

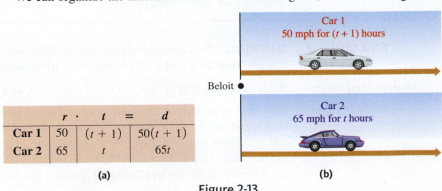

(b)

Figure 2-13

The distance the cars travel can be expressed in two ways: as $50(t + 1)$ miles, and as $65t$ miles.

Since these distances are equal when car 2 overtakes car 1, we can form the equation:

The distance that car 1 goes	equals	the distance that car 2 goes.
$50(t + 1)$	$=$	$65t$

Solve the equation   We can solve this equation as follows.

$$50(t + 1) = 65t \qquad \text{\color{red}{This is the equation to solve.}}$$

$$50t + 50 = 65t \qquad \text{\color{red}{Use the distributive property to remove parentheses.}}$$

$$50 = 15t \qquad \text{\color{red}{Subtract } 50t \text{ from both sides.}}$$

$$\frac{50}{15} = t \qquad \text{\color{red}{Divide both sides by 15.}}$$

$$t = \frac{10}{3} \qquad \text{\color{red}{Simplify the fraction: } \frac{50}{15} = \frac{10 \cdot \overset{1}{5}}{3 \cdot \underset{1}{5}} = \frac{10}{3}}$$

State the conclusion   Recall that t represents the time car 2 travels. Car 2 will overtake car 1 in $\frac{10}{3}$, or $3\frac{1}{3}$ hours.

Check the result   In $3\frac{1}{3}$ hours, car 2 will have traveled $65\left(\frac{10}{3}\right)$, or $\frac{650}{3}$, miles. With a 1-hour head start, car 1 will have traveled $50\left(\frac{10}{3} + 1\right) = 50\left(\frac{13}{3}\right)$, or $\frac{650}{3}$, miles. Since these distances are equal, the solution checks.

 **SELF CHECK 3**   A car leaves Plano, heading east at 55 mph. One hour later, a second car leaves Plano, heading east along the same route at 65 mph. How long will it take for the second car to overtake the first car?

COMMENT   In the previous example, we could let t represent the time traveled by the first car. Then $(t - 1)$ would represent the time traveled by the second car.

2   Solve a liquid mixture application using a linear equation in one variable.

EXAMPLE 4   **MIXING ACID**   A chemist has one solution that is 50% sulfuric acid and another that is 20% sulfuric acid. How much of each should she use to make 12 liters of a solution that is 30% sulfuric acid?

Analyze the problem   We will let x represent the number of liters of the 50% sulfuric acid solution. Since there must be 12 liters of the final mixture, $(12 - x)$ represents the number of liters of 20% sulfuric acid solution to use.

Form an equation   Liquid mixture applications are based on the relationship $rb = a$, where b is the base, r is the rate, and a is the amount.

If x represents the number of liters of 50% solution to use, the amount of sulfuric acid in the solution will be $0.50x$ liters. The amount of sulfuric acid in the 20% solution will be $0.20(12 - x)$ liters. The amount of sulfuric acid in the final mixture will be $0.30(12)$ liters. We can organize this information in a chart or a diagram, as shown in Figure 2-14 on the next page.

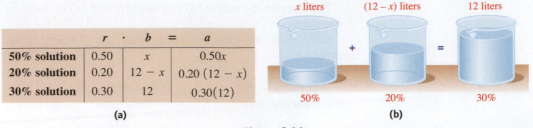

	r	·	b	=	a
50% solution	0.50		x		$0.50x$
20% solution	0.20		$12 - x$		$0.20(12 - x)$
30% solution	0.30		12		$0.30(12)$

(a)   (b)

Figure 2-14

Since the number of liters of sulfuric acid in the 50% solution plus the number of liters of sulfuric acid in the 20% solution will equal the number of liters of sulfuric acid in the mixture, we can form the equation.

The amount of sulfuric acid in the 50% solution	plus	the amount of sulfuric acid in the 20% solution	equals	the amout of sulfuric acid in the final mixture.
50% of x	+	20% of $(12 - x)$	=	30% of 12

Solve the equation   We can solve this equation as follows.

$0.5x + 0.2(12 - x) = 0.3(12)$ This is the equation to solve written in decimal form.

$5x + 2(12 - x) = 3(12)$ Multiply both sides by 10 to clear the equation of decimals.

$5x + 24 - 2x = 36$ Use the distributive property to remove parentheses.

$3x + 24 = 36$ Combine like terms.

$3x = 12$ Subtract 24 from both sides.

$x = 4$ Divide both sides by 3.

State the conclusion   Recall that we let x represent the number of liters of the 50% sulfuric acid solution. The chemist must mix 4 liters of the 50% solution and $12 - 4 = 8$ liters of the 20% solution.

Check the result   The amount of acid in 4 liters of 50% solution is $4(0.50) = 2$ liters.

The amount of acid in 8 liters of 20% solution is $8(0.20) = 1.6$ liters.

The amount of acid in 12 liters of 30% solution is $12(0.30) = 3.6$ liters.

Since $2 + 1.6 = 3.6$, the results check.

SELF CHECK 4   A chemist has one solution that is 40% sulfuric acid and another that is 10% sulfuric acid. How much of each should she use to make 20 liters of a solution that is 28% sulfuric acid?

3   Solve a dry mixture application using a linear equation in one variable.

EXAMPLE 5   **MIXING NUTS**   Fancy cashews are not selling at \$9 per pound. However, filberts are selling well at \$6 per pound. How many pounds of filberts should be combined with 50 pounds of cashews to obtain a mixture that can be sold at \$7 per pound?

Analyze the problem   We will let x represent the number of pounds of filberts in the mixture. Since we will be adding the filberts to 50 pounds of cashews, the total number of pounds of the mixture will be $(50 + x)$.

Form an equation   Dry mixture applications are based on the relationship $v = pn$, where v is the value of the mixture, p is the price per pound, and n is the number of pounds. At \$6 per pound, x pounds of the filberts are worth \$6x. At \$9 per pound, the 50 pounds of cashews are worth $9 \cdot 50$, or $450. The mixture will weigh $(50 + x)$ pounds, and at $7 per pound, it will be worth $7(50 + x)$. The *value* of the filberts (in dollars), $6x$, plus the *value* of the cashews (in dollars), 450, is equal to the *value* of the mixture (in dollars), $7(50 + x)$. We can organize this information in a table or a diagram, as shown in Figure 2-15.

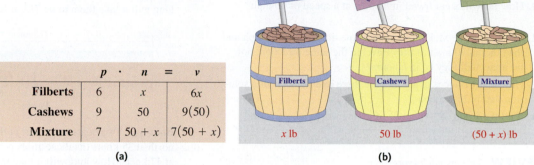

	p	$\cdot$	n	$=$	v
Filberts	6		x		$6x$
Cashews	9		50		$9(50)$
Mixture	7		$50 + x$		$7(50 + x)$

(a)   (b)

Figure 2-15

We can form the equation:

The value of the filberts	plus	the value of the cashews	equals	the value of the mixture.
$6x$	$+$	$9(50)$	$=$	$7(50 + x)$

Solve the equation   We can solve this equation as follows.

$6x + 9(50) = 7(50 + x)$ This is the equation to solve.

$6x + 450 = 350 + 7x$ Use the distributive property to remove parentheses and simplify.

$100 = x$ Subtract $6x$ and 350 from both sides.

State the conclusion   Recall that we let x represent the number of pounds of filberts to use. There should be 100 pounds of filberts in the mixture.

Check the result   The value of 100 pounds of filberts at $6 per pound is $ 600

The value of 50 pounds of cashews at $9 per pound is $ 450

The value of the mixture is $ 1,050

The value of 150 pounds of mixture at $7 per pound is also $1,050.

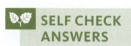 **SELF CHECK 5**   Cashews are selling at $8 per pound, and filberts sell at $5 per pound. How many pounds of filberts should be combined with 40 pounds of cashews to obtain a mixture that can be sold at $6 per pound?

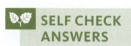 **SELF CHECK ANSWERS**   **1.** 1.6 hours   **2.** 2 hours   **3.** $5\frac{1}{2}$ hours   **4.** 12 liters of 40% sulfuric acid solution, 8 liters of 10% sulfuric acid solution   **5.** 80 pounds of filberts

---

NOW TRY THIS

1. A nurse has 5 ml of a 10% solution of benzalkonium chloride. If a doctor orders a 40% solution, how much pure benzalkonium chloride must he add to the solution to obtain the desired strength?

2. A paramedic has 5 ml of a 5% saline solution. If she needs a 4% saline solution, how much distilled water must she add to obtain the desired strength?

2.6 Exercises ✽

WARM-UPS

1. How far will a car travel in h hours at a speed of 60 mph?

2. Two cars leave McKinney at the same time, one at 60 mph and the other at 70 mph. If they travel in the same direction, how far apart will they be in h hours?

3. How many ounces of alcohol are there in 20 ounces of a solution that is 25% alcohol?

4. Find the value of 10 pounds of coffee worth $\$d$ per pound.

REVIEW *Simplify each expression.*

5. $5 + 2(-7)$

6. $\dfrac{-5(3) - 2(-2)}{6 - (-5)}$

7. $3^3 - 5^2$

8. $4^2 + 2(5) - (-3)$

Solve each equation.

9. $-6x + 4 = -8$

10. $\dfrac{1}{4}y - 3 = 6$

11. $\dfrac{2}{3}p + 1 = 5$

12. $3(z - 4) = 5(z + 4)$

VOCABULARY AND CONCEPTS *Fill in the blanks.*

13. Motion problems are based on the formula _____.

14. Liquid mixture problems are based on the formula _____.

15. Dry mixture problems are based on the formula _____.

16. The information in motion and mixture problems can be organized in the form of a ____ or a _____.

APPLICATIONS *Solve.* SEE EXAMPLES 1–2. (OBJECTIVE 1)

17. Travel times Ashford and Bartlett are 315 miles apart. A car leaves Ashford bound for Bartlett at 50 mph. At the same time, another car leaves Bartlett bound for Ashford at 55 mph. How long will it take them to meet?

315 mi
Ashford   Bartlett
50 mph   55 mph

18. Travel times Granville and Preston are 535 miles apart. A car leaves Preston bound for Granville at 47 mph. At the same time, another car leaves Granville bound for Preston at 60 mph. How long will it take them to meet?

19. Paving highways Two crews working toward each other are 9.45 miles apart. One crew paves 1.5 miles of highway per day, and the other paves 1.2 miles per day. How long will it take them to meet?

20. Biking Two friends who live 20 miles apart ride bikes toward each other. One averages 11 mph, and the other averages 9 mph. How long will it take for them to meet?

21. Travel times Two cars leave Peoria at the same time, one heading east at 60 mph and the other west at 50 mph. How long will it take them to be 715 miles apart?

715 mi
50 mph   60 mph
Peoria

22. Boating Two boats leave port at the same time, one heading north at 35 knots (nautical miles per hour) and the other south at 47 knots. How long will it take them to be 738 nautical miles apart?

23. Hiking Two boys with two-way radios that have a range of 2 miles leave camp and walk in opposite directions. If one boy walks 3 mph and the other walks 4 mph, how long will it take before they lose radio contact?

24. Biking Two cyclists leave a park and ride in opposite directions, one averaging 9 mph and the other 6 mph. If they have two-way radios with a 5-mile range, for how many minutes will they remain in radio contact?

Solve. SEE EXAMPLE 3. (OBJECTIVE 1)

25. Chasing a bus Complete the table and compute how long it will take the car to overtake the bus if the bus had a 2-hour head start.

	r	$\cdot$	t	$=$	d
Car	60 mph		t		
Bus	50 mph		$t + 2$		

26. Hot pursuit Two crooks rob a bank and flee to the east at 66 mph. In 30 minutes, the police follow them in a helicopter, flying at 132 mph. How long will it take for the police to overtake the robbers?

27. Travel times Two cars start together and head east, one averaging 42 mph and the other averaging 53 mph. See the illustration. In how many hours will the cars be 82.5 miles apart?

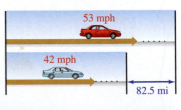

53 mph
42 mph
82.5 mi

28. Aviation A plane leaves an airport and flies south at 180 mph. Later, a second plane leaves the same airport and flies south at 450 mph. If the second plane overtakes the first one in $1\frac{1}{2}$ hours, how much of a head start did the first plane have?

29. Speed of trains Two trains are 330 miles apart, and their speeds differ by 20 mph. They travel toward each other and meet in 3 hours. Find the speed of each train.

30. Speed of airplanes Two planes are 6,000 miles apart, and their speeds differ by 200 mph. They travel toward each other and meet in 5 hours. Find the speed of the slower plane.

31. Average speeds An automobile averaged 40 mph for part of a trip and 50 mph for the remainder. If the 5-hour trip covered 210 miles, for how long did the car average 40 mph?

32. Vacation driving A family drove to the Grand Canyon, averaging 45 mph. They returned using the same route, averaging 60 mph. If they spent a total of 7 hours of driving time, how far is their home from the Grand Canyon?

Solve. SEE EXAMPLE 4. (OBJECTIVE 2)

33. Chemistry A solution contains 0.3 liters of sulfuric acid. If this represents 12% of the total amount, find the total amount.

34. Medicine A laboratory has a solution that contains 3 ounces of benzalkonium chloride. If this is 15% of the total solution, how many ounces of solution does the lab have?

35. Mixing fuels How many gallons of fuel costing $3.35 per gallon must be mixed with 20 gallons of a fuel costing $3.85 per gallon to obtain a mixture costing $3.55 per gallon?

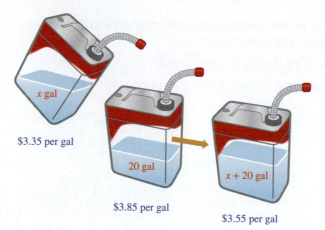

$3.35 per gal

20 gal

$3.85 per gal

$x + 20$ gal

$3.55 per gal

36. Mixing paint Paint costing $19 per gallon is to be mixed with 5 gallons of paint thinner costing $3 per gallon to make a paint that can be sold for $14 per gallon. Refer to the table and compute how much paint will be produced.

	p	$\cdot$	n	$=$	r
Paint	$19		x gal		$19x$
Thinner	$3		5 gal		$3(5)$
Mixture	$14		$(x + 5)$ gal		$14(x + 5)$

37. Brine solutions How many gallons of a 3% salt solution must be mixed with 50 gallons of a 7% solution to obtain a 5% solution?

38. Making cottage cheese To make low-fat cottage cheese, milk containing 4% butterfat is mixed with 10 gallons of milk containing 1% butterfat to obtain a mixture containing 2% butterfat. How many gallons of the fattier milk must be used?

39. Antiseptic solutions A nurse wants to add water to 30 ounces of a 10% solution of benzalkonium chloride to dilute it to an 8% solution. How much water must she add?

40. Mixing photographic chemicals A photographer wants to mix 2 liters of a 5% acetic acid solution with a 10% solution to get a 7% solution. How many liters of 10% solution must be added?

Solve. SEE EXAMPLE 5. (OBJECTIVE 3)

41. Mixing candy Lemon drops are to be mixed with jelly beans to make 100 pounds of mixture. Refer to the illustration and compute how many pounds of each candy should be used.

Lemon Drops $1.90/lb

Jelly Beans $1.20/lb

Mixture $1.48/lb

42. Blending gourmet tea One grade of tea, worth $3.20 per pound, is to be mixed with another grade worth $2 per pound to make 20 pounds that will sell for $2.72 per pound. How much of each grade of tea must be used?

43. Mixing nuts A bag of peanuts is worth 30¢ less than a bag of cashews. Equal amounts of peanuts and cashews are used to make 40 bags of a mixture that sells for $1.05 per bag. How much would a bag of cashews be worth?

44. Mixing candy Twenty pounds of lemon drops are to be mixed with cherry chews to make a mixture that will sell for $1.80 per pound. How much of the more expensive candy should be used? See the table.

Candy	Price per pound
Peppermint patties	$1.35
Lemon drops	$1.70
Licorice lumps	$1.95
Cherry chews	$2.00

45. Coffee blends A store sells regular coffee for $7 a pound and a gourmet coffee for $12 a pound. To reduce the inventory by 30 pounds of the gourmet coffee, the shopkeeper plans to make a gourmet blend to sell for $9 a pound. How many pounds of regular coffee should be used?

46. Lawn seed blends A garden store sells Kentucky bluegrass seed for $6 per pound and ryegrass seed for $3 per pound. How much rye must be mixed with 100 pounds of bluegrass to obtain a blend that will sell for $5 per pound?

47. Mixing coffee A shopkeeper sells chocolate coffee beans for $7 per pound. A customer asks the shopkeeper to mix 2 pounds of chocolate beans with 5 pounds of hazelnut coffee beans. If the customer paid $6 per pound for the mixture, what is the price per pound of the hazelnut beans?

48. Trail mix Fifteen pounds of trail mix are made by mixing 2 pounds of raisins worth $3 per pound with peanuts worth $4 per pound and M&Ms worth $5 per pound. How many pounds of peanuts must be used if the mixture is to be worth $4.20 per pound?

WRITING ABOUT MATH

49. Describe the steps you would use to analyze and solve a problem.

50. Create a mixture problem that could be solved by using the equation $4x + 6(12 - x) = 5(12)$.

51. Create a mixture problem of your own, and solve it.

52. In mixture problems, explain why it is important to distinguish between the quantity and the value (or strength) of the materials being combined.

SOMETHING TO THINK ABOUT

53. Is it possible for the equation of a problem to have a solution, but for the problem to have no solution? For example, is it possible to find two consecutive even integers whose sum is 16?

54. Invent a motion problem that leads to an equation that has a solution, although the problem does not.

55. Consider the problem: How many gallons of a 10% and a 20% solution should be mixed to obtain a 30% solution? Without solving it, how do you know that the problem has no solution?

56. What happens if you try to solve Exercise 55?

Section 2.7

Solving Linear Inequalities in One Variable

Objectives

1 Solve a linear inequality in one variable using the properties of inequality and graph the solution on a number line.

2 Solve a compound linear inequality in one variable.

3 Solve an application involving a linear inequality in one variable.

Vocabulary

inequality
solution of an inequality
addition property of
    inequality

multiplication property of
    inequality
double inequality

compound inequality
interval

Getting Ready

Graph each set on the number line.

1. All real numbers greater than -1

2. All real numbers less than or equal to 5

3. All real numbers between -2 and 4

4. All real numbers less than -2 or greater than or equal to 4

Many times, we will encounter mathematical statements indicating that two quantities are not necessarily equal. These statements are called *inequalities*.

1 Solve a linear inequality in one variable using the properties of inequality and graph the solution on a number line.

Recall the meaning of the following symbols.

INEQUALITY SYMBOLS		
$<$	means	"is less than"
$>$	means	"is greater than"
$\leq$	means	"is less than or equal to"
$\geq$	means	"is greater than or equal to"

An **inequality** is a statement that indicates that two quantities are not necessarily equal. A **solution of an inequality** is any number that makes the inequality true. The number 2 is a solution of the inequality

$$x \leq 3$$

because $2 \leq 3$.

This inequality has many more solutions, because any real number that is less than or equal to 3 will satisfy it. We can use a graph on the number line to represent the solutions of the inequality. The red arrow in Figure 2-16 indicates all those points with coordinates that satisfy the inequality $x \leq 3$.

Figure 2-16

The bracket at the point with coordinate 3 indicates that the number 3 is a solution of the inequality $x \leq 3$.

The graph of the inequality $x > 1$ appears in Figure 2-17. The red arrow indicates all those points whose coordinates satisfy the inequality. The parenthesis at the point with coordinate 1 indicates that 1 is not a solution of the inequality $x > 1$.

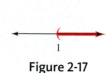

Figure 2-17

To solve more complicated inequalities, we need to use the addition, subtraction, multiplication, and division properties of inequalities. When we use any of these properties, the resulting inequality will have the same solutions as the original one.

ADDITION PROPERTY OF INEQUALITY	Suppose a, b, and c are real numbers. If $a < b$, then $a + c < b + c$.

SUBTRACTION PROPERTY OF INEQUALITY	Suppose a, b, and c are real numbers. If $a < b$, then $a - c < b - c$.

Similar statements can be made for the symbols $>$, $\leq$, and $\geq$.

The **addition property of inequality** can be stated this way: *If any quantity is added to both sides of an inequality, the resulting inequality has the same direction as the original inequality.*

The subtraction property of inequality can be stated this way: *If any quantity is subtracted from both sides of an inequality, the resulting inequality has the same direction as the original inequality.*

COMMENT The subtraction property of inequality is included in the addition property: To *subtract* a number a from both sides of an inequality, we *add* the *negative* of a to both sides.

EXAMPLE 1 Solve $2x + 5 > x - 4$ and graph the solution on a number line.

Solution To isolate the x on the left side of the $>$ sign, we proceed as if we were solving an equation.

$$2x + 5 > x - 4$$
$$2x + 5 \, \mathbf{- \, 5} > x - 4 \, \mathbf{- \, 5} \quad \text{Subtract 5 from both sides.}$$
$$2x > x - 9 \quad \text{Combine like terms.}$$
$$2x \, \mathbf{- \, x} > x - 9 \, \mathbf{- \, x} \quad \text{Subtract } x \text{ from both sides.}$$
$$x > -9 \quad \text{Combine like terms.}$$

Figure 2-18

The graph of the solution (see Figure 2-18) includes all points to the right of -9 but does not include -9 itself. For this reason, we use a parenthesis at -9.

 **SELF CHECK 1** Graph the solution of $2x - 2 < x + 1$.

If both sides of the true inequality $6 < 9$ are multiplied or divided by a *positive* number, such as 3, another true inequality results.

$$6 < 9$$
$$\mathbf{3 \cdot 6} < \mathbf{3 \cdot 9} \qquad \text{Multiply both sides by 3.}$$
$$18 < 27 \qquad \text{True.}$$

$$6 < 9$$
$$\frac{6}{\mathbf{3}} < \frac{9}{\mathbf{3}} \qquad \text{Divide both sides by 3.}$$
$$2 < 3 \qquad \text{True.}$$

The inequalities $18 < 27$ and $2 < 3$ are true.

However, if both sides of $6 < 9$ are multiplied or divided by a negative number, such as -3, the direction of the inequality symbol must be reversed to produce another true inequality.

$$6 < 9$$
$$\mathbf{-3 \cdot 6} > \mathbf{-3 \cdot 9} \qquad \begin{array}{l}\text{Multiply both sides} \\ \text{by } -3 \text{ and reverse} \\ \text{the direction of the} \\ \text{inequality.}\end{array}$$
$$-18 > -27 \qquad \text{True.}$$

$$6 < 9$$
$$\frac{6}{\mathbf{-3}} > \frac{9}{\mathbf{-3}} \qquad \begin{array}{l}\text{Divide both sides by } -3 \\ \text{and reverse the direction of} \\ \text{the inequality.}\end{array}$$
$$-2 > -3 \qquad \text{True.}$$

The inequality $-18 > -27$ is true, because -18 lies to the right of -27 on the number line. The inequality $-2 > -3$ is true, because -2 lies to the right of -3 on the number line. This example suggests the multiplication and division properties of inequality.

MULTIPLICATION PROPERTY OF INEQUALITY	Suppose a, b, and c are real numbers.
	If $a < b$ and $c > 0$, then $ac < bc$.
	If $a < b$ and $c < 0$, then $ac > bc$.

DIVISION PROPERTY OF INEQUALITY	Suppose a, b, and c are real numbers.
	If $a < b$ and $c > 0$, then $\dfrac{a}{c} < \dfrac{b}{c}$.
	If $a < b$ and $c < 0$, then $\dfrac{a}{c} > \dfrac{b}{c}$.

COMMENT In the previous definitions, we did not consider the case of $c = 0$. If $a < b$ and $c = 0$, then $ac = bc$, but $\frac{a}{c}$ and $\frac{b}{c}$ are not defined.

Similar statements can be made for the symbols $>$, $\leq$, and $\geq$.

The **multiplication property of inequality** can be stated this way:

If unequal quantities are multiplied by the same positive quantity, the results will be unequal and in the same direction as the original inequality.

If unequal quantities are multiplied by the same negative quantity, the results will be unequal but in the opposite direction of the original inequality.

The division property of inequality can be stated this way:

If unequal quantities are divided by the same positive quantity, the results will be unequal and in the same direction as the original inequality.

If unequal quantities are divided by the same negative quantity, the results will be unequal but in the opposite direction of the original inequality.

To *divide* both sides of an inequality by a nonzero number c, we could instead *multiply* both sides by $\frac{1}{c}$.

COMMENT Remember that if both sides of an inequality are multiplied by a *positive* number, the direction of the resulting inequality remains the same. However, if both sides of an inequality are multiplied by a *negative* number, the direction of the resulting inequality must be reversed.

Note that the procedures for solving inequalities are the same as for solving equations, except that we must reverse the inequality symbol whenever we multiply or divide by a negative number.

EXAMPLE 2   Solve $3x + 7 \leq -5$ and graph the solution on the number line.

Solution   To isolate x on the left side, we proceed as if we were solving an equation.

$$3x + 7 \leq -5$$
$$3x + 7 \mathbf{- 7} \leq -5 \mathbf{- 7} \qquad \text{Subtract 7 from both sides.}$$
$$3x \leq -12 \qquad \text{Combine like terms.}$$
$$\frac{3x}{\mathbf{3}} \leq \frac{-12}{\mathbf{3}} \qquad \text{Divide both sides by 3.}$$
$$x \leq -4$$

Figure 2-19

The solution consists of all real numbers that are less than or equal to -4. The bracket at -4 in the graph of Figure 2-19 indicates that -4 is one of the solutions.

 SELF CHECK 2   Graph the solution of $2x - 5 \geq -3$ on the number line.

EXAMPLE 3   Solve $5 - 3x \leq 14$ and graph the solution on the number line.

Solution   To isolate x on the left side, we proceed as if we were solving an equation. This time, we will have to reverse the inequality symbol.

$$5 - 3x \leq 14$$
$$5 - 3x \mathbf{- 5} \leq 14 \mathbf{- 5} \qquad \text{Subtract 5 from both sides.}$$
$$-3x \leq 9 \qquad \text{Combine like terms.}$$
$$\frac{-3x}{\mathbf{-3}} \geq \frac{9}{\mathbf{-3}} \qquad \text{Divide both sides by } -3 \text{ and reverse the direction of the } \leq \text{ symbol.}$$
$$x \geq -3$$

Figure 2-20

Since both sides of the inequality were divided by -3, the direction of the inequality was *reversed*. The graph of the solution appears in Figure 2-20. The bracket at -3 indicates that -3 is one of the solutions.

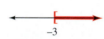 **SELF CHECK 3**   Graph the solution of $6 - 7x \geq -15$ on the number line.

2   Solve a compound linear inequality in one variable.

Two inequalities often can be combined into a **double inequality** or **compound inequality** to indicate that numbers lie *between* two fixed values. For example, the inequality $2 < x < 5$ indicates that x is greater than 2 and that x is also less than 5. The solution of $2 < x < 5$ consists of all numbers that lie *between* 2 and 5. The graph of this set (called an **interval**) appears in Figure 2-21.

Figure 2-21

EXAMPLE 4   Solve $-4 < 2(x - 1) \leq 4$ and graph the solution on the number line.

Solution   To isolate x in the center, we proceed as if we were solving an equation with three parts: a left side, a center, and a right side.

$$-4 < 2(x - 1) \leq 4$$
$$-4 < 2x - 2 \leq 4 \qquad \text{Use the distributive property to remove parentheses.}$$
$$-2 < 2x \leq 6 \qquad \text{Add 2 to all three parts.}$$
$$-1 < x \leq 3 \qquad \text{Divide all three parts by 2.}$$

The graph of the solution appears in Figure 2-22.

 SELF CHECK 4   Graph the solution of $0 \leq 4(x + 5) < 26$ on the number line.

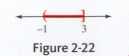

Figure 2-22

3   Solve an application involving a linear inequality in one variable.

When solving applications, there are certain words that help us translate a sentence into a mathematical inequality.

Words	Sentence	Inequality
at least	To earn a grade of A, you must score at least 90%.	$S \geq 90\%$
is less than	The perimeter is less than 30 feet.	$P < 30$ ft
is no less than	The perimeter is no less than 100 centimeters.	$P \geq 100$ cm
is more than	The area is more than 30 square inches.	$A > 30$ sq in.
exceeds	The car's speed exceeded the limit of 45 mph.	$S > 45$ mph
cannot exceed	The salary cannot exceed $50,000.	$S \leq \$50,000$
at most	The perimeter is at most 75 feet.	$P \leq 75$ ft
is between	The altitude is between 10,000 and 15,000 feet.	$10,000 < A < 15,000$

EXAMPLE 5   **GRADES**   A student has scores of 72, 74, and 78 points on three mathematics examinations. How many points does he need on his last exam to earn a B or better, an average of at least 80 points?

Solution   We can let x represent the score on the fourth (last) exam. To find the average grade, we add the four scores and divide by 4. To earn a B, this average must be greater than or equal to 80 points.

The average of the four grades	is greater than or equal to	80.
$\dfrac{72 + 74 + 78 + x}{4}$	$\geq$	80

We can solve this inequality for x.

$$\frac{224 + x}{4} \geq 80 \qquad \text{Add.}$$

$$224 + x \geq 320 \qquad \text{Multiply both sides by 4.}$$

$$x \geq 96 \qquad \text{Subtract 224 from both sides.}$$

To earn a B, the student must score at least 96 points.

 SELF CHECK 5   A student has scores of 70, 75, and 77 points on three mathematics examinations. How many points does he need on his last exam to earn a B or better, an average of at least 80 points?

EXAMPLE 6   **EQUILATERAL TRIANGLES**   If the perimeter of an equilateral triangle is less than 15 feet, how long could each side be?

Solution   Recall that each side of an equilateral triangle is the same length and that the perimeter of a triangle is the sum of the lengths of its three sides. If we let x represent the length of one of the sides in feet, then $(x + x + x)$ represents the perimeter. Since the perimeter is to be less than 15 feet, we have the following inequality:

$$x + x + x < 15$$
$$3x < 15 \quad \textcolor{red}{\text{Combine like terms.}}$$
$$x < 5 \quad \textcolor{red}{\text{Divide both sides by 3.}}$$

Each side of the triangle must be less than 5 feet long.

SELF CHECK 6   If the perimeter of an equilateral triangle is less than 21 feet, how long could each side be?

SELF CHECK ANSWERS

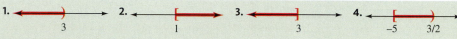

1. 3   **2.** 1   **3.** 3   **4.** −5   3/2

5. at least 98 points   **6.** less than 7 feet long

NOW TRY THIS

Solve each inequality and graph the solution.

1. $2(x - 3) \le 2x - 1$

2. $-5x - 7 > 5(3 - x)$

3. A person's body-mass index (BMI) determines the amount of body fat. BMI is represented by the formula $B = 703\frac{w}{h^2}$, where w is weight (in pounds) and h is height (in inches). A 5-foot 8-inch gymnast must maintain a normal body-mass index. If the normal range for men is represented by $18.5 < 703\frac{w}{h^2} < 25$, within what range should the gymnast maintain his weight? Give the answer to the nearest tenth of a pound.

2.7   Exercises

WARM-UPS *Determine whether the sign would stay the same or need to be reversed if the variable remains on the left side of each inequality when solving.*

1. $2x < 4$

2. $x + 5 \ge 6$

3. $-3x \le -6$

4. $-x > 2$

5. $2x - 5 < 7$

6. $5 - 2x < 7$

REVIEW *Simplify each expression.*

7. $3x^2 - 2(y^2 - x^2)$

8. $5(xy + 2) - 3xy - 8$

9. $\frac{1}{3}(x + 6) - \frac{4}{3}(x - 9)$

10. $\frac{4}{5}x(y + 1) - \frac{9}{5}y(x - 1)$

VOCABULARY AND CONCEPTS *Fill in the blanks.*

11. The symbol $<$ means _____. The symbol $>$ means _____.

12. The symbol __ means "is greater than or equal to." The symbol ___ means "is less than or equal to."

13. Two inequalities often can be combined into a _____ _____ or *compound inequality*.

14. The graph of the solution of $2 < x < 5$ on the number line is called an _____.

15. An _____ is a statement indicating that two quantities are not necessarily equal.

16. A _____ of an inequality is any number that makes the inequality true.

GUIDED PRACTICE *Solve each inequality and graph the solution on the number line.* SEE EXAMPLE 1. (OBJECTIVE 1)

17. $x + 5 > 8$

18. $x + 5 \ge 2$

19. $3x - 6 \le 2x - 7$

20. $3 + x < 2$

Solve each inequality and graph the solution on the number line. SEE EXAMPLE 2. (OBJECTIVE 1)

21. $3x - 10 \le 2$

22. $9x + 13 \ge 8x$

23. $6x - 2 > 3x - 11$ **24.** $7x + 6 \geq 4x$

25. $9x + 4 > 4x - 1$ **26.** $5x + 7 < 2x + 1$

27. $\frac{5}{2}(7x - 15) + x \geq \frac{13}{2}x - \frac{3}{2}$

28. $\frac{5}{3}(x + 1) \leq -x + \frac{2}{3}$

Solve each inequality and graph the solution on the number line.
SEE EXAMPLE 3. (OBJECTIVE 1)

29. $-x - 3 \leq 7$ **30.** $-x - 9 > 3$

31. $-3x - 5 < 4$ **32.** $3x + 7 \leq 4x - 2$

33. $-5x + 17 > 37$ **34.** $7x - 9 > 5$

35. $-3x - 7 > -1$ **36.** $-2x - 5 \leq 4x + 1$

37. $9 - 2x > 24 - 7x$ **38.** $13 - 17x < 34 - 10x$

39. $2(x + 7) \leq 4x - 6$ **40.** $9(x - 11) > 13 + 7x$

Solve each inequality and graph the solution on the number line.
SEE EXAMPLE 4. (OBJECTIVE 2)

41. $2 < x - 5 < 5$ **42.** $3 < x - 2 < 7$

43. $-4 < 2(x + 1) \leq 12$ **44.** $-9 \leq 3(x - 2) < 6$

45. $0 \leq x + 10 \leq 10$ **46.** $-8 < x - 8 < 8$

47. $-6 < 3(x + 2) < 9$ **48.** $-18 \leq 9(x - 5) < 27$

ADDITIONAL PRACTICE *Solve each inequality and graph the solution on the number line.*

49. $8 + x > 7$ **50.** $7x - 16 < 6x$

51. $7 - x \leq 3x - 1$ **52.** $2 - 3x \geq 6 + x$

53. $8(5 - x) \leq 10(8 - x)$ **54.** $17(3 - x) \geq 3 - 13x$

55. $\frac{3x - 3}{2} < 2x + 2$ **56.** $\frac{x + 7}{3} \geq x - 3$

57. $\frac{2(x + 5)}{3} \leq 3x - 6$ **58.** $\frac{3(x - 1)}{4} > x + 1$

59. $9 < -3x < 15$ **60.** $-4 \leq -4x < 12$

61. $-3 \leq \frac{x}{2} \leq 5$ **62.** $-12 \leq \frac{x}{3} < 0$

63. $3 \leq 2x - 1 < 5$ **64.** $4 < 3x - 5 \leq 7$

65. $0 < 10 - 5x \leq 15$ **66.** $1 \leq -7x + 8 \leq 15$

67. $-4 < \frac{x - 2}{2} < 6$ **68.** $-1 \leq \frac{x + 1}{3} \leq 3$

APPLICATIONS *Express each solution as an inequality.* SEE *EXAMPLES 5–6. (OBJECTIVE 3)*

69. Calculating grades A student has test scores of 68, 75, and 79 points. What must she score on the fourth exam to have an average score of at least 80 points?

70. Calculating grades A student has test scores of 84, 89, and 93 points. What must he score on the fourth exam to have an average score of at least 90 points?

71. Geometry The perimeter of a square is no less than 68 centimeters. How long can a side be?

72. Geometry The perimeter of an equilateral triangle is at most 57 feet. What could be the length of a side? (*Hint*: All three sides of an equilateral triangle are equal.)

Express each solution as an inequality.

73. Fleet averages An automobile manufacturer produces three light trucks in equal quantities. One model has an economy rating of 17 miles per gallon, and the second model is rated for 19 mpg. If the manufacturer is required to have a fleet average of at least 21 mpg, what economy rating is required for the third model?

74. Avoiding service charges When the average daily balance of a customer's checking account is less than $500 in any business week, the bank assesses a $5 service charge. Bill's account balances for the week were as shown in the table. What must Friday's balance be to avoid the service charge?

Monday	$540.00
Tuesday	$435.50
Wednesday	$345.30
Thursday	$310.00

75. Land elevations The land elevations in Nevada fall from the 13,143-foot height of Boundary Peak to the Colorado River at 470 feet. To the nearest tenth, what is the range of these elevations in miles? (*Hint*: 1 mile is 5,280 feet.)

76. Homework A teacher requires that students do homework at least 2 hours a day. How many minutes should a student work each week?

77. Plane altitudes A pilot plans to fly at an altitude of between 17,500 and 21,700 feet. To the nearest tenth, what will be the range of altitudes in miles? (*Hint*: There are 5,280 feet in 1 mile.)

78. Getting exercise A certain exercise program recommends that your daily exercise period should exceed 15 minutes but should not exceed 30 minutes per day. In hours, find the range of exercise time for one week.

79. Comparing temperatures To hold the temperature of a room between 23° and 26° Celsius, what Fahrenheit temperatures must be maintained? (*Hint:* Fahrenheit temperature (F) and Celsius temperature (C) are related by the formula $C = \frac{5}{9}(F - 32)$.)

80. Melting iron To melt iron, the temperature of a furnace must be at least 1,540°C but at most 1,650°C. What range of Fahrenheit temperatures must be maintained?

81. Phonograph records The radii of old phonograph records lie between 5.9 and 6.1 inches. What variation in circumference can occur? (*Hint*: The circumference of a circle is given by the formula $C = 2\pi r$, where r is the radius. Use 3.14 to approximate π.)

82. Pythons A large snake, the African Rock Python, can grow to a length of 25 feet. To the nearest hundredth, find the snake's range of lengths in meters. (*Hint*: There are about 3.281 feet in 1 meter.)

83. Comparing weights The normal weight of a 6 foot 2 inch man is between 150 and 190 pounds. To the nearest hundredth, what would such a person weigh in kilograms? (*Hint*: There are approximately 2.2 pounds in 1 kilogram.)

84. Manufacturing The time required to assemble a television set at the factory is 2 hours. A stereo receiver requires only 1 hour. The labor force at the factory can supply at least 644 and at most 805 hours of assembly time per week. When the factory is producing 3 times as many television sets as stereos, how many stereos could be manufactured in 1 week?

85. Geometry A rectangle's length is 3 feet less than twice its width, and its perimeter is between 24 and 48 feet. What might be its width?

86. Geometry A rectangle's width is 8 feet less than 3 times its length, and its perimeter is between 8 and 16 feet. What might be its length?

WRITING ABOUT MATH

87. Explain why multiplying both sides of an inequality by a negative constant reverses the direction of the inequality.

88. Explain the use of parentheses and brackets in the graphing of the solution of an inequality.

SOMETHING TO THINK ABOUT

89. To solve the inequality $1 < \frac{1}{x}$, one student multiplies both sides by x to get $x < 1$. Why is this not correct?

90. Find the solution of $1 < \frac{1}{x}$. (*Hint*: Will any negative values of x work?)

Projects

PROJECT 1

Build a scale similar to the one shown in Figure 2-1. Demonstrate to your class how you would use the scale to solve the following equations.

a. $x - 4 = 6$ **b.** $x + 3 = 2$ **c.** $2x = 6$

d. $\frac{x}{2} = 3$ **e.** $3x - 2 = 5$ **f.** $\frac{x}{3} + 1 = 2$

PROJECT 2

Use a calculator to determine whether the following statements are true or false.

a. $7^5 = 5^7$ **b.** $2^3 + 7^3 = (2 + 7)^3$

c. $(-4)^4 = -4^4$ **d.** $\frac{10^3}{5^3} = 2^3$

e. $8^4 \cdot 9^4 = (8 \cdot 9)^4$ **f.** $2^3 \cdot 3^3 = 6^3$

g. $\frac{3^{10}}{3^2} = 3^5$ **h.** $[(1.2)^3]^2 = [(1.2)^2]^3$

i. $(7.2)^2 - (5.1)^2 = (7.2 - 5.1)^2$

Reach for Success
EXTENSION OF READING THIS MATHEMATICS TEXTBOOK

READING THIS MATHEMATICS TEXTBOOK

Now that we have discussed the overall structure of this textbook, let's move on to some special features designed to help you work through each section.

Prior to the homework exercise set, you will find the following features:	
• Getting Ready	What does this mean to you? _____ _____
• Examples All examples are cross-referenced with the objectives. Thus, if your instructor does not cover a particular objective, you may omit that example.	How many objectives are in this section? _____ How many of these are you required to know? _____ Which examples may you omit? _____
• Self Checks After each numbered example, you will find a Self Check practice exercise so that you may immediately determine your reading comprehension.	Explain how the Self Checks might help you understand and retain the text material. _____ _____
• Now Try This Created to provide collaborative exercises to be discussed in small groups, these problems are slightly more difficult than the examples and encourage you to deepen your understanding or provide transition to a future topic.	Explain why the Now Try This might be considered transitional group-work exercises. _____ _____ _____

Match each exercise set feature with its description.	
_____ 1. Additional Practice	A. Gets you ready for homework
_____ 2. Applications	B. Keeps previously learned skills current
_____ 3. Guided Practice	C. Helps you speak the language
_____ 4. Review	D. Provides references for assistance
_____ 5. Something to Think About	E. Begins to duplicate the exam format
_____ 6. Vocabulary	F. Shows relevance of the topics to the world
_____ 7. Warm-up	G. Increases communication skills
_____ 8. Writing About Math	H. Transitions students' concepts
When beginning the exercises, be sure to *read the instructions* carefully.	What is the importance of understanding the instructions in the following two exercises? 1. Find the perimeter of the figure. 2. Find the area of the figure. 7 in. [ ] 21 in.

Don't forget the computer software, Enhanced WebAssign. The exercises there can be done on the computer with additional help options, an excellent resource particularly during late evening when your other resources are not available!

2 Review

SECTION 2.1 Solving Basic Linear Equations in One Variable

DEFINITIONS AND CONCEPTS	EXAMPLES
An **equation** is a statement indicating that two quantities are equal. An **expression** is a mathematical statement that does not compare two quantities.	**Equations:** $3x = 5$ $3x - 4 = 10$ $8x - 7 = -2x$ **Expressions:** $5x + 1$ $5x^2 + 3x - 2$ $-8(2x - 4)$
A number is said to *satisfy* an equation if it makes the equation true when substituted for the variable.	To determine whether 3 is a solution of the equation $2x + 5 = 11$, substitute 3 for x and determine whether the result is a true statement. $$2x + 5 = 11$$ $$2(3) + 5 \stackrel{?}{=} 11$$ $$6 + 5 \stackrel{?}{=} 11$$ $$11 = 11$$ Since the result is a true statement, 3 satisfies the equation.
Addition and subtraction properties of equality: Any real number can be added to (or subtracted from) both sides of an equation to form another equation with the same solutions as the original equation.	To solve $x - 3 = 8$, add 3 to both sides. To solve $x + 3 = 8$, subtract 3 from both sides. $$x - 3 = 8 \qquad\qquad x + 3 = 8$$ $$x - 3 + 3 = 8 + 3 \qquad x - 3 - 3 = 8 - 3$$ $$x = 11 \qquad\qquad\quad x = 5$$ Verify that each result satisfies its corresponding equation.
Two equations are **equivalent equations** when they have the same solutions.	$3x + 4 = 10$ and $3x = 6$ are equivalent equations because 2 is the only solution of each equation.
Multiplication and division properties of equality: Both sides of an equation can be multiplied (or divided) by any *nonzero* real number to form another equation with the same solutions as the original equation.	To solve $\frac{x}{3} = 4$, multiply both sides by 3. To solve $3x = 12$, divide both sides by 3. $$\frac{x}{3} = 4 \qquad\qquad 3x = 12$$ $$3\left(\frac{x}{3}\right) = 3(4) \qquad \frac{3x}{3} = \frac{12}{3}$$ $$x = 12 \qquad\qquad x = 4$$ Verify that each result satisfies its corresponding equation.
Sale price = regular price − markdown	If a coat regularly costs $150 and is marked down $25, its selling price is $150 − $25 = $125.
Retail price = wholesale cost + markup	If the wholesale cost of a TV is $500 and it is marked up $200, its retail price is $500 + $200 = $700.
A **percent** is the numerator of a fraction with a denominator of 100. Amount = rate · base	$6\% = \dfrac{6}{100} = 0.06 \qquad 8\% = \dfrac{8}{100} = 0.08$ An amount of $150 will be earned when a base of $3,000 is invested at a rate of 5%. $$a = rb$$ $$150 = 0.05 \cdot 3{,}000$$ $$150 = 150$$

REVIEW EXERCISES

Determine whether each statement is an expression or an equation.

1. $3(x + 4)$ **2.** $x + 4 = 3$

Determine whether the given number is a solution of the equation.

3. $6 - 4x = 2;\ -1$ **4.** $2(x + 3) = 3(x - 4);\ 18$

Solve each equation and check all solutions.

5. $x - 6 = -7$ **6.** $-y - 3 = 10$

7. $p + 5 = 9$ **8.** $c + 9 = -6$

9. $p + \dfrac{1}{2} = -\dfrac{1}{2}$ **10.** $x + \dfrac{5}{7} = \dfrac{5}{7}$

11. $z + \dfrac{2}{3} = \dfrac{1}{3}$ **12.** $b - \dfrac{1}{4} = -\dfrac{3}{4}$

13. Retail sales   A necklace is on sale for $69.95. If it has been marked down $35.45, what is its regular price?

14. Retail sales   A suit that has been marked up $115.25 sells for $212.95. Find its wholesale price.

Solve each equation and check all solutions.

15. $3x = 15$ **16.** $8r = -16$

17. $-10z = 5$ **18.** $14q = 21$

19. $\dfrac{y}{3} = 6$ **20.** $\dfrac{w}{7} = -5$

21. $\dfrac{a}{-7} = \dfrac{1}{14}$ **22.** $\dfrac{p}{12} = \dfrac{1}{2}$

Solve.

23. What number is 35% of 700?

24. 72% of what number is 936?

25. What percent of 2,300 is 851?

26. 72 is what percent of 576?

SECTION 2.2   Solving More Linear Equations in One Variable

DEFINITIONS AND CONCEPTS	EXAMPLES
Solving a linear equation may require the use of several properties of equality. If the equation contains any fractions, consider clearing the fractions first.	To solve $\dfrac{x}{3} - 4 = -8$, proceed as follows: $\dfrac{x}{3} - 4 + 4 = -8 + 4$ To undo the subtraction of 4, add 4 to both sides. $\dfrac{x}{3} = -4$ Simplify. $3\left(\dfrac{x}{3}\right) = 3(-4)$ To undo the division of 3, multiply both sides by 3. $x = -12$
Retail price $=$ cost $+\ \dfrac{\text{percent of}}{\text{markup}} \cdot$ cost Markup $=$ percent of markup $\cdot$ cost	A wholesale cost of a necklace is $125. If its retail price is $150, find the percent of markup. $150 = 125 + p \cdot 125$ $25 = 125p$ Subtract 125 from both sides. $0.20 = p$ Divide both sides by 125. The percent of markup is 20%.
Sale price $=$ regular price $-\ \dfrac{\text{percent of}}{\text{markdown}} \cdot$ regular price Markdown (discount) $=\ \dfrac{\text{percent of}}{\text{markdown}} \cdot$ regular price	A used textbook that was originally priced at $95 is now priced at $57. Find the percent of markdown. $57 = 95 - p \cdot 95$ $-38 = -95p$ Subtract 95 from both sides. $0.40 = p$ Divide both sides by -95. The used textbook has a markdown of 40%.

REVIEW EXERCISES

Solve each equation and check all solutions.

27. $5y + 6 = 21$ **28.** $5y - 9 = 1$

29. $-12z + 4 = -8$ **30.** $17z + 3 = 20$

31. $13 - 13p = 0$ **32.** $10 + 7p = -4$

33. $23a - 43 = 3$ **34.** $84 - 21a = -63$

35. $3x + 7 = 1$ **36.** $7 - 9x = 16$

37. $\dfrac{b + 3}{4} = 2$ **38.** $\dfrac{b - 7}{2} = -2$

39. $\dfrac{3y - 2}{4} = -5$ **40.** $\dfrac{3x + 10}{2} = -1$

41. $\dfrac{x}{2} + 7 = 11$ **42.** $\dfrac{r}{3} - 3 = 7$

43. $\dfrac{2x}{3} - 4 = 6$ **44.** $\dfrac{y}{4} - \dfrac{6}{5} = -\dfrac{1}{5}$

45. $\dfrac{a}{2} + \dfrac{3}{4} = 6$ **46.** $\dfrac{x}{8} - 2.3 = 3.2$

47. Retail sales   An iPhone is on sale for $240, a 25% savings from the regular price. Find the regular price.

48. Tax rates   A $38 dictionary costs $40.47 with sales tax. Find the tax rate.

49. Percent of increase   A Turkish rug was purchased for $560. If it is now worth $1,100, find the percent of increase to the nearest 10th.

50. Percent of decrease   A clock on sale for $221.84 was regularly priced at $470. Find the percent of decrease.

SECTION 2.3   Simplifying Expressions to Solve Linear Equations in One Variable

DEFINITIONS AND CONCEPTS	EXAMPLES
Like terms are terms with the same variables having the same exponents. They can be combined by adding their numerical coefficients and using the same variables and exponents.	Combine like terms. $4(x + 3) + 6(x - 5)$ $= 4x + 12 + 6x - 30$ Use the distributive property to remove parentheses. $= 10x - 18$ Combine like terms: $4x + 6x = 10x$, $12 - 30 = -18$.
An **identity** is an equation that is true for all values of its variable.	Show that the following equation is an identity. $2(x - 5) + 6x = 8(x - 1) - 2$ $2x - 10 + 6x = 8x - 8 - 2$ Use the distributive property to remove parentheses. $8x - 10 = 8x - 10$ Combine like terms. $-10 = -10$ Subtract $8x$ from both sides. Since the final result is always true, the equation is an identity and its solution set is $\mathbb{R}$.
A **contradiction** is an equation that is true for no values of its variable.	Show that the following equation is a contradiction. $6x - 2(x + 5) = 4x - 1$ $6x - 2x - 10 = 4x - 1$ Use the distributive property to remove parentheses. $4x - 10 = 4x - 1$ Combine like terms. $-10 = -1$ Subtract $4x$ from both sides. Since the final result is false, the equation is a contradiction and its solution set is $\varnothing$.

REVIEW EXERCISES

Simplify each expression, if possible.

51. $5x + 9x$ **52.** $7a + 12a$

53. $18b - 13b$ **54.** $21x - 23x$

55. $5y - 7y$ **56.** $19x - 19$

57. $7(x + 2) + 2(x - 7)$ **58.** $2(3 - x) + x - 6x$

Solve each equation and check all solutions.

59. $10y - 14 = 3y$ **60.** $6(a + 4) = 3a$

61. $2x - 19 = 2 - x$ **62.** $5b - 19 = 2b + 20$

63. $3x + 20 = 5 - 2x$ **64.** $0.9x + 10 = 0.7x + 1.8$

65. $10(p - 3) = 3(p + 11)$

66. $2(5x - 7) = 2(x - 35)$

67. $\dfrac{3u - 6}{5} = 3$ **68.** $\dfrac{5v - 35}{3} = -5$

69. $\dfrac{2(b + 4)}{3} = b - 2$

70. $\dfrac{3(x - 1)}{6} - 5 = \dfrac{2(x + 3)}{6}$

Classify each equation as an identity or a contradiction and give the solution.

71. $4x - 2 = x + 3(x + 1) - 5$

72. $-3(a + 1) - a = -4a + 3$

73. $2(x - 1) + 4 = 4(1 + x) - (2x + 2)$

74. $3(2x + 1) + 3 = 9(x + 2) + 9 - 3x$

SECTION 2.4   Formulas

DEFINITIONS AND CONCEPTS	EXAMPLES
A **literal equation** or **formula** often can be solved for any of its variables.	Solve $2x + 3y = 6$ for y. $2x + 3y = 6$ $3y = -2x + 6$ Subtract $2x$ from both sides. $y = -\dfrac{2}{3}x + 2$ Divide both sides by 3.

REVIEW EXERCISES

Solve each equation for the indicated variable.

75. $E = IR$ for R **76.** $i = prt$ for t

77. $P = I^2R$ for R **78.** $V = \dfrac{1}{3}Bh$ for B

79. $p = a + b + c$ for c

80. $y = mx + b$ for m

81. $V = \pi r^2 h$ for h

82. $A = \dfrac{3}{2}(B + 4)$ for B

83. $F = \dfrac{GMm}{d^2}$ for G

84. $P = \dfrac{RT}{mV}$ for m

SECTION 2.5   Introduction to Problem Solving

DEFINITIONS AND CONCEPTS	EXAMPLES
To solve applications, follow these steps: **1.** Analyze the situation and choose a variable. **2.** Form an equation. **3.** Solve the equation. **4.** State the conclusion. **5.** Check the result to be certain it satisfies the given conditions.	The length of a rectangular frame is 4 in. longer than twice the width. If the perimeter is 38 in., find the width of the frame. **1.** Let w represent the width of the frame in inches. **2.** The width of the frame is w and because the length is 4 in. longer than twice the width, the length is $(2w + 4)$. Since the frame is a rectangle, its perimeter is the sum of two widths and two lengths. This perimeter is 38. So $$2w + 2(2w + 4) = 38$$ **3.** To solve the equation, proceed as follows: $$2w + 2(2w + 4) = 38$$ $$2w + 4w + 8 = 38$$ $$6w + 8 = 38$$ $$6w = 30$$ $$w = 5$$ **4.** The frame is 5 in. wide. **5.** If the width is 5 in., the length is $2 \cdot 5 + 4 = 14$ in. The perimeter is $2 \cdot 5 + 2 \cdot 14 = 10 + 28 = 38$. The result checks.
If the sum of the measures of two angles is 90°, the angles are called **complementary angles**.	Find the complement of an angle measuring 42°. $$x + 42 = 90$$ $$x + 42 - 42 = 90 - 42$$ $$x = 58°$$
If the sum of the measures of two angles is 180°, the angles are called **supplementary angles**.	Find the supplement of an angle measuring 42°. $$x + 42 = 180$$ $$x + 42 - 42 = 180 - 42$$ $$x = 138°$$

REVIEW EXERCISES

85. Carpentry   A carpenter wants to cut an 8-foot board into two pieces so that one piece is 7 feet shorter than twice the longer piece. Where should he make the cut?

86. Find the value of x.   **87.** Find the value of x.

       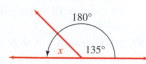

84 in.

88. Find the complement of an angle that measures 12°.

89. Find the supplement of an angle that measures 75°.

90. Rectangles   If the length of the rectangular painting in the illustration is 3 inches more than twice the width, how wide is the rectangle?

91. Investing   A woman has $27,000. Part is invested for 1 year in a certificate of deposit paying 7% interest, and the remaining amount in a cash management fund paying 9%. The total interest on the two investments is $2,110. How much does she invest at each rate?

SECTION 2.6   Motion and Mixture Applications

DEFINITIONS AND CONCEPTS	**EXAMPLES**
Distance = rate · time $$d = rt$$	Two cars leave McKinney at the same time traveling in opposite directions. If the speed of one car is 62 mph and the cars are 268 miles apart after two hours, what is the speed of the second car? Let x represent the speed, in mph, of the second car. $$2(62) + 2x = 268$$ $$124 + 2x = 268$$ $$2x = 268 - 124$$ $$2x = 144$$ $$x = 72 \text{ mph}$$ The second car is traveling at 72 mph.
Value = price · number $$v = pn$$	How many pounds of candy priced at $2.50 per pound must be mixed with 30 pounds of another candy priced at $3.75 per pound to make a mixture that would sell for $3.25 per pound? Let x represent the number of pounds of candy priced at $2.50 per pound. $$2.50x + 3.75(30) = 3.25(x + 30)$$ $$2.50x + 112.50 = 3.25x + 97.50$$ $$2.50x - 2.50x + 112.50 = 3.25x - 2.50x + 97.50$$ $$112.50 = 0.75x + 97.50$$ $$112.50 - 97.50 = 0.75x + 97.50 - 97.50$$ $$15 = 0.75x$$ $$20 = x$$ There should be 20 pounds of the $2.50 per pound candy.

REVIEW EXERCISES

92. Riding bicycles   A bicycle path is 7 miles long. A man walks from one end at the rate of 4 mph. At the same time, a friend bicycles from the other end, traveling at 10 mph. In how many minutes will they meet?

93. Tornadoes   During a storm, two teams of scientists leave a university at the same time in specially designed vans to search for tornadoes. The first team travels east at 20 mph and the second travels west at 25 mph. If their radios have a range of up to 90 miles, how long will it be before they lose radio contact?

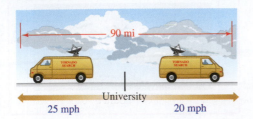

90 mi

University

25 mph   20 mph

94. Band trips   A bus carrying the members of a marching band and a truck carrying their instruments leave a high school at the same time and travel in the same direction.

The bus travels at 60 mph and the truck at 50 mph. In how many hours will they be 90 miles apart?

95. Mixing milk   A container is partly filled with 12 liters of whole milk containing 4% butterfat. How much 1% milk must be added to get a mixture that is 2% butterfat?

96. Photography   A photographer wants to mix 2 liters of a 6% acetic acid solution with a 12% solution to get an 8% solution. How many liters of 12% solution must be added?

97. Mixing candy   A store manager mixes candy worth 90¢ per pound with gumdrops worth $1.50 per pound to make 20 pounds of a mixture worth $1.20 per pound. How many pounds of each kind of candy must he use?

SECTION 2.7   Solving Linear Inequalities in One Variable

DEFINITIONS AND CONCEPTS	EXAMPLES
Inequalities are solved by techniques similar to those used to solve equations, with this exception: *If both sides of an inequality are multiplied or divided by a negative number, the direction of the inequality must be reversed.* The solution of an inequality can be graphed on the number line.	To solve the inequality $-3x - 8 < 7$, proceed as follows: $$-3x - 8 < 7$$ $$-3x < 15$$ $$x > -5 \quad \text{Divide both sides by } -3 \text{ and reverse the inequality symbol.}$$ The graph of $x > -5$ is $\xleftarrow{\hspace{3cm}}\overset{\textstyle(}{\underset{-5}{\rule{0pt}{0pt}}}\xrightarrow{\hspace{3cm}}$

REVIEW EXERCISES

Graph the solution to each inequality on a number line.

98. $3x + 2 < 5$ **99.** $-5x - 8 > 7$

100. $5x - 3 \geq 2x + 9$ **101.** $7x + 1 \leq 8x - 5$

102. $5(3 - x) \leq 3(x - 3)$ **103.** $8(2 - x) > 4 - 2x$

104. $8 < x + 2 < 13$ **105.** $4 \leq 2 - 2x < 8$

106. Swimming pools   By city ordinance, the perimeter of a rectangular swimming pool cannot exceed 68 feet. The width is 6 feet shorter than the length. What possible lengths will meet these conditions?

2   Test

Determine whether the given number is a solution of the equation.

1. $4x + 5 = -3; -2$ **2.** $3(x + 2) = 2x + 4; 2$

Solve each equation.

3. $x + 17 = -19$ **4.** $a - 15 = 32$

5. $12x = -144$ **6.** $\dfrac{x}{7} = -1$

7. $8x + 2 = -14$ **8.** $3 = 5 - 2x$

9. $\dfrac{2x - 5}{3} = 3$ **10.** $23 - 5(x + 10) = -12$

Simplify each expression.

11. $x + 5(x - 3)$ **12.** $3x - 5(2 - x)$

13. $-3(x + 3) + 3(x - 3)$ **14.** $-4(2x - 5) - 7(4x + 1)$

Solve each equation.

15. $8x + 6 = 2(4x + 3)$

16. $2(x + 6) = 2(x - 2)$

17. $\dfrac{3x - 18}{2} = 6x$

18. $\dfrac{7}{8}(x - 4) = 5x - \dfrac{7}{2}$

Solve each equation for the variable indicated.

19. $d = rt$ for t

20. $S = 2\pi rh + 2\pi r^2$ for h

21. $A = 2\pi rh$ for h

22. $x + 2y = 5$ for y

23. Find the value of x.

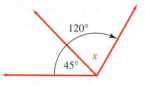

24. Find the supplement of a 79° angle.

25. Investing A student invests part of $10,000 at 6% annual interest and the rest at 5%. If the annual income from these investments is $560, how much was invested at each rate?

26. Traveling A car leaves Rockford at the rate of 65 mph, bound for Madison. At the same time, a truck leaves Madison at the rate of 55 mph, bound for Rockford. If the cities are 72 miles apart, how long will it take for the car and the truck to meet?

27. Mixing solutions How many liters of water must be added to 30 liters of a 10% brine solution to dilute it to an 8% solution?

28. Mixing nuts Twenty pounds of cashews are to be mixed with peanuts to make a mixture that will sell for $4 per pound. How many pounds of peanuts should be used?

Nut	Price per pound
Cashews	$6
Peanuts	$3

Graph the solution of each inequality.

29. $8x - 20 \geq 4$

30. $x - 2(x + 7) > 14$

31. $-4 \leq 2(x + 1) < 10$

32. $-10 < 2(4 - x) \leq 12$

⌒ Cumulative Review ⌒

Classify each number as an integer, a rational number, an irrational number, and/or a real number. Each number may be in several classifications.

1. $\dfrac{27}{9}$

2. -0.25

Graph each set of numbers on the number line.

3. The natural numbers between 2 and 7

4. The real numbers between -2 and 5

Simplify each expression.

5. $\dfrac{|-3| - |3|}{|-3 - 3|}$

6. $\dfrac{14}{15} \cdot \dfrac{3}{4}$

7. $2\dfrac{3}{5} + 5\dfrac{1}{2}$

8. $35.7 - 0.05$

Let $x = -5$, $y = 3$, and $z = 0$, and evaluate each expression.

9. $(2z - 3x)y$

10. $\dfrac{x - 3y + |z|}{2 - x}$

11. $x^2 - y^2 + z^2$

12. $\dfrac{x}{y} + \dfrac{y + 2}{3 - z}$

13. What is $4\dfrac{1}{2}\%$ of 220?

14. 1,688 is 32% of what number?

Consider the algebraic expression $3x^3 + 5x^2y + 37y$.

15. Identify the coefficient of the second term.

16. List the factors of the third term.

Simplify each expression.

17. $4x + 7y - 9x$

18. $3(x - 7) + 2(8 - x)$

19. $2x^2y^3 - 4x^2y^3$

20. $3(5 - x) - 6(x + 2)$

Solve each equation and check the result.

21. $2(x - 7) + 5 = 3x$

22. $\dfrac{x - 5}{3} - 5 = 7$

23. $\dfrac{2x - 1}{5} = \dfrac{1}{2}$

24. $2(a - 3) - 3(a - 2) = -a$

Solve each formula for the variable indicated.

25. $A = \frac{1}{2}h(b + B)$ for h

26. $y = mx + b$ for x

27. Auto sales An auto dealer's promotional ad appears in the illustration. One car is selling for $23,499. What was the dealer's invoice?

700 cars to choose from!
Buy at
3%
over dealer invoice!

28. Furniture pricing A sofa and a $300 chair are discounted 35%, and are priced at $780 for both. Find the original price of the sofa.

29. Cost of a car The total cost of a new car, including an 8.5% sales tax, is $24,618.65. Find the cost before tax.

30. Manufacturing concrete Concrete contains 3 times as much gravel as cement. How many pounds of cement are in 500 pounds of dry concrete mix?

31. Building construction A 35-foot beam, 1 foot wide and 2 inches thick, is cut into three sections. One section is 14 feet long. Of the remaining two sections, one is twice as long as the other. Will the shortest section span an 8-foot-wide doorway?

32. Installing solar heating One solar panel in the illustration is 3.4 feet wider than the other. Find the width of each panel.

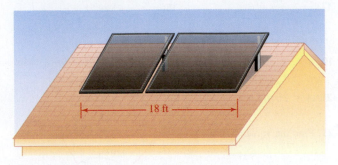

18 ft

33. Electric bills An electric company charges $52.50 per month, plus 28¢ for every kWh of energy used. One resident's bill was $203.70. How many kWh were used that month?

34. Installing gutters A contractor charges $47.75 for the installation of rain gutters, plus $2.05 per foot. If one installation cost $427, how many feet of gutter were required?

Evaluate each expression.

35. $4^2 - 5^2$ **36.** $(4 - 5)^2$

37. $5(4^3 - 2^3)$ **38.** $-2(5^4 - 7^3)$

Graph the solutions of each inequality.

39. $8(4 + x) > 10(6 + x)$ **40.** $-9 < 3(x + 2) \le 3$

Graphs; Equations of Lines; Functions; Variation

© Jess Alford/Jupiterimages

Careers and Mathematics

PSYCHOLOGIST

Psychologists study the human mind and human behavior. Research psychologists investigate the physical, cognitive, emotional, or social aspects of human behavior. Psychologists in health service fields provide mental health care in hospitals, clinics, schools, or private settings. Psychologists employed in applied settings, such as business, industry, or government, provide training, conduct research, and design organizational systems.

Job Outlook:
Overall employment of psychologists is expected to grow faster than the average for all occupations through 2016.

Annual Earnings:
$45,300–$77,750

For More Information:
http://www.bls.gov/oco/ocos056.htm

For a Sample Application:
See Problem 66 in Section 3.2.

In this chapter

In this chapter, we will discuss equations with two variables. We will see that the relationships set up by these equations also can be expressed in tables or graphs.

We begin the chapter by using the rectangular coordinate system to plot ordered pairs and graph linear equations. After explaining how to find the slope of a line, we will show how to write the equation of a line that passes through two known points. We then discuss functions, one of the most important ideas in mathematics.

Reach for Success   Understanding Your Syllabus

When you go on a trip, you have a plan, an itinerary. You might even have a travel representative along on the trip as a guide to handle the details and any unexpected situations that might occur. Your course syllabus is similar to your itinerary and your instructor is similar to the travel guide.

A syllabus is mandatory for all college courses and serves as a contract between the student and the instructor. Examine the syllabus you received for this course to identify important information.

<table>
<tr><td colspan="2">Course Syllabus</td><td>Tests</td></tr>
<tr><td colspan="2">Course Title: Beginning Algebra
Instructor's Office: B232 University Building
Office Hours: MTWR: 8:30am - 10:00am plus other times by appointment
Course Resources: The college provides a Math Lab at no charge to support student success.
Supplies: Textbook: Beginning and Intermediate Algebra: A Guided Approach, 7th edition by Karr/Massey/Gustafson
A graphing calculator is recommended. The TI 84 and Casio fx-9760GII are each supported by a tear-out card in the back of the textbook which provide keystrokes for calculator concepts covered in the text.
Access to homework will be through WebAssign. The course code required to register for this site will be provided the first day of the semester.</td><td>Tests count 40% of your final grade.
Tests are given in class on the date indicated in your schedule. There are no make-ups. However, if you miss a test, your Final Exam grade will be used to replace it. If you take all tests, the Final Exam grade, if higher, can be used to replace a lower test grade.</td></tr>
<tr><td colspan="2">Method of Evaluation:</td><td>Homework
Homework counts 10% of your final grade.
You have a 3 calendar day "extension" after the deadline in which to submit homework. There is, however, a 10% penalty each day but ONLY for the problems that were not completed by the deadline.</td></tr>
<tr><td>4 Tests (10% each)
20 Homework Assignments
10 Online Labs
Midterm
Comprehensive Final Exam</td><td>40%
10%
10%
20%
20%</td><td>LABS
Labs count 10% of your final grade.
Late labs are not accepted, i.e., there is no 3-day "extension."
MIDTERM
The Midterm counts 20% of your final grade. If you miss the midterm, your Final Exam grade will be used to replace it.
FINAL EXAM</td></tr>
<tr><td>90 - 100   80 - 89   70 - 79   60 - 69   Below 60
A   B   C   D   F</td><td></td><td>The Final Exam counts 20% of your final grade.
The final exam is cumulative, meaning that it includes all topics covered during the semester.</td></tr>
</table>

Your syllabus will include your instructor's contact information, office location, and e-mail address. It typically contains the instructor's office hours that are used to meet with and help students.

Write the information regarding your instructor:

Office location: _____

E-mail address: _____

Office hours: _____

Write down at least one listed office hour during which you could meet with your instructor if necessary. _____

Does your instructor include a statement such as "and other times by appointment" with the office hours?

If your schedule absolutely prohibits you from meeting with your instructor during office hours, write down at least two times during the week that you are available and could request an appointment.

_____

Most instructors prefer that students be proactive. For example, it is important to contact your instructor prior to any known life event that may interfere with meeting due dates.

Write down any issues that might interfere with your work this semester and that you might wish to discuss with your instructor.

_____

Your syllabus will include requirements of the course and how each will be used to determine your grade. Knowing exactly how your grade will be computed, you can estimate your average at any time.

Write down how the following will be used to determine your grade.

Tests (how many?) _____ _____ points or _____ percentage

Homework/daily work _____ points or _____ percentage

Attendance _____ points or _____ percentage

Quizzes _____ points or _____ percentage

Other _____ points or _____ percentage

How would you use this information to calculate your grade?

_____

A Successful Study Strategy . . .

 Know your instructor's office hours so you can have a plan of action if it becomes necessary to meet. In addition, knowing how your grade will be determined the first day of class can eliminate surprises at the end of the semester.

At the end of the chapter you will find additional exercises to guide you to planning for a successful semester.

Section 3.1

The Rectangular Coordinate System

Objectives

1. Graph ordered pairs and mathematical relationships.
2. Interpret the meaning of graphed data.
3. Interpret information from a step graph.

Vocabulary

rectangular coordinate
   system
Cartesian coordinate system
perpendicular lines
x-axis

y-axis
origin
coordinate plane
Cartesian plane
quadrants

ordered pairs
x-coordinate
y-coordinate
coordinates

Getting Ready

Graph each set of numbers on the number line.

1. −2, 1, 3

2. All numbers greater than −2

3. All numbers less than or equal to 3

4. All numbers between −3 and 2

It is often said, "A picture is worth a thousand words." In this section, we will show how numerical relationships can be described by using mathematical pictures called *graphs*. We also will show how we can obtain important information by reading graphs.

1 ## Graph ordered pairs and mathematical relationships.

When designing the Gateway Arch in St. Louis, shown in Figure 3-1(a) on the next page, architects created a mathematical model of the arch called a *graph*. This graph, shown in Figure 3-1(b), is drawn on a grid called the **rectangular coordinate system**. This coordinate system is sometimes called a **Cartesian coordinate system** after the 17th-century French mathematician René Descartes.

A rectangular coordinate system (see Figure 3-2 on the next page) is formed by two perpendicular number lines. Recall that **perpendicular lines** are lines that meet at a 90° angle.

- The horizontal number line is called the **x-axis**.
- The vertical number line is called the **y-axis**.

The positive direction on the x-axis is to the right, and the positive direction on the y-axis is upward. The scale on each axis should fit the data. For example, the axes of the graph of the arch shown in Figure 3-1(b) are scaled in units of 100 feet. If no scale is indicated on the axes, we assume that the axes are scaled in units of 1.

René Descartes
1596–1650
Descartes is famous for his work in philosophy as well as for his work in mathematics. His philosophy is expressed in the words "I think, therefore I am." He is best known in mathematics for his invention of a coordinate system and his work with conic sections.

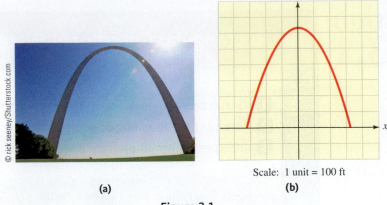

Scale: 1 unit = 100 ft

(a)   **(b)**

Figure 3-1

The point where the axes cross is called the **origin**. This is the 0 point on each axis. The two axes form a **coordinate plane** (often referred to as the **Cartesian plane**) and divide it into four regions called **quadrants**, which are numbered as shown in Figure 3-2.

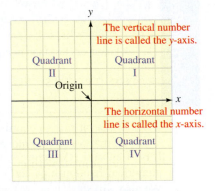

Figure 3-2

Each point in a coordinate plane can be identified by a pair of real numbers x and y, written as (x, y) and called **ordered pairs**. The first number in the pair is the **x-coordinate**, and the second number is the **y-coordinate**. The numbers are called the **coordinates** of the point. Some examples of ordered pairs are $(3, -4)$, $\left(-1, -\frac{3}{2}\right)$, $(0, 2.5)$, and the origin $(0, 0)$.

$$(3, -4)$$
↑   ↑

In an ordered pair, the   The y-coordinate
x-coordinate is listed first.   is listed second.

The process of locating a point in the coordinate plane is called *graphing* or *plotting* the point. In Figure 3-3(a), we show how to graph the point A with coordinates of $(3, -4)$. Since the x-coordinate is positive, we start at the origin and move 3 units to the right along the x-axis. Since the y-coordinate is negative, we then move down 4 units to locate point A. Point A is the *graph* of $(3, -4)$ and lies in quadrant IV.

To plot the point $B(-4, 3)$, we start at the origin, move 4 units to the left along the x-axis, and then move up 3 units to locate point B. Point B lies in quadrant II.

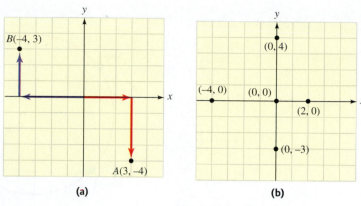

Figure 3-3

COMMENT   Note that point A with coordinates of $(3, -4)$ is not the same as point B with coordinates $(-4, 3)$. Since the order of the coordinates of a point is important, we call them ordered pairs.

In Figure 3-3(b), we see that the points $(-4, 0), (0, 0)$, and $(2, 0)$ lie on the x-axis. In fact, all points with a y-coordinate of 0 will lie on the x-axis.

From Figure 3-3(b), we also see that the points $(0, -3), (0, 0)$, and $(0, 4)$ lie on the y-axis. All points with an x-coordinate of 0 lie on the y-axis. From the figure, we also can see that the coordinates of the origin are $(0, 0)$ and will lie on both the x- and y-axes.

EXAMPLE 1   **GRAPHING POINTS**   Plot the points.

   a. $A(-2, 3)$ **b.** $B\left(-1, -\frac{3}{2}\right)$ **c.** $C(0, 2.5)$ **d.** $D(4, 2)$

Solution   **a.** To plot point A with coordinates $(-2, 3)$, we start at the origin, move 2 units to the *left* on the x-axis, and move 3 units *up*. Point A lies in quadrant II. (See Figure 3-4.)

   b. To plot point B with coordinates of $\left(-1, -\frac{3}{2}\right)$, we start at the origin and move 1 unit to the *left* and $\frac{3}{2}$ (or $1\frac{1}{2}$) units *down*. Point B lies in quadrant III, as shown in Figure 3-4.

   c. To graph point C with coordinates of $(0, 2.5)$, we start at the origin and move 0 units on the x-axis and 2.5 units *up*. Point C lies on the y-axis, as shown in Figure 3-4.

   d. To graph point D with coordinates of $(4, 2)$, we start at the origin and move 4 units to the *right* and 2 units *up*. Point D lies in quadrant I, as shown in Figure 3-4.

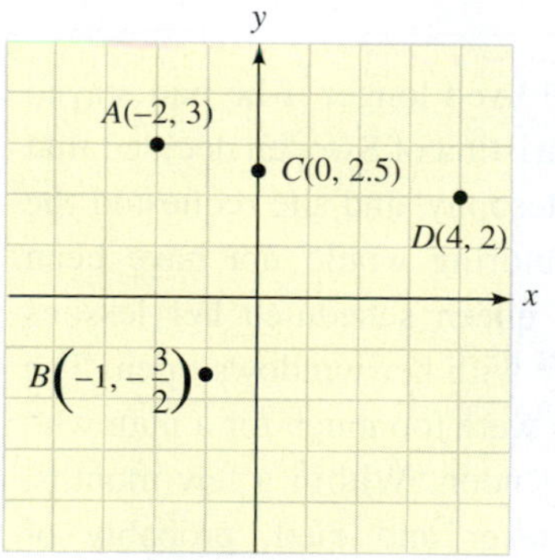

Figure 3-4

SELF CHECK 1   Plot the points.   **a.** $E(2, -2)$ **b.** $F(-4, 0)$ **c.** $G\left(1.5, \frac{5}{2}\right)$ **d.** $H(0, 5)$

EXAMPLE 2   **ORBITS**   The circle shown in Figure 3-5 is an approximate graph of the orbit of Earth. The graph is made up of infinitely many points, each with its own x- and y-coordinates. Use the graph to find the approximate coordinates of Earth's position during the months of February, May, and August.

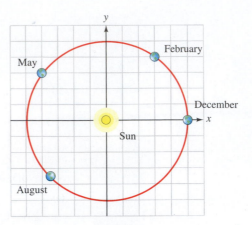

Figure 3-5

Solution   To find the coordinates of each position, we start at the origin and move left or right along the *x*-axis to find the *x*-coordinate and then up or down to find the *y*-coordinate. See Table 3-1.

TABLE 3-1		
Month	**Position of Earth on the graph**	**Coordinates**
February	3 units to the *right*, then 4 units *up*	$(3, 4)$
May	4 units to the *left*, then 3 units *up*	$(-4, 3)$
August	3.5 units to the *left*, then 3.5 units *down*	$(-3.5, -3.5)$

SELF CHECK 2   From Figure 3-5, find the coordinates of Earth's position in December.

Perspective

As a child, René Descartes was frail and often sick. To improve his health, eight-year-old René was sent to a Jesuit school. The headmaster encouraged him to sleep in the morning as long as he wished. As a young man, Descartes spent several years as a soldier and world traveler, but his interests included mathematics and philosophy, as well as science, literature, writing, and taking it easy. The habit of sleeping late continued throughout his life. He claimed that his most productive thinking occurred when he was lying in bed. According to one story, Descartes first thought of analytic geometry as he watched a fly walking on his bedroom ceiling.

Descartes might have lived longer if he had stayed in bed. In 1649, Queen Christina of Sweden decided that she needed a tutor in philosophy, and she requested the services of Descartes. Tutoring would not have been difficult, except that the queen scheduled her lessons before dawn in her library with her windows open. The cold Stockholm mornings were too much for a man who was used to sleeping past noon. Within a few months, Descartes developed a fever and died, probably of pneumonia.

Every day, we deal with quantities that are related.

- The distance that we travel depends on how fast we are going.
- Our weight depends on how much we eat.
- The amount of water in a tank depends on how long the water has been running.

We often can use graphs to visualize relationships between two quantities. For example, suppose that we know the number of gallons of water that are in a tank at several time intervals after the water has been turned on. We can list that information in a *table of values*. (See Figure 3-6.)

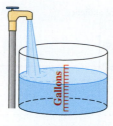

Time (minutes)	Water in tank (gallons)	
0	0	→ (0, 0)
1	3	→ (1, 3)
3	9	→ (3, 9)
4	12	→ (4, 12)
↑	↑	↑
x-coordinate	y-coordinate	The data in the table can be expressed as ordered pairs (x, y).

At various times, the amount of water in the tank was measured and recorded in the table of values.

Figure 3-6

The information in the table can be used to construct a graph that shows the relationship between the amount of water in the tank and the time the water has been running. Since the amount of water in the tank *depends* on the time, we will associate *time* with the x-axis and the *amount of water* with the y-axis.

To construct the graph in Figure 3-7, we plot the four ordered pairs and draw a line through the resulting data points.

COMMENT Note that the scale for the gallons of water (y-axis) is $1\frac{1}{2}$ units while the scale for minutes (x-axis) is 1 unit. The scales on both axes do not have to be the same, but remember to label them!

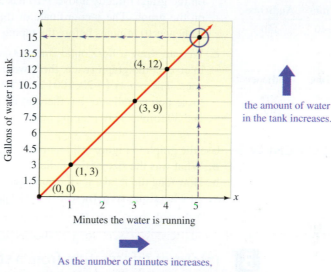

the amount of water in the tank increases.

As the number of minutes increases,

Figure 3-7

From the graph, we can see that the amount of water in the tank increases as the water is allowed to run. We also can use the graph to make observations about the amount of water in the tank at other times. For example, the dashed line on the graph shows that in 5 minutes, the tank will contain 15 gallons of water.

2 Interpret the meaning of graphed data.

In the next example, we show that valuable information can be obtained from reading a graph.

EXAMPLE 3 **READING GRAPHS** The graph in Figure 3-8 on the next page shows the number of people in an audience before, during, and after the taping of a television show. On the x-axis, 0 represents the time when taping began. Use the graph to answer the following questions, and record each result in a table of values.

a. How many people were in the audience when taping began?

b. What was the size of the audience 10 minutes before taping began?

c. At what times were there exactly 100 people in the audience?

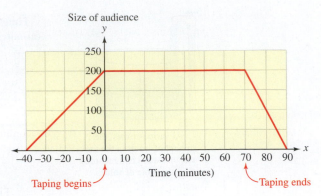

Figure 3-8

Solution   **a.** The time when taping began is represented by 0 on the *x*-axis. Since the point on the graph directly above 0 has a *y*-coordinate of 200, the point $(0, 200)$ is on the graph. The *y*-coordinate of this point indicates that 200 people were in the audience when the taping began.

Time	Audience
0	200

b. Ten minutes before taping began is represented by -10 on the *x*-axis. Since the point on the graph directly above -10 has a *y*-coordinate of 150, the point $(-10, 150)$ is on the graph. The *y*-coordinate of this point indicates that 150 people were in the audience 10 minutes before the taping began.

Time	Audience
-10	150

c. We can draw a horizontal line passing through 100 on the *y*-axis. Since this line intersects the graph twice, there were two times when 100 people were in the audience. The points $(-20, 100)$ and $(80, 100)$ are on the graph. The *y*-coordinates of these points indicate that there were 100 people in the audience 20 minutes before and 80 minutes after taping began.

Time	Audience
-20	100
80	100

 **SELF CHECK 3**   Use the graph in Figure 3-8 to answer the following questions.
a. What was the size of the audience that watched the taping?
b. How long did it take for the audience to leave the studio after taping ended?
c. At what times were there exactly 50 people in the audience?

---

3   Interpret information from a step graph.

The graph in Figure 3-9 shows the cost of renting a trailer for different periods of time. For example, the cost of renting the trailer for 4 days is $60, which is the *y*-coordinate of the point with coordinates of $(4, 60)$. For renting the trailer for a period lasting over 4 and up to 5 days, the cost jumps to $70. Since the jumps in cost form steps in the graph, we call the graph a *step graph*.

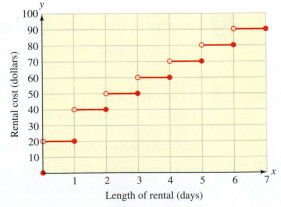

Figure 3-9

EXAMPLE 4 **STEP GRAPHS** Use the information in Figure 3-9 to answer the following questions. Write the results in a table of values.

  a. Find the cost of renting the trailer for 2 days.

  b. Find the cost of renting the trailer for $5\frac{1}{2}$ days.

  c. How long can you rent the trailer if you have $50?

  d. Is the rental cost per day the same?

Solution **a.** We locate 2 days on the *x*-axis and move up to locate the point on the graph directly above the 2. Since the point has coordinates $(2, 40)$, a two-day rental would cost $40. We enter this ordered pair in Table 3-2.

TABLE 3-2

Length of rental (days)	Cost (dollars)
2	40
$5\frac{1}{2}$	80
3	50

  b. We locate $5\frac{1}{2}$ days on the *x*-axis and move straight up to locate the point on the graph with coordinates $\left(5\frac{1}{2}, 80\right)$, which indicates that a $5\frac{1}{2}$-day rental would cost $80. We enter this ordered pair in Table 3-2.

  c. We draw a horizontal line through the point labeled 50 on the *y*-axis. Since this line intersects one step of the graph, we can look down to the *x*-axis to find the *x*-values that correspond to a *y*-value of 50. From the graph, we see that the trailer can be rented for more than 2 and up to 3 days for $50. We write $(3, 50)$ in Table 3-2.

  d. No, the cost per day is not the same. If we look at the *y*-coordinates, we see that for the first day, the rental fee is $20. For the second day, the cost jumps another $20. For the third day, and all subsequent days, the cost jumps only $10.

 **SELF CHECK 4** Use the information in Figure 3-9 to answer the following questions. Write the results in a table of values.

  a. Find the cost of renting the trailer for 4 days.

  b. Find the cost of renting the trailer for $2\frac{1}{2}$ days.

  c. How long can you rent the trailer if you have $80?

SELF CHECK ANSWERS

1.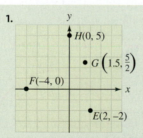

2. $(5, 0)$ 3. **a.** 200 people **b.** 20 min **c.** 30 min before and about 85 min after taping began

4.

Length of rental (days)	Cost (dollars)
4	60
$2\frac{1}{2}$	50
6	80

NOW TRY THIS

1. Find three ordered pairs that represent the information stated below.

   Damon paid $1,150 for 2 airline tickets. Javier paid $1,250 for 3 tickets, and Caroline paid $1,400 for 4 tickets.

2. Because the size of some data is large, we sometimes insert a // (break) symbol on the *x*- and/or *y*-axis of the rectangular coordinate system near the origin to indicate that the designated scale does not begin until the first value is listed.

   Plot the points from Problem 1 on a single set of coordinate axes with an appropriate scale.

3.1 Exercises

WARM-UPS

1. In which quadrant does the graph of $(-2, -6)$ lie?
2. At what point do the coordinate axes intersect?
3. In which quadrant does the graph of $(3, -5)$ lie?
4. On which axis does the point $(2, 0)$ lie?

REVIEW

5. Evaluate: $-5 - 5(-4)$
6. Evaluate: $(-3)^2 + (-10)$
7. What is the opposite of -8?
8. Simplify: $-|-2 - 10|$
9. Solve: $-4x + 7 = -21$
10. Solve $P = 2l + 2w$ for w.
11. Evaluate $(x + 1)(x + y)^2$ for $x = -2$ and $y = -5$.
12. Simplify: $-5(x + 2) - 3(4 - x)$

VOCABULARY AND CONCEPTS   *Fill in the blanks.*

13. The pair of numbers $(-1, -5)$ is called an _____.
14. In the _____ $\left(-\frac{3}{2}, -5\right)$, $-\frac{3}{2}$ is called the __coordinate and -5 is called the __coordinate.
15. The point with coordinates $(0, 0)$ is the _____.
16. The x-and y-axes divide the _____ into four regions called _____.
17. The point with coordinates $(4, 2)$ can be graphed on a _____ or _____ coordinate system.
18. The rectangular coordinate system is formed by two _____ number lines called the __ and __axes.
19. The values x and y in the ordered pair (x, y) are called the _____ of its corresponding point.
20. The process of locating the position of a point on a coordinate plane is called _____ the point.

Answer the question or fill in the blanks.

21. Do $(3, 2)$ and $(2, 3)$ represent the same point?
22. In the ordered pair $(2, 3)$, is 3 associated with the horizontal or the vertical axis?
23. To plot the point with coordinates $(-5, 4.5)$, we start at the _____, move 5 units to the ___, and then move 4.5 units ___.
24. To plot the point with coordinates $\left(-\frac{3}{2}, -4\right)$, we start at the _____, move $\frac{3}{2}$ units to the ___, and then move 4 units _____.
25. In which quadrant do points with a negative x-coordinate and a positive y-coordinate lie?
26. In which quadrant do points with a positive x-coordinate and a negative y-coordinate lie?

GUIDED PRACTICE   *Plot each point on a coordinate grid. SEE EXAMPLE 1. (OBJECTIVE 1)*

27. $A(-3, 4)$, $B(4, 3.5)$, $C\left(-2, -\frac{5}{2}\right)$, $D(0, -4)$, $E\left(\frac{3}{2}, 0\right)$, $F(3, -4)$

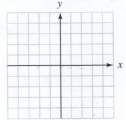

28. $G(4, 4)$, $H(0.5, -3)$, $I(-4, -4)$, $J(0, -1)$, $K(0, 0)$, $L(0, 3)$, $M(-2, 0)$

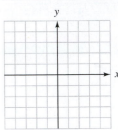

Use each graph to complete the table. SEE EXAMPLE 2. (OBJECTIVE 1)

29.

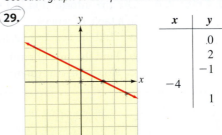

x	y
	0
	2
	-1
-4	
	1

30.

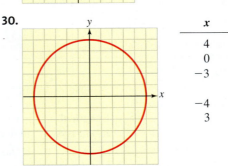

x	y
4	
0	
-3	
	0
-4	
3	

The graph gives the heart rate of a woman before, during, and after an aerobic workout. Use the graph to answer the following questions. SEE EXAMPLE 3. (OBJECTIVE 2)

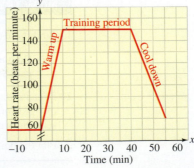

31. What information does the point $(-10, 60)$ give us?

32. After beginning the workout, how long did it take the woman to reach her training-zone heart rate?

33. What was her heart rate one-half hour after beginning the workout?

34. For how long did she work out at the training-zone level?

35. At what times was her heart rate 100 beats per minute?

36. How long was her cooldown period?

37. What was the difference in her heart rate before the workout and after the cooldown period?

38. What was her approximate heart rate 5 minutes after beginning her cooldown?

Use the graph to answer the questions. SEE EXAMPLE 4. (OBJECTIVE 3)

DVD rentals *The charges for renting a movie are shown in the graph.*

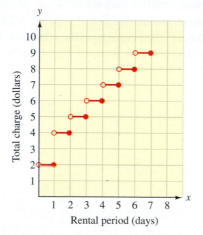

39. Find the charge for a 1-day rental.

40. Find the charge for a 3-day rental.

41. Find the charge if the DVD is kept for 5 days.

42. Find the charge if the DVD is kept for a week.

ADDITIONAL PRACTICE *Write the coordinates of each point.*

43.

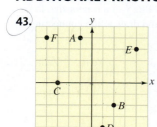

44.

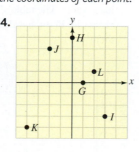

Postage rates *The graph gives the first-class postage rates for mailing letters weighing up to 3.5 ounces.*

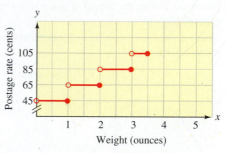

45. Find the cost of postage to mail each of the following letters first class: 1-ounce; $2\frac{1}{2}$-ounce.

46. Find the cost of postage to mail each of the following letters first class: 1.5-ounce; 3.25-ounce.

47. Find the difference in postage for a 0.75-ounce letter and a 2.75-ounce letter.

48. What is the heaviest letter that can be mailed for $1.05?

APPLICATIONS *Use the graphs to answer the questions. SEE EXAMPLES 3 AND 4. (OBJECTIVES 2–3)*

49. **Road maps** Road maps usually have a coordinate system to help locate cities. Use the map to locate Carbondale, Champaign, Chicago, Peoria, and Rockford. Express each answer in the form (number, letter).

50. **Battling ships** In a computer game of battling ships, players use coordinates to drop depth charges from a battleship to hit a hidden submarine. What coordinates should be used to make three hits on the exposed submarine shown in the illustration? Express each answer in the form (letter, number).

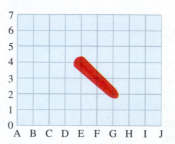

51. Water pressure   The graphs show the paths of two streams of water from the same hose held at two different angles.

   a. At which angle does the stream of water shoot higher? How much higher?

   b. At which angle does the stream of water shoot out farther? How much farther?

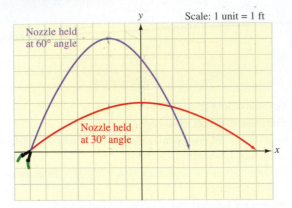

52. Golf swings   To correct her swing, a golfer was videotaped and then had her image displayed on a computer monitor so that it could be analyzed by a golf pro. See the illustration. Give the coordinates of the points that are highlighted on the arc of her swing.

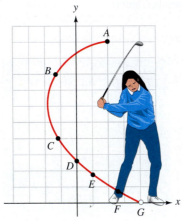

53. Gas mileage   The table gives the number of miles (*y*) that a truck can be driven on *x* gallons of gasoline. Plot the ordered pairs and draw a line connecting the points.

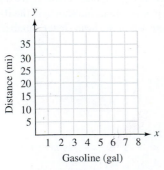

Gallons	Miles
2	10
3	15
5	25

   a. Estimate how far the truck can go on 6 gallons of gasoline.

   b. How many gallons of gas are needed to travel a distance of 40 miles?

   c. Estimate how far the truck can go on 6.5 gallons of gasoline.

54. Wages   The table gives the amount *y* (in dollars) that a student can earn by working *x* hours. Plot the ordered pairs and draw a line connecting the points.

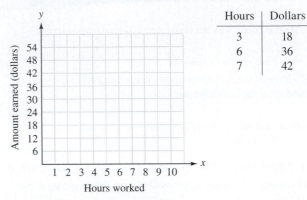

Hours	Dollars
3	18
6	36
7	42

   a. How much will the student earn in 5 hours?

   b. How long would the student have to work to earn $12?

   c. Estimate how much the student will earn in 3.5 hours.

55. Value of a car   The table shows the value *y* (in thousands of dollars) of a car that is *x* years old. Plot the ordered pairs and draw a line connecting the points.

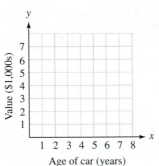

Years	Value (in thousands)
3	7
4	5.5
5	4

   a. What does the point $(3, 7)$ on the graph tell you?

   b. Estimate the value of the car when it is 7 years old.

   c. After how many years will the car be worth $2,500?

56. Depreciation   As a piece of farm machinery gets older, it loses value. The table shows the value *y* of a tractor that is *x* years old. Plot the ordered pairs and draw a line connecting them.

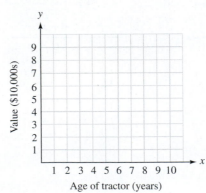

Years	Value
0	9
6	5
9	3

a. What does the point $(0, 9)$ on the graph tell you?

b. Estimate the value of the tractor in 3 years.

c. When will the tractor's value fall below $30,000?

WRITING ABOUT MATH

57. Explain why the point with coordinates $(-3, 3)$ is not the same as the point with coordinates $(3, -3)$.

58. Explain what is meant when we say that the rectangular coordinate graph of the St. Louis Arch is made up of *infinitely many* points.

59. Explain how to plot the point with coordinates $(-2, 5)$.

60. Explain why the coordinates of the origin are $(0, 0)$.

SOMETHING TO THINK ABOUT

61. Could you have a coordinate system in which the coordinate axes were not perpendicular? How would it be different?

62. René Descartes is famous for saying, "I think, therefore I am." What do you think he meant by that?

Section 3.2

Graphs of Linear Equations in Two Variables

Objectives

1. Determine whether an ordered pair satisfies an equation in two variables.
2. Construct a table of values given an equation.
3. Graph a linear equation in two variables by constructing a table of values.
4. Graph a linear equation in two variables using the intercept method.
5. Graph a horizontal line and a vertical line given an equation.
6. Write a linear equation in two variables from given information, graph the equation, and interpret the graphed data.

Vocabulary

input value	dependent variable	y-intercept
output value	independent variable	intercept method
linear equation in two variables	x-intercept	general form

Getting Ready

In Problems 1–4, let $y = 2x + 1$.

1. Find the value of y when $x = 0$.

2. Find the value of y when $x = 2$.

3. Find the value of y when $x = -2$.

4. Find the value of y when $x = \frac{1}{2}$.

5. Find five pairs of numbers with a sum of 8.

6. Find five pairs of numbers with a difference of 5.

In this section, we will discuss how to graph linear equations in two variables. We will then show how to create tables and graphs using a graphing calculator.

1 Determine whether an ordered pair satisfies an equation in two variables.

The equation $x + 2y = 5$ contains the two variables x and y. The solutions of such equations are ordered pairs of numbers. For example, the ordered pair $(1, 2)$ is a solution, because the equation is satisfied when $x = 1$ and $y = 2$.

$$x + 2y = 5$$
$$1 + 2(2) = 5 \quad \text{Substitute 1 for } x \text{ and 2 for } y.$$
$$1 + 4 = 5 \quad \text{Multiply}$$
$$5 = 5$$

EXAMPLE 1 Is the pair $(-2, 4)$ a solution of $y = 3x + 9$?

Solution We substitute -2 for x and 4 for y and determine whether the resulting equation is true.

$$y = 3x + 9 \qquad \text{This is the original equation.}$$
$$4 \stackrel{?}{=} 3(-2) + 9 \qquad \text{Substitute } -2 \text{ for } x \text{ and 4 for } y.$$
$$4 \stackrel{?}{=} -6 + 9 \qquad \text{Do the multiplication: } 3(-2) = -6.$$
$$4 = 3 \qquad \text{Do the addition: } -6 + 9 = 3.$$

Since the equation $4 = 3$ is false, the pair $(-2, 4)$ is not a solution.

SELF CHECK 1 Is $(-1, -5)$ a solution of $y = 5x$?

2 Construct a table of values given an equation.

To find solutions of equations in x and y, we can pick numbers at random, substitute them for x, and find the corresponding values of y. For example, to find some ordered pairs that satisfy $y = 5 - x$, we can let $x = 1$ (called the **input value**), substitute 1 for x, and solve for y (called the **output value**).

(1)
$$y = 5 - x$$

x	y	(x, y)
1	4	$(1, 4)$

$$y = 5 - x \qquad \text{This is the original equation.}$$
$$y = 5 - 1 \qquad \text{Substitute the input value of 1 for } x.$$
$$y = 4 \qquad \text{The output is 4.}$$

The ordered pair $(1, 4)$ is a solution. As we find solutions, we will list them in a *table of values* like Table (1) at the left.

If $x = 2$, we have

(2)
$$y = 5 - x$$

x	y	(x, y)
1	4	$(1, 4)$
2	3	$(2, 3)$

$$y = 5 - x \qquad \text{This is the original equation.}$$
$$y = 5 - 2 \qquad \text{Substitute the input value of 2 for } x.$$
$$y = 3 \qquad \text{The output is 3.}$$

A second solution is $(2, 3)$. We list it in Table (2) at the left.

If $x = 5$, we have

(3)
$$y = 5 - x$$

x	y	(x, y)
1	4	$(1, 4)$
2	3	$(2, 3)$
5	0	$(5, 0)$

$$y = 5 - x \qquad \text{This is the original equation.}$$
$$y = 5 - 5 \qquad \text{Substitute the input value of 5 for } x.$$
$$y = 0 \qquad \text{The output is 0.}$$

A third solution is $(5, 0)$. We list it in Table (3) at the left.

If $x = -1$, we have

(4)	$y = 5 - x$		
	x	y	(x, y)
	1	4	$(1, 4)$
	2	3	$(2, 3)$
	5	0	$(5, 0)$
	−1	**6**	**(−1, 6)**

$y = 5 - x$			This is the original equation.
$y = 5 - (-1)$			Substitute the input value of −1 for x.
$y = 6$			The output is 6.

A fourth solution is $(-1, 6)$. We list it in Table (4) at the left.
    If $x = 6$, we have

(5)	$y = 5 - x$		
	x	y	(x, y)
	1	4	$(1, 4)$
	2	3	$(2, 3)$
	5	0	$(5, 0)$
	−1	6	$(-1, 6)$
	6	**−1**	**(6, −1)**

$y = 5 - x$	This is the original equation.
$y = 5 - 6$	Substitute the input value of 6 for x.
$y = -1$	The output is −1.

A fifth solution is $(6, -1)$. We list it in Table (5) at the left.
    Since we can choose any real number for x, and since any choice of x will give a corresponding value of y, we can see that the equation $y = 5 - x$ has *infinitely many solutions*.

3 Graph a linear equation in two variables by constructing a table of values.

A *linear equation* is any equation that can be written in the form $Ax + By = C$, where A, B, and C are real numbers and A and B are not both 0. To graph the equation $y = 5 - x$, we plot the ordered pairs listed in the table on a rectangular coordinate system, as in Figure 3-10. From the figure, we can see that the five points lie on a line.

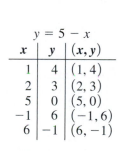

$y = 5 - x$		
x	y	(x, y)
1	4	$(1, 4)$
2	3	$(2, 3)$
5	0	$(5, 0)$
−1	6	$(-1, 6)$
6	−1	$(6, -1)$

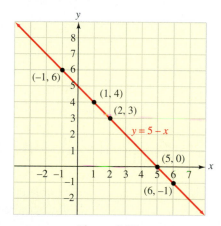

Figure 3-10

We draw a line through the points. The arrowheads on the line show that the graph continues forever in both directions. Since the graph of any solution of $y = 5 - x$ will lie on this line, the line is a picture of all of the solutions of the equation. The line is said to be the *graph* of the equation.
    Any equation, such as $y = 5 - x$, whose graph is a line is called a **linear equation in two variables**. Any point on the line has coordinates that satisfy the equation, and the graph of any pair (x, y) that satisfies the equation is a point on the line.
    Since we usually will choose a number for x first and then find the corresponding value of y, the value of y depends on x. For this reason, we call y the **dependent variable** and x the **independent variable**. The value of the independent variable is the input value, and the value of the dependent variable is the output value.
    Although only two points are needed to graph a linear equation, we often plot a third point as a check. If the three points do not lie on a line, at least one of them is in error.

COMMENT   The equation $y = 5 - x$ can be written as $x + y = 5$, which is in the form $Ax + By = C$.

GRAPHING LINEAR EQUATIONS IN TWO VARIABLES

1. Find two ordered pairs (x, y) that satisfy the equation by choosing arbitrary input values for x and solving for the corresponding output values of y. A third point may be used to provide a check.

2. Plot each resulting pair (x, y) on a rectangular coordinate system. If they do not lie on a line, check your calculations.

3. Draw the line passing through the points.

EXAMPLE 2   Graph by constructing a table of values and plotting points: $y = 3x - 4$

Solution   We find three ordered pairs that satisfy the equation.

If $x = 1$	If $x = 2$	If $x = 3$
$y = 3\textbf{\textit{x}} - 4$	$y = 3\textbf{\textit{x}} - 4$	$y = 3\textbf{\textit{x}} - 4$
$y = 3(\textbf{1}) - 4$	$y = 3(\textbf{2}) - 4$	$y = 3(\textbf{3}) - 4$
$y = -1$	$y = 2$	$y = 5$

We enter the results in a table of values, plot the points, and draw a line through the points. The graph appears in Figure 3-11.

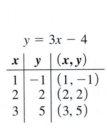

$y = 3x - 4$

x	y	(x, y)
1	−1	$(1, -1)$
2	2	$(2, 2)$
3	5	$(3, 5)$

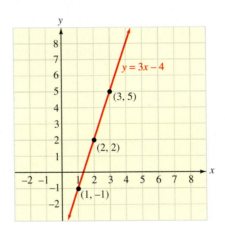

Figure 3-11

SELF CHECK 2   Graph: $y = 3x$

EXAMPLE 3   Graph by constructing a table of values and plotting points: $y = -0.4x + 2$

Solution   We find three ordered pairs that satisfy the equation.

If $x = -5$	If $x = 0$	If $x = 5$
$y = -0.4\textbf{\textit{x}} + 2$	$y = -0.4\textbf{\textit{x}} + 2$	$y = -0.4\textbf{\textit{x}} + 2$
$y = -0.4(\textbf{−5}) + 2$	$y = -0.4(\textbf{0}) + 2$	$y = -0.4(\textbf{5}) + 2$
$y = 2 + 2$	$y = 2$	$y = -2 + 2$
$y = 4$		$y = 0$

We enter the results in a table of values, plot the points, and draw a line through the points. The graph appears in Figure 3-12.

$$y = -0.4x + 2$$

x	y	(x, y)
-5	4	$(-5, 4)$
0	2	$(0, 2)$
5	0	$(5, 0)$

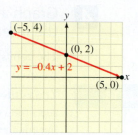

Figure 3-12

SELF CHECK 3   Graph: $y = 1.5x - 2$

EXAMPLE 4   Graph by constructing a table of values and plotting points: $y - 4 = \dfrac{1}{2}(x - 8)$

Solution   We first solve for y and simplify.

$$y - 4 = \frac{1}{2}(x - 8)$$

$$y - 4 = \frac{1}{2}x - 4 \qquad \text{Use the distributive property to remove parentheses.}$$

$$y = \frac{1}{2}x \qquad \text{Add 4 to both sides.}$$

We now find three ordered pairs that satisfy the equation.

If $x = 0$ *If* $x = 2$ *If* $x = -4$

$$y = \frac{1}{2}\textbf{\textit{x}} \qquad\qquad y = \frac{1}{2}\textbf{\textit{x}} \qquad\qquad y = \frac{1}{2}\textbf{\textit{x}}$$

$$y = \frac{1}{2}(\textbf{0}) \qquad\quad y = \frac{1}{2}(\textbf{2}) \qquad\quad y = \frac{1}{2}(\textbf{-4})$$

$$y = 0 \qquad\qquad\quad y = 1 \qquad\qquad\quad y = -2$$

We enter the results in a table of values, plot the points, and draw a line through the points. The graph appears in Figure 3-13.

$$y - 4 = \frac{1}{2}(x - 8)$$

x	y	(x, y)
0	0	$(0, 0)$
2	1	$(2, 1)$
-4	-2	$(-4, -2)$

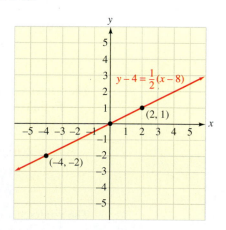

Figure 3-13

SELF CHECK 4   Graph: $y + 3 = \frac{1}{3}(x - 6)$

Graph a linear equation in two variables using the intercept method.

The points where a line intersects the x- and y-axes are called **intercepts** of the line.

x- AND y-INTERCEPTS

The **x-intercept** of a line is the point $(a, 0)$ where the line intersects the x-axis. (See Figure 3-14.) To find a, substitute 0 for y in the equation of the line and solve for x.

A **y-intercept** of a line is the point $(0, b)$ where the line intersects the y-axis. To find b, substitute 0 for x in the equation of the line and solve for y.

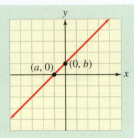

Figure 3-14

Plotting the x- and y-intercepts and drawing a line through them is called the **intercept method** of graphing a line. This method is useful for graphing equations written in *general form*.

GENERAL FORM OF THE EQUATION OF A LINE

If A, B, and C are real numbers and A and B are not both 0, then the equation

$$Ax + By = C$$

is called the **general form** of the equation of a line.

COMMENT   Whenever possible, we will write the general form $Ax + By = C$ so that A, B, and C are integers and $A \geq 0$. We also will make A, B, and C as small as possible. For example, the equation $6x + 12y = 24$ can be written as $x + 2y = 4$ by dividing both sides by 6.

EXAMPLE 5   Graph using the intercept method: $3x + 2y = 6$

Solution   To find the y-intercept, we let $x = 0$ and solve for y.

$$3x + 2y = 6$$
$$3(0) + 2y = 6 \qquad \text{Substitute 0 for } x.$$
$$2y = 6 \qquad \text{Simplify.}$$
$$y = 3 \qquad \text{Divide both sides by 2.}$$

The y-intercept is the point with coordinates $(0, 3)$.
To find the x-intercept, we let $y = 0$ and solve for x.

$$3x + 2y = 6$$
$$3x + 2(0) = 6 \qquad \text{Substitute 0 for } y.$$
$$3x = 6 \qquad \text{Simplify.}$$
$$x = 2 \qquad \text{Divide both sides by 3.}$$

The x-intercept is the point with coordinates $(2, 0)$.
As a check, we plot one more point. If $x = 4$, then

$$3x + 2y = 6$$
$$3(4) + 2y = 6 \qquad \text{Substitute 4 for } x.$$
$$12 + 2y = 6 \qquad \text{Simplify.}$$
$$2y = -6 \qquad \text{Subtract 12 from both sides.}$$
$$y = -3 \qquad \text{Divide both sides by 2.}$$

The point $(4, -3)$ is on the graph. We plot these three points and join them with a line. The graph of $3x + 2y = 6$ is shown in Figure 3-15.

$3x + 2y = 6$

x	y	(x, y)
0	3	$(0, 3)$
2	0	$(2, 0)$
4	−3	$(4, -3)$

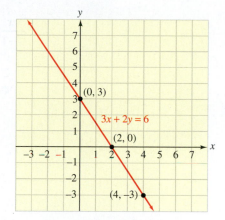

Figure 3-15

 SELF CHECK 5 Graph: $4x + 3y = 6$

---

5 Graph a horizontal line and a vertical line given an equation.

Equations such as $y = 3$ and $x = -2$ are linear equations, because they can be written in the general form $Ax + By = C$.

$y = 3$	is equivalent to	$0x + 1y = 3$
$x = -2$	is equivalent to	$1x + 0y = -2$

Next, we discuss how to graph these types of linear equations.

EXAMPLE 6 Graph: **a.** $y = 3$ **b.** $x = -2$

Solution **a.** We can write the equation $y = 3$ in general form as $\mathbf{0}x + y = 3$. Since the coefficient of x is 0, the numbers chosen for x have no effect on y. The value of y is always 3. For example, if we substitute -3 for x, we get

$$0\mathbf{x} + y = 3$$
$$0(\mathbf{-3}) + y = 3$$
$$0 + y = 3$$
$$y = 3$$

The table in Figure 3-16(a) on the next page gives several pairs that satisfy the equation $y = 3$. After plotting these pairs and joining them with a line, we see that the graph of $y = 3$ is a horizontal line that intersects the y-axis at 3. The y-intercept is $(0, 3)$. There is no x-intercept.

b. We can write $x = -2$ in general form as $x + \mathbf{0}y = -2$. Since the coefficient of y is 0, the values of y have no effect on x. The value of x is always -2. A table of values and the graph are shown in Figure 3-16(b). The graph of $x = -2$ is a vertical line that intersects the x-axis at -2. The x-intercept is $(-2, 0)$. There is no y-intercept.

$y = 3$

x	y	(x, y)
-3	3	$(-3, 3)$
0	3	$(0, 3)$
2	3	$(2, 3)$
4	3	$(4, 3)$

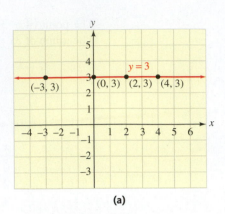

(a)

$x = -2$

x	y	(x, y)
-2	-2	$(-2, -2)$
-2	0	$(-2, 0)$
-2	2	$(-2, 2)$
-2	3	$(-2, 3)$

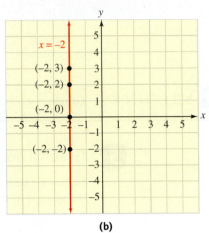

(b)

Figure 3-16

 **SELF CHECK 6**   Identify the graph of each equation as a horizontal or a vertical line.
a. $x = 5$ **b.** $y = -3$ **c.** $x = 0$

From the results of Example 6, we have the following facts.

EQUATIONS OF HORIZONTAL AND VERTICAL LINES

Suppose a and b are real numbers.

The equation $y = b$ represents a horizontal line that intersects the y-axis at $(0, b)$.

If $b = 0$, the line is the x-axis.

The equation $x = a$ represents a vertical line that intersects the x-axis at $(a, 0)$.

If $a = 0$, the line is the y-axis.

6   Write a linear equation in two variables from given information, graph the equation, and interpret the graphed data.

In Chapter 2, we solved applications using one variable. In the next example, we will write an equation containing two variables to describe an application and then graph the equation.

EXAMPLE 7   **BIRTHDAY PARTIES**   A restaurant offers a party package that includes food, drinks, cake, and party favors for a cost of $25 plus $3 per child. Write a linear equation that will give the cost for a party of any size. Graph the equation and determine the meaning of the y-intercept in the context of this problem.

Solution   We can let c represent the cost of the party and n represent the number of children attending. Then c will be the sum of the basic charge of $25 and the cost per child times the number of children attending.

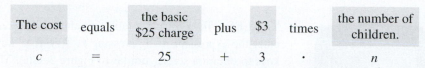

The cost	equals	the basic $25 charge	plus	$3	times	the number of children.
c	$=$	25	$+$	3	$\cdot$	n

For the equation $c = 25 + 3n$, the independent variable (input) is n, the number of children. The dependent variable (output) is c, the cost of the party. We will find three points on the graph of the equation by choosing n-values of 0, 5, and 10 and finding the corresponding c-values. The results are recorded in the table.

$c = 25 + 3n$		**If $n = 0$**	**If $n = 5$**	**If $n = 10$**
n	c	$c = 25 + 3(\textbf{\textcolor{red}{0}})$	$c = 25 + 3(\textbf{\textcolor{red}{5}})$	$c = 25 + 3(\textbf{\textcolor{red}{10}})$
0	25	$c = 25$	$c = 25 + 15$	$c = 25 + 30$
5	40		$c = 40$	$c = 55$
10	55			

Next, we graph the points in Figure 3-17 and draw a line through them. We don't draw an arrowhead on the left, because it doesn't make sense to have a *negative* number of children attend a party.

From the graph, we can determine the y-intercept is $(0, 25)$. The $25 represents the setup cost for a party with no attendees.

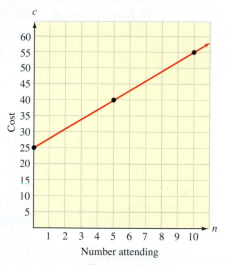

Figure 3-17

COMMENT   The scale for the cost (y-axis) is 5 units and the scale for the number attending (x-axis) is 1. Since the scales on the x- and y-axes are not the same, you must label them!

 **SELF CHECK 7**   Write a linear equation that will give the cost for a party of any size if the party package costs $30 plus $4 per child. Then graph the equation.

Accent on technology

▸ Creating Tables and Graphs

So far, we have graphed equations by creating tables and plotting points. This method is often tedious and time-consuming. Fortunately, creating tables and graphing equations is easier when we use a graphing calculator.

Although we will use calculators to generate tables and graph equations, we will show complete keystrokes only for the TI 83-84 family of calculators. For keystrokes of other calculators, please consult your owner's manual.

All graphing calculators have a *viewing window* that is used to display graphs. We will first discuss how to generate tables and then discuss how to draw graphs.

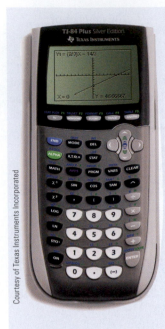

Courtesy of Texas Instruments Incorporated

TABLES　To construct a table of values for the equation $y = x^2$, press the **Y =** key, enter the expression x^2 by pressing **x, T, θ, n** **x²** keys, and press the **2nd** and **GRAPH** keys to obtain a screen similar to Figure 3-18(a). You can use the up and down keys to scroll through the table to obtain a screen like Figure 3-18(b).

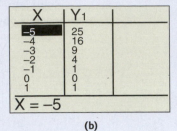

X	Y₁	
0	0	
1	1	
2	4	
3	9	
4	16	
5	25	
6	36	
X = 0		

(a)

X	Y₁	
−5	25	
−4	16	
−3	9	
−2	4	
−1	1	
0	0	
1	1	
X = −5		

(b)

Figure 3-18

GRAPHS　To see an accurate graph, we must often set the minimum and maximum values for the x- and y-coordinates. The standard window settings of

$$\text{Xmin} = -10 \qquad \text{Xmax} = 10 \qquad \text{Ymin} = -10 \qquad \text{Ymax} = -10$$

indicate that -10 is the minimum x- and y-coordinate to be used in the graph, and that 10 is the maximum x- and y-coordinate to be used. We will usually express window values in interval notation. In this notation, the standard settings are

$$X = [-10, 10] \quad Y = [-10, 10]$$

To graph the equation $2x - 3y = 14$ with a calculator, we must first solve the equation for y.

$$2x - 3y = 14$$

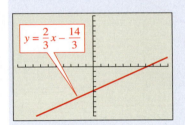

$y = \dfrac{2}{3}x - \dfrac{14}{3}$

Figure 3-19

$$-3y = -2x + 14 \qquad \textcolor{red}{\text{Subtract } 2x \text{ from both sides.}}$$

$$y = \frac{2}{3}x - \frac{14}{3} \qquad \textcolor{red}{\text{Divide both sides by } -3.}$$

Press the **Y =** key and enter the equation as $(2/3)x - 14/3$ by pressing **(2 ÷ 3) x, T, θ, n** **− 1 4 ÷ 3**, and press **GRAPH** to obtain the line shown in Figure 3-19.

COMMENT　To graph an equation with a graphing calculator, the equation must be solved for y.

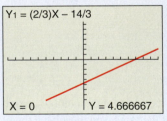

Y₁ = (2/3)X − 14/3

X = 0　　　Y = 4.666667

Figure 3-20

　We can find the y-intercept by using the **TRACE** feature of the calculator. From the graph, press **TRACE**. The y-intercept will be highlighted and its coordinates will be displayed at the bottom of the screen. See Figure 3-20.

For instructions regarding the use of a Casio graphing calculator, please refer to the Casio Keystroke Guide in the back of the book.

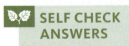

SELF CHECK ANSWERS

1. yes　**2.**

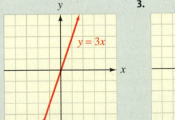

$y = 3x$

3.

$y = 1.5x - 2$

4.

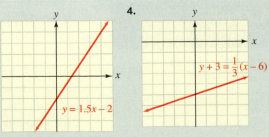

$y + 3 = \frac{1}{3}(x - 6)$

5.

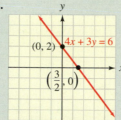

(0, 2) $4x + 3y = 6$

$\left(\frac{3}{2}, 0\right)$

6. a. vertical
b. horizontal
c. vertical

7. $c = 30 + 4n$

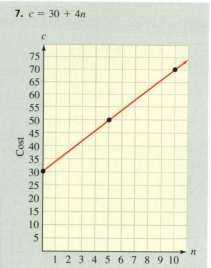

Cost

Number attending

NOW TRY THIS

1. Given $8x - 7y = 12$, complete the ordered pair $(-2, \quad)$ that satisfies the equation.

2. Graph: $y - 5 = 0$

3. Graph: $y = x$

4. Identify the x-intercept and the y-intercept of $y = \frac{2}{3}x + 8$.

3.2 Exercises

WARM-UPS

1. How many points determine a line?

2. Define the intercepts of a line.

3. If $y = 3x + 2$, find the value of y when $x = 1$.

4. Find three pairs (x, y) with a sum of 7.

5. Which lines have no y-intercepts?

6. Which lines have no x-intercepts?

REVIEW

7. Solve: $\frac{2}{3}x = -12$

8. Combine like terms: $3t - 4T + 5T - 6t$

9. Is $\dfrac{x + 5}{6}$ an expression or an equation?

10. Write the formula used to find the perimeter of a rectangle.

11. What number is 0.5% of 250?

12. Solve: $6 - 4x > 22$

13. Subtract: $-2.5 - (-2.6)$

14. Evaluate: $(-5)^3$

VOCABULARY AND CONCEPTS   *Fill in the blanks.*

15. The equation $y = x + 1$ is a _____ equation in ___ variables.

16. An ordered pair is a _____ of an equation if the numbers in the ordered pair satisfy the equation.

17. In equations containing the variables x and y, x is called the _____ variable and y is called the _____ variable.

18. When constructing a table of values, the values of _ are the input values and the values of _ are the output values.

19. An equation whose graph is a line and whose variables are to the first power is called a _____ equation.

20. The equation $Ax + By = C$ is the _____ form of the equation of a line.

21. The _____ of a line is the point $(0, b)$ where the line intersects the y-axis.

22. The _____ of a line is the point $(a, 0)$ where the line intersects the x-axis.

GUIDED PRACTICE *Determine whether the ordered pair satisfies the equation.* SEE EXAMPLE 1. (OBJECTIVE 1)

23. $x - 2y = -4$; $(4, 4)$

24. $y = -7x - 3$; $(4, -30)$

25. $y = \frac{2}{3}x + 5$; $(6, 12)$

26. $y = -\frac{2}{5}x - 1$; $(5, -3)$

Complete each table of values. Check your work with a graphing calculator. (OBJECTIVE 2)

27. $y = x - 3$

x	y	(x, y)
0		
1		
-2		
-4		

28. $y = x - 2$

x	y	(x, y)
0		
-1		
-2		
3		

29. $y = -2x$

x	y	(x, y)
0		
1		
3		
-2		

30. $y = -1.7x + 2$

x	y	(x, y)
-3		
-1		
0		
3		

Graph each equation by constructing a table of values and then plotting the points. SEE EXAMPLE 2. (OBJECTIVE 3)

31. $y = 2x$

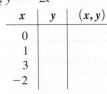

32. $y = -\frac{1}{2}x$

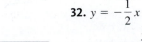

33. $y = 2x - 1$

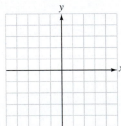

34. $y = 3x + 1$

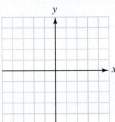

Graph each equation by constructing a table of values and then plotting the points. SEE EXAMPLE 3. (OBJECTIVE 3)

35. $y = 1.2x - 2$

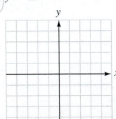
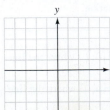

36. $y = -2.4x + 1$

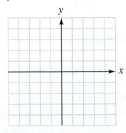
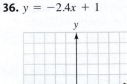

37. $y = 2.5x - 5$

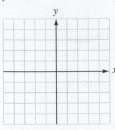

38. $y = x$

Graph each equation by constructing a table of values and then plotting the points. SEE EXAMPLE 4. (OBJECTIVE 3)

39. $y = \frac{x}{2} - 2$

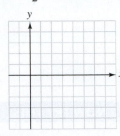

40. $y = \frac{x}{3} - 3$

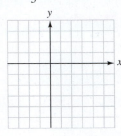

41. $y - 3 = -\frac{1}{2}(2x + 4)$

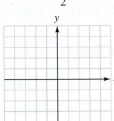

42. $y + 1 = 3(x - 1)$

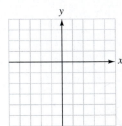
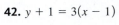

Graph each equation using the intercept method. Write the equation in general form, if necessary. SEE EXAMPLE 5. (OBJECTIVE 4)

43. $x + y = 7$

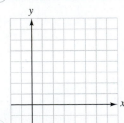

44. $x + y = -2$

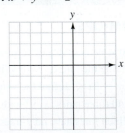

45. $2x + 3y = 12$

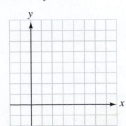

46. $3x - 2y = 6$

Graph each equation. SEE EXAMPLE 6. (OBJECTIVE 5)

47. $y = -5$

48. $x = 4$

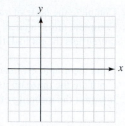

49. $x = 5$

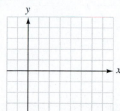

50. $y = 4$

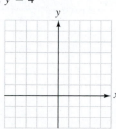

59. $y = -3x$

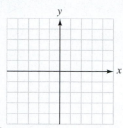

60. $x + y = -2$

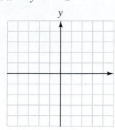

61. $y + 2 = \dfrac{3}{4}(4x + 8)$

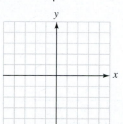

62. $y = 4.5x + 2$

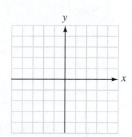

ADDITIONAL PRACTICE *Graph each equation using any method.*

51. $y = -3x - 1$

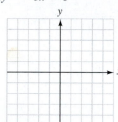

52. $2x = 5$

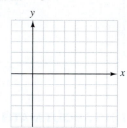

53. $x - y = -2$

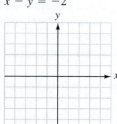

54. $y = 0$

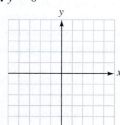

55. $3y = 7$

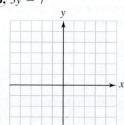

56. $x = 0$

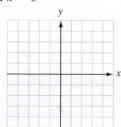

57. $x - y = 7$

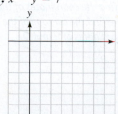

58. $y = -2x + 5$

APPLICATIONS SEE EXAMPLE 7. (OBJECTIVE 6)

63. Educational costs Each semester, a college charges a service fee of $50 plus $25 for each unit taken by a student.

  a. Write a linear equation that gives the total enrollment cost *c* for a student taking *u* units.

  b. Complete the table of values and graph the equation. See the illustration.

  c. What does the *y*-intercept of the line tell you?

  d. Use the graph to find the total cost for a student taking 18 units the first semester and 12 units the second semester.

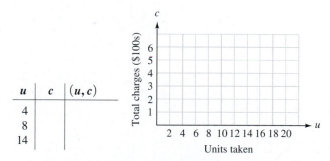

64. Group rates To promote the sale of tickets for a cruise to Alaska, a travel agency reduces the regular ticket price of $3,000 by $5 for each individual traveling in the group.

  a. Write a linear equation that would find the ticket price *T* for the cruise if a group of *p* people travel together.

  b. Complete the table of values and then graph the equation. See the illustration on the following page.

  c. As the size of the group increases, what happens to the ticket price?

  d. Use the graph to determine the cost of an individual ticket if a group of 40 will be traveling together.

p	T	(p, T)
10		
30		
60		

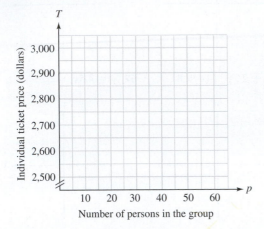

the number of trials n the rat had been given by the equation $t = 25 - 0.25n$.

a. Complete the table of values and then graph the equation.

b. Complete this sentence: From the graph, we see that the more trials the rat had, the . . .

c. From the graph, estimate the time it will take the rat to complete the maze on its 32nd trial.

d. Interpret the meaning of the y-intercept.

n	t	(n, t)
4		
12		
16		

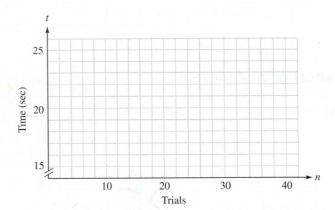

65. Physiology   Physiologists have found that a woman's height h in inches can be approximated using the linear equation $h = 3.9r + 28.9$, where r represents the length of her radius bone in inches.

a. Complete the table of values (round to the nearest tenth), and then graph the equation.

b. Complete this sentence: From the graph, we see that the longer the radius bone, the . . .

c. From the graph, estimate the height of a girl whose radius bone is 10 inches long.

r	h	(r, h)
7		(7,)
8.5		(8.5,)
9		(9,)

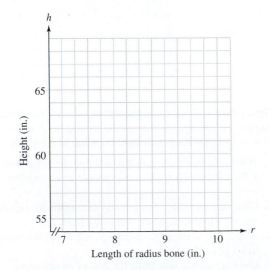

66. Research   A psychology major found that the time t in seconds that it took a white rat to complete a maze was related to

WRITING ABOUT MATH

67. From geometry, we know that two points determine a line. Explain why it is good practice when graphing linear equations to find and plot three points instead of just two.

68. Explain the process used to find the x- and y-intercepts of the graph of a line.

69. What is a table of values? Why is it often called a table of solutions?

70. When graphing an equation in two variables, how many solutions of the equation must be found?

71. Give examples of an equation in one variable and an equation in two variables. How do their solutions differ?

72. What does it mean when we say that an equation in two variables has infinitely many solutions?

SOMETHING TO THINK ABOUT

If points $P(a, b)$ and $Q(c, d)$ are two points on a rectangular coordinate system and point M is midway between them, then point M is called the midpoint of the line segment joining P and Q. (See the illustration on the following page.) To find the coordinates of the midpoint $M(x_M, y_M)$ of the segment PQ, we find the average of the x-coordinates and the average of the y-coordinates of P and Q.

$$x_M = \frac{a + c}{2} \quad \text{and} \quad y_M = \frac{b + d}{2}$$

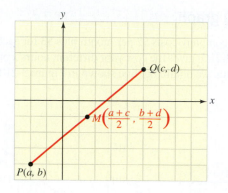

Find the coordinates of the midpoint of the line segment with the given endpoints.

73. $P(5, 3)$ and $Q(7, 9)$

74. $P(3, 8)$ and $Q(9, -2)$

75. $P(2, -7)$ and $Q(-3, 12)$

76. $P(-8, 12)$ and $Q(3, -9)$

77. $A(4, 6)$ and $B(10, 6)$

78. $A(8, -6)$ and the origin

79. $A(x, 3)$ and $B(x - 1, -4)$

80. $A(-2, y + 1)$ and $B(6, y - 1)$

Section 3.3

Slope of a Line

Objectives

1. Find the slope of a line given a graph.
2. Find the slope of a line passing through two specified points.
3. Find the slope of a line given an equation.
4. Identify the slope of a horizontal and vertical line.
5. Determine whether two lines are parallel, perpendicular, or neither parallel nor perpendicular.
6. Interpret slope in an application.

Vocabulary

slope
subscript notation
rise

run
hypotenuse

parallel lines
negative reciprocals

Getting Ready

Simplify each expression, if possible.

1. $\dfrac{6 - 2}{12 - 8}$ **2.** $\dfrac{-12 - 3}{11 - 8}$ **3.** $\dfrac{16 - 16}{6 - 2}$ **4.** $\dfrac{2 - 9}{7 - 7}$

We have seen that two points can be used to graph a line. We can also graph a line if we know the coordinates of only one point and the slant (or steepness) of the line. A measure of this slant is called the **slope** of the line. Slope will be discussed in this section.

1 Find the slope of a line given a graph.

A research service offered by an Internet company costs \$2 per month plus \$3 for each hour of connect time. The table shown in Figure 3-21(a) gives the cost y for different hours x of connect time. If we construct a graph from this data, we obtain the line shown in Figure 3-21(b).

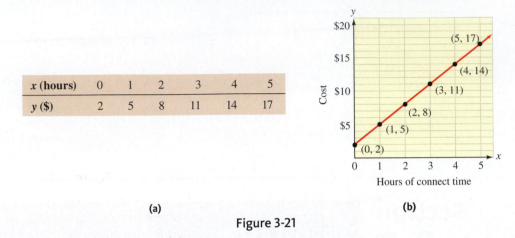

x (hours)	0	1	2	3	4	5
y (\$)	2	5	8	11	14	17

(a)   (b)

Figure 3-21

From the graph, we can see that if x changes from 0 to 1, y changes from 2 to 5. As x changes from 1 to 2, y changes from 5 to 8, and so on. The ratio of the change in y divided by the change in x is the constant 3.

$$\frac{\text{Change in } y}{\text{Change in } x} = \frac{5-2}{1-0} = \frac{8-5}{2-1} = \frac{11-8}{3-2} = \frac{14-11}{4-3} = \frac{17-14}{5-4} = \frac{3}{1} = 3$$

The ratio of the change in y divided by the change in x between any two points on any line is always a constant. This constant *rate of change* is called the slope of the line and usually is denoted by the letter m.

To distinguish between the coordinates of points P and Q in Figure 3-22, we use **subscript notation**. Point P is denoted as $P(x_1, y_1)$ and is read as "point P with coordinates of x sub 1 and y sub 1." Point Q is denoted as $Q(x_2, y_2)$ and is read as "point Q with coordinates of x sub 2 and y sub 2."

As a point on the line in Figure 3-22 moves from P to Q, its y-coordinate changes by the amount $y_2 - y_1$, and its x-coordinate changes by $x_2 - x_1$. The change in y is often called the **rise** of the line between points P and Q, and the change in x is often called the **run**.

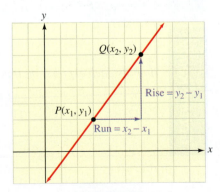

Figure 3-22

EXAMPLE 1   Find the slope of the line shown in Figure 3-23(a).

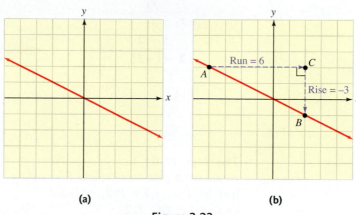

(a)   (b)

Figure 3-23

Solution   In Figure 3-23(b), we choose two points on the line and call them A and B. Then we draw right triangle ABC, having a horizontal leg and a vertical leg. The longest side AB of the right triangle is called the **hypotenuse**. As we move from A to B, we move to the right, a run of 6, and then down, a rise of -3. To find the slope of the line, we write a ratio.

$$m = \frac{\text{rise}}{\text{run}}$$ The slope of a line is the ratio of the rise to the run.

$$m = \frac{-3}{6}$$ From Figure 3-23(b), the rise is -3 and the run is 6.

$$m = -\frac{1}{2}$$ Simplify the fraction.

The slope of the line is $-\frac{1}{2}$.

SELF CHECK 1   Find the slope of the line shown in Figure 3-23(a) using two points different from those used in Example 1.

2   Find the slope of a line passing through two specified points.

Once we know the coordinates of two points on a line, we can substitute those coordinates into the *slope formula*.

SLOPE OF A NONVERTICAL LINE

The slope of the nonvertical line passing through points $P(x_1, y_1)$ and $Q(x_2, y_2)$ is

$$m = \frac{\text{change in } y}{\text{change in } x} = \frac{rise}{run} = \frac{y_2 - y_1}{x_2 - x_1} \quad (x_2 \neq x_1)$$

COMMENT   You can use the coordinates of any two points on a line to compute the slope of the line and obtain the same result.

EXAMPLE 2   Use the two points shown in Figure 3-24 on the next page to find the slope of the line passing through the points $P(-3, 2)$ and $Q(2, -5)$.

Solution   We can let $P(x_1, y_1) = P(-3, 2)$ and $Q(x_2, y_2) = Q(2, -5)$. Then $x_1 = -3$, $y_1 = 2$, $x_2 = 2$, and $y_2 = -5$. To find the slope, we substitute these values into the slope formula and simplify.

$$m = \frac{\text{change in } y}{\text{change in } x}$$

$$= \frac{y_2 - y_1}{x_2 - x_1}$$

$$= \frac{-5 - 2}{2 - (-3)} \quad \text{Substitute } -5 \text{ for } y_2, 2 \text{ for } y_1, 2 \text{ for } x_2, \text{ and } -3 \text{ for } x_1.$$

$$= \frac{-7}{5} \quad \text{Simplify.}$$

$$= -\frac{7}{5}$$

The slope of the line is $-\frac{7}{5}$. We would obtain the same result if we had let $P(x_1, y_1) = P(2, -5)$ and $Q(x_2, y_2) = Q(-3, 2)$.

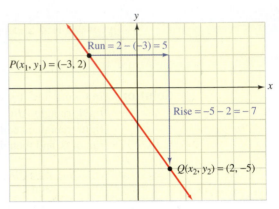

Figure 3-24

SELF CHECK 2   Find the slope of the line passing through points $P(-2, 4)$ and $Q(3, -4)$.

COMMENT   When calculating slope, always subtract the y-values and the x-values *in the same order.*

$$m = \frac{y_2 - y_1}{x_2 - x_1} \qquad m = \frac{y_1 - y_2}{x_1 - x_2} \quad \text{True.}$$

However, the following are not true:

$$m = \frac{y_2 - y_1}{x_1 - x_2} \qquad m = \frac{y_1 - y_2}{x_2 - x_1}$$

Everyday connections
Wind Power

The chart on the right reflects the capacity of California and Texas to produce megawatts of electricity from wind turbines. Although the states have the capacity to produce large amounts of electricity from wind power, the actual output is much less.

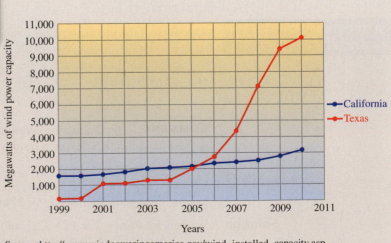

Source: http://www.windpoweringamerica.gov/wind_installed_capacity.asp

Use the data from the graph to answer the questions.

1. About how many megawatts of wind power could California have produced in 2003? Texas?

2. Compute the rate of change in the capacity of megawatts power in Texas between 2000 and 2001.

3. Between what two years was there the greatest rate of change in Texas? Estimate this rate.

4. What was the overall rate of change for Texas between 1999 and 2010?

5. If the actual production of megawatts is roughly 30% of the capacity, how many megawatts of power were produced by California and Texas in 2010?

3 **Find the slope of a line given an equation.**

If we need to find the slope of a line from a given equation, we could graph the line and count squares to determine the rise and the run. Instead, we could find the x- and y-intercepts and then use the slope formula.

In Example 3, we will find the slope of the line determined by the equation $3x - 4y = 12$. To do so, we will find the x- and y-intercepts of the line and use the slope formula.

EXAMPLE 3   Find the slope of the line determined by $3x - 4y = 12$.

Solution   We first find the coordinates of the intercepts of the line.

- If $x = 0$, then $y = -3$, and the point $(0, -3)$ is on the line.
- If $y = 0$, then $x = 4$, and the point $(4, 0)$ is on the line.

We then refer to Figure 3-25 and use the slope formula to find the slope of the line passing through $(0, -3)$ and $(4, 0)$.

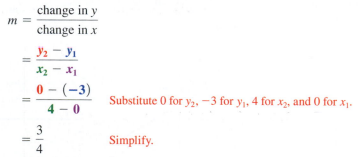

$$m = \frac{\text{change in } y}{\text{change in } x}$$

$$= \frac{y_2 - y_1}{x_2 - x_1}$$

$$= \frac{0 - (-3)}{4 - 0} \qquad \text{Substitute 0 for } y_2, -3 \text{ for } y_1, 4 \text{ for } x_2, \text{ and 0 for } x_1.$$

$$= \frac{3}{4} \qquad \text{Simplify.}$$

Figure 3-25

The slope of the line is $\frac{3}{4}$.

 SELF CHECK 3   Find the slope of the line determined by $4x + 3y = 12$.

4 **Identify the slope of a horizontal and vertical line.**

If $P(x_1, y_1)$ and $Q(x_2, y_2)$ are points on the horizontal line shown in Figure 3-26(a) on the next page, then $y_1 = y_2$, and the numerator of the slope formula

$$\frac{y_2 - y_1}{x_2 - x_1} \qquad \text{On a horizontal line, } x_2 \neq x_1.$$

is 0. Thus, the value of the fraction is 0, and the slope of the horizontal line is 0.

If $P(x_1, y_1)$ and $Q(x_2, y_2)$ are two points on the vertical line shown in Figure 3-26(b), then $x_1 = x_2$, and the denominator of the slope formula

$$\frac{y_2 - y_1}{x_2 - x_1} \qquad \text{On a vertical line, } y_2 \neq y_1.$$

is 0. Since the denominator cannot be 0, a vertical line has an undefined slope.

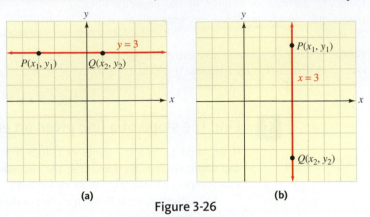

(a)   (b)

Figure 3-26

SLOPES OF HORIZONTAL AND VERTICAL LINES

All horizontal lines (lines with equations of the form $y = b$) have a slope of 0.

Vertical lines (lines with equations of the form $x = a$) have an undefined slope.

COMMENT   It is also true that if a line has a slope of 0, it is a horizontal line with an equation of the form $y = b$. Also, if a line has an undefined slope, it is a vertical line with an equation of the form $x = a$.

If a line rises as we follow it from left to right as in Figure 3-27(a), the line is said to be *increasing* and its slope is positive. If a line drops as we follow it from left to right, as in Figure 3-27(b), it is said to be *decreasing* and its slope is negative. If a line is horizontal, as in Figure 3-27(c), it is said to be *constant* and its slope is 0. If a line is vertical, as in Figure 3-27(d), it has an undefined slope.

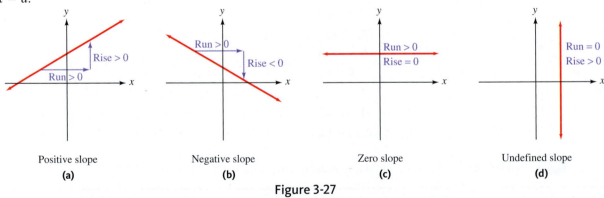

Positive slope   Negative slope   Zero slope   Undefined slope
(a)   (b)   (c)   (d)

Figure 3-27

5   Determine whether two lines are parallel, perpendicular, or neither parallel nor perpendicular.

To see a relationship between parallel lines and their slopes, we refer to the parallel lines l_1 and l_2 shown in Figure 3-28, with slopes of m_1 and m_2, respectively. Because right triangles *ABC* and *DEF* are similar, it follows that

COMMENT   The triangle is an uppercase delta in the Greek alphabet.

$$m_1 = \frac{\Delta y \text{ of } l_1}{\Delta x \text{ of } l_1} \qquad \begin{array}{l}\text{Read } \Delta y \text{ as "the change in } y\text{."}\\\text{Read } \Delta x \text{ as "the change in } x\text{."}\end{array}$$

$$= \frac{\Delta y \text{ of } l_2}{\Delta x \text{ of } l_2}$$

$$= m_2$$

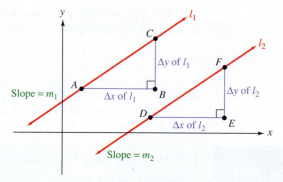

Figure 3-28

This illustrates that if two nonvertical lines are parallel, they have the same slope. It is also true that when two distinct nonvertical lines have the same slope, they are **parallel lines**.

SLOPES OF PARALLEL LINES	Nonvertical parallel lines have the same slope.

Nonvertical parallel lines have the same slope.

Two distinct nonvertical lines having the same slope are parallel.

Since vertical lines are parallel, two distinct lines each with an undefined slope are parallel.

EXAMPLE 4   The lines in Figure 3-29 are parallel. Find the slope of line l_2.

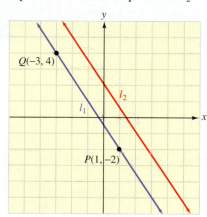

Figure 3-29

Solution   From the information in the figure, we can find the slope of line l_1. Since the lines are parallel, they will have equal slopes. Therefore, the slope of line l_2 will be equal to the slope of line l_1.

$$m = \frac{y_2 - y_1}{x_2 - x_1}$$

$$m = \frac{-2 - 4}{1 - (-3)}$$

$$= \frac{-6}{4}$$

$$= -\frac{3}{2}$$

The slope of line l_1 is $-\frac{3}{2}$. Because the lines are parallel, the slope of line l_2 is also equal to $-\frac{3}{2}$.

SELF CHECK 4   Find the slope of any line parallel to a line with a slope of 6.

Two real numbers a and b are called **negative reciprocals** or opposite reciprocals if $a \cdot b = -1$. For example,

$$-\frac{4}{3} \quad \text{and} \quad \frac{3}{4}$$

are negative reciprocals, because $-\frac{4}{3}\left(\frac{3}{4}\right) = -1$.

The following relates perpendicular lines and their slopes.

SLOPES OF PERPENDICULAR LINES	If two nonvertical lines are perpendicular, their slopes are negative reciprocals.
	If the slopes of two lines are negative reciprocals, the lines are perpendicular.

Because a horizontal line is perpendicular to a vertical line, a line with a slope of 0 is perpendicular to a line with an undefined slope.

EXAMPLE 5   Are the lines shown in Figure 3-30 perpendicular?

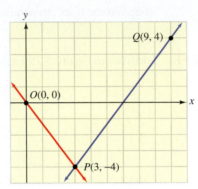

Figure 3-30

Solution   We find the slopes of the lines to see whether they are negative reciprocals.

$$\text{Slope of } OP = \frac{\Delta y}{\Delta x} \qquad\qquad \text{Slope of } PQ = \frac{\Delta y}{\Delta x}$$

$$= \frac{y_2 - y_1}{x_2 - x_1} \qquad\qquad\qquad = \frac{y_2 - y_1}{x_2 - x_1}$$

$$= \frac{-4 - 0}{3 - 0} \qquad\qquad\qquad = \frac{4 - (-4)}{9 - 3}$$

$$= -\frac{4}{3} \qquad\qquad\qquad\qquad = \frac{8}{6}$$

$$\qquad\qquad\qquad\qquad\qquad\qquad = \frac{4}{3}$$

Since their slopes are not negative reciprocals, the lines are not perpendicular.

SELF CHECK 5   In Figure 3-30, is the line PQ perpendicular to a line passing through the points $(3, -4)$ and $(0, -1)$?

6   Interpret slope in an application.

Many applications involve equations of lines and their slopes.

EXAMPLE 6   **COST OF CARPET**   If carpet costs $25 per square yard, the total cost c of n square yards is the price per square yard times the number of square yards purchased.

c	equals	the cost per square yard	times	the number of square yards.
c	$=$	25	$\cdot$	n

Graph the equation $c = 25n$ and interpret the slope of the line.

Solution   We can graph the equation on a coordinate system with a vertical c-axis and a horizontal n-axis. Figure 3-31 shows a table of ordered pairs and the graph.

$c = 25n$

n	c	(n, c)
10	250	$(10, 250)$
20	500	$(20, 500)$
30	750	$(30, 750)$
40	1,000	$(40, 1,000)$
50	1,250	$(50, 1,250)$

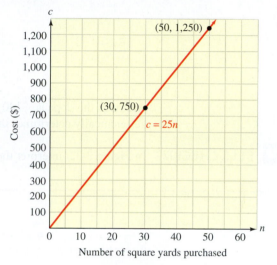

Figure 3-31

COMMENT   Recall that any two points on a line can be used to find the slope. In this example, we chose $(30, 750)$ and $(50, 1{,}250)$.

If we choose the points $(30, 750)$ and $(50, 1{,}250)$ and substitute into the slope formula, we have

$$m = \frac{\text{change in } c}{\text{change in } n}$$

$$= \frac{c_2 - c_1}{n_2 - n_1}$$

$$= \frac{1{,}250 - 750}{50 - 30} \qquad \text{Substitute 1,250 for } c_2, \text{ 750 for } c_1, \text{ 50 for } n_2, \text{ and 30 for } n_1.$$

$$= \frac{500}{20} \qquad \text{Subtract.}$$

$$= 25$$

The slope of 25 is the ratio of the change in the cost to the change in the number of square yards purchased. As a rate of change, the cost of carpet is 25 dollars per square yard.

SELF CHECK 6   If carpet costs $30 per square yard, the total cost c of n square yards is $c = 30n$. Graph and interpret the slope of the line.

EXAMPLE 7   **RATE OF DESCENT**   It takes a skier 25 minutes to complete the course shown in Figure 3-32 on the next page. Find his average rate of descent in feet per minute.

Solution   To find the average rate of descent, we must find the ratio of the change in altitude to the change in time. To find this ratio, we calculate the slope of the line passing through the points $(0, 12{,}000)$ and $(25, 8{,}500)$.

$$\text{Slope} = \frac{12,000 - 8,500}{0 - 25} = \frac{3,500}{-25} = -140$$

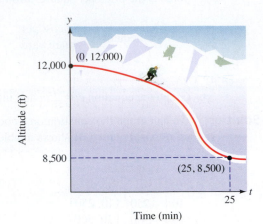

Figure 3-32

Although the slope is -140, the rate of change of descent is 140 ft/min. The term *descent* automatically determines the direction (negative).

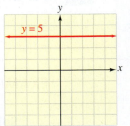

SELF CHECK 7   In Figure 3-32, interpret the ordered pair $(0, 12,000)$.

SELF CHECK ANSWERS

1. $-\frac{1}{2}$ **2.** $-\frac{8}{5}$ **3.** $-\frac{4}{3}$ **4.** 6   **5.** no   **6.** The slope of 30 is the ratio of the change in the cost to the change in the number of square yards purchased. As a rate of change, the cost of carpet is 30 dollars per square yard.

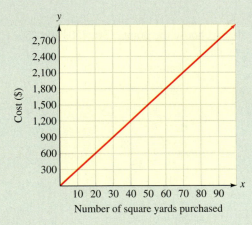

7. The skier begins at an elevation of 12,000 feet.

NOW TRY THIS

1. Interpret the sign of the slope of the three sections of the graph shown at right.

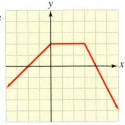

2. Find the slope of any line parallel to the line shown.

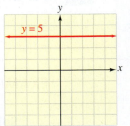

3. Find the slope, if any, of any line perpendicular to the graph shown in Problem 2.

4. The lines in the graph are parallel. Find the value of y.

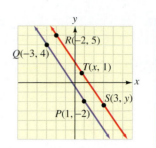

5. Graph the line passing through the points $(2.1, -3)$ and $(4.7, 1.6)$. Find the slope of this line.

3.3   Exercises

WARM-UPS   *Simplify each expression.*

1. $\dfrac{3 - 7}{-6 - 4}$

2. $\dfrac{8 - (-1)}{-5 - (-2)}$

3. $\dfrac{9 - (-8)}{-2 - (-2)}$

4. $\dfrac{-7 + 7}{12 - 8}$

REVIEW   *Simplify each expression.*

5. $4(a - 3) + 2a$

6. $5(y - 8) - 2y$

7. $4z - 6(z + w) + 2w$

8. $7(b - 3) - (b + 2)$

9. $3(a - b) - 2(a + b)$

10. $2m - 4(m - n) + 2n$

VOCABULARY AND CONCEPTS   *Fill in the blanks.*

11. The slope of a line is the change in _ divided by the change in _.

12. The point (x_1, y_1) is read as "the ordered pair x ___ 1, y ___ 1."

13. The vertical change between two points is called the ___. The horizontal change between two points is called the ___. Slope is sometimes defined as ___ over ___.

14. The slope of a _____ line is 0. The slope of a vertical line is _____.

15. The longest side of a right triangle is called the _____.

16. _____ lines are lines that have the same slope.

17. Slopes of nonvertical _____ lines are negative reciprocals.

18. If a line rises as x gets larger, the line has a _____ slope.

19. If a line has a positive slope, the line is said to be _____. If a line has a negative slope, the line is said to be _____.

20. If the product of two numbers is -1, the numbers are said to be _____ _____.

GUIDED PRACTICE

Find the slope of each line. **SEE EXAMPLE 1. (OBJECTIVE 1)**

21.

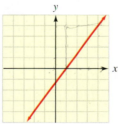

22.

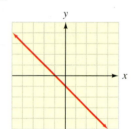

23.

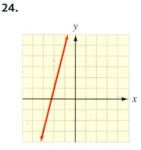

24.

25.

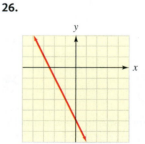

26.

27.

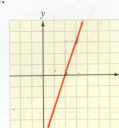

28.

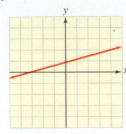

Find the slope of the line that passes through the given points.
SEE EXAMPLE 2. *(OBJECTIVE 2)*

29. $(0, 0), (3, 9)$

30. $(9, 6), (0, 0)$

31. $(2, -5), (4, -3)$

32. $(-2, -8), (3, 2)$

33. $(3, -1), (-6, 2)$

34. $(0, -8), (-5, 0)$

35. $(4, 15), (-6, -11)$

36. $(-6, -2), (-1, -9)$

Find the slope of the line determined by each equation. SEE EXAMPLE
3. *(OBJECTIVE 3)*

37. $3x + 2y = 12$

38. $2x - y = 6$

39. $2x = 5y - 10$

40. $x = y$

41. $y = \dfrac{x - 4}{2}$

42. $x = \dfrac{3 - y}{4}$

43. $y = 5x - 4$

44. $y = -\dfrac{2}{3}x + 6$

Find the slope of each vertical or horizontal line. (OBJECTIVE 4)

45.

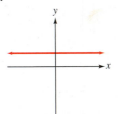

46.

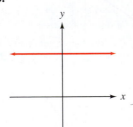

47. passing through $(-7, -5)$ and $(-7, -2)$

48. passing through $(3, -5)$ and $(3, 14)$

49. passing through $(6, -2)$ and $(-4, -2)$

50. passing through $(-3, 5)$ and $(9, 5)$

51. $4x - 2 = 12$

52. $-8y + 7 = 15$

53. $2(y + 1) = 3y$

54. $5x + 2 = 6(x + 1)$

55. $x - y = \dfrac{7 - 5y}{5}$

56. $x + y = \dfrac{2 + 3x}{3}$

*Determine whether the lines with the given slopes are parallel,
perpendicular, or neither parallel nor perpendicular.* SEE EXAMPLES 4–5.
(OBJECTIVE 5)

57. $m_1 = 3, m_2 = -\dfrac{1}{3}$

58. $m_1 = \dfrac{1}{4}, m_2 = 4$

59. $m_1 = 4, m_2 = 0.25$

60. $m_1 = -5, m_2 = \dfrac{1}{-0.2}$

61. $m_1 = \dfrac{2}{3}, m_2 = \dfrac{4}{6}$

62. $m_1 = \dfrac{5}{6}, m_2 = \dfrac{6}{5}$

63. $m_1 = 0, m_2$ is undefined

64. $m_1 = 0, m_2 = 0$

ADDITIONAL PRACTICE *Find the slope of the line that
passes through the given points.*

65. $(2, -1), (-3, 2)$

66. $(2, -8), (3, -8)$

67. $(2, -3), (-3, 2)$

68. $(-4, 1), (-1, 4)$

*Determine whether the slope of the line in each graph is positive,
negative, 0, or undefined.*

69.

70.

71.

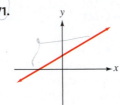

72.

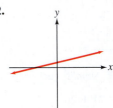

73.

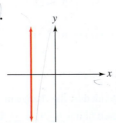

74.

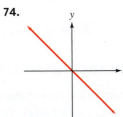

*Find the slopes of lines PQ and PR and determine whether the points
P, Q, and R lie on the same line. (Hint: Two lines with the same slope
and a point in common must be the same line.)*

75. $P(-2, 4), Q(4, 8), R(8, 12)$

76. $P(6, 10), Q(0, 6), R(3, 8)$

77. $P(-4, 10), Q(-6, 0), R(-1, 5)$

78. $P(-10, -13), Q(-8, -10), R(-12, -16)$

79. $P(-2, 4), Q(0, 8), R(2, 12)$

80. $P(8, -4), Q(0, -12), R(8, -20)$

*Determine whether the line PQ is parallel, perpendicular, or neither
to a line with a slope of -2.*

81. $P(-2, -3), Q(-4, 1)$

82. $P(-5, 3), Q(-7, 2)$

83. $P(-2, 1), Q(6, 5)$

84. $P(3, 4), Q(-3, -5)$

85. $P(4, -3), Q(-2, 0)$

86. $P(-2, 3), Q(4, -9)$

87. Find the equation of the x-axis and its slope.

88. Find the equation of the y-axis and its slope, if any.

APPLICATIONS *SEE EXAMPLES 6–7. (OBJECTIVE 6)*

89. **Grade of a road**
If the vertical rise of
the road shown in
the illustration is
24 feet for a hori-
zontal run of 1 mile,
find the slope of the
road. (*Hint*: 1 mile =
5,280 feet.)

90. **Pitch of a roof**   If
the rise of the roof shown
in the illustration is 5 feet
for a run of 12 feet, find
the pitch of the roof.

91. **Slope of a ramp**   If a ramp rises 3 feet over a run of 15 feet, find its slope.

92. **Slope of a ladder**   A ladder leans against a building and reaches a height of 24 feet. If its base is 10 feet from the building, find the slope of the ladder.

93. **Rate of growth**   When a college started an aviation pro-gram, the administration agreed to predict enrollments using a straight-line method. If the enrollment during the first year was 12, and the enrollment during the fifth year was 26, find the rate of growth per year (the slope of the line). See the illustration.

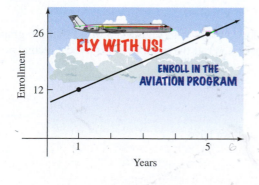

94. **Rate of growth**   A small business predicts sales accord-ing to a straight-line method. If sales were $50,000 in the first year and $110,000 in the third year, find the rate of growth in sales per year (the slope of the line).

95. **Rate of decrease**   The price of computer equipment has been dropping steadily for the past ten years. If a desktop PC cost $6,700 ten years ago, and the same computing power cost $2,200 three years ago, find the rate of decrease per year. (Assume a straight-line model).

96. **Hospital costs**   The table shows the changing mean daily cost for a hospital room. Find the rate of change per year of the portion of the room cost that was absorbed by the hospi-tal between 2000 and 2010.

	Cost passed on to the patient	Total cost to the hospital
2000	$130	$245
2005	214	459
2010	295	670

WRITING ABOUT MATH

97. Explain why a vertical line has no defined slope.

98. Explain how to determine from their slopes whether two lines are parallel, perpendicular, or neither.

SOMETHING TO THINK ABOUT

99. The points $(3, a)$, $(5, 7)$, and $(7, 10)$ lie on a line. Find the value of a.

100. The line passing through points $A(1, 3)$ and $B(-2, 7)$ is perpendicular to the line passing through points $C(4, b)$ and $D(8, -1)$. Find the value of b.

Section 3.4

Point-Slope Form

Objectives

1 Write the point-slope form of the equation of a line with a given slope that passes through a specified point.

2 Determine the slope of a line from a graph and use it with a specified point to write an equation for the line.

3 Graph a line given the point-slope form of an equation.

4 Find an equation of the line representing real-world data.

Vocabulary

point-slope form

Getting Ready

Solve each equation for y.

1. $2x + y = 12$

2. $3x - y = 7$

3. $4x + 2y = 9$

4. $5x - 4y = 12$

Earlier in the chapter, we were given a linear equation and saw how to construct its graph. In this section, we will start with a graph and see how to write its equation.

1 **Write the point-slope form of the equation of a line with a given slope that passes through a specified point.**

Consider the line shown in Figure 3-33(a). From the graph, we can determine that the slope is 2 and passes through the point $(3, 1)$. If the point (x, y) is to be a second point on the line, it must satisfy the equation

$$\frac{y - 1}{x - 3} = 2 \qquad \text{By formula, the slope of the line is } \tfrac{y-1}{x-3}, \text{ which is given to be 2.}$$

After multiplying both sides of this equation by $(x - 3)$, we have

(1) $y - 1 = 2(x - 3)$

We now consider the line with slope m shown in Figure 3-33(b). From the graph, we can see that it passes through the point (x_1, y_1). If the point (x, y) is to be a second point on the line, it must satisfy the equation

$$\frac{y - y_1}{x - x_1} = m \qquad \text{This is the slope formula.}$$

After multiplying both sides by $(x - x_1)$, we have

(2) $y - y_1 = m(x - x_1)$

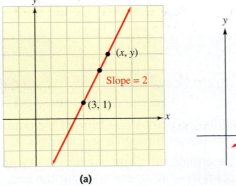

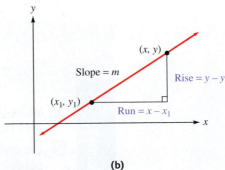

(a)   **(b)**

Figure 3-33

Because Equations 1 and 2 display the coordinates of a point on a line and the slope of the line, they are written in a form called **point-slope form**.

POINT-SLOPE FORM OF THE EQUATION OF A LINE

The equation of the line that has a slope of m and passes through the point (x_1, y_1) is
$$y - y_1 = m(x - x_1)$$

EXAMPLE 1   Find the point-slope equation of the line that has a slope of $-\frac{3}{5}$ and passes through the point $(3, -6)$.

Solution   We substitute $-\frac{3}{5}$ for m, 3 for x_1, and -6 for y_1 into the point-slope equation.

$$y - \boldsymbol{y_1} = m(x - \boldsymbol{x_1}) \qquad \text{This is the formula for point-slope form.}$$

$$y - (\boldsymbol{-6}) = -\frac{\boldsymbol{3}}{\boldsymbol{5}}(x - \boldsymbol{3}) \qquad \text{Substitute } -6 \text{ for } y_1, -\tfrac{3}{5} \text{ for } m, \text{ and 3 for } x_1.$$

$$y + 6 = -\frac{3}{5}(x - 3) \qquad -(-6) = 6$$

 SELF CHECK 1   Find the point-slope equation of the line that has a slope of 3 and passes through the point $(-2, 8)$.

EXAMPLE 2   Find the point-slope equation of the line that has a slope of $\frac{2}{3}$ and passes through the point $(-4, 2)$. Then solve it for y.

Solution   We substitute $\frac{2}{3}$ for m, -4 for x_1, and 2 for y_1 into the point-slope form.

$$y - \boldsymbol{y_1} = \boldsymbol{m}(x - \boldsymbol{x_1}) \qquad \text{This is the point-slope equation.}$$

$$y - \boldsymbol{2} = \frac{\boldsymbol{2}}{\boldsymbol{3}}[x - (\boldsymbol{-4})] \qquad \text{Substitute } -4 \text{ for } x_1, 2 \text{ for } y_1, \text{ and } \tfrac{2}{3} \text{ for } m.$$

$$y - 2 = \frac{2}{3}(x + 4) \qquad -(-4) = 4$$

To solve the equation for y, we proceed as follows.

$$y - 2 = \frac{2}{3}x + \frac{8}{3} \qquad \text{Use the distributive property to remove parentheses.}$$

$$y = \frac{2}{3}x + \frac{14}{3} \qquad \text{Add 2 to both sides and simplify.}$$

The equation is $y = \frac{2}{3}x + \frac{14}{3}$.

 SELF CHECK 2   Find the point-slope equation of the line that has a slope of $\frac{3}{4}$ and passes through the point $(-6, -2)$. Then solve it for y.

2   Determine the slope of a line from a graph and use it with a specified point to write an equation for the line.

In the next example, we are given a graph and asked to write its equation.

EXAMPLE 3   Find the point-slope equation of the line shown in Figure 3-34 on the next page. Then solve it for y.

Solution   To find the equation of the line that passes through two known points, we must find the slope by substituting -6 for y_2, 4 for y_1, 8 for x_2, and -5 for x_1.

$$m = \frac{y_2 - y_1}{x_2 - x_1}$$ This is the slope formula.

$$= \frac{-6 - 4}{8 - (-5)}$$ Substitute -6 for y_2, 4 for y_1, 8 for x_2, and -5 for x_1.

$$= -\frac{10}{13}$$

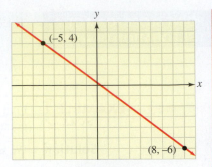

Figure 3-34

Since the line passes through both points, we can choose either one and substitute its coordinates into the point-slope equation. If we choose $(-5, 4)$, we substitute -5 for x_1, 4 for y_1, and $-\frac{10}{13}$ for m.

$$y - y_1 = m(x - x_1)$$ This is point-slope form.

$$y - 4 = -\frac{10}{13}[x - (-5)]$$ Substitute.

$$y - 4 = -\frac{10}{13}(x + 5)$$ $-(-5) = 5$

To solve for y, we proceed as follows:

$$y - 4 = -\frac{10}{13}x - \frac{50}{13}$$ Use the distributive property to remove parentheses.

$$y = -\frac{10}{13}x + \frac{2}{13}$$ Add 4 to both sides and simplify.

The equation is $y = -\frac{10}{13}x + \frac{2}{13}$.

 **SELF CHECK 3**   Find the point-slope equation of the line that passes through $(-2, 5)$ and $(5, -2)$. Then solve it for y.

---

3   ## Graph a line given the point-slope form of an equation.

Graphing a linear equation written in point-slope form does not require solving for one of the variables. For example, to graph

$$y - 2 = \frac{3}{4}(x - 1)$$

we compare the equation to point-slope form

$$y - y_1 = m(x - x_1)$$

and note that the slope of the line is $m = \frac{3}{4}$ and that it passes through the point $(x_1, y_1) = (1, 2)$. Because the slope is $\frac{3}{4}$, we can start at the point $(1, 2)$ and locate another point on the line by counting 4 units to the right and 3 units up as shown in Figure 3-35. The change in x from point P to point Q is 4, and the corresponding change in y is 3. The line joining these points is the graph of the equation.

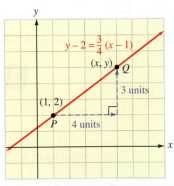

Figure 3-35

4   Find an equation of the line representing real-world data.

EXAMPLE 4   **WATER BILLING**   In a city, the monthly cost for water is related to the number of gallons used by a linear equation. If a customer is charged $12 for using 1,000 gallons and $16 for using 1,800 gallons, find the cost of 2,000 gallons.

Solution   When a customer uses 1,000 gallons, the charge is $12. When he uses 1,800 gallons, the charge is $16. Since y is related to x by a linear equation, the points $(1,000, 12)$ and $(1,800, 16)$ will lie on a line as shown in Figure 3-36. To write the equation of the line passing through these points, we first find the slope of the line.

$$m = \frac{y_2 - y_1}{x_2 - x_1}$$ This is the slope formula.

$$= \frac{16 - 12}{1,800 - 1,000}$$ Substitute.

$$= \frac{4}{800}$$

$$= \frac{1}{200}$$

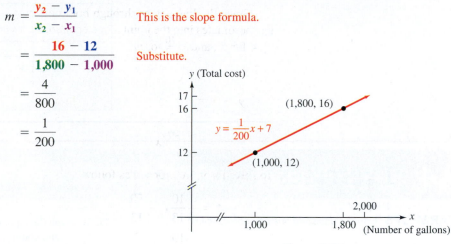

Figure 3-36

We then substitute $\frac{1}{200}$ for m and the coordinates of one of the known points, say $(1,000, 12)$ into the point-slope equation of a line and solve for y.

$$y - y_1 = m(x - x_1)$$ This is point-slope form.

$$y - 12 = \frac{1}{200}(x - 1,000)$$ Substitute.

$$y - 12 = \frac{1}{200}x - \frac{1,000}{200}$$ Use the distributive property to remove parentheses.

$$y - 12 = \frac{1}{200}x - 5$$ $\frac{1,000}{200} = 5$

(3) $$y = \frac{1}{200}x + 7$$ Add 12 to both sides and simplify.

To find the charge for 2,000 gallons, we substitute 2,000 for x in Equation 3 and find the value of y.

$$y = \frac{1}{200}(2,000) + 7$$ Substitute 2,000 for x.

$$y = 10 + 7$$ Multiply.

$$y = 17$$ Simplify.

The charge for 2,000 gallons of water is $17.

SELF CHECK 4   Find the cost for 5,000 gallons of water.

SELF CHECK ANSWERS   1. $y - 8 = 3(x + 2)$ 2. $y = \frac{3}{4}x + \frac{5}{2}$ 3. $y = -x + 3$ 4. $32

NOW TRY THIS

1. Find the point-slope form of the equation of the line passing through the point $(-3, 2)$ that is parallel to the line with slope of 2. Solve this equation for y.

2. Find the point-slope form of the equation of the line perpendicular to a line with a slope of $\frac{1}{4}$ passing through the point $(-1, -4)$.

3. Write the point-slope form of the equation of the line shown.

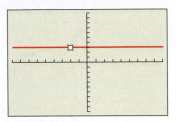

3.4   Exercises

WARM-UPS *Find the slope and one point on each line.*

1. $y - 5 = 2(x - 3)$
2. $y - (-2) = -4[x - (-1)]$
3. $y - (-5) = \frac{7}{8}[x - (-4)]$
4. $y - (-8) = -\frac{4}{5}(x - 3)$

REVIEW *Solve each equation.*

5. $3(x + 2) + x = 5x$
6. $12b + 6(3 - b) = b + 3$
7. $\frac{5(2 - x)}{3} - 1 = x + 5$
8. $r - 1 = \frac{r + 2}{2} + 6$

9. **Junk mail** According to the U.S. Postal Service, people open 53% of the junk mail they receive. If a carpet cleaning business sends out 15,000 promotional ads, how many are likely to be read?

10. **Mixing coffee** To make a mixture of 80 pounds of coffee worth $272, a grocer mixes coffee worth $3.25 a pound with coffee worth $3.85 a pound. How many pounds of cheaper coffee should the grocer use?

VOCABULARY AND CONCEPTS *Fill in the blanks.*

11. Write the point-slope form of the equation of a line. _____

12. The graph of the equation $y - 3 = 7[x - (-2)]$ has a slope of _ and passes through _____.

13. To graph the equation $y - 2 = \frac{3}{2}(x - 1)$, we start at the point _____ and count _ units to the right and _ units up to locate a second point on the line. The graph is the line joining the two points.

14. To graph the equation $y - 4 = -\frac{2}{3}(x - 2)$, we start at the point _____ and count _ units to the right and _ units down to locate a second point on the line. The graph is the line joining the two points.

GUIDED PRACTICE *Write the point-slope equation of the line with the given slope that passes through the given point.* SEE EXAMPLE 1. (OBJECTIVE 1)

15. $m = 3, (0, 0)$
16. $m = -5, (2, 1)$
17. $m = -7, (-1, -2)$
18. $m = -4, (0, 0)$
19. $m = 2, (-5, 3)$
20. $m = -\frac{5}{6}, (12, 7)$
21. $m = -\frac{6}{7}, (6, 5)$
22. $m = -3, (0, 7)$

Write the point-slope equation of the line with the given slope that passes through the given point. Solve each equation for y. SEE EXAMPLE 2. (OBJECTIVE 1)

23. $m = -5, (2, -3)$ 24. $m = 7, (1, 4)$

25. $m = 5, (0, 7)$ **26.** $m = -8, (0, -2)$

27. $m = -3, (2, 0)$ **28.** $m = 4, (-5, 0)$

29. $m = \dfrac{1}{3}, (6, -2)$ **30.** $m = -\dfrac{3}{5}, (5, 7)$

Use point-slope form to write an equation of the given line. Solve each equation for y. SEE EXAMPLE 3. *(OBJECTIVE 2)*

31.

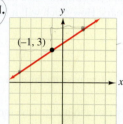

32.

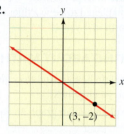

33.

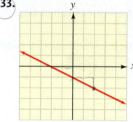

34.

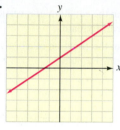

Graph each equation. On the graph, label the ordered pair and the slope identified in the given point-slope equation. (OBJECTIVE 3)

35. $y - 3 = 2(x - 1)$ **36.** $y - 1 = -2(x - 3)$

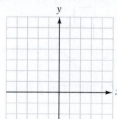

37. $y - 4 = -\dfrac{2}{3}(x - 2)$ **38.** $y - (-2) = \dfrac{3}{2}(x - 1)$

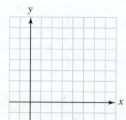

ADDITIONAL PRACTICE *Write the point-slope equation of the line with the given slope that passes through the given point.*

39. $m = 0.5, (-1, -8)$ **40.** $m = \dfrac{8}{7}, (5, -3)$

41. $m = -4, (-3, 2)$ **42.** $m = 1.5, (-3, -9)$

Use point-slope form to write an equation of the given line. Solve each equation for y.

43. $(0, 0), (4, 4)$ **44.** $(-2, -5), (4, -2)$

45. $(3, 4), (0, -3)$ **46.** $(4, 0), (6, -8)$

Graph each equation. On the graph, label the ordered pair and the slope identified in the given point-slope equation.

47. $y - 1 = -\dfrac{1}{2}(x + 2)$ **48.** $y - (-2) = \dfrac{3}{2}(x + 3)$

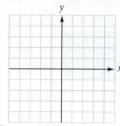

49. $y + 2 = 3(x - 2)$ **50.** $y + 3 = -2(x - 1)$

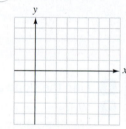

Write the point-slope equation of the line with the given properties. Solve each equation for y.

51. $m = -\dfrac{2}{3}, (3, -1)$ **52.** $m = -\dfrac{5}{3}, (6, -7)$

53. $m = 0.5, (10, 8)$ **54.** $m = -0.25, (8, -7)$

55. $(1, 2), (-3, -4)$ **56.** $(3, -4), (5, -1)$

APPLICATIONS *Assume that x and y are related by a linear equation.* SEE EXAMPLE 4. *(OBJECTIVE 4)*

57. Electric bills One month, a customer received an electric bill for $200 when she used 1,450 kilowatt hours (kWh) of electricity. The next month, she received a bill for $250 when she used 1,850 kWh. Compute her bill for the month when she used 1,500 kWh.

58. Telephone costs One month, a customer received a phone bill for $30.37 when he made 67 calls. The next month, he received a bill for $31.69 when he made 79 calls. Compute his bill for a month when he made 47 calls.

59. Value of antiques   An antique table was purchased for $370 and is expected to be worth $450 in 2 years. What will it be worth in 13 years?

60. Value of antiques   An antique clock is expected to be worth $350 after 2 years and $530 after 5 years. What will the clock be worth after 7 years?

61. Computer repairs   A repair service charges a fixed amount to repair a computer, plus an hourly rate. Use the information in the table to find the hourly rate.

62. Predicting fires   A local fire department recognizes that city growth and the number of reported fires are related by a linear equation. City records show that 300 fires were reported in a year when the local population was 57,000 people, and 325 fires were reported in a year when the population was 59,000 people. How many fires can be expected when the population reaches 100,000 people?

63. Real estate   Three years after a cottage was purchased it was appraised at $147,700. The property is now 10 years old and is worth $172,200. Find its original purchase price.

64. Rate of depreciation   A truck that cost $27,600 when new is expected to be worthless in 12 years. How much will it be worth in 9 years?

WRITING ABOUT MATH

65. Explain how to find the equation of a line passing through two given points.

66. Explain how to graph an equation written in point-slope form.

SOMETHING TO THINK ABOUT

67. Can the equation of a vertical line be written in point-slope form? Explain.

68. Can the equation of a horizontal line be written in point-slope form? Explain.

Section 3.5

Slope-Intercept Form

Objectives

1 Find the slope and *y*-intercept given a linear equation.

2 Write an equation of the line that has a given slope and passes through a specified point.

3 Use the slope-intercept form to write an equation of the line that passes through two given points.

4 Graph a linear equation using the slope and *y*-intercept.

5 Determine whether two linear equations define lines that are parallel, perpendicular, or neither parallel nor perpendicular.

6 Write an equation of the line passing through a specified point and parallel or perpendicular to a given line.

7 Write an equation of a line representing real-world data.

Vocabulary

slope-intercept form   linear depreciation   annual depreciation rate

Solve each equation for y.

1. $4x + 8y = 12$

2. $6x - 5y = 7$

In this section, we will discuss another form of an equation of a line called *slope-intercept form*.

1 Find the slope and *y*-intercept given a linear equation.

Since the *y*-intercept of the line shown in Figure 3-37 is the point $(0, b)$, we can write its equation by substituting 0 for x_1 and b for y_1 in the point-slope equation and simplifying.

$$y - y_1 = m(x - x_1)$$
$$y - b = m(x - 0)$$
$$y - b = mx$$
$$(1) \qquad y = mx + b$$

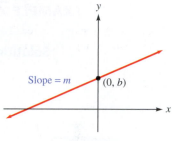

Figure 3-37

Because Equation 1 displays the slope m and the y-coordinate b of the y-intercept, it is called the **slope-intercept form** of the equation of a line.

SLOPE-INTERCEPT FORM OF THE EQUATION OF A LINE

The equation of the line with slope m and y-intercept $(0, b)$ is

$$y = mx + b$$

EXAMPLE 1   Find the slope and y-intercept of the line with the given equation.

a. $y = \frac{2}{3}x + 5$ **b.** $3x - 4y = 9$

Solution   **a.** Since the equation is solved for y, we can identify the slope and the y-intercept from the equation $y = \frac{2}{3}x + 5$.

$$y = \frac{2}{3}x + 5$$

The slope is $\frac{2}{3}$, and the y-intercept is $(0, 5)$.

b. To write the equation $3x - 4y = 9$ in slope-intercept form, we solve it for y.

$$3x - 4y = 9$$

$$-4y = -3x + 9 \qquad \text{Subtract } 3x \text{ from both sides.}$$

$$y = \frac{3}{4}x - \frac{9}{4} \qquad \text{Divide both sides by } -4.$$

We can read the slope and the y-intercept from the equation because it is now in $y = mx + b$ form.

$$y = \frac{3}{4}x + \left(-\frac{9}{4}\right)$$

The slope is $\frac{3}{4}$. The y-intercept is $\left(0, -\frac{9}{4}\right)$.

🌿 SELF CHECK 1   Find the slope and y-intercept of $2x + 5y = 12$.

---

2   **Write an equation of the line that has a given slope and passes through a specified point.**

If we know the slope of a line and the coordinates of one point on the line, we can write the equation of the line.

EXAMPLE 2   Write the slope-intercept equation of the line that has a slope of 5 and passes through the point $(-2, 9)$.

Solution   We are given that $m = 5$ and that the ordered pair $(-2, 9)$ satisfies the equation. To find b, we can substitute -2 for x, 9 for y, and 5 for m in the equation $y = mx + b$.

$$y = mx + b$$
$$9 = 5(-2) + b \qquad \text{Substitute 9 for } y, \text{ 5 for } m, \text{ and } -2 \text{ for } x.$$
$$9 = -10 + b \qquad \text{Multiply.}$$
$$19 = b \qquad \text{Add 10 to both sides.}$$

Since $m = 5$ and $b = 19$, the equation in slope-intercept form is $y = 5x + 19$.

🌿 SELF CHECK 2   Write an equation of the line that has a slope of -2 and passes through the point $(2, 3)$.

COMMENT   Note that Example 2 could be written using the point-slope form, since the given information contained a point and the slope.

3   **Use the slope-intercept form to write an equation of the line that passes through two given points.**

EXAMPLE 3   Use the slope-intercept form to write an equation of the line that passes through the points $(-3, 4)$ and $(5, -8)$.

Solution   We first find the slope of the line.

$$m = \frac{y_2 - y_1}{x_2 - x_1} \qquad \text{This is the formula for slope.}$$
$$= \frac{-8 - 4}{5 - (-3)} \qquad \text{Substitute } -8 \text{ for } y_2, \text{ 4 for } y_1, \text{ 5 for } x_2, \text{ and } -3 \text{ for } x_1.$$
$$= \frac{-12}{8} \qquad -8 - 4 = -12, 5 - (-3) = 8$$
$$= -\frac{3}{2} \qquad \text{Simplify.}$$

Then we write an equation of the line that has a slope of $-\frac{3}{2}$ and passes through either of the given points, say $(-3, 4)$. To do so, we substitute $-\frac{3}{2}$ for m, -3 for x, 4 for y, and solve the equation $y = mx + b$ for b.

$$y = mx + b$$
$$4 = -\frac{3}{2}(-3) + b \qquad \text{Substitute.}$$
$$4 = \frac{9}{2} + b \qquad -\frac{3}{2}(-3) = \frac{9}{2}$$
$$-\frac{1}{2} = b \qquad \text{Subtract } \frac{9}{2} \text{ from both sides and simplify.}$$

Since $m = -\frac{3}{2}$ and $b = -\frac{1}{2}$, an equation is $y = -\frac{3}{2}x + \left(-\frac{1}{2}\right)$, or more simply, $y = -\frac{3}{2}x - \frac{1}{2}$.

 **SELF CHECK 3** Use the slope-intercept form to write an equation of the line that passes through the points $(2, -1)$ and $(-3, 7)$.

4 **Graph a linear equation using the slope and *y*-intercept.**

We can graph linear equations when they are written in the slope-intercept form. For example, to graph $y = \frac{4}{3}x - 2$, we note that $b = -2$ and that the y-intercept is $(0, b) = (0, -2)$. (See Figure 3-38.)

Since the slope of the line is $\frac{4}{3}$, we can locate another point on the line by starting at the point $(0, -2)$ and counting 3 units to the right and 4 units up. The change in x is 3, and the corresponding change in y is 4. The line joining the two points is the graph of the equation.

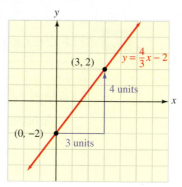

Figure 3-38

EXAMPLE 4 Find the slope and the y-intercept of the line with the equation $2(x - 3) = -3(y + 5)$. Then graph the line.

Solution We write the equation in the form $y = mx + b$ to find the slope m and the y-intercept $(0, b)$.

$$2(x - 3) = -3(y + 5)$$
$$2x - 6 = -3y - 15 \qquad \text{Use the distributive property to remove parentheses.}$$
$$2x + 3y - 6 = -15 \qquad \text{Add } 3y \text{ to both sides.}$$
$$3y - 6 = -2x - 15 \qquad \text{Subtract } 2x \text{ from both sides.}$$
$$3y = -2x - 9 \qquad \text{Add 6 to both sides.}$$
$$y = -\frac{2}{3}x - 3 \qquad \text{Divide both sides by 3.}$$

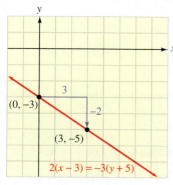

Figure 3-39

The slope is $-\frac{2}{3}$ and the y-intercept is $(0, -3)$. To draw the graph, we plot the y-intercept $(0, -3)$ and then locate a second point on the line by moving 3 units to the right and 2 units down. We draw a line through the two points to obtain the graph shown in Figure 3-39.

 **SELF CHECK 4** Find the slope and the y-intercept of the line with the equation $-3(x + 2) = 2(y - 3)$.

5 Determine whether two linear equations define lines that are parallel, perpendicular, or neither parallel nor perpendicular.

Recall from an earlier section that different lines having the same slope are parallel. Also recall that if the slopes of two lines are negative reciprocals, the lines are perpendicular. We will use these facts in the next examples.

EXAMPLE 5 Show that the lines represented by $4x + 8y = 16$ and $x = 6 - 2y$ are parallel.

Solution We solve each equation for y to see whether their slopes are equal and then determine if the lines are different.

$$4x + 8y = 16 \qquad\qquad x = 6 - 2y$$
$$8y = -4x + 16 \qquad\qquad 2y = -x + 6$$
$$y = -\frac{1}{2}x + 2 \qquad\qquad y = -\frac{1}{2}x + 3$$

Since the slope of each line is $-\frac{1}{2}$, the lines are possibly parallel, but we need to know whether the lines are distinct. Since the values of b in these equations are different, the lines are distinct and, therefore, parallel.

 SELF CHECK 5 Are the lines represented by $3x + y = 2$ and $2y = 6x - 3$ parallel?

EXAMPLE 6 Show that the lines represented by $4x + 8y = 16$ and $4x - 2y = 22$ are perpendicular.

Solution We solve each equation for y to see whether the slopes of their graphs are negative reciprocals.

$$4x + 8y = 16 \qquad\qquad 4x - 2y = 22$$
$$8y = -4x + 16 \qquad\qquad -2y = -4x + 22$$
$$y = -\frac{1}{2}x + 2 \qquad\qquad y = 2x - 11$$

Since the slopes of $-\frac{1}{2}$ and 2 are negative reciprocals $\left(-\frac{1}{2} \cdot 2 = -1\right)$, the lines are perpendicular.

 SELF CHECK 6 Are the lines represented by $y = 2x + 3$ and $2x + y = 7$ perpendicular?

6 Write an equation of the line passing through a specified point and parallel or perpendicular to a given line.

We will now use the slope properties of parallel and perpendicular lines to write more equations of lines.

EXAMPLE 7 Write an equation of the line passing through $(-3, 2)$ and parallel to the line $y = 8x - 5$.

Solution Since $y = 8x - 5$ is written in slope-intercept form, the slope of its graph is the coefficient of x, which is 8. The desired equation is to have a graph that is parallel to the line, thus its slope must be 8 as well.

To find the equation of the line, we substitute -3 for x, 2 for y, and 8 for m in the slope-intercept equation and solve for b.

$$y = mx + b$$
$$2 = 8(-3) + b \quad \text{Substitute.}$$
$$2 = -24 + b \quad \text{Multiply.}$$
$$26 = b \quad \text{Add 24 to both sides.}$$

Since the slope of the desired line is 8 and the y-intercept is $(0, 26)$, the equation is $y = 8x + 26$.

 SELF CHECK 7   Write an equation of the line passing through $(0, 0)$ and parallel to the line $y = 8x - 3$.

EXAMPLE 8   Write an equation of the line passing through $(-2, 5)$ and perpendicular to the line $y = 8x - 3$.

Solution   Since the slope of the given line is 8, the slope of the desired line must be $-\frac{1}{8}$, which is the negative reciprocal of 8.

To find the equation of the desired line, we substitute -2 for x, 5 for y, and $-\frac{1}{8}$ for m in the slope-intercept form and solve for b.

$$y = mx + b$$
$$5 = -\frac{1}{8}(-2) + b \quad \text{Substitute.}$$
$$5 = \frac{1}{4} + b \quad \text{Multiply.}$$
$$\frac{19}{4} = b \quad \text{Subtract } \tfrac{1}{4} \text{ from both sides.}$$

Since the slope of the desired line is $-\frac{1}{8}$ and the y-intercept is $\left(0, \frac{19}{4}\right)$, the equation is $y = -\frac{1}{8}x + \frac{19}{4}$.

 SELF CHECK 8   Write the equation of the line passing through $(0, 0)$ and perpendicular to the line $y = 8x - 3$.

7   Write an equation of a line representing real-world data.

As machinery wears out, it becomes worth less. Accountants often estimate the decreasing value of aging equipment with **linear depreciation**, a method based on linear equations.

EXAMPLE 9   **LINEAR DEPRECIATION**   A company buys a $12,500 computer with an estimated life of 6 years. The computer can then be sold as scrap for an estimated *salvage value* of $500. If y represents the value of the computer after x years of use and y and x are related by the equation of a line,

a. find an equation of the line and graph it.

b. find the value of the computer after 2 years.

c. find the economic meaning of the y-intercept of the line.

d. find the economic meaning of the slope of the line.

Solution   **a.** To find an equation of the line, we calculate its slope and then use the slope-intercept form to find its equation.

When the computer is new, its age x is 0 and its value y is the purchase price of $12,500. We can represent this information as the point with coordinates $(0, 12{,}500)$. When it is six years old, $x = 6$ and $y = 500$, its *salvage value*. We can represent this information as the point with coordinates $(6, 500)$. Since the line passes through these two points as shown in Figure 3-40, the slope of the line is

$$m = \frac{y_2 - y_1}{x_2 - x_1}$$

$$= \frac{500 - 12,500}{6 - 0}$$

$$= \frac{-12,000}{6}$$

$$= -2,000$$

The slope is $-2,000$.

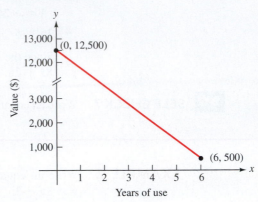

Figure 3-40

To find an equation of the line, we substitute $-2,000$ for m, 0 for x, and $12,500$ for y in the slope-intercept form and solve for b.

$$y = mx + b$$

$$12,500 = -2,000(0) + b$$

$$12,500 = b$$

The current value of the computer is related to its age by the equation $y = -2,000x + 12,500$.

b. To find the value of the computer after 2 years, we substitute 2 for x in the equation $y = -2,000x + 12,500$.

$$y = -2,000x + 12,500$$

$$y = -2,000(2) + 12,500$$

$$= -4,000 + 12,500$$

$$= 8,500$$

In 2 years, the computer will be worth $8,500.

c. Since the y-intercept of the graph is $(0, 12,500)$, the y-coordinate of the y-intercept is the computer's original purchase price.

d. Each year, the value decreases by $2,000, shown by the slope of the line, $-2,000$. The slope of the depreciation line is called the **annual depreciation rate**.

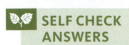 **SELF CHECK 9**   Find the equation of the line if the computer was sold as scrap for a value of $350. Find the economic meaning of the slope of the line.

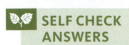 **SELF CHECK ANSWERS**   **1.** $-\frac{2}{5}, \left(0, \frac{12}{5}\right)$ **2.** $y = -2x + 7$ **3.** $y = -\frac{8}{5}x + \frac{11}{5}$ **4.** $-\frac{3}{2}, (0,0)$ **5.** no   **6.** no   **7.** $y = 8x$
8. $y = -\frac{1}{8}x$ **9.** $y = -2,025x + 12,500$; each year, the value decreases by $2,025, shown by the slope of the line, $-2,025$

NOW TRY THIS

1. Find the slope and y-intercept of the line $y = x$.
2. Write an equation of the line passing through $(2, -6)$ and perpendicular to $2x + 3y = 12$.

3. In one state, approximately 82,000 fires were reported in a year when the population was 20,000,000 and 94,000 were reported in a year when the population was 23,000,000. Interpret the slope in the context of this problem. Predict the number of fires that will occur when the population of the state is 27,000,000.

3.5 Exercises

WARM-UPS *Solve each equation for y.*

1. $2x + 4y = 12$ 2. $4x - 2y = 5$

Solve each equation for b.

3. $2 = \dfrac{3}{5}(10) + b$ 4. $-4 = \dfrac{3}{2}(-5) + b$

REVIEW *Solve each equation.*

5. $2x + 3 = 7$ 6. $3x - 2 = 7$
7. $3(y - 2) = y + 1$ 8. $-4z - 6 = 2(z + 3)$

VOCABULARY AND CONCEPTS *Fill in the blanks.*

9. The formula for the slope-intercept form of the equation of a line is _____. The slope of the graph of $y = -3x + 7$ is ___ and the y-intercept is ____.
10. In the straight-line method, the declining value of aging equipment is called _____. The slope of the depreciation line is called the annual depreciation ___.
11. The numbers 3 and $-\dfrac{1}{3}$ are called negative _____.
12. If two distinct lines have the same slope, they are _____.

GUIDED PRACTICE *Find the slope and the y-intercept of the line with the given equation. SEE EXAMPLE 1. (OBJECTIVE 1)*

13. $y = 7x - 5$ 14. $y = -2x + 5$

15. $y = -\dfrac{2}{5}x + 6$ 16. $y = \dfrac{3}{7}x - 8$

17. $3x - 2y = 8$ 18. $x - 2y = -6$

19. $-2x - 6y = 5$ 20. $10x - 15y = 4$

Write the slope-intercept equation of the line that has the given slope and passes through the given point. SEE EXAMPLE 2. (OBJECTIVE 2)

21. $m = 12, (0, 0)$ 22. $m = -6, (0, 0)$

23. $m = -5, (0, -4)$ 24. $m = -2, (0, 11)$

25. $m = -7, (7, 5)$ 26. $m = 3, (-2, -5)$

27. $m = 0, (3, -5)$ 28. $m = -7, (1, -10)$

Write the slope-intercept equation of the line that passes through the given points. SEE EXAMPLE 3. (OBJECTIVE 3)

29. $(1, -5), (3, -11)$ 30. $(0, 7), (-2, 3)$

31. $(-8, 10), (8, 2)$ 32. $(-4, 5), (2, -6)$

Write each equation in slope-intercept form to find the slope and the y-intercept. Then use the slope and y-intercept to graph the line. SEE EXAMPLE 4. (OBJECTIVE 4)

33. $x - y = 1$ 34. $x + y = 2$

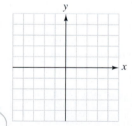   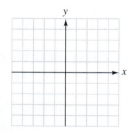

35. $2x = 3y - 6$ 36. $5x = -4y + 10$

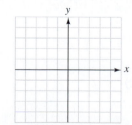   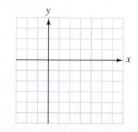

37. $3y = -2x + 18$ **38.** $-8x = 9y + 36$

 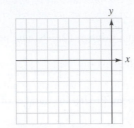

Determine whether the graphs represented by each pair of equations are parallel, perpendicular, or neither.
SEE EXAMPLES 5–6. *(OBJECTIVE 5)*

39. $y = -6x + 3, y = -6x - 4$

40. $y = 5x - 13, y = \frac{1}{5}x + 28$

41. $x + y = 7, y = x - 3$

42. $x = y + 5, y = x + 8$

43. $y = 3x + 9, 2y = 6x - 10$

44. $2x + 3y = 11, 3x - 2y = 15$

45. $x = 3y + 9, y = -3x + 2$

46. $3x + 6y = 7, y = \frac{1}{2}x$

Write the slope-intercept equation of the line that passes through the given point and is parallel to the given line. SEE EXAMPLE 7.
(OBJECTIVE 6)

47. $(0, 0), y = 4x - 9$

48. $(-3, 2), x = -3y + 8$

49. $(2, 5), 4x - y = 7$

50. $(-6, 3), y + 3x = -12$

Write the slope-intercept equation of the line that passes through the given point and is perpendicular to the given line.
SEE EXAMPLE 8. *(OBJECTIVE 6)*

51. $(0, 0), y = 4x - 9$

52. $(0, -4), x = -3y - 2$

53. $(2, 5), 4x - y = 7$

54. $(-6, 3), y + 3x = -12$

ADDITIONAL PRACTICE *Find the slope and the y-intercept of the line determined by the given equation.*

55. $7x = 2y - 4$ **56.** $15x = -2y - 14$

Write an equation of the line with the following properties. Write the equation in slope-intercept form.

57. $m = \frac{2}{3}$, passing through $(-3, 4)$

58. $m = 0$, passing through $(5, 3)$

59. $m = -\frac{4}{3}$, passing through $(6, -2)$

60. $m = \frac{3}{4}$, passing through $(8, 5)$

61. passing through $(3, -1), (-3, -9)$

62. passing through $(9, 8), (-6, -2)$

63. passing through $(4, -2)$ and parallel to $x = \frac{5}{4}y - 2$

64. passing through $(4, -2)$ and perpendicular to $x = \frac{5}{4}y - 2$

65. passing through $(1, -5)$ and perpendicular to $x = -\frac{3}{4}y + 5$

66. passing through $(1, -5)$ and parallel to $x = -\frac{3}{4}y + 5$

67. perpendicular to the line $y = 5$ and passing through $(-2, 7)$.

68. parallel to the line $y = 7$ and passing through $(-2.5, 3)$.

69. parallel to the line $x = 8$ and passing through $(5, 2)$.

70. perpendicular to the line $x = 5$ and passing through $(1, -3)$.

For Exercises 71–78, determine whether the graphs represented by each pair of equations are parallel, perpendicular, or neither.

71. $y = 8, x = 4$

72. $x = -3, x = -7$

73. $3x = y - 2, 3(y - 3) + x = 0$

74. $2y = 8, x = y$

75. $4x + 5y = 20, 5x - 4y = 20$

76. $6x - 15y = 11, 2x - 5y = 12$

77. $2x + 3y = 12, 6x + 9y = 32$

78. $5x + 6y = 30, 6x + 5y = 24$

APPLICATIONS *For Exercises 79–85, assume straight-line depreciation or straight-line appreciation.* SEE EXAMPLE 9.
(OBJECTIVE 7)

79. Depreciation   A taxicab was purchased for $24,300. Its salvage value at the end of its 7-year useful life is expected to be $1,900. Find the depreciation equation.

80. Depreciation   A small business purchases the computer shown in the illustration. It will be depreciated over a 4-year period, when its salvage value will be $300. Find the depreciation equation.

81. Appreciation   An apartment building was purchased for $450,000. The owners expect the property to double in value in 12 years. Find the appreciation equation.

82. Appreciation A house purchased for $112,000 is expected to double in value in 12 years. Find its appreciation equation.

83. Depreciation Find the depreciation equation for the TV in the want ad.

> **For Sale:** 3-year-old 54-inch TV, $1,900 new. Asking $1,190. Call 875-5555. Ask for Mike.

84. Depreciating word processors A word processor cost $555 when new and is expected to be worth $80 after 5 years. What will it be worth after 3 years?

85. Salvage values A copier cost $1,050 when new and will have a salvage value of $90 when it is replaced in 8 years. Find its annual depreciation rate.

86. Car repairs An auto shop charges an hourly rate for repairs. If it costs $69 for a $1\frac{1}{2}$-hour radiator repair and $230 for a 5-hour transmission overhaul, what is the hourly rate?

87. Printer charges To print advertising brochures, a printer charges a fixed setup cost, plus $50 for every 100 brochures. If 700 brochures cost $375, and 1,000 brochures cost $525, what is the setup cost?

88. Cost of rain gutters An installer of rain gutters charges a service charge, plus a dollar amount per foot. If one neighbor installed 250 feet and paid $435, and another neighbor installed 300 feet and paid $510, what is the service charge?

89. Teacher pensions Average pensions for Illinois teachers nearly doubled from 1993 to 2002, as shown in the illustration. If pensions grew in a linear fashion over the next 10 years, what was the average pension in 2012? (*Hint:* Label 1993 as year 0, 1994 as year 1, etc.)

ANNUAL PENSION
State average, scale in thousands

$22,176

$42,144 (*Cook County's average is $58,841*)

Source: Chicago Tribune

90. Controlling exotic plants Eurasian water milfoil, an undesirable aquatic plant, has infested the waters of northern Wisconsin. To control the weed, lakes are seeded with a chemical that kills it. Four lakes on the Eagle River chain that were treated for milfoil are shown in the following table. If there is a cost per acre to treat a lake, and a state permit costs $800, complete the table.

Lake	Acres treated	Cost
Catfish	36	$22,400
Cranberry	42	$26,000
Eagle	53	
Yellow Birch	17	

Source: Eagle River Chain of Lakes Association

An investor bought an office building for $465,000 excluding the value of the land. At that time, a real-estate appraiser estimated that the building would retain 80% of its value after 40 years. For tax purposes, the investor used linear depreciation to depreciate the building over a period of 40 years.

91. Real estate Find the estimated value of the property after 40 years. Then find the slope of the straight-line depreciation graph of the building by finding the slope of the line passing through $(0, 465,000)$ and $(40, ?)$. Write the equation of the depreciation graph.

92. Real estate See Exercise 91. Find the estimated value of the property after 35 years. If the investor sells the building for $400,000 at that time, find the taxable capital gain.

WRITING ABOUT MATH

93. Explain how to use the slope-intercept form to graph the equation $y = -\frac{3}{5}x + 2$.

94. In straight-line depreciation, explain why the slope of the line is called the *rate of depreciation*.

SOMETHING TO THINK ABOUT

95. Solve $Ax + By = C$ for y and thereby show that the slope of its graph is $-\frac{A}{B}$ and its y-intercept is $\left(0, \frac{C}{B}\right)$.

96. Show that the x-intercept of the graph of $Ax + By = C$ is $\left(\frac{C}{A}, 0\right)$.

97. Can the equation of a vertical line be written in slope-intercept form? Explain.

98. Can the equation of a horizontal line be written in slope-intercept form? Explain.

99. If the graph of $y = ax + b$ passes through quadrants I, II, and IV, what can be known about the constants a and b?

100. The graph of $Ax + By = C$ passes through the quadrants I and IV, only. What is known about the constants A, B, and C?

Investigate the properties of slope and y-intercept by experimenting with the following problems.

101. Graph $y = mx + 2$ for several positive values of m. What do you notice?

102. Graph $y = mx + 2$ for several negative values of m. What do you notice?

103. Graph $y = 2x + b$ for several increasing positive values of b. What do you notice?

104. Graph $y = 2x + b$ for several decreasing negative values of b. What do you notice?

105. How will the graph of $y = \frac{1}{2}x + 5$ compare to the graph of $y = \frac{1}{2}x - 5$?

106. How will the graph of $y = \frac{1}{2}x - 5$ compare to the graph of $y = \frac{1}{2}x$?

Section 3.6   Functions

Objectives

1. Find the domain and range of a set of ordered pairs.
2. Determine whether a given equation defines y to be a function of x.
3. Evaluate a function written in function notation.
4. Graph a function and determine its domain and range.
5. Determine whether a graph represents a function.

Vocabulary

relation   range   linear function
domain   function   vertical line test

Getting Ready

Let $y = 2x - 1$. Find the value of y when

1. $x = 0$ 2. $x = 2$ 3. $x = -1$ 4. $x = -2$

Let $y = -3x + 2$. Find the value of y when

5. $x = 0$ 6. $x = 2$ 7. $x = -1$ 8. $x = -2$

In this section, we will discuss *relations* and *functions*. We include these concepts in this chapter because they involve ordered pairs.

1   Find the domain and range of a set of ordered pairs.

Table 3-3 shows the number of medals won by United States athletes during five Winter Olympics.

TABLE 3-3

USA Winter Olympic Medal Count

Year	1992	1994	1998	2002	2006	2010
Medals	11	13	13	34	25	37

We can display the data in the table as a set of ordered pairs, where the *first component* represents the year and the *second component* represents the number of medals won by U.S. athletes.

$$\{(1992, 11), (1994, 13), (1998, 13), (2002, 34), (2006, 25), (2010, 37)\}$$

A set of ordered pairs, such as this, is called a **relation**. The set of all first components is called the **domain** of the relation, and the set of all second components is called the **range** of the relation.

EXAMPLE 1   Find the domain and range of the relation $\{(-2, -5), (4, 7), (8, 9)\}$.

Solution   The domain is the set of first components of the ordered pairs:

$$\{-2, 4, 8\}$$

The range is the set of second components of the ordered pairs:

$$\{-5, 7, 9\}$$

 SELF CHECK 1   Find the domain and range of the relation $\{(-3, -2), (-1, 3), (4, 5)\}$.

When to each first component in a relation there corresponds exactly one second component, the relation is called a **function**.

FUNCTION   A function is a set of ordered pairs (a relation) in which to each first component, there corresponds exactly one second component.

2   **Determine whether a given equation defines y to be a function of x.**

Earlier in the chapter, we constructed the following table of ordered pairs for the equation $y = 3x - 4$ by substituting specific values for x and computing the corresponding values of y. Since the equation determines a set of ordered pairs, the equation determines a relation.

$$y = 3x - 4$$

x	y	(x, y)
-2	-10	$(-2, -10)$
-1	-7	$(-1, -7)$
0	-4	$(0, -4)$
1	-1	$(1, -1)$
2	2	$(2, 2)$

If $x = -2$, then $y = 3(-2) - 4 = -10$.
If $x = -1$, then $y = 3(-1) - 4 = -7$.
If $x = 0$, then $y = 3(0) - 4 = -4$.
If $x = 1$, then $y = 3(1) - 4 = -1$.
If $x = 2$, then $y = 3(2) - 4 = 2$.

↑   ↑
Input values   Output values

From the table or equation, we can see that each *input value x* determines exactly one *output value y.* Because this is true, the equation also defines y to be a *function* of x. This leads to the following definition.

y IS A FUNCTION OF x   Any equation in x and y where each value of x (the *input*) determines exactly one value of y (the *output*) is called a function. In this case, we say that **y is a function of x.**

The set of all input values x is called the domain of the function, and the set of all output values y is called the range of the function.

COMMENT   A function is always a relation, but a relation is not necessarily a function. For example, the relation $\{(2, 1), (2, 3)\}$ is not a function because the input 2 determines two different values in the range: 1 and 3.

Since each value of y in a function depends on a specific value of x, y is the *dependent variable* and x is the *independent variable*. The graph of a function is the graph of the equation that defines the function. In Example 2, we graph a linear equation. Any function that is defined by a linear equation is called a **linear function**.

EXAMPLE 2   Determine whether the equations define y to be a function of x.
a. $y = x^2$ **b.** $x = y^2$

Solution   **a.** We construct a table of ordered pairs for the equation $y = x^2$ by substituting values for x and computing the corresponding values.

$$y = x^2$$

x	y	(x, y)	
-2	4	$(-2, 4)$	If $x = -2$, then $y = (-2)^2 = 4$.
-1	1	$(-1, 1)$	If $x = -1$, then $y = (-1)^2 = 1$.
0	0	$(0, 0)$	If $x = 0$, then $y = 0$.
1	1	$(1, 1)$	If $x = 1$, then $y = (1)^2 = 1$.
2	4	$(2, 4)$	If $x = 2$, then $y = (2)^2 = 4$.
3	9	$(3, 9)$	If $x = 3$, then $y = (3)^2 = 9$.

↑ Input values   ↑ Output values

From the table we can see that each input value x determines exactly one output value y. The relation is a function.

b. We construct a table of ordered pairs for the equation $x = y^2$. Because y is squared, it will be more convenient to substitute values for y and compute the corresponding values for x.

$$x = y^2$$

x	y	(x, y)	
4	-2	$(4, -2)$	If $y = -2$, then $x = (-2)^2 = 4$.
1	-1	$(1, -1)$	If $y = -1$, then $x = (-1)^2 = 1$.
0	0	$(0, 0)$	If $y = 0$, then $x = 0$.
1	1	$(1, 1)$	If $y = 1$, then $x = (1)^2 = 1$.
4	2	$(4, 2)$	If $y = 2$, then $x = (2)^2 = 4$.
9	3	$(9, 3)$	If $y = 3$, then $x = (3)^2 = 9$.

↑ Input values   ↑ Output values

From the table we can see that each input value x does not determine exactly one output value y. The relation is not a function.

 **SELF CHECK 2**   Determine whether the equations define y to be a function of x.
a. $y = -x^2$ **b.** $x = -y^2$

3   **Evaluate a function written in function notation.**

There is a special notation for functions that uses the symbol $f(x)$, read as "f of x."

FUNCTION NOTATION   The notation $y = f(x)$ denotes that the variable y is a function of x.

COMMENT The notation $f(x)$ does not mean "f times x."

The notation $y = f(x)$ provides a way to denote the values of y in a function that correspond to individual values of x. For example, if $y = f(x)$, the value of y that is determined by $x = 3$ is denoted as $f(3)$. Similarly, $f(-1)$ represents the value of y that corresponds to $x = -1$.

EXAMPLE 3 Let $f(x) = 2x - 3$ and find **a.** $f(3)$ **b.** $f(-1)$ **c.** $f(0)$
d. the value of x for which $f(x) = 5$.

Solution **a.** We replace x with 3.

$$f(\boldsymbol{x}) = 2\boldsymbol{x} - 3$$
$$f(\boldsymbol{3}) = 2(\boldsymbol{3}) - 3$$
$$= 6 - 3$$
$$= 3$$

b. We replace x with -1.

$$f(\boldsymbol{x}) = 2\boldsymbol{x} - 3$$
$$f(\boldsymbol{-1}) = 2(\boldsymbol{-1}) - 3$$
$$= -2 - 3$$
$$= -5$$

c. We replace x with 0.

$$f(\boldsymbol{x}) = 2\boldsymbol{x} - 3$$
$$f(\boldsymbol{0}) = 2(\boldsymbol{0}) - 3$$
$$= 0 - 3$$
$$= -3$$

d. We replace $f(x)$ with 5 and solve for x.

$$f(x) = 2x - 3$$
$$\boldsymbol{5} = 2x - 3$$
$$8 = 2x \qquad \text{Add 3 to both sides.}$$
$$4 = x \qquad \text{Divide both sides by 2.}$$

 **SELF CHECK 3** Use the function of Example 3 and find **a.** $f(-2)$ **b.** $f\left(\frac{3}{2}\right)$ **c.** $f(-5)$
d. the value for which $f(x) = 11$.

---

4 **Graph a function and determine its domain and range.**

For the graph of the function shown in Figure 3-41, the domain (the set of input values x) is shown on the x-axis, and the range (the set of output values y) is shown on the y-axis. The domain of this function is the set of real numbers $\mathbb{R}$. The range is also the set of real numbers $\mathbb{R}$.

Sonya Kovalevskaya
1850–1891
This talented young Russian woman hoped to study mathematics at the University of Berlin, but strict rules prohibited women from attending lectures. Undaunted, she studied privately with the great mathematician Karl Weierstrauss and published several important papers.

Figure 3-41

Not all functions are linear functions. For example, the graph of the *absolute value function* defined by the equation $y = |x|$ is not a line, as the next example will show.

EXAMPLE 4 Graph $f(x) = |x|$ and determine the domain and range.

Solution We begin by setting up a table of values.

$y = f(x) = |x|$

x	y	(x, y)
-2	2	$(-2, 2)$
-1	1	$(-1, 1)$
0	0	$(0, 0)$
1	1	$(1, 1)$
2	2	$(2, 2)$

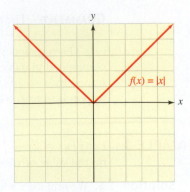

Figure 3-42

From Figure 3-42, we can determine that the domain is the set of all real numbers $\mathbb{R}$. This makes sense because we can find the absolute value of any real number.

From the graph, we can see that the range is the set of values where $y \geq 0$. We write this as $\{y \mid y$ is a real number and $y \geq 0\}$. The absolute value of any real number is always positive or 0.

SELF CHECK 4   Graph $f(x) = -|x|$ and determine the domain and range.

5 Determine whether a graph represents a function.

A **vertical line test** can be used to determine whether the graph of an equation represents a function. If any vertical line intersects a graph more than once, the graph cannot represent a function, because to one number x, there would correspond more than one value of y.

The graph in Figure 3-43(a) represents a function, because every vertical line that intersects the graph does so exactly once. The graph in Figure 3-43(b) does not represent a function, because some vertical lines intersect the graph more than once.

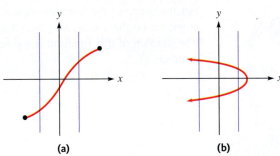

(a)   (b)

Figure 3-43

EXAMPLE 5   Determine whether each graph represents a function.

a. 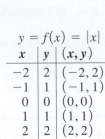   b. 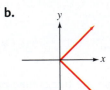   c. 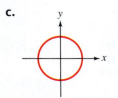   d.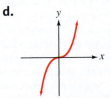

Solution   a. Since any vertical line that intersects the graph does so only once, the graph represents a function.

b. Since some vertical lines that intersect the graph do so twice, the graph does not represent a function.

c. Since some vertical lines that intersect the graph do so twice, the graph does not represent a function.

d. Since any vertical line that intersects the graph does so only once, the graph represents a function.

 SELF CHECK 5 Determine whether each graph represents a function.

a.

b.

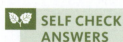 **SELF CHECK ANSWERS**

1. domain: $\{-3, -1, 4\}$; range: $\{-2, 3, 5\}$ 2. a. yes   b. no   3. a. -7 b. 0   c. -13 d. $x = 7$
4. domain: $\mathbb{R}$; range: $\{y \mid y$ is a real number and $y \leq 0\}$

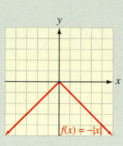

5. a. yes   b. no

NOW TRY THIS

Refer to the graph and find:

1. $f(1)$
2. x when $f(x) = -3$
3. the domain
4. the range.

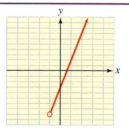

3.6 Exercises

WARM-UPS *Let $y = 2x + 1$. Find the value of y when*

1. $x = 0$
2. $x = 1$
3. $x = -1$
4. $x = -2$

REVIEW *Solve each equation.*

5. $4x = 3(x + 2)$
6. $6y + 3 = 7(y - 1)$
7. $5(2 - a) = 3(a + 6)$
8. $3x = 1.5(5 - x) - 12$

VOCABULARY AND CONCEPTS *Fill in the blanks.*

9. A _____ is a set of ordered pairs.
10. In a relation, the set of first components is called the _____ of the relation. The set of second components is called the ____.
11. Any equation in x and y where each ____ x determines exactly one output y is called a _____.
12. In a function, the set of all inputs is called the _____ of the function.

13. In a function, the set of all outputs is called the _____ of the function.

14. In the function $y = f(x)$, y is called the _____ variable.

15. In the function $y = f(x)$, x is called the _____ variable.

16. The function $y = f(x) = mx + b$ is a _____ function and $y = f(x) = |x|$ is called the _____ value function.

17. If a vertical line intersects a graph more than once, the graph _____ represent a function.

18. If $y = f(x)$, the value of y that is determined by $x = 4$ is denoted as ____.

GUIDED PRACTICE *Find the domain and range of each relation. SEE EXAMPLE 1. (OBJECTIVE 1)*

19. $\{(-3, -1), (1, 2), (3, 7)\}$

20. $\{(-5, -2), (0, 3), (3, 9), (5, 12)\}$

21. $\{(0, 5), (2, 7), (-3, -8), (4, 0)\}$

22. $\{(3, 7), (-2, -5), (3, -1), (2, -5)\}$

Determine whether the equation defines y to be a function of x. SEE EXAMPLE 2. (OBJECTIVE 2)

23. $y = -x + 1$

24. $y = \dfrac{1}{2}x - 3$

25. $x = -y^2$

26. $x = |y - 1|$

Find $f(3)$, $f(0)$, $f(-1)$, and the value of x for which $f(x) = 3$. SEE EXAMPLE 3. (OBJECTIVE 3)

27. $f(x) = -3x$

28. $f(x) = -4x$

29. $f(x) = 2x - 3$

30. $f(x) = 3x - 5$

31. $f(x) = 7 + 5x$

32. $f(x) = 3 - 3x$

33. $f(x) = 9 - 2x$

34. $f(x) = 12 + 3x$

35. $f(x) = \dfrac{1}{2}x + \dfrac{3}{2}$

36. $f(x) = -\dfrac{1}{3}x + \dfrac{1}{3}$

Graph each function and state its domain and range. SEE EXAMPLE 4. (OBJECTIVE 4)

37. $y = x + 2$

38. $y = -x + 1$

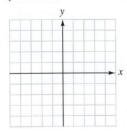 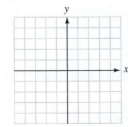

39. $f(x) = \dfrac{1}{2}|x|$

40. $f(x) = -2|x|$

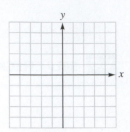 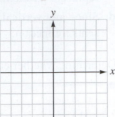

Determine whether each graph represents a function. SEE EXAMPLE 5. (OBJECTIVE 5)

41.

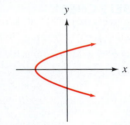

42.

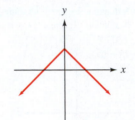

43.

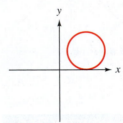

44.

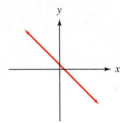

ADDITIONAL PRACTICE *Find $f(1)$, $f(-2)$, and $f(3)$.*

45. $f(x) = x^2 + 1$

46. $f(x) = x^2 - 2$

47. $f(x) = x^3 - 1$

48. $f(x) = x^3$

49. $f(x) = (x - 1)^2$

50. $f(x) = (x + 3)^2$

51. $f(x) = 3x^2 - 2x$

52. $f(x) = 5x^2 + 2x - 1$

53. $f(x) = \dfrac{1}{2}x^2 - 2x + 3$

54. $f(x) = \dfrac{3}{4}x^2 + 4x - 5$

Graph each function and state the domain and range.

55. $y = -\dfrac{1}{2}x + 2$

56. $y = \dfrac{1}{2}x - 3$

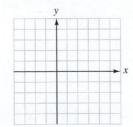 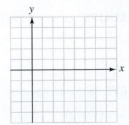

57. $f(x) = |x| - 1$

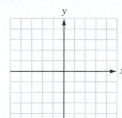

58. $f(x) = -|x| + 2$

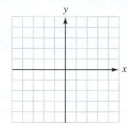

In a certain city, the monthly cost of residential electric service is given by the function $C = 0.08n + 17$, where n is the number of kilowatt hours (kWh) used, 8¢ is the cost per kWh, and $17 is a fixed charge.

63. Find the monthly bill if 500 kWh were used.

64. Find the monthly bill if 600 kWh were used.

65. Find the monthly bill if 800 kWh were used.

66. Find the monthly bill if 1,000 kWh were used.

WRITING ABOUT MATH

67. Define the domain and range of a function.

68. Explain why the vertical line test works.

APPLICATIONS *In a certain city, the monthly cost of telephone service is given by the function $C = 0.10n + 12$, where n is the number of calls, 10¢ is the cost per call, and $12 is a fixed charge.*

59. Find the cost of making 20 calls in a month.

60. Find the cost of making 60 calls in a month.

61. Find the cost of making 100 calls in a month.

62. Find the cost of making 400 calls in a month.

SOMETHING TO THINK ABOUT *Let $f(x) = 2x + 1$ and $g(x) = x$. Assume $f(x) \neq 0$ and $g(x) \neq 0$.*

69. Is $f(x) + g(x) = g(x) + f(x)$?

70. Is $f(x) - g(x) = g(x) - f(x)$?

71. Is $f(x) \cdot g(x) = g(x) \cdot f(x)$?

72. Is $f(x) + g(x) = f(x) \cdot g(x)$?

Section 3.7

Variation

Objectives

1 Solve a direct variation application.
2 Solve an inverse variation application.
3 Solve a joint variation application.
4 Solve a combined variation application.

Vocabulary

direct variation   inverse variation   joint variation
constant of variation   rational function   combined variation

Getting Ready

Solve for k.

1. $8 = 2k$ **2.** $8 = \dfrac{k}{2}$ **3.** $A = kbh$ **4.** $P = \dfrac{kT}{V}$

We now introduce some special terminology that scientists use to describe special functions.

1 Solve a direct variation application.

The most basic type of variation is called **direct variation**.

DIRECT VARIATION

The words *y varies directly with x* mean that

$$y = kx$$

for some constant k. The constant k is called the **constant of variation**.

The more force that is applied to a spring, the more it will stretch. Scientists call this fact Hooke's law: *The distance a spring will stretch varies directly with the force applied.* If d represents distance and f represents force, this relationship can be expressed by the equation

(1) $d = kf$

COMMENT Since k is a constant, the equation $y = kx$ defines a linear function. In the examples and problems in this section, k will always be a positive number.

where k is the constant of variation. If a spring stretches 5 inches when a weight of 2 pounds is attached, we can find the constant of variation by substituting 5 for d and 2 for f in Equation 1 and solving for k.

$$d = kf$$
$$5 = k(2)$$
$$\frac{5}{2} = k$$

To find the distance that the spring will stretch when a weight of 6 pounds is attached, we substitute $\frac{5}{2}$ for k and 6 for f in Equation 1 and solve for d.

$$d = \frac{5}{2}f$$
$$d = \frac{5}{2}(6)$$
$$d = 15$$

The spring will stretch 15 inches when a weight of 6 pounds is attached.

EXAMPLE 1 At a constant speed, the distance traveled varies directly with time. If a bus driver can drive 105 miles in 3 hours, how far would he drive in 5 hours?

Solution We let d represent the distance traveled and let t represent time. We then translate the words *distance varies directly with time* into the equation

(2) $d = kt$

To find the constant of variation, k, we substitute 105 for d and 3 for t in Equation 2 and solve for k.

$$d = kt$$
$$105 = k(3)$$
$$35 = k \qquad \text{Divide both sides by 3.}$$

We can now substitute 35 for k in Equation 2 to obtain Equation 3.

(3) $d = 35t$

To find the distance traveled in 5 hours, we substitute 5 for t in Equation 3.

$$d = 35(5)$$
$$d = 175$$

In 5 hours, the bus driver would travel 175 miles.

COMMENT Remember to label the answer with the appropriate units.

 SELF CHECK 1 How far would the bus driver travel in 8 hours?

2 Solve an inverse variation application.

A second common type of variation is called **inverse variation**.

INVERSE VARIATION

The words *y varies inversely with x* mean that

$$y = \frac{k}{x}$$

for some constant *k*. The constant *k* is the **constant of variation**.

COMMENT Since *k* is a constant, the equation $y = \frac{k}{x}$ defines a special function called a **rational function**. We will discuss these functions later in the book. In this section, *k* will always be a positive number.

Under constant temperature, the volume occupied by a gas varies inversely with its pressure. If *V* represents volume and *p* represents pressure, this relationship is expressed by the equation

(4) $\qquad V = \dfrac{k}{p}$

EXAMPLE 2 A gas occupies a volume of 15 cubic inches when placed under 4 pounds per square inch of pressure. How much pressure is needed to compress the gas into a volume of 10 cubic inches?

Solution To find the constant of variation, we substitute 15 for *V* and 4 for *p* in Equation 4 and solve for *k*.

$$V = \frac{k}{p}$$

$$15 = \frac{k}{4}$$

$$60 = k \qquad \text{Multiply both sides by 4.}$$

To find the pressure needed to compress the gas into a volume of 10 cubic inches, we substitute 60 for *k* and 10 for *V* in Equation 4 and solve for *p*.

$$V = \frac{k}{p}$$

$$10 = \frac{60}{p}$$

$$10p = 60 \qquad \text{Multiply both sides by } p.$$

$$p = 6 \qquad \text{Divide both sides by 10.}$$

It will take 6 pounds per square inch of pressure to compress the gas into a volume of 10 cubic inches.

 SELF CHECK 2 How much pressure is needed to compress the gas into a volume of 8 cubic inches?

---

3 Solve a joint variation application.

A third type of variation is called **joint variation**.

JOINT VARIATION

The words *y varies jointly with x and z* mean that

$$y = kxz$$

for some constant k. The constant k is the **constant of variation**.

The area A of a rectangle depends on its length l and its width w by the formula

$$A = lw$$

We could say that the area of the rectangle varies jointly with its length and its width. In this example, the constant of variation is $k = 1$.

EXAMPLE 3 The area of a triangle varies jointly with the length of its base and its height. If a triangle with an area of 63 square inches has a base of 18 inches and a height of 7 inches, find the area of a triangle with a base of 12 inches and a height of 10 inches.

Solution We let A represent the area of the triangle, b represent the length of the base, and h represent the height. We translate the words *area varies jointly with the length of the base and the height* into the formula

(5) $A = kbh$

We are given that $A = 63$ when $b = 18$ and $h = 7$. To find k, we substitute these values in Equation 5 and solve for k.

$$A = kbh$$
$$63 = k(18)(7)$$
$$63 = k(126)$$
$$\frac{63}{126} = k \qquad \text{Divide both sides by 126.}$$
$$\frac{1}{2} = k \qquad \text{Simplify.}$$

Thus, $k = \frac{1}{2}$, and the formula for finding the area is

(6) $A = \dfrac{1}{2}bh$

To find the area of a triangle with a base of 12 inches and a height of 10 inches, we substitute 12 for b and 10 for h in Equation 6.

$$A = \frac{1}{2}bh$$
$$A = \frac{1}{2}(12)(10)$$
$$A = 60$$

The area is 60 square inches.

SELF CHECK 3 Find the area of a triangle with a base of 14 inches and a height of 8 inches.

4 Solve a combined variation application.

The final type of variation, called **combined variation**, involves a combination of direct and inverse variation.

EXAMPLE 4 The pressure of a fixed amount of gas varies directly with its temperature and inversely with its volume. A sample of gas at a pressure of 1 atmosphere occupies a volume of 3 cubic meters when its temperature is 273 Kelvin (about 0° Celsius). Find the pressure after the gas is heated to 364 K and compressed to 1 cubic meter.

Solution We let P represent the pressure of the gas, T represent its temperature, and V represent its volume. The words *the pressure varies directly with temperature and inversely with volume* translate into the equation

$$(7) \quad P = \frac{kT}{V}$$

To find k, we substitute 1 for P, 273 for T, and 3 for V in Equation 7.

$$P = \frac{kT}{V}$$

$$1 = \frac{k(273)}{3}$$

$$1 = 91k \qquad \frac{273}{3} = 91$$

$$\frac{1}{91} = k$$

Since $k = \frac{1}{91}$, the formula is

$$P = \frac{1}{91} \cdot \frac{T}{V} \quad \text{or} \quad P = \frac{T}{91V}$$

To find the pressure under the new conditions, we substitute 364 for T and 1 for V in the previous equation and solve for P.

$$P = \frac{T}{91V}$$

$$P = \frac{364}{91(1)}$$

$$= 4$$

The pressure of the heated and compressed gas is 4 atmospheres.

 **SELF CHECK 4** Find the pressure after the gas is heated to 373.1 K and compressed into 1 cubic meter.

SELF CHECK ANSWERS

1. 280 mi **2.** 7.5 lb/in.^{2} **3.** 56 in.^{2} **4.** 4.1 atmospheres

NOW TRY THIS

The deflection (the perpendicular distance from horizontal) of a beam is inversely proportional to its width and the cube of its depth. If the deflection is 0.125 inch when the width is 4 inches and the depth is 2 inches, find the deflection when the width is 8 inches and the depth is 1 inch.

3.7   Exercises

WARM-UPS *Determine whether each equation indicates direct variation, inverse variation, joint variation, or combined variation.*

1. $y = \dfrac{7}{x}$

2. $y = 5x$

3. $y = \dfrac{6x}{z}$

4. $y = 3xz$

5. $y = \dfrac{1}{2}x$

6. $y = \dfrac{2x}{z}$

7. $y = \dfrac{1}{2}xz$

8. $y = \dfrac{4}{x}$

REVIEW *Solve each inequality and graph the solution.*

9. $x - 2 > 5$

10. $-x + 2 \le 5$

11. $-5x + 4 \ge -6$

12. $7(x - 3) + 7 \le 9x$

VOCABULARY AND CONCEPTS *Fill in the blanks. Assume that k is a constant.*

13. The equation $y = kx$ represents _____ variation.

14. The equation _____ represents inverse variation. This equation determines a _____ function.

15. In the equation $y = kx$, k is called the _____ of variation.

16. Hooke's law is an example of _____ variation.

17. The equation $y = kxz$ represents _____ variation.

18. The equation $y = \dfrac{kx}{z}$ represents combined variation and means that y varies _____ with x and _____ with z.

Express each sentence as a formula.

19. The distance d a car can travel while moving at a constant speed varies directly with n, the number of gallons of gasoline it consumes.

20. A farmer's harvest h varies directly with a, the number of acres he plants.

21. For a fixed area, the length l of a rectangle varies inversely with its width w.

22. The value v of a boat varies inversely with its age a.

23. The area A of a circle varies directly with the square of its radius r.

24. The distance s that a body falls varies directly with the square of the time t.

25. The distance d traveled varies jointly with the speed s and the time t.

26. The simple interest i on a savings account that contains a fixed amount of money varies jointly with the rate r and the time t.

27. The current I varies directly with the voltage V and inversely with the resistance R.

28. The force of gravity F varies directly with the product of the masses m_1 and m_2, and inversely with the square of the distance between them.

GUIDED PRACTICE *Express each direct variation as an equation. Then find the requested value. Assume that all variables represent positive numbers.* SEE EXAMPLE 1. *(OBJECTIVE 1)*

29. y varies directly with x. If $y = 24$ when $x = 8$, find y when $x = 11$.

30. A varies directly with z. If $A = 30$ when $z = 5$, find A when $z = 9$.

31. r varies directly with s. If $r = 21$ when $s = 6$, find r when $s = 12$.

32. d varies directly with t. If $d = 15$ when $t = 3$, find t when $d = 3$.

Express each inverse variation as an equation. Then find the requested value. Assume that all variables represent positive numbers. SEE EXAMPLE 2. *(OBJECTIVE 2)*

33. y varies inversely with x. If $y = 6$ when $x = 2$, find y when $x = 4$.

34. V varies inversely with p. If $V = 60$ when $p = 12$, find V when $p = 9$.

35. r varies inversely with s. If $r = 40$ when $s = 10$, find r when $s = 15$.

36. J varies inversely with v. If $J = 90$ when $v = 5$, find J when $v = 45$.

Express each joint variation as an equation. Then find the requested value. Assume that all variables represent positive numbers. SEE EXAMPLE 3. *(OBJECTIVE 3)*

37. y varies jointly with r and s. If $y = 4$ when $r = 2$ and $s = 6$, find y when $r = 3$ and $s = 4$.

38. A varies jointly with x and y. If $A = 18$ when $x = 3$ and $y = 3$, find A when $x = 7$ and $y = 9$.

39. D varies jointly with p and q. If $D = 16$ when p and q are both 8, find D when p and q are both 12.

40. z varies jointly with r and the square of s. If $z = 24$ when r and s are 2, find z when $r = 3$ and $s = 4$.

Express each combined variation as an equation. Then find the requested value. Assume that all variables represent positive numbers. SEE EXAMPLE 4. *(OBJECTIVE 4)*

41. y varies directly with a and inversely with b. If $y = 3$ when $a = 4$ and $b = 12$, find y when $a = 10$ and $b = 18$.

42. y varies directly with the square of x and inversely with z. If $y = 1$ when $x = 2$ and $z = 10$, find y when $x = 4$ and $z = 5$.

43. y varies directly with x and inversely with z. If $y = 1$ when $x = 3$ and $z = 7$, find y when $x = 8$ and $z = 10$.

44. p varies directly with q and inversely with r. If $p = 5$ when $q = 1$ and $r = 6$, find p when $q = 5$ and $r = 10$.

ADDITIONAL PRACTICE *Express each variation as an equation. Then find the requested value. Assume that all variables represent positive numbers.*

45. s varies directly with t^2. If $s = 20$ when $t = 5$, find s when $t = 15$.

46. y varies directly with x^3. If $y = 16$ when $x = 2$, find y when $x = 3$.

47. y varies inversely with x^2. If $y = 6$ when $x = 4$, find y when $x = 2$.

48. i varies inversely with d^2. If $i = 8$ when $d = 3$, find i when $d = 6$.

49. y varies jointly with x and w^2. If $y = 12$ when $x = 2$ and $w = 3$, find y when $x = 3$ and $w = 2$.

50. I varies jointly with r^2 and q. If $I = 28$ when $r = 2$ and $q = 7$, find I when $r = 4$ and $q = 6$.

51. b varies directly with c and inversely with d^2. If $b = 5$ when $c = 2$ and $d = 4$, find b when $c = 36$ and $d = 2$.

52. x varies directly with c^2 and inversely with f. If $x = 16$ when $c = 8$ and $f = 12$, find x when $c = 10$ and $f = 5$.

53. c varies jointly with d and h. If $c = 196$ when $d = 6$ and $h = 4$, find d when $c = 705.6$ and $h = 16$.

54. y varies inversely with the square of x. If $y = 16$ when $x = 10$, find x when $y = 6,400$.

55. q varies jointly with the quotient of p and r. If $q = \frac{2}{3}$ when $p = 3$ and $r = 6$, find p when $q = 9$ and $r = 15$.

56. F varies directly with the product of m_1 and m_2 and inversely with d^2. If $F = 1,250$ when $m_1 = 400$ and $m_2 = 500$ and $d = 100$, find m_1 when $F = 1,550$ and all other values remain constant.

APPLICATIONS *Set up a variation equation and solve for the requested value.*

57. Objects in free fall The distance traveled by an object in free fall varies directly with the square of the time that it falls. If the object falls 256 feet in 4 seconds, how far will it fall in 6 seconds?

58. Traveling range The distance that a car can travel without refueling varies directly with the number of gallons of gasoline in the tank. If a car can go 360 miles on 12 gallons of gas, how far can it go on 7 gallons?

59. Computing interest For a fixed rate and principal, the interest earned in a bank account paying simple interest varies directly with the length of time the principal is left on deposit. If an investment of $5,000 earns $700 in 2 years, how much will it earn in 7 years?

60. Computing forces The force of gravity acting on an object varies directly with the mass of the object. The force on a mass of 5 kilograms is 49 newtons. What is the force acting on a mass of 12 kilograms?

61. Commuting time The time it takes a car to travel a certain distance varies inversely with its rate of speed. If a certain trip takes 3 hours when the driver travels at 50 mph, how long will the trip take when the driver travels at 60 mph?

62. Geometry For a fixed area, the length of a rectangle is inversely proportional to its width. A rectangle has a width of 8 feet and a length of 10 feet. If the length is increased to 16 feet, find the width of the rectangle.

63. Computing pressures If the temperature of a gas is constant, the volume occupied varies inversely with the pressure. If a gas occupies a volume of 40 cubic meters under a pressure of 8 atmospheres, find the volume when the pressure is changed to 6 atmospheres.

64. Computing depreciation Assume that the value of a machine varies inversely with its age. If a drill press is worth $300 when it is 2 years old, find its value when it is 6 years old. How much has the machine depreciated in those 4 years?

65. Computing interest The interest earned on a fixed amount of money varies jointly with the annual interest rate and the time that the money is left on deposit. If an account earns $120 at 8% annual interest when left on deposit for 2 years, how much interest would be earned in 3 years at an annual rate of 12%?

66. Cost of a well The cost of drilling a water well is jointly proportional to the length and diameter of the steel casing. If a 30-foot well using 4-inch casing costs $1,200, find the cost of a 35-foot well using 6-inch casing.

67. Electronics The current in a circuit varies directly with the voltage and inversely with the resistance. If a current of 4 amperes flows when 36 volts is applied to a 9-ohm resistance, find the current when the voltage is 42 volts and the resistance is 11 ohms.

68. Road construction The time it takes to build a highway varies directly with the length of the highway and inversely with the number of workers. If it takes 100 workers 4 weeks to build 2 miles of highway, how long will it take 80 workers to build 10 miles of highway?

WRITING ABOUT MATH

69. Explain why the words *y varies jointly with x and z* mean the same as the words *y varies directly with the product of x and z*.

70. Explain the meaning of combined variation.

SOMETHING TO THINK ABOUT

71. Can direct variation be defined as $\frac{y}{x} = k$, rather than $y = kx$?

72. Can inverse variation be defined as $xy = k$, rather than $y = \frac{k}{x}$?

Projects

PROJECT 1

In Section 3.6, we defined functions in terms of an equation in x and y, where each input number x determines a single output number y. In this case, all functions were functions whose domains and ranges were subsets of the real numbers. The concept of function can be defined in a more general way.

> A **function** is a correspondence between a set of input values x (called the **domain**) and a set of output values y (called the **range**), where to each x-value in the domain there corresponds exactly one y-value in the range.

Using this definition, the domain and range of a function do not have to be sets of real numbers. For example, the correspondence formed by the set of states in the U.S. and the set of governors determines a function because:

> *To every state, there corresponds exactly one governor.*

In this function, the domain is the set of states and the range is the set of governors.

However, many correspondences in the real world do not determine functions. For example, the correspondence formed by mothers and their children does not determine a function because:

> *To every mother, there corresponds one or more children.*

a. Think of five correspondences in the real world that determine functions.

b. Think of five correspondences in the real world that do not determine functions.

PROJECT 2

Graphs are often used in newspapers and magazines to convey complex information at a glance. Unfortunately, it is easy to use graphs to convey misleading information. For example, the profit percents of a company for several years are given in the table and two graphs in Figure 3-44.

The first graph in the figure accurately indicates the company's steady performance over five years. But because the vertical axis of the second graph does not start at zero, the performance appears deceptively erratic.

Year	Profit
2006	6.2%
2007	6.0%
2008	6.2%
2009	6.1%
2010	6.3%
2011	6.6%

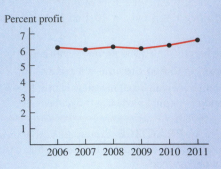

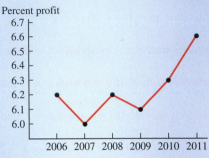

Figure 3-44

As your college's head librarian, you spend much of your time writing reports, either trying to make the school library look good (for college promotional literature) or bad (to encourage greater funding). In 2006, the library held a collection of 17,000 volumes. Over the years, the library has acquired many new books and has retired several old books. The details appear in the table.

Using the data in the table,

■ Draw a misleading graph that makes the library look good.

■ Draw a misleading graph that makes the library look bad.

■ Draw a graph that accurately reflects the library's condition.

Year	Volumes acquired	Volumes removed
2006	215	137
2007	217	145
2008	235	185
2009	257	210
2010	270	200
2011	275	180

Reach for Success
EXTENSION OF UNDERSTANDING YOUR SYLLABUS

Now that you know where your instructor's office is, how to get in touch with him or her, and how your grade will be determined, take a look at other information in the syllabus that may be important to know before you get too far into the semester.

On your syllabus, locate your instructor's make-up policy.	Does your instructor accept late work? _____
	If so, under what conditions? State the work that is accepted late and any penalty assessed.
	Test make-up policy _____
	Penalty assessed _____
	Other requirements make-up policy _____
	Penalty assessed _____
If you know of any conflicts in meeting established deadlines, contact your instructor *prior* to the due date.	If not, try to reschedule any personal obligations.
On your syllabus, locate your instructor's attendance policy.	What is your instructor's attendance policy?
	_____
	What does the syllabus indicate about arriving late or leaving class early?
	_____

Understanding any penalties for late work can make a difference in the successful completion of the course. Do not hesitate to contact your instructor if you are unclear regarding any of the policies stated on the syllabus.

3   Review

SECTION 3.1   The Rectangular Coordinate System

DEFINITIONS AND CONCEPTS	EXAMPLES
Any **ordered pair of real numbers** represents a point on the rectangular coordinate system.	Plot $(2, 6)$, $(-2, 6)$, $(-2, -6)$, $(2, -6)$, and $(0, 0)$. 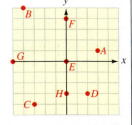
The point where the axes cross, $(0, 0)$, is called the **origin**. The four regions of a coordinate plane are called **quadrants**.	The origin is represented by the ordered pair $(0, 0)$. The ordered pair $(2, 6)$ is found in quadrant I. The ordered pair $(-2, 6)$ is found in quadrant II. The ordered pair $(-2, -6)$ is found in quadrant III. The ordered pair $(2, -6)$ is found in quadrant IV.

REVIEW EXERCISES

Plot each point on the rectangular coordinate system in the illustration.

1. $A(1, 3)$ **2.** $B(1, -3)$
3. $C(-3, 1)$ **4.** $D(-3, -1)$
5. $E(0, 5)$ **6.** $F(-5, 0)$

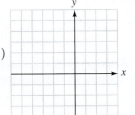

Find the coordinates of each point in the illustration.

7. A **8.** B
9. C **10.** D
11. E **12.** F
13. G **14.** H

SECTION 3.2   Graphs of Linear Equations in Two Variables

DEFINITIONS AND CONCEPTS	EXAMPLES
An ordered pair of real numbers is a **solution** to an equation in two variables if it satisfies the equation.	The ordered pair $(-1, 5)$ satisfies the equation $x - 2y = -11$. $(-1) - 2(5) \stackrel{?}{=} -11$ Substitute -1 for x and 5 for y. $-1 - 10 \stackrel{?}{=} -11$ $-11 = -11$ True. Since the results are equal, $(-1, 5)$ is a solution.
To graph a linear equation, **1.** Find two ordered pairs (x, y) that satisfy the equation. **2.** Plot each pair on the rectangular coordinate system. **3.** Draw a line passing through the two points. General form of an equation of a line: $\quad Ax + By = C$ $\quad$ (A and B are not both 0.)	Graph: $x - y = -2$. Create a table of values by either using the original equation or by solving for y first. $\quad x - y = -2$ This is the original equation. $\quad\quad -y = -x - 2$ Subtract x from both sides. $\quad\quad\quad y = x + 2$ Divide both sides by -1.

Then find two ordered pairs that satisfy the equation. Use a third as a check.

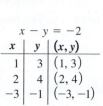

$$x - y = -2$$

x	y	(x, y)
1	3	$(1, 3)$
2	4	$(2, 4)$
−3	−1	$(-3, -1)$

We then plot the points and draw a line passing through them.

The equation $y = b$ represents a horizontal line that intersects the y-axis at $(0, b)$.	The graph of $y = 5$ is a horizontal line passing through $(0, 5)$.
The equation $x = a$ represents a vertical line that intersects the x-axis at $(a, 0)$.	The graph of $x = 3$ is a vertical line passing through $(3, 0)$.

REVIEW EXERCISES

Determine whether each pair satisfies the equation $3x - 4y = 12$.

15. $(0, -3)$ **16.** $\left(\frac{4}{3}, 2\right)$

Graph each equation on a rectangular coordinate system.

17. $y = x - 5$ **18.** $y = 2x + 1$

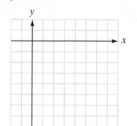 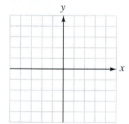

19. $y = \frac{x}{2} + 2$ **20.** $y = 3$

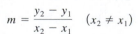 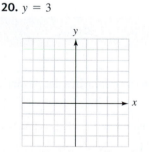

21. $x + y = 4$ **22.** $x - y = -3$

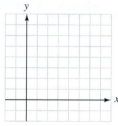 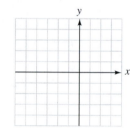

23. $3x + 5y = 15$ **24.** $7x - 4y = 28$

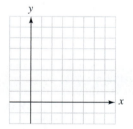 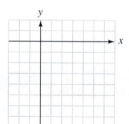

SECTION 3.3   Slope of a Line

DEFINITIONS AND CONCEPTS	**EXAMPLES**
The **slope** of a line passing through (x_1, y_1) and (x_2, y_2) is given by the formula $$m = \frac{y_2 - y_1}{x_2 - x_1} \quad (x_2 \neq x_1)$$	The slope of a line passing through $(-1, 4)$ and $(5, -3)$ is given by $$m = \frac{y_2 - y_1}{x_2 - x_1} = \frac{-3 - 4}{5 - (-1)} = \frac{-7}{6} = -\frac{7}{6}$$

Horizontal lines have a slope of 0.	$y = 5$ is a horizontal line with slope 0.
Vertical lines have an undefined slope.	$x = 5$ is a vertical line with an undefined slope.
Parallel lines have the same slope.	Two lines with slopes $\frac{2}{3}$ and $\frac{2}{3}$ are parallel.
The product of the slopes of nonvertical perpendicular lines is -1.	Two lines with slopes 3 and $-\frac{1}{3}$ are perpendicular.

REVIEW EXERCISES

Find the slope of the line passing through the given points.

25. $(3, 8), (7, 2)$ **26.** $(-1, 3), (3, -2)$

27. $(-2, -5), (-4, 9)$ **28.** $(-8, 2), (3, 2)$

Find the slope of the line determined by each equation.

29. $5x - 2y = 10$ **30.** $x = -4$

Determine whether the slope of each line is positive, negative, 0, or undefined.

31.

32.

33.

34.

Determine whether lines with the given slopes are parallel, perpendicular, or neither parallel nor perpendicular.

35. 4 and $-\frac{1}{4}$ **36.** 0.2 and $\frac{1}{5}$

37. -5 and $-\frac{1}{5}$ **38.** $\frac{1}{2}$ and 2

39. -7 and $\frac{1}{7}$ **40.** 0.25 and $\frac{1}{4}$

41. Find the slope of a roof if it rises 4 feet for every run of 12 feet.

42. Find the average rate of growth of a business if sales were \$25,000 the first year and \$66,000 the third year.

SECTION 3.4   Point-Slope Form

DEFINITIONS AND CONCEPTS	**EXAMPLES**
Point-slope form of a line: $$y - y_1 = m(x - x_1)$$ where m is the slope and (x_1, y_1) is a point on the line.	To write the point-slope form of a line passing through $(6, 5)$ with slope $\frac{1}{2}$, proceed as follows: $y - y_1 = m(x - x_1)$ This is point-slope form. $y - 5 = \frac{1}{2}(x - 6)$ Substitute. $y - 5 = \frac{1}{2}(x - 6)$

REVIEW EXERCISES

Use the point-slope form of a linear equation to find the equation of each line. Then solve the equation for y.

43. $m = 5$ and passing through $(-2, 3)$

44. $m = -\frac{1}{3}$ and passing through $\left(1, \frac{2}{3}\right)$

45. $m = \frac{1}{9}$ and passing through $(-27, -2)$

46. $m = -\frac{3}{5}$ and passing through $\left(1, -\frac{1}{5}\right)$

SECTION 3.5   Slope-Intercept Form

DEFINITIONS AND CONCEPTS	EXAMPLES
Slope-intercept form of a line: $$y = mx + b$$ where m is the slope and $(0, b)$ is the y-intercept.	To find the slope and y-intercept of the graph of $5x - 2y = 8$, we solve the equation for y. $$5x - 2y = 8$$ $$-2y = -5x + 8 \qquad \text{Subtract } 5x \text{ from both sides.}$$ $$y = \tfrac{5}{2}x - 4 \qquad \text{Divide both sides by } -2.$$ The slope is $\frac{5}{2}$ and the y-intercept is $(0, -4)$. To find the slope-intercept form of a line with slope $\frac{1}{2}$ and y-intercept $(0, 5)$, proceed as follows: $$y = \boldsymbol{m}x + \boldsymbol{b} \qquad \text{This is slope-intercept form.}$$ $$y = \tfrac{1}{2}x + \boldsymbol{5} \qquad \text{Substitute.}$$ $$y = \tfrac{1}{2}x + 5$$

REVIEW EXERCISES

Find the slope and the y-intercept of the line defined by each equation.

47. $y = -\frac{x}{3} + 6$ **48.** $3x - 6y = 9$

49. $2x + 5y = 1$ **50.** $x + 3y = 1$

Use the slope-intercept form to find the equation of each line.

51. $m = -3$, y-intercept $(0, 2)$
52. $m = 0$, y-intercept $(0, -7)$
53. $m = 7$, y-intercept $(0, 0)$
54. $m = \frac{1}{2}$, y-intercept $\left(0, -\frac{3}{2}\right)$

Graph the line that passes through the given point and has the given slope.

55. $(-1, 3)$, $m = -2$ **56.** $(1, -2)$, $m = 2$

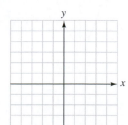   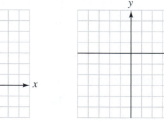

57. $\left(0, \frac{1}{2}\right)$, $m = \frac{3}{2}$ **58.** $(-3, 0)$, $m = -\frac{5}{2}$

    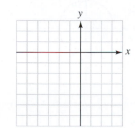

Determine whether the graphs of the equations are parallel, perpendicular, or neither parallel nor perpendicular.

59. $y = 3x$, $x = 3y$
60. $3x = y$, $x = -3y$
61. $x + 2y = y - x$, $2x + y = 3$
62. $3x + 2y = 7$, $2x - 3y = 8$

Write the equation of each line in slope-intercept form.

63. parallel to $y = 7x - 18$, passing through $(2, 5)$

64. parallel to $3x + 2y = 7$, passing through $(-3, 5)$

65. perpendicular to $2x - 5y = 12$, passing through $(0, 0)$

66. perpendicular to $y = \dfrac{x}{3} + 17$, y-intercept at $(0, -4)$

Find the slope-intercept equation of a line that passes through the given points.

67. $(-2, 5)$, $(1, -1)$

68. $(10, 8)$, $(2, 4)$

69. Depreciation   A company buys a fax machine for $2,700 and will sell it for $200 at the end of its useful life. If the company depreciates the machine linearly over a 5-year period, what will the machine be worth after 3 years?

SECTION 3.6   Functions

DEFINITIONS AND CONCEPTS	EXAMPLES
A **relation** is any set of ordered pairs.	The equation $y = |x - 1|$ represents a relation because it determines a set of ordered pairs (x, y).
Any equation in x and y where each value of x determines one value of y is called a **function**.	The equation $y = |x - 1|$ also represents a function because each value of x determines one value of y.
The set of all inputs into a function is called the **domain**. The set of all outputs is called the **range**.	The graph of $y = 4x - 1$ is a line with slope 4 and y-intercept $(0, -1)$. The domain is the set of real numbers $\mathbb{R}$, and the range is the set of real numbers $\mathbb{R}$.
The notation $y = f(x)$ denotes that the variable y (**dependent variable**) is a function of x (**independent variable**).	If $y = f(x) = -7x + 4$, find $f(2)$. $f(\mathbf{2}) = -7(\mathbf{2}) + 4$ Substitute 2 for x. $ = -14 + 4$ $ = -10$ In ordered pair form, we can write $(2, -10)$.
If any vertical line intersects a graph more than once, the graph does not represent a function. This is called the vertical line test.	Determine whether each graph represents a function. **a.** 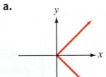 **b.** The graph in part **a** does not represent a function because it does not pass the vertical line test. The graph in part **b** represents a function because it passes the vertical line test.

REVIEW EXERCISES

70. Find the domain and range of the relation
$\{(-3, -2), (0, -1), (-3, 5)\}$.

71. Determine whether the equation $y = \dfrac{x}{2} - 1$ defines y to be a function of x.

Let $f(x) = -2x + 5$ and find each value.

72. $f(0)$

73. $f(-5)$

74. $f\left(-\dfrac{1}{2}\right)$

75. $f(6)$

76. Graph $f(x) = |x| - 3$ and state its domain and range.

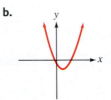

Determine whether each graph represents a function.

77.

78.

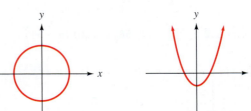

SECTION 3.7   Variation

DEFINITIONS AND CONCEPTS	EXAMPLES
$y = kx$ represents **direct variation**.	$y = 5x$ represents direct variation with a constant of variation, $k = 5$.
$y = \frac{k}{x}$ represents **inverse variation**.	$y = \frac{5}{x}$ represents inverse variation with a constant of variation, $k = 5$.
$y = kxz$ represents **joint variation**.	$y = 5xz$ represents joint variation with a constant of variation, $k = 5$.
Direct and inverse variation are used together in **combined variation**.	$y = \frac{5x}{z}$ represents combined variation with a constant of variation, $k = 5$.

REVIEW EXERCISES

Express each variation as an equation. Then find the requested value.

79. s varies directly with the square of t. Find s when $t = 6$ if $s = 32$ when $t = 2$.

80. l varies inversely with w. Find the constant of variation if $l = 30$ when $w = 20$.

81. R varies jointly with b and c. If $R = 72$ when $b = 4$ and $c = 24$, find R when $b = 6$ and $c = 18$.

82. s varies directly with w and inversely with the square of m. If $s = \frac{7}{4}$ when w and m are 4, find s when $w = 5$ and $m = 7$.

3   Test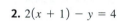

Graph each equation.

1. $y = \dfrac{x}{2} + 1$

2. $2(x + 1) - y = 4$

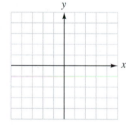
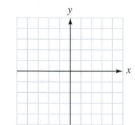

3. $x = 1$

4. $2y = 8$

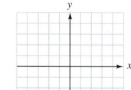
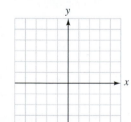

5. Find the slope of the line passing through $(-1, 3)$ and $(5, 8)$.

6. Find the slope of the line passing through $(-1, 3)$ and $(3, -1)$.

7. Find the slope of the line determined by $3x - 4y = 5$.

8. Find the y-intercept of the line determined by $2y - 7(x + 5) = 7$.

9. Find the slope of the line $y = 5$, if any.

10. Find the slope of the line $x = -2$, if any.

11. If two lines are parallel, their slopes are _____.

12. If two lines are nonvertical perpendicular lines, the product of their slopes is ___.

Determine whether lines with the given slopes are parallel, perpendicular, or neither.

13. 0.5 and -2

14. $\dfrac{20}{5}$ and 4

15. If a ramp rises 3 feet over a run of 12 feet, find the slope of the ramp.

16. If a business had sales of \$50,000 the second year in business and \$100,000 in sales the fifth year, find the annual dollar growth in sales.

17. Find the slope of a line parallel to a line with a slope of 2.

18. Find the slope of a line perpendicular to a line with a slope of 2.

19. Write the equation of a line that has a slope of 7 and passes through the point $(-2, 5)$. Give the result in point-slope form.

20. Write the equation of a line that has a slope of $\frac{3}{4}$ and a y-intercept of $(0, -5)$. Give the result in slope-intercept form.

21. Write the equation of a line that is perpendicular to the x-axis and passes through $(-7, 10)$.

22. Write the equation of a line that passes through $(3, -5)$ and is perpendicular to the line with the equation $y = \frac{1}{3}x + 11$. Give the result in slope-intercept form.

Suppose that $f(x) = 3x - 2$ and find each value.

23. $f(2)$

24. $f(-3)$

Graph the function and state its domain and range.

25. $f(x) = -|x| + 4$

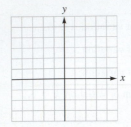

Determine whether each graph represents a function.

26.

27.

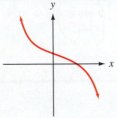

28. If y varies directly with x and $y = 32$ when $x = 8$, find x when $y = 4$.

29. If i varies inversely with the square of d, find the constant of variation if $i = 100$ when $d = 2$.

30. If y varies jointly with x and z, and $y = 16$ when $x = 4$ and $z = \frac{1}{2}$, find y when $x = 6$ and $z = 3$.

Polynomials

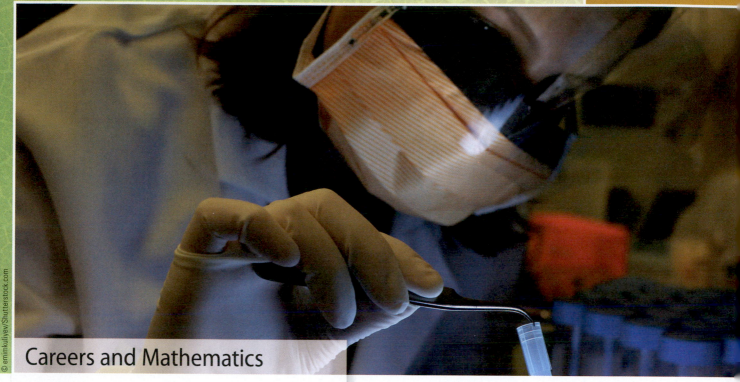

© emimkuliyev/Shutterstock.com

Careers and Mathematics

MEDICAL SCIENTISTS

Medical scientists research human diseases to improve human health. Most conduct biomedical research to gain knowledge of the life processes of living organisms, including viruses, bacteria, and other infectious agents. They study biological systems to understand the causes of disease and develop treatments. Medical scientists try to identify changes in cells or chromosomes that signal the development of medical problems, such as various types of cancer.

Job Outlook:
Employment of medical scientists is expected to increase 36 percent by 2020. This is faster than the average for all occupations.

Annual Earnings:
$44,830–$142,800

For More Information:
http://www.bls.gov/ooh/life-physical-and-social-science/medical-scientists.htm

For a Sample Application:
See Problem 73 in Section 4.3.

In this chapter

In this chapter, we will recognize patterns for integer exponents and express very large and small numbers in scientific notation. We then discuss special algebraic expressions, called polynomials, and show how to add, subtract, multiply, and divide them.

Reach for Success Setting Course Goals

Two professional football teams do not show up on a Sunday afternoon as if they were playing a game of sandlot football. Instead, before the game they've studied films, they've practiced the plays—they have a "game plan."

In this exercise you will set some realistic goals for this course and then establish a "game plan" that will help you achieve your goals.

© Drew Hallowell/Getty Images

We're going to assume one of your goals is to be successful in this course.

The grade I am willing to work to achieve is a/an _____.

Most college classes recommend 2–3 hours of outside time per week for every 1 hour in class if students want to be successful. With mathematics, you may find it takes more time.

Considering other course commitments as well as any work and family commitments, state the number of hours *outside of class* you realistically believe you can devote each week to this one course. _____

To help yourself be successful in this course,

- complete all homework and other assignments on time.

- take good notes in class.

- ask questions in class.

- study on a daily basis.

- participate in a study group.

- utilize the mathematics or tutor center.

- attend instructor reviews (if available) or office hours.

- access online tutoring services (if available).

List at least three things you are willing to add to your "game plan" and commit to, that will support your success in this course.

1. _____

2. _____

3. _____

Can you think of anything else you can do to improve your performance in this class?

_____

Consider your answers to the questions above. Is your plan sufficient to achieve your goal?

YES _____   NO _____

Please explain. _____

A Successful Study Strategy . . .

Set a goal, outline a plan of action, and stick to the plan to achieve that goal.

At the end of the chapter you will find an additional exercise to guide you through a successful semester.

Section 4.1

Natural-Number Exponents

Objectives

1. Write an exponential expression without exponents.
2. Write a repeated multiplication expression using exponents.
3. Simplify an expression by using the product rule for exponents.
4. Simplify an expression by using the power rules for exponents.
5. Simplify an expression by using the quotient rule for exponents.

Vocabulary

base   exponent   power

Getting Ready

Evaluate each expression.

1. 2^3 **2.** 3^2 **3.** $3(2)$ **4.** $2(3)$

5. $2^3 + 2^2$ **6.** $2^3 \cdot 2^2$ **7.** $3^3 - 3^2$ **8.** $\dfrac{3^3}{3^2}$

In this section, we will revisit the topic of exponents. This time we will develop the basic properties used to manipulate exponential expressions.

1. Write an exponential expression without exponents.

We have used natural-number exponents to indicate repeated multiplication. For example,

$$2^5 = 2 \cdot 2 \cdot 2 \cdot 2 \cdot 2 = 32 \quad (-7)^3 = (-7)(-7)(-7) = -343$$
$$x^4 = x \cdot x \cdot x \cdot x \qquad -y^5 = -1 \cdot y \cdot y \cdot y \cdot y \cdot y$$

These examples suggest a definition for x^n, where n is a natural number.

NATURAL-NUMBER EXPONENTS

If n is a natural number, then

$$x^n = \overbrace{x \cdot x \cdot x \cdot \cdots \cdot x}^{n \text{ factors of } x}$$

In the exponential expression x^n, x is called the **base** and n is called the **exponent**. The entire expression is called a **power** of x.

Base $\rightarrow x^n \leftarrow$ Exponent

If an exponent is a natural number, it tells how many times its base is to be used as a factor. An exponent of 1 indicates that its base is to be used one time as a factor, an exponent of 2 indicates that its base is to be used two times as a factor, and so on.

$$3^1 = 3 \qquad (-y)^1 = -y \qquad (-4z)^2 = (-4z)(-4z) \qquad (t^2)^3 = t^2 \cdot t^2 \cdot t^2$$

EXAMPLE 1 Find each value to show that **a.** $(-2)^4$ and **b.** -2^4 have different values.

Solution We find each power and show that the results are different.

a. $(-2)^4 = (-2)(-2)(-2)(-2)$ **b.** $-2^4 = -(2^4)$
$$= 16$$
$$= -(2 \cdot 2 \cdot 2 \cdot 2)$$
$$= -16$$

Since $16 \neq -16$, it follows that $(-2)^4 \neq -2^4$.
Note that in part a the base is -2 but in part b the base is 2.

SELF CHECK 1 Show that $(-4)^3$ and -4^3 have the same value.

COMMENT There is a pattern regarding even and odd exponents. If the exponent of a base is even, the result is positive. If the exponent of a base is odd, the result will be the same sign as the original base.

EXAMPLE 2 Write each expression without exponents.

a. r^3 **b.** $(-2s)^4$ **c.** $\left(\dfrac{1}{3}ab\right)^5$

Solution **a.** $r^3 = r \cdot r \cdot r$

b. $(-2s)^4 = (-2s)(-2s)(-2s)(-2s)$

c. $\left(\dfrac{1}{3}ab\right)^5 = \left(\dfrac{1}{3}ab\right)\left(\dfrac{1}{3}ab\right)\left(\dfrac{1}{3}ab\right)\left(\dfrac{1}{3}ab\right)\left(\dfrac{1}{3}ab\right)$

SELF CHECK 2 Write each expression without exponents.   **a.** x^4
b. $(3z)^4$ **c.** $\left(-\dfrac{1}{2}xy\right)^3$

2   Write a repeated multiplication expression using exponents.

Many expressions can be written more compactly by using exponents.

EXAMPLE 3 Write each expression using one exponent.
a. $3 \cdot 3 \cdot 3 \cdot 3 \cdot 3$ **b.** $(5z)(5z)(5z)$

Solution **a.** Since 3 is used as a factor five times,

$$3 \cdot 3 \cdot 3 \cdot 3 \cdot 3 = 3^5$$

b. Since $5z$ is used as a factor three times,

$$(5z)(5z)(5z) = (5z)^3$$

SELF CHECK 3 Write each expression using one exponent.

a. $7 \cdot 7 \cdot 7 \cdot 7$ **b.** $\left(\dfrac{1}{3}xy\right)\left(\dfrac{1}{3}xy\right)$

3 Simplify an expression by using the product rule for exponents.

To develop a pattern for multiplying exponential expressions with the same base, we consider the product $x^2 \cdot x^3$. Since the expression x^2 means that x is to be used as a factor two times and the expression x^3 means that x is to be used as a factor three times, we have

$$x^2 x^3 = \overbrace{x \cdot x}^{2 \text{ factors of } x} \cdot \overbrace{x \cdot x \cdot x}^{3 \text{ factors of } x}$$

$$= \overbrace{x \cdot x \cdot x \cdot x \cdot x}^{5 \text{ factors of } x}$$

$$= x^5$$

In general,

$$x^m \cdot x^n = \overbrace{x \cdot x \cdot x \cdots x}^{m \text{ factors of } x} \overbrace{x \cdot x \cdot x \cdot x \cdots x}^{n \text{ factors of } x}$$

$$= \overbrace{x \cdot x \cdot x \cdot x \cdot x \cdot x \cdots x \cdot x \cdot x}^{m+n \text{ factors of } x}$$

$$= x^{m+n}$$

This discussion suggests the following pattern: *To multiply two exponential expressions with the same base, keep the base and add the exponents.*

PRODUCT RULE FOR EXPONENTS	If m and n are natural numbers, then $$x^m x^n = x^{m+n}$$

EXAMPLE 4   Simplify each expression.

   a. $x^3 x^4 = x^{3+4}$ Keep the base and add the exponents.

          $= x^7$ $3 + 4 = 7$

   b. $y^2 y^4 y = (\mathbf{y^2 y^4})y$ Use the associative property of multiplication to group y^2 and y^4 together.

          $= (\mathbf{y^{2+4}})y$ Keep the base and add the exponents.

          $= \mathbf{y^6} y$ $2 + 4 = 6$

          $= y^{6+1}$ Keep the base and add the exponents: $y = y^1$.

          $= y^7$ $6 + 1 = 7$

 SELF CHECK 4   Simplify each expression.   **a.** zz^3 **b.** $x^2 x^3 x^6$

EXAMPLE 5   Simplify: $(2y^3)(3y^2)$

Solution $(2y^3)(3y^2) = 2(3)y^3 y^2$ Use the commutative and associative properties of multiplication to group the coefficients together and the variables together.

          $= 6y^{3+2}$ Multiply the coefficients. Keep the base and add the exponents.

          $= 6y^5$ $3 + 2 = 5$

 SELF CHECK 5   Simplify: $(4x)(-3x^2)$

COMMENT   The product rule for exponents applies only to exponential expressions with the same base. An expression such as $x^2 y^3$ cannot be simplified, because x^2 and y^3 have different bases.

4 Simplify an expression by using the power rules for exponents.

To find another pattern of exponents, we consider the expression $(x^3)^4$, which can be written as $x^3 \cdot x^3 \cdot x^3 \cdot x^3$. Because each of the four factors of x^3 contains three factors of x, there are $4 \cdot 3$ (or 12) factors of x. Thus, the expression can be written as x^{12}.

$$(x^3)^4 = x^3 \cdot x^3 \cdot x^3 \cdot x^3$$

$$= \overbrace{x \cdot x \cdot x \cdot x \cdot x \cdot x \cdot x \cdot x \cdot x \cdot x \cdot x \cdot x}^{12 \text{ factors of } x}$$
$$\underbrace{}_{x^3} \underbrace{}_{x^3} \underbrace{}_{x^3} \underbrace{}_{x^3}$$

$$= x^{12}$$

In general,

$$(x^m)^n = \overbrace{x^m \cdot x^m \cdot x^m \cdot \cdots \cdot x^m}^{n \text{ factors of } x^m}$$

$$= \overbrace{x \cdot x \cdot x \cdot x \cdot x \cdot x \cdot x \cdots \cdot x}^{m \cdot n \text{ factors of } x}$$

$$= x^{m \cdot n}$$

The previous discussion suggests the following pattern: *To raise an exponential expression to a power, keep the base and multiply the exponents.*

POWER RULE FOR EXPONENTS

If m and n are natural numbers, then

$$(x^m)^n = x^{mn}$$

EXAMPLE 6   Simplify:

a. $(2^3)^7 = 2^{3 \cdot 7}$ Keep the base and multiply the exponents.
    $= 2^{21}$ $3 \cdot 7 = 21$

b. $(z^7)^7 = z^{7 \cdot 7}$ Keep the base and multiply the exponents.
    $= z^{49}$ $7 \cdot 7 = 49$

SELF CHECK 6   Simplify:
a. $(y^5)^2$ **b.** $(u^x)^y$

In the next example, the product and power rules of exponents are both used.

EXAMPLE 7   Simplify:

a. $(x^2x^5)^2 = (x^7)^2$ **b.** $(y^6y^2)^3 = (y^8)^3$
    $= x^{14}$ $= y^{24}$

c. $(z^2)^4(z^3)^3 = z^8z^9$ **d.** $(x^3)^2(x^5x^2)^3 = x^6(x^7)^3$
    $= z^{17}$ $= x^6x^{21}$
                    $= x^{27}$

SELF CHECK 7   Simplify:
a. $(a^4a^3)^3$ **b.** $(k^9k^4)^2$ **c.** $(a^3)^3(a^4)^2$ **d.** $(y^5)^3(y^4y)^5$

To find more patterns for exponents, we consider the expressions $(2x)^3$ and $\left(\frac{2}{x}\right)^3$.

$$(2x)^3 = (2x)(2x)(2x) \qquad\qquad \left(\frac{2}{x}\right)^3 = \left(\frac{2}{x}\right)\left(\frac{2}{x}\right)\left(\frac{2}{x}\right) \quad (x \neq 0)$$

$$= (2 \cdot 2 \cdot 2)(x \cdot x \cdot x) \qquad\qquad = \frac{2 \cdot 2 \cdot 2}{x \cdot x \cdot x}$$

$$= 2^3 x^3 \qquad\qquad\qquad\qquad = \frac{2^3}{x^3}$$

$$= 8x^3 \qquad\qquad\qquad\qquad\quad = \frac{8}{x^3}$$

These examples suggest the following patterns: *To raise a product to a power, we raise each factor of the product to that power, and to raise a quotient to a power, we raise both the numerator and denominator to that power.*

PRODUCT TO A POWER RULE FOR EXPONENTS	If n is a natural number, then $$(xy)^n = x^n y^n$$
QUOTIENT TO A POWER RULE FOR EXPONENTS	If n is a natural number, and if $y \neq 0$, then $$\left(\frac{x}{y}\right)^n = \frac{x^n}{y^n}$$

EXAMPLE 8   Simplify. Assume no division by zero.

a. $(ab)^4 = a^4 b^4$

b. $(3c)^3 = 3^3 c^3$
$= 27c^3$

c. $(x^2 y^3)^5 = (x^2)^5 (y^3)^5$
$= x^{10} y^{15}$

d. $(-2x^3 y)^2 = (-2)^2 (x^3)^2 y^2$
$= 4x^6 y^2$

e. $\left(\frac{4}{k}\right)^3 = \frac{4^3}{k^3}$
$= \frac{64}{k^3}$

f. $\left(\frac{3x^2}{2y^3}\right)^5 = \frac{3^5 (x^2)^5}{2^5 (y^3)^5}$
$= \frac{243x^{10}}{32y^{15}}$

SELF CHECK 8   Simplify. Assume no division by zero.

a. $(cd)^5$ **b.** $(4x)^3$ **c.** $(a^5 b)^3$

d. $(3x^2 y)^2$ **e.** $\left(\frac{z}{8}\right)^3$ **f.** $\left(\frac{2x^3}{3y^2}\right)^4$

5 **Simplify an expression by using the quotient rule for exponents.**

To find a pattern for dividing exponential expressions, we consider the fraction $\frac{4^5}{4^2}$, where the exponent in the numerator is greater than the exponent in the denominator. We can simplify the fraction as follows:

$$\frac{4^5}{4^2} = \frac{4 \cdot 4 \cdot 4 \cdot 4 \cdot 4}{4 \cdot 4}$$

$$= \frac{\overset{1}{\cancel{4}} \cdot \overset{1}{\cancel{4}} \cdot 4 \cdot 4 \cdot 4}{\underset{1}{\cancel{4}} \cdot \underset{1}{\cancel{4}}}$$

$$= 4^3$$

The result of 4^3 has a base of 4 and an exponent of $5 - 2$ (or 3). This suggests that *to divide exponential expressions with the same base, we keep the base and subtract the exponents.*

QUOTIENT RULE FOR EXPONENTS	If m and n are natural numbers, $m > n$ and $x \neq 0$, then
	$$\frac{x^m}{x^n} = x^{m-n}$$

EXAMPLE 9   Simplify. Assume no division by zero.

Solution   **a.** $\dfrac{x^4}{x^3} = x^{4-3}$

$= x^1$

$= x$

b. $\dfrac{8\mathbf{y^2 y^6}}{6y^3} = \dfrac{8\mathbf{y^8}}{6y^3}$

$= \dfrac{8y^{8-3}}{6}$

$= \dfrac{4y^5}{3}$

c. $\dfrac{a^3 a^5 a^7}{a^4 a} = \dfrac{a^{15}}{a^5}$

$= a^{15-5}$

$= a^{10}$

d. $\dfrac{(a^3 b^4)^2}{ab^5} = \dfrac{a^6 b^8}{ab^5}$

$= a^{6-1} b^{8-5}$

$= a^5 b^3$

SELF CHECK 9   Simplify. Assume no division by zero.

a. $\dfrac{a^5}{a^3}$ **b.** $\dfrac{6b^2 b^3}{4b^4}$ **c.** $\dfrac{a^4 a^7 a}{a^3 a^5}$ **d.** $\dfrac{(x^2 y^3)^2}{x^3 y^4}$

SELF CHECK ANSWERS

1. both are -64 **2. a.** $x \cdot x \cdot x \cdot x$ **b.** $(3z)(3z)(3z)(3z)$ **c.** $\left(-\frac{1}{2}xy\right)\left(-\frac{1}{2}xy\right)\left(-\frac{1}{2}xy\right)$
3. a. 7^4 **b.** $\left(\frac{1}{3}xy\right)^2$ **4. a.** z^4 **b.** x^{11} **5.** $-12x^3$ **6. a.** y^{10} **b.** u^{xy} **7. a.** a^{21} **b.** k^{26} **c.** a^{17}
d. y^{40} **8. a.** $c^5 d^5$ **b.** $64x^3$ **c.** $a^{15} b^3$ **d.** $9x^4 y^2$ **e.** $\frac{z^3}{512}$ **f.** $\frac{16x^{12}}{81y^8}$ **9. a.** a^2 **b.** $\frac{3b}{2}$ **c.** a^4 **d.** xy^2

NOW TRY THIS

Simplify each expression. Assume no division by zero.

1. If $x^{1/2}$ has meaning, find $(x^{1/2})^2$.

2. $-3^2(x^2 - 2^2)$

3. a. $x^{p+1} x^p$ **b.** $(x^{p+1})^2$ **c.** $\dfrac{x^{2p+1}}{x^p}$

4.1   Exercises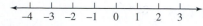

WARM-UPS *Perform the operations.*

1. $4 \cdot 4 \cdot 4$

2. $2 \cdot 2 \cdot 2 \cdot 2$

3. $(-7)(-7)(-7)$

4. $(-3)(-3)(-3)(-3)$

Evaluate each expression.

5. 6^2

6. $(-6)^2$

7. $3(4)^2$

8. $(3 \cdot 4)^2$

REVIEW

9. Graph the real numbers $-3, 0, 2,$ and $-\frac{3}{2}$ on a number line.

$$\begin{array}{ccccccccc} \leftarrow & | & | & | & | & | & | & | & \rightarrow \\ -4 & -3 & -2 & -1 & 0 & 1 & 2 & 3 \end{array}$$

10. Graph the real numbers $-2 < x \le 3$ on a number line.

$$\begin{array}{ccccccc} \leftarrow & | & | & | & | & | & | & \rightarrow \\ -3 & -2 & -1 & 0 & 1 & 2 & 3 \end{array}$$

Write each algebraic expression as an English phrase.

11. $3(x + y)$

12. $3x + y$

Write each English phrase as an algebraic expression.

13. Three greater than the absolute value of twice x

14. The sum of the numbers y and z decreased by the sum of their squares

VOCABULARY AND CONCEPTS *Fill in the blanks.*
Assume no division by zero.

15. The ____ of the exponential expression $(-5)^3$ is __. The exponent is _.

16. The base of the exponential expression -5^3 is _. The _____ is 3.

17. $(4y)^3$ means _____.

18. Write $(-5x)(-5x)(-5x)(-5x)$ as a power. _____

19. $y^5 =$ _____

20. $x^m x^n =$ ____

21. $(xy)^n =$ ___

22. $\left(\dfrac{a}{b}\right)^n =$ ___

23. $(a^b)^c =$ ___

24. $\dfrac{x^m}{x^n} =$ ____

25. The area of the square is $s \cdot s$. Why do you think the symbol s^2 is called "s squared"?

26. The volume of the cube is $s \cdot s \cdot s$. Why do you think the symbol s^3 is called "s cubed"?

Identify the base and the exponent in each expression.

27. x^4

28. 4^x

29. 7^2

30. $(-7)^2$

31. $(2y)^3$

32. $(-3x)^2$

33. $-x^4$

34. $(-x)^4$

35. x

36. $(xy)^3$

37. $2x^3$

38. $-3y^6$

GUIDED PRACTICE *Evaluate each expression. SEE EXAMPLE 1.*
(OBJECTIVE 1)

39. -5^2

40. $(-5)^2$

41. $2^2 + 3^2$

42. $2^3 - 2^2$

43. $-3(6^2 - 2^3)$

44. $2(4^3 + 3^2)$

45. $(-4)^3 + (-3)^2$

46. $-4^3 - 3^2$

Write each expression without using exponents. SEE EXAMPLE 2.
(OBJECTIVE 1)

47. 5^3

48. -4^5

49. $-5x^6$

50. $3x^3$

51. $-2y^4$

52. $(-2y)^4$

53. $(3t)^5$

54. $a^3 b^2$

Write each expression using one exponent. SEE EXAMPLE 3.
(OBJECTIVE 2)

55. $2 \cdot 2 \cdot 2$

56. $5 \cdot 5$

57. $x \cdot x \cdot x \cdot x$

58. $y \cdot y \cdot y \cdot y \cdot y \cdot y$

59. $(2x)(2x)(2x)$

60. $(-4y)(-4y)$

61. $-4 \cdot t \cdot t \cdot t \cdot t$

62. $-8 \cdot a \cdot a \cdot a \cdot a \cdot a$

Simplify. SEE EXAMPLE 4. (OBJECTIVE 3)

63. $x^4 x^3$

64. $y^5 y^2$

65. $x^5 x^5$

66. yy^3

67. $a^3 a^4 a^5$

68. $b^2 b^3 b^5$

69. $y^3 (y^2 y^4)$

70. $(y^4 y)y^6$

Simplify. SEE EXAMPLE 5. (OBJECTIVE 3)

71. $4x^2 (3x^5)$

72. $-2y(y^3)$

73. $(5x^4)(-2x)$

74. $(-3y^6)(-6y^5)$

Simplify. SEE EXAMPLE 6. (OBJECTIVE 4)

75. $(3^2)^4$

76. $(4^3)^3$

77. $(y^5)^3$

78. $(b^3)^6$

Simplify. *SEE EXAMPLE 7. (OBJECTIVE 4)*

79. $(x^2x^3)^5$ **80.** $(y^3y^4)^4$

81. $(a^2a^7)^3$ **82.** $(q^3q^3)^5$

83. $(x^5)^2(x^7)^3$ **84.** $(y^3y)^2(y^2)^2$

85. $(r^3r^2)^4(r^3r^5)^2$ **86.** $(yy^3)^3(y^2y^3)^4(y^3y^3)^2$

Simplify. Assume no division by 0. *SEE EXAMPLE 8. (OBJECTIVE 4)*

87. $(xy)^3$ **88.** $(uv^2)^4$

89. $(r^3s^2)^2$ **90.** $(a^3b^2)^3$

91. $(4ab^2)^2$ **92.** $(3x^2y)^3$

93. $(-2r^2s^3t)^3$ **94.** $(-3x^2y^4z)^2$

95. $\left(\dfrac{a}{b}\right)^3$ **96.** $\left(\dfrac{r^2}{s}\right)^4$

97. $\left(\dfrac{2x}{3y^2}\right)^4$ **98.** $\left(\dfrac{4u^2}{5v^4}\right)^3$

Simplify. Assume no division by 0. *SEE EXAMPLE 9. (OBJECTIVE 5)*

99. $\dfrac{x^5}{x^3}$ **100.** $\dfrac{a^6}{a^3}$

101. $\dfrac{y^3y^4}{yy^2}$ **102.** $\dfrac{b^4b^5}{b^2b^3}$

103. $\dfrac{12a^2a^3a^4}{4(a^4)^2}$ **104.** $\dfrac{16(aa^2)^3}{2a^2a^3}$

105. $\dfrac{(ab^2)^3}{(ab)^2}$ **106.** $\dfrac{(m^3n^4)^3}{(mn^2)^3}$

ADDITIONAL PRACTICE Simplify. Assume no division by 0.

107. tt^2 **108.** w^3w^5

109. $6x^3(-x^2)(-x^4)$ **110.** $-2x(-x^2)(-3x)$

111. $(-2a^5)^3$ **112.** $(-3b^7)^4$

113. $(3zz^2z^3)^5$ **114.** $(4t^3t^6t^2)^2$

115. $(s^3)^3(s^2)^2(s^5)^4$ **116.** $(s^2)^3(s^3)^2(s^4)^4$

117. $\left(\dfrac{-2a}{b}\right)^5$ **118.** $\left(\dfrac{2t}{3}\right)^4$

119. $\left(\dfrac{b^2}{3a}\right)^3$ **120.** $\left(\dfrac{a^3b}{c^4}\right)^5$

121. $\dfrac{17(x^4y^3)^8}{34(x^5y^2)^4}$ **122.** $\dfrac{35(r^3s^2)^2}{49r^2s^3}$

123. $\left(\dfrac{y^3y}{2yy^2}\right)^3$ **124.** $\left(\dfrac{3t^3t^4t^5}{4t^2t^6}\right)^3$

125. $\left(\dfrac{-2r^3r^3}{3r^4r}\right)^3$ **126.** $\left(\dfrac{-6y^4y^5}{5y^3y^5}\right)^2$

127. $\dfrac{20(r^4s^3)^4}{6(rs^3)^3}$ **128.** $\dfrac{15(x^2y^5)^5}{21(x^3y)^2}$

APPLICATIONS

129. Bouncing balls   When a certain ball is dropped, it always rebounds to one-half of its previous height. If the ball is dropped from a height of 32 feet, explain why the expression $32\left(\frac{1}{2}\right)^4$ represents the height of the ball on the fourth bounce. Find the height of the fourth bounce.

130. Having babies   The probability that a couple will have n baby boys in a row is given by the formula $\left(\frac{1}{2}\right)^n$. Find the probability that a couple will have four baby boys in a row.

131. Investing   If an investment of \$1,000 doubles every seven years, find the value of the investment after 28 years.

If P dollars are invested at a rate r, compounded annually, it will grow to A dollars in t years according to the formula

$$A = P(1 + r)^t$$

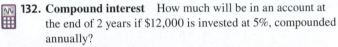 **132. Compound interest**   How much will be in an account at the end of 2 years if \$12,000 is invested at 5%, compounded annually?

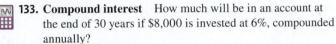 **133. Compound interest**   How much will be in an account at the end of 30 years if \$8,000 is invested at 6%, compounded annually?

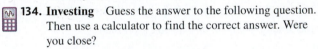 **134. Investing**   Guess the answer to the following question. Then use a calculator to find the correct answer. Were you close?

If the value of 1¢ is to double every day, what will the penny be worth after 31 days?

WRITING ABOUT MATH

135. Describe how you would multiply two exponential expressions with like bases.

136. Describe how you would divide two exponential expressions with like bases.

SOMETHING TO THINK ABOUT

137. Is the operation of raising to a power commutative? That is, is $a^b = b^a$? Explain.

138. Is the operation of raising to a power associative? That is, is $(a^b)^c = a^{(bc)}$? Explain.

Section 4.2

Zero and Negative-Integer Exponents

Objectives

1. Simplify an expression containing an exponent of zero.
2. Simplify an expression containing a negative-integer exponent.
3. Simplify an expression containing a variable exponent.

Getting Ready

Simplify by dividing out common factors. Assume that $y \neq 0$, $x \neq 0$.

1. $\dfrac{3 \cdot 3 \cdot 3}{3 \cdot 3 \cdot 3 \cdot 3}$

2. $\dfrac{2yy}{2yyy}$

3. $\dfrac{3xx}{3xx}$

4. $\dfrac{xxy}{xxxyy}$

In the previous section, we discussed natural-number exponents. We now continue the discussion to include 0 and negative-integer exponents.

1   Simplify an expression containing an exponent of zero.

When we discussed the quotient rule for exponents in the previous section, the exponent in the numerator was always greater than the exponent in the denominator. We now consider what happens when the exponents are equal.

If we apply the quotient rule to the fraction $\dfrac{5^3}{5^3}$, where the exponents in the numerator and denominator are equal, we obtain 5^0. However, because any nonzero number divided by itself equals 1, we also obtain 1.

$$\frac{5^3}{5^3} = 5^{3-3} = \mathbf{5^0} \qquad \frac{5^3}{5^3} = \frac{\overset{1}{\cancel{5}} \cdot \overset{1}{\cancel{5}} \cdot \overset{1}{\cancel{5}}}{\underset{1}{\cancel{5}} \cdot \underset{1}{\cancel{5}} \cdot \underset{1}{\cancel{5}}} = \mathbf{1}$$

These are equal.

For this reason, we define 5^0 to be equal to 1. In general, the following is true.

ZERO EXPONENTS

If x is any nonzero real number, then

$$x^0 = 1$$

EXAMPLE 1   Write each expression without exponents.

a. $\left(\dfrac{1}{13}\right)^0 = 1$

b. $\dfrac{x^5}{x^5} = x^{5-5} \quad (x \neq 0)$

$\qquad = x^0$

$\qquad = 1$

c. $3x^0 = 3(\mathbf{1})$

$\qquad = 3$

d. $(3x)^0 = 1$

e. $\dfrac{6^n}{6^n} = 6^{n-n}$ \qquad\qquad **f.** $\dfrac{y^m}{y^m} = y^{m-m}$ $(y \neq 0)$

$\qquad\qquad = 6^0$ \qquad\qquad\qquad\qquad $= y^0$

$\qquad\qquad = 1$ \qquad\qquad\qquad\qquad\quad $= 1$

Parts c and d illustrate that $3x^0 \neq (3x)^0$.

SELF CHECK 1 Write each expression without exponents.   **a.** $(-0.115)^0$ **b.** $\dfrac{4^2}{4^2}$
c. $-2x^0$ **d.** $(-9y)^0$ **e.** $\dfrac{-2^x}{-2^x}$ **f.** $\dfrac{x^m}{x^m}$ $(x \neq 0)$

---

2 **Simplify an expression containing a negative-integer exponent.**

If we apply the quotient rule to $\dfrac{6^2}{6^5}$, where the exponent in the numerator is less than the exponent in the denominator, we obtain 6^{-3}. However, by dividing out two factors of 6, we also obtain $\dfrac{1}{6^3}$.

$$\dfrac{6^2}{6^5} = 6^{2-5} = \mathbf{6^{-3}} \qquad \dfrac{6^2}{6^5} = \dfrac{\overset{1}{\cancel{6}} \cdot \overset{1}{\cancel{6}}}{\underset{1}{\cancel{6}} \cdot \underset{1}{\cancel{6}} \cdot 6 \cdot 6 \cdot 6} = \dfrac{\mathbf{1}}{\mathbf{6^3}}$$

— These are equal. —

For these reasons, we define 6^{-3} to be $\dfrac{1}{6^3}$. In general, the following is true.

NEGATIVE EXPONENTS   If x is any nonzero number and n is a natural number, then

$$x^{-n} = \dfrac{1}{x^n} \qquad \text{and} \qquad \dfrac{1}{x^{-n}} = x^n$$

COMMENT   A negative exponent can be viewed as "taking the reciprocal."

EXAMPLE 2   Express each quantity without negative exponents or parentheses. Assume no variables are 0.

a. $3^{-5} = \dfrac{1}{3^5}$ \qquad\qquad\qquad\qquad **b.** $x^{-4} = \dfrac{1}{x^4}$

$\qquad\quad = \dfrac{1}{243}$

c. $(2x)^{-2} = \dfrac{1}{(2x)^2}$ \qquad\qquad\qquad **d.** $2x^{-2} = 2\left(\dfrac{1}{x^2}\right)$

$\qquad\qquad\quad = \dfrac{1}{4x^2}$ \qquad\qquad\qquad\qquad\qquad $= \dfrac{2}{x^2}$

e. $(-3a)^{-4} = \dfrac{1}{(-3a)^4}$ \qquad\qquad **f.** $\dfrac{1}{(3x)^{-2}} = \dfrac{(3x)^2}{1}$

$\qquad\qquad\qquad = \dfrac{1}{81a^4}$ \qquad\qquad\qquad\qquad\quad $= 9x^2$

SELF CHECK 2 Write each expression without negative exponents or parentheses. Assume no variables are 0.   **a.** 4^{-3} **b.** a^{-5} **c.** $(3y)^{-3}$ **d.** $-5x^{-3}$
e. $(-2x)^{-3}$ \qquad **f.** $\dfrac{1}{(2x)^{-3}}$

Because of the definitions of negative and zero exponents, the product, power, and quotient rules are true for all integer exponents.

PROPERTIES OF INTEGER EXPONENTS	If m and n are integers and no base is 0 $(x \neq 0, y \neq 0)$, then

$$x^m x^n = x^{m+n} \qquad (x^m)^n = x^{mn} \qquad (xy)^n = x^n y^n \qquad \left(\frac{x}{y}\right)^n = \frac{x^n}{y^n}$$

$$x^0 = 1 \qquad x^{-n} = \frac{1}{x^n} \qquad \frac{1}{x^{-n}} = x^n \qquad \frac{x^m}{x^n} = x^{m-n}$$

EXAMPLE 3   Simplify and write the result without negative exponents. Assume no variables are 0.

COMMENT   If m and n are natural numbers, $m < n$ and $x \neq 0$, then $\dfrac{x^m}{x^n} = \dfrac{1}{x^{n-m}}$.

a. $(x^{-3})^2 = x^{-6}$
$= \dfrac{1}{x^6}$

b. $\dfrac{x^3}{x^7} = x^{3-7}$
$= x^{-4}$
$= \dfrac{1}{x^4}$

c. $\dfrac{y^{-4}y^{-3}}{y^{-20}} = \dfrac{y^{-7}}{y^{-20}}$
$= y^{-7-(-20)}$
$= y^{-7+20}$
$= y^{13}$

d. $\dfrac{12a^3b^4}{4a^5b^2} = 3a^{3-5}b^{4-2}$
$= 3a^{-2}b^2$
$= \dfrac{3b^2}{a^2}$

e. $\left(-\dfrac{x^3 y^2}{xy^{-3}}\right)^{-2} = (-x^{3-1}y^{2-(-3)})^{-2}$
$= (-x^2 y^5)^{-2}$
$= \dfrac{1}{(-x^2 y^5)^2}$
$= \dfrac{1}{x^4 y^{10}}$

SELF CHECK 3   Simplify and write the result without negative exponents. Assume no variables are 0.

a. $(x^4)^{-3}$ b. $\dfrac{a^4}{a^8}$ c. $\dfrac{a^{-4}a^{-5}}{a^{-3}}$ d. $\dfrac{20x^5 y^3}{5x^3 y^6}$ e. $\left(\dfrac{x^4 y^5}{x^{-7} y^9}\right)^{-3}$

3   Simplify an expression containing a variable exponent.

The properties of exponents are also true when the exponents are algebraic expressions.

EXAMPLE 4   Simplify. Assume no base is 0.

a. $x^{2m}x^{3m} = x^{2m+3m}$
$= x^{5m}$

b. $\dfrac{y^{2m}}{y^{4m}} = y^{2m-4m}$
$= y^{-2m}$
$= \dfrac{1}{y^{2m}}$

c. $a^{2m-1}a^{2m} = a^{2m-1+2m}$
$= a^{4m-1}$

d. $(b^{m+1})^2 = b^{(m+1)2}$
$= b^{2m+2}$

SELF CHECK 4   Simplify. Assume no base is 0.

a. $z^{3n}z^{2n}$ b. $\dfrac{z^{3n}}{z^{5n}}$ c. $x^{3m+2}x^m$ d. $(x^{m+2})^3$

Accent on technology

▸ Finding Present Value

To find out how much money P (called the *present value*) must be invested at an annual rate i (expressed as a decimal) to have A in n years, we use the formula $P = A(1 + i)^{-n}$. To find out how much we must invest at 6% to have $50,000 in 10 years, we substitute 50,000 for A, 0.06 (6%) for i, and 10 for n to get

$$P = A(1 + i)^{-n}$$
$$P = 50,000(1 + 0.06)^{-10}$$

To evaluate P with a calculator, we enter these numbers and press these keys:

(1 + .06) y^x 10 +/− × 50000

Using a calculator with a y^x and a +/− key.

50000 (1 + .06) ∧ (−) 10 **ENTER**

Using a TI84 graphing calculator.

Either way, we see that we must invest $27,919.74 to have $50,000 in 10 years.

For instructions regarding the use of a Casio graphing calculator, please refer to the Casio Keystroke Guide in the back of the book.

SELF CHECK ANSWERS

1. a. 1 b. 1 c. −2 d. 1 e. 1 f. 1 2. a. $\frac{1}{64}$ b. $\frac{1}{a^5}$ c. $\frac{1}{27y^3}$ d. $-\frac{5}{x^3}$ e. $-\frac{1}{8x^3}$ f. $8x^3$ 3. a. $\frac{1}{x^{12}}$
b. $\frac{1}{a^4}$ c. $\frac{1}{a^6}$ d. $\frac{4x^2}{y^3}$ e. $\frac{y^{12}}{x^{33}}$ 4. a. z^{5n} b. $\frac{1}{z^{2n}}$ c. x^{4m+2} d. x^{3m+6}

NOW TRY THIS

Simplify each expression. Write your answer with positive exponents only. Assume no variables are 0.

1. $-2(x^2y^5)^0$

2. $-3x^{-2}$

3. $9^2 - 9^0$

4. Explain why the instructions above include the statement "Assume no variables are 0."

4.2   Exercises

WARM-UPS *Identify the base in each expression.*

1. $3x^{-2}$

2. $(3x)^{-2}$

3. $(5x)^0$

4. $5x^0$

Simplify by dividing out common factors. Assume no variable is 0.

5. $\frac{3 \cdot a \cdot a}{3 \cdot a \cdot a \cdot a}$

6. $\frac{-2 \cdot x \cdot x \cdot x}{-2 \cdot x \cdot x}$

7. $\frac{a \cdot a \cdot b}{a \cdot b \cdot b}$

8. $\frac{x \cdot x \cdot y \cdot y}{x \cdot x \cdot x \cdot y \cdot y}$

REVIEW

9. If $a = -2$ and $b = 3$, evaluate $\frac{3a^2 + 4b + 8}{a + 2b^2}$.

10. Evaluate: $|-3 + 5 \cdot 2|$

Solve each equation.

11. $5\left(x - \frac{1}{2}\right) = \frac{7}{2}$

12. $\frac{5(2 - x)}{6} = \frac{x + 6}{2}$

13. Solve $P = L + \frac{s}{f}i$ for s.

14. Solve $P = L + \frac{s}{f}i$ for i.

VOCABULARY AND CONCEPTS *Fill in the blanks.*

15. If x is any nonzero real number, then $x^0 = $ __. If x is any nonzero real number, then $x^{-n} = $ ___.

16. Since $\frac{6^4}{6^4} = 6^{4-4} = 6^0$ and $\frac{6^4}{6^4} = 1$, we define 6^0 to be __.

17. Since $\dfrac{8^3}{8^5} = 8^{3-5} = 8^{-2}$ and $\dfrac{8^3}{8^5} = \dfrac{8 \cdot 8 \cdot 8}{8 \cdot 8 \cdot 8 \cdot 8 \cdot 8} = \dfrac{1}{8^2}$, we define 8^{-2} to be ___.

18. The amount P that must be deposited now to have A dollars in the future is called the _____.

GUIDED PRACTICE
Write each expression without exponents. Assume no variable is 0. SEE EXAMPLE 1. (OBJECTIVE 1)

19. 9^0

20. $\dfrac{a^6}{a^6}$

21. $5x^0$

22. $(5x)^0$

23. $\left(\dfrac{a^2b^3}{ab^4}\right)^0$

24. $\dfrac{2}{3}\left(\dfrac{xyz}{x^2y}\right)^0$

25. $\dfrac{8^y}{8^y}$

26. $\dfrac{a^n}{a^n}$

27. $(-x)^0$

28. $-x^0$

29. $\dfrac{x^0 - 5x^0}{2x^0}$

30. $\dfrac{4a^0 + 2a^0}{3a^0}$

Simplify each expression by writing it as an expression without negative exponents or parentheses. Assume no variables are 0. SEE EXAMPLE 2. (OBJECTIVE 2)

31. 5^{-4}

32. 9^{-2}

33. a^{-5}

34. y^{-3}

35. $(2y)^{-4}$

36. $(-3x)^{-1}$

37. $(-5p)^{-3}$

38. $(-4z)^{-2}$

39. $(y^2y^4)^{-2}$

40. $(x^3x^2)^{-3}$

41. $-5x^{-4}$

42. $-7y^{-2}$

Simplify and write the result without negative exponents. Assume no variables are 0. SEE EXAMPLE 3. (OBJECTIVE 2)

43. $\dfrac{y^4}{y^5}$

44. $\dfrac{x^5}{x^9}$

45. $\dfrac{a^4}{a^9}$

46. $\dfrac{z^5}{z^8}$

47. $\dfrac{x^{-2}x^{-3}}{x^{-10}}$

48. $\dfrac{a^{-4}a^{-2}}{a^{-12}}$

49. $\dfrac{15a^3b^8}{3a^4b^4}$

50. $\dfrac{14b^5c^4}{21b^3c^5}$

51. $(a^{-6})^4$

52. $(-b^{-3})^9$

53. $\left(-\dfrac{b^5}{b^{-3}}\right)^{-4}$

54. $\left(\dfrac{a^4}{a^{-3}}\right)^3$

Simplify each expression. Assume no base is 0. SEE EXAMPLE 4. (OBJECTIVE 3)

55. $x^{2m}x^m$

56. $y^{5m}y^{4m}$

57. $\dfrac{x^{3n}}{x^{6n}}$

58. $\dfrac{y^m}{y^{7m}}$

59. $y^{3m+2}y^{-m}$

60. $x^{m+1}x^m$

61. $(x^{n+4})^3$

62. $(y^{m-3})^4$

63. $u^{2m}v^{3n}u^{3m}v^{-3n}$

64. $r^{2m}s^{-3}r^{3m}s^3$

65. $(y^{2-n})^{-4}$

66. $(x^{3-4n})^{-2}$

ADDITIONAL PRACTICE
Simplify each expression and write the result without using parentheses or negative exponents. Assume no variable base is 0.

67. $2^5 \cdot 2^{-2}$

68. $10^2 \cdot 10^{-4} \cdot 10^5$

69. $5^{-2} \cdot 5^5 \cdot 5^{-3}$

70. $3^{-4} \cdot 3^5 \cdot 3^{-3}$

71. $\dfrac{8^5 \cdot 8^{-3}}{8}$

72. $\dfrac{6^2 \cdot 6^{-3}}{6^{-2}}$

73. $\dfrac{2^5 \cdot 2^7}{2^6 \cdot 2^{-3}}$

74. $\dfrac{5^{-2} \cdot 5^{-4}}{5^{-6}}$

75. a^{-9}

76. c^{-4}

77. $\dfrac{y^{3m}}{y^{2m}}$

78. $\dfrac{z^{4m}}{z^{2m}}$

79. $(4t)^{-3}$

80. $(-6r)^{-2}$

81. $(ab^2)^{-3}$

82. $(m^2n^3)^{-2}$

83. $(x^2y)^{-2}$

84. $(x^{-1}y^2)^{-3}$

85. $\dfrac{b^0b^3}{b^{-3}b^4}$

86. $\dfrac{(r^2)^3}{(r^3)^4}$

87. $(m^3n^4)^{-3}$

88. $(c^2d^3)^{-2}$

89. $\dfrac{x^{12}x^{-7}}{x^3x^4}$

90. $\dfrac{(b^3)^4}{(b^5)^4}$

91. $(ab^2)^{-2}$

92. $(x^2y)^{-3}$

93. $(-2x^3y^{-2})^{-5}$

94. $(-3u^{-2}v^3)^{-3}$

95. $(a^{-2}b^{-3})^{-4}$

96. $(y^{-3}z^5)^{-6}$

97. $\left(\dfrac{b^5}{b^{-2}}\right)^{-2}$

98. $\left(\dfrac{b^{-2}}{b^3}\right)^{-3}$

99. $\left(\dfrac{6a^2b^3}{2ab^2}\right)^{-2}$

100. $\left(\dfrac{15r^2s^{-2}t}{3r^{-3}s^3}\right)^{-3}$

101. $\left(\dfrac{18a^2b^3c^{-4}}{3a^{-1}b^2c}\right)^{-3}$

102. $\left(\dfrac{21x^{-2}y^2z^{-2}}{7x^3y^{-1}}\right)^{-2}$

103. $\left(\dfrac{-3r^4r^{-3}}{r^{-3}r^7}\right)^3$

104. $\left(\dfrac{12y^3z^{-2}}{3y^{-4}z^3}\right)^2$

105. $\left(\dfrac{14u^{-2}v^3}{21u^{-3}v}\right)^4$

106. $\dfrac{(17x^5y^{-5}z)^{-3}}{(17x^{-5}y^3z^2)^{-4}}$

107. $\dfrac{x^{3n}}{x^{6n}}$

108. $(y^2)^{m+1}$

APPLICATIONS
For 109–111, see Accent on Technology.

109. Present value How much money must be invested at 7% to have $100,000 in 40 years?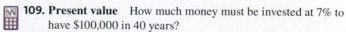

110. Present value How much money must be invested at 9% to have $100,000 in 40 years?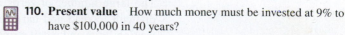

111. Present value   How much must be invested at 4% annual interest to have $1,000,000 in 60 years?

112. Biology   During bacterial reproduction, the time required for a population to double is called the **generation time**. If b bacteria are introduced into a medium, then after the generation time has elapsed, there will be $2b$ bacteria. After n generations, there will be $b \cdot 2^n$ bacteria. Give the meaning of this expression when $n = 0$.

WRITING ABOUT MATH

113. Explain how you would help a friend understand that 2^{-3} is not equal to -8.

114. Describe how you would verify on a calculator that
$$2^{-3} = \frac{1}{2^3}$$

SOMETHING TO THINK ABOUT

115. If a positive number x is raised to a negative power, is the result greater than, equal to, or less than x? Explore the possibilities.

116. We know that $x^{-n} = \frac{1}{x^n}$. Is it also true that $x^n = \frac{1}{x^{-n}}$? Explain.

Section 4.3

Scientific Notation

Objectives

1 Convert a number from standard notation to scientific notation.
2 Convert a number from scientific notation to standard notation.
3 Use scientific notation to simplify an expression.

Vocabulary

standard notation   scientific notation

Getting Ready

Evaluate each expression.

1. 10^2 **2.** 10^3 **3.** 10^1 **4.** 10^{-2}
5. $5(10^2)$ **6.** $8(10^3)$ **7.** $3(10^1)$ **8.** $7(10^{-2})$

We now use exponents to express very large and very small numbers that are written in **standard notation** in a compact form called **scientific notation**. In science, almost all large and small numbers are written in scientific notation.

1 Convert a number from standard notation to scientific notation.

Scientists often deal with extremely large and extremely small numbers. For example,

- The distance from Earth to the Sun is approximately 150,000,000 kilometers.
- Ultraviolet light emitted from a mercury arc has a wavelength of approximately 0.000025 centimeter.

The large number of zeros in these numbers written in standard notation makes them difficult to read and hard to remember. Scientific notation provides a compact way of writing large and small numbers.

SCIENTIFIC NOTATION A number is written in scientific notation if it is written as the product of a number between 1 (including 1) and 10 and an integer power of 10.

Each of the following numbers is written in scientific notation.

$$3.67 \times 10^6 \qquad 2.24 \times 10^{-4} \qquad 9.875 \times 10^{22}$$

Every number that is written in scientific notation has the following form:

An integer exponent.

$$\blacksquare.\blacksquare \times 10^{\blacksquare}$$

A number between 1 and 10.

EXAMPLE 1 Write 150,000,000 in scientific notation.

Solution We note that 1.5 lies between 1 and 10. To obtain 150,000,000, the decimal point in 1.5 must be moved eight places to the right. Because multiplying a number by 10 moves the decimal point one place to the right, we can accomplish this by multiplying 1.5 by 10 eight times.

$$1.5\,0\,0\,0\,0\,0\,0\,0$$

8 places to the right.

150,000,000 written in scientific notation is 1.5×10^8.

 SELF CHECK 1 Write 93,000,000 in scientific notation.

EXAMPLE 2 Write 0.000025 in scientific notation.

Solution We note that 2.5 is between 1 and 10. To obtain 0.000025, the decimal point in 2.5 must be moved five places to the left. We can accomplish this by dividing 2.5 by 10^5, which is equivalent to multiplying 2.5 by $\frac{1}{10^5}$ (or by 10^{-5}).

$$0\,0\,0\,0\,2.5$$

5 places to the left.

In scientific notation, 0.000025 is written 2.5×10^{-5}.

 SELF CHECK 2 Write 0.0012 in scientific notation.

EXAMPLE 3 Write **a.** 235,000 and **b.** 0.00000235 in scientific notation.

Solution **a.** $235,000 = 2.35 \times 10^5$, because 2.35 is between 1 and 10 and the decimal point must be moved 5 places to the right.

b. $0.00000235 = 2.35 \times 10^{-6}$, because 2.35 is between 1 and 10 and the decimal point must be moved 6 places to the left.

 SELF CHECK 3 Write the following in scientific notation: **a.** 17,500 **b.** 0.657

Perspective

THE METRIC SYSTEM

A common metric unit of length is the kilometer, which is 1,000 meters. Because 1,000 is 10^3, we can write 1 km = 10^3 m. Similarly, 1 centimeter is one-hundredth of a meter: 1 cm = 10^{-2} m. In the metric system, prefixes such as *kilo* and *centi* refer to powers of 10. Other prefixes are used in the metric system, as shown in the table.

To appreciate the magnitudes involved, consider these facts: Light, which travels 186,000 miles every second, will travel about one foot in one nanosecond. The distance to the nearest star (except for the Sun) is 43 petameters, and the diameter of an atom is about 10 nanometers. To measure some quantities, however, even these units are inadequate. The Sun, for example, radiates 5×10^{26} watts. That's a lot of light bulbs!

Prefix	Symbol	Meaning
peta	P	$10^{15} = $ 1,000,000,000,000,000.
tera	T	$10^{12} = $ 1,000,000,000,000.
giga	G	$10^{9} = $ 1,000,000,000.
mega	M	$10^{6} = $ 1,000,000.
kilo	k	$10^{3} = $ 1,000.
hecta	h	$10^{2} = $ 100.
deca	da	$10^{1} = $ 10.
deci	d	$10^{-1} = $ 0.1
centi	c	$10^{-2} = $ 0.01
milli	m	$10^{-3} = $ 0.001
micro	μ	$10^{-6} = $ 0.000 001
nano	n	$10^{-9} = $ 0.000 000 001
pico	p	$10^{-12} = $ 0.000 000 000 001

EXAMPLE 4   Write 432.0×10^{-5} in scientific notation.

Solution   The number 432.0×10^{-5} is not written in scientific notation, because 432.0 is not a number between 1 and 10. To write the number in scientific notation, we proceed as follows:

$$432.0 \times 10^{-5} = 4.32 \times 10^{2} \times 10^{-5} \qquad \text{Write 432.0 in scientific notation.}$$
$$= 4.32 \times 10^{-3} \qquad 10^{2} \times 10^{-5} = 10^{-3}$$

 SELF CHECK 4   Write 85×10^{-3} in scientific notation.

2   Convert a number from scientific notation to standard notation.

We can convert a number written in scientific notation to standard notation by reversing the process of converting standard notation to scientific notation. To convert a number to standard notation, move the decimal point the number of places indicated by the exponent. If the exponent is positive, this represents a large number and the decimal point will move to the right. If the exponent is negative, this represents a small number and the decimal point will move to the left. For example, to write 9.3×10^{7} in standard notation, we move the decimal point seven places to the right. Since we already have one number to the right of the decimal, we will need to insert 6 zeros for place value.

$$9.3 \times 10^7 = 9.3 \times 10,000,000$$
$$= 93,000,000 \qquad \text{Move the decimal point 7 places to the right.}$$

EXAMPLE 5   Write   **a.** 3.4×10^5 and   **b.** 2.1×10^{-4} in standard notation.

Solution   **a.** $3.4 \times 10^5 = 3.4 \times 100,000$
$$= 340,000 \qquad \text{Move the decimal point 5 places to the right.}$$

b. $2.1 \times 10^{-4} = 2.1 \times \dfrac{1}{10^4}$

$$= 2.1 \times \dfrac{1}{10,000}$$

$$= 0.00021 \qquad \text{Move the decimal point 4 places to the left.}$$

 SELF CHECK 5   Write the following in standard notation: **a.** 4.76×10^5 **b.** 9.8×10^{-3}

Each of the following numbers is written in both scientific and standard notation. In each case, the exponent gives the number of places that the decimal point moves, and the sign of the exponent indicates the direction that it moves.

$2.37 \times 10^6 = 2\,3\,7\,0\,0\,0\,0.$ Move the decimal point 6 places to the right.

$8.375 \times 10^{-3} = 0\,.\,0\,0\,8\,3\,7\,5$ Move the decimal point 3 places to the left.

$9.77 \times 10^0 = 9.77$ No movement of the decimal point.

3   Use scientific notation to simplify an expression.

Another advantage of scientific notation becomes apparent when we simplify fractions such as

$$\frac{(0.0032)(25,000)}{0.00040}$$

that contain very large or very small numbers. Although we can simplify this fraction by using arithmetic or a calculator, scientific notation provides an alternative way. First, we write each number in scientific notation; then we do the arithmetic on the numbers and the exponential expressions separately. Finally, we write the result in standard form, if desired.

$$\frac{(0.0032)(25,000)}{0.00040} = \frac{(3.2 \times 10^{-3})(2.5 \times 10^4)}{4.0 \times 10^{-4}}$$

$$= \frac{(3.2)(2.5)}{4.0} \times \frac{10^{-3}10^4}{10^{-4}}$$

$$= \frac{8.0}{4.0} \times 10^{-3+4-(-4)}$$

$$= 2.0 \times 10^5$$

$$= 200,000$$

EXAMPLE 6   **SPEED OF LIGHT**   In a vacuum, light travels 1 meter in approximately 0.000000003 second. How long does it take for light to travel 500 kilometers?

Solution   Since 1 kilometer = 1,000 meters, the length of time for light to travel 500 kilometers $(500 \cdot 1,000 \text{ meters})$ is given by

$$(0.000000003)(500)(1,000) = (3 \times 10^{-9})(5 \times 10^2)(1 \times 10^3)$$
$$= 3(5) \times 10^{-9+2+3}$$
$$= \mathbf{15} \times 10^{-4}$$
$$= \mathbf{1.5 \times 10^1} \times 10^{-4}$$
$$= 1.5 \times 10^{-3}$$
$$= 0.0015$$

Light travels 500 kilometers in approximately 0.0015 second (or 1.5 millisecond).

 SELF CHECK 6   How long does it take for light to travel 700 kilometers?

Accent
on technology

▶ Finding Powers
  of Decimals

To find the value of $(453.46)^5$, we can use a calculator and enter these numbers and press these keys:

453.46 **y^x** 5 **=**   *Using a calculator with a* **y^x** *key.*

453.46 **∧** 5 **ENTER**   *Using a TI84 graphing calculator.*

Either way, we have $(453.46)^5 = 1.917321395 \times 10^{13}$. Since this number is too large to show on the display, the calculator gives the result as **1.917321395 E13**.

For instructions regarding the use of a Casio graphing calculator, please refer to the Casio Keystroke Guide in the back of the book.

 **SELF CHECK ANSWERS**

1. 9.3×10^7 **2.** 1.2×10^{-3} **3. a.** 1.75×10^4 **b.** 6.57×10^{-1} **4.** 8.5×10^{-2} **5. a.** 476,000 **b.** 0.0098 **6.** Light travels 700 kilometers in approximately 0.0021 second (or 2.1 milliseconds).

NOW TRY THIS

1. Write the result shown on the graphing calculator screen in
   a. scientific notation
   b. standard notation

2. Write the result shown on the graphing calculator screen in
   a. scientific notation
   b. standard notation

3. There were approximately 1.45728×10^7 inches of wiring in one space shuttle. How many miles of wiring is this?   (*Hint*: Recall that 5,280 feet = 1 mile.)

4.3   Exercises

WARM-UPS *Identify the number between 1 and 10 to be used to write the number in scientific notation.*

1. 39,000,000
2. 0.000000028
3. 83700
4. 921,400,000,000
5. 0.0000000001052
6. 0.000625

REVIEW

7. If $a = -1$, find the value of $-3a^{33}$.
8. Evaluate $\dfrac{3a^2 - 2b}{2a + 2b}$ if $a = 4$ and $b = 3$.

Determine which property of real numbers justifies each statement.

9. $a \cdot c = c \cdot a$
10. $7(u + 3) = 7u + 7 \cdot 3$

Solve each equation.

11. $6(x - 5) + 9 = 3x$
12. $8(3x - 5) - 4(2x + 3) = 12$

VOCABULARY AND CONCEPTS *Fill in the blanks.*

13. A number is written in _____ when it is written as the product of a number between 1 (including 1) and 10 and an integer power of 10.
14. The number 125,000 is written in _____ notation.

GUIDED PRACTICE *Write each number in scientific notation.* SEE EXAMPLES 1–3. (OBJECTIVE 1)

15. 450,000
16. 4,750
17. 1,700,000
18. 290,000
19. 0.0059
20. 0.00083
21. 0.00000275
22. 0.000000055

Write each number in scientific notation. SEE EXAMPLE 4. (OBJECTIVE 1)

23. 42.5×10^2
24. 0.07×10^4
25. 0.37×10^{-4}
26. 25.2×10^{-3}

Write each number in standard notation. SEE EXAMPLE 5. (OBJECTIVE 2)

27. 2.3×10^2
28. 4.25×10^4
29. 8.12×10^5
30. 1.2×10^3
31. 1.15×10^{-3}
32. 8.16×10^{-5}
33. 9.76×10^{-4}
34. 6.52×10^{-3}

Use scientific notation to simplify each expression. Give all answers in standard notation. SEE EXAMPLE 6. (OBJECTIVE 3)

35. $(3.4 \times 10^2)(2.1 \times 10^3)$
36. $(4.1 \times 10^{-3})(3.4 \times 10^4)$
37. $\dfrac{9.3 \times 10^2}{3.1 \times 10^{-2}}$
38. $\dfrac{7.2 \times 10^6}{1.2 \times 10^8}$
39. $\dfrac{96,000}{(12,000)(0.00004)}$
40. $\dfrac{(0.48)(14,400,000)}{96,000,000}$
41. $\dfrac{2,475}{(132,000,000)(0.25)}$
42. $\dfrac{147,000,000,000,000}{25(0.000049)}$

ADDITIONAL PRACTICE *Write each number in scientific notation.*

43. 0.0000051
44. 0.04
45. 863,000,000
46. 514,000
47. $\dfrac{2.4 \times 10^2}{6 \times 10^{23}}$
48. $\dfrac{1.98 \times 10^2}{6 \times 10^{23}}$

Write each number in standard notation.

49. 37×10^7
50. 0.07×10^3
51. 0.32×10^{-4}
52. 617×10^{-2}

Determine which number of each pair is the larger.

53. 37.2 or 3.72×10^2
54. 37.2 or 3.72×10^{-1}
55. 3.72×10^3 or 4.72×10^3
56. 3.72×10^3 or 4.72×10^2
57. 3.72×10^{-1} or 4.72×10^{-2}
58. 3.72×10^{-3} or 2.72×10^{-2}

APPLICATIONS

59. **Distance to Alpha Centauri** The distance from Earth to the nearest star outside our solar system is approximately 25,700,000,000,000 miles. Write this number in scientific notation.

60. **Speed of sound** The speed of sound in air is 33,100 centimeters per second. Write this number in scientific notation.

61. **Distance to Mars** The distance from Mars to the Sun is approximately 1.14×10^8. Write this number in standard notation.

62. **Distance to Venus** The distance from Venus to the Sun is approximately 6.7×10^7. Write this number in standard notation.

63. **Length of one meter** One meter is approximately 0.00622 mile. Write this number in scientific notation.

64. Angstroms   One angstrom is 1×10^{-7} millimeter. Write this number in standard notation.

65. Distance between Mercury and the Sun   The distance from Mercury to the Sun is approximately 3.6×10^{7} miles. Use scientific notation to express this distance in feet. (*Hint:* 5,280 feet = 1 mile.)

66. Oil reserves   Recently, Venezuela was believed to have crude oil reserves of about 2.965×10^{11} barrels. A barrel contains 42 gallons of oil. Use scientific notation to express its oil reserves in gallons.

67. National debt   The U.S. national debt in September 2011 was approximately $1,645.12 billion. Write this number in scientific notation, and then translate that to standard notation. (*Hint:* 1 billion = 1.0×10^{9}.)

68. National debt   If the U.S. national debt continued to grow at a rate of $600,000.00 per day, calculate the debt in September 2012. Compare this debt to that of today. Has it increased or decreased?

69. Speed of sound   The speed of sound in air is approximately 3.3×10^{4} centimeters per second. Use scientific notation to express this speed in kilometers per second. (*Hint:* 100 centimeters = 1 meter and 1,000 meters = 1 kilometer.)

70. Light year   One light year is approximately 5.87×10^{12} miles. Use scientific notation to express this distance in feet. (*Hint:* 5,280 feet = 1 mile.)

71. Wavelengths   Some common types of electromagnetic waves are given in the table. List the wavelengths in order from shortest to longest.

Type	Use	Wavelength (m)
Visible light	Lighting	9.3×10^{-6}
Infrared	Photography	3.7×10^{-5}
X-rays	Medical	2.3×10^{-11}

72. Wavelengths   More common types of electromagnetic waves are given in the table. List the wavelengths in order from longest to shortest.

Type	Use	Wavelength (m)
Radio waves	Communication	3.0×10^{2}
Microwaves	Cooking	1.1×10^{-2}
Ultraviolet	Sun lamp	6.1×10^{-8}

The bulk of the surface area of the red blood cell shown in the illustration is contained on its top and bottom. That area is $2\pi r^2$, twice the area of one circle. If there are N discs, their total surface area T will be N times the surface area of a single disc: $T = 2N\pi r^2$.

73. Red blood cells   The red cells in human blood pick up oxygen in the lungs and carry it to all parts of the body. Each cell is a tiny circular disc with a radius of about 0.00015 in. Because the amount of oxygen carried depends on the surface area of the cells, and the cells are so tiny, a great number are needed—about 25 trillion in an average adult. Write these two numbers in scientific notation.

74. Red blood cells   Find the total surface area of all the red blood cells in the body of an average adult. See Exercise 73.

WRITING ABOUT MATH

75. In what situations would scientific notation be more convenient than standard notation?

76. To multiply a number by a power of 10, we move the decimal point. Which way, and how far? Explain.

SOMETHING TO THINK ABOUT

77. Two positive numbers are written in scientific notation. How could you decide which is larger, without converting either to standard notation?

 78. The product $1 \cdot 2 \cdot 3 \cdot 4 \cdot 5$, or 120, is called **5 factorial**, written 5!. Similarly, the number $6! = 6 \cdot 5 \cdot 4 \cdot 3 \cdot 2 \cdot 1 = 720$. Factorials get large very quickly. Calculate 30!, and write the number in standard notation. How large a factorial can you compute with a calculator?

Section 4.4

Polynomials

Objectives

1. Determine whether an expression is a polynomial.
2. Classify a polynomial as a monomial, binomial, or trinomial, if applicable.
3. Find the degree of a polynomial.
4. Evaluate a polynomial.
5. Evaluate a polynomial function.
6. Graph a linear, quadratic, and cubic polynomial function.

Vocabulary

polynomial	degree of a polynomial	polynomial function
monomial	descending powers of a	quadratic function
binomial	variable	parabola
trinomial	ascending powers of a	cubic function
degree of a monomial	variable	

Getting Ready

Write each expression using exponents.

1. $2xxyyy$
2. $3xyyy$
3. $2xx + 3yy$
4. $xxx + yyy$
5. $(3xxy)(2xyy)$
6. $(5xyzzz)(xyz)$
7. $3(5xy)\left(\frac{1}{3}xy\right)$
8. $(xy)(xz)(yz)(xyz)$

In algebra, exponential expressions may be combined to form **polynomials**. In this section, we will introduce the topic of polynomials and graph some basic polynomial functions.

1 Determine whether an expression is a polynomial.

Recall that expressions such as

$$3x \qquad 4y^2 \qquad -8x^2y^3 \qquad \text{and} \qquad 25$$

with constant and/or variable factors are called *algebraic terms*. The coefficients of the first three of these terms are 3, 4, and -8, respectively. Because $25 = 25x^0$, 25 is referred to as a constant.

POLYNOMIALS

A polynomial is an algebraic expression that is a single term or the sum of several terms containing whole-number exponents on the variables.

Here are some examples of polynomials:

$$8xy^2t \qquad 3x + 2 \qquad 4y^2 - 2y + 3 \qquad 3a - 4b - 4c + 8d$$

COMMENT The expression $2x^3 - 3y^{-2}$ is not a polynomial, because the second term contains a negative exponent on a variable base.

EXAMPLE 1   Determine whether each expression is a polynomial.

    a. $x^2 + 2x + 1$ A polynomial.

    b. $3x^{-1} - 2x - 3$ No. The first term has a negative exponent on a variable base.

    c. $\dfrac{1}{2}x^3 - 2.3x + 5$ A polynomial.

    d. $-2x + 3x^{1/2}$ No. The second term has a fractional exponent on a variable base.

 **SELF CHECK 1**   Determine whether each expression is a polynomial.

    a. $3x^{-4} + 2x^2 - 3$ **b.** $7.5x^3 - 4x^2 - 3x$

    c. $6x^2 - 9x^{1/3}$ **d.** $\dfrac{2}{3}x^4 + \dfrac{7}{9}x^3 - 7x$

2   Classify a polynomial as a monomial, binomial, or trinomial, if applicable.

A polynomial with one term is called a **monomial**. A polynomial with two terms is called a **binomial**. A polynomial with three terms is called a **trinomial**. Here are some examples.

Monomials	Binomials	Trinomials
$5x^2y$	$3u^3 - 4u^2$	$-5t^2 + 4t + 3$
$-6x$	$18a^2b + 4ab$	$27x^3 - 6x - 2$
29	$-29z^{17} - 1$	$-32r^6 + 7y^3 - z$

EXAMPLE 2   Classify each polynomial as a monomial, a binomial, or a trinomial, if applicable.

    a. $5x^4 + 3x$ Since the polynomial has two terms, it is a binomial.

    b. $7x^4 - 5x^3 - 2$ Since the polynomial has three terms, it is a trinomial.

    c. $-5x^2y^3$ Since the polynomial has one term, it is a monomial.

    d. $9x^5 - 5x^2 + 8x - 7$ Since the polynomial has four terms, it has no special name. It is none of these.

 **SELF CHECK 2**   Classify each polynomial as a monomial, a binomial, or a trinomial, if applicable.

    a. $5x$ **b.** $-5x^2 + 2x - 5$

    c. $16x^2 - 9y^2$ **d.** $x^9 + 7x^4 - x^2 + 6x - 1$

3   Find the degree of a polynomial.

The monomial $7x^6$ is called a monomial of sixth degree or a monomial of degree 6, because the variable x occurs as a factor six times. The monomial $3x^3y^4$ is a monomial of the seventh degree, because the variables x and y occur as factors a total of seven times. Other examples are

    $-2x^3$ is a monomial of degree 3.

    $47x^2y^3$ is a monomial of degree 5.

    $18x^4y^2z^8$ is a monomial of degree 14.

    8 is a monomial of degree 0, because $8 = 8x^0$.

These examples illustrate the following definition.

DEGREE OF A MONOMIAL	If a is a nonzero coefficient, the **degree of the monomial** ax^n is n. The degree of a monomial with several variables is the sum of the exponents on those variables.

COMMENT   Note that the degree of ax^n is not defined when $a = 0$. Since $ax^n = 0$ when $a = 0$, the constant 0 has no defined degree.

Because each term of a polynomial is a monomial, we define the degree of a polynomial by considering the degree of each of its terms.

DEGREE OF A POLYNOMIAL	The **degree of a polynomial** is the degree of its term with largest degree.

For example,

- $x^2 + 2x$ is a binomial of degree 2, because the degree of its first term is 2 and the degree of its other term is less than 2.
- $3x^3y^2 + 4x^4y^4 - 3x^3$ is a trinomial of degree 8, because the degree of its second term is 8 and the degree of each of its other terms is less than 8.
- $25x^4y^3z^7 - 15xy^8z^{10} - 32x^8y^8z^3 + 4$ is a polynomial of degree 19, because its second and third terms are of degree 19. Its other terms have degrees less than 19.

EXAMPLE 3   Find the degree of each polynomial.

a. $-4x^3 - 5x^2 + 3x$ 3, the degree of the first term because it has largest degree.

b. $5x^4y^2 + 7xy^2 - 16x^3y^5$ 8, the degree of the last term.

c. $-17a^2b^3c^4 + 12a^3b^4c$ 9, the degree of the first term.

 SELF CHECK 3   Find the degree of each polynomial.

a. $15p^3q^4 - 25p^4q^2$ b. $-14rs^3t^4 + 12r^3s^3t^3$

c. $16mn^6 - 9m^2n^2 + 3m^3n^3$

If the polynomial contains a single variable, we usually write it with its exponents in **descending order** where the term with the highest degree is listed first, followed by the term with the next highest degree, and so on. If we reverse the order, the polynomial is said to be written with its exponents in **ascending order**.

4   Evaluate a polynomial.

When a number is substituted for the variable in a polynomial, the polynomial takes on a numerical value. Finding that value is called *evaluating the polynomial*.

EXAMPLE 4   Evaluate the polynomial $3x^2 + 2$ when

a. $x = 0$ b. $x = 2$ c. $x = -3$ d. $x = -\frac{1}{5}$.

Solution   a. $3x^2 + 2 = 3(\mathbf{0})^2 + 2$ b. $3x^2 + 2 = 3(\mathbf{2})^2 + 2$

$= 3(0) + 2$ $= 3(4) + 2$

$= 0 + 2$ $= 12 + 2$

$= 2$ $= 14$

c. $3x^2 + 2 = 3(-3)^2 + 2$
$= 3(9) + 2$
$= 27 + 2$
$= 29$

d. $3x^2 + 2 = 3\left(-\dfrac{1}{5}\right)^2 + 2$
$= 3\left(\dfrac{1}{25}\right) + 2$
$= \dfrac{3}{25} + \dfrac{50}{25}$
$= \dfrac{53}{25}$

 SELF CHECK 4   Evaluate $3x^2 + x - 2$ when **a.** $x = 2$ **b.** $x = -1$
c. $x = -2$ **d.** $x = \dfrac{1}{2}$

When we evaluate a polynomial for several values of its variable, we often write the results in a table.

EXAMPLE 5   Evaluate the polynomial $x^3 + 1$ for the following values and write the results in a table.
a. $x = -2$ **b.** $x = -1$ **c.** $x = 0$ **d.** $x = 1$ **e.** $x = 2$

Solution

x	$x^3 + 1$	
a. -2	-7	$x^3 + 1 = (-2)^3 + 1 = -7$
b. -1	0	$x^3 + 1 = (-1)^3 + 1 = 0$
c. 0	1	$x^3 + 1 = (0)^3 + 1 = 1$
d. 1	2	$x^3 + 1 = (1)^3 + 1 = 2$
e. 2	9	$x^3 + 1 = (2)^3 + 1 = 9$

SELF CHECK 5   Complete the following table.

x	$-x^3 + 1$
-2	
-1	
0	
1	
2	

5   Evaluate a polynomial function.

Since the right sides of the functions $f(x) = 2x - 3$, $f(x) = x^2$, and $f(t) = -16t^2 + 64t$ are polynomials, they are called **polynomial functions**. We can evaluate these functions at specific values of the variable by evaluating the polynomial on the right side.

EXAMPLE 6   Given $f(x) = 2x - 3$, find $f(-2)$.

Solution   To find $f(-2)$, we substitute -2 for x and evaluate the function.

$f(x) = 2x - 3$
$f(-2) = 2(-2) - 3$
$= -4 - 3$
$= -7$

Thus, $f(-2) = -7$.

SELF CHECK 6   Given $f(x) = 2x - 3$, find $f(5)$.

Accent on technology

▸ Height of a Rocket

The height h (in feet) of a toy rocket launched straight up into the air with an initial velocity of 64 feet per second is given by the polynomial function

$$h = f(t) = -16t^2 + 64t$$

In this case, the height h is the dependent variable, and the time t is the independent variable because the height depends on the time elapsed. To find the height of the rocket 3.5 seconds after launch, we substitute 3.5 for t and evaluate h.

$$h = -16t^2 + 64t$$
$$h = -16(\mathbf{3.5})^2 + 64(\mathbf{3.5})$$

To evaluate h with a calculator, we enter these numbers and press these keys:

16 +/− × 3.5 x² + (64 × 3.5) =   Using a calculator with a +/− key.

(−) 16 × 3.5 x² + 64 × 3.5 **ENTER**   Using a TI84 graphing calculator.

Either way, the display reads 28. After 3.5 seconds, the rocket will be 28 feet above the ground.

For instructions regarding the use of a Casio graphing calculator, please refer to the Casio Keystroke Guide in the back of the book.

6   Graph a linear, quadratic, and cubic polynomial function.

We can graph polynomial functions as we graphed equations in Section 3.2. We make a table of values, plot points, and draw the line or curve that passes through those points.

In the next example, we graph the function $f(x) = 2x - 3$. Since its graph is a line, recall that it is a linear function.

COMMENT   The ordered pair (x, y) can be written as $(x, f(x))$.

EXAMPLE 7   Graph: $f(x) = 2x - 3$

Solution   We substitute numbers for x, compute the corresponding values of $f(x)$, and list the results in a table, as in Figure 4-1. We then plot the pairs $(x, f(x))$ and draw a line through the points, as shown in the figure. From the graph, we can see that x can be any value. This confirms that the domain is the set of real numbers $\mathbb{R}$. We also can see that $f(x)$ can be any value. This confirms the range is also the set of real numbers $\mathbb{R}$.

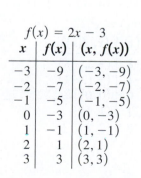

$$f(x) = 2x - 3$$

x	$f(x)$	$(x, f(x))$
-3	-9	$(-3, -9)$
-2	-7	$(-2, -7)$
-1	-5	$(-1, -5)$
0	-3	$(0, -3)$
1	-1	$(1, -1)$
2	1	$(2, 1)$
3	3	$(3, 3)$

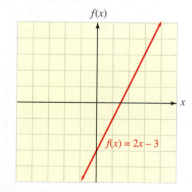

Figure 4-1

SELF CHECK 7   Graph: $f(x) = \frac{1}{2}x + 3$

In the next example, we graph the function $f(x) = x^2$, referred to as the squaring function. Since the polynomial on the right side is of second degree, we call this function a **quadratic function**.

EXAMPLE 8 Graph $f(x) = x^2$. State the domain and the range.

Solution We substitute numbers for x, compute the corresponding values of $f(x)$, and list the results in a table, as in Figure 4-2. We then plot the pairs $(x, f(x))$ and draw a smooth curve through the points, as shown in the figure. This curve is called a **parabola**. From the graph, we can see that x can be any value. This confirms that the domain is the set of real numbers $\mathbb{R}$. We can also see that $f(x)$ is always a positive number or 0. This confirms that the range is $\{f(x) \mid f(x) \text{ is a real number and } f(x) \geq 0\}$. In interval notation, this is $[0, \infty)$.

Amalie Noether
1882–1935
Albert Einstein described Noether as the most creative female mathematical genius since the beginning of higher education for women. Her work was in the area of abstract algebra. Although she received a doctoral degree in mathematics, she was denied a mathematics position in Germany because she was a woman.

$$f(x) = x^2$$

x	$f(x)$	$(x, f(x))$
-3	9	$(-3, 9)$
-2	4	$(-2, 4)$
-1	1	$(-1, 1)$
0	0	$(0, 0)$
1	1	$(1, 1)$
2	4	$(2, 4)$
3	9	$(3, 9)$

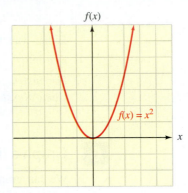

Figure 4-2

 SELF CHECK 8 Graph $f(x) = x^2 - 3$. State the domain and the range.

In the next example, we graph the function $f(x) = x^3$, referred to as the cubing function. Since the polynomial on the right side is of third degree, we call this function a **cubic function**.

EXAMPLE 9 Graph $f(x) = x^3$. State the domain and the range.

Solution We substitute numbers for x, compute the corresponding values of $f(x)$, and list the results in a table, as in Figure 4-3. We then plot the pairs $(x, f(x))$ and draw a smooth curve through the points, as shown in the figure. From the figure, we can see that the domain and the range are both $\mathbb{R}$.

$$f(x) = x^3$$

x	$f(x)$	$(x, f(x))$
-2	-8	$(-2, -8)$
-1	-1	$(-1, -1)$
0	0	$(0, 0)$
1	1	$(1, 1)$
2	8	$(2, 8)$

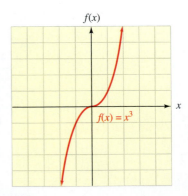

Figure 4-3

 SELF CHECK 9 Graph $f(x) = x^3 + 3$. State the domain and the range.

Accent on technology

▸Graphing Polynomial Functions

It is possible to use a graphing calculator to generate tables and graphs for polynomial functions. For example, Figure 4-4 shows calculator tables and the graphs of $f(x) = 2x - 3$, $f(x) = x^2$, and $f(x) = x^3$. (*Note*: Although you are graphing the function $f(x)$, the calculator will represent it as y.)

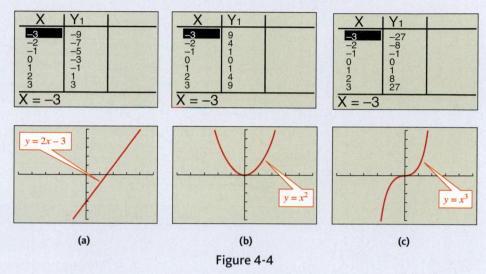

X	Y₁
−3	−9
−2	−7
−1	−5
0	−3
1	−1
2	1
3	3

X = −3

X	Y₁
−3	9
−2	4
−1	1
0	0
1	1
2	4
3	9

X = −3

X	Y₁
−3	−27
−2	−8
−1	−1
0	0
1	1
2	8
3	27

X = −3

$y = 2x - 3$

$y = x^2$

$y = x^3$

(a)   (b)   (c)

Figure 4-4

For instructions regarding the use of a Casio graphing calculator, please refer to the Casio Keystroke Guide in the back of the book.

EXAMPLE 10 Graph $f(x) = x^2 - 2x$. State the domain and the range.

Solution We substitute numbers for x, compute the corresponding values of $f(x)$, and list the results in a table, as in Figure 4-5. We then plot the pairs $(x, f(x))$ and draw a smooth curve through the points, as shown in the figure. From the graph, we can see the domain is $\mathbb{R}$ and the range is $[-1, \infty)$.

$f(x) = x^2 - 2x$

x	$f(x)$	$(x, f(x))$
−2	8	$(-2, 8)$
−1	3	$(-1, 3)$
0	0	$(0, 0)$
1	−1	$(1, -1)$
2	0	$(2, 0)$
3	3	$(3, 3)$
4	8	$(4, 8)$

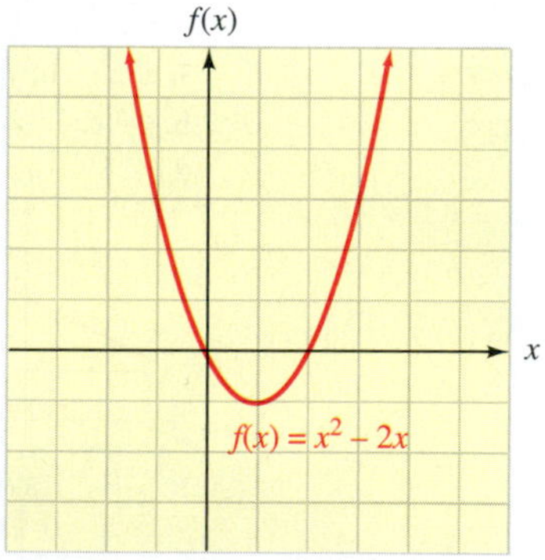

$f(x) = x^2 - 2x$

Figure 4-5

SELF CHECK 10 Graph $f(x) = x^2 - 2x$. State the domain and the range.

Everyday connections
NBA Salaries

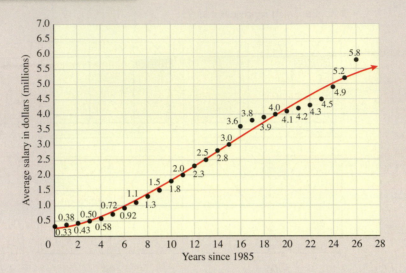

$$f(t) = 0.2457 + 0.0563t + 0.0127t^2 - 0.0002802t^3$$

The polynomial function shown above models average player salary in the National Basketball Association during the time period 1985–2011, where t equals the number of years since 1985. The black dots represent average salaries and points on the red graph represent predicted salaries. Use the graph to answer the following questions.

1. a. What was the average player salary in 2002?

   b. What was the average player salary predicted by the function f in 2002?

2. What is the predicted average salary for a player in 2017?

3. Does this function yield a realistic prediction of the average player salary in 2022?

SELF CHECK ANSWERS

1. a. no   **b.** yes   **c.** no   **d.** yes   **2. a.** monomial   **b.** trinomial   **c.** binomial   **d.** none of these   **3. a.** 7
b. 9   **c.** 7   **4. a.** 12   **b.** 0   **c.** 8   **d.** $-\frac{3}{4}$ **5.** 9, 2, 1, 0, -7 **6.** 7

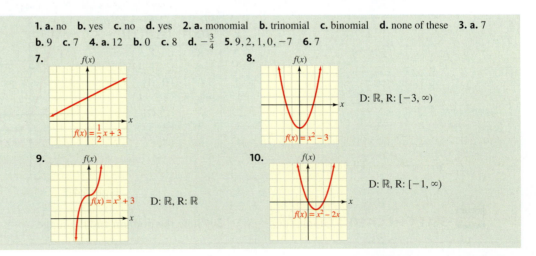

7. $f(x) = \frac{1}{2}x + 3$

8.   D: $\mathbb{R}$, R: $[-3, \infty)$ $f(x) = x^2 - 3$

9. $f(x) = x^3 + 3$ D: $\mathbb{R}$, R: $\mathbb{R}$

10.   D: $\mathbb{R}$, R: $[-1, \infty)$ $f(x) = x^2 - 2x$

NOW TRY THIS

1. Classify each polynomial and state its degree.

   a. $9x^2 - 6x^8$

   b. 1

2. If $f(x) = -3x^2 - 2x$,

   a. find $f(-4)$.

   b. find $f(3p)$.

3. 📈 Use a graphing calculator to graph $f(x) = \frac{1}{2}x^3 - x^2 + 3x + 1$.

4.4   Exercises

WARM-UPS *Write each expression using exponents.*

1. $5 \cdot a \cdot a + 2 \cdot b \cdot b \cdot b$

2. $7 \cdot x \cdot x \cdot x + 8 \cdot y \cdot y$

3. $4 \cdot a \cdot a \cdot a \cdot b \cdot b$

4. $9 \cdot x \cdot x \cdot y \cdot y \cdot y \cdot y$

5. $x \cdot x \cdot x + 3 \cdot x \cdot x + x$

6. $5 \cdot a + 4 \cdot a \cdot a + a \cdot a \cdot a$

7. $a \cdot a \cdot a + b \cdot b \cdot b$

8. $2 \cdot x \cdot x + 5 \cdot y \cdot y$

REVIEW *Solve each equation.*

9. $5(u - 5) + 9 = 2(u + 4)$

10. $8(3a - 5) - 12 = 4(2a + 3)$

Solve each inequality and graph the solution set.

11. $-4(3y + 2) \le 28$

12. $-5 < 3t + 4 \le 13$

Write each expression without using parentheses or negative exponents. Assume no variable is zero.

13. $(x^3 x^5)^4$

14. $(b^4)^3(b^2)^3$

15. $\left(\dfrac{y^2 y^5}{y^4}\right)^3$

16. $\left(\dfrac{2t^3}{t}\right)^{-4}$

VOCABULARY AND CONCEPTS *Fill in the blanks.*

17. An expression such as $3t^4$ with a constant and/or a variable is called an _____ term.

18. A _____ is an algebraic expression that is one term, or the sum of two or more terms, containing whole-number exponents on the variables.

19. A _____ is a polynomial with one term. A _____ is a polynomial with two terms. A _____ is a polynomial with three terms.

20. If $a \neq 0$, the _____ of ax^n is n.

21. The degree of a monomial with several variables is the ____ of the exponents on those variables.

22. The graph of a _____ function is a line.

23. A function of the form $y = f(x)$ where $f(x)$ is a polynomial is called a _____ function.

24. The function $f(x) = x^2$ is called the squaring or _____ function.

25. The function $f(x) = x^3$ is called the cubing or _____ function.

26. The graph of a quadratic function is called a _____.

27. The polynomial $8x^5 - 3x^3 + 6x^2 - 1$ is written with its exponents in _____ order. Its degree is _.

28. The polynomial $-2x + x^2 - 5x^3 + 7x^4$ is written with its exponents in _____ order. Its degree is _.

29. Any equation in x and y where each input value x determines exactly one output value y is called a _____.

30. $f(x)$ is read as ____.

GUIDED PRACTICE *Determine whether each expression is a polynomial. SEE EXAMPLE 1. (OBJECTIVE 1)*

31. $x^4 + 3x^2 - 5$

32. $2x^{-2} - 3x + 7$

33. $3x^{1/2} - 4$

34. $0.5x^5 - 0.25x^2$

Classify each polynomial as a monomial, a binomial, a trinomial, or none of these. SEE EXAMPLE 2. (OBJECTIVE 2)

35. $3x + 7$

36. $5x^2 + 2x$

37. $3y^2 + 4y + 3$

38. $3xy$

39. $3z^2$

40. $3x^4 - 2x^3 + 3x - 1$

41. $5t - 32$

42. $-8x^3y^2z^5$

Give the degree of each polynomial. SEE EXAMPLE 3. (OBJECTIVE 3)

43. $4x^7 - 5x^2$

44. $3x^5 - 4x^2$

45. $-2x^2 + 3x^3$

46. $-6x^4 + 3x^2 - 5x$

47. $3x^2y^3 + 5x^3y^5$

48. $-2x^2y^3 + 4x^3y^2z$

49. $-5r^2s^2t - 3r^3st^2 + 3$

50. $8r^3s^2t^4 - 3r^5t^6 + 2$

Evaluate $-x^2 - 4$ *for each value.* SEE EXAMPLE 4. (OBJECTIVE 4)

51. $x = 0$ **52.** $x = 1$

53. $x = -1$ **54.** $x = -2$

Complete each table. SEE EXAMPLE 5. (OBJECTIVE 4)

55.

x	$x^2 - 3$
-2	
-1	
0	
1	
2	

56.

x	$-x^2 + 3$
-2	
-1	
0	
1	
2	

57.

x	$x^3 + 2$
-2	
-1	
0	
1	
2	

58.

x	$-x^3 + 2$
-2	
-1	
0	
1	
2	

If $f(x) = 5x + 1$, *find each value.* SEE EXAMPLE 6. (OBJECTIVE 5)

59. $f(-3)$ **60.** $f(4)$

61. $f\left(-\dfrac{1}{2}\right)$ **62.** $f\left(\dfrac{2}{5}\right)$

Graph each polynomial function. State the domain and range. SEE EXAMPLES 7–10. (OBJECTIVE 6)

63. $f(x) = x^2 - 1$ **64.** $f(x) = x^2 + 2$

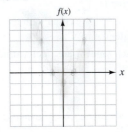   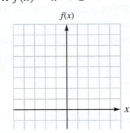

65. $f(x) = x^3 + 2$ **66.** $f(x) = x^3 - 2$

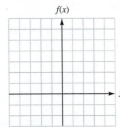   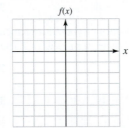

ADDITIONAL PRACTICE *Evaluate* $5x - 3$ *for each value.*

67. $x = 2$ **68.** $x = 0$

69. $x = -1$ **70.** $x = -2$

Classify each polynomial as a monomial, a binomial, a trinomial, or none of these.

71. $s^2 - 23s + 31$ **72.** $12x^3 - 12x^2 + 36x - 3$

73. $3x^5 - 2x^4 - 3x^3 + 17$ **74.** x^3

75. $\dfrac{1}{2}x^3 + 3$ **76.** $x^3 - 1$

Give an example of a polynomial that is . . .

77. a binomial   **78.** a monomial

79. a trinomial   **80.** not a monomial, a binomial, or a trinomial

81. of degree 3   **82.** of degree 1

83. of degree 0   **84.** of no defined degree

Give the degree of each polynomial.

85. $x^{12} + 3x^2y^3z^4$ **86.** 17^2x

87. 38   **88.** -25

If $f(x) = x^2 - 2x + 3$, *find each value.*

89. $f(5)$ **90.** $f(3)$

91. $f(-2)$ **92.** $f(-1)$

93. $f(0.5)$ **94.** $f(1.2)$

APPLICATIONS *Use a calculator to help solve each.*

95. Height of a rocket See the Accent on Technology section on page 253. Find the height of the rocket 2 seconds after launch.

96. Height of a rocket Again referring to page 253, make a table of values to find the rocket's height at various times. For what values of t will the height of the rocket be 0?

97. Computing revenue The revenue r (in dollars) that a manufacturer of desk chairs receives is given by the polynomial function

$$r = f(d) = -0.08d^2 + 100d$$

where d is the number of chairs manufactured. Find the revenue received when 815 chairs are manufactured.

98. Falling balloons Some students threw balloons filled with water from a dormitory window. The height h (in feet) of the balloons t seconds after being thrown is given by the polynomial function

$$h = f(t) = -16t^2 + 12t + 20$$

How far above the ground is a balloon 1.5 seconds after being thrown?

99. Stopping distance The number of feet that a car travels before stopping depends on the driver's reaction time and the braking distance. For one driver, the stopping distance d is given by the function $d = f(v) = 0.04v^2 + 0.9v$, where v is the velocity of the car. Find the stopping distance when the driver is traveling at 30 mph.

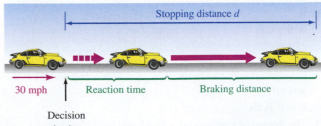

100. Stopping distance Find the stopping distance of the car discussed in Exercise 99 when the driver is going 70 mph.

WRITING ABOUT MATH

101. Describe how to determine the degree of a polynomial.
102. Describe how to classify a polynomial as a monomial, a binomial, a trinomial, or none of these.

SOMETHING TO THINK ABOUT

103. Find a polynomial whose value will be 1 if you substitute $\frac{3}{2}$ for x.
104. Graph the function $f(x) = -x^2$. What do you discover?

Section 4.5

Adding and Subtracting Polynomials

Objectives

1. Add two or more monomials.
2. Subtract two monomials.
3. Add two polynomials.
4. Subtract two polynomials.
5. Simplify an expression using the order of operations and combining like terms.
6. Solve an application requiring operations with polynomials.

Vocabulary

subtrahend   minuend

Getting Ready

Combine like terms and simplify, if possible.

1. $3x + 2x$
2. $5y - 3y$
3. $19x + 6x$
4. $8z - 3z$
5. $9r + 3r$
6. $4r - 3s$
7. $7r - 7r$
8. $17r - 17r^2$

In this section, we will discuss how to add and subtract polynomials.

---

1 ## Add two or more monomials.

Recall that like terms have the same variables with the same exponents. For example,

$3xyz^2$ and $-2xyz^2$ are like terms.   The variables and their exponents are the same.

$\frac{1}{2}ab^2c$ and $\frac{1}{3}a^2bc^2$ are unlike terms.   The variables and their corresponding exponents are NOT the same.

Also recall that to combine like terms, we add (or subtract) their coefficients and keep the same variables with the same exponents. For example,

$$2y + 5y = (2 + 5)y \qquad\qquad -3x^2 + 7x^2 = (-3 + 7)x^2$$
$$= 7y \qquad\qquad\qquad\qquad\qquad = 4x^2$$

Likewise,

$$4x^3y^2 + 9x^3y^2 = 13x^3y^2 \qquad\qquad 4r^2s^3t^4 + 7r^2s^3t^4 = 11r^2s^3t^4$$

EXAMPLE 1   Perform the following additions.

a. $5xy^3 + 7xy^3 = 12xy^3$

b. $-7x^2y^2 + 6x^2y^2 + 3x^2y^2 = -x^2y^2 + 3x^2y^2$
$$= 2x^2y^2$$

c. $(2x^2)^2 + 81x^4 = 4x^4 + 81x^4 \qquad (2x^2)^2 = (2x^2)(2x^2) = 4x^4$
$$= 85x^4$$

SELF CHECK 1   Perform the following additions.   **a.** $6a^3b^2 + 5a^3b^2$
b. $-2pq^2 + 5pq^2 + 8pq^2$ **c.** $27x^6 + (2x^2)^3$

2 Subtract two monomials.

To subtract one monomial from another, we add the opposite of the monomial that is to be subtracted. In symbols, $x - y = x + (-y)$.

EXAMPLE 2   Find each difference.

a. $8x^2 - 3x^2 = 8x^2 + (-3x^2)$
$$= 5x^2$$

b. $6x^3y^2 - 9x^3y^2 = 6x^3y^2 + (-9x^3y^2)$
$$= -3x^3y^2$$

c. $-3r^2st^3 - 5r^2st^3 = -3r^2st^3 + (-5r^2st^3)$
$$= -8r^2st^3$$

SELF CHECK 2   Find each difference.   **a.** $12m^3 - 7m^3$ **b.** $-4p^3q^2 - 8p^3q^2$
c. $-8x^3y^2z - 12x^3y^2z$

3 Add two polynomials.

Because of the distributive property, we can remove parentheses enclosing several terms when the sign preceding the parentheses is $+$.

$$+(3x^2 + 3x - 2) = \mathbf{+1}(3x^2 + 3x - 2)$$
$$= \mathbf{1}(3x^2) + \mathbf{1}(3x) + \mathbf{1}(-2)$$
$$= 3x^2 + 3x + (-2)$$
$$= 3x^2 + 3x - 2$$

We can add polynomials by removing parentheses, if necessary, and then combining any like terms that are contained within the polynomials.

EXAMPLE 3 Add: $(3x^2 - 3x + 2) + (2x^2 + 7x - 4)$

Solution
$(3x^2 - 3x + 2) + (2x^2 + 7x - 4)$
$= 3x^2 - 3x + 2 + 2x^2 + 7x - 4$ Remove parentheses.
$= 3x^2 + 2x^2 - 3x + 7x + 2 - 4$ Use the commutative property of addition.
$= 5x^2 + 4x - 2$ Combine like terms.

SELF CHECK 3 Add: $(2a^2 - a + 4) + (5a^2 + 6a - 5)$

Additions such as Example 3 often are written with like terms aligned vertically. We then can add the polynomials column by column.

$$\begin{array}{r} 3x^2 - 3x + 2 \\ + 2x^2 + 7x - 4 \\ \hline 5x^2 + 4x - 2 \end{array}$$

EXAMPLE 4 Add:
$$\begin{array}{r} 4x^2y + 8x^2y^2 - 3x^2y^3 \\ + 3x^2y - 8x^2y^2 + 8x^2y^3 \\ \hline 7x^2y \qquad\qquad + 5x^2y^3 \end{array}$$

SELF CHECK 4 Add:
$$\begin{array}{r} 4pq^2 + 6pq^3 - 7pq^4 \\ + 2pq^2 - 8pq^3 + 9pq^4 \end{array}$$

4 Subtract two polynomials.

We can remove parentheses enclosing several terms when the sign preceding the parentheses is negative by distributing a -1 to each term *within the parentheses*.

$$\begin{aligned} -(3x^2 + 3x - 2) &= \mathbf{-1}(3x^2 + 3x - 2) \\ &= \mathbf{-1}(3x^2) + \mathbf{(-1)}(3x) + \mathbf{(-1)}(-2) \\ &= -3x^2 + (-3x) + 2 \\ &= -3x^2 - 3x + 2 \end{aligned}$$

This suggests that the way to subtract polynomials is to remove parentheses by changing the sign of all terms being subtracted and then combine like terms.

EXAMPLE 5 Subtract:

a. $(3x - 4) - (5x + 7) = 3x - 4 - 5x - 7$
$$= -2x - 11$$

b. $(3x^2 - 4x - 6) - (2x^2 - 6x + 12) = 3x^2 - 4x - 6 - 2x^2 + 6x - 12$
$$= x^2 + 2x - 18$$

c. $(-4rt^3 + 2r^2t^2) - (-3rt^3 + 2r^2t^2) = -4rt^3 + 2r^2t^2 + 3rt^3 - 2r^2t^2$
$$= -rt^3$$

SELF CHECK 5 Subtract: **a.** $(4x - 1) - (x - 5)$ **b.** $(6x^2 - 5x + 4) - (-3x^2 + 6x + 1)$
c. $(-2a^2b + 5ab^2) - (-5a^2b - 7ab^2)$

To subtract polynomials in vertical form, we add the negative of the **subtrahend** (the bottom polynomial) to the **minuend** (the top polynomial) to obtain the difference.

EXAMPLE 6   Subtract $(3x^2y - 2xy^2)$ from $(2x^2y + 4xy^2)$.

Solution   We write the subtraction in vertical form, change the signs of the terms of the subtrahend, and add.

$$
\begin{array}{r}
2x^2y + 4xy^2 \\
- \quad 3x^2y - 2xy^2 \\
\hline
\end{array}
\quad \rightarrow \quad
\begin{array}{r}
2x^2y + 4xy^2 \\
+ \; - 3x^2y + 2xy^2 \\
\hline
- \; x^2y + 6xy^2
\end{array}
$$

In horizontal form, the result is the same.

$$2x^2y + 4xy^2 - (3x^2y - 2xy^2) = 2x^2y + 4xy^2 - 3x^2y + 2xy^2$$
$$= -x^2y + 6xy^2$$

 SELF CHECK 6   Subtract:
$$
\begin{array}{r}
5p^2q - 6pq + 7q \\
- \quad 2p^2q + 2pq - 8q \\
\hline
\end{array}
$$

EXAMPLE 7   Subtract $(6xy^2 + 4x^2y^2 - x^3y^2)$ from $(-2xy^2 - 3x^3y^2)$.

Solution
$$
\begin{array}{r}
-2xy^2 \qquad\quad - 3x^3y^2 \\
- \quad 6xy^2 + 4x^2y^2 - \; x^3y^2 \\
\hline
\end{array}
\quad \rightarrow \quad
\begin{array}{r}
-2xy^2 \qquad\quad - 3x^3y^2 \\
+ \; -6xy^2 - 4x^2y^2 + \; x^3y^2 \\
\hline
-8xy^2 - 4x^2y^2 - 2x^3y^2
\end{array}
$$

COMMENT   Be careful of the order of subtration. "Subtract x from y" means $y - x$.

In horizontal form, the result is the same.

$$-2xy^2 - 3x^3y^2 - (6xy^2 + 4x^2y^2 - x^3y^2)$$
$$= -2xy^2 - 3x^3y^2 - 6xy^2 - 4x^2y^2 + x^3y^2$$
$$= -8xy^2 - 4x^2y^2 - 2x^3y^2$$

 SELF CHECK 7   Subtract $(-2pq^2 - 2p^2q^2 + 3p^3q^2)$ from $(5pq^2 + 3p^2q^2 - p^3q^2)$.

5   **Simplify an expression using the order of operations and combining like terms.**

Because of the distributive property, we can remove parentheses enclosing several terms that are multiplied by a monomial by multiplying every term within the parentheses by that monomial. For example, to add $3(2x + 5)$ and $2(4x - 3)$, we proceed as follows:

$$3(2x + 5) + 2(4x - 3) = 6x + 15 + 8x - 6$$
$$= 6x + 8x + 15 - 6 \qquad \text{Use the commutative property of addition.}$$
$$= 14x + 9 \qquad \text{Combine like terms.}$$

EXAMPLE 8   Simplify:

a. $3(x^2 + 4x) + 2(x^2 - 4) = 3x^2 + 12x + 2x^2 - 8$
$$= 5x^2 + 12x - 8$$

b. $8(y^2 - 2y + 3) - 4(2y^2 + y - 3) = 8y^2 - 16y + 24 - 8y^2 - 4y + 12$
$$= -20y + 36$$

c. $-4(x^2y^2 - x^2y + 3x) - (x^2y^2 - 2x) + 3(x^2y^2 + 2x^2y)$

$\quad = -4x^2y^2 + 4x^2y - 12x - x^2y^2 + 2x + 3x^2y^2 + 6x^2y$

$\quad = -2x^2y^2 + 10x^2y - 10x$

SELF CHECK 8 Simplify: **a.** $2(a^3 - 3a) + 5(a^3 + 2a)$
b. $5(m^2 + 6m - 9) - (8m^2 - 2m + 5)$
c. $5(x^2y + 2x^2) - (x^2y - 3x^2)$

---

6 Solve an application requiring operations with polynomials.

EXAMPLE 9 **PROPERTY VALUES** A house purchased for \$95,000 is expected to appreciate according to the formula $y_1 = 2{,}500x + 95{,}000$, where y_1 is the value of the house after x years. A second house purchased for \$125,000 is expected to appreciate according to the formula $y_2 = 4{,}500x + 125{,}000$. Find one formula that will give the value of both properties after x years.

Solution The value of the first house after x years is given by the polynomial $2{,}500x + 95{,}000$. The value of the second house after x years is given by the polynomial $4{,}500x + 125{,}000$. The value of both houses will be the sum of these two polynomials.

$$2{,}500x + 95{,}000 + 4{,}500x + 125{,}000 = 7{,}000x + 220{,}000$$

The total value y of the properties is given by $y = 7{,}000x + 220{,}000$.

SELF CHECK 9 Find the total values of the properties after 20 years.

SELF CHECK ANSWERS **1. a.** $11a^3b^2$ **b.** $11pq^2$ **c.** $35x^6$ **2. a.** $5m^3$ **b.** $-12p^3q^2$ **c.** $-20x^3y^2z$ **3.** $7a^2 + 5a - 1$
4. $6pq^2 - 2pq^3 + 2pq^4$ **5. a.** $3x + 4$ **b.** $9x^2 - 11x + 3$ **c.** $3a^2b + 12ab^2$ **6.** $3p^2q - 8pq + 15q$
7. $7pq^2 + 5p^2q^2 - 4p^3q^2$ **8. a.** $7a^3 + 4a$ **b.** $-3m^2 + 32m - 50$ **c.** $4x^2y + 13x^2$ **9.** \$360,000

NOW TRY THIS

1. If the lengths of the sides of a triangle represent consecutive even integers, find a polynomial that represents the perimeter of the triangle.

2. If the length of a rectangle is $(15x - 3)$ ft and the width is $(8x + 17)$ ft, find a polynomial that represents the perimeter.

3. If the length of one side of a rectangle is represented by the polynomial $(4x - 18)$ cm, and the perimeter is $(12x - 36)$ cm, find a polynomial that represents the width.

4.5 Exercises

WARM-UPS *Simplify.*

1. $x(7 + 2)$

2. $y(3 + 9)$

3. $a(16 - 4)$

4. $b(12 - 8)$

Determine whether the terms are like or unlike.

5. $6x^2, 6x$

6. $5a, 7b$

7. $4x^3, 5x^3$

8. $-2a^2b, 6a^2b$

REVIEW *Let $a = 3$, $b = -2$, $c = -1$, and $d = 2$. Evaluate each expression.*

9. $ab + cd$

10. $ac - bd$

11. $a(b - c)$

12. $d(b + a)$

13. Solve the inequality $-4(2x - 9) \geq 12$ and graph the solution set.

14. The kinetic energy of a moving object is given by the formula

$$K = \frac{mv^2}{2}$$

Solve the formula for m.

VOCABULARY AND CONCEPTS *Fill in the blanks.*

15. A _____ is a polynomial with one term.

16. If two polynomials are subtracted in vertical form, the bottom polynomial is called the _____, and the top polynomial is called the _____.

17. To add like monomials, add the numerical _____ and keep the _____.

18. $a - b = a +$ ____

19. To add two polynomials, combine any _____ contained in the polynomials.

20. To subtract polynomials, use the distributive property to remove parentheses and combine _____.

If the terms are like terms, add them. If they are unlike terms, state unlike terms.

21. $3y$, $4y$

22. $3x^2$, $5x^2$

23. $3x$, $3y$

24. $3x^2$, $6x$

25. $3x^3$, $4x^3$, $6x^3$

26. $-2y^4$, $-6y^4$, $10y^4$

27. $-5x^3y^2$, $13x^3y^2$

28. 23, $12x$

29. $15x^4y^2$, $-9x^4y^2$, $4x^4y^2$

30. $32x^5y^3$, $-21x^5y^3$, $-11x^5y^3$

31. $-x^2y$, xy, $3xy^2$

32. $4x^3y^2z$, $-6x^3y^2z$, $2x^3y^2z$

GUIDED PRACTICE *Simplify. SEE EXAMPLE 1. (OBJECTIVE 1)*

33. $4y + 5y$

34. $3t + 6t$

35. $15x^2 + 10x^2$

36. $25r^4 + 15r^4$

37. $-7t^6 + 3(t^2)^3$

38. $-6(p^2)^4 + 10p^8$

39. $26x^2y^4 + 3x^2y^4$

40. $-16a^4b^2 + 10a^4b^2$

Simplify. SEE EXAMPLE 2. (OBJECTIVE 2)

41. $-18a - 3a$

42. $20b - 15b$

43. $32u^3 - 16u^3$

44. $25xy^2 - 7xy^2$

45. $18x^5y^2 - 11x^5y^2$

46. $17x^6y - 22x^6y$

47. $-14ab^3 - 6ab^3$

48. $17m^2n - 20m^2n$

Add. SEE EXAMPLE 3. (OBJECTIVE 3)

49. $(3x + 7) + (4x - 3)$

50. $(5y - 8) + (2y + 6)$

51. $(6y^2 - 2y + 5) + (2y^2 + 5y - 8)$

52. $(3x^2 - 3x - 2) + (3x^2 + 4x - 3)$

Perform the operation. SEE EXAMPLE 4. (OBJECTIVE 3)

53.
$$\begin{aligned} 3x^2 + 4x + 5 \\ + \underline{2x^2 - 3x + 6} \end{aligned}$$

54.
$$\begin{aligned} 2x^3 + 2x^2 - 3x + 5 \\ + \underline{3x^3 - 4x^2 - \ x - 7} \end{aligned}$$

55.
$$\begin{aligned} 2x^3 - 3x^2 + 4x - 7 \\ + \underline{-9x^3 - 4x^2 - 5x + 6} \end{aligned}$$

56.
$$\begin{aligned} -3x^3 + 4x^2 - 4x + 9 \\ + \underline{\ \ 2x^3 \qquad\quad + 9x - 3} \end{aligned}$$

Subtract. SEE EXAMPLE 5. (OBJECTIVE 4)

57. $(4a + 3) - (2a - 4)$

58. $(5b - 7) - (3b + 5)$

59. $(2a^2 - 6a + 3) - (-3a^2 - 4a + 5)$

60. $(4b^2 + 5b - 1) - (-4b^2 - 3b - 6)$

Perform the operation. SEE EXAMPLE 6. (OBJECTIVE 4)

61.
$$\begin{aligned} 3x^2 + 4x - 5 \\ - \underline{-2x^2 - 2x + 3} \end{aligned}$$

62.
$$\begin{aligned} 3y^2 - 4y + \ 7 \\ - \underline{6y^2 - 6y - 13} \end{aligned}$$

63.
$$\begin{aligned} 4x^3 + 4x^2 - 3x + 10 \\ - \underline{5x^3 - 2x^2 - 4x - \ 4} \end{aligned}$$

64.
$$\begin{aligned} 3x^3 + 4x^2 + 7x + 12 \\ - \underline{-4x^3 + 6x^2 + 9x - \ 3} \end{aligned}$$

Perform the operation. SEE EXAMPLE 7. (OBJECTIVE 4)

65. Subtract $(8x + 2y)$ from $(-3x - 7y)$.

66. Subtract $(2x + 5y)$ from $(5x - 8y)$.

67. Subtract $(4x^2 - 3x + 2)$ from $(2x^2 - 3x + 1)$.

68. Subtract $(-4a + b)$ from $(6a^2 + 5a - b)$.

Simplify. SEE EXAMPLE 8. (OBJECTIVE 5)

69. $2(x + 3) + 4(x - 2)$

70. $5(y - 3) - 7(y + 4)$

71. $2(x^2 - 5x - 4) - 3(x^2 - 5x - 4) + 6(x^2 - 5x - 4)$

72. $7(x^2 + 3x + 1) + 9(x^2 + 3x + 1) - 5(x^2 + 3x + 1)$

ADDITIONAL PRACTICE *Perform the operations.*

73. $3rst + 4rst + 7rst$
74. $-2ab + 7ab - 3ab$
75. $-4a^2bc + 5a^2bc - 7a^2bc$
76. $(3x)^2 - 4x^2 + 10x^2$
77. $-3x^3y^6 + 2(xy^2)^3 - (3x)^3y^6$
78. $(-3x^2y)^4 + (4x^4y^2)^2 - 2x^8y^4$
79. $5x^5y^{10} - (2xy^2)^5 + (3x)^5y^{10}$
80. $5(x + y) + 7(x + y)$
81. $-8(x - y) + 11(x - y)$
82. $(4c^2 + 3c - 2) + (3c^2 + 4c + 2)$
83. $(-3z^2 - 4z + 7) + (2z^2 + 2z - 1) - (2z^2 - 3z + 7)$

84. $\quad -3x^2y + 4xy + 25y^2$
$\quad + \underline{\quad 5x^2y - 3xy - 12y^2}$

85. $\quad -6x^3z - 4x^2z^2 + 7z^3$
$\quad + \underline{\; -7x^3z + 9x^2z^2 - 21z^3}$

86. $\quad -2x^2y^2 - 4xy + 12y^2$
$\quad - \underline{\; 10x^2y^2 + 9xy - 24y^2}$

87. $\quad 25x^3 - 45x^2z + 31xz^2$
$\quad - \underline{\; 12x^3 + 27x^2z - 17xz^2}$

88. $2(a^2b^2 - ab) - 3(ab + 2ab^2) + (b^2 - ab + a^2b^2)$

89. $3(xy^2 + y^2) - 2(xy^2 - 4y^2 + y^3) + 2(y^3 + y^2)$
90. $-4(x^2y^2 + xy^3 + xy^2z) - 2(x^2y^2 - 4xy^2z) - 2(8xy^3 - y)$

91. Find the sum when $(x^2 + x - 3)$ is added to the sum of $(2x^2 - 3x + 4)$ and $(3x^2 - 2)$.
92. Find the difference when $(t^3 - 2t^2 + 2)$ is subtracted from the sum of $(3t^3 + t^2)$ and $(-t^3 + 6t - 3)$.
93. Find the difference when $(-3z^3 - 4z + 7)$ is subtracted from the sum of $(2z^2 + 3z - 7)$ and $(-4z^3 - 2z - 3)$.
94. Find the sum when $(3x^2 + 4x - 7)$ is added to the sum of $(-2x^2 - 7x + 1)$ and $(-4x^2 + 8x - 1)$.

APPLICATIONS *Consider the following information: If a house was purchased for $105,000 and is expected to appreciate $900 per year, its value y after x years is given by the formula*
$y = 900x + 105,000.$ *SEE EXAMPLE 9. (OBJECTIVE 6)*

95. **Value of a house** Find the expected value of the house in 10 years.
96. **Value of a house** A second house was purchased for $120,000 and was expected to appreciate $1,000 per year. Find a polynomial equation that will give the value of the house in x years.
97. **Value of a house** Find one polynomial equation that will give the combined value y of both houses after x years.

98. **Value of two houses** Find the value of the two houses after 25 years.

Consider the following information: A business bought two computers, one for $6,600 and the other for $9,200. The first computer is expected to depreciate $1,100 per year and the second $1,700 per year.

99. **Value of a computer** Write a polynomial equation that will give the value of
   a. the first computer after x years.
   b. the second computer after x years.
100. **Value of two computers**
   a. Find one polynomial equation that will give the value of both computers after x years.
   b. Find the value of the computers after 3 years.

WRITING ABOUT MATH

101. How do you recognize like terms?
102. How do you add like terms?

SOMETHING TO THINK ABOUT *Let $P(x) = 3x - 5$. Find each value.*

103. $P(x + h) + P(x)$
104. $P(x + h) - P(x)$
105. If $P(x) = x^{23} + 5x^2 + 73$ and $Q(x) = x^{23} + 4x^2 + 73$, find $P(7) - Q(7)$.
106. If two numbers written in scientific notation have the same power of 10, they can be added as similar terms:

$$2 \times 10^3 + 3 \times 10^3 = 5 \times 10^3$$

Without converting to standard form, how could you add

$$2 \times 10^3 + 3 \times 10^4$$

Section 4.6

Multiplying Polynomials

Objectives

1. Multiply two or more monomials.
2. Multiply a polynomial by a monomial.
3. Multiply a binomial by a binomial.
4. Multiply a polynomial by a binomial.
5. Solve an equation that simplifies to a linear equation.
6. Solve an application involving multiplication of polynomials.

Vocabulary

FOIL method　　　　　　　　squaring a binomial　　　　　　conjugate binomials

Getting Ready

Simplify.

1. $(2x)(3)$　　　　2. $(3xxx)(x)$　　　　3. $5x^2 \cdot x$　　　　4. $8x^2x^3$

Use the distributive property to remove parentheses.

5. $3(x + 5)$　　　6. $-2(x + 5)$　　　7. $4(y - 3)$　　　8. $-2(y^2 - 3)$

We now discuss how to multiply polynomials by beginning with a review of multiplying two monomials. We will then introduce multiplying polynomials with more than one term. The section concludes with a discussion of how these techniques can be used to solve linear equations.

1 Multiply two or more monomials.

We have previously multiplied monomials by other monomials. For example, to multiply $4x^2$ by $-2x^3$, we use the commutative and associative properties of multiplication to group the numerical factors together and the variable factors together. Then we multiply the numerical factors and multiply the variable factors.

$$4x^2(-2x^3) = 4(-2)x^2x^3$$
$$= -8x^5$$

This example suggests the following strategy.

MULTIPLYING MONOMIALS

To multiply two simplified monomials, multiply the numerical factors and then multiply the variable factors.

EXAMPLE 1 Multiply: **a.** $3x^5(2x^5)$ **b.** $-2a^2b^3(5ab^2)$

Solution **a.** $3x^5(2x^5) = 3(2)x^5x^5$

$= 6x^{10}$

b. $-2a^2b^3(5ab^2) = -2(5)a^2ab^3b^2$

$= -10a^3b^5$

SELF CHECK 1 Multiply: **a.** $(5a^2b^3)(6a^3b^4)$ **b.** $(-15p^3q^2)(5p^3q^2)$

---

2 Multiply a polynomial by a monomial.

To find the product of a monomial and a polynomial with more than one term, we use the distributive property. To multiply $2x + 4$ by $5x$, for example, we proceed as follows:

$5x(2x + 4) = 5x \cdot 2x + 5x \cdot 4$ Use the distributive property to remove parentheses.

$= 10x^2 + 20x$ Multiply the monomials $5x \cdot 2x = 10x^2$ and $5x \cdot 4 = 20x$.

This example suggests the following process.

MULTIPLYING POLYNOMIALS BY MONOMIALS	To multiply a polynomial with more than one term by a monomial, use the distributive property to remove parentheses and simplify.

EXAMPLE 2 Multiply: **a.** $3a^2(3a^2 - 5a)$ **b.** $-2xz^2(2x - 3z + 2z^2)$

Solution **a.** $3a^2(3a^2 - 5a) = 3a^2 \cdot 3a^2 - 3a^2 \cdot 5a$ Use the distributive property to remove parentheses.

$= 9a^4 - 15a^3$ Multiply.

b. $-2xz^2(2x - 3z + 2z^2)$

$= -2xz^2 \cdot 2x + (-2xz^2) \cdot (-3z) + (-2xz^2) \cdot 2z^2$ Use the distributive property to remove parentheses.

$= -4x^2z^2 + 6xz^3 + (-4xz^4)$ Multiply.

$= -4x^2z^2 + 6xz^3 - 4xz^4$

Recall that subtracting is equivalent to adding the opposite.

SELF CHECK 2 Multiply:
a. $2p^3(3p^2 - 5p)$
b. $-5a^2b(3a + 2b - 4ab)$

---

3 Multiply a binomial by a binomial.

To multiply two binomials, we must use the distributive property more than once. For example, to multiply $(2a - 4)$ by $(3a + 5)$, we proceed as follows.

$(2a - 4)(3a + 5) = (2a - 4) \cdot 3a + (2a - 4) \cdot 5$ Use the distributive property to remove parentheses.

$= 3a(2a - 4) + 5(2a - 4)$ Use the commutative property of multiplication.

$$= 3a \cdot 2a + 3a \cdot (-4) + 5 \cdot 2a + 5 \cdot (-4) \qquad \text{Use the distributive property.}$$

$$= 6a^2 - 12a + 10a - 20 \qquad \text{Do the multiplications.}$$

$$= 6a^2 - 2a - 20 \qquad \text{Combine like terms.}$$

This example suggests the following strategy.

MULTIPLYING TWO BINOMIALS

To multiply two binomials, multiply each term of one binomial by each term of the other binomial and combine like terms.

EXAMPLE 3   Find each product.
(Using the FOIL method)

a. $(3x + 4)(2x - 3) = 3x(2x) + 3x(-3) + 4(2x) + 4(-3)$

$$= 6x^2 - 9x + 8x - 12$$

$$= 6x^2 - x - 12$$

b. $(2y - 7)(5y - 4) = 2y(5y) + 2y(-4) + (-7)(5y) + (-7)(-4)$

$$= 10y^2 - 8y - 35y + 28$$

$$= 10y^2 - 43y + 28$$

c. $(2r - 3s)(2r + t) = 2r(2r) + 2r(t) - 3s(2r) - 3s(t)$

$$= 4r^2 + 2rt - 6sr - 3st$$

$$= 4r^2 + 2rt - 6rs - 3st$$

SELF CHECK 3   Find each product.
a. $(5x + 3)(6x + 7)$
b. $(2a - 1)(3a + 2)$
c. $(5y - 2z)(2y + 3z)$

To multiply binomials, we can apply the distributive property using a mnemonic device, called the **FOIL method**. FOIL is an acronym for **F**irst terms, **O**uter terms, **I**nner terms, and **L**ast terms. To use this method to multiply $(2a - 4)$ by $(3a + 5)$, we

1. multiply the **F**irst terms $2a$ and $3a$ to obtain $6a^2$,

2. multiply the **O**uter terms $2a$ and 5 to obtain $10a$,

3. multiply the **I**nner terms -4 and $3a$ to obtain $-12a$, and

4. multiply the **L**ast terms -4 and 5 to obtain -20.

Then we simplify the resulting polynomial, if possible.

COMMENT   FOIL is simply a mnemonic for applying the distributive property to multiply two binomials in a given order.

$$(2a - 4)(3a + 5) = 2a(3a) + 2a(5) + (-4)(3a) + (-4)(5)$$

$$= 6a^2 + 10a - 12a - 20 \qquad \text{Simplify.}$$

$$= 6a^2 - 2a - 20 \qquad \text{Combine like terms.}$$

EXAMPLE 4   Simplify each expression.

a. $3(2x - 3)(x + 1)$

$$= 3(2x^2 + 2x - 3x - 3) \quad \text{Multiply the binomials.}$$
$$= 3(2x^2 - x - 3) \quad \text{Combine like terms.}$$
$$= 6x^2 - 3x - 9 \quad \text{Use the distributive property to remove parentheses.}$$

b. $(x + 1)(x - 2) - 3x(x + 3)$

$$= x^2 - 2x + x - 2 - 3x^2 - 9x \quad \text{Use the distributive property to remove parentheses.}$$
$$= -2x^2 - 10x - 2 \quad \text{Combine like terms.}$$

SELF CHECK 4   Simplify each expression.
a. $-3(6x - 5)(2x - 3)$
b. $(x + 3)(2x - 1) + 2x(x - 1)$

The products discussed in Example 5 are sometimes called special products. These include **squaring a binomial** and multiplying conjugate binomials. Binomials that have the same terms, but with opposite signs between the terms, are called **conjugate binomials**.

EXAMPLE 5   Find each product.

a. $(x + y)^2 = (x + y)(x + y) \quad \text{Square the binomial.}$

$$= x^2 + xy + xy + y^2 \quad \text{Distribute.}$$
$$= x^2 + 2xy + y^2 \quad \text{Combine like terms.}$$

The square of the sum of two quantities has three terms: *the square of the first quantity, plus twice the product of the quantities, plus the square of the second quantity.*

COMMENT   Note that
$$(x + y)^2 \neq x^2 + y^2$$
and
$$(x - y)^2 \neq x^2 - y^2$$

b. $(x - y)^2 = (x - y)(x - y) \quad \text{Square the binomial.}$

$$= x^2 - xy - xy + y^2 \quad \text{Distribute.}$$
$$= x^2 - 2xy + y^2 \quad \text{Combine like terms.}$$

The square of the difference of two quantities has three terms: *the square of the first quantity, minus twice the product of the quantities, plus the square of the second quantity.*

c. $(x + y)(x - y) = x^2 - xy + xy - y^2 \quad \text{Distribute.}$

$$= x^2 - y^2 \quad \text{Combine like terms.}$$

The product of the sum and the difference of two quantities is a binomial. *It is the product of the first quantities minus the product of the second quantities.*

SELF CHECK 5   Find each product.
a. $(p + 2)^2$
b. $(p - 2)^2$
c. $(p + 2q)(p - 2q)$

Because the products discussed in Example 5 occur so often, it may be helpful to recognize their forms.

However, you should multiply these as binomials by applying the distributive property until you discover the pattern for yourself.

SPECIAL PRODUCTS

$$(x + y)^2 = x^2 + 2xy + y^2$$
$$(x - y)^2 = x^2 - 2xy + y^2$$
$$(x + y)(x - y) = x^2 - y^2$$

4 ## Multiply a polynomial by a binomial.

We must use the distributive property more than once to multiply a polynomial by a binomial. For example, to multiply $(3x^2 + 3x - 5)$ by $(2x + 3)$, we proceed as follows:

$$\mathbf{(2x + 3)}(3x^2 + 3x - 5) = \mathbf{(2x + 3)}\,3x^2 + \mathbf{(2x + 3)}\,3x + \mathbf{(2x + 3)(-5)}$$

$$= 3x^2(2x + 3) + 3x(2x + 3) - 5(2x + 3)$$
$$= 6x^3 + 9x^2 + 6x^2 + 9x - 10x - 15$$
$$= 6x^3 + 15x^2 - x - 15$$

This example suggests the following process.

MULTIPLYING POLYNOMIALS

To multiply one polynomial by another, multiply each term of one polynomial by each term of the other polynomial and combine like terms.

It is often convenient to organize the work vertically.

EXAMPLE 6   **a.** Multiply:

$$
\begin{array}{r}
3a^2 - 4a\ + 7 \\
\times \quad\quad\quad 2a\ + 5 \\
\hline
\end{array}
$$

$5(3a^2 - 4a + 7) \rightarrow$ $15a^2 - 20a + 35$
$2a(3a^2 - 4a + 7) \rightarrow$ $6a^3 - 8a^2 + 14a$
$$\overline{\ 6a^3 + 7a^2 - 6a + 35}$$

b. Multiply:

$$
\begin{array}{r}
3y^2 - 5y\ + 4 \\
\times \quad\quad - 4y^2 - 3 \\
\hline
\end{array}
$$

$-3(3y^2 - 5y + 4) \rightarrow$ $- 9y^2 + 15y - 12$
$-4y^2(3y^2 - 5y + 4) \rightarrow$ $-12y^4 + 20y^3 - 16y^2$
$$\overline{-12y^4 + 20y^3 - 25y^2 + 15y\ - 12}$$

🌿 **SELF CHECK 6**   Multiply:
a. $(3x + 2)(2x^2 - 4x + 5)$
b. $(-2x^2 + 3)(2x^2 - 4x - 1)$

COMMENT   An expression can be simplified by combining its like terms. An equation (two quantities set equal) can be solved. Remember that

Expressions are to be simplified.

Equations are to be solved.

5   Solve an equation that simplifies to a linear equation.

To solve an equation such as $(x + 2)(x + 3) = x(x + 7)$, we can use the distributive property to remove the parentheses on the left and the right sides and proceed as follows:

$$(x + 2)(x + 3) = x(x + 7)$$

$x^2 + 3x + 2x + 6 = x^2 + 7x$	Use the distributive property to remove parentheses.
$x^2 + 5x + 6 = x^2 + 7x$	Combine like terms.
$5x + 6 = 7x$	Subtract x^2 from both sides.
$6 = 2x$	Subtract $5x$ from both sides.
$3 = x$	Divide both sides by 2.

Check:

$$(x + 2)(x + 3) = x(x + 7)$$

$(3 + 2)(3 + 3) \stackrel{?}{=} 3(3 + 7)$	Replace x with 3.
$5(6) \stackrel{?}{=} 3(10)$	Do the additions within parentheses.
$30 = 30$	

Since the answer checks, the solution is 3.

EXAMPLE 7   Solve: $(x + 5)(x + 4) = (x + 9)(x + 10)$

Solution   We remove parentheses on both sides of the equation and proceed as follows:

$$(x + 5)(x + 4) = (x + 9)(x + 10)$$

$x^2 + 4x + 5x + 20 = x^2 + 10x + 9x + 90$	Use the distributive property to remove parentheses.
$x^2 + 9x + 20 = x^2 + 19x + 90$	Combine like terms.
$9x + 20 = 19x + 90$	Subtract x^2 from both sides.
$20 = 10x + 90$	Subtract $9x$ from both sides.
$-70 = 10x$	Subtract 90 from both sides.
$-7 = x$	Divide both sides by 10.

Check:

$$(x + 5)(x + 4) = (x + 9)(x + 10)$$

$(-7 + 5)(-7 + 4) \stackrel{?}{=} (-7 + 9)(-7 + 10)$	Replace x with -7.
$(-2)(-3) \stackrel{?}{=} (2)(3)$	Do the additions within parentheses.
$6 = 6$	

Since the result checks, the solution is -7.

SELF CHECK 7   Solve: $(x + 2)(x - 4) = (x + 6)(x - 3)$

6   Solve an application involving multiplication of polynomials.

EXAMPLE 8   **DIMENSIONS OF A PAINTING**   A square paint-ing is surrounded by a border 2 inches wide. If the area of the border is 96 square inches, find the dimensions of the painting.

Solution

Analyze the problem   Refer to Figure 4-6, which shows a square painting surrounded by a border 2 inches wide. We can let x represent the length in inches of each side of the square painting. The outer rectangle is also a square, and one length is $(x + 2 + 2)$ or $(x + 4)$ inches.

Figure 4-6
© Shutterstock.com/Olga Lyubkina

Form an equation   We know that the area of the border is 96 square inches, the area of the larger square is $(x + 4)(x + 4)$, and the area of the painting is $x \cdot x$. If we subtract the area of the paint-ing from the area of the larger square, the difference is 96 (the area of the border).

The area of the large square	minus	the area of the square painting	equals	the area of the border.
$(x + 4)(x + 4)$	$-$	$x \cdot x$	$=$	96

Solve the equation

$$(x + 4)(x + 4) - x^2 = 96$$

$x^2 + 8x + 16 - x^2 = 96$ Use the distributive property to remove parentheses.

$8x + 16 = 96$ Combine like terms.

$8x = 80$ Subtract 16 from both sides.

$x = 10$ Divide both sides by 8.

State the conclusion   The dimensions of the painting are 10 inches by 10 inches.

Check the result   Check the result.

🌿 **SELF CHECK 8**   If the area of the border is 112 square inches, find the dimensions of the painting.

🌿 **SELF CHECK ANSWERS**   **1. a.** $30a^5b^7$ **b.** $-75p^6q^4$ **2. a.** $6p^5 - 10p^4$ **b.** $-15a^3b - 10a^2b^2 + 20a^3b^2$ **3. a.** $30x^2 + 53x + 21$
b. $6a^2 + a - 2$ **c.** $10y^2 + 11yz - 6z^2$ **4. a.** $-36x^2 + 84x - 45$ **b.** $4x^2 + 3x - 3$ **5. a.** $p^2 + 4p + 4$
b. $p^2 - 4p + 4$ **c.** $p^2 - 4q^2$ **6. a.** $6x^3 - 8x^2 + 7x + 10$ **b.** $-4x^4 + 8x^3 + 8x^2 - 12x - 3$ **7.** 2
8. The dimensions are 12 inches by 12 inches.

NOW TRY THIS

Simplify or solve as appropriate.

1. $-\dfrac{1}{2}x(8x^2 - 16x + 2)$

2. $(2x - 3)(4x^2 + 6x + 9)$

3. $(x - 2)(x + 5) = (x - 1)(x + 8)$

4. Find a representation of the area of a square with one side represented by $(3x + 5)$ ft.

4.6 Exercises

WARM-UPS *Simplify.*

1. $3x^2(5x)$

2. $6y(2y^2)$

3. $-4x(2y)$

4. $7x(-3y)$

Use the distributive property to remove parentheses.

5. $-5(x - 4)$

6. $-3(y^2 + 2)$

7. $2(4x^2 - 9)$

8. $7(z^2 + z)$

REVIEW *For 9–12, determine which property of real numbers justifies each statement.*

9. $4x + 5x^2 = 5x^2 + 4x$

10. $(x + 3) + y = x + (3 + y)$

11. $3(ab) = (ab)3$

12. $a + 0 = a$

13. Solve: $\frac{5}{3}(5y + 6) - 10 = 0$

14. Solve: $F = \frac{GMm}{d^2}$ for m

VOCABULARY AND CONCEPTS *Fill in the blanks.*

15. A polynomial with one term is called a _____.

16. A binomial is a polynomial with ___ terms. The binomials $(a + b)$ and $(a - b)$ are called _____ binomials.

17. Products in the form $(a + b)^2$, $(a - b)^2$, or $(a + b)(a - b)$ are called _____.

18. In the acronym FOIL, F stands for _____, O stands for _____, I stands for _____ and L stands for _____.

Consider the product $(2x + 5)(3x - 4)$.

19. The product of the first terms is ___.

20. The product of the outer terms is ___.

21. The product of the inner terms is ___.

22. The product of the last terms is ___.

GUIDED PRACTICE *Multiply.* SEE EXAMPLE 1. (OBJECTIVE 1)

23. $(3x^2)(4x^3)$

24. $(-2a^3)(3a^2)$

25. $(-5t^3)(2t^4)$

26. $(-6a^2)(-3a^5)$

27. $(2x^2y^3)(3x^3y^2)$

28. $(-x^3y^6z)(x^2y^2z^7)$

29. $(3b^2)(-2b)(4b^3)$

30. $(3y)(2y^2)(-y^4)$

Multiply. SEE EXAMPLE 2. (OBJECTIVE 2)

31. $3(x + 4)$

32. $-3(a - 2)$

33. $-4(t + 7)$

34. $6(s^2 - 3)$

35. $3x(x - 2)$

36. $-5y(y + 3)$

37. $-2x^2(3x^2 - x)$

38. $4b^3(2 - 2b)$

39. $3xy(x + y)$

40. $-4x^2(3x^2 - x)$

41. $-6x^2(2x^2 + 3x - 5)$

42. $3y^3(2y^2 - 7y - 8)$

43. $\frac{1}{4}x^2(8x^5 - 4)$

44. $\frac{4}{3}a^2b(6a - 5b)$

45. $-\frac{2}{3}r^2t^2(9r - 3t)$

46. $-\frac{4}{5}p^2q(10p + 15q)$

Find each product. SEE EXAMPLE 3. (OBJECTIVE 3)

47. $(a + 4)(a + 5)$

48. $(y - 3)(y + 5)$

49. $(3x - 2)(x + 4)$

50. $(t + 4)(2t - 3)$

51. $(4a - 2)(2a - 3)$

52. $(2b - 1)(3b + 4)$

53. $(3x - 5)(2x + 1)$

54. $(5y - 7)(2y - 5)$

55. $(2s + 3t)(3s - t)$

56. $(3a - 2b)(4a + b)$

57. $(u + v)(u + 2t)$

58. $(x - 5y)(a + 2y)$

Simplify each expression. SEE EXAMPLE 4. (OBJECTIVE 3)

59. $2(x - 4)(x + 1)$

60. $-3(2x + 3y)(3x - 4y)$

61. $3a(a + b)(a - b)$

62. $-2r(r + s)(r + s)$

63. $(3x - 2y)(x + y)$

64. $(2a - 3b)(3a - 2b)$

65. $(2x - 3)(x + 1) - 5x(x + 2)$

66. $(x + 2)(3x - 1) + 3x(x - 2)$

Find each product. SEE EXAMPLE 5. (OBJECTIVE 3)

67. $(x + 5)^2$

68. $(y - 9)^2$

69. $(x - 4)^2$

70. $(a + 3)^2$

71. $(4t + 3)^2$

72. $(3t - 2)^2$

73. $(x - 2y)^2$

74. $(3a + 2b)^2$

75. $(r + 4)(r - 4)$

76. $(y + 6)(y - 6)$

77. $(4x + 5)(4x - 5)$

78. $(5z + 1)(5z - 1)$

Multiply. SEE EXAMPLE 6. (OBJECTIVE 4)

79. $(2x + 3)(x^2 + 4x - 1)$ **80.** $(3x - 2)(2x^2 - x + 2)$

81. $(4t + 3)(t^2 + 2t + 3)$

82. $(3x + y)(2x^2 - 3xy + y^2)$

83. $\begin{array}{r} 4x + 3 \\ \underline{x + 2} \end{array}$ **84.** $\begin{array}{r} 5r + 6 \\ \underline{2r - 1} \end{array}$

85. $\begin{array}{r} 4x - 2y \\ \underline{3x + 5y} \end{array}$ **86.** $\begin{array}{r} x^2 + x + 1 \\ \underline{x - 1} \end{array}$

Solve each equation. SEE EXAMPLE 7. (OBJECTIVE 5)

87. $(s - 4)(s + 1) = s^2 + 5$

88. $(y - 5)(y - 2) = y^2 - 4$

89. $z(z + 2) = (z + 4)(z - 4)$

90. $(z + 3)(z - 3) = z(z - 3)$

91. $(x + 4)(x - 4) = (x - 2)(x + 6)$

92. $(y - 1)(y + 6) = (y - 3)(y - 2) + 8$

93. $(a - 3)^2 = (a + 3)^2$

94. $(b + 2)^2 = (b - 1)^2$

ADDITIONAL PRACTICE *Simplify.*

95. $(x^2y^5)(x^2z^5)(-3y^2z^3)$ **96.** $(-r^4st^2)(2r^2st)(rst)$

97. $(x + 3)(2x - 3)$ **98.** $(2x + 3)(2x - 5)$

99. $(t - 3)(t - 3)$ **100.** $(z - 5)(z - 5)$

101. $(-2r - 3s)(2r + 7s)$ **102.** $(2a - 3b)^2$

103. $(3x - 2)^2$ **104.** $(x - 2y)(x^2 + 2xy + 4y^2)$

105. $(3x - y)(x^2 + 3xy - y^2)$
106. $(xyz^3)(xy^2z^2)^3$

Simplify or solve as appropriate.

107. $3xy(x + y) - 2x(xy - x)$
108. $(a + b)(a - b) - (a + b)(a + b)$
109. $(2x - 1)(2x + 1) = x(4x + 1)$
110. $7s^2 + (s - 3)(2s + 1) = (3s - 1)^2$
111. $(x + 2)^2 = (x - 2)^2$
112. $(2s - 3)(s + 2) = (2s + 1)(s - 3)$
113. $(x + y)(x - y) + x(x + y)$
114. $(x - 3)^2 - (x + 3)^2$
115. $(3x + 4)(2x - 2) - (2x + 1)(x + 3)$
116. $4 + (2y - 3)^2 = (2y - 1)(2y + 3)$
117. $3y(y + 2) = 3(y + 1)(y - 1)$
118. $(b + 2)(b - 2) + 2b(b + 1)$

119. Millstones   The radius of one millstone in the illustration is 3 meters greater than the radius of the other, and their areas differ by 15π square meters. Find the radius of the larger millstone.

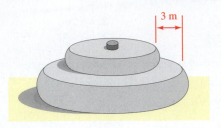

3 m

120. Bookbinding   Two square sheets of cardboard used for making book covers differ in area by 44 square inches. An edge of the larger square is 2 inches greater than an edge of the smaller square. Find the length of an edge of the smaller square.

121. Baseball   In major league baseball, the distance between bases is 30 feet greater than it is in softball. The bases in major league baseball mark the corners of a square that has an area 4,500 square feet greater than for softball. Find the distance between the bases in baseball.

122. Pulley designs   The radius of one pulley in the illustration is 1 inch greater than the radius of the second pulley, and their areas differ by 4π square inches. Find the radius of the smaller pulley.

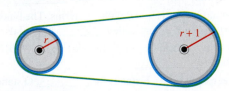

r $r + 1$

WRITING ABOUT MATH

123. Describe the steps involved in finding the product of a binomial and its conjugate.

124. Writing the expression $(x + y)^2$ as $x^2 + y^2$ illustrates a common error. Explain.

SOMETHING TO THINK ABOUT

125. The area of the square in the illustration is the total of the areas of the four smaller regions. The picture illustrates the product $(x + y)^2$. Explain.

126. The illustration represents the product of two binomials. Explain.

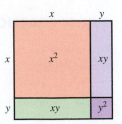

Section 4.7

Dividing Polynomials by Monomials

Objectives

1. Divide a monomial by a monomial.
2. Divide a polynomial by a monomial.
3. Solve a formula for a specified variable.

Getting Ready

Simplify each fraction.

1. $\dfrac{4x^2y^3}{2xy}$

2. $\dfrac{9xyz}{9xz}$

3. $\dfrac{15x^2y}{10x}$

4. $\dfrac{6x^2y}{6xy^2}$

5. $\dfrac{(2x^2)(5y^2)}{10xy}$

6. $\dfrac{(5x^3y)(6xy^3)}{10x^4y^4}$

In this section, we will show how to divide polynomials by monomials. We will discuss how to divide polynomials by polynomials and introduce all vocabulary associated with long division in the next section.

1 Divide a monomial by a monomial.

We have seen that dividing by a number is equivalent to multiplying by its reciprocal. For example, dividing the number 8 by 2 gives the same answer as multiplying 8 by $\frac{1}{2}$.

$$\frac{8}{2} = 4 \qquad \text{and} \qquad 8 \cdot \frac{1}{2} = 4$$

In general, the following is true.

DIVISION

$$\frac{a}{b} = a \cdot \frac{1}{b} \quad (b \neq 0)$$

Recall that to simplify a fraction, we write both its numerator and its denominator as the product of several factors and then divide out all common factors. For example,

$$\frac{20}{25} = \frac{4 \cdot 5}{5 \cdot 5} \qquad \text{Factor: } 20 = 4 \cdot 5 \text{ and } 25 = 5 \cdot 5$$

$$= \frac{4 \cdot \overset{1}{\cancel{5}}}{\underset{1}{\cancel{5}} \cdot 5} \qquad \text{Divide out the common factor of 5.}$$

$$= \frac{4}{5}$$

We can use the same method to simplify algebraic fractions that contain variables. We must assume, however, that no variable is 0.

$$\frac{3p^2q}{6pq^3} = \frac{3 \cdot p \cdot p \cdot q}{2 \cdot 3 \cdot p \cdot q \cdot q \cdot q}$$ Factor: $p^2 = p \cdot p, 6 = 2 \cdot 3,$ and $q^3 = q \cdot q \cdot q.$

$$= \frac{\overset{1}{\cancel{3}} \cdot \overset{1}{\cancel{p}} \cdot p \cdot \overset{1}{\cancel{q}}}{2 \cdot \cancel{3} \cdot \cancel{p} \cdot \cancel{q} \cdot q \cdot q}$$ Divide out the common factors of 3, p, and q.

$$= \frac{p}{2q^2}$$

To divide monomials, we can either use the previous method or use the rules of exponents.

COMMENT In all examples and exercises in this section, we will assume that no variables are 0 to avoid the possibility of division by 0. Recall that division by 0 is undefined.

EXAMPLE 1 Simplify: **a.** $\dfrac{x^2y}{xy^2}$ **b.** $\dfrac{-8a^3b^2}{4ab^3}$

Solution   ***Using Fractions***   ***Using the Properties of Exponents***

a. $\dfrac{x^2y}{xy^2} = \dfrac{x \cdot x \cdot y}{x \cdot y \cdot y}$ $\dfrac{x^2y}{xy^2} = x^{2-1}y^{1-2}$

$$= \frac{\overset{1}{\cancel{x}} \cdot x \cdot \overset{1}{\cancel{y}}}{\underset{1}{\cancel{x}} \cdot y \cdot \underset{1}{\cancel{y}}}$$ $= x^1 y^{-1}$

                                            $= x \cdot \dfrac{1}{y}$

$$= \frac{x}{y}$$ $= \dfrac{x}{y}$

b. $\dfrac{-8a^3b^2}{4ab^3} = \dfrac{-2 \cdot 4 \cdot a \cdot a \cdot a \cdot b \cdot b}{4 \cdot a \cdot b \cdot b \cdot b}$ $\dfrac{-8a^3b^2}{4ab^3} = \dfrac{(-1)2^3a^3b^2}{2^2ab^3}$

$$= \frac{-2 \cdot \overset{1}{\cancel{4}} \cdot \overset{1}{\cancel{a}} \cdot a \cdot a \cdot \overset{1}{\cancel{b}} \cdot \overset{1}{\cancel{b}}}{\underset{1}{\cancel{4}} \cdot \underset{1}{\cancel{a}} \cdot \underset{1}{\cancel{b}} \cdot \underset{1}{\cancel{b}} \cdot b}$$ $= (-1)2^{3-2}a^{3-1}b^{2-3}$

                                          $= (-1)2^1a^2b^{-1}$

                                          $= -2a^2 \cdot \dfrac{1}{b}$

$$= \frac{-2a^2}{b}$$ $= \dfrac{-2a^2}{b}$

 **SELF CHECK 1** Simplify: **a.** $\dfrac{r^5s^3}{rs^4}$ **b.** $\dfrac{-5p^2q^3}{10pq^4}$

2   Divide a polynomial by a monomial.

In Chapter 1, we saw that

$$\frac{a}{d} + \frac{b}{d} = \frac{a+b}{d}$$

Since this is true, we also have

$$\frac{a+b}{d} = \frac{a}{d} + \frac{b}{d}$$

This suggests that, to divide a polynomial by a monomial, we can divide each term of the polynomial in the numerator by the monomial in the denominator.

EXAMPLE 2   Simplify: $\dfrac{9x + 6y}{3xy}$

Solution $\dfrac{9x + 6y}{3xy} = \dfrac{9x}{3xy} + \dfrac{6y}{3xy}$ Divide each term in the numerator by the monomial.

$= \dfrac{3}{y} + \dfrac{2}{x}$ Simplify each fraction.

SELF CHECK 2   Simplify: $\dfrac{4a - 8b}{4ab}$

EXAMPLE 3   Simplify: $\dfrac{6x^2y^2 + 4x^2y - 2xy}{2xy}$

Solution $\dfrac{6x^2y^2 + 4x^2y - 2xy}{2xy}$

COMMENT   Remember that any nonzero value divided by itself is 1.

$= \dfrac{6x^2y^2}{2xy} + \dfrac{4x^2y}{2xy} - \dfrac{2xy}{2xy}$ Divide each term in the numerator by the monomial.

$= 3xy + 2x - 1$ Simplify each fraction.

SELF CHECK 3   Simplify: $\dfrac{9a^2b - 6ab^2 + 3ab}{3ab}$

EXAMPLE 4   Simplify: $\dfrac{12a^3b^2 - 4a^2b + a}{6a^2b^2}$

Solution $\dfrac{12a^3b^2 - 4a^2b + a}{6a^2b^2}$

$= \dfrac{12a^3b^2}{6a^2b^2} - \dfrac{4a^2b}{6a^2b^2} + \dfrac{a}{6a^2b^2}$ Divide each term in the numerator by the monomial.

$= 2a - \dfrac{2}{3b} + \dfrac{1}{6ab^2}$ Simplify each fraction.

SELF CHECK 4   Simplify: $\dfrac{14p^3q + pq^2 - p}{7p^2q}$

EXAMPLE 5   Simplify: $\dfrac{(x - y)^2 - (x + y)^2}{xy}$

Solution $\dfrac{(x - y)^2 - (x + y)^2}{xy}$

$= \dfrac{x^2 - 2xy + y^2 - (x^2 + 2xy + y^2)}{xy}$ Square the binomials in the numerator.

$= \dfrac{x^2 - 2xy + y^2 - x^2 - 2xy - y^2}{xy}$ Use the distributive property to remove parentheses.

$= \dfrac{-4xy}{xy}$ Combine like terms.

$= -4$ Simplify.

Simplify: $\dfrac{(x + y)^2 - (x - y)^2}{2xy}$

3 Solve a formula for a specified variable.

The cross-sectional area of the trapezoidal drainage ditch shown in Figure 4-7 is given by the formula $A = \frac{1}{2}h(B + b)$, where B and b represent its bases and h represents its height. To solve the formula for b, we proceed as follows.

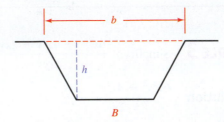

Figure 4-7

$$A = \frac{1}{2}h(B + b)$$

$$\mathbf{2}(A) = \mathbf{2}\left[\frac{1}{2}h(B + b)\right] \qquad \text{Multiply both sides by 2 to clear fractions.}$$

$$2A = h(B + b) \qquad \text{Simplify.}$$

$$2A = hB + hb \qquad \text{Use the distributive property to remove parentheses.}$$

$$2A - \mathbf{hB} = hB - \mathbf{hB} + hb \qquad \text{Subtract } hB \text{ from both sides.}$$

$$2A - hB = hb \qquad \text{Combine like terms.}$$

$$\frac{2A - hB}{\mathbf{h}} = \frac{hb}{\mathbf{h}} \qquad \text{Divide both sides by } h.$$

$$\frac{2A - hB}{h} = b$$

EXAMPLE 6 Another student worked the previous problem in a different way and got a result of $b = \frac{2A}{h} - B$. Is this also correct?

Solution To show that this result is correct, we must show that $\frac{2A - hB}{h} = \frac{2A}{h} - B$. We can do this by dividing $(2A - hB)$ by h.

$$\frac{2A - hB}{h} = \frac{2A}{h} - \frac{hB}{h} \qquad \text{Divide each term in the numerator by the monomial.}$$

$$= \frac{2A}{h} - B \qquad \text{Simplify: } \frac{hB}{h} = B.$$

The results are the same.

Suppose another student got $2A - B$. Is this result correct?

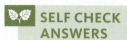

SELF CHECK ANSWERS   **1. a.** $\frac{r^4}{s}$ **b.** $-\frac{p}{2q}$ **2.** $\frac{1}{b} - \frac{2}{a}$ **3.** $3a - 2b + 1$ **4.** $2p + \frac{q}{7p} - \frac{1}{7pq}$ **5.** 2   **6.** no

---

NOW TRY THIS

Perform each division.

1. $\dfrac{6 - 2i}{3}$

2. **a.** $\dfrac{2x^{p+1}}{6x^{p-1}}$ **b.** $\dfrac{x^{m-1}}{x^{1-m}}$

3. $\dfrac{(x + 3)^4 - (x + 3)^2}{(x + 3)^2}$

4.7 Exercises

WARM-UPS *Simplify each fraction.*

1. $\dfrac{3}{21}$ 2. $\dfrac{27}{81}$

3. $\dfrac{-64}{72}$ 4. $\dfrac{-125}{50}$

5. $\dfrac{70}{420}$ 6. $\dfrac{-3{,}612}{-3{,}612}$

7. $\dfrac{8{,}423}{-8{,}423}$ 8. $\dfrac{-288}{-112}$

REVIEW *Identify each polynomial as a monomial, a binomial, a trinomial, or none of these.*

9. $5a^2b + 2ab^2$

10. $-3x^3y$

11. $-2x^3 + 3x^2 - 4x + 12$

12. $17t^2 - 15t + 27$

13. Find the degree of the trinomial $3x^2 - 2x + 4$.

14. What is the numerical coefficient of the second term of the trinomial $-7t^2 - 5t + 17$?

VOCABULARY AND CONCEPTS *Fill in the blanks.*

15. A _____ is an algebraic expression in which the exponents on the variables are whole numbers.

16. A _____ is a polynomial with one algebraic term.

17. Any nonzero value divided by itself is __.

18. Division by __ is undefined.

19. $\dfrac{1}{b} \cdot a = $ _____

20. $\dfrac{15x - 6y}{6xy} = \dfrac{15x}{\_\_\_} - \dfrac{6y}{6xy}$

GUIDED PRACTICE *In all fractions, assume that no denominators are 0. Perform each division by simplifying each fraction. Write all answers without using negative or zero exponents.* SEE EXAMPLE 1. *(OBJECTIVE 1)*

21. $\dfrac{xy}{yz}$ 22. $\dfrac{a^2b}{ab^2}$

23. $\dfrac{r^3s^2}{rs^3}$ 24. $\dfrac{y^4z^3}{y^2z^2}$

25. $\dfrac{8x^3y^2}{4xy^3}$ 26. $\dfrac{-3y^3z}{6yz^2}$

27. $\dfrac{12u^5v}{-4u^2v^3}$ 28. $\dfrac{16rst^2}{-8rst^3}$

Simplify. SEE EXAMPLE 2. *(OBJECTIVE 2)*

29. $\dfrac{6x + 9y}{3xy}$ 30. $\dfrac{8x + 12y}{4xy}$

31. $\dfrac{xy + 6}{3y}$ 32. $\dfrac{ab + 10}{2b}$

33. $\dfrac{5x - 10y}{25xy}$ 34. $\dfrac{2x - 32}{16x}$

35. $\dfrac{3x^2 + 6y^3}{3x^2y^2}$ 36. $\dfrac{4a^2 - 9b^2}{12ab}$

Simplify. SEE EXAMPLES 3–4. *(OBJECTIVE 2)*

37. $\dfrac{4x - 2y + 8z}{4xy}$ 38. $\dfrac{5a^2 + 10b^2 - 15ab}{5ab}$

39. $\dfrac{12x^3y^2 - 8x^2y - 4x}{4xy}$ 40. $\dfrac{12a^2b^2 - 8a^2b - 4ab}{4ab}$

41. $\dfrac{-25x^2y + 30xy^2 - 5xy}{-5xy}$

42. $\dfrac{-30a^2b^2 - 15a^2b - 10ab^2}{-10ab}$

43. $\dfrac{15a^3b^2 - 10a^2b^3}{5a^2b^2}$

44. $\dfrac{9a^4b^3 - 16a^3b^4}{12a^2b}$

Simplify each numerator and perform the division. SEE EXAMPLE 5.
(OBJECTIVE 2)

45. $\dfrac{5x(4x - 2y)}{2y}$

46. $\dfrac{9y^2(x^2 - 3xy)}{3x^2}$

47. $\dfrac{(-2x)^3 + (3x^2)^2}{6x^2}$

48. $\dfrac{(-3x^2y)^3 + (3xy^2)^3}{27x^3y^4}$

49. $\dfrac{4x^2y^2 - 2(x^2y^2 + xy)}{2xy}$

50. $\dfrac{-5a^3b - 5a(ab^2 - a^2b)}{10a^2b^2}$

51. $\dfrac{(a + b)^2 - (a - b)^2}{2ab}$

52. $\dfrac{(x - y)^2 + (x + y)^2}{2x^2y^2}$

Determine whether the two formulas are the same. SEE EXAMPLE 6.
(OBJECTIVE 3)

53. $l = \dfrac{P - 2w}{2}$ and $l = \dfrac{P}{2} - w$.

54. $r = \dfrac{G + 2b}{2b}$ and $r = \dfrac{G}{2b} + b$.

55. Phone bills   On a phone bill, the following formulas are given to compute the average cost per minute of x minutes of phone usage. Are they equivalent?

$$C = \dfrac{0.15x + 12}{x} \quad \text{and} \quad C = 0.15 + \dfrac{12}{x}$$

56. Electric bills   On an electric bill, the following formulas are given to compute the average cost of x kwh of electricity. Are they equivalent?

$$C = \dfrac{0.08x + 5}{x} \quad \text{and} \quad C = 0.08x + \dfrac{5}{x}$$

ADDITIONAL PRACTICE   *In all fractions, assume that no denominators are 0. Simplify each expression.*

57. $\dfrac{120}{160}$

58. $\dfrac{-90}{360}$

59. $\dfrac{5,880}{2,660}$

60. $\dfrac{-762}{366}$

61. $\dfrac{-16r^3y^2}{-4r^2y^4}$

62. $\dfrac{35xyz^2}{-7x^2yz}$

63. $\dfrac{-65rs^2t}{15r^2s^3t}$

64. $\dfrac{112u^3z^6}{-42u^3z^6}$

65. $\dfrac{x^2x^3}{xy^6}$

66. $\dfrac{(xy)^2}{x^2y^3}$

67. $\dfrac{(a^3b^4)^3}{ab^4}$

68. $\dfrac{(a^2b^3)^3}{a^6b^6}$

69. $\dfrac{12a + 2b}{6ab}$

70. $\dfrac{5ab + 30a^2}{10a}$

71. $\dfrac{16x - 8y}{24xy}$

72. $\dfrac{30xy^2 - 24x^2y + 12xy}{6xy}$

73. $\dfrac{2x(8x - 3y)}{4x}$

74. $\dfrac{3a^2b - 6(ab + a^2b^2)}{3ab}$

75. $\dfrac{8x^3 + 16x^2 - 4x}{4x}$

76. $\dfrac{12y^4 - 9y^3 + 6y^2}{-6y^2}$

77. $\dfrac{-(3x^3y^4)^3}{-(9x^4y^5)^2}$

78. $\dfrac{-15b^5 + 12b^3 - 18b^2 - 3b}{3b}$

79. $\dfrac{(a^2a^3)^4}{(a^4)^3}$

80. $\dfrac{(t^{-3}t^5)}{(t^2)^{-3}}$

81. $\dfrac{(3x - y)(2x - 3y)}{6xy}$

82. $\dfrac{(2m - n)(3m - 2n)}{-3m^2n^2}$

WRITING ABOUT MATH

83. Describe how you would simplify the fraction

$$\dfrac{4x^2y + 8xy^2}{4xy}$$

84. A student incorrectly attempts to simplify the fraction $\dfrac{3x + 5}{x + 5}$ as follows:

$$\dfrac{3x + 5}{x + 5} = \dfrac{3x + 5}{x + 5} = 3$$

How would you explain the error?

SOMETHING TO THINK ABOUT

85. If $x = 501$, evaluate $\dfrac{x^{500} - x^{499}}{x^{499}}$.

86. An exercise reads as follows:

Simplify: $\dfrac{3x^3y + 6xy^2}{3xy^3}$

It contains a misprint: one mistyped letter or digit. The correct answer is $\dfrac{x^2}{y} + 2$. Correct the exercise.

Section 4.8

Dividing Polynomials by Polynomials

Objectives

1. Divide a polynomial by a binomial.
2. Divide a polynomial by a binomial by first writing exponents in descending order.
3. Divide a polynomial with one or more missing terms by a binomial.

Vocabulary

divisor   quotient   remainder
dividend

Getting Ready

Divide.

1. $12\overline{)156}$ 2. $17\overline{)357}$ 3. $13\overline{)247}$ 4. $19\overline{)247}$

We now complete our work of operations on polynomials by considering how to divide one polynomial by another.

1 Divide a polynomial by a binomial.

To divide one polynomial by another, we use a method similar to long division in arithmetic. Recall that the parts of a division problem are defined as

$$divisor\overline{)dividend}^{\,quotient\,+\,remainder}$$

Recall that division by zero is undefined. Therefore, the divisor cannot be 0. We must exclude any value of the variable that will result in a divisor of zero.

EXAMPLE 1   Divide $(x^2 + 5x + 6)$ by $(x + 2)$.   Assume no division by 0.

Solution   Here the **divisor** is $x + 2$ and the **dividend** is $x^2 + 5x + 6$. We proceed as follows:

Step 1:
$$\begin{array}{r} x \\ x + 2\overline{)x^2 + 5x + 6} \end{array}$$

How many times does x divide x^2? $\frac{x^2}{x} = x$
Write x above the division symbol.

Step 2:
$$\begin{array}{r} x \\ x + 2\overline{)x^2 + 5x + 6} \\ x^2 + 2x \end{array}$$

Multiply each item in the divisor by x.
Write the product under $x^2 + 5x$ and draw a line.

Step 3:
$$\begin{array}{r} x \\ x + 2\overline{)x^2 + 5x + 6} \\ -\underline{x^2 - 2x} \\ 3x + 6 \end{array}$$

Subtract $(x^2 + 2x)$ from $(x^2 + 5x)$ by adding the negative of $(x^2 + 2x)$ to $(x^2 + 5x)$.
Bring down the 6.

Step 4:

$$\begin{array}{r} x + 3 \\ x + 2\overline{)x^2 + 5x + 6} \\ -x^2 - 2x \\ \hline 3x + 6 \end{array}$$

How many times does x divide $3x$? $\dfrac{3x}{x} = +3$

Write $+3$ above the division symbol.

Step 5:

$$\begin{array}{r} x + 3 \\ x + 2\overline{)x^2 + 5x + 6} \\ -x^2 - 2x \\ \hline 3x + 6 \\ -3x - 6 \end{array}$$

Multiply each term in the divisor by 3. Write the product under the $3x + 6$ and draw a line.

Step 6:

$$\begin{array}{r} x + 3 \\ x + 2\overline{)x^2 + 5x + 6} \\ -x^2 - 2x \\ \hline 3x + 6 \\ -3x - 6 \\ \hline 0 \end{array}$$

Subtract $(3x + 6)$ from $(3x + 6)$ by adding the negative of $(3x + 6)$.

The **quotient** is $x + 3$, and the **remainder** is 0.

Step 7: Check by verifying that $x + 2$ times $x + 3$ is $x^2 + 5x + 6$.

$$(x + 2)(x + 3) = x^2 + 3x + 2x + 6$$
$$= x^2 + 5x + 6$$

SELF CHECK 1 Divide: $(x^2 + 7x + 12)$ by $(x + 3)$ Assume no division by 0.

We need to consider division problems with a remainder other than 0 such as in the next example.

EXAMPLE 2 Divide: $\dfrac{6x^2 - 7x - 2}{2x - 1}$ Assume no division by 0.

Solution Here the divisor is $2x - 1$ and the dividend is $6x^2 - 7x - 2$.

Step 1:

$$\begin{array}{r} 3x \\ 2x - 1\overline{)6x^2 - 7x - 2} \end{array}$$

How many times does $2x$ divide $6x^2$? $\dfrac{6x^2}{2x} = 3x$

Write $3x$ above the division symbol.

Step 2:

$$\begin{array}{r} 3x \\ 2x - 1\overline{)6x^2 - 7x - 2} \\ 6x^2 - 3x \end{array}$$

Multiply each term in the divisor by $3x$. Write the product under $6x^2 - 7x$ and draw a line.

Step 3:

$$\begin{array}{r} 3x \\ 2x - 1\overline{)6x^2 - 7x - 2} \\ -6x^2 + 3x \\ \hline -4x - 2 \end{array}$$

Subtract $(6x^2 - 3x)$ from $(6x^2 - 7x)$ by adding the negative of $(6x^2 - 3x)$ to $(6x^2 - 7x)$.

Bring down the -2.

Step 4:

$$\begin{array}{r} 3x - 2 \\ 2x - 1\overline{)6x^2 - 7x - 2} \\ -6x^2 + 3x \\ \hline -4x - 2 \end{array}$$

How many times does $2x$ divide $-4x$? $\dfrac{-4x}{2x} = -2$

Write -2 above the division symbol.

Step 5:

$$\begin{array}{r} 3x - 2 \\ 2x - 1\overline{)6x^2 - 7x - 2} \\ -6x^2 + 3x \\ \hline -4x - 2 \\ +4x + 2 \end{array}$$

Multiply each term in the divisor by -2. Write the product under $-4x - 2$ and draw a line.

Step 6:

$$
\begin{array}{r}
3x - 2 \\
2x - 1 \overline{\smash{\big)}\ 6x^2 - 7x - 2} \\
\underline{-\ 6x^2 + 3x} \\
-4x - 2 \\
\underline{+4x - 2} \\
-4
\end{array}
$$

Subtract $(-4x + 2)$ from $(-4x - 2)$ by adding the negative of $(-4x + 2)$.

COMMENT The division process ends when the degree of the remainder is less than the degree of the divisor.

Here the quotient is $3x - 2$, and the remainder is -4. It is common to write the answer in quotient $+ \frac{\text{remainder}}{\text{divisor}}$ form:

$$
3x - 2 + \frac{-4}{2x - 1}
$$

where the fraction $\frac{-4}{2x - 1}$ is formed by dividing the remainder by the divisor.

Step 7: To check the answer, we multiply $(3x - 2)$ by $(2x - 1)$ and add (-4). The result should be the dividend.

$$
(2x - 1)(3x - 2) - 4 = 6x^2 - 4x - 3x + 2 - 4
$$
$$
= 6x^2 - 7x - 2
$$

 SELF CHECK 2   Divide: $\dfrac{8x^2 + 6x - 3}{2x + 3}$ Assume no division by 0.

Division by a binomial may include more than one variable as illustrated in the next example.

EXAMPLE 3   Divide: $\dfrac{6x^2 - xy - y^2}{3x + y}$ Assume no division by 0.

Solution   Here the divisor is $3x + y$ and the dividend is $6x^2 - xy - y^2$.

Step 1:

$$
\begin{array}{r}
2x \\
3x + y \overline{\smash{\big)}\ 6x^2 -\ xy - y^2}
\end{array}
$$

How many times does $3x$ divide $6x^2$? $\frac{6x^2}{3x} = 2x$ Write $2x$ above the division symbol.

Step 2:

$$
\begin{array}{r}
2x \\
3x + y \overline{\smash{\big)}\ 6x^2 -\ xy - y^2} \\
\underline{6x^2 + 2xy}
\end{array}
$$

Multiply each term in the divisor by $2x$. Write the product under $6x^2 - xy$ and draw a line.

Step 3:

$$
\begin{array}{r}
2x \\
3x + y \overline{\smash{\big)}\ 6x^2 -\ xy - y^2} \\
\underline{-6x^2 - 2xy} \\
-3xy - y^2
\end{array}
$$

Subtract $(6x^2 + 2xy)$ from $(6x^2 - xy)$ by adding the negative of $(6x^2 + 2xy)$ to $(6x^2 - xy)$.

Bring down the $-y^2$.

Step 4:

$$
\begin{array}{r}
2x - y \\
3x + y \overline{\smash{\big)}\ 6x^2 -\ xy - y^2} \\
\underline{-6x^2 - 2xy} \\
-3xy - y^2
\end{array}
$$

How many times does $3x$ divide $-3xy$? $\frac{-3xy}{3x} = -y$ Write $-y$ above the division symbol.

Step 5:

$$
\begin{array}{r}
2x - y \\
3x + y \overline{\smash{\big)}\ 6x^2 -\ xy - y^2} \\
\underline{-6x^2 - 2xy} \\
-3xy - y^2 \\
\underline{+3xy + y^2}
\end{array}
$$

Multiply each term in the divisor by $-y$. Write the product under the $-3x - y^2$ and draw a line.

Step 6:

$$
\begin{array}{r}
2x - y \\
3x + y\overline{)6x^2 -\ xy - y^2} \\
\underline{-6x^2 - 2xy} \\
-3xy - y^2 \\
\underline{+3xy + y^2} \\
0
\end{array}
$$

Subtract $(-3xy - y^2)$ from $(-3xy - y^2)$ by adding the negative of $(-3xy - y^2)$.

The quotient is $2x - y$ and the remainder is 0.

SELF CHECK 3   Divide $(6x^2 - xy - y^2)$ by $(2x - y)$. Assume no division by 0.

2 Divide a polynomial by a binomial by first writing exponents in descending order.

The division method works best when exponents of the terms in the divisor and the dividend are written in descending order. This means that the term involving the highest power of x appears first, the term involving the second-highest power of x appears second, and so on. For example, the terms in

$$3x^3 + 2x^2 - 7x + 5 \qquad 5 = 5x^0$$

have their exponents written in descending order.

If the powers in the dividend or divisor are not in descending order, we can use the commutative property of addition to write them that way.

EXAMPLE 4   Divide: $\dfrac{4x^2 + 2x^3 + 12 - 2x}{x + 3}$ Assume no division by 0.

Solution   We write the dividend so that the exponents are in descending order and divide.

$$
\begin{array}{r}
2x^2 - 2x + 4 \\
x + 3\overline{)2x^3 + 4x^2 - 2x + 12} \\
\underline{-2x^3 - 6x^2} \\
-2x^2 - 2x \\
\underline{+2x^2 + 6x} \\
+4x + 12 \\
\underline{-4x - 12} \\
0
\end{array}
$$

Check: $(x + 3)(2x^2 - 2x + 4) = 2x^3 - 2x^2 + 4x + 6x^2 - 6x + 12$

$$= 2x^3 + 4x^2 - 2x + 12$$

SELF CHECK 4   Divide: $\dfrac{x^2 - 10x + 6x^3 + 4}{2x - 1}$ Assume no division by 0.

3 Divide a polynomial with one or more missing terms by a binomial.

When we write the terms of a dividend in descending powers of x, we may notice that some powers of x are missing. For example, if the dividend is $3x^4 - 7x^2 - 3x + 15$, the term involving x^3 is missing. When this happens, we should either write the term with a coefficient of 0 or leave a blank space for it. In this case, we would write the dividend as

$$3x^4 + 0x^3 - 7x^2 - 3x + 15 \qquad \text{or} \qquad 3x^4 \qquad -7x^2 - 3x + 15$$

EXAMPLE 5 Divide: $\dfrac{x^2 - 4}{x + 2}$ Assume no division by 0.

Solution Since $x^2 - 4$ does not have a term involving x, we must either include the term $0x$ or leave a space for it.

$$
\begin{array}{r}
x - 2 \\
x + 2 \overline{)x^2 + 0x - 4} \\
\underline{-x^2 - 2x} \\
-2x - 4 \\
\underline{+2x + 4} \\
0
\end{array}
$$

Check: $(x + 2)(x - 2) = x^2 - 2x + 2x - 4$
$$= x^2 - 4$$

SELF CHECK 5 Divide: $\dfrac{x^2 - 9}{x - 3}$ Assume no division by 0.

SELF CHECK ANSWERS

1. $x + 4$ **2.** $4x - 3 + \dfrac{6}{2x + 3}$ **3.** $3x + y$ **4.** $3x^2 + 2x - 4$ **5.** $x + 3$

NOW TRY THIS

Assume no division by 0.

1. Identify the missing term(s): $8x^3 - 7x + 2x^5 - x^2$

2. Perform the division: $\dfrac{x^2 + 3x - 5}{x + 3}$

3. $(8x^2 - 2x + 3) \div (1 + 2x)$

4. The area of a rectangle is represented by $(3x^2 + 17x - 6)$ m^2 and the width is represented by $(3x - 1)$ m. Find a polynomial representation of the length.

4.8 Exercises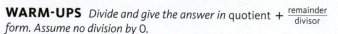

WARM-UPS *Divide and give the answer in* quotient $+ \dfrac{\text{remainder}}{\text{divisor}}$ *form. Assume no division by 0.*

1. $16\overline{)384}$ **2.** $26\overline{)806}$

3. $19\overline{)271}$ **4.** $15\overline{)241}$

If each of the expressions is a divisor, determine what value must be excluded. Write your answer as $x \ne$ value.

5. $x + 2$ **6.** $3 - x$

7. $2x - 7$ **8.** $4x + 5$

REVIEW

9. List the composite numbers between 20 and 30.

10. Graph the set of prime numbers between 10 and 20 on a number line.

10 11 12 13 14 15 16 17 18 19 20

Let $a = -2$ and $b = 3$. Evaluate each expression.

11. $|a - b|$ **12.** $|a + b|$

13. $-|a^2 - b^2|$ **14.** $a - |-b|$

Simplify each expression.

15. $4(3x^2 - 5x + 2) + 3(2x^2 + 6x - 3)$

16. $-2(y^3 + 2y^2 - y) - 3(3y^3 + y)$

VOCABULARY AND CONCEPTS   *Fill in the blanks.*

17. In the long division $x + 1\overline{)x^2 + 2x + 1}$, $x + 1$ is called the
_____, and $x^2 + 2x + 1$ is called the _____.

18. The answer to a division problem is called the _____.

19. If a division does not come out even, the leftover part is called
a _____.

20. The exponents in $2x^4 + 3x^3 + 4x^2 - 7x - 2$ are said to be
written in _____ order.

Write each polynomial with the powers in descending order.

21. $8x^2 + 2x^3 - 5 + 3x$

22. $5x^2 + 7x^3 - 3x - 9$

23. $7x + 6x^3 - 4x^2 + 5x^4$

24. $7x^5 + x^3 - x^2 + 2x^4$

Identify the missing terms in each polynomial.

25. $5x^4 + 2x^2 - 1$

26. $-3x^5 - 2x^3 + 4x - 6$

GUIDED PRACTICE   *Perform each division. Assume no division by 0. SEE EXAMPLE 1. (OBJECTIVE 1)*

27. Divide $(x^2 + 5x + 4)$ by $(x + 1)$.

28. Divide $(y^2 + 13y + 12)$ by $(y + 1)$.

29. $x + 5\overline{)x^2 + 7x + 10}$

30. $x + 6\overline{)x^2 + 5x - 6}$

31. $\dfrac{x^2 - 5x + 6}{x - 2}$ **32.** $\dfrac{z^2 - 7z + 12}{z - 3}$

33. $a - 7\overline{)a^2 - 11a + 28}$ **34.** $t - 6\overline{)t^2 - 3t - 18}$

Perform each division. Assume no division by 0. SEE EXAMPLE 2. (OBJECTIVE 1)

35. $\dfrac{8a^2 - 10a + 3}{4a - 3}$ **36.** $\dfrac{9a^2 - 9a - 4}{3a - 4}$

37. $\dfrac{3b^2 + 11b + 6}{3b + 2}$ **38.** $\dfrac{8a^2 + 2a - 3}{2a - 1}$

39. $\dfrac{2x^2 + 5x + 2}{2x + 3}$ **40.** $\dfrac{3x^2 - 8x + 3}{3x - 2}$

41. $\dfrac{4x^2 + 6x - 1}{2x + 1}$ **42.** $\dfrac{6x^2 - 11x + 2}{3x - 1}$

Perform each division. Assume no division by 0. SEE EXAMPLE 3. (OBJECTIVE 1)

43. Divide $(a^2 + 2ab + b^2)$ by $(a + b)$.

44. Divide $(a^2 - 2ab + b^2)$ by $(a - b)$.

45. $x + 2y\overline{)2x^2 + 3xy - 2y^2}$

46. $x + 3y\overline{)2x^2 + 5xy - 3y^2}$

47. $\dfrac{2x^2 - 7xy + 3y^2}{2x - y}$ **48.** $\dfrac{3x^2 + 5xy - 2y^2}{x + 2y}$

49. $\dfrac{12a^2 - ab - b^2}{3a - b}$ **50.** $\dfrac{2m^2 + 7mn - 4n^2}{2m - n}$

*Write the powers of x in descending order (if necessary) and perform
each division. Assume no division by 0. SEE EXAMPLE 4. (OBJECTIVE 2)*

51. $5x + 3\overline{)11x + 10x^2 + 3}$

52. $2x - 7\overline{)-x - 21 + 2x^2}$

53. $4 + 2x\overline{)-10x - 28 + 2x^2}$

54. $1 + 3x\overline{)9x^2 + 1 + 6x}$

*Perform each division. Assume no division by 0. SEE EXAMPLE 5.
(OBJECTIVE 3)*

55. $\dfrac{x^2 - 16}{x - 4}$ **56.** $\dfrac{x^2 - 25}{x + 5}$

57. $\dfrac{4x^2 - 9}{2x + 3}$ **58.** $\dfrac{25x^2 - 16}{5x - 4}$

59. $\dfrac{x^3 - 8}{x - 2}$ **60.** $\dfrac{x^3 + 27}{x + 3}$

ADDITIONAL PRACTICE   *Perform each division. If there is
a remainder, leave the answer in* quotient $+ \frac{remainder}{divisor}$ *form. Assume
no division by 0.*

61. $2x + 3\overline{)2x^3 + 7x^2 + 4x - 3}$

62. $2x - 1\overline{)2x^3 - 3x^2 + 5x - 2}$

63. $\dfrac{x^3 + 3x^2 + 3x + 1}{x + 1}$

64. $\dfrac{x^3 + 6x^2 + 12x + 8}{x + 2}$

65. $3x - 4\overline{)15x^3 - 23x^2 + 16x}$

66. $2y + 3\overline{)21y^2 + 6y^3 - 20}$

67. $3x + 2\overline{)6x^3 + 10x^2 + 7x + 2}$

68. $4x + 3\overline{)4x^3 - 5x^2 - 2x + 3}$

69. $\dfrac{2x^3 + 7x^2 + 4x - 4}{2x + 3}$

70. $\dfrac{6x^3 + x^2 + 2x - 2}{3x - 1}$

71. $\dfrac{2x^3 + 4x^2 - 2x + 3}{x - 2}$

72. $3x - 2y\overline{)-10y^2 + 13xy + 3x^2}$

73. $2x - y\overline{)xy - 2y^2 + 6x^2}$

74. $2x + y\overline{)2x^3 + 3x^2y + 3xy^2 + y^3}$

WRITING ABOUT MATH

75. Distinguish among *dividend, divisor, quotient,* and
remainder.

76. How would you check the results of a division?

SOMETHING TO THINK ABOUT

77. Find the error in the following work.

$$
\begin{array}{r}
x + 1 \\
x - 2\overline{)x^2 + 3x - 2} \\
\underline{x^2 - 2x} \\
x - 2 \\
\underline{x - 2} \\
0
\end{array}
$$

78. Find the error in the following work.

$$
\begin{array}{r}
3x \\
x + 2\overline{)3x^2 + 10x + 7} \\
\underline{3x^2 + 6x} \\
4x + 7
\end{array}
= 3x + \dfrac{4x + 7}{x + 2}
$$

79. Divide: $\dfrac{a^3 + a}{a + 3}$

80. Divide: $\dfrac{x^3 + y^3}{x + y}$

Projects

PROJECT 1

Let $f(x) = 3x^2 + 3x - 2$, $g(x) = 2x^2 - 5$, and $t(x) = x + 2$. Perform each operation.

a. $f(x) + g(x)$

b. $g(x) - t(x)$

c. $f(x) \cdot g(x)$

d. $\dfrac{f(x)}{t(x)}$

e. $f(x) + g(x) + t(x)$

f. $f(x) \cdot g(x) \cdot t(x)$

g. $\dfrac{f(x) - t(x) + g(x)}{t(x)}$

h. $\dfrac{[g(x)]^2}{t(x)}$

PROJECT 2

To discover a pattern in the behavior of polynomials, consider the polynomial $2x^2 - 3x - 5$. First, evaluate the polynomial at $x = 1$ and $x = 3$. Then divide the polynomial by $(x - 1)$ and again by $(x - 3)$.

a. What do you notice about the remainders of these divisions?

b. Try others. For example, evaluate the polynomial at $x = 2$ and then divide by $(x - 2)$.

c. Can you make the pattern hold when you evaluate the polynomial at $x = -2$?

d. Does the pattern hold for other polynomials? Try some polynomials of your own, experiment, and report your conclusions.

Reach for Success

EXTENSION OF STUDY STRATEGIES

Let's consider how you can increase test preparedness *prior* to an exam. What would you do differently if you only knew then what you know now?

There is a known learning pattern called the Forgetting Curve, first recognized in 1885 by Hermann Ebbinghaus. The basic premise is that by reviewing material from a one-hour lecture for ten minutes the next day you are firing the synapses that make connections in your brain. Then, reviewing the material just a few minutes every day after that significantly increases your ability to recall that information. When you repeat this information, you send a message to your brain, "Here it is again; it must be important! I need to know this."

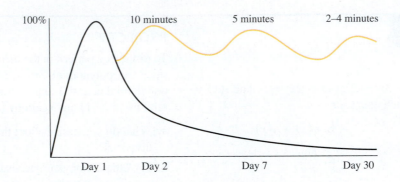

According to the graph, by Day 7, how much time would you need to spend to remember the material?_____

By Day 30, how much time would you need to review the material?_____

Can you think of factors that might influence the speed of forgetting?_____

_____

Do you feel overwhelmed by this amount of review? Does it seem to take too much time?_____

All-night study sessions rarely help with exam performance. Would it surprise you to know that it actually takes *less* time to review daily than it does to relearn the material the night before a test?_____

Give this a try and share your experience with your instructor.

You might not be surprised to learn that difficulty of the material, its connectedness to a previously learned topic, stress, and lack of sleep could all influence how well you can remember the material. While studying mathematics, work toward understanding the material. That will help increase the connections to other material and make it easier to remember.

4 Review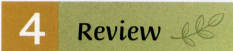

SECTION 4.1   Natural-Number Exponents

DEFINITIONS AND CONCEPTS	EXAMPLES
If n is a natural number, then $$x^n = \overbrace{x \cdot x \cdot x \cdot \cdots \cdot x}^{n \text{ factors of } x}$$	$$x^5 = \overbrace{x \cdot x \cdot x \cdot x \cdot x}^{5 \text{ factors of } x} \qquad x^7 = \overbrace{x \cdot x \cdot x \cdot x \cdot x \cdot x \cdot x}^{7 \text{ factors of } x}$$
If m and n are integers, then $$x^m x^n = x^{m+n}$$ $$(x^m)^n = x^{m \cdot n}$$ $$(xy)^n = x^n y^n$$ $$\left(\frac{x}{y}\right)^n = \frac{x^n}{y^n} \quad (y \neq 0)$$ $$\frac{x^m}{x^n} = x^{m-n} \quad (x \neq 0)$$	$$x^2 \cdot x^7 = x^{2+7} = x^9$$ $$(x^2)^7 = x^{2 \cdot 7} = x^{14}$$ $$(xy)^3 = x^3 y^3$$ $$\left(\frac{x}{y}\right)^3 = \frac{x^3}{y^3} \quad (y \neq 0)$$ $$\frac{x^7}{x^2} = x^{7-2} = x^5 \quad (x \neq 0)$$

REVIEW EXERCISES

Write each expression without exponents.

1. $(-3x)^4$

2. $\left(\dfrac{1}{2}pq\right)^3$

Evaluate each expression.

3. 5^3 **4.** 3^5

5. $(-6)^2$ **6.** -6^2

7. $3^2 + 2^2$ **8.** $(3 + 2)^2$

Perform the operations and simplify.

9. $x^4 x^6$ **10.** $x^2 x^7$

11. $(y^7)^3$ **12.** $(x^{21})^2$

13. $(ab)^3$ **14.** $(3x)^4$

15. $b^3 b^4 b^5$ **16.** $-z^2(z^3 y^2)$

17. $(16s)^2 s$ **18.** $-3y(y^5)$

19. $(2x^4 y^2)^3$ **20.** $(5x^3 y)^2$

21. $\dfrac{x^7}{x^3}$ **22.** $\left(\dfrac{x^2 y}{xy^2}\right)^2$

23. $\dfrac{8(y^2 x)^2}{4(yx^2)^2}$ **24.** $\dfrac{(5y^2 z^3)^3}{25(yz)^5}$

SECTION 4.2   Zero and Negative-Integer Exponents

DEFINITIONS AND CONCEPTS	EXAMPLES
$$x^0 = 1 \quad (x \neq 0)$$ $$x^{-n} = \frac{1}{x^n} \quad (x \neq 0)$$ $$\frac{1}{x^{-n}} = x^n \quad (x \neq 0)$$	$$(2x)^0 = 1 \quad (x \neq 0)$$ $$x^{-3} = \frac{1}{x^3} \quad (x \neq 0)$$ $$\frac{1}{x^{-3}} = x^3 \quad (x \neq 0)$$

REVIEW EXERCISES

Write each expression without negative exponents or parentheses.

25. x^0 **26.** $(3x^2 y^2)^0$

27. $(3x^0)^2$ **28.** $(3x^2 y^0)^2$

29. x^{-3} **30.** $x^{-2} x^3$

31. $y^4 y^{-3}$ **32.** $\dfrac{x^3}{x^{-7}}$

33. $(x^{-3} x^4)^{-2}$ **34.** $(a^{-2} b)^{-3}$

35. $\left(\dfrac{x^2}{x}\right)^{-5}$ **36.** $\left(\dfrac{15z^4}{5z^3}\right)^{-2}$

SECTION 4.3   Scientific Notation

DEFINITIONS AND CONCEPTS	EXAMPLES
A number is written in scientific notation if it is written as the product of a number between 1 (including 1) and 10 and an integer power of 10.	4,582,000,000 is written as 4.582×10^9 in scientific notation. 0.00035 is written as 3.5×10^{-4} in scientific notation.

REVIEW EXERCISES

Write each number in scientific notation.

37. 728   **38.** 6,230

39. 0.0275   **40.** 0.00942

41. 7.73   **42.** 753×10^3

43. 0.018×10^{-2} **44.** 600×10^2

Write each number in standard notation.

45. 3.87×10^4 **46.** 7.98×10^{-5}

47. 2.68×10^0 **48.** 5.76×10^1

49. 739×10^{-2} **50.** 0.437×10^{-3}

51. $\dfrac{(0.00012)(0.00004)}{0.00000016}$ **52.** $\dfrac{(4,800)(20,000)}{600,000}$

SECTION 4.4   Polynomials

DEFINITIONS AND CONCEPTS	EXAMPLES
A polynomial is an algebraic expression that is one term or the sum of terms containing whole-number exponents on the variables.	Polynomials: $9xy$, $5x^2 + 9x - 1$, and $11x - 5y$

If a is a nonzero coefficient, the degree of the monomial ax^n is n.
 	
 The degree of a polynomial is the same as the degree of its term with largest degree.
	Find the degree of each term and the degree of the polynomial $8x^2 - 5x + 3$.
	    The degree of the first term is 2.
	    The degree of the second term is 1.
	    The degree of the third term is 0.
	    The degree of the polynomial is 2.

When a number is substituted for the variable in a polynomial, the polynomial takes on a numerical value.	Evaluate $5x - 4$ when $x = -3$. $5x - 4 = 5(-3) - 4$ Substitute -3 for x. $\qquad = -15 - 4$ Simplify. $\qquad = -19$
Finding a function value for a polynomial uses the same process as evaluating a polynomial for a specified value.	If $f(x) = x^2 - 8x + 3$, find $f(-3)$. $f(x) = x^2 - 8x + 3$ $f(-3) = (-3)^2 - 8(-3) + 3$ Substitute -3 for x. $\qquad = 9 + 24 + 3$ Simplify. $\qquad = 36$ Since the result is 36, $f(-3) = 36$.
To graph a polynomial function, create a table of values, plot the ordered pairs $(x, f(x))$, and draw a smooth curve through those points. Determine the domain and the range from the graph.	Graph the polynomial function $f(x) = x^2 - 8x + 3$ and state the domain and range: <table><tr><td>x</td><td>$f(x) = x^2 - 8x + 3$</td><td>$(x, f(x))$</td></tr><tr><td>-1</td><td>$f(-1) = (-1)^2 - 8(-1) + 3 = 12$</td><td>$(-1, 12)$</td></tr><tr><td>$0$</td><td>$f(0) = (0)^2 - 8(0) + 3 = 3$</td><td>$(0, 3)$</td></tr><tr><td>$1$</td><td>$f(1) = (1)^2 - 8(1) + 3 = -4$</td><td>$(1, -4)$</td></tr><tr><td>$2$</td><td>$f(2) = (2)^2 - 8(2) + 3 = -9$</td><td>$(2, -9)$</td></tr><tr><td>$4$</td><td>$f(4) = (4)^2 - 8(4) + 3 = -13$</td><td>$(4, -13)$</td></tr><tr><td>$5$</td><td>$f(5) = (5)^2 - 8(5) + 3 = -12$</td><td>$(5, -12)$</td></tr></table>

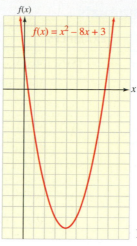

$$f(x) = x^2 - 8x + 3$$

D: $\mathbb{R}$, R: $[-13, \infty)$

REVIEW EXERCISES

Find the degree of each polynomial and classify it as a monomial, a binomial, or a trinomial.

53. $29x^8$

54. $5^3x + x^2$

55. $-3x^5 + x - 1$

56. $9xy + 21x^3y^2$

Evaluate $3x + 2$ for each value of x.

57. $x = 7$

58. $x = -4$

59. $x = -2$

60. $x = \dfrac{2}{3}$

Evaluate $5x^4 - x$ for each value of x.

61. $x = 3$

62. $x = 0$

63. $x = -2$

64. $x = -0.3$

If $f(x) = x^2 - 4$, find each value.

65. $f(0)$

66. $f(-4)$

67. $f(-2)$

68. $f\left(\dfrac{1}{2}\right)$

Graph each polynomial function. State the domain and range.

69. $f(x) = x^2 - 5$

70. $f(x) = x^3 - 2$

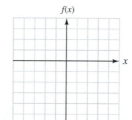

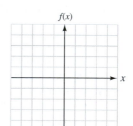

SECTION 4.5   Adding and Subtracting Polynomials

DEFINITIONS AND CONCEPTS	EXAMPLES
We can add polynomials by removing parentheses, if necessary, and then combining any like terms that are contained within the polynomials.	$(8x^3 - 6x + 13) + (9x - 7)$ $= 8x^3 - 6x + 13 + 9x - 7$ Remove parentheses. $= 8x^3 + 3x + 6$ Combine like terms.
We can subtract polynomials by dropping the negative sign and the parentheses, and *changing the sign of every term within the second set of parentheses.*	$(8x^3 - 6x + 13) - (9x - 7)$ $= 8x^3 - 6x + 13 - 9x + 7$ Change the sign of each term in the second set of parentheses. $= 8x^3 - 15x + 20$ Combine like terms.

REVIEW EXERCISES

Simplify each expression, if possible.

71. $6x - 4x + x$

72. $5x + 4y$

73. $(xy)^2 + 3x^2y^2$

74. $-2x^2yz + 3yx^2z$

75. $(3x^2 + 2x) + (5x^2 - 8x)$

76. $(7a^2 + 2a - 5) - (3a^2 - 2a + 1)$

77. $3(9x^2 + 3x + 7) - 2(11x^2 - 5x + 9)$

78. $4(4x^3 + 2x^2 - 3x - 8) - 5(2x^3 - 3x + 8)$

SECTION 4.6   Multiplying Polynomials

DEFINITIONS AND CONCEPTS	EXAMPLES
To multiply two monomials, first multiply the numerical factors and then multiply the variable factors using the properties of exponents.	$(5x^2y^3)(4xy^2)$ $= 5(4)x^2xy^3y^2$ Use the commutative property of multiplication. $= 20x^3y^5$ Use multiplication and the properties of exponents.
To multiply a polynomial with more than one term by a monomial, multiply each term of the polynomial by the monomial and simplify.	$4x(3x^2 + 2x)$ $= \mathbf{4x} \cdot 3x^2 + \mathbf{4x} \cdot 2x$ Use the distributive property. $= 12x^3 + 8x^2$ Multiply.
To multiply two binomials, use the distributive property.	$(2x - 5)(x + 3)$ $= \mathbf{2x}(x) + \mathbf{2x}(3) + (-5)(x) + (-5)(3)$ $= 2x^2 + 6x - 5x - 15$ $= 2x^2 + x - 15$
Special products: Squaring a binomial: $(x + y)^2 = (x + y)(x + y) = x^2 + 2xy + y^2$ $(x - y)^2 = (x - y)(x - y) = x^2 - 2xy + y^2$ Multiplying conjugate binomials: $(x + y)(x - y) = x^2 - y^2$	$(x + 7)^2 = (x + 7)(x + 7)$ $= x^2 + 14x + 49$ $(x - 7)^2 = (x - 7)(x - 7)$ $= x^2 - 14x + 49$ $(2x + 3)(2x - 3) = 4x^2 - 6x + 6x - 9$ $= 4x^2 - 9$
To multiply one polynomial by another, multiply each term of one polynomial by each term of the other polynomial, and simplify.	$(x + 2)(4x^2 - x + 3)$ $4x^3 - x^2 + 3x + 8x^2 - 2x + 6$ $4x^3 + 7x^2 + x + 6$
To solve an equation that simplifies to a linear equation, use the distributive property and order of operations and proceed to solve.	Solve: $(x + 3)(x - 5) = x^2 - 3(x + 1)$ $x^2 - 5x + 3x - 15 = x^2 - 3x - 3$ $x^2 - 2x - 15 = x^2 - 3x - 3$ $-2x - 15 = -3x - 3$ $x - 15 = -3$ $x = 12$

REVIEW EXERCISES

Find each product.

79. $(4x^3y^5)(3x^2y)$

80. $(xyz^3)(x^3z)^2$

Find each product.

81. $4(2x + 3)$

82. $3(2x + 4)$

83. $x^2(3x^2 - 5)$

84. $2y^2(y^2 + 5y)$

85. $-x^2y(y^2 - xy)$

86. $-3xy(xy - x)$

Find each product.

87. $(x + 5)(x + 4)$

88. $(2x + 1)(x - 1)$

89. $(3a - 3)(2a + 2)$

90. $6(a - 1)(a + 1)$

91. $(a - b)(2a + b)$

92. $(3x - y)(2x + y)$

Find each product.

93. $(x + 6)(x + 6)$

94. $(x + 5)(x - 5)$

95. $(y - 7)(y + 7)$

96. $(x + 4)^2$

97. $(x - 3)^2$

98. $(y - 2)^2$

99. $(3y + 2)^2$

100. $(y^2 + 1)(y^2 - 1)$

Find each product.

101. $(3x + 1)(x^2 + 2x + 1)$

102. $(2a - 3)(4a^2 + 6a + 9)$

Solve each equation.

103. $x^2 + 3 = x(x + 3)$

104. $x^2 + x = (x + 1)(x + 2)$

105. $(x + 2)(x - 5) = (x - 4)(x - 1)$

106. $(x - 1)(x - 2) = (x - 3)(x + 1)$

107. $x^2 + x(x + 2) = x(2x + 1) + 1$

108. $(x + 5)(3x + 1) = x^2 + (2x - 1)(x - 5)$

SECTION 4.7   Dividing Polynomials by Monomials

DEFINITIONS AND CONCEPTS

To divide a polynomial by a monomial, divide each term in the numerator by the monomial in the denominator.

EXAMPLES

Divide: $\dfrac{12x^6 - 8x^4 + 2x}{2x}$ Assume no division by 0.

$$\dfrac{12x^6 - 8x^4 + 2x}{2x}$$

$$= \dfrac{12x^6}{2x} - \dfrac{8x^4}{2x} + \dfrac{2x}{2x} \quad \text{Divide each term in the numerator by the monomial in the denominator.}$$

$$= 6x^5 - 4x^3 + 1$$

REVIEW EXERCISES

Perform each division. Assume no variable is 0.

109. $\dfrac{3x + 6y}{2xy}$

110. $\dfrac{21x^2y^2 - 7xy}{7xy}$

111. $\dfrac{15a^2bc + 20ab^2c - 25abc^2}{-5abc}$

112. $\dfrac{(x + y)^2 + (x - y)^2}{-2xy}$

SECTION 4.8   Dividing Polynomials by Polynomials

DEFINITIONS AND CONCEPTS

Use long division to divide one polynomial by another. Answers are written in $quotient + \frac{remainder}{divisor}$ form.

EXAMPLES

Divide: $\dfrac{6x^2 - 3x + 5}{x - 3}$ Assume no division by 0.

$$\begin{array}{r} 6x + 15 \\ x - 3 \overline{) 6x^2 - 3x + 5} \\ \underline{-6x^2 + 18x} \\ 15x + 5 \\ \underline{-15x + 45} \\ 50 \end{array}$$

The result is $6x + 15 + \dfrac{50}{x - 3}$.

REVIEW EXERCISES

Perform each division. Assume no division by 0.

113. $x + 2 \overline{) x^2 + 3x + 5}$

114. $x - 1 \overline{) x^2 - 6x + 5}$

115. $x - 4 \overline{) 3x^2 - 11x - 4}$

116. $3x - 1 \overline{) 3x^2 + 14x - 2}$

117. $2x - 1 \overline{) 6x^3 + x^2 + 1}$

118. $3x + 1 \overline{) -13x - 4 + 9x^3}$

4 Test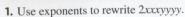

1. Use exponents to rewrite $2xxxyyyy$.
2. Evaluate: $3^2 + 5^3$

Write each expression as an expression containing only one exponent.

3. $y^3(y^5y)$

4. $(-3b^2)(2b^3)(-b^2)$

5. $(2x^3)^5(x^2)^3$

6. $(2rr^2r^3)^3$

Simplify each expression. Write answers without using parentheses or negative exponents. Assume no variable is 0.

7. $-7x^0$

8. $5y^{-6}y^3$

9. $\dfrac{y^2}{yy^{-2}}$

10. $\left(\dfrac{a^2b^{-1}}{4a^3b^{-2}}\right)^{-3}$

11. Write 540,000 in scientific notation.
12. Write 0.0025 in scientific notation.
13. Write 7.4×10^3 in standard notation.
14. Write 6.7×10^{-4} in standard notation.
15. Classify $3x^2 + 2$ as a monomial, a binomial, or a trinomial.
16. Find the degree of the polynomial $3x^2y^3z^4 + 2x^3y^2z - 5x^3y^3z^5$.
17. Evaluate $x^2 + x - 2$ when $x = -2$.
18. Graph the polynomial function $f(x) = x^2 + 2$. State the domain and range.

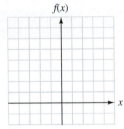

19. Simplify: $-6(x - y) + 2(x + y) - 3(x + 2y)$
20. Simplify: $-2(x^2 + 3x - 1) - 3(x^2 - x + 2) + 5(x^2 + 2)$

21. $\begin{array}{r} 3x^3 + 4x^2 - x - 7 \\ + 2x^3 - 2x^2 + 3x + 2 \\ \hline \end{array}$

22. $\begin{array}{r} 2x^2 - 7x + 3 \\ - 3x^2 - 2x - 1 \\ \hline \end{array}$

Find each product.

23. $(-2x^3)(2x^2y)$
24. $(-5x^4)(4x^5y)$
25. $(2x - 5)(3x + 4)$
26. $(2x - 3)(x^2 - 2x + 4)$

Simplify each expression. Assume no division by 0.

27. Simplify: $\dfrac{8x^2y^3z^4}{16x^3y^2z^4}$

28. Simplify: $\dfrac{6a^2 - 12b^2}{24ab}$

29. Divide: $2x + 3 \overline{)2x^2 - x - 6}$
30. Solve: $(a + 2)^2 = (a - 3)^2$

Cumulative Review

Evaluate each expression. Let $x = 2$ and $y = -5$.

1. $4 + 5x$

2. $3y^2 - 4$

3. $\dfrac{3x - y}{xy}$

4. $\dfrac{x^2 - y^2}{x + y}$

Solve each equation.

5. $\dfrac{4}{5}x + 6 = 18$

6. $x - 2 = \dfrac{x + 2}{3}$

7. $2(5x + 2) = 3(3x - 2)$

8. $4(y + 1) = -2(4 - y)$

Graph the solution of each inequality.

9. $3x - 4 > 2$

10. $7x - 9 < 5$

11. $-2 < -x + 3 < 5$

12. $0 \le \dfrac{4 - x}{3} \le 2$

Solve each formula for the indicated variable.

13. $A = p + prt$ for r **14.** $A = \dfrac{1}{2}bh$ for h

Graph each equation.

15. $3x - 4y = 12$ **16.** $y - 2 = \dfrac{1}{2}(x - 4)$

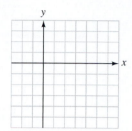   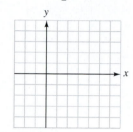

Let $f(x) = 5x - 2$ and find each value.

17. $f(4)$ **18.** $f(-1)$

19. $f(-2)$ **20.** $f\left(\dfrac{1}{5}\right)$

Write each expression as an expression using only one exponent. Assume no division by 0.

21. $y^4(y^2y^8)$ **22.** $\dfrac{x^5y^3}{x^4y^4}$

23. $\dfrac{a^4b^{-3}}{a^{-3}b^3}$ **24.** $\left(\dfrac{-x^{-2}y^3}{x^{-3}y^2}\right)^2$

Perform each operation. Assume no division by 0.

25. $(3x^2 + 2x - 7) - (2x^2 - 2x + 7)$
26. $(4x - 5)(3x + 2)$
27. $(x - 2)(x^2 + 2x + 4)$

28. $x - 3 \overline{)2x^2 - 5x - 3}$

29. Astronomy   The parsec, a unit of distance used in astronomy, is 3×10^{16} meters. The distance from Earth to Betelgeuse, a star in the constellation Orion, is 1.6×10^2 parsecs. Use scientific notation to express this distance in meters.

30. Surface area   The total surface area A of a box with dimensions l, w, and d is given by the formula

$$A = 2lw + 2wd + 2ld$$

If $A = 202$, $l = 9$, and $w = 5$, find d.

31. Concentric circles   The area of the ring between the two concentric circles of radius r and R is given by the formula

$$A = \pi(R + r)(R - r)$$

If $r = 3$ and $R = 17$, find A to the nearest tenth.

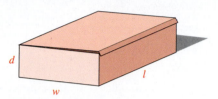

32. Employee discounts   Employees at an appliance store can purchase merchandise at 25% less than the regular price. An employee buys a TV set for $414.72, including 8% sales tax. Find the regular price of the TV.

Factoring Polynomials and Solving Equations by Factoring

5

© iStockphoto.com/elkor

Careers and Mathematics

ATHLETES, COACHES, AND SCOUTS

We are a nation of sports fans and sports players. Some of those who participate in amateur sports dream of becoming paid professional athletes, coaches, or sports officials, but very few beat the long odds of making a full-time living from professional sports. Nearly 42% of athletes and coaches work part time. Education and training for coaches and athletes vary greatly by the level and type of sport.

Job Outlook:
People who are state-certified to teach academic subjects in addition to physical education will have the best prospects for obtaining coaching and instructor jobs. Employment of athletes, coaches, and related workers is expected to increase faster than average for all occupations through 2020.

Annual Earnings:
Athletes: $28,340
Coaches and scouts: $26,950

For More Information:
http://www.bls.gov/oco/ocos251.htm

For a Sample Application:
See Problem 21 in Section 5.8.

REACH FOR SUCCESS

In this chapter

In this chapter, we will reverse the operation of multiplying polynomials and show which polynomials were used to find a given product. We will use this skill to solve many equations and applications.

Reach for Success Organizing Your Time

Life during your college years should be fun but it also requires time-management skills if you want to be academically successful. Although you previously examined a typical 24-hour day and a week, looking at a longer period will help you identify potential time conflicts. The popularity of electronic calendars has increased significantly. Regardless of the format, it is important to find a calendar that works for you.

© iStockphoto.com/Ryan Balderas

Populate a two-week calendar with all the graded assignments from your syllabus in each of your classes.

Begin by filling in the dates in the chart. Then

1. Record all exams in the boxes.
2. Include due dates for any papers.
3. Add any labs/homework assignments.
4. Record employment hours including any scheduled overtime.

Sun _____	Mon _____	Tue _____	Wed _____	Thu _____	Fri _____	Sat _____
Sun _____	Mon _____	Tue _____	Wed _____	Thu _____	Fri _____	Sat _____

Do you know of any other time requirements for this month? If so, record them in your chart. Be sure to include any personal appointments and life events (e.g., birthdays, weddings).

Your calendar might look like this:

Sun 1	Mon 2 English outline due	Tue 3 Dentist 1 p.m. Work	Wed 4 Work	Thu 5 Mathematics test	Fri 6 RJ's school play	Sat 7 Volunteer work Wedding anniversary
Sun 8 Sociology group project meeting 2 p.m.	Mon 9 Work Kenley's Dr. appt. 2:30 p.m.	Tue 10 Turn in history paper Math lab due	Wed 11 Read history chapters 13–15	Thu 12 Sociology group project due Work	Fri 13 English paper due Fred's birthday	Sat 14 Work

A Successful Study Strategy . . .

Use a planner to schedule your entire semester. By seeing even a week at a time, you will be able to avoid staying up all night to finish a paper or prepare for an exam. All-nighters rarely benefit anyone!

At the end of the chapter you will find an additional exercise to guide you in planning for a successful semester.

Section 5.1
Factoring Out the Greatest Common Factor; Factoring by Grouping

Objectives

1. Identify the greatest common factor of two or more monomials.
2. Factor a polynomial containing a greatest common factor.
3. Factor a polynomial containing a negative greatest common factor.
4. Factor a polynomial containing a binomial greatest common factor.
5. Factor a four-term polynomial using grouping.

Vocabulary

fundamental theorem of arithmetic   greatest common factor (GCF)   factoring by grouping

Getting Ready

Simplify each expression.

1. $5(x + 3)$
2. $7(y - 8)$
3. $x(3x - 2)$
4. $y(5y + 9)$
5. $3(x + y) + a(x + y)$
6. $x(y + 1) + 5(y + 1)$
7. $5(x + 1) - y(x + 1)$
8. $x(x + 2) - y(x + 2)$

In this chapter, we will reverse the operation of multiplication and show how to find the factors of a known product. The process of finding the individual factors of a product is called *factoring*. We will limit our discussion of factoring polynomials to those that factor using only rational numbers.

1 Identify the greatest common factor of two or more monomials.

Recall that a natural number greater than 1 whose only factors are 1 and the number itself is called a prime number.

The prime numbers less than 50 are

2, 3, 5, 7, 11, 13, 17, 19, 23, 29, 31, 37, 41, 43, and 47

A natural number is said to be in prime-factored form if it is written as the product of factors that are prime numbers.

To find the prime-factored form of a natural number, we can use a factoring tree. For example, to find the prime-factored form of 60, we proceed as follows:

Solution 1		*Solution 2*	
1. Start with 60.	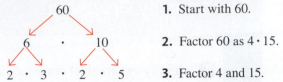	**1.** Start with 60.	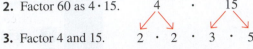
2. Factor 60 as $6 \cdot 10$.		**2.** Factor 60 as $4 \cdot 15$.	
3. Factor 6 and 10.		**3.** Factor 4 and 15.	

We stop when only prime numbers appear. In either case, the prime factorization of 60 is $2 \cdot 2 \cdot 3 \cdot 5$. Thus, the prime-factored form of 60 is $2^2 \cdot 3 \cdot 5$. This illustrates the **fundamental theorem of arithmetic**, which states that there is exactly one prime factorization for any natural number greater than 1.

The largest natural number that divides each group of natural numbers is called their **greatest common factor (GCF)**. The GCF of 42, 60, and 90 is 6, because 6 is the largest natural number that divides each of these numbers:

COMMENT We found the GCF, 6, by using the prime factorizations of 42, 60, and 90.

$$\frac{42}{6} = 7 \qquad \frac{60}{6} = 10 \qquad \text{and} \qquad \frac{90}{6} = 15$$

$$42 = \mathbf{2 \cdot 3} \cdot 7$$
$$60 = \mathbf{2} \cdot 2 \cdot \mathbf{3} \cdot 5$$
$$90 = \mathbf{2 \cdot 3} \cdot 3 \cdot 5$$

Algebraic monomials also can have a greatest common factor. The right sides of the equations show the prime factorizations of $6a^2b^3$, $4a^3b^2$, and $18a^2b$.

$$6a^2b^3 = \mathbf{2} \cdot 3 \cdot \mathbf{a \cdot a} \cdot \mathbf{b} \cdot b \cdot b$$
$$4a^3b^2 = \mathbf{2} \cdot 2 \cdot \mathbf{a \cdot a} \cdot a \cdot \mathbf{b} \cdot b$$
$$18a^2b = \mathbf{2} \cdot 3 \cdot 3 \cdot \mathbf{a \cdot a} \cdot \mathbf{b}$$

Since all three of these monomials have one factor of 2, two factors of a, and one factor of b, the GCF is

$$2 \cdot a \cdot a \cdot b \qquad \text{or} \qquad 2a^2b$$

Perspective

Much of the mathematics that we have inherited from earlier times is the result of teamwork. In a battle early in the 12th century, control of the Spanish city of Toledo was taken from the Mohammedans, who had ruled there for four centuries. Libraries in this great city contained many books written in Arabic, full of knowledge that was unknown in Europe.

The Archbishop of Toledo wanted to share this knowledge with the rest of the world. He knew that these books should be translated into Latin, the universal language of scholarship. But what European scholar could read Arabic? The citizens of Toledo knew both Arabic and Spanish, and most scholars of Europe could read Spanish.

Teamwork saved the day. A citizen of Toledo read the Arabic text aloud, in Spanish. The scholars listened to the Spanish version and wrote it down in Latin. One of these scholars was an Englishman, Robert of Chester. It was he who translated al-Khowarazmi's book, *Ihm al-jabr wa'l muqabalah*, the beginning of the subject we now know as algebra.

EXAMPLE 1 Find the GCF of $10x^3y^2$, $60x^2y$, and $30xy^2$.

Solution Find the prime factorization of each of the three monomials.

$$10x^3y^2 = \mathbf{2} \cdot \mathbf{5} \cdot \mathbf{x} \cdot x \cdot x \cdot \mathbf{y} \cdot y$$
$$60x^2y = \mathbf{2} \cdot 2 \cdot 3 \cdot \mathbf{5} \cdot \mathbf{x} \cdot x \cdot \mathbf{y}$$
$$30xy^2 = \mathbf{2} \cdot 3 \cdot \mathbf{5} \cdot \mathbf{x} \cdot \mathbf{y} \cdot y$$

List each common factor the least number of times it appears in any one monomial: 2, 5, x, and y. Find the product of the factors in the list:

$$2 \cdot 5 \cdot x \cdot y = 10xy$$

 SELF CHECK 1 Find the GCF of $20a^2b^3$, $12ab^4$, and $8a^3b^2$.

The process of factoring will be used to convert an addition or subtraction problem to a multiplication problem. This technique will then be used to solve equations later in this chapter and to simplify rational expressions in the next chapter.

2 **Factor a polynomial containing a greatest common factor.**

Recall that the distributive property provides a way to multiply a polynomial by a monomial. For example,

$$\mathbf{3x^2}(2x - 3y) = \mathbf{3x^2} \cdot 2x + \mathbf{3x^2} \cdot (-3y)$$
$$= 6x^3 - 9x^2y$$

To reverse this process and factor the product $6x^3 - 9x^2y$, we can find the GCF of each term (which is $3x^2$) and then use the distributive property.

$$6x^3 - 9x^2y = \mathbf{3x^2} \cdot 2x + \mathbf{3x^2} \cdot (-3y)$$
$$= \mathbf{3x^2}(2x - 3y)$$

This process is called factoring out the greatest common factor.

FINDING THE GREATEST COMMON FACTOR (GCF)	
	1. Identify the number of terms.
	2. Find the prime factorization of each term.
	3. List each common factor the least number of times it appears in any one term.
	4. Find the product of the factors found in the list to obtain the GCF.

EXAMPLE 2 Factor: $12y^2 + 20y$

Solution To find the GCF, we find the prime factorization of $12y^2$ and $20y$.

$$\left. \begin{array}{l} 12y^2 = \mathbf{2} \cdot \mathbf{2} \cdot 3 \cdot \mathbf{y} \cdot y \\ 20y = \mathbf{2} \cdot \mathbf{2} \cdot 5 \cdot \mathbf{y} \end{array} \right\} \quad \text{GCF} = 4y$$

We can use the distributive property to factor out the GCF of $4y$.

$$12y^2 + 20y = \mathbf{4y} \cdot 3y + \mathbf{4y} \cdot 5$$
$$= \mathbf{4y}(3y + 5)$$

Check by verifying that $4y(3y + 5) = 12y^2 + 20y$.

 SELF CHECK 2 Factor: $15x^3 - 20x^2$

EXAMPLE 3   Factor: $35a^3b^2 - 14a^2b^3$

Solution   To find the GCF, we find the prime factorization of $35a^3b^2$ and $-14a^2b^3$.

$$\left.\begin{array}{l}35a^3b^2 = 5 \cdot \mathbf{7} \cdot \mathbf{a} \cdot \mathbf{a} \cdot a \cdot \mathbf{b} \cdot \mathbf{b} \\ -14\mathbf{a^2b^3} = -\mathbf{2} \cdot \mathbf{7} \cdot \mathbf{a} \cdot \mathbf{a} \cdot \mathbf{b} \cdot \mathbf{b} \cdot b\end{array}\right\} \quad \text{GCF} = 7a^2b^2$$

We factor out the GCF of $7a^2b^2$.

$$35a^3b^2 - 14a^2b^3 = \mathbf{7a^2b^2} \cdot 5a + \mathbf{7a^2b^2} \cdot (-2b)$$
$$= \mathbf{7a^2b^2}(5a - 2b)$$

Check: $7a^2b^2(5a - 2b) = 35a^3b^2 - 14a^2b^3$

 SELF CHECK 3   Factor: $40x^2y^3 + 15x^3y^2$

EXAMPLE 4   Factor: $a^2b^2 - ab$

Solution   We factor out the GCF, which is ab.

$$a^2b^2 - ab = \mathbf{ab} \cdot ab + \mathbf{ab}(-1)$$
$$= \mathbf{ab}(ab - 1)$$

COMMENT The last term of $a^2b^2 - ab$ has an implied coefficient of -1. When ab is factored out, we must write the coefficient of -1.

Check: $ab(ab - 1) = a^2b^2 - ab$

SELF CHECK 4   Factor: $x^3y^5 - x^2y^3$

EXAMPLE 5   Factor: $12x^3y^2z + 6x^2yz - 3xz$

Solution   We factor out the GCF, which is $3xz$.

$$12x^3y^2z + 6x^2yz - 3xz = \mathbf{3xz} \cdot 4x^2y^2 + \mathbf{3xz} \cdot 2xy + \mathbf{3xz}(-1)$$
$$= \mathbf{3xz}(4x^2y^2 + 2xy - 1)$$

Check: $3xz(4x^2y^2 + 2xy - 1) = 12x^3y^2z + 6x^2yz - 3xz$

SELF CHECK 5   Factor: $6ab^2c - 12a^2bc + 3ab$

---

3 **Factor a polynomial containing a negative greatest common factor.**

It is often useful to factor -1 out of a polynomial, especially if the leading coefficient is negative.

EXAMPLE 6   Factor -1 out of $-a^3 + 2a^2 - 4$.

Solution $-a^3 + 2a^2 - 4$

$$= (\mathbf{-1})a^3 + (\mathbf{-1})(-2a^2) + (\mathbf{-1})4 \quad \text{Write each term with a factor of } -1.$$
$$= \mathbf{-1}(a^3 - 2a^2 + 4) \quad \text{Factor out the GCF, } -1.$$
$$= -(a^3 - 2a^2 + 4)$$

Check: $-(a^3 - 2a^2 + 4) = -a^3 + 2a^2 - 4$

SELF CHECK 6   Factor -1 out of $-b^4 - 3b^2 + 2$.

EXAMPLE 7   Factor out the negative of the GCF: $-18a^2b + 6ab^2 - 12a^2b^2$

Solution   The GCF is $6ab$. To factor out its negative, we factor out $-6ab$.

$$-18a^2b + 6ab^2 - 12a^2b^2 = (\mathbf{-6ab})3a + (\mathbf{-6ab})(-b) + (\mathbf{-6ab})2ab$$
$$= \mathbf{-6ab}(3a - b + 2ab)$$

Check: $-6ab(3a - b + 2ab) = -18a^2b + 6ab^2 - 12a^2b^2$

 SELF CHECK 7   Factor out the negative of the GCF: $-25xy^2 - 15x^2y + 30x^2y^2$

Sometimes a polynomial does not contain a GCF that is a monomial, but rather a polynomial. The next objective will discuss greatest common factors that are binomials or a product of a monomial and a binomial.

4   Factor a polynomial containing a binomial greatest common factor.

If the GCF of several terms is a polynomial, we can factor out the common polynomial factor. For example, since $(a + b)$ is a common factor of $(a + b)x$ and $(a + b)y$, we can factor out the $(a + b)$.

$$(\mathbf{a + b})x + (\mathbf{a + b})y = (\mathbf{a + b})(x + y)$$

We can check by verifying that $(a + b)(x + y) = (a + b)x + (a + b)y$.

EXAMPLE 8   Factor $(a + 3)$ out of $(a + 3) + (a + 3)^2$.

Solution   Recall that $a + 3$ is equal to $(a + 3)^1$ and that $(a + 3)^2$ is equal to $(a + 3)(a + 3)$. We can factor out $(a + 3)$ and simplify.

$$(a + 3) + (a + 3)^2 = (\mathbf{a + 3}) \cdot 1 + (\mathbf{a + 3}) \cdot (a + 3)$$
$$= (\mathbf{a + 3})[1 + (a + 3)] \qquad \text{Factor out } a + 3, \text{ the GCF.}$$
$$= (a + 3)(a + 4) \qquad \text{Combine like terms.}$$

 SELF CHECK 8   Factor out $(y + 2)$: $(y + 2)^2 - 3(y + 2)$

EXAMPLE 9   Factor: $6a^2b^2(x + 2y) - 9ab(x + 2y)$

Solution   The GCF of $6a^2b^2$ and $9ab$ is $3ab$. We can factor out this GCF as well as $(x + 2y)$.

$$6a^2b^2(x + 2y) - 9ab(x + 2y)$$
$$= \mathbf{3ab} \cdot 2ab(\mathbf{x + 2y}) - \mathbf{3ab} \cdot 3(\mathbf{x + 2y})$$
$$= \mathbf{3ab}(\mathbf{x + 2y})(2ab - 3) \qquad \text{Factor out } 3ab(x + 2y), \text{ the GCF.}$$

 SELF CHECK 9   Factor: $4p^3q^2(2a + b) + 8p^2q^3(2a + b)$

5   Factor a four-term polynomial using grouping.

Suppose we want to factor

$$ax + ay + cx + cy$$

Although no factor is common to all four terms, there is a common factor of a in $ax + ay$ and a common factor of c in $cx + cy$. In this case, we group the first two terms and group the last two terms. We can factor out the a from the first two terms and the c from the last two terms to obtain

$$ax + ay + cx + cy = a\mathbf{(x + y)} + c\mathbf{(x + y)}$$
$$= \mathbf{(x + y)}(a + c) \qquad \text{Factor out } (x + y).$$

We can check the result by multiplication.

$$(x + y)(a + c) = ax + cx + ay + cy$$
$$= ax + ay + cx + cy$$

Thus, $ax + ay + cx + cy$ factors as $(x + y)(a + c)$. This type of factoring is called **factoring by grouping**.

EXAMPLE 10 Factor: $2c + 2d - cd - d^2$

Solution $2c + 2d - cd - d^2 = 2\mathbf{(c + d)} - d\mathbf{(c + d)}$ Factor out 2 from $(2c + 2d)$ and $-d$ from $(-cd - d^2)$.

$$= \mathbf{(c + d)}(2 - d) \qquad \text{Factor out } (c + d).$$

Check: $(c + d)(2 - d) = 2c - cd + 2d - d^2$
$$= 2c + 2d - cd - d^2$$

SELF CHECK 10 Factor: $3a + 3b - ac - bc$

EXAMPLE 11 Factor: $5x^3 - 10x^2 + 4x - 8$

Solution $5x^3 - 10x^2 + 4x - 8 = 5x^2\mathbf{(x - 2)} + 4\mathbf{(x - 2)}$ Factor out $5x^2$ from $(5x^3 - 10x^2)$ and 4 from $(4x - 8)$.

$$= (x - 2)(5x^2 + 4) \qquad \text{Factor out } (x - 2).$$

SELF CHECK 11 Factor: $6y^3 + 3y^2 - 14y - 7$

COMMENT When factoring expressions, the final result must be a product. Expressions such as $2(c + d) - d(c + d)$ and $x(xy - a) - 1(xy - a)$ are *not* in factored form.

EXAMPLE 12 Factor: **a.** $a(c - d) + b(d - c)$ **b.** $ac + bd - ad - bc$

Solution   **a.** $a(c - d) + b(d - c) = a(c - d) - b(-d + c)$ Factor -1 from $(d - c)$.
$$= a(c - d) - b(c - d) \qquad -d + c = c - d$$
$$= (c - d)(a - b) \qquad \text{Factor out } (c - d).$$

b. In this example, we cannot factor anything from the first two terms or the last two terms. However, if we rearrange the terms, we can factor by grouping.

$$ac \mathbf{+ bd - ad} - bc = ac \mathbf{- ad + bd} - bc \qquad bd - ad = -ad + bd$$
$$= a(c - d) + b(d - c) \qquad \text{Factor } a \text{ from } (ac - ad) \text{ and } b \text{ from } (bd - bc).$$
$$= a(c - d) - b(c - d) \qquad d - c = -1(c - d)$$
$$= (c - d)(a - b) \qquad \text{Factor out } (c - d).$$

SELF CHECK 12 Factor: **a.** $2(x - y) - z(y - x)$
  b. $ax - by - ay + bx$

COMMENT In Example 12(b), we also could have factored the polynomial if we had rearranged the terms as $ac - bc - ad + bd$.

NOW TRY THIS

Factor.

1. $(x + y)(x^2 - 3) + (x + y)$

2. a. $x^{2n} + x^n$
   b. $x^3 + x^{-1}$

3. Which of the following is equivalent to $\frac{3 - x}{x + 2}$? There may be more than one answer.

   a. $\dfrac{-(x - 3)}{x + 2}$ **b.** $\dfrac{x - 3}{x + 2}$ **c.** $\dfrac{-x + 3}{x + 2}$ **d.** $-\dfrac{x - 3}{x + 2}$

5.1   Exercises

WARM-UPS *Find the greatest common factor.*

1. 3, 6, and 9
2. 2, 6, 10
3. 4, 16, 32
4. 5, 15, 20
5. 4, 6, 10
6. 6, 12, 21
7. 12, 18, 24
8. 30, 45, 60

REVIEW *Solve each equation and check all solutions.*

9. $3x - 2(x + 1) = 5$
10. $5(y - 1) + 1 = y$
11. $\dfrac{2x - 7}{5} = 3$
12. $2x - \dfrac{x}{2} = 5x$

VOCABULARY AND CONCEPTS *Fill in the blanks.*

13. If a natural number is written as the product of prime numbers, it is written in _____ form.

14. The _____ states that each natural number greater than 1 has exactly one prime factorization.

15. The GCF of several natural numbers is the _____ number that divides each of the numbers.

16. To find the prime factorization of a natural number, you can use a _____.

17. To factor a four-term polynomial, it is often necessary to factor by _____.

18. Check the results of a factoring problem by _____.

Find the prime factorization of each number.

19. 12
20. 24
21. 40
22. 98
23. 225
24. 144
25. 288
26. 968

GUIDED PRACTICE *Find the GCF of the given monomials.*
SEE EXAMPLE 1. (OBJECTIVE 1)

27. $3x, 6xy$
28. $5xy^2, 10xy$
29. $5x^2, 10x$
30. $8y^3, 12y^2x$
31. $4ab, 18b$
32. $7a^2b, 14ab^2$
33. $6x^2y^2, 12xyz, 18xy^2z^3$
34. $4a^3b^2c, 12ab^2c^2, 20ab^2c^2$

Complete each factorization. SEE EXAMPLE 2. (OBJECTIVE 2)

35. $9a + 15 = 3(\blacksquare + 5)$
36. $3t - 27 = 3(t - \blacksquare)$
37. $4a + 12 = \blacksquare(a + 3)$
38. $5b - 15 = \blacksquare(b - 3)$
39. $8x + 12 = \blacksquare(2x + 3)$
40. $6x - 15 = \blacksquare(2x - 5)$
41. $4y^2 + 8y - 2xy = 2y(2y + \blacksquare - \blacksquare)$
42. $3b^2 - 9b - 6ab = 3b(\blacksquare - \blacksquare - 2a)$
43. $3x^2 - 6xy + 9xy^2 = \blacksquare(\blacksquare - 2y + 3y^2)$
44. $10x^3 + 8xy - 4xy^2 = 2x(\blacksquare + 4y - \blacksquare)$
45. $r^4 + r^2 = r^2(\blacksquare + 1)$ **46.** $a^3 - a^2 = \blacksquare(a - 1)$

Factor each polynomial by factoring out the GCF. SEE EXAMPLE 2. (OBJECTIVE 2)

47. $3x + 6$
48. $2y - 10$
49. $4x - 8$
50. $4t + 12$
51. $6x^2 - 9x$
52. $15a^2 + 3a$
53. $4b^3 - 10b^2$
54. $8y^3 - 12y^2$

Factor each polynomial by factoring out the GCF. SEE EXAMPLE 3.
(OBJECTIVE 2)

55. $t^3 + 2t^2$ **56.** $b^3 - 3b^2$

57. $10x^2y^3 + 15xy^4$ **58.** $16a^4b^2 - 24a^2b^3$

Factor each polynomial by factoring out the GCF. SEE EXAMPLE 4.
(OBJECTIVE 2)

59. $a^3b^3z^3 - a^2b^3z^2$ **60.** $r^3s^6t^9 + r^2s^2t^2$

61. $24x^2y^3z^4 + 8xy^2z^3$ **62.** $3x^2y^3 - 9x^4y^3z$

Factor each polynomial by factoring out the GCF. SEE EXAMPLE 5.
(OBJECTIVE 2)

63. $3x + 3y - 6z$ **64.** $2x - 4y + 8z$

65. $ab + ac - ad$ **66.** $4y^2 + 8y - 2xy$

67. $rs - rt + ru$ **68.** $3x^2 - 6xy + 9xy^2$

69. $a^2b^2x^2 + a^3b^2x^2 - a^3b^3x^3$

70. $4x^2y^2z^2 - 6xy^2z^2 + 12xyz^2$

Factor out -1 from each polynomial. SEE EXAMPLE 6. (OBJECTIVE 3)

71. $-x - 2$ **72.** $-y + 3$

73. $-a - b$ **74.** $-x - 2y$

75. $-2x + 5y$ **76.** $-3x + 8z$

77. $-3ab - 5ac + 9bc$ **78.** $-6yz + 12xz - 5xy$

Factor out the negative of the GCF. SEE EXAMPLE 7. (OBJECTIVE 3)

79. $-3x^2y - 6xy^2$

80. $-4a^2b^2 + 6ab^2$

81. $-4a^2b^2c^2 + 14a^2b^2c - 10ab^2c^2$

82. $-25x^4y^3z^2 + 30x^2y^3z^4$

Complete each factorization. (OBJECTIVE 4)

83. $a(x + y) + b(x + y) = (x + y)$ ▮

84. $x(a + b) + p(a + b) =$ ▮$(x + p)$

85. $p(m - n) - q(m - n) =$ ▮$(p - q)$

86. $(r - s)p - (r - s)q = (r - s)$ ▮

87. $3(r - 2s) - x(r - 2s) =$ ▮$(3 - x)$

88. $x(a + 2b) + y(a + 2b) = (a + 2b)$ ▮

89. $(x + 3)(x + 1) - y(x + 1) = (x + 1)$ ▮

90. $x(x^2 + 2) - y(x^2 + 2) =$ ▮$(x - y)$

Factor each expression. SEE EXAMPLE 8. (OBJECTIVE 4)

91. $x(y + 1) - 5(y + 1)$

92. $3(x + y) - a(x + y)$

93. $(3x - y)(x^2 - 2) + (x^2 - 2)$

94. $(x - 5y)(a + 2) - (x - 5y)$

95. $(x + y)^2 + b(x + y)$ **96.** $(a - b)c + (a - b)d$

97. $(x - 3)^2 + (x - 3)$

98. $(3t + 5)^2 - (3t + 5)$

Factor each expression. SEE EXAMPLE 9. (OBJECTIVE 4)

99. $5a(2a - 1) - 10b(2a - 1)$

100. $3x^2(r + 3s) - 6y^2(r + 3s)$

101. $3x(c - 3d) + 6y(c - 3d)$

102. $9a^2b^2(3x - 2y) - 6ab(3x - 2y)$

Factor each polynomial by grouping. SEE EXAMPLE 10. (OBJECTIVE 5)

103. $2x + 2y + ax + ay$ **104.** $bx + bz + 5x + 5z$

105. $9p - 9q + mp - mq$ **106.** $7r + 7s - kr - ks$

Factor each expression. SEE EXAMPLE 11. (OBJECTIVE 5)

107. $9x^3 + 3x^2 + 12x + 4$ **108.** $6y^3 - 12y^2 - 5y + 10$

109. $8a^3 - 2a^2 - 4a + 1$ **110.** $7b^3 + 14b^2 + 2b + 4$

Factor each expression. SEE EXAMPLE 12. (OBJECTIVE 5)

111. $ax + bx - a - b$ **112.** $mp - np - m + n$

113. $x(a - b) + y(b - a)$ **114.** $p(m - n) - q(n - m)$

ADDITIONAL PRACTICE *Factor each expression completely.*

115. $r^4 + r^2$ **116.** $a^3 + a^2$

117. $12uvw^3 - 18uv^2w^2$

118. $14xyz - 16x^2y^2z$

119. $-14a^6b^6 + 49a^2b^3 - 21ab$

120. $-5a^2b^3c + 15a^3b^4c^2 - 25a^4b^3c$

121. $3x(a + b + c) - 2y(a + b + c)$

122. $2m(a - 2b + 3c) - 21xy(a - 2b + 3c)$

123. $14x^2y(r + 2s - t) - 21xy(r + 2s - t)$

124. $5xy^3(2x - y + 3z) + 25xy^2(2x - y + 3z)$

125. $3tv - 9tw + uv - 3uw$

126. $ce - 2cf + 3de - 6df$

127. $-4abc - 4ac^2 + 2bc + 2c^2$

128. $2x^3z - 4x^2z + 32xz - 64z$

129. $ax^3 + bx^3 + 2ax^2y + 2bx^2y$

130. $4a^2b + 12a^2 - 8ab - 24a$

131. $y^3 - 3y^2 - 5y + 15$

132. $3ab + 9a - 2b - 6$

133. $2r - bs - 2s + br$

134. $xy + 7 + y + 7x$

135. $ar^2 - brs + ars - br^2$

136. $a^2bc + a^2c + abc + ac$

WRITING ABOUT MATH

137. When we add $5x$ and $7x$, we combine like terms: $5x + 7x = 12x$. Explain how this is related to factoring out a common factor.

138. Explain how you would factor $x(a - b) + y(b - a)$.

SOMETHING TO THINK ABOUT

139. Think of two positive integers. Divide their product by their greatest common factor. Why do you think the result is called the least common multiple of the two integers? (*Hint*: The multiples of an integer such as 5 are 5, 10, 15, 20, 25, 30, and so on.)

140. Two integers are called *relatively prime* if their greatest common factor is 1. For example, 6 and 25 are relatively prime, but 6 and 15 are not. If the greatest common factor of three integers is 1, must any two of them be relatively prime? Explain.

Section 5.2

Factoring the Difference of Two Squares

Objectives

1 Factor the difference of two squares.
2 Completely factor a polynomial.

Vocabulary

difference of two squares   sum of two squares   prime polynomial

Getting Ready

Multiply the binomials.

1. $(a + b)(a - b)$

2. $(2r + s)(2r - s)$

3. $(3x + 2y)(3x - 2y)$

4. $(4x^2 + 3)(4x^2 - 3)$

Whenever we multiply binomial conjugates, binomials of the form $(x + y)$ and $(x - y)$, we obtain a binomial of the form $x^2 - y^2$.

$$(x + y)(x - y) = x^2 - xy + xy - y^2$$
$$= x^2 - y^2$$

In this section, we will show how to reverse the multiplication process and factor binomials such as $x^2 - y^2$ into binomial conjugates.

1 Factor the difference of two squares.

The binomial $x^2 - y^2$ is called the **difference of two squares**, because x^2 is the square of x and y^2 is the square of y. The difference of the squares of two quantities always factors into binomial conjugates.

FACTORING THE DIFFERENCE OF TWO SQUARES	$x^2 - y^2 = (x + y)(x - y)$

COMMENT The factorization of $x^2 - y^2$ also can be expressed as $(x - y)(x + y)$.

To factor $x^2 - 9$, we note that it can be written in the form $x^2 - 3^2$.

$$x^2 - 3^2 = (x + 3)(x - 3)$$

We can check by verifying that $(x + 3)(x - 3) = x^2 - 9$.

To factor the difference of two squares, it is helpful to know the integers that are perfect squares. The number 400, for example, is a perfect square, because $20^2 = 400$. The integer squares less than 400 are

$$1, 4, 9, 16, 25, 36, 49, 64, 81, 100, 121, 144, 169, 196, 225, 256, 289, 324, 361$$

Expressions containing variables such as x^4y^2 are also perfect squares, because they can be written as the square of a quantity:

$$x^4y^2 = (x^2y)^2$$

EXAMPLE 1   Factor: $25x^2 - 49$

Solution   We can write $25x^2 - 49$ in the form $(5x)^2 - 7^2$.

$$25x^2 - 49 = (5x)^2 - 7^2$$
$$= (5x + 7)(5x - 7)$$

We can check by multiplying $(5x + 7)$ and $(5x - 7)$.

$$(5x + 7)(5x - 7) = 25x^2 - 35x + 35x - 49$$
$$= 25x^2 - 49$$

 SELF CHECK 1   Factor: $16a^2 - 81$

EXAMPLE 2   Factor: $4y^4 - 25z^2$

Solution   We can write $4y^4 - 25z^2$ in the form $(2y^2)^2 - (5z)^2$.

$$4y^4 - 25z^2 = (2y^2)^2 - (5z)^2$$
$$= (2y^2 + 5z)(2y^2 - 5z)$$

Check by multiplication.

 SELF CHECK 2   Factor: $9m^2 - 64n^4$

2   Completely factor a polynomial.

We often must factor out a greatest common factor before factoring the difference of two squares. To factor $8x^2 - 32$, for example, we factor out the GCF of 8 and then factor the resulting difference of two squares.

$$8x^2 - 32 = 8(x^2 - 4) \qquad \text{Factor out 8, the GCF.}$$
$$= 8(x^2 - 2^2) \qquad \text{Write 4 as } 2^2.$$
$$= 8(x + 2)(x - 2) \qquad \text{Factor the difference of two squares.}$$

We can check by multiplication:

$$8(x + 2)(x - 2) = 8(x^2 - 4)$$
$$= 8x^2 - 32$$

EXAMPLE 3   Factor completely: $2a^2x^3y - 8b^2xy$

Solution   We factor out the GCF of $2xy$ and then factor the resulting difference of two squares.

$2a^2x^3y - 8b^2xy$

$= \mathbf{2xy} \cdot a^2x^2 - \mathbf{2xy} \cdot 4b^2$ The GCF is $2xy$.

$= \mathbf{2xy}(a^2x^2 - 4b^2)$ Factor out $2xy$.

$= 2xy[(ax)^2 - (2b)^2]$ Write a^2x^2 as $(ax)^2$ and $4b^2$ as $(2b)^2$.

$= 2xy(ax + 2b)(ax - 2b)$ Factor the difference of two squares.

Check by multiplication.

 SELF CHECK 3   Factor completely: $2p^2q^2s - 18r^2s$

Sometimes we must factor a difference of two squares more than once to completely factor a polynomial. For example, the binomial $625a^4 - 81b^4$ can be written in the form $(25a^2)^2 - (9b^2)^2$, which factors as

$$625a^4 - 81b^4 = (25a^2)^2 - (9b^2)^2$$
$$= (25a^2 + 9b^2)(\mathbf{25a^2 - 9b^2})$$

Since the factor $25a^2 - 9b^2$ can be written in the form $(5a)^2 - (3b)^2$, it is the difference of two squares and can be factored as $(5a + 3b)(5a - 3b)$. Thus,

$$625a^4 - 81b^4 = (25a^2 + 9b^2)(5a + 3b)(5a - 3b)$$

The binomial $25a^2 + 9b^2$ is the **sum of two squares**, because it can be written in the form $(5a)^2 + (3b)^2$. If we are limited to rational coefficients, binomials that are the sum of two squares cannot be factored unless they contain a GCF. Polynomials that do not factor are called **prime polynomials**.

EXAMPLE 4   Factor completely: $2x^4y - 32y$

Solution $2x^4y - 32y = \mathbf{2y} \cdot x^4 - \mathbf{2y} \cdot 16$ The GCF is $2y$.

$= \mathbf{2y}(x^4 - 16)$ Factor out $2y$.

$= 2y(x^2 + 4)(\mathbf{x^2 - 4})$ Factor $x^4 - 16$.

$= 2y(x^2 + 4)(\mathbf{x + 2})(\mathbf{x - 2})$ Factor $x^2 - 4$. Note that $x^2 + 4$ is the sum of two squares and does not factor using rational coefficients.

Check by multiplication.

 SELF CHECK 4   Factor completely: $48a^5 - 3ab^4$

Example 5 requires the techniques of factoring out a common factor, factoring by grouping, and factoring the difference of two squares.

EXAMPLE 5 Factor completely: $2x^3 - 8x + 2yx^2 - 8y$

Solution $2x^3 - 8x + 2yx^2 - 8y = 2(x^3 - 4x + yx^2 - 4y)$ Factor out 2, the GCF.

$\qquad\qquad\qquad\qquad = 2[x(x^2 - 4) + y(x^2 - 4)]$ Factor out x from $(x^3 - 4x)$ and y from $(yx^2 - 4y)$.

$\qquad\qquad\qquad\qquad = 2[(x^2 - 4)(x + y)]$ Factor out $x^2 - 4$.

$\qquad\qquad\qquad\qquad = 2(x + 2)(x - 2)(x + y)$ Factor $x^2 - 4$.

Check by multiplication.

 SELF CHECK 5 Factor completely: $3a^3 - 12a + 3a^2b - 12b$

 **SELF CHECK ANSWERS**

   1. $(4a + 9)(4a - 9)$ **2.** $(3m + 8n^2)(3m - 8n^2)$ **3.** $2s(pq + 3r)(pq - 3r)$
   4. $3a(4a^2 + b^2)(2a + b)(2a - b)$ **5.** $3(a + 2)(a - 2)(a + b)$

NOW TRY THIS

Factor completely.

1.   **a.** $x^2 - \dfrac{1}{9}$

     b. $2x^2 - 0.72$

     c. $16 - x^2$

2. $(x + y)^2 - 25$

3. $x^{2n} - 9$

5.2   Exercises

WARM-UPS *Complete each factorization.*

1. $x^2 - 9 = (x + 3)$ ▨

2. $y^2 - 36 = (y + 6)$ ▨

3. $z^2 - 4 =$ ▨ $(z - 2)$

4. $p^2 - q^2 = (p + q)$ ▨

5. $25 - t^2 = (5 + t)$ ▨

6. $36 - r^2 =$ ▨ $(6 - r)$

7. $81 - y^2 =$ ▨ $(9 - y)$

8. $49 - y^4 = (7 + y^2)$ ▨

9. $4m^2 - 9n^2 = (2m + 3n)$ ▨

10. $16x^4 - 49y^2 = (4x^2 + 7y)$ ▨

11. $25y^6 - 64x^4 =$ ▨ $(5y^3 - 8x^2)$

12. $16p^2 - 25q^2 =$ ▨ $(4p - 5q)$

REVIEW

13. In the study of the flow of fluids, Bernoulli's law is given by the following equation. Solve it for p.

$$\frac{p}{w} + \frac{v^2}{2g} + h = k$$

14. Solve Bernoulli's law for h. (See Exercise 13.)

VOCABULARY AND CONCEPTS *Fill in the blanks.*

15. A binomial of the form $a^2 - b^2$ is called the _____ _____.

16. A binomial of the form _____ is called the sum of two squares.

17. A polynomial that cannot be factored over the rational numbers is said to be a _____ polynomial.

18. The ____ of two squares cannot be factored unless it has a GCF.

GUIDED PRACTICE *Factor each polynomial.* **SEE EXAMPLE 1.** **(OBJECTIVE 1)**

19. $x^2 - 36$ **20.** $x^2 - 25$

21. $y^2 - 49$ **22.** $y^2 - 81$

23. $4y^2 - 49$ **24.** $16z^2 - 9$

25. $25x^4 - 81$ **26.** $64x^4 - 9$

Factor each polynomial. SEE EXAMPLE 2. (OBJECTIVE 1)

27. $9x^2 - y^2$ **28.** $4x^2 - z^2$

29. $25t^2 - 36u^2$ **30.** $49u^2 - 64v^2$

31. $100a^2 - 49b^2$ **32.** $36a^2 - 121b^2$

33. $x^4 - 9y^2$ **34.** $121a^2 - 144b^2$

Factor each polynomial completely. SEE EXAMPLE 3. (OBJECTIVE 2)

35. $8x^2 - 32y^2$ **36.** $2a^2 - 200b^2$

37. $2a^2 - 8y^2$ **38.** $3x^2 - 27y^2$

39. $100x^2 - 16y^2$ **40.** $45u^2 - 20v^2$

41. $x^3 - xy^2$ **42.** $a^2b - b^3$

43. $4a^2x - 9b^2x$ **44.** $4b^2y - 16c^2y$

45. $3m^3 - 3mn^2$ **46.** $5x^3 - 5xy^2$

Factor each polynomial completely. SEE EXAMPLE 4. (OBJECTIVE 2)

47. $a^4 - 16$

48. $b^4 - 256$

49. $a^4 - b^4$

50. $m^4 - 16n^4$

Factor each polynomial completely. SEE EXAMPLE 5. (OBJECTIVE 2)

51. $2x^4 - 2y^4$

52. $a^5 - ab^4$

53. $a^4b - b^5$

54. $m^5 - 16mn^4$

55. $2x^4y - 512y^5$

56. $2x^8y^2 - 32y^6$

57. $a^3 - 9a + 3a^2 - 27$

58. $2x^3 - 18x - 6x^2 + 54$

ADDITIONAL PRACTICE *Factor each polynomial completely. If a polynomial is prime, so indicate.*

59. $49y^2 - 225z^4$

60. $196x^4 - 169y^2$

61. $4x^4 - x^2y^2$

62. $9xy^2 - 4xy^4$

63. $x^4 - 81$

64. $y^4 - 625$

65. $16y^8 - 81z^4$

66. $x^8 - y^8$

67. $a^2 + b^2$

68. $36a^2 + 49b^2$

69. $x^8y^8 - 1$

70. $a^3 - 49a + 2a^2 - 98$

71. $2a^3b - 242ab^3$

72. $25x^2 + 36y^2$

73. $3a^{10} - 3a^2b^4$

74. $2p^{10}q - 32p^2q^5$

75. $3a^8 - 243a^4b^8$

76. $a^6b^2 - a^2b^6c^4$

77. $a^2b^7 - 625a^2b^3$

78. $16x^3y^4z - 81x^3y^4z^5$

79. $3a^5y + 6ay^5$

80. $81r^4 - 256s^4$

81. $144a^4 + 169b^4$

82. $2m^3n^2 - 32mn^2 + 8m^2 - 128$

83. $2x^9y + 2xy^9$

84. $25(a - b)^2 - 16$

85. $49(y + 1)^2 - x^2$

86. $b^3 - 25b - 2b^2 + 50$

87. $y^3 - 16y - 3y^2 + 48$

88. $243r^5s - 48rs^5$

89. $50c^4d^2 - 8c^2d^4$

WRITING ABOUT MATH

90. Explain how to factor the difference of two squares.

91. Explain why $x^4 - y^4$ is not completely factored as $(x^2 + y^2)(x^2 - y^2)$.

SOMETHING TO THINK ABOUT

92. It is easy to multiply 399 by 401 without a calculator: The product is $400^2 - 1$, or 159,999. Explain.

93. Use the method in the previous exercise to find $498 \cdot 502$ without a calculator.

Section 5.3

Factoring Trinomials with a Leading Coefficient of 1

Objectives

1. Factor a trinomial of the form $ax^2 + bx + c$, $a = 1$, by trial and error.
2. Factor a trinomial containing a negative greatest common factor.
3. Identify a prime trinomial.
4. Factor a polynomial completely.
5. Factor a trinomial of the form $ax^2 + bx + c$, $a = 1$, by grouping (*ac* method).
6. Factor a perfect-square trinomial.

Vocabulary

key number   perfect-square trinomial

Getting Ready

Multiply the binomials.

1. $(x + 6)(x + 6)$ **2.** $(y - 7)(y - 7)$ **3.** $(a - 3)(a - 3)$

4. $(x + 4)(x + 5)$ **5.** $(r - 2)(r - 5)$ **6.** $(m + 3)(m - 7)$

7. $(a - 3b)(a + 4b)$ **8.** $(u - 3v)(u - 5v)$ **9.** $(x + 4y)(x - 6y)$

We now discuss how to factor trinomials of the form $x^2 + bx + c$, where the coefficient of x^2 is 1 and there are no common factors.

1 Factor a trinomial of the form $ax^2 + bx + c$, $a = 1$ by trial and error.

The product of two binomials is often a trinomial. For example,

$$(x + 3)(x + 3) = x^2 + 6x + 9 \quad \text{and} \quad (x - 3y)(x + 4y) = x^2 + xy - 12y^2$$

For this reason, we should not be surprised that many trinomials factor into the product of two binomials. To develop a method for factoring trinomials, we multiply $(x + a)$ and $(x + b)$.

$$
\begin{aligned}
(x + a)(x + b) &= x^2 + bx + ax + ab && \text{Use the distributive property.} \\
&= x^2 + ax + bx + ab && \text{Apply the commutative property of addition.} \\
&= x^2 + (a + b)x + ab && \text{Factor the GCF, } x\text{, out of } ax + bx.
\end{aligned}
$$

From the result, we can see that

- the first term is the product of x and x.
- the coefficient of the middle term is the sum of a and b, and
- the last term is the product of a and b.

We can use these facts to factor trinomials with leading coefficients of 1.

EXAMPLE 1   Factor: $x^2 + 5x + 6$

Solution   To factor this trinomial, we will write it as the product of two binomials. Since the first term of the trinomial is x^2, the first term of each binomial factor must be x because $x \cdot x = x^2$. To fill in the following blanks, we must find two integers whose product is $+6$ and whose sum is $+5$.

$$x^2 + 5x + 6 = (x \quad)(x \quad)$$

The positive factorizations of 6 and the sums of the factors are shown in the following table.

Product of the factors	Sum of the factors
$1(6) = 6$	$1 + 6 = 7$
$2(3) = 6$	$2 + 3 = 5$

The last row contains the integers $+2$ and $+3$, whose product is $+6$ and whose sum is $+5$. So we can fill in the blanks with $+2$ and $+3$.

$$x^2 + 5x + 6 = (x + 2)(x + 3)$$

To check the result, we verify that $(x + 2)$ times $(x + 3)$ is $x^2 + 5x + 6$.

$$(x + 2)(x + 3) = x^2 + 3x + 2x + 2 \cdot 3$$
$$= x^2 + 5x + 6$$

SELF CHECK 1   Factor: $y^2 + 5y + 4$

COMMENT   In Example 1, the factors can be written in either order due to the commutative property of multiplication. An equivalent factorization is $x^2 + 5x + 6 = (x + 3)(x + 2)$.

EXAMPLE 2   Factor: $y^2 - 7y + 12$

Solution   Since the first term of the trinomial is y^2, the first term of each binomial factor must be y. To fill in the following blanks, we must find two integers whose product is $+12$ and whose sum is -7.

$$y^2 - 7y + 12 = (y \quad)(y \quad)$$

The factorizations of 12 and the sums of the factors are shown in the table.

Product of the factors	Sum of the factors
$1(12) = 12$	$1 + 12 = 13$
$2(6) = 12$	$2 + 6 = 8$
$3(4) = 12$	$3 + 4 = 7$
$-1(-12) = 12$	$-1 + (-12) = -13$
$-2(-6) = 12$	$-2 + (-6) = -8$
$-3(-4) = 12$	$-3 + (-4) = -7$

The last row contains the integers -3 and -4, whose product is $+12$ and whose sum is -7. So we can fill in the blanks with -3 and -4.

$$y^2 - 7y + 12 = (y - 3)(y - 4)$$

To check the result, we verify that $(y - 3)$ times $(y - 4)$ is $y^2 - 7y + 12$.

$$(y - 3)(y - 4) = y^2 - 3y - 4y + 12$$
$$= y^2 - 7y + 12$$

 SELF CHECK 2   Factor: $p^2 - 5p + 6$

EXAMPLE 3   Factor: $a^2 + 2a - 15$

Solution   Since the first term is a^2, the first term of each binomial factor must be a. To fill in the blanks, we must find two integers whose product is -15 and whose sum is $+2$.

$$a^2 + 2a - 15 = (a \quad)(a \quad)$$

The factorizations of -15 and the sums of the factors are shown in the table.

Product of the factors	Sum of the factors
$1(-15) = -15$	$1 + (-15) = -14$
$3(-5) = -15$	$3 + (-5) = -2$
$5(-3) = -15$	$5 + (-3) = 2$
$15(-1) = -15$	$15 + (-1) = 14$

The third row contains the integers $+5$ and -3, whose product is -15 and whose sum is $+2$. So we can fill in the blanks with $+5$ and -3.

$$a^2 + 2a - 15 = (a + 5)(a - 3)$$

Check: $(a + 5)(a - 3) = a^2 - 3a + 5a - 15$
$$= a^2 + 2a - 15$$

 SELF CHECK 3   Factor: $p^2 + 3p - 18$

EXAMPLE 4   Factor: $z^2 - 4z - 21$

Solution   Since the first term is z^2, the first term of each binomial factor must be z. To fill in the blanks, we must find two integers whose product is -21 and whose sum is -4.

$$z^2 - 4z - 21 = (z \quad)(z \quad)$$

The factorizations of -21 and the sums of the factors are shown in the table.

Product of the factors	Sum of the factors
$1(-21) = -21$	$1 + (-21) = -20$
$3(-7) = -21$	$3 + (-7) = -4$
$7(-3) = -21$	$7 + (-3) = 4$
$21(-1) = -21$	$21 + (-1) = 20$

The second row contains the integers $+3$ and -7, whose product is -21 and whose sum is -4. So we can fill in the blanks with $+3$ and -7.

$$z^2 - 4z - 21 = (z + 3)(z - 7)$$

Check: $(z + 3)(z - 7) = z^2 - 7z + 3z - 21$
$$= z^2 - 4z - 21$$

SELF CHECK 4   Factor: $q^2 - 2q - 24$

COMMENT When factoring a trinomial written in descending order, if the last term is positive the binomial factors will have the same sign as the middle term. If the last term is negative the binomial factors will have different signs.

The next example is a trinomial containing two variables.

EXAMPLE 5 Factor: $x^2 + xy - 6y^2$

Solution Since the first term is x^2, the first term of each binomial factor must be x. Since the last term is $-6y^2$, the second term of each binomial factor has a factor of y. To fill in the blanks, we must find coefficients whose product is -6 that will give a middle coefficient of 1.

$$x^2 + xy - 6y^2 = (x \,\boxed{}\, y)(x \,\boxed{}\, y)$$

The factorizations of -6 and the sums of the factors are shown in the table.

Product of the factors	Sum of the factors
$1(-6) = -6$	$1 + (-6) = -5$
$2(-3) = -6$	$2 + (-3) = -1$
$3(-2) = -6$	$3 + (-2) = 1$
$6(-1) = -6$	$6 + (-1) = 5$

The third row contains the integers 3 and -2. These are the only integers whose product is -6 and will give the correct middle coefficient of 1. So we can fill in the blanks with 3 and -2.

$$x^2 + xy - 6y^2 = (x + 3y)(x - 2y)$$

Check: $(x + 3y)(x - 2y) = x^2 - 2xy + 3xy - 6y^2$
$$= x^2 + xy - 6y^2$$

 SELF CHECK 5 Factor: $a^2 + ab - 12b^2$

2 **Factor a trinomial containing a negative greatest common factor.**

When the coefficient of the first term is -1, we begin by factoring out -1.

EXAMPLE 6 Factor: $-x^2 + 2x + 15$

Solution We factor out -1 and then factor the trinomial.

$$-x^2 + 2x + 15 = -(x^2 - 2x - 15) \quad \text{Factor out } -1.$$
$$= -(x - 5)(x + 3) \quad \text{Factor } x^2 - 2x - 15.$$

COMMENT In Example 6, it is not necessary to factor out the -1, but by doing so, it usually will be easier to factor the remaining trinomial.

Check: $-(x - 5)(x + 3) = -(x^2 + 3x - 5x - 15)$
$$= -(x^2 - 2x - 15)$$
$$= -x^2 + 2x + 15$$

 SELF CHECK 6 Factor: $-x^2 + 11x - 18$

3 **Identify a prime trinomial.**

If a trinomial cannot be factored using only rational coefficients, it is called a prime polynomial over the set of rational numbers.

EXAMPLE 7   Factor: $x^2 + 2x + 3$

Solution   To factor the trinomial, we must find two integers whose product is $+3$ and whose sum is 2. The possible factorizations of 3 and the sums of the factors are shown in the table.

Product of the factors	Sum of the factors
$1(3) = 3$	$1 + 3 = 4$
$-1(-3) = 3$	$-1 + (-3) = -4$

Since two integers whose product is $+3$ and whose sum is $+2$ do not exist, $x^2 + 2x + 3$ cannot be factored. It is a prime trinomial.

 SELF CHECK 7   Factor: $x^2 - 4x + 6$

---

4   Factor a polynomial completely.

To *factor* an expression means to factor the expression *completely*. The following examples require more than one type of factoring.

EXAMPLE 8   Factor: $-3ax^2 + 9a - 6ax$

Solution   We first write the trinomial in descending powers of x and factor out the common factor of $-3a$.

$$-3ax^2 + 9a - 6ax = -3ax^2 - 6ax + 9a$$
$$= -3a\,(\boldsymbol{x^2 + 2x - 3})$$

Then we factor the trinomial $x^2 + 2x - 3$.

$$-3ax^2 + 9a - 6ax = -3a\,(\boldsymbol{x + 3})(\boldsymbol{x - 1})$$

Check: $-3a(x + 3)(x - 1) = -3a(x^2 + 2x - 3)$
$$= -3ax^2 - 6ax + 9a$$
$$= -3ax^2 + 9a - 6ax$$

 SELF CHECK 8   Factor: $-2pq^2 + 6p - 4pq$

We have factored four-term polynomials by grouping two terms and two terms. The next example will require grouping three terms.

EXAMPLE 9   Factor: $m^2 - 2mn + n^2 - 64a^2$

Solution   We group the first three terms together and factor the resulting trinomial.

$$m^2 - 2mn + n^2 - 64a^2 = (m - n)(m - n) - 64a^2$$
$$= (m - n)^2 - (8a)^2$$

Then we factor the resulting difference of two squares.

$$m^2 - 2mn + n^2 - 64a^2 = (m - n)^2 - (8a)^2$$
$$= (m - n + 8a)(m - n - 8a)$$

 SELF CHECK 9   Factor: $p^2 + 4pq + 4q^2 - 25y^2$

5 Factor a trinomial of the form $ax^2 + bx + c$, $a = 1$, by grouping (*ac* method).

An alternative way of factoring trinomials of the form $ax^2 + bx + c$, $a = 1$, uses the technique of factoring by grouping, sometimes referred to as the *ac method*. For example, to factor $x^2 + x - 12$ by grouping, we proceed as follows:

1. Determine the values of a and c ($a = 1$ and $c = -12$) and find ac:

$$(1)(-12) = -12$$

This number is called the **key number**.

2. Find two factors of the key number -12 whose sum is $b = 1$. Two such factors are $+4$ and -3.

$$+4(-3) = -12 \qquad \text{and} \qquad +4 + (-3) = 1$$

3. Use the factors $+4$ and -3 as the coefficients of two terms whose sum is x. Write these two terms to replace x.

$$x^2 + x - 12 = x^2 + 4x - 3x - 12 \qquad x = +4x - 3x$$

4. Factor the right side of the previous equation by grouping.

$$x^2 + 4x - 3x - 12 = x(x + 4) - 3(x + 4) \qquad \text{Factor } x \text{ out of } (x^2 + 4x) \text{ and } -3 \text{ out of } (-3x - 12).$$

$$= (x + 4)(x - 3) \qquad \text{Factor out } (x + 4).$$

Check this factorization by multiplication.

EXAMPLE 10 Factor $y^2 + 7y + 10$ by grouping.

Solution We note that this equation is in the form $y^2 + by + c$, with $a = 1$, $b = 7$, and $c = 10$. First, we determine the key number ac:

$$ac = 1(10) = 10$$

Then we find two factors of 10 whose sum is $b = 7$. Two such factors are $+2$ and $+5$. We use these factors as the coefficients of two terms whose sum is $7y$.

$$y^2 + 7y + 10 = y^2 + 2y + 5y + 10 \qquad 7y = +2y + 5y$$

Finally, we factor the right side of the previous equation by grouping.

$$y^2 + 2y + 5y + 10 = y(y + 2) + 5(y + 2) \qquad \text{Factor out } y \text{ from } (y^2 + 2y) \text{ and factor out 5 from } (5y + 10).$$

$$= (y + 2)(y + 5) \qquad \text{Factor out } (y + 2).$$

 SELF CHECK 10 Use grouping to factor $p^2 - 7p + 12$.

EXAMPLE 11 Factor: $z^2 - 4z - 21$

Solution This is the trinomial of Example 4. To factor it by grouping, we note that the trinomial is in the form $z^2 + bz + c$, with $a = 1$, $b = -4$, and $c = -21$. First, we determine the key number ac:

$$ac = 1(-21) = -21$$

Then we find two factors of -21 whose sum is $b = -4$. Two such factors are $+3$ and -7. We use these factors as the coefficients of two terms whose sum is $-4z$. We write these two terms to replace $-4z$.

$$z^2 - 4z - 21 = z^2 + 3z - 7z - 21 \qquad -4z = +3z - 7z$$

Carl Friedrich Gauss
1777–1855
Many people consider Gauss to be the greatest mathematician of all time. He made contributions in the areas of number theory, solutions of equations, geometry of curved surfaces, and statistics. For his efforts, he has earned the title "Prince of the Mathematicians."

Finally, we factor the right side of the previous equation by grouping.

$$z^2 + 3z - 7z - 21 = z(z + 3) - 7(z + 3)$$ Factor out z from $(z^2 + 3z)$ and factor out -7 from $(-7z - 21)$.

$$= (z + 3)(z - 7)$$ Factor out $(z + 3)$.

 SELF CHECK 11   Use grouping to factor $a^2 + 2a - 15$. This is the trinomial of Example 3.

---

6   Factor a perfect-square trinomial.

We have discussed the following special-product relationships used to square binomials.

1. $(x + y)^2 = x^2 + 2xy + y^2$

2. $(x - y)^2 = x^2 - 2xy + y^2$

These relationships can be used in reverse order to factor special trinomials called **perfect-square trinomials**.

PERFECT-SQUARE TRINOMIALS	**1.** $x^2 + 2xy + y^2 = (x + y)^2$ **2.** $x^2 - 2xy + y^2 = (x - y)^2$

In words, Formula 1 states that *if a trinomial is the square of one quantity, plus twice the product of the two quantities, plus the square of the second quantity, it factors into the square of the sum of the quantities.*

Formula 2 states that *if a trinomial is the square of one quantity, minus twice the product of the two quantities, plus the square of the second quantity, it factors into the square of the difference of the quantities.*

The trinomials on the left sides of the previous equations are perfect-square trinomials because they are the results of squaring a binomial. Although we can factor perfect-square trinomials by using the techniques discussed earlier in this section, we usually can factor them by inspecting their terms. For example, $x^2 + 8x + 16$ is a perfect-square trinomial, because

- The first term x^2 is the square of x.
- The last term 16 is the square of 4.
- The middle term $8x$ is twice the product of x and 4.

Thus,

$$x^2 + 8x + 16 = x^2 + 2(x)(4) + 4^2$$
$$= (x + 4)^2$$

EXAMPLE 12   Factor: $x^2 - 10x + 25$

Solution $x^2 - 10x + 25$ is a perfect-square trinomial, because

- The first term x^2 is the square of x.
- The last term 25 is the square of 5.
- The middle term $-10x$ is the negative of twice the product of x and 5.

Thus,

$$x^2 - 10x + 25 = x^2 - 2(x)(5) + 5^2$$
$$= (x - 5)^2$$

 SELF CHECK 12   Factor: $x^2 + 10x + 25$

 **SELF CHECK ANSWERS**

1. $(y + 1)(y + 4)$ **2.** $(p - 3)(p - 2)$ **3.** $(p + 6)(p - 3)$ **4.** $(q + 4)(q - 6)$
5. $(a - 3b)(a + 4b)$ **6.** $-(x - 9)(x - 2)$ **7.** prime **8.** $-2p(q + 3)(q - 1)$
9. $(p + 2q + 5y)(p + 2q - 5y)$ **10.** $(p - 4)(p - 3)$ **11.** $(a + 5)(a - 3)$ **12.** $(x + 5)^2$

NOW TRY THIS

Factor.

1. $18 + 3x - x^2$

2. $x^2 + \dfrac{2}{5}x + \dfrac{1}{25}$

3. $x^{2n} + x^n - 2$

5.3 Exercises

WARM-UPS *Find two numbers whose product is (a) and whose sum is (b).*

1. (a) 4 (b) 5
2. (a) 6 (b) −5
3. (a) −6 (b) 1
4. (a) −6 (b) −1
5. (a) 4 (b) −5
6. (a) −12 (b) 4
7. (a) −18 (b) 7
8. (a) 24 (b) 11

REVIEW *Graph the solution of each inequality on a number line.*

9. $x - 3 > 5$
10. $x + 4 \leq 3$

11. $-3x - 5 \geq 4$
12. $2x - 3 < 7$

13. $\dfrac{3(x - 1)}{4} < 12$
14. $\dfrac{-2(x + 3)}{3} \geq 9$

15. $-2 < x \leq 4$
16. $-5 \leq x + 1 < 5$

VOCABULARY AND CONCEPTS *Complete each relationship for a perfect-square trinomial.*

17. $x^2 + 2xy + y^2 = $ _____
18. $x^2 - 2xy + y^2 = $ _____

Complete each factorization.

19. $x^2 + 5x + 6 = (x + 2)(x + \boxed{})$
20. $x^2 - 5x + 6 = (x \boxed{} 2)(x \boxed{} 3)$
21. $x^2 + x - 6 = (x \boxed{} 2)(x + \boxed{})$
22. $x^2 - x - 6 = (x \boxed{} 3)(x + \boxed{})$
23. $x^2 + 5x - 6 = (x + \boxed{})(x - \boxed{})$
24. $x^2 - 7x + 6 = (x - \boxed{})(x - \boxed{})$

25. $y^2 + 6y + 8 = (y + \boxed{})(y + \boxed{})$
26. $z^2 - 3z - 10 = (z + \boxed{})(z - \boxed{})$
27. $x^2 - xy - 2y^2 = (x + \boxed{})(x - \boxed{})$
28. $a^2 + ab - 6b^2 = (a + \boxed{})(a - \boxed{})$

GUIDED PRACTICE *Factor. SEE EXAMPLE 1. (OBJECTIVE 1)*

29. $x^2 + 5x + 4$
30. $y^2 + 4y + 3$

31. $z^2 + 6z + 8$
32. $x^2 + 7x + 10$

33. $x^2 + 8x + 15$
34. $z^2 + 8z + 16$

35. $x^2 + 12x + 20$
36. $y^2 + 9y + 20$

Factor. SEE EXAMPLE 2. (OBJECTIVE 1)

37. $t^2 - 9t + 14$
38. $c^2 - 9c + 8$

39. $x^2 - 8x + 12$
40. $p^2 - 10p + 16$

41. $x^2 - 9x + 20$
42. $t^2 - 3t + 2$

43. $r^2 - 8r + 7$
44. $y^2 - 9y + 18$

Factor. SEE EXAMPLE 3. (OBJECTIVE 1)

45. $q^2 + 8q - 9$
46. $x^2 + 5x - 24$

47. $s^2 + 11s - 26$
48. $b^2 + 6b - 7$

49. $c^2 + 4c - 5$
50. $x^2 + 8x - 20$

51. $y^2 + 4y - 12$
52. $a^2 + 9a - 36$

Factor. SEE EXAMPLE 4. (OBJECTIVE 1)

53. $b^2 - 5b - 6$ **54.** $t^2 - 5t - 50$

55. $a^2 - 10a - 39$ **56.** $a^2 - 4a - 5$

57. $m^2 - 3m - 10$ **58.** $y^2 - 2y - 35$

59. $x^2 - 3x - 40$ **60.** $x^2 - 6x - 16$

Factor. SEE EXAMPLE 5. (OBJECTIVE 1)

61. $m^2 + 5mn - 14n^2$ **62.** $m^2 - mn - 12n^2$

63. $a^2 - 4ab - 12b^2$ **64.** $a^2 + 7ab - 18b^2$

65. $a^2 + 10ab + 9b^2$ **66.** $u^2 + 2uv - 15v^2$

67. $m^2 - 11mn + 10n^2$ **68.** $x^2 + 6xy + 9y^2$

Factor each trinomial. Factor out -1 *first.* SEE EXAMPLE 6. (OBJECTIVE 2)

69. $-x^2 - 7x - 10$ **70.** $-x^2 + 9x - 20$

71. $-y^2 - 2y + 15$ **72.** $-y^2 - 3y + 18$

73. $-t^2 - 4t + 32$ **74.** $-t^2 - t + 30$

75. $-r^2 + 14r - 40$ **76.** $-r^2 + 14r - 45$

Factor each trinomial. If prime, so indicate. SEE EXAMPLE 7. (OBJECTIVE 3)

77. $a^2 + 3a + 10$ **78.** $v^2 + 9v + 15$

79. $r^2 - 9r - 12$ **80.** $b^2 + 6b - 18$

Factor. SEE EXAMPLE 8. (OBJECTIVE 4)

81. $2x^2 + 20x + 42$ **82.** $-2b^2 + 20b - 18$

83. $3y^3 - 21y^2 + 18y$ **84.** $-5a^3 + 25a^2 - 30a$

85. $3z^2 - 15tz + 12t^2$ **86.** $5m^2 + 45mn - 50n^2$

87. $-4x^2y - 4x^3 + 24xy^2$ **88.** $3x^2y^3 + 3x^3y^2 - 6xy^4$

Factor. SEE EXAMPLE 9. (OBJECTIVE 4)

89. $x^2 + 4x + 4 - y^2$

90. $p^2 - 2p + 1 - q^2$

91. $b^2 - 6b + 9 - c^2$

92. $m^2 + 8m + 16 - n^2$

Factor. SEE EXAMPLES 10–11. (OBJECTIVE 5)

93. $x^2 + 6x + 5$ **94.** $y^2 + 4y + 3$

95. $t^2 - 9t + 14$ **96.** $c^2 - 9c + 8$

97. $a^2 + 6a - 16$ **98.** $x^2 + 5x - 24$

99. $y^2 - y - 30$ **100.** $a^2 - 4a - 32$

Factor. SEE EXAMPLE 12. (OBJECTIVE 6)

101. $x^2 + 6x + 9$ **102.** $x^2 + 10x + 25$

103. $y^2 - 8y + 16$ **104.** $z^2 - 2z + 1$

105. $u^2 - 18u + 81$ **106.** $v^2 - 14v + 49$

107. $x^2 + 4xy + 4y^2$ **108.** $a^2 + 12ab + 36b^2$

ADDITIONAL PRACTICE *Factor. Write each trinomial in descending powers of one variable, if necessary. If a polynomial is prime, so indicate.*

109. $4 - 5x + x^2$ **110.** $y^2 + 7 + 8y$

111. $10y + 9 + y^2$ **112.** $x^2 - 13 - 12x$

113. $-r^2 + 2s^2 + rs$ **114.** $u^2 - 3v^2 + 2uv$

115. $x^2 - 9x - 27$ **116.** $-a^2 + 5b^2 + 4ab$

117. $-a^2 - 6ab - 5b^2$ **118.** $3y^3 + 6y^2 + 3y$

119. $12xy + 4x^2y - 72y$ **120.** $y^2 + 2yz + z^2$

121. $r^2 - 2rs + 4s^2$ **122.** $r^2 + 24r + 144$

123. $r^2 - 10rs + 25s^2$

124. $a^2 + 2ab + b^2 - 4$

125. $a^2 + 6a + 9 - b^2$

126. $5y^3 + 10y^2 + 5y$

127. $t^2 + 18t + 81$

128. $c^2 - a^2 + 8a - 16$

WRITING ABOUT MATH

129. Explain how you would write a trinomial in descending order.

130. Explain how to use the distributive property to check the factoring of a trinomial.

SOMETHING TO THINK ABOUT

131. Two students factor $2x^2 + 20x + 42$ and get two different answers: $(2x + 6)(x + 7)$ and $(x + 3)(2x + 14)$. Do both answers check? Why don't they agree? Is either completely correct?

132. Find the error:

$x = y$	
$x^2 = xy$	Multiply both sides by x.
$x^2 - y^2 = xy - y^2$	Subtract y^2 from both sides.
$(x + y)(x - y) = y(x - y)$	Factor.
$x + y = y$	Divide both sides by $(x - y)$.
$y + y = y$	Substitute y for its equal, x.
$2y = y$	Combine like terms.
$2 = 1$	Divide both sides by y.

Section 5.4

Factoring Trinomials with a Leading Coefficient Other Than 1

Objectives

1. Factor a trinomial of the form $ax^2 + bx + c$ using trial and error.
2. Factor a trinomial of the form $ax^2 + bx + c$ by grouping (ac method).
3. Factor a polynomial involving a perfect-square trinomial.

Getting Ready

Multiply and combine like terms.

1. $(2x + 1)(3x + 2)$
2. $(3y - 2)(2y - 5)$
3. $(4t - 3)(2t + 3)$
4. $(2r + 5)(2r - 3)$
5. $(2m - 3)(3m - 2)$
6. $(4a + 3)(4a + 1)$

In the previous section, we saw how to factor trinomials whose leading coefficients are 1. We now show how to factor trinomials whose leading coefficients are other than 1.

1 ## Factor a trinomial of the form $ax^2 + bx + c$ using trial and error.

In the previous section, we only considered factors of c. We must now consider more combinations of factors when we factor trinomials with leading coefficients other than 1.

EXAMPLE 1   Factor: $2x^2 + 5x + 3$

Solution   Since the first term is $2x^2$, the first terms of the binomial factors must be $2x$ and x. To fill in the blanks, we must find two factors of $+3$ that will give a middle term of $+5x$.

$$\left(2x \;\rule{1cm}{0.4pt}\;\right)\left(x \;\rule{1cm}{0.4pt}\;\right)$$

Since the sign of each term of the trinomial is $+$, we need to consider only positive factors of the last term (3). Since the positive factors of 3 are 1 and 3, there are two possible factorizations.

$$(2x + 1)(x + 3) \qquad \text{or} \qquad (2x + 3)(x + 1)$$

The first possibility is incorrect, because it gives a middle term of $7x$. The second possibility is correct, because it gives a middle term of $5x$. Thus,

$$2x^2 + 5x + 3 = (2x + 3)(x + 1)$$

Check by multiplication.

 SELF CHECK 1   Factor: $3x^2 + 7x + 2$

EXAMPLE 2   Factor: $6x^2 - 17x + 5$

Solution   Since the first term is $6x^2$, the first terms of the binomial factors must be $6x$ and x or $3x$ and $2x$. To fill in the blanks, we must find two factors of $+5$ that will give a middle term of $-17x$.

$$\left(6x \quad\right)\left(x \quad\right) \qquad \text{or} \qquad \left(3x \quad\right)\left(2x \quad\right)$$

Since the sign of the third term is $+$ and the sign of the middle term is $-$, we need to consider only negative factors of the last term (5). Since the negative factors of 5 are -1 and -5, there are four possible factorizations.

$$(6x - 1)(x - 5) \qquad\qquad (6x - 5)(x - 1)$$

The one to choose $\longrightarrow$ $(3x - 1)(2x - 5)$ $\qquad (3x - 5)(2x - 1)$

Only the possibility printed in red gives the correct middle term of $-17x$. Thus,

$$6x^2 - 17x + 5 = (3x - 1)(2x - 5)$$

Check by multiplication.

 SELF CHECK 2   Factor: $6x^2 - 7x + 2$

EXAMPLE 3   Factor: $3y^2 - 5y - 12$

Solution   Since the sign of the third term of $3y^2 - 5y - 12$ is $-$, the signs between the binomial factors will be opposites. Because the first term is $3y^2$, the first terms of the binomial factors must be $3y$ and y.

Since $1(-12), 2(-6), 3(-4), 12(-1), 6(-2)$, and $4(-3)$ all give a product of -12, there are 12 possible combinations to consider.

$$(3y + 1)(y - 12) \qquad\qquad \mathbf{(3y - 12)(y + 1)}$$
$$(3y + 2)(y - 6) \qquad\qquad \mathbf{(3y - 6)(y + 2)}$$
$$\mathbf{(3y + 3)(y - 4)} \qquad\qquad (3y - 4)(y + 3)$$
$$\mathbf{(3y + 12)(y - 1)} \qquad\qquad (3y - 1)(y + 12)$$
$$\mathbf{(3y + 6)(y - 2)} \qquad\qquad (3y - 2)(y + 6)$$

The one to choose $\longrightarrow$ $(3x + 4)(y - 3)$ $\qquad \mathbf{(3y - 3)(y + 4)}$

The combinations printed in blue cannot work, because one of the factors has a common factor. This implies that $3y^2 - 5y - 12$ would have a common factor, which it does not.

After mentally trying the remaining factors, we see that only $(3y + 4)(y - 3)$ gives the correct middle term of $-5x$. Thus,

$$3y^2 - 5y - 12 = (3y + 4)(y - 3)$$

Check by multiplication.

 SELF CHECK 3   Factor: $5a^2 - 7a - 6$

EXAMPLE 4 Factor: $6b^2 + 7b - 20$

Solution Since the first term is $6b^2$, the first terms of the binomial factors must be $6b$ and b or $3b$ and $2b$. To fill in the blanks, we must find two factors of -20 that will give a middle term of $+7b$.

$$(6b \quad)(b \quad) \qquad \text{or} \qquad (3b \quad)(2b \quad)$$

Since the sign of the third term is $-$, the signs inside the binomial factors will be different. Because the factors of the last term (20) are 1, 2, 4, 5, 10, and 20, there are many possible combinations for the last terms. We must try to find a combination that will give a last term of -20 and a sum of the products of the outer terms and inner terms of $+7b$.

If we choose factors of $6b$ and b for the first terms and -5 and 4 for the last terms, we have

$$(6b - 5)(b + 4)$$

$$-5b$$
$$\underline{24b}$$
$$19b$$

which gives an incorrect middle term of $19b$.

If we choose factors of $3b$ and $2b$ for the first terms and -4 and $+5$ for the last terms, we have

$$(3b - 4)(2b + 5)$$

$$-8b$$
$$\underline{15b}$$
$$7b$$

which gives the correct middle term of $+7b$ and the correct last term of -20. Thus,

$$6b^2 + 7b - 20 = (3b - 4)(2b + 5)$$

Check by multiplication.

SELF CHECK 4 Factor: $4x^2 + 4x - 3$

Although, the next example has two variables, the process remains the same. When two variables are involved, first write the polynomial in descending powers of one of the variables. Then, proceed as previously discussed.

EXAMPLE 5 Factor: $2x^2 + 7xy + 6y^2$

Solution Notice the polynomial is written in descending powers of x. Since the first term is $2x^2$, the first terms of the binomial factors must be $2x$ and x. To fill in the blanks, we must find two factors of $6y^2$ that will give a middle term of $+7xy$.

$$(2x \quad)(x \quad)$$

Since the sign of each term is $+$, the signs inside the binomial factors will be $+$. The possible factors of the last term, $6y^2$, are

$$y \text{ and } 6y \qquad \text{or} \qquad 3y \text{ and } 2y$$

We must try to find a combination that will give a last term of $+6y^2$ and a middle term of $+7xy$.

If we choose y and $6y$ to be the factors of the last term, we have

$$(2x + y)(x + 6y)$$

$$\begin{array}{r} xy \\ 12xy \\ \hline 13xy \end{array}$$

which gives an incorrect middle term of $13xy$.

If we choose $3y$ and $2y$ to be the factors of the last term, we have

$$(2x + 3y)(x + 2y)$$

$$\begin{array}{r} 3xy \\ 4xy \\ \hline 7xy \end{array}$$

which gives a correct middle term of $7xy$. Thus,

$$2x^2 + 7xy + 6y^2 = (2x + 3y)(x + 2y)$$

Check by multiplication.

🍃 **SELF CHECK 5**   Factor: $4x^2 + 8xy + 3y^2$

Because some guesswork is often necessary, it is difficult to give specific rules for factoring trinomials. However, the following hints are often helpful.

FACTORING GENERAL TRINOMIALS USING TRIAL AND ERROR

1. Write the trinomial in descending powers of one variable.
2. Factor out any GCF (including -1 if that is necessary to make the coefficient of the first term positive).
3. If the sign of the third term is $+$, the signs between the terms of the binomial factors are the same as the sign of the middle term. If the sign of the third term is $-$, the signs between the terms of the binomial factors are opposites.
4. Try combinations of first terms and last terms until you find one that works, or until you exhaust all the possibilities. If no combination works, the trinomial is prime.
5. Check the factorization by multiplication.

EXAMPLE 6   Factor: $2x^2y - 8x^3 + 3xy^2$

Solution   **Step 1:**   Write the trinomial in descending powers of x.

$$-8x^3 + 2x^2y + 3xy^2$$

Step 2:   Factor out the negative of the GCF, which is $-x$.

$$-8x^3 + 2x^2y + 3xy^2 = -x(8x^2 - 2xy - 3y^2)$$

Step 3:   Because the sign of the third term of the trinomial factor is $-$, the signs within its binomial factors will be opposites.

Step 4:   Find the binomial factors of the trinomial.

$$-8x^3 + 2x^2y + 3xy^2 = -x(\mathbf{8x^2 - 2xy - 3y^2})$$
$$= -x(\mathbf{2x + y})(\mathbf{4x - 3y})$$

Step 5: Check by multiplication.

$$-x(2x + y)(4x - 3y) = -x(8x^2 - 6xy + 4xy - 3y^2)$$
$$= -x(8x^2 - 2xy - 3y^2)$$
$$= -8x^3 + 2x^2y + 3xy^2$$
$$= 2x^2y - 8x^3 + 3xy^2$$

SELF CHECK 6   Factor: $12y - 2y^3 - 2y^2$

When there are a larger number of possible combinations, some people prefer to use an alternative to trial and error, the *ac* method. A visual of the *ac* method, referred to as the Box method, will also be discussed.

2   Factor a trinomial of the form $ax^2 + bx + c$ by grouping (*ac* method).

Another way to factor trinomials of the form $ax^2 + bx + c$ uses the grouping (*ac* method), first discussed in the previous section. For example, to factor $6x^2 - 17x + 5$ (Example 2) by grouping, we note that $a = 6$, $b = -17$, and $c = 5$ and proceed as follows:

1. Determine the product *ac*: $6(+5) = 30$. This is the *key number*.

2. Find two factors of the key number 30 whose sum is -17. Two such factors are -15 and -2.

$$-15(-2) = 30 \qquad \text{and} \qquad -15 + (-2) = -17$$

3. Use -15 and -2 as coefficients of two terms to be placed between $6x^2$ and 5 to replace $-17x$.

$$6x^2 - 17x + 5 = 6x^2 - 15x - 2x + 5$$

4. Factor the right side of the previous equation by grouping.

$$6x^2 - 15x - 2x + 5 = 3x(2x - 5) - 1(2x - 5) \qquad \text{Factor the GCF, } 3x, \text{ from } (6x^2 - 15x) \text{ and } -1 \text{ from } (-2x + 5).$$
$$= (2x - 5)(3x - 1) \qquad \text{Factor out the GCF, } (2x - 5).$$

Verify this factorization by multiplication.

EXAMPLE 7   Factor $4y^2 + 12y + 5$ by grouping.

Solution   To factor this trinomial by grouping, we note that it is written in the form $ay^2 + by + c$, with $a = 4$, $b = 12$, and $c = 5$. Since $a = 4$ and $c = 5$, we have $ac = 20$.

We now find two factors of 20 whose sum is 12. Two such factors are 10 and 2. We use these factors as coefficients of two terms to be placed between $4y^2$ and 5 to replace $+12y$.

$$4y^2 + 12y + 5 = 4y^2 + 10y + 2y + 5$$

Finally, we factor the right side of the previous equation by grouping.

$$4y^2 + 10y + 2y + 5 = 2y(2y + 5) + (2y + 5) \qquad \text{Factor the GCF, } 2y, \text{ from } (4y^2 + 10y).$$
$$= 2y(2y + 5) + 1 \cdot (2y + 5) \qquad \text{Write the coefficient of 1.}$$
$$= (2y + 5)(2y + 1) \qquad \text{Factor out the GCF, } (2y + 5).$$

Check by multiplication.

SELF CHECK 7   Use grouping to factor $2p^2 - 7p + 3$.

EXAMPLE 8   Factor: $6b^2 + 7b - 20$

Solution   This is the trinomial of Example 4. Since $a = 6$ and $c = -20$ in the trinomial, $ac = -120$. We now find two factors of -120 whose sum is $+7$. Two such factors are 15 and -8. We use these factors as coefficients of two terms to be placed between $6b^2$ and -20 to replace $+7b$.

$$6b^2 + 7b - 20 = 6b^2 + 15b - 8b - 20$$

COMMENT   When using the grouping method, if no pair of factors of ac produces the desired value b, the trinomial is prime over the rationals.

Finally, we factor the right side of the previous equation by grouping.

$$6b^2 + 15b - 8b - 20 = 3b(2b + 5) - 4(2b + 5) \qquad \text{Factor the GCF, } 3b, \text{ from} \\ (6b^2 + 15b) \text{ and } -4 \text{ from} \\ (-8b - 20).$$

$$= (2b + 5)(3b - 4) \qquad \text{Factor out the GCF, } (2b + 5).$$

Check by multiplication.

 **SELF CHECK 8**   Factor: $3y^2 - 4y - 4$

Another option for factoring general trinomials is to use a *box method*, which is a visual of the *ac* method. To illustrate, we will factor $6x^2 + 5x - 4$.

1. Draw a 2 × 2 box.

2. In the top left corner, write the first term of $6x^2$ and in the bottom right corner write the last term, -4.

$6x^2$	
	-4

3. Multiply these two terms to get $-24x^2$. Find two factors of $-24x^2$ that will add to give the middle term of the trinomial, $5x$. These will be $8x$ and $-3x$. Write these two values in the remaining boxes. Their placement does not matter.

$6x^2$	$-3x$
$8x$	-4

4. Factor the greatest common factor from each row and each column.

	$2x$	-1
$3x$	$6x^2$	$-3x$
4	$8x$	-4

The sums of the factors of the rows and columns are the factors of the original trinomial, $6x^2 + 5x - 4$.

$$6x^2 + 5x - 4 = (3x + 4)(2x - 1)$$

3   Factor a polynomial involving a perfect-square trinomial.

As before, we can factor perfect-square trinomials by inspection.

EXAMPLE 9   Factor: $4x^2 - 20x + 25$

Solution $4x^2 - 20x + 25$ is a perfect-square trinomial, because

COMMENT   It is recommended that you factor a perfect-square trinomial as you would any other trinomial until you discover the pattern for yourself.

- The first term $4x^2$ is the square of $2x$: $(2x)^2 = 4x^2$.
- The last term 25 is the square of 5: $5^2 = 25$.
- The middle term $-20x$ is the negative of twice the product of $2x$ and 5.

Thus,

$$4x^2 - 20x + 25 = (2x)^2 - 2(2x)(5) + 5^2$$
$$= (2x - 5)^2$$

Check by multiplication.

 SELF CHECK 9   Factor: $9x^2 - 12x + 4$

The next examples combine several factoring techniques. Although factoring a polynomial with four terms typically involves grouping two terms and two terms, there are times we need to group three terms. It may be helpful to look for a perfect-square trinomial in either the first three or last three terms.

EXAMPLE 10   Factor: $4x^2 - 4xy + y^2 - 9$

Solution
$$4x^2 - 4xy + y^2 - 9$$
$$= (4x^2 - 4xy + y^2) - 9 \qquad \text{Group the first three terms.}$$
$$= (2x - y)(2x - y) - 9 \qquad \text{Factor the perfect-square trinomial.}$$
$$= (2x - y)^2 - 9 \qquad \text{Write } (2x - y)(2x - y) \text{ as } (2x - y)^2.$$
$$= [(2x - y) + 3][(2x - y) - 3] \qquad \text{Factor the difference of two squares.}$$
$$= (2x - y + 3)(2x - y - 3) \qquad \text{Simplify each factor.}$$

Check by multiplication.

 SELF CHECK 10   Factor: $x^2 + 4x + 4 - y^2$

EXAMPLE 11   Factor: $9 - 4x^2 - 4xy - y^2$

Solution
$$9 - 4x^2 - 4xy - y^2 = 9 - (4x^2 + 4xy + y^2) \qquad \text{Factor } -1 \text{ from the last three terms.}$$

$$= 9 - (2x + y)(2x + y) \qquad \text{Factor the perfect-square trinomial.}$$

$$= 9 - (2x + y)^2 \qquad \text{Write } (2x + y)(2x + y) \text{ as } (2x + y)^2.$$

$$= [3 + (2x + y)][3 - (2x + y)] \qquad \text{Factor the difference of two squares.}$$

$$= (3 + 2x + y)(3 - 2x - y) \qquad \text{Simplify each factor.}$$

Check by multiplication.

 SELF CHECK 11   Factor: $16 - a^2 - 2ab - b^2$

NOW TRY THIS

1. If the area of a rectangle can be expressed by the polynomial $(35x^2 - 31x + 6)$ cm^2, find polynomial expressions for the length and width.

2. Factor, if possible:
   a. $-8x^2 - 15 + 22x$
   b. $15x^2 - 28x - 12$

5.4 Exercises

WARM-UPS *Fill in the blanks.*

1. $6x^2 + 7x + 2 = (2x + 1)(3x + \blacksquare)$
2. $3t^2 + t - 2 = (3t - \blacksquare)(t + 1)$
3. $6x^2 + x - 2 = (3x + \blacksquare)(2x - \blacksquare)$
4. $6x^2 + 5x + 1 = (\blacksquare x + 1)(3x + 1)$
5. $15x^2 - 7x - 4 = (5x - \blacksquare)(3x + 1)$
6. $2x^2 + 5x + 3 = (2x + \blacksquare)(x + 1)$

REVIEW

7. The nth term l of an arithmetic sequence is

   $$l = f + (n - 1)d$$

   where f is the first term and d is the common difference. Remove the parentheses and solve for n.

8. The sum S of n consecutive terms of an arithmetic sequence is

   $$S = \frac{n}{2}(f + l)$$

   where f is the first term and l is the nth term. Solve for f.

VOCABULARY AND CONCEPTS *Fill in the blanks.*

9. If the sign of the first and third terms of a trinomial are $+$, the signs within the binomial factors are _____ the sign of the middle term.

10. If the sign of the first term of a trinomial is $+$ and the sign of the third term is $-$, the signs within the binomial factors are _____.

11. An alternative method to trial and error is the _____ or grouping.

12. A visual of the ac method is the _____.

Complete each factorization.

13. $6x^2 + 5x - 1 = (x \,\blacksquare\, 1)(6x \,\blacksquare\, 1)$
14. $6x^2 + x - 1 = (2x \,\blacksquare\, 1)(3x \,\blacksquare\, 1)$

15. $4x^2 + 4x - 3 = (2x + \blacksquare)(2x - \blacksquare)$
16. $4x^2 - x - 3 = (4x + \blacksquare)(x - \blacksquare)$
17. $12x^2 - 7xy + y^2 = (3x - \blacksquare)(4x - \blacksquare)$
18. $6x^2 + 5xy - 6y^2 = (2x + \blacksquare)(3x - \blacksquare)$

GUIDED PRACTICE *Factor.* SEE EXAMPLE 1. (OBJECTIVE 1)

19. $3a^2 + 7a + 2$ 20. $6y^2 + 7y + 2$

21. $3a^2 + 10a + 3$ 22. $2b^2 + 13b + 15$

23. $5t^2 + 13t + 6$ 24. $16y^2 + 10y + 1$

25. $16x^2 + 16x + 3$ 26. $4z^2 + 13z + 3$

Factor. SEE EXAMPLE 2. (OBJECTIVE 1)

27. $5y^2 - 23y + 12$ 28. $10x^2 - 9x + 2$

29. $2y^2 - 7y + 3$ 30. $7z^2 - 26z + 15$

31. $16m^2 - 14m + 3$ 32. $4t^2 - 4t + 1$

33. $6x^2 - 7x + 2$ 34. $20x^2 - 23x + 6$

Factor. SEE EXAMPLE 3. (OBJECTIVE 1)

35. $3a^2 - 4a - 4$ 36. $2x^2 - 3x - 2$

37. $12y^2 - y - 1$ 38. $8u^2 - 2u - 15$

39. $12y^2 - 5y - 2$ 40. $10x^2 - x - 2$

41. $10y^2 - 3y - 1$ 42. $14x^2 - 3x - 2$

Factor each trinomial. *SEE EXAMPLE 4. (OBJECTIVE 1)*

43. $8q^2 + 10q - 3$ **44.** $2m^2 + 5m - 12$

45. $10x^2 + 21x - 10$ **46.** $4y^2 + 5y - 6$

47. $30x^2 - 23x - 14$ **48.** $18x^2 - 15x - 25$

49. $8x^2 - 14x - 15$ **50.** $6x^2 - 7x - 20$

Factor. *SEE EXAMPLE 5. (OBJECTIVE 1)*

51. $2x^2 + 3xy + y^2$ **52.** $5m^2 + 8mn + 3n^2$

53. $3x^2 - 4xy + y^2$ **54.** $2b^2 - 5bc + 2c^2$

55. $10p^2 - 11pq - 6q^2$ **56.** $2u^2 + 3uv - 2v^2$

57. $6p^2 - pq - 2q^2$ **58.** $8r^2 - 10rs - 25s^2$

Write the terms of each trinomial in descending powers of one variable. Then factor. *SEE EXAMPLE 6. (OBJECTIVE 1)*

59. $15 + 8a^2 - 26a$ **60.** $16 - 40a + 25a^2$

61. $12x^2 + 10y^2 - 23xy$ **62.** $3ab + 20a^2 - 2b^2$

Factor. *SEE EXAMPLES 7–8. (OBJECTIVE 2)*

63. $6x^2 + 7x + 2$ **64.** $6m^2 + 19m + 3$

65. $-26x + 6x^2 - 20$ **66.** $-42 + 9a^2 - 3a$

Factor. *SEE EXAMPLE 9. (OBJECTIVE 3)*

67. $9x^2 - 12x + 4$ **68.** $9x^2 + 6x + 1$

69. $25x^2 + 30x + 9$ **70.** $16y^2 - 24y + 9$

71. $9a^2 + 24a + 16$ **72.** $4x^2 - 4x + 1$

73. $16x^2 - 40x + 25$ **74.** $25x^2 - 20x + 4$

Factor. *SEE EXAMPLES 10–11. (OBJECTIVE 3)*

75. $16x^2 + 8xy + y^2 - 9$

76. $9x^2 - 6x + 1 - d^2$

77. $9 - a^2 - 4ab - 4b^2$

78. $16m^2 - 24m - n^2 + 9$

ADDITIONAL PRACTICE *Factor. If the polynomial is prime, so indicate.*

79. $-12y^2 - 12 + 25y$ **80.** $-12t^2 + 1 + 4t$

81. $5x^2 + 2 + x$ **82.** $25 + 2u^2 + 3u$

83. $4x^2 + 10x - 6$ **84.** $9x^2 + 21x - 18$

85. $y^3 + 13y^2 + 12y$ **86.** $6x^3 - 15x^2 - 9x$

87. $9y^3 + 3y^2 - 6y$ **88.** $30r^5 + 63r^4 - 30r^3$

89. $6s^5 - 26s^4 - 20s^3$ **90.** $4a^2 - 15ab + 9b^2$

91. $12x^2 + 5xy - 3y^2$ **92.** $b^2 + 4a^2 + 16ab$

93. $3b^2 + 3a^2 - ab$ **94.** $4a^2 - 4ab - 8b^2$

95. $25x^2 + 20xy + 4y^2$ **96.** $6r^2 + rs - 2s^2$

97. $-16x^4y^3 + 30x^3y^4 + 4x^2y^5$

98. $9p^2 + 1 + 6p - q^2$

99. $24a^2 + 14ab + 2b^2$ **100.** $9a^2 + 6ac - c^2$

WRITING ABOUT MATH

101. Describe an organized approach to finding all of the possibilities when you attempt to factor $12x^2 - 4x + 9$.

102. Explain how to determine whether a trinomial is prime.

SOMETHING TO THINK ABOUT

103. For what values of b will the trinomial $6x^2 + bx + 6$ be factorable?

104. For what values of b will the trinomial $5y^2 - by - 3$ be factorable?

Section 5.5

Factoring the Sum and Difference of Two Cubes

Objectives

1. Factor the sum of two cubes.
2. Factor the difference of two cubes.
3. Factor a polynomial involving the sum or difference of two cubes.

Vocabulary

sum of two cubes   difference of two cubes

Getting Ready

Find each product.

1. $(x - 3)(x^2 + 3x + 9)$ **2.** $(x + 2)(x^2 - 2x + 4)$

3. $(y + 4)(y^2 - 4y + 16)$ **4.** $(r - 5)(r^2 + 5r + 25)$

5. $(a - b)(a^2 + ab + b^2)$ **6.** $(a + b)(a^2 - ab + b^2)$

Recall that the difference of the squares of two quantities factors into the product of two binomials. One binomial is the sum of the quantities, and the other is the difference of the quantities.

$$x^2 - y^2 = (x + y)(x - y) \qquad \text{or} \qquad F^2 - L^2 = (F + L)(F - L)$$

In this section, we will discuss formulas for factoring the *sum of two cubes* and the *difference of two cubes*.

1 Factor the sum of two cubes.

To discover the pattern for factoring the sum of two cubes, we find the following product:

$$(x + y)(x^2 - xy + y^2) = x^3 - x^2y + xy^2 + x^2y - xy^2 + y^3 \qquad \text{Use the distributive property.}$$

$$= x^3 + y^3 \qquad \text{Combine like terms.}$$

This result justifies the formula for factoring the **sum of two cubes**.

FACTORING THE SUM OF TWO CUBES

$$x^3 + y^3 = (x + y)(x^2 - xy + y^2)$$

To factor the sum of two cubes, it is helpful to know the cubes of the numbers from 1 to 10:

1, 8, 27, 64, 125, 216, 343, 512, 729, 1,000

Expressions containing variables such as x^6y^3 are also perfect cubes, because they can be written as the cube of a quantity:

$$x^6y^3 = (x^2y)^3$$

EXAMPLE 1   Factor: $x^3 + 8$

Solution   The binomial $x^3 + 8$ is the sum of two cubes, because

$$x^3 + 8 = x^3 + 2^3$$

Thus, $x^3 + 8$ factors as $(x + 2)$ times the trinomial $(x^2 - 2 \cdot x + 2^2)$.

$$x^3 + \mathbf{8} = x^3 + \mathbf{2}^3$$
$$= (\mathbf{x} + \mathbf{2})(x^2 - \mathbf{x} \cdot \mathbf{2} + \mathbf{2}^2)$$
$$= (x + 2)(x^2 - 2x + 4)$$

To check, we can use the distributive property and combine like terms.

$$(x + 2)(x^2 - 2x + 4) = x^3 - 2x^2 + 4x + 2x^2 - 4x + 8$$
$$= x^3 - 8$$

 SELF CHECK 1   Factor: $p^3 + 64$

EXAMPLE 2   Factor: $8b^3 + 27c^3$

Solution   The binomial $8b^3 + 27c^3$ is the sum of two cubes, because

$$8b^3 + 27c^3 = (2b)^3 + (3c)^3$$

Thus, the binomial $8b^3 + 27c^3$ factors as $(2b + 3c)$ times the trinomial $(2b)^2 - (2b)(3c) + (3c)^2$.

$$8b^3 + 27c^3 = (2b)^3 + (3c)^3$$
$$= (2b + 3c)[(2b)^2 - (2b)(3c) + (3c)^2]$$
$$= (2b + 3c)(4b^2 - 6bc + 9c^2)$$

To check, we can use the distributive property and combine like terms.

$$(2b + 3c)(4b^2 - 6bc + 9c^2)$$
$$= 8b^3 - 12b^2c + 18bc^2 + 12b^2c - 18bc^2 + 27c^3$$
$$= 8b^3 + 27c^3$$

 SELF CHECK 2   Factor: $1{,}000p^3 + q^3$

2   Factor the difference of two cubes.

To discover the pattern for factoring the difference of two cubes, we find the following product:

$$(x - y)(x^2 + xy + y^2) = x^3 + x^2y + xy^2 - x^2y - xy^2 - y^3 \quad \text{Use the distributive property.}$$
$$= x^3 - y^3 \quad \text{Combine like terms.}$$

This result justifies the formula for factoring the **difference of two cubes**.

FACTORING THE DIFFERENCE OF TWO CUBES

$$x^3 - y^3 = (x - y)(x^2 + xy + y^2)$$

EXAMPLE 3   Factor: $a^3 - 64b^3$

Solution   The binomial $a^3 - 64b^3$ is the difference of two cubes.

$$a^3 - 64b^3 = a^3 - (4b)^3$$

Thus, its factors are the difference $a - 4b$ and the trinomial $a^2 + a(4b) + (4b)^2$.

$$\begin{aligned}
\mathbf{a^3 - 64b^3} &= \mathbf{a^3 - (4b)^3} \\
&= \mathbf{(a - 4b)[a^2 + a(4b) + (4b)^2]} \\
&= (a - 4b)(a^2 + 4ab + 16b^2)
\end{aligned}$$

To check, we can use the distributive property and combine like terms.

$$\begin{aligned}
(a - 4b)&(a^2 + 4ab + 16b^2) \\
&= a^3 + 4a^2b + 16ab^2 - 4a^2b - 16ab^2 - 64b^3 \\
&= a^3 - 64b^3
\end{aligned}$$

 SELF CHECK 3   Factor: $27p^3 - 8$

3   Factor a polynomial involving the sum or difference of two cubes.

Sometimes we must factor out a greatest common factor before factoring a sum or difference of two cubes.

EXAMPLE 4   Factor: $-2t^5 + 128t^2$

Solution
$$\begin{aligned}
-2t^5 + 128t^2 &= -2t^2(t^3 - 64) \\
&= -2t^2(t^3 - 4^3) \\
&= -2t^2(t - 4)(t^2 + 4t + 16)
\end{aligned}$$

Factor the GCF, $-2t^2$, from $(-2t^5 + 128t^2)$.

Write $t^3 - 64$ as $t^3 - 4^3$.

Factor $t^3 - 4^3$.

We can check by multiplication.

 SELF CHECK 4   Factor: $-3p^4 + 81p$

If a binomial is both the difference of two squares and the difference of two cubes, we will factor the difference of two squares first.

EXAMPLE 5   Factor: $x^6 - 64$

Solution   If we consider the polynomial to be the difference of two squares, we can factor it as follows:

$$\begin{aligned}
x^6 - 64 &= (x^3)^2 - 8^2 \\
&= (x^3 + 8)(x^3 - 8)
\end{aligned}$$

Because $x^3 + 8$ is the sum of two cubes and $x^3 - 8$ is the difference of two cubes, each of these binomials can be factored.

$$\begin{aligned}
x^6 - 64 &= \mathbf{(x^3 + 8)(x^3 - 8)} \\
&= \mathbf{(x + 2)(x^2 - 2x + 4)(x - 2)(x^2 + 2x + 4)}
\end{aligned}$$

We can check by multiplication.

 SELF CHECK 5   Factor: $a^6 - 1$

SELF CHECK ANSWERS

1. $(p + 4)(p^2 - 4p + 16)$ 2. $(10p + q)(100p^2 - 10pq + q^2)$ 3. $(3p - 2)(9p^2 + 6p + 4)$ 4. $-3p(p - 3)(p^2 + 3p + 9)$ 5. $(a + 1)(a^2 - a + 1)(a - 1)(a^2 + a + 1)$

NOW TRY THIS

Factor.

1. $x^3 - \frac{1}{8}$
2. $x^3 - y^{12}$
3. $64x^3 - 8$
4. $x^3(x^2 - 9) - 8(x^2 - 9)$

5.5 Exercises

WARM-UPS *Write each expression as a cube.*

1. 8
2. x^6
3. -27
4. 125
5. y^{12}
6. $27x^3$
7. $-y^9$
8. $-64y^6$

Square each expression.

9. 3
10. x^2
11. $4y^2$
12. $3x$
13. $-2x$
14. $-3y^2$
15. $-x^5$
16. $-5y^3$

REVIEW

17. The length of one fermi is 1×10^{-13} centimeter, approximately the radius of a proton. Express this number in standard notation.
18. In the 14th century, the Black Plague killed about 25,000,000 people, which was 25% of the population of Europe. Find the population at that time, expressed in scientific notation.

VOCABULARY AND CONCEPTS *Fill in the blanks.*

19. A polynomial in the form of $a^3 + b^3$ is called a ________ ____.
20. A polynomial in the form of $a^3 - b^3$ is called a ________ ________.

Complete each formula.

21. $x^3 + y^3 = (x + y)$
22. $x^3 - y^3 = (x - y)$

GUIDED PRACTICE *Factor.* SEE EXAMPLE 1. (OBJECTIVE 1)

23. $a^3 + 8$
24. $b^3 + 125$
25. $125x^3 + 8$
26. $8 + x^3$
27. $y^3 + 1$
28. $1 + 8x^3$
29. $125 + a^3$
30. $64 + b^3$

Factor. SEE EXAMPLE 2. (OBJECTIVE 1)

31. $m^3 + n^3$
32. $a^3 + 8b^3$
33. $x^3 + y^3$
34. $27x^3 + y^3$
35. $8u^3 + w^3$
36. $x^3y^3 + 1$

Factor. SEE EXAMPLE 3. (OBJECTIVE 2)

37. $x^3 - y^3$
38. $b^3 - 27$
39. $x^3 - 8$
40. $a^3 - 64$
41. $s^3 - t^3$
42. $27 - y^3$
43. $125p^3 - q^3$
44. $x^3 - 27y^3$
45. $27a^3 - b^3$
46. $64x^3 - 27$

Factor. SEE EXAMPLE 4. (OBJECTIVE 3)

47. $2x^3 + 54$

48. $5x^3 - 5$

49. $-x^3 + 216$

50. $-x^3 - 125$

51. $64m^3x - 8n^3x$

52. $16r^4 + 128rs^3$

53. $x^4y + 216xy^4$

54. $16a^5 - 54a^2b^3$

Factor each polynomial completely. Factor a difference of two squares first. SEE EXAMPLE 5. (OBJECTIVE 3)

55. $x^6 - 1$

56. $x^6 - y^6$

57. $x^{12} - y^6$

58. $a^{12} - 64$

ADDITIONAL PRACTICE   *Factor.*

59. $y^3 + 8$

60. $x^3 - y^9$

61. $27x^3 + 125$

62. $81r^4s^2 - 24rs^5$

63. $64 - z^3$

64. $27x^3 - 125y^3$

65. $64x^3 + 27y^3$

66. $216a^4b^4 - 1{,}000ab^7$

67. $3(x^3 + y^3) - z(x^3 + y^3)$

68. $y^7z - yz^4$

69. $x(27y^3 - z^3) + 5(27y^3 - z^3)$

70. $(m^3 + 8n^3) + (m^3x + 8n^3x)$

71. $(a^3x + b^3x) - (a^3y + b^3y)$

72. $(a^4 + 27a) - (a^3b + 27b)$

73. $y^3(y^2 - 1) - 27(y^2 - 1)$

74. $z^3(y^2 - 4) + 8(y^2 - 4)$

WRITING ABOUT MATH

75. Explain how to factor $a^3 + b^3$.

76. Explain the difference between $x^3 - y^3$ and $(x - y)^3$.

SOMETHING TO THINK ABOUT

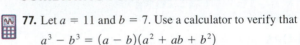 **77.** Let $a = 11$ and $b = 7$. Use a calculator to verify that
$$a^3 - b^3 = (a - b)(a^2 + ab + b^2)$$

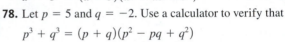 **78.** Let $p = 5$ and $q = -2$. Use a calculator to verify that
$$p^3 + q^3 = (p + q)(p^2 - pq + q^2)$$

Section 5.6

Summary of Factoring Techniques

Objectives

1 Factor a polynomial.

Getting Ready

Multiply.

1. $3ax(x + a)$

2. $(x + 3y)(x - 3y)$

3. $(x - 2)(x^2 + 2x + 4)$

4. $2(x + 2)(x - 2)$

5. $(x - 5)(x + 2)$

6. $(2x - 3)(3x - 2)$

7. $2(3x - 1)(x - 2)$

8. $(a + b)(x + y)(x - y)$

In this section, we will discuss ways to approach any factoring problem.

1 **Factor a polynomial.**

Suppose we want to factor the trinomial

$$x^4y + 7x^3y - 18x^2y$$

Since it is written in descending powers of x, we begin by identifying the problem type. The first type we look for is one that contains a common factor. Because the trinomial has the greatest common factor of x^2y, we factor it out first:

$$x^4y + 7x^3y - 18x^2y = x^2y(x^2 + 7x - 18)$$

We can factor the remaining trinomial $x^2 + 7x - 18$ as $(x + 9)(x - 2)$. Thus,

$$x^4y + 7x^3y - 18x^2y = x^2y(x^2 + 7x - 18)$$
$$= x^2y(x + 9)(x - 2)$$

To identify the type of factoring problem, we follow these steps.

FACTORING A POLYNOMIAL

1. Write the polynomial in descending powers of one variable.

2. Factor out all common factors.

3. If an expression has two terms, check to see if the problem type is
   a. the difference of two squares: $x^2 - y^2 = (x + y)(x - y)$
   b. the sum of two cubes: $x^3 + y^3 = (x + y)(x^2 - xy + y^2)$
   c. the difference of two cubes: $x^3 - y^3 = (x - y)(x^2 + xy + y^2)$

4. If an expression has three terms, check to see if it is a perfect-square trinomial:

$$x^2 + 2xy + y^2 = (x + y)(x + y)$$
$$x^2 - 2xy + y^2 = (x - y)(x - y)$$

   If the trinomial is not a perfect-square trinomial, attempt to factor the trinomial as a general trinomial.

5. If an expression has four terms, try to factor the expression by grouping. It may be necessary to rearrange the terms.

6. Continue factoring until each nonmonomial factor is prime.

7. If the polynomial does not factor, the polynomial is *prime* over the set of rational numbers.

8. Check the results by multiplying.

EXAMPLE 1 Factor: $x^5y^2 - xy^6$

Solution Since the polynomial is written in descending powers of x, we begin by factoring out the greatest common factor of xy^2.

$$x^5y^2 - xy^6 = xy^2(x^4 - y^4)$$

The expression $x^4 - y^4$ has two terms. We check to see whether it is the difference of two squares, which it is. As the difference of two squares, it factors as $(x^2 + y^2)(x^2 - y^2)$.

$$x^5y^2 - xy^6 = xy^2(x^4 - y^4)$$
$$= xy^2(x^2 + y^2)(x^2 - y^2)$$

The binomial $x^2 + y^2$ is the sum of two squares and cannot be factored. However, $x^2 - y^2$ is the difference of two squares and factors as $(x + y)(x - y)$.

$$x^5y^2 - xy^6 = xy^2(x^4 - y^4)$$
$$= xy^2(x^2 + y^2)(x^2 - y^2)$$
$$= xy^2(x^2 + y^2)(x + y)(x - y)$$

Since each individual factor is prime, the given expression is in completely factored form.

SELF CHECK 1   Factor: $-a^5b + ab^5$

EXAMPLE 2   Factor: $x^6 - x^4y^2 - x^3y^3 + xy^5$

Solution   Factor out the greatest common factor of x.

$$x^6 - x^4y^2 - x^3y^3 + xy^5 = x(x^5 - x^3y^2 - x^2y^3 + y^5)$$

Since $x^5 - x^3y^2 - x^2y^3 + y^5$ has four terms, we try factoring it by grouping:

$$x^6 - x^4y^2 - x^3y^3 + xy^5 = x(x^5 - x^3y^2 - x^2y^3 + y^5)$$
$$= x[x^3(x^2 - y^2) - y^3(x^2 - y^2)]$$
$$= x(x^2 - y^2)(x^3 - y^3) \qquad \text{Factor out } x^2 - y^2.$$

Finally, we factor the difference of two squares and the difference of two cubes:

$$x^6 - x^4y^2 - x^3y^3 + xy^5 = x(x^2 - y^2)(x^3 - y^3)$$

$$= x(x + y)(x - y)(x - y)(x^2 + xy + y^2)$$
$$= x(x + y)(x - y)^2(x^2 + xy + y^2) \qquad \begin{array}{l}(x - y)(x - y) \\ = (x - y)^2\end{array}$$

Since each factor is prime, the given expression is in completely factored form.

SELF CHECK 2   Factor: $2a^5 - 2a^2b^3 - 8a^3 + 8b^3$

SELF CHECK ANSWERS   **1.** $-ab(a^2 + b^2)(a + b)(a - b)$ **2.** $2(a + 2)(a - 2)(a - b)(a^2 + ab + b^2)$

NOW TRY THIS

Factor.

1. $4x^2 + 16$
2. $ax^2 + bx^2 - 36a - 36b$
3. $9x^2 - 9x$
4. $64 - x^6$

5.6   Exercises

WARM-UPS   *Indicate which factoring technique you would use first, if any.*

1. $3x^2 - 9x$

2. $49 - 16a^2$

3. $125 + r^3s^3$

4. $ax + ay - x - y$

5. $x^2 + 36$

6. $16x^2 - 24$

7. $25r^2 - s^4$

8. $8a^3 - 27b^3$

REVIEW *Solve each equation, if possible.*

9. $2(t + 5) + t = 3(t + 2) + 4$
10. $5 + 3(2x - 1) = 2(4 + 3x) - 24$
11. $6 + 2(t + 3) = t + 3$
12. $4m - 3 = -2(m + 1) - 3$

VOCABULARY AND CONCEPTS *Fill in the blanks.*

13. The first step in any factoring problem is to factor out all common _____, if possible.
14. If a polynomial has two terms, check to see if it is the _____, the sum of two cubes, or the _____ of two cubes.
15. If a polynomial has three terms, try to factor it as the product of two _____.
16. If a polynomial has four or more terms, try factoring by _____.

PRACTICE *Factor.*

17. $6x + 3$
18. $x^2 - 36$
19. $x^2 + 10x + 9$
20. $a^3 - 27$
21. $8t^2 - 6t - 9$
22. $4x^2 - 25$
23. $t^2 - 2t + 1$
24. $6p^2 - 3p - 2$
25. $2x^2 - 50$
26. $t^4 - 16$
27. $x^2 + 7x + 1$
28. $10r^2 - 13r - 4$
29. $-2x^5 + 128x^2$
30. $49 - 28z + 4z^2$
31. $14t^3 - 40t^2 + 6t^4$
32. $6x^2 + 7x - 20$
33. $6x^2 - x - 16$
34. $30a^4 + 5a^3 - 200a^2$
35. $6a^3 + 35a^2 - 6a$
36. $21r^3 - 10t^2 + t$
37. $16x^2 - 40x^3 + 25x^4$
38. $25a^2 - 60a + 36$
39. $-84x^2 - 147x - 12x^3$
40. $x^3 - 5x^2 - 25x + 125$
41. $8x^6 - 8$
42. $16x^2 + 64$
43. $5x^3 - 5x^5 + 25x^2$
44. $12y^3 - 27y$
45. $9x^2 + 12x + 16$
46. $70p^4q^3 - 35p^4q^2 + 49p^5q^2$
47. $2ab^2 + 8ab - 24a$
48. $2x^2y - 4xy^2$
49. $-8p^3q^7 - 4p^2q^3$
50. $8m^2n^3 - 24mn^4$
51. $4a^2 - 4ab + b^2 - 9$
52. $3rs + 6r^2 - 18s^2$
53. $8a^3 - b^3$
54. $ac + ad + bc + bd$
55. $x^2y^2 - 2x^2 - y^2 + 2$
56. $a^2c + a^2d^2 + bc + bd^2$
57. $a^2 + 2ab + b^2 - y^2$
58. $2x^3 + 54y^3$
59. $a^2(x - 3) - b^2(x - 3)$
60. $5x^3y^3z^4 + 25x^2y^3z^2 - 35x^3y^2z^5$
61. $8p^6 - 27q^6$
62. $3c^2 - 11cd - 4d^2$
63. $125p^3 - 64y^3$
64. $8a^2x^3y - 2b^2xy$
65. $-16x^4y^2z + 24x^5y^3z^4 - 15x^2y^3z^7$

66. $2ac + 4ad + bc + 2bd$
67. $81p^4 - 16q^4$
68. $4x^2 + 9y^2$
69. $54x^3 + 250y^6$
70. $4x^2 + 4x + 1 - y^2$
71. $x^5 - x^3y^2 + x^2y^3 - y^5$

72. $a^3x^3 - a^3y^3 + b^3x^3 - b^3y^3$

73. $2a^2c - 2b^2c + 4a^2d - 4b^2d$
74. $3a^2x^2 + 6a^2x + 3a^2 - 3b^2$

WRITING ABOUT MATH

75. Explain how to identify the type of factoring required to factor a polynomial.
76. Which factoring technique do you find most difficult? Why?

SOMETHING TO THINK ABOUT

77. Write $x^6 - y^6$ as $(x^3)^2 - (y^3)^2$, factor it as the difference of two squares, and show that you get

$$(x + y)(x^2 - xy + y^2)(x - y)(x^2 + xy + y^2)$$

Write $x^6 - y^6$ as $(x^2)^3 - (y^2)^3$, factor it as the difference of two cubes, and show that you get

$$(x + y)(x - y)(x^4 + x^2y^2 + y^4)$$

78. Verify that the results of Exercise 77 agree by showing the parts in color agree. Which do you think is completely factored?

Section 5.7 Solving Equations by Factoring

Objectives

1. Solve a quadratic equation in one variable using the zero-factor property.
2. Solve a higher-order polynomial equation in one variable.

Vocabulary

quadratic equation　　　　zero-factor property　　　　higher-order polynomial
quadratic form　　　　　　　　　　　　　　　　　　　　equation

Getting Ready

Solve each equation.

1. $x + 3 = 4$　　**2.** $y - 8 = 5$　　**3.** $3x - 2 = 7$　　**4.** $5y + 9 = 19$

In this section, we will learn how to use factoring to solve many equations that contain second-degree polynomials in one variable. These equations are called *quadratic equations*.

Equations such as

$$3x + 2 = 0 \quad \text{and} \quad 9x - 6 = 0$$

that contain first-degree polynomials are *linear equations*. Equations such as

$$9x^2 - 6x = 0 \quad \text{and} \quad 3x^2 + 4x - 7 = 0$$

that contain second-degree polynomials are called **quadratic equations**.

QUADRATIC EQUATIONS　A quadratic equation in one variable is an equation of the form

$$ax^2 + bx + c = 0 \quad \text{(This is called \textbf{quadratic form}.)}$$

where a, b, and c are real numbers, and $a \neq 0$.

1 Solve a quadratic equation in one variable using the zero-factor property.

Many quadratic equations can be solved by factoring. For example, to solve the quadratic equation

$$x^2 + 5x - 6 = 0$$

which is already in quadratic form, we begin by factoring the trinomial and writing the equation as

$$(1) \quad (x + 6)(x - 1) = 0$$

This equation indicates that the product of two quantities is 0. However, if the product of two quantities is 0, then at least one of those quantities must be 0. This fact is called the **zero-factor property**.

ZERO-FACTOR PROPERTY	Suppose a and b represent two real numbers. If $ab = 0$, then $a = 0$ or $b = 0$.

COMMENT In mathematics, when we use the word "or" it is understood to mean one or the other or both. Thus, if $a \cdot b = 0$, both a and b could be equal to 0.

By applying the zero-factor property to Equation 1, we have

$$x + 6 = 0 \qquad \text{or} \qquad x - 1 = 0$$

We can solve each of these linear equations to get

$$x = -6 \qquad \text{or} \qquad x = 1$$

To check, we substitute -6 for x, and then 1 for x in the original equation and simplify.

For $x = -6$	*For $x = 1$*
$x^2 + 5x - 6 = 0$	$x^2 + 5x - 6 = 0$
$(-6)^2 + 5(-6) - 6 \stackrel{?}{=} 0$	$(1)^2 + 5(1) - 6 \stackrel{?}{=} 0$
$36 - 30 - 6 \stackrel{?}{=} 0$	$1 + 5 - 6 \stackrel{?}{=} 0$
$6 - 6 \stackrel{?}{=} 0$	$6 - 6 \stackrel{?}{=} 0$
$0 = 0$	$0 = 0$

Both solutions check.

The quadratic equations $9x^2 - 6x = 0$ and $4x^2 - 25 = 0$ are each missing a term. The first equation is missing the constant term, and the second equation is missing the term involving x. These types of equations often can be solved by factoring.

EXAMPLE 1 Solve: $9x^2 - 6x = 0$

Solution We begin by factoring the left side of the equation.

$$9x^2 - 6x = 0$$
$$3x(3x - 2) = 0 \qquad \text{\color{red}Factor out the common factor of } 3x.$$

By the zero-factor property, we have

$$3x = 0 \qquad \text{or} \qquad 3x - 2 = 0$$

Solve each linear equation.

$$
\begin{array}{c|cl}
3x = 0 \quad \text{or} & 3x - 2 = 0 & \\
& 3x = 2 & \text{\color{red}Add 2 to both sides.} \\
\dfrac{3x}{3} = \dfrac{0}{3} & \dfrac{3x}{3} = \dfrac{2}{3} & \text{\color{red}Divide both sides of each equation by 3.} \\
x = 0 & x = \dfrac{2}{3} &
\end{array}
$$

Check: We substitute these results for x in the original equation and simplify.

$$\begin{array}{ll} \textbf{\textit{For x = 0}} & \textbf{\textit{For x = }}\dfrac{\textbf{2}}{\textbf{3}} \\ 9x^2 - 6x = 0 & 9x^2 - 6x = 0 \\ 9(0)^2 - 6(0) \stackrel{?}{=} 0 & 9\left(\dfrac{2}{3}\right)^2 - 6\left(\dfrac{2}{3}\right) \stackrel{?}{=} 0 \\ 0 - 0 \stackrel{?}{=} 0 & 9\left(\dfrac{4}{9}\right) - 6\left(\dfrac{2}{3}\right) \stackrel{?}{=} 0 \\ 0 = 0 & 4 - 4 \stackrel{?}{=} 0 \\ & 0 = 0 \end{array}$$

Both solutions check.

SELF CHECK 1   Solve: $5y^2 + 10y = 0$

EXAMPLE 2   Solve: $4x^2 - 25 = 0$

Solution   We proceed as follows:

$$\begin{array}{lll} 4x^2 - 25 = 0 & & \\ (2x + 5)(2x - 5) = 0 & & \text{Factor } 4x^2 - 25. \\ 2x + 5 = 0 \quad \text{or} \quad 2x - 5 = 0 & & \text{Apply the zero-factor property.} \\ 2x = -5 \qquad\qquad 2x = 5 & & \text{Isolate the variable term.} \\ x = -\dfrac{5}{2} \qquad\qquad x = \dfrac{5}{2} & & \text{Divide both sides of each equation by 2.} \end{array}$$

Check each solution.

$$\begin{array}{ll} \textbf{\textit{For x = }}-\dfrac{\textbf{5}}{\textbf{2}} & \textbf{\textit{For x = }}\dfrac{\textbf{5}}{\textbf{2}} \\ 4x^2 - 25 = 0 & 4x^2 - 25 = 0 \\ 4\left(-\dfrac{5}{2}\right)^2 - 25 \stackrel{?}{=} 0 & 4\left(\dfrac{5}{2}\right)^2 - 25 \stackrel{?}{=} 0 \\ 4\left(\dfrac{25}{4}\right) - 25 \stackrel{?}{=} 0 & 4\left(\dfrac{25}{4}\right) - 25 \stackrel{?}{=} 0 \\ 0 = 0 & 0 = 0 \end{array}$$

Both solutions check.

SELF CHECK 2   Solve: $9p^2 - 64 = 0$

In the next example, we solve an equation whose polynomial is a trinomial.

EXAMPLE 3   Solve: $x^2 - 3x - 18 = 0$

Solution

$$\begin{array}{lll} x^2 - 3x - 18 = 0 & & \\ (x + 3)(x - 6) = 0 & & \text{Factor } x^2 - 3x - 18. \\ x + 3 = 0 \quad \text{or} \quad x - 6 = 0 & & \text{Apply the zero-factor property.} \\ x = -3 \qquad\qquad x = 6 & & \text{Solve each linear equation.} \end{array}$$

Check each solution.

SELF CHECK 3   Solve: $x^2 + 3x - 18 = 0$

EXAMPLE 4   Solve: $2x^2 + 3x = 2$

Solution   We write the equation in the form $ax^2 + bx + c = 0$ and solve for x.

COMMENT   To apply the zero-factor property, the equation must be set equal to 0 prior to factoring.

$$2x^2 + 3x = 2$$
$$2x^2 + 3x - 2 = 0 \qquad \text{Subtract 2 from both sides.}$$
$$(2x - 1)(x + 2) = 0 \qquad \text{Factor } 2x^2 + 3x - 2.$$

$2x - 1 = 0$ or $x + 2 = 0$ Apply the zero-factor property.

$\quad 2x = 1 \qquad\qquad\quad x = -2$ Solve each linear equation.

$$x = \frac{1}{2}$$

Check each solution.

 SELF CHECK 4   Solve: $3x^2 - 5x = 2$

Some equations must be simplified before we write them in quadratic form. Many times this requires the distributive property as illustrated in the next example.

EXAMPLE 5   Solve: $(x - 2)(x - 6) = -3$

Solution   We must write the equation in the form $ax^2 + bx + c = 0$ before we can solve for x. We first multiply the binomials and then set the result equal to zero.

$$(x - 2)(x - 6) = -3$$
$$x^2 - 6x - 2x + 12 = -3 \qquad \text{Multiply.}$$
$$x^2 - 8x + 12 = -3 \qquad \text{Combine like terms.}$$
$$x^2 - 8x + 15 = 0 \qquad \text{Add 3 to both sides.}$$
$$(x - 3)(x - 5) = 0 \qquad \text{Factor the trinomial.}$$

$(x - 3) = 0$ or $x - 5 = 0$ Apply the zero-factor property.

$\quad x = 3 \qquad\qquad\quad x = 5$ Solve each linear equation.

Check each solution.

 SELF CHECK 5   Solve: $(x + 1)(x - 5) = 7$

2 Solve a higher-order polynomial equation in one variable.

A **higher-order polynomial equation** is any equation in one variable with a degree of 3 or larger.

EXAMPLE 6   Solve: $x^3 - 2x^2 - 63x = 0$

Solution   We begin by completely factoring the left side.

$$x^3 - 2x^2 - 63x = 0$$
$$x(x^2 - 2x - 63) = 0 \qquad \text{Factor out } x, \text{ the GCF.}$$
$$x(x + 7)(x - 9) = 0 \qquad \text{Factor the trinomial.}$$

$x = 0$ or $x + 7 = 0$ or $x - 9 = 0$ Set each factor equal to 0.

$\qquad\qquad\quad x = -7 \qquad\qquad x = 9$ Solve each linear equation.

Check each solution.

 SELF CHECK 6   Solve: $x^3 - x^2 - 2x = 0$

As with quadratic equations, higher-order equations in one variable must be set equal to zero to solve by factoring.

EXAMPLE 7   Solve: $6x^3 + 12x = 17x^2$

Solution   To set the equation equal to 0, we subtract $17x^2$ from both sides. Then we proceed as follows:

$$6x^3 + 12x = 17x^2$$

$$6x^3 - 17x^2 + 12x = 0 \qquad \text{\color{red}Subtract } 17x^2 \text{ from both sides.}$$

$$x(6x^2 - 17x + 12) = 0 \qquad \text{\color{red}Factor out } x, \text{ the GCF.}$$

$$x(2x - 3)(3x - 4) = 0 \qquad \text{\color{red}Factor } 6x^2 - 17x + 12.$$

$$x = 0 \quad \text{or} \quad 2x - 3 = 0 \quad \text{or} \quad 3x - 4 = 0 \qquad \text{\color{red}Set each factor equal to 0.}$$

$$2x = 3 \qquad\qquad 3x = 4 \qquad \text{\color{red}Solve the linear equations.}$$

$$x = \frac{3}{2} \qquad\qquad x = \frac{4}{3}$$

Check each solution.

 SELF CHECK 7   Solve: $6x^3 + 7x^2 = 5x$

Everyday connections
Selling Calendars

A bookshop is selling calendars at a price of \$4 each. At this price, the store can sell 12 calendars per day. The manager estimates that for each \$1 increase in the selling price, the store will sell 3 fewer calendars per day. Each calendar costs the store \$2. We can represent the store's total daily profit from calendar sales by the function $p(x) = -3x^2 + 30x - 48$, where x represents the selling price, in dollars, of a calendar. Find the selling price at which the profit $p(x)$ equals zero.

SELF CHECK ANSWERS   **1.** $0, -2$ **2.** $\frac{8}{3}, -\frac{8}{3}$ **3.** $3, -6$ **4.** $2, -\frac{1}{3}$ **5.** $6, -2$ **6.** $0, 2, -1$ **7.** $0, \frac{1}{2}, -\frac{5}{3}$

NOW TRY THIS

Solve each equation.

1. $8x^2 - 8 = 0$

2. $x^2 = 3x$

3. $x(x + 10) = -25$

5.7   Exercises

WARM-UPS *Solve.*

1. $(x - 8)(x - 7) = 0$ **2.** $(x + 9)(x - 2) = 0$

3. $(x - 2)(x + 3) = 0$ **4.** $(x - 3)(x - 2) = 0$

5. $(x - 4)(x + 1) = 0$ **6.** $(x + 5)(x + 2) = 0$

7. $(2x - 5)(3x + 6) = 0$ **8.** $(3x - 4)(x + 1) = 0$

REVIEW *Simplify each expression and write all results without using negative exponents.*

9. $u^3 u^2 u^4$

10. $\dfrac{y^6}{y^8}$

11. $\dfrac{a^3 b^4}{a^2 b^5}$

12. $(-2x^6)^0$

VOCABULARY AND CONCEPTS *Fill in the blanks.*

13. An equation of the form $ax^2 + bx + c = 0$, where $a \neq 0$, is called a _____ equation.

14. The property "If $ab = 0$, then $a = \_$ or $b = \_$" is called the _____ property.

15. A quadratic equation contains a _____-degree polynomial in one variable.

16. If the product of three factors is 0, then at least one of the numbers must be $\_$.

GUIDED PRACTICE *Solve. SEE EXAMPLE 1. (OBJECTIVE 1)*

17. $x^2 + 7x = 0$ **18.** $x^2 - 12x = 0$

19. $x^2 - 2x + 1 = 0$ **20.** $x^2 + x - 20 = 0$

21. $x^2 - 3x = 0$ **22.** $x^2 + 5x = 0$

23. $5x^2 + 7x = 0$ **24.** $2x^2 - 5x = 0$

25. $x^2 - 7x = 0$ **26.** $2x^2 + 10x = 0$

27. $3x^2 + 8x = 0$ **28.** $5x^2 - x = 0$

Solve. SEE EXAMPLE 2. (OBJECTIVE 1)

29. $x^2 - 25 = 0$ **30.** $x^2 - 36 = 0$

31. $9y^2 - 4 = 0$ **32.** $16z^2 - 25 = 0$

Solve. SEE EXAMPLE 3. (OBJECTIVE 1)

33. $x^2 - 13x + 12 = 0$ **34.** $x^2 + 7x + 6 = 0$

35. $x^2 - 2x - 15 = 0$ **36.** $x^2 + x - 20 = 0$

37. $x^2 - 3x - 18 = 0$ **38.** $x^2 + 3x - 10 = 0$

39. $x^2 - x - 20 = 0$ **40.** $x^2 - 10x + 24 = 0$

Solve. SEE EXAMPLE 4. (OBJECTIVE 1)

41. $6x^2 + x = 2$ **42.** $12x^2 + 5x = 3$

43. $2x^2 - 5x = -2$ **44.** $5p^2 - 6p = -1$

45. $x^2 = 49$ **46.** $z^2 = 25$

47. $4x^2 = 81$ **48.** $9y^2 = 64$

Solve. SEE EXAMPLE 5. (OBJECTIVE 1)

49. $x(6x + 5) = 6$ **50.** $x(2x - 3) = 14$

51. $(x + 1)(8x + 1) = 18x$ **52.** $4x(3x + 2) = x + 12$

Solve. SEE EXAMPLE 6. (OBJECTIVE 2)

53. $(x + 4)(x - 5)(x - 7) = 0$

54. $(x + 2)(x + 3)(x - 4) = 0$

55. $(x - 1)(x^2 + 5x + 6) = 0$

56. $(x - 2)(x^2 - 8x + 7) = 0$

57. $x^3 + 3x^2 + 2x = 0$ **58.** $x^3 - 7x^2 + 10x = 0$

59. $x^3 - 27x - 6x^2 = 0$ **60.** $x^3 - 22x - 9x^2 = 0$

Solve. SEE EXAMPLE 7. (OBJECTIVE 2)

61. $6x^3 + 20x^2 = -6x$ **62.** $2x^3 - 2x^2 = 4x$

63. $x^3 + 7x^2 = x^2 - 9x$ **64.** $x^3 + 10x^2 = 2x^2 - 16x$

ADDITIONAL PRACTICE *Solve.*

65. $x^2 - 4x = 0$ **66.** $15x^2 - 20x = 0$

67. $9x^2 + 5x = 0$ **68.** $5x^2 + x = 0$

69. $(x + 3)(x^2 + 2x - 15) = 0$

70. $(x + 4)(x^2 - 2x - 15)$

71. $x^2 - 4x - 21 = 0$ **72.** $x^2 + 2x - 15 = 0$

73. $2y - 8 = -y^2$ **74.** $-3y + 18 = y^2$

75. $(p^2 - 81)(p + 2) = 0$ **76.** $(4q^2 - 49)(q - 7) = 0$

77. $15x^2 - 2 = 7x$ **78.** $8x^2 + 10x = 3$

79. $x^2 + 8 - 9x = 0$ **80.** $45 + x^2 - 14x = 0$

81. $a^2 + 8a = -15$ **82.** $a^2 - a = 56$

83. $3x^2 - 8x = 3$ **84.** $2x^2 - 11x = 21$

85. $2x^2 + x - 3 = 0$ **86.** $6q^2 - 5q + 1 = 0$

87. $14m^2 + 23m + 3 = 0$ **88.** $35n^2 - 34n + 8 = 0$

89. $(x + 2)(x^2 + x - 20) = 0$
90. $(x + 1)(x^2 - 9x + 8) = 0$
91. $z^2 - 81 = 0$ **92.** $x^2 - 16 = 0$
93. $4x^2 - 1 = 0$ **94.** $9y^2 - 1 = 0$
95. $x^3 + 1.3x^2 - 0.3x = 0$ **96.** $2.4x^3 - x^2 - 0.4x = 0$

98. Explain the error in this solution.

$$5x^2 + 2x = 10$$
$$x(5x + 2) = 10$$
$$x = 10 \quad \text{or} \quad 5x + 2 = 10$$
$$5x = 8$$
$$x = \frac{8}{5}$$

WRITING ABOUT MATH

97. If the product of several numbers is 0, at least one of the numbers is 0. Explain why.

SOMETHING TO THINK ABOUT

99. Solve in two ways: $3a^2 + 9a - 2a - 6 = 0$
100. Solve in two ways: $p^2 - 2p + p - 2 = 0$

Section 5.8   Solving Applications

Objectives

1. Solve an integer application using a quadratic equation.
2. Solve a motion application using a quadratic equation.
3. Solve a geometric application using a quadratic equation.

Getting Ready

1. One side of a square is s inches long. Find an expression that represents its area.

2. The length of a rectangle is 4 centimeters more than twice the width. If w represents the width, find an expression that represents the length.

3. If x represents the smaller of two consecutive integers, find an expression that represents their product.

4. The length of a rectangle is 3 inches greater than the width. If w represents the width of the rectangle, find an expression that represents the area.

Finally, we can use the methods for solving quadratic equations discussed in the previous section to solve applications.

 Solve an integer application using a quadratic equation.

EXAMPLE 1   One integer is 5 less than another and their product is 84. Find the integers.

Analyze the problem   We are asked to find two integers. Let x represent the larger number. Then $x - 5$ represents the smaller number.

Form an equation   We know that the product of the integers is 84. Since a product refers to multiplication, we can form the equation $x(x - 5) = 84$.

Solve the equation   To solve the equation, we proceed as follows.

$$x(x - 5) = 84$$
$$x^2 - 5x = 84 \qquad \text{Use the distributive property to remove parentheses.}$$
$$x^2 - 5x - 84 = 0 \qquad \text{Subtract 84 from both sides.}$$
$$(x - 12)(x + 7) = 0 \qquad \text{Factor.}$$
$$x - 12 = 0 \quad \text{or} \quad x + 7 = 0 \qquad \text{Apply the zero-factor property.}$$
$$x = 12 \qquad \qquad x = -7 \qquad \text{Solve each linear equation.}$$

We have two different values for the first integer,

$$x = 12 \quad \text{or} \quad x = -7$$

and two different values for the second integer,

$$x - 5 = 7 \quad \text{or} \quad x - 5 = -12$$

State the conclusion   There are two pairs of integers: 12 and 7, and -7 and -12.

Check the result   The number 7 is five less than 12 and $12 \cdot 7 = 84$. The number -12 is five less than -7 and $-7 \cdot -12 = 84$. Both pairs of integers check.

 SELF CHECK 1   One integer is 7 more than another and their product is 60. Find the integers.

COMMENT   In Example 1, we could have let x represent the smaller number, in which case the larger number would be described as $(x + 5)$. The results would be the same.

2   Solve a motion application using a quadratic equation.

EXAMPLE 2   **FLYING OBJECTS**   If an object is launched straight up into the air with an initial velocity of 112 feet per second, its height after t seconds is given by the formula

$$h = 112t - 16t^2$$

where h represents the height of the object in feet. After this object has been launched, in how many seconds will it hit the ground?

Analyze the problem   We are asked to find the number of seconds it will take for an object to hit the ground. When the object is launched, it will go up and then come down. When it hits the ground, its height will be 0. So, we let $h = 0$.

Form an equation   If we substitute 0 for h in the formula $h = 112t - 16t^2$, the new equation will be $0 = 112t - 16t^2$ and we will solve for t.

$$\boldsymbol{h} = 112t - 16t^2$$
$$\boldsymbol{0} = 112t - 16t^2$$

Solve the equation   We solve the equation as follows.

$$0 = 112t - 16t^2$$
$$0 = 16t(7 - t) \qquad \text{Factor out } 16t, \text{ the GCF.}$$
$$16t = 0 \quad \text{or} \quad 7 - t = 0 \qquad \text{Set each factor equal to 0.}$$
$$t = 0 \qquad \qquad t = 7 \qquad \text{Solve each linear equation.}$$

When $t = 0$, the object's height above the ground is 0 feet, because it has not been released. When $t = 7$, the height is again 0 feet. The object has hit the ground.

State the conclusion   The object hits the ground in 7 seconds.

Check the result   When $t = 7$,

$$h = 112(\mathbf{7}) - 16(\mathbf{7})^2$$
$$= 184 - 16(49)$$
$$= 0$$

Since the height is 0 feet, the object has hit the ground after 7 seconds.

SELF CHECK 2   If this object is launched with an initial velocity of 96 feet per second, how many seconds will it take to hit the ground?

---

3   ## Solve a geometric application using a quadratic equation.

Recall that the area of a rectangle is given by the formula

$$A = lw$$

where A represents the area, l the length, and w the width of the rectangle. The perimeter of a rectangle is given by the formula

$$P = 2l + 2w$$

where P represents the perimeter of the rectangle, l the length, and w the width.

EXAMPLE 3   **RECTANGLES**   Assume that the rectangle in Figure 5-1 has an area of 52 square centimeters and that its length is 1 centimeter more than 3 times its width. Find the perimeter of the rectangle.

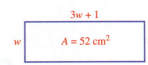

Figure 5-1

Analyze the problem   We are asked to find the perimeter of the rectangle. To do so, we must know both the length and the width. If we let w represent the width of the rectangle in centimeters, then $(3w + 1)$ represents its length in centimeters.

Form and solve an equation   We are given that the area of the rectangle is 52 square centimeters. We can use this fact to find the values of the width and length and then find the perimeter. To find the width, we can substitute 52 for A and $(3w + 1)$ for l in the formula $A = lw$ and solve for w.

$$\mathbf{A} = \mathbf{l}w$$
$$\mathbf{52} = (\mathbf{3w + 1})w$$

$$52 = 3w^2 + w \qquad \text{\color{red}Use the distributive property}$$
$$\text{\color{red}to remove parentheses.}$$

$$0 = 3w^2 + w - 52 \qquad \text{\color{red}Subtract 52 from both sides.}$$
$$0 = (3w + 13)(w - 4) \qquad \text{\color{red}Factor.}$$

$$3w + 13 = 0 \qquad \text{or} \qquad w - 4 = 0 \qquad \text{\color{red}Apply the zero-factor property.}$$
$$3w = -13 \qquad \qquad \qquad w = 4 \qquad \text{\color{red}Solve each linear equation.}$$
$$w = -\frac{13}{3}$$

Because the width of a rectangle cannot be negative, we discard the result $w = -\frac{13}{3}$. Thus, the width of the rectangle is 4 centimeters, and the length is given by

$$3\mathbf{w} + 1 = 3(\mathbf{4}) + 1$$
$$= 12 + 1$$
$$= 13$$

The dimensions of the rectangle are 4 centimeters by 13 centimeters. We find the perimeter by substituting 13 for l and 4 for w in the formula for the perimeter.

$$P = 2l + 2w$$
$$= 2(13) + 2(4)$$
$$= 26 + 8$$
$$= 34$$

State the conclusion   The perimeter of the rectangle is 34 centimeters.

Check the result   A rectangle with dimensions of 13 centimeters by 4 centimeters does have an area of 52 square centimeters, and the length is 1 centimeter more than 3 times the width. A rectangle with these dimensions has a perimeter of 34 centimeters.

 SELF CHECK 3   If the rectangle has an area of 80 square centimeters, use the same relationship for the length and the width to find the perimeter.

EXAMPLE 4   **TRIANGLES**   The triangle in Figure 5-2 has an area of 10 square centimeters and a height that is 3 centimeters less than twice the length of its base. Find the length of the base and the height of the triangle.

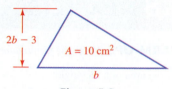

Figure 5-2

Analyze the problem   We are asked to find the length of the base and the height of the triangle, so we will let b represent the length of the base of the triangle in centimeters. Then $(2b - 3)$ represents the height in centimeters.

Form and solve an equation   Because the area is 10 square centimeters, we can substitute 10 for A and $(2b - 3)$ for h in the formula $A = \frac{1}{2}bh$ and solve for b.

$$A = \frac{1}{2}bh$$

$$10 = \frac{1}{2}b(2b - 3)$$

$$20 = b(2b - 3) \qquad \text{Multiply both sides by 2 to clear fractions.}$$
$$20 = 2b^2 - 3b \qquad \text{Use the distributive property to remove parentheses.}$$
$$0 = 2b^2 - 3b - 20 \qquad \text{Subtract 20 from both sides.}$$
$$0 = (2b + 5)(b - 4) \qquad \text{Factor.}$$

$$2b + 5 = 0 \qquad \text{or} \qquad b - 4 = 0 \qquad \text{Set both factors equal to 0.}$$
$$2b = -5 \qquad\qquad\quad b = 4 \qquad \text{Solve each linear equation.}$$
$$b = -\frac{5}{2}$$

Because a triangle cannot have a negative number for the length of its base, we discard the result $b = -\frac{5}{2}$. The length of the base of the triangle is 4 centimeters.

State the conclusion   Its height is $2(4) - 3$, or 5 centimeters.

Check the result   If the base of the triangle has a length of 4 centimeters and the height of the triangle is 5 centimeters, its height is 3 centimeters less than twice the length of its base. Its area is 10 square centimeters.

$$A = \frac{1}{2}bh = \frac{1}{2}(4)(5) = 2(5) = 10$$

 **SELF CHECK 4** If the triangle in the figure had an area of 52 square centimeters, use the same relation-ship for the height and base to find the length of the base and the height of the triangle.

 **SELF CHECK ANSWERS** **1.** 5 and 12; −5 and −12 **2.** It will hit the ground after 6 seconds. **3.** The perimeter of the rectangle is 42 cm. **4.** The length of the base is 8 cm and the height is 13 cm.

NOW TRY THIS

A cell phone is shaped like a rectangle and its longer edge is one inch shorter than twice its shorter edge. If the area of the phone is 10 square inches, find the dimensions of the phone.

5.8 Exercises

WARM-UPS *Write the formula for . . .*

1. the area of a rectangle.
2. the area of a triangle.
3. the area of a square.
4. the volume of a rectangular solid.
5. the perimeter of a rectangle.
6. the perimeter of a square.

REVIEW *Solve each equation.*

7. $-2(5x + 2) = 3(2 - 3x)$
8. $4(3a - 2) - 12 = 2a$
9. **Rectangles** A rectangle is 5 times as long as it is wide, and its perimeter is 132 feet. Find its area.
10. **Investing** A woman invested $15,000, part at 7% simple annual interest and part at 8% annual interest. If she receives $1,100 interest per year, how much did she invest at 7%?

VOCABULARY AND CONCEPTS *Fill in the blanks.*

11. The first step in the problem-solving process is to _____ the problem.
12. The last step in the problem-solving process is to _____ _____.

APPLICATIONS

13. **Integer** One integer is 4 more than another. Their product is 32. Find the integers.
14. **Integer** One integer is 5 less than 4 times another. Their product is 21. Find the integers.

15. **Integer** If 4 is added to the square of an integer, the result is 5 less than 10 times that integer. Find the integer(s).
16. **Integer** If 5 times the square of an integer is added to 3 times the integer, the result is 2. Find the integer.

An object has been launched straight up into the air. The formula $h = vt - 16t^2$ gives the height h of the object above the ground after t seconds when it is launched upward with an initial velocity v.

17. **Time of flight** After how many seconds will an object hit the ground if it was launched with a velocity of 144 feet per second?
18. **Time of flight** After how many seconds will an object hit the ground if it was launched with a velocity of 160 feet per second?
19. **Ballistics** If a cannonball is fired with an upward velocity of 224 feet per second, at what times will it be at a height of 640 feet?
20. **Ballistics** A cannonball's initial upward velocity is 128 feet per second. At what times will it be 192 feet above the ground?
21. **Exhibition diving** At a resort, tourists watch swim-mers dive from a cliff to the water 64 feet below. A diver's height h above the water t seconds after diving is given by $h = -16t^2 + 64$. How long does a dive last?
22. **Forensic medicine** The kinetic energy E of a moving object is given by $E = \frac{1}{2}mv^2$, where m is the mass of the object (in kilograms) and v is the object's velocity (in meters per second). Kinetic energy is measured in joules. By the damage done to a victim, a police pathologist determines that the energy of a 3-kilogram mass at impact was 54 joules. Find the velocity at impact.

In Exercises 23–24, note that in the triangle $y^2 = h^2 + x^2$. (Pythagorean theorem)

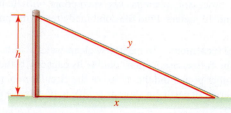

23. Ropes courses A camper slides down the cable of a high-adventure ropes course to the ground as shown in the illustration. At what height did the camper start his slide?

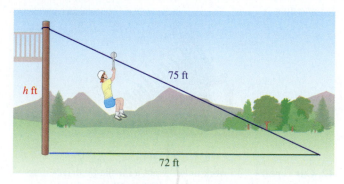

24. Ropes courses If the pole and the landing area discussed in Exercise 23 are 48 feet apart and the high end of the cable is 36 feet, how long is the cable?

25. Insulation The area of the rectangular slab of foam insulation is 36 square meters. Find the dimensions of the slab.

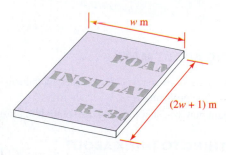

26. Shipping pallets The length of a rectangular shipping pallet is 2 feet less than 3 times its width. Its area is 21 square feet. Find the dimensions of the pallet.

27. Carpentry A rectangular room containing 143 square feet is 2 feet longer than it is wide. How long a crown molding is needed to trim the perimeter of the ceiling?

28. Designing tents The length of the base of the triangular sheet of canvas above the door of the tent shown is 2 feet more than twice its height. The area is 30 square feet. Find the height and the length of the base of the triangle.

29. Dimensions of a triangle The height of a triangle is 2 inches less than 5 times the length of its base. The area is 36 square inches. Find the length of the base and the height of the triangle.

30. Area of a triangle The base of a triangle is numerically 3 less than its area, and the height is numerically 6 less than its area. Find the area of the triangle.

31. Area of a triangle The length of the base and the height of a triangle are numerically equal. Their sum is 6 less than the number of units in the area of the triangle. Find the area of the triangle.

32. Dimensions of a parallelogram The formula for the area of a parallelogram is $A = bh$. The area of the parallelogram in the illustration is 200 square centimeters. If its base is twice its height, how long is the base?

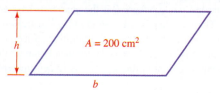

33. Swimming pool borders The owners of the rectangular swimming pool want to surround the pool with a crushed-stone border of uniform width. They have enough stone to cover 74 square meters. How wide should they make the border? (*Hint:* The area of the larger rectangle minus the area of the smaller is the area of the border.)

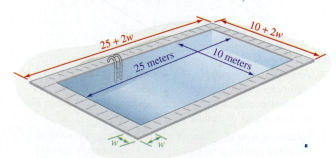

34. House construction   The formula for the area of a trapezoid is $A = \frac{h(B + b)}{2}$. The area of the trapezoidal truss in the illustration is 24 square meters. Find the height of the trapezoid if one base is 8 meters and the other base is the same as the height.

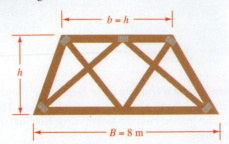

35. Volume of a solid   The volume of a rectangular solid is given by the formula $V = lwh$, where l is the length, w is the width, and h is the height. The volume of the rectangular solid in the illustration is 210 cubic centimeters. Find the width of the rectangular solid if its length is 10 centimeters and its height is 1 centimeter longer than twice its width.

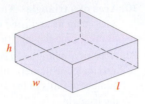

36. Volume of a pyramid   The volume of a pyramid is given by the formula $V = \frac{Bh}{3}$, where B is the area of its base and h is its height. The volume of the pyramid in the illustration is 192 cubic centimeters. Find the dimensions of its rectangular base if one edge of the base is 2 centimeters longer than the other, and the height of the pyramid is 12 centimeters.

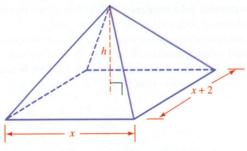

37. Volume of a pyramid   The volume of a pyramid is 84 cubic centimeters. Its height is 9 centimeters, and one side of its rectangular base is 3 centimeters shorter than the other. Find the dimensions of its base. (See Exercise 36.)

38. Volume of a solid   The volume of a rectangular solid is 180 cubic centimeters. Its height is 3 centimeters, and its width is 4 centimeters shorter than its length. Find the sum of its length and width. (See Exercise 35.)

39. Sewage treatment   In one step in waste treatment, sewage is exposed to air by placing it in circular aeration pools. One sewage processing plant has two such pools, with diameters of 38 and 44 meters. Find the combined area of the pools.

40. Sewage treatment   To meet new clean-water standards, the plant in Exercise 39 must double its capacity by building another pool. Find the radius of the circular pool that the engineering department should specify to double the plant's capacity.

In Exercises 41–42, $a^2 + b^2 = c^2$.

41. Tornado damage   The tree shown below was blown down in a tornado. Find x and the height of the tree when it was standing.

42. Car repairs   To work under a car, a mechanic drives it up steel ramps like the ones shown below. Find the length of each side of the ramp.

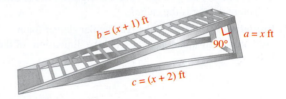

WRITING ABOUT MATH

43. Explain the steps you would use to set up and solve an application problem.

44. Explain how you should check the solution to an application.

SOMETHING TO THINK ABOUT

45. Here is an easy-sounding problem:

> *The length of a rectangle is 2 feet greater than the width, and the area is 18 square feet. Find the width of the rectangle.*

Set up the equation. Can you solve it? Why not?

46. Does the equation in Exercise 45 have a solution, even if you can't find it? If it does, find an estimate of the solution.

Projects

Because the length of each side of the largest square in Figure 5-3 is $x + y$, its area is $(x + y)^2$. This area is also the sum of four smaller areas, which illustrates the factorization

$$x^2 + 2xy + y^2 = (x + y)^2$$

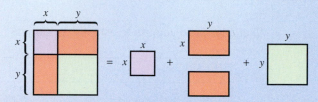

Figure 5-3

What factorization is illustrated by each of the following figures?

a.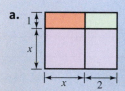

b.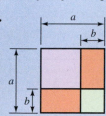

c.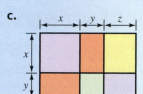

d.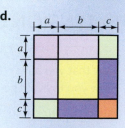

e. Factor the expression

$$a^2 + ac + 2a + ab + bc + 2b$$

and draw a figure that illustrates the factorization.

f. Verify the factorization

$$x^3 + 3x^2y + 3xy^2 + y^3 = (x + y)^3$$

Hint: Expand the right side:

$$(x + y)^3 = (x + y)(x + y)(x + y)$$

Then draw a figure that illustrates the factorization.

Reach for Success
EXTENSION OF ORGANIZING YOUR TIME

Now that you have created a calendar of all your responsibilities, let's look at how to establish *priorities*. This will allow you to avoid potential conflicts that could create unnecessary stress. Planning is the key to academic success!

Establish *your* criteria for dealing with scheduling conflicts. You cannot make a decision without knowing what criteria to use.

You might consider:	What criteria do you use to establish priorities?
• Financial: Is it going to cost a lot to do it?	1. _____
• Time: Do you have the time to do it?	2. _____
• Importance: Are there reasons to do it?	3. _____
• Risk: What happens if you do not do it or if you are late?	4. _____
• Ability to complete: Do you have the skills needed?	

Now, look at the two-week calendar below. Do you see any potential scheduling conflicts? _____

Sun 1	Mon 2	Tue 3	Wed 4	Thu 5	Fri 6	Sat 7
	English outline due	Dentist 1 p.m. Work	Work	Mathematics test	RJ's school play	Volunteer work Wedding anniversary
Sun 8	Mon 9	Tue 10	Wed 11	Thu 12	Fri 13	Sat 14
Sociology group project meeting 2 p.m.	Work Kenley's Dr. appt. 2:30 p.m.	Turn in history paper Math lab due	Read history chapters 13–15	Sociology group project due Work	English paper due Fred's birthday	Work

If you have to work on Tuesday and Wednesday the 3rd and 4th, when do you study for the mathematics test due on the 5th?

Your plan: _____

_____

As you can see, the second week is busy with many assignments. To avoid time conflicts, consider completing some of the assignments the first week rather than waiting until the second week. What other strategies could you use?

_____

_____

_____

Most instructors are willing to help you with study skills in addition to mathematical skills. Do not hesitate to ask them!

5 Review

SECTION 5.1   Factoring Out the Greatest Common Factor; Factoring by Grouping

DEFINITIONS AND CONCEPTS	EXAMPLES
A natural number is in **prime-factored form** if it is written as the product of prime-number factors.	$42 = \mathbf{6} \cdot 7 = \mathbf{2} \cdot \mathbf{3} \cdot 7$ $56 = \mathbf{8} \cdot 7 = \mathbf{2} \cdot \mathbf{4} \cdot 7 = \mathbf{2} \cdot \mathbf{2} \cdot \mathbf{2} \cdot 7 = 2^3 \cdot 7$
The **greatest common factor (GCF)** of several monomials is found by taking each common prime factor the fewest number of times it appears in any one monomial.	Find the GCF of $12x^3y$, $42x^2y^2$, and $32x^2y^3$. $\left.\begin{array}{l} 12x^3y = 2 \cdot \mathbf{2} \cdot 3 \cdot \mathbf{x} \cdot \mathbf{x} \cdot x \cdot \mathbf{y} \\ 42x^2y^2 = \mathbf{2} \cdot 3 \cdot 7 \cdot \mathbf{x} \cdot \mathbf{x} \cdot \mathbf{y} \cdot y \\ 32x^2y^3 = \mathbf{2} \cdot 2 \cdot 2 \cdot 2 \cdot 2 \cdot \mathbf{x} \cdot \mathbf{x} \cdot \mathbf{y} \cdot y \cdot y \end{array}\right\}$ GCF $= 2x^2y$
If the leading coefficient of a polynomial is negative, it is often useful to factor out -1.	Factor completely: $-x^3 + 3x^2 - 5$ $-x^3 + 3x^2 - 5$ $= (\mathbf{-1})x^3 + (\mathbf{-1})(-3x^2) + (\mathbf{-1})5$ $= \mathbf{-1}(x^3 - 3x^2 + 5)$ Factor out -1. $= -(x^3 - 3x^2 + 5)$ The coefficient of 1 need not be written.
If a polynomial has four terms, consider factoring it by grouping.	Factor completely: $x^2 + xy + 3x + 3y$ Factor x from $(x^2 + xy)$ and 3 from $(3x + 3y)$ and proceed as follows: $x^2 + xy + 3x + 3y = x(\mathbf{x + y}) + 3(\mathbf{x + y})$ $= (\mathbf{x + y})(x + 3)$ Factor out $(x + y)$.

REVIEW EXERCISES

Find the prime factorization of each number.

1. 24

2. 45

3. 96

4. 102

5. 87

6. 99

7. 2,050

8. 4,096

Factor.

9. $4x + 12y$

10. $5ax^2 + 15a$

11. $7x^2 + 14x$

12. $9x^2 - 3x$

13. $2x^3 + 4x^2 - 8x$

14. $-ax - ay + az$

15. $ax + ay - a$

16. $x^2yz + xy^2z$

17. $(a - b)x + (a - b)y$

18. $(x + y)^2 + (x + y)$

19. $2x^2(x + 2) + 6x(x + 2)$

20. $5x(a + b)^2 - 10x(a + b)$

21. $3p + 9q + ap + 3aq$

22. $ar - 2as + 7r - 14s$

23. $x^2 + ax + bx + ab$

24. $xy + 2x - 2y - 4$

25. $xa + yb + ya + xb$

26. $x^3 - 4x^2 + 3x - 12$

SECTION 5.2 Factoring the Difference of Two Squares

DEFINITIONS AND CONCEPTS	EXAMPLES
To factor the difference of two squares, use the pattern $x^2 - y^2 = (x + y)(x - y)$	$x^2 - 36 = x^2 - 6^2 = (x + 6)(x - 6)$
If we limited to rational coefficients, binomials that are the sum of two squares cannot be factored over the real numbers unless they contain a GCF.	$9x^2 + 36 = 9(x^2 + 4)$ Factor out 9, the GCF. $(x^2 + 4)$ does not factor.

REVIEW EXERCISES

Factor.

27. $x^2 - 25$

28. $x^2 y^2 - 16$

29. $(x + 2)^2 - y^2$

30. $z^2 - (x + y)^2$

31. $2x^2 y + 18y^3$

32. $(x + y)^2 - z^2$

SECTIONS 5.3–5.4 Factoring Trinomials

DEFINITIONS AND CONCEPTS	EXAMPLES
Factor trinomials using these steps (trial and error): **1.** Write the trinomial with the exponents of one variable in descending order. **2.** Factor out any greatest common factor (including -1 if that is necessary to make the coefficient of the first term positive). **3.** If the sign of the third term is $+$, the signs between the terms of the binomial factors are the same as the sign of the trinomial's second term. If the sign of the third term is $-$, the signs between the terms of the binomials are opposites. **4.** Try various combinations of first terms and last terms until you find the one that works. If none work, the trinomial is prime. **5.** Check by multiplication.	Factor completely: $12 - x^2 - x$ **1.** We will begin by writing the exponents of x in descending order. $12 - x^2 - x = -x^2 - x + 12$ **2.** Factor out -1 to get $= -(x^2 + x - 12)$ **3.** Since the sign of the third term is $-$, the signs between the binomials are opposites. **4.** We find the combination that works. $= -(x + 4)(x - 3)$ **5.** Since $-(x + 4)(x - 3) = 12 - x^2 - x$, the factorization is correct.
Factor trinomials by grouping (*ac* method) using these steps: **1.** Write the trinomial in $ax^2 + bx + c$ form. **2.** Find the key number ac. **3.** Find two factors of the key number whose sum is b. **4.** Use the factors as the coefficients of two terms whose sum is bx to replace bx in the trinomial. **5.** Factor the polynomial by grouping.	$2x^2 - x - 10$ $a = 2, b = -1, c = -10$ Determine the key number $ac = 2(-10) = -20$. $= 2x^2 \mathbf{- 5x + 4x} - 10$ Replace $-x$ with $-5x + 4x$. (The two factors whose product is -20 and difference is -1 are -5 and 4.) $= x(\mathbf{2x - 5}) + 2(\mathbf{2x - 5})$ Factor by grouping. $= (\mathbf{2x - 5})(x + 2)$

REVIEW EXERCISES

Factor.

33. $x^2 + 7x + 10$ **34.** $x^2 - 8x + 15$

35. $x^2 + 2x - 24$ **36.** $x^2 - 4x - 12$

37. $2x^2 - 5x - 3$ **38.** $3x^2 - 14x - 5$

39. $15x^2 + x - 2$ **40.** $6x^2 + 3x - 3$

41. $6x^3 + 17x^2 - 3x$ **42.** $4x^3 - 5x^2 - 6x$

43. $12x - 4x^3 - 2x^2$ **44.** $-4a^3 + 4a^2b + 24ab^2$

SECTION 5.5   Factoring the Sum and Difference of Two Cubes

DEFINITIONS AND CONCEPTS	EXAMPLES
The sum and difference of two cubes factor according to the patterns $$x^3 + y^3 = (x + y)(x^2 - xy + y^2)$$ $$x^3 - y^3 = (x - y)(x^2 + xy + y^2)$$	$$x^3 + 64 = x^3 + 4^3 = (x + 4)(x^2 - x \cdot 4 + 4^2)$$ $$= (x + 4)(x^2 - 4x + 16)$$ $$x^3 - 64 = x^3 - 4^3 = (x - 4)(x^2 - x(-4) + (-4)^2)$$ $$= (x - 4)(x^2 + 4x + 16)$$

REVIEW EXERCISES

Factor.

45. $c^3 - 125$

46. $d^3 + 8$

47. $2x^3 + 54$

48. $2ab^4 - 2ab$

SECTION 5.6   Summary of Factoring Techniques

DEFINITIONS AND CONCEPTS	EXAMPLES
Factoring polynomials: **1.** Write the polynomial in descending powers of one variable. **2.** Factor out all common factors. **3.** If an expression has two terms, check to see if it is **a.** the **difference of two squares**: $$a^2 - b^2 = (a + b)(a - b)$$ **b.** the **sum of two cubes**: $$a^3 + b^3 = (a + b)(a^2 - ab + b^2)$$ **c.** the **difference of two cubes**: $$a^3 - b^3 = (a - b)(a^2 + ab + b^2)$$	Factor: $-2x^6 + 2y^6$ Factor out the common factor of -2. $$-2x^6 + 2y^6 = -2 \cdot x^6 - (-2)y^6 = -2(x^6 - y^6)$$ Identify $x^6 - y^6$ as the difference of two squares and factor it: $$-2(x^6 - y^6) = -2[(x^3)^2 - (y^3)^2] = -2(x^3 + y^3)(x^3 - y^3)$$ Then identify $(x^3 + y^3)$ as the sum of two cubes and $(x^3 - y^3)$ as the difference of two cubes and factor each binomial: $$-2(x^6 - y^6) = -2[(x^3)^2 - (y^3)^2] = -2(x^3 + y^3)(x^3 - y^3)$$ $$= -2(x + y)(x^2 - xy + y^2)(x - y)(x^2 + xy + y^2)$$

4. If an expression has three terms, check to see if it is a **perfect-square trinomial square**:

$$a^2 + 2ab + b^2 = (a + b)(a + b)$$
$$a^2 - 2ab + b^2 = (a - b)(a - b)$$

If the trinomial is not a perfect-square trinomial, attempt to factor it as a **general trinomial**.

5. If an expression has four or more terms, factor it by **grouping**.

6. Continue factoring until each individual factor is prime, except possibly a monomial factor.

7. If the polynomial does not factor, the polynomial is **prime** over the set of rational numbers.

8. Check the results by multiplying.

Factor: $4x^2 - 12x + 9$
This is a perfect-square trinomial because it has the form $(2x)^2 - 2(2x)(3) + (3)^2$. It factors as $(2x - 3)^2$.

Factor: $ax - bx + ay - by$
Since the expression has four terms, use factoring by grouping:
$$ax - bx + ay - by = x(a - b) + y(a - b)$$
$$= (a - b)(x + y)$$

REVIEW EXERCISES

Factor.

49. $3x^2y - xy^2 - 6xy + 2y^2$
50. $5x^2 - 5x - 30$
51. $2a^2x + 2abx + a^3 + a^2b$

52. $2x^2 - 8x - 11$
53. $x^2 - 9 + ax + 3a$
54. $10x^3 - 80y^3$

SECTION 5.7   Solving Equations by Factoring

DEFINITIONS AND CONCEPTS	EXAMPLES
A **quadratic equation** is an equation of the form $ax^2 + bx + c = 0$, where a, b, and c are real numbers and $a \neq 0$.	$2x^2 + 5x = 8$ and $x^2 - 5x = 0$ are quadratic equations.

Zero-factor property:

If a and b represent two real numbers and if $ab = 0$, then $a = 0$ or $b = 0$.

The zero-factor property can be extended to any number of factors.

To solve the quadratic equation $x^2 - 3x = 4$, proceed as follows:

$$x^2 - 3x = 4$$
$$x^2 - 3x - 4 = 0 \qquad \text{Subtract 4 from both sides to write the equation in quadratic form.}$$
$$(x + 1)(x - 4) = 0 \qquad \text{Factor } x^2 - 3x - 4.$$
$$x + 1 = 0 \quad \text{or} \quad x - 4 = 0 \qquad \text{Apply the zero-factor property.}$$
$$x = -1 \quad \mid \quad x = 4 \qquad \text{Solve each linear equation.}$$

REVIEW EXERCISES

Solve.

55. $x^2 + 5x = 0$
56. $2x^2 - 6x = 0$
57. $3x^2 = 2x$
58. $5x^2 + 25x = 0$
59. $y^2 - 49 = 0$
60. $x^2 - 25 = 0$
61. $a^2 - 9a + 20 = 0$
62. $(x - 1)(x + 4) = 6$
63. $2x - x^2 + 24 = 0$
64. $16 + x^2 - 10x = 0$

65. $2x^2 - 5x - 3 = 0$
66. $2x^2 + x - 3 = 0$
67. $16x^2 = 9$
68. $9x^2 = 4$
69. $x^3 - 7x^2 + 12x = 0$
70. $x^3 + 5x^2 + 6x = 0$
71. $2x^3 + 5x^2 = 3x$
72. $3x^3 - 2x = x^2$

SECTION 5.8   Solving Applications

DEFINITIONS AND CONCEPTS	EXAMPLES
Use the methods for solving quadratic equations discussed in Section 5.7 to solve applications.	Assume that the area of a rectangle is 240 square inches and that its length is 4 inches less than twice its width. Find the perimeter of the rectangle.
1. Analyze the problem.	Let w represent the width of the rectangle in inches. Then $(2w - 4)$ represents its length. We can find the length and width by substituting into the formula for the area: $A = l \cdot w$.
2. Form an equation.	
3. Solve the equation.	$240 = (2w - 4)w$

$$240 = 2w^2 - 4w \qquad \text{Use the distributive property to remove parentheses.}$$
$$0 = 2w^2 - 4w - 240 \qquad \text{Subtract 240 from both sides.}$$
$$0 = w^2 - 2w - 120 \qquad \text{Divide both sides by 2.}$$
$$0 = (w - 12)(w + 10) \qquad \text{Factor.}$$
$$w - 12 = 0 \quad \text{or} \quad w + 10 = 0 \qquad \text{Apply the zero-factor property.}$$
$$w = 12 \qquad \qquad w = -10 \qquad \text{Solve each linear equation.}$$

Because the width cannot be negative, we discard the result $w = -10$. Thus, the width of the rectangle is 12 inches, and the length is given by

$$2w - 4 = 2(12) - 4$$
$$= 24 - 4$$
$$= 20$$

The dimensions of the rectangle are 12 in. by 20 in. We find the perimeter by substituting 20 for l and 12 for w in the formula for perimeter.

$$P = 2l + 2w = 2(20) + 2(12) = 40 + 24 = 64$$

4. State the conclusion.   The perimeter of the rectangle is 64 inches.

5. Check the result.   A rectangle with dimensions 12 in. by 20 in. does have an area of 240 square inches, and the length is 4 inches less than twice the width.

REVIEW EXERCISES

73. Numbers   The sum of two numbers is 12, and their product is 35. Find the numbers.

74. Numbers   If 3 times the square of a positive number is added to 5 times the number, the result is 2. Find the number.

75. Dimensions of a rectangle   A rectangle is 2 feet longer than it is wide, and its area is 48 square feet. Find its dimensions.

76. Gardening   A rectangular flower bed is 3 feet longer than twice its width, and its area is 27 square feet. Find its dimensions.

77. Geometry   A rectangle is 3 feet longer than it is wide. Its area is numerically equal to its perimeter. Find its dimensions.

78. Geometry   A triangle has a height 1 foot longer than its base. If its area is 36 square feet, find its height.

5 Test

1. Find the prime factorization of 120.
2. Find the prime factorization of 108.

Factor.

3. $60ab^2c^3 + 30a^3b^2c - 25a$
4. $3x^2(a + b) - 6xy(a + b)$
5. $ax + ay + bx + by$
6. $x^2 - 64$
7. $2a^2 - 32b^2$
8. $16x^4 - 81y^4$
9. $x^2 + 5x - 6$
10. $x^2 - 9x - 22$
11. $-x^2 - 10xy - 9y^2$
12. $6x^2 - 30xy + 24y^2$
13. $3x^2 + 13x + 4$
14. $2a^2 + 5a - 12$
15. $2x^2 + 3x - 1$
16. $12 - 25x + 12x^2$
17. $12a^2 + 6ab - 36b^2$
18. $x^3 - 8y^3$

19. $216 + 8a^3$
20. $x^9z^3 - y^3z^6$

Solve each equation.

21. $x^2 = -10x$
22. $2x^2 + 5x + 3 = 0$
23. $16y^2 - 64 = 0$
24. $-3(y - 6) + 2 = y^2 + 2$
25. $10x^2 - 13x = 9$
26. $10x^2 - x = 9$
27. $10x^2 + 43x = 9$
28. $10x^2 - 89x = 9$

29. **Cannon fire** A cannonball is fired straight up into the air with a velocity of 192 feet per second. In how many seconds will it hit the ground? (Its height above the ground is given by the formula $h = vt - 16t^2$, where v is the velocity and t is the time in seconds.)

30. **Base of a triangle** The base of a triangle with an area of 40 square meters is 2 meters longer than it is high. Find the base of the triangle.

Rational Expressions and Equations; Ratio and Proportion

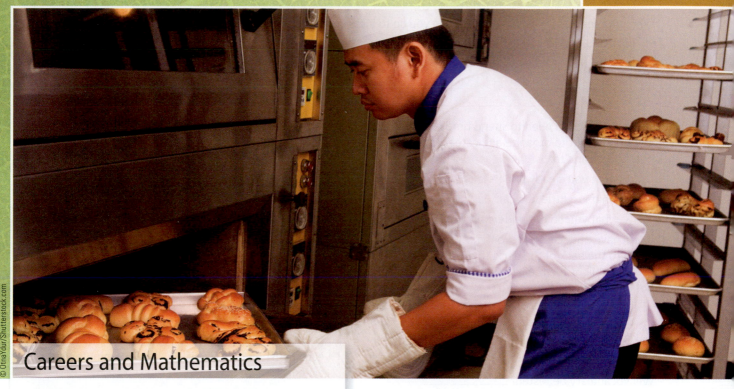

© OtnaYdur/Shutterstock.com

Careers and Mathematics

FOOD-PROCESSING OCCUPATIONS; BAKER

Bakers mix and bake ingredients in accordance to recipes to produce varying quantities of breads, pastries, and other baked goods. Bakers commonly are employed in grocery stores and specialty shops that produce small quantities of baked goods. In manufacturing, bakers produce goods in large quantities. Bakers make up 21% of all food-processing workers. Training varies widely among the food-processing occupations. Bakers often start as apprentices or as trainees.

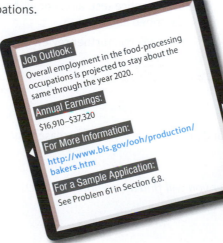

Job Outlook:
Overall employment in the food-processing occupations is projected to stay about the same through the year 2020.

Annual Earnings:
$16,910–$37,320

For More Information:
http://www.bls.gov/ooh/production/bakers.htm

For a Sample Application:
See Problem 61 in Section 6.8.

REACH FOR SUCCESS

6.1 Simplifying Rational Expressions

6.2 Multiplying and Dividing Rational Expressions

6.3 Adding and Subtracting Rational Expressions

6.4 Simplifying Complex Fractions

6.5 Solving Equations That Contain Rational Expressions

6.6 Solving Applications of Equations That Contain Rational Expressions

6.7 Ratios

6.8 Proportions and Similar Triangles

■ *Projects*
REACH FOR SUCCESS EXTENSION
CHAPTER REVIEW
CHAPTER TEST
CUMULATIVE REVIEW

In this chapter

In Chapter 6, we will discuss rational expressions, the fractions of algebra. After learning how to simplify, add, subtract, multiply, and divide them, we will solve equations and applications that involve rational expressions. We then will conclude by discussing ratio and proportion.

Reach for Success Preparing for a Test

Preparation is necessary to be successful in any event. You have practiced and worked out the kinks. You are well rested, and you are motivated. Now is the time. Are you ready?

When preparing for class tests, what can you do to increase your confidence and feel less anxious?

© bikeriderlondon/Shutterstock.com

To know what you need to spend time reviewing, you might want to organize and categorize your notes according to:

- what you clearly understand.
- what is a little unclear.
- what you do not understand at all.

If you have note cards, move the cards that you understand to the back of the deck. If you do not have note cards, highlight the topics in your notebook that you do not understand. Continue to work on these. Seek help.

List the objectives that are giving you the most difficulty.

1. _____
2. _____
3. _____
4. _____
5. _____
6. _____
7. _____
8. _____

How many questions will be on the exam?
_____

Will vocabulary be tested? _____

Will the formulas be provided? _____

Determine how much time you should allow yourself for each question on a test if you have 60 minutes to take a 20-question test. _____

Write down the formulas that you will need for the exam.

_____

Use the Chapter Review at the end of the chapter to prepare for the exam. After completing the review, take the Chapter Test without notes, grade it, and mark the objectives you missed.

If the test includes completion, multiple choice, true/false, essay, or matching, be certain to practice those formats.

Did your instructor provide a written or computerized review for the test? _____ If so, work it as if it were the test—in a single sitting and without notes. How did you do? _____

Note any objectives you missed.

_____

If your instructor did not have a review, make up your own practice test, alone or with a group. Then, take it as described above.

How will you further review any objectives you missed?

_____

A Successful Study Strategy . . .

Begin your review early and practice every day to increase your confidence and thus reduce anxiety.

At the end of the chapter you will find an additional exercise to guide you in planning for a successful semester.

Section 6.1

Simplifying Rational Expressions

Objectives

1. Identify all values of a variable for which a rational expression is undefined.
2. Find the domain of a rational function.
3. Write a rational expression in simplest form.
4. Simplify a rational expression containing factors that are negatives.

Vocabulary

rational expression   simplest form

Getting Ready

Simplify.

1. $\dfrac{12}{16}$ **2.** $\dfrac{16}{8}$ **3.** $\dfrac{25}{55}$ **4.** $\dfrac{36}{72}$

Fractions such as $\frac{1}{2}$ and $\frac{3}{4}$ that are the quotient of two integers are *rational numbers*. Expressions such as

$$\frac{a}{a+2} \qquad \text{and} \qquad \frac{5x^2+3}{x^2+x-12}$$

where the denominators and/or numerators are polynomials, are called **rational expressions**. Since rational expressions indicate division, we must exclude any values of the variable that will make the denominator equal to 0. For example, a cannot be -2 in the rational expression

$$\frac{a}{a+2}$$

because the denominator will be 0:

$$\frac{a}{a+2} = \frac{-2}{-2+2} = \frac{-2}{0}$$

When the denominator of a rational expression is 0, we say that the expression is undefined.

1 Identify all values of a variable for which a rational expression is undefined.

EXAMPLE 1   Identify all values of x such that the following rational expression is undefined.

$$\frac{5x^2+3}{x^2+x-12}$$

Solution   To find the values of x that make the rational expression undefined, we set its denominator equal to 0 and solve for x.

$$x^2 + x - 12 = 0$$
$$(x + 4)(x - 3) = 0 \qquad \text{Factor the trinomial.}$$
$$x + 4 = 0 \quad \text{or} \quad x - 3 = 0 \quad \text{Apply the zero-factor property.}$$
$$x = -4 \qquad \qquad x = 3 \quad \text{Solve each equation.}$$

We can check by substituting 3 and -4 for x and verifying that these values make the denominator of the rational expression equal to 0.

$$\textbf{\textit{For x = 3}} \qquad\qquad\qquad \textbf{\textit{For x = -4}}$$

$$\frac{5x^2 + 3}{x^2 + x - 12} = \frac{5(3)^2 + 3}{3^2 + 3 - 12} \qquad\qquad \frac{5x^2 + 3}{x^2 + x - 12} = \frac{5(-4)^2 + 3}{(-4)^2 + (-4) - 12}$$

$$= \frac{5(9) + 3}{9 + 3 - 12} \qquad\qquad\qquad = \frac{5(16) + 3}{16 - 4 - 12}$$

$$= \frac{45 + 3}{12 - 12} \qquad\qquad\qquad = \frac{80 + 3}{12 - 12}$$

$$= \frac{48}{0} \qquad\qquad\qquad\qquad = \frac{83}{0}$$

Since the denominator is 0 when $x = 3$ or $x = -4$, the rational expression $\dfrac{5x^2 + 3}{x^2 + x - 12}$ is undefined at these values.

SELF CHECK 1   Find all values of x such that the following rational expression is undefined.

$$\frac{3x^2 - 2}{x^2 - 2x - 3}$$

---

2   Find the domain of a rational function.

The same process is used to find the domain of a rational function. Recall that the domain is the set of all values that can be substituted for the variable. We learned that the domain for the linear function $f(x) = x + 3$ was $\mathbb{R}$, all real numbers.

EXAMPLE 2   Find the domain for the rational function:

$$f(x) = \frac{5x^2 + 3}{x^2 + x - 12}$$

Solution   We follow the steps in Example 1 to find the values that would make the expression undefined. Those values are 3 and -4. To write the domain, we must identify all values that can be substituted. Therefore, in set-builder notation, the domain of the rational function

$$f(x) = \frac{5x^2 + 3}{x^2 + x - 12}$$

is $\{x \mid x \in \mathbb{R}, x \neq 3, -4\}$. This is read as "the set of all values of x such that x is a real number where $x \neq 3$ and $x \neq -4$."

SELF CHECK 2   Find the domain for the rational function: $f(x) = \dfrac{6x + 1}{x^2 + 5x - 6}$

3 Write a rational expression in simplest form.

We have seen that a fraction can be simplified by dividing out common factors shared by its numerator and denominator. For example,

$$\frac{18}{30} = \frac{3\cdot 6}{5\cdot 6} = \frac{3\cdot \overset{1}{\cancel{6}}}{5\cdot \cancel{6}} = \frac{3}{5} \qquad -\frac{6}{15} = -\frac{3\cdot 2}{3\cdot 5} = -\frac{\overset{1}{\cancel{3}}\cdot 2}{\cancel{3}\cdot 5} = -\frac{2}{5}$$

These examples illustrate the *fundamental property of fractions*, first discussed in Chapter 1.

THE FUNDAMENTAL PROPERTY OF FRACTIONS

If a, b, and x are real numbers, then

$$\frac{a\cdot x}{b\cdot x} = \frac{a}{b} \quad (b \neq 0 \text{ and } x \neq 0)$$

Since rational expressions are fractions, we can use the fundamental property of fractions to simplify rational expressions. We factor the numerator and denominator of the rational expression and divide out all common factors. When all common factors have been divided out, we say that the rational expression has been written in **simplest form**.

EXAMPLE 3 Simplify $\dfrac{21x^2y}{14xy^2}$. Assume that the denominator is not 0.

Solution We will factor the numerator and the denominator and then divide out any common factors, if possible.

$$\frac{21x^2y}{14xy^2} = \frac{3\cdot 7\cdot x\cdot x\cdot y}{2\cdot 7\cdot x\cdot y\cdot y} \qquad \text{Factor the numerator and denominator.}$$

$$= \frac{3\cdot \overset{1}{\cancel{7}}\cdot \overset{1}{\cancel{x}}\cdot x\cdot \overset{1}{\cancel{y}}}{2\cdot \underset{1}{\cancel{7}}\cdot \underset{1}{\cancel{x}}\cdot y\cdot \underset{1}{\cancel{y}}} \qquad \text{Divide out the common factors of 7, } x\text{, and } y.$$

$$= \frac{3x}{2y}$$

This rational expression also can be simplified by using the rules of exponents.

$$\frac{21x^2y}{14xy^2} = \frac{3\cdot 7}{2\cdot 7} x^{2-1}y^{1-2} \qquad \frac{x^2}{x} = x^{2-1}; \frac{y}{y^2} = y^{1-2}$$

$$= \frac{3}{2} xy^{-1} \qquad 2-1=1; 1-2=-1$$

$$= \frac{3}{2}\cdot \frac{x}{y} \qquad y^{-1} = \frac{1}{y}$$

$$= \frac{3x}{2y} \qquad \text{Multiply.}$$

SELF CHECK 3 Simplify $\dfrac{32a^3b^2}{24ab^4}$. Assume that the denominator is not 0.

EXAMPLE 4   Simplify $\dfrac{x^2 + 3x}{3x + 9}$. Assume that the denominator is not 0.

Solution   We will factor the numerator and the denominator and then divide out any common factors, if possible.

$$\frac{x^2 + 3x}{3x + 9} = \frac{x(x + 3)}{3(x + 3)} \qquad \text{Factor the numerator and the denominator.}$$

$$= \frac{x\cancel{(x + 3)}^{1}}{3\cancel{(x + 3)}_{1}} \qquad \text{Divide out the common factor of } x + 3.$$

$$= \frac{x}{3}$$

SELF CHECK 4   Simplify $\dfrac{x^2 - 5x}{5x - 25}$. Assume that the denominator is not 0.

Perspective

The fraction $\frac{8}{4}$ is equal to 2, because $4 \cdot 2 = 8$. The expression $\frac{8}{0}$ is undefined, because there is no number x for which $0 \cdot x = 8$. The expression $\frac{0}{0}$ presents a different problem, however, because $\frac{0}{0}$ seems to equal any number. For example, $\frac{0}{0} = 17$, because $0 \cdot 17 = 0$. Similarly, $\frac{0}{0} = \pi$, because $0 \cdot \pi = 0$. Since "no answer" and "any answer" are both unacceptable, division by 0 is not allowed.

Although $\frac{0}{0}$ represents many numbers, there is often one best answer. In the 17th century, mathematicians such as Sir Isaac Newton (1642–1727) and Gottfried Wilhelm von Leibniz (1646–1716) began to look more closely at expressions related to the fraction $\frac{0}{0}$. One of these expressions, called a **derivative**, is the foundation of **calculus**, an important area of mathematics discovered independently by both Newton and Leibniz. They discovered that under certain conditions, there was one best answer. Expressions related to $\frac{0}{0}$ are called **indeterminate forms**.

Any number divided by 1 remains unchanged. For example,

$$\frac{37}{1} = 37, \qquad \frac{5x}{1} = 5x, \qquad \text{and} \qquad \frac{3x + y}{1} = 3x + y$$

In general, for any real number a, the following is true.

DIVISION BY 1
$$\frac{a}{1} = a$$

EXAMPLE 5   Simplify $\dfrac{x^3 + x^2}{1 + x}$. Assume that the denominator is not 0.

Solution   We will factor the numerator and then divide out any common factors, if possible.

$$\frac{x^3 + x^2}{1 + x} = \frac{x^2(x + 1)}{1 + x} \qquad \text{Factor the numerator.}$$

$$= \frac{x^2\cancel{(x + 1)}^{1}}{\cancel{1 + x}_{1}} \qquad \text{Divide out the common factor of } x + 1.$$

$$= \frac{x^2}{1}$$

$$= x^2 \qquad \text{Denominators of 1 need not be written.}$$

SELF CHECK 5   Simplify $\dfrac{x^2 - x}{x - 1}$. Assume that the denominator is not 0.

EXAMPLE 6   Simplify $\dfrac{x^2 + 13x + 12}{x^2 - 144}$. Assume that the denominator is not 0.

Solution   We will factor the numerator and the denominator and then divide out any common factors, if possible.

$$\frac{x^2 + 13x + 12}{x^2 - 144} = \frac{(x + 1)(x + 12)}{(x + 12)(x - 12)} \qquad \text{Factor the numerator and denominator.}$$

$$= \frac{(x + 1)\cancel{(x + 12)}^{1}}{\cancel{(x + 12)}_{1}(x - 12)} \qquad \text{Divide out the common factor of } (x + 12).$$

$$= \frac{x + 1}{x - 12}$$

 SELF CHECK 6   Simplify $\dfrac{x^2 - 9}{x^3 - 3x^2}$. Assume that the denominator is not 0.

COMMENT   Remember that only *factors* common to the *entire numerator* and *entire denominator* can be divided out. *Terms* that are common to the numerator and denominator *cannot* be divided out.

$$\frac{5 + 8}{5} = \frac{13}{5} \qquad\qquad \frac{5 + 8}{5} = \frac{\cancel{5}^{1} + 8}{\cancel{5}_{1}} = \frac{1 + 8}{1} = 9$$

EXAMPLE 7   Simplify $\dfrac{5(x + 3) - 5}{7(x + 3) - 7}$. Assume that the denominator is not 0.

Solution   We cannot divide out $(x + 3)$, because it is not a factor of the entire numerator, nor is it a factor of the entire denominator. Instead, we simplify the numerator and denominator, factor them, and divide out all common factors, if any.

$$\frac{5(x + 3) - 5}{7(x + 3) - 7} = \frac{5x + 15 - 5}{7x + 21 - 7} \qquad \text{Use the distributive property.}$$

$$= \frac{5x + 10}{7x + 14} \qquad \text{Combine like terms.}$$

$$= \frac{5(x + 2)}{7(x + 2)} \qquad \text{Factor the numerator and denominator.}$$

$$= \frac{5\cancel{(x + 2)}^{1}}{7\cancel{(x + 2)}_{1}} \qquad \text{Divide out the common factor of } (x + 2).$$

$$= \frac{5}{7}$$

 SELF CHECK 7   Simplify $\dfrac{4(x - 2) + 4}{3(x - 2) + 3}$. Assume that the denominator is not 0.

EXAMPLE 8   Simplify: $\dfrac{x(x + 3) - 3(x - 1)}{x^2 + 3}$

Solution   Since the denominator $x^2 + 3$ is always positive, there are no restrictions on x. To simplify the fraction, we will simplify the numerator and then divide out any common factors, if possible.

$$\frac{x(x + 3) - 3(x - 1)}{x^2 + 3} = \frac{x^2 + 3x - 3x + 3}{x^2 + 3} \qquad \text{Use the distributive property.}$$

$$= \frac{x^2 + 3}{x^2 + 3} \qquad \text{Combine like terms.}$$

$$= \frac{\overset{1}{\cancel{(x^2 + 3)}}}{\underset{1}{\cancel{(x^2 + 3)}}} \qquad \text{Divide out the common factor of } (x^2 + 3).$$

$$= 1$$

SELF CHECK 8   Simplify: $\dfrac{a(a + 2) - 2(a - 1)}{a^2 + 2}$

Sometimes rational expressions do not simplify. For example, to attempt to simplify

$$\frac{x^2 + x - 2}{x^2 + x}$$

we factor the numerator and denominator.

$$\frac{x^2 + x - 2}{x^2 + x} = \frac{(x + 2)(x - 1)}{x(x + 1)}$$

Because there are no factors common to the numerator and denominator, this rational expression is already in simplest form.

EXAMPLE 9   Simplify $\dfrac{x^3 + 8}{x^2 + ax + 2x + 2a}$. Assume that the denominator is not 0.

Solution   We will factor the numerator and the denominator and then divide out any common factors, if possible.

$$\frac{x^3 + 8}{x^2 + ax + 2x + 2a} = \frac{(x + 2)(x^2 - 2x + 4)}{x(x + a) + 2(x + a)} \qquad \begin{array}{l}\text{Factor the numerator and begin to} \\ \text{factor the denominator.}\end{array}$$

$$= \frac{(x + 2)(x^2 - 2x + 4)}{(x + a)(x + 2)} \qquad \text{Finish factoring the denominator.}$$

$$= \frac{\overset{1}{\cancel{(x + 2)}}(x^2 - 2x + 4)}{(x + a)\underset{1}{\cancel{(x + 2)}}} \qquad \begin{array}{l}\text{Divide out the common factor of} \\ (x + 2).\end{array}$$

$$= \frac{x^2 - 2x + 4}{x + a}$$

SELF CHECK 9   Simplify $\dfrac{ab + 3a - 2b - 6}{a^3 - 8}$. Assume that the denominator is not 0.

4 Simplify a rational expression containing factors that are negatives.

If the terms of two polynomials are the same, except for signs, the polynomials are called *negatives* or *opposites* of each other. For example,

$x - y$ and $y - x$ are negatives (opposites),

$2a - 1$ and $1 - 2a$ are negatives (opposites), and

$3x^2 - 2x + 5$ and $-3x^2 + 2x - 5$ are negatives (opposites).

Example 10 shows why the quotient of two polynomials that are negatives is always -1.

EXAMPLE 10 Simplify **a.** $\dfrac{x - y}{y - x}$ **b.** $\dfrac{2a - 1}{1 - 2a}$. Assume that no denominators are 0.

Solution We can rearrange terms in each numerator, factor out -1, and proceed as follows:

a. $\dfrac{x - y}{y - x} = \dfrac{-y + x}{y - x}$ **b.** $\dfrac{2a - 1}{1 - 2a} = \dfrac{-1 + 2a}{1 - 2a}$

$= \dfrac{-(y - x)}{y - x}$ $= \dfrac{-(1 - 2a)}{1 - 2a}$

$= \dfrac{-\overset{1}{\cancel{(y - x)}}}{\underset{1}{\cancel{y - x}}}$ $= \dfrac{-\overset{1}{\cancel{(1 - 2a)}}}{\underset{1}{\cancel{1 - 2a}}}$

$= -1$ $= -1$

SELF CHECK 10 Simplify $\dfrac{3p - 2q}{2q - 3p}$. Assume that the denominator is not 0.

The previous example suggests this important result.

DIVISION OF NEGATIVES The quotient of any nonzero expression and its negative is -1. In symbols, we have

if $a \neq b$, then $\dfrac{a - b}{b - a} = -1$.

SELF CHECK ANSWERS 1. $3, -1$ 2. $\{x \mid x \in \mathbb{R}, x \neq -6, 1\}$ 3. $\dfrac{4a^2}{3b^2}$ 4. $\dfrac{x}{5}$ 5. x 6. $\dfrac{x + 3}{x^2}$ 7. $\dfrac{4}{3}$ 8. 1 9. $\dfrac{b + 3}{a^2 + 2a + 4}$ 10. -1

NOW TRY THIS

1. Evaluate $\dfrac{x - 3}{x + 4}$ for **a.** $x = 3$ **b.** $x = 0$ **c.** $x = -4$

2. Simplify $\dfrac{4x + 20}{4x - 12}$. Assume $x \neq 3$.

3. **a.** Find all value(s) of x for which $\dfrac{x + 1}{9x^2 - x}$ is undefined.

   b. Find the domain of $f(x) = \dfrac{x + 1}{9x^2 - x}$.

6.1   Exercises

WARM-UPS *Simplify each rational expression.*

1. $\dfrac{14}{21}$

2. $\dfrac{34}{17}$

3. $\dfrac{12}{16}$

4. $\dfrac{100}{25}$

5. $\dfrac{15}{35}$

6. $\dfrac{28}{35}$

7. $\dfrac{-18}{54}$

8. $-\dfrac{20}{12}$

REVIEW

9. State the associative property of addition.

10. State the distributive property.

11. What is the additive identity?

12. What is the multiplicative identity?

13. Find the additive inverse of $-\dfrac{7}{5}$.

14. Find the multiplicative inverse of $-\dfrac{7}{5}$.

VOCABULARY AND CONCEPTS   *Fill in the blanks.*

15. In a fraction, the part above the fraction bar is called the _____.

16. In a fraction, the part below the fraction bar is called the _____.

17. The denominator of a fraction cannot be __.

18. A fraction that has polynomials in its numerator and denominator is called a _____ expression.

19. $x - 2$ and $2 - x$ are called _____ of each other.

20. To *simplify* a rational expression means to write it in _____ terms.

21. The fundamental property of fractions states that $\dfrac{ac}{bc} = $ ___.

22. Any number x divided by 1 is __.

23. To simplify a rational expression, we _____ the numerator and denominator and divide out _____ factors.

24. A rational expression cannot be simplified when it is written in _____.

GUIDED PRACTICE   *Find all values of the variable for which the following rational expressions are undefined.* SEE EXAMPLE 1. (OBJECTIVE 1)

25. $\dfrac{4y + 1}{y + 4}$

26. $\dfrac{5x - 2}{x - 6}$

27. $\dfrac{3x - 13}{x^2 - x - 2}$

28. $\dfrac{3p^2 + 7p}{8p^2 + 2p - 1}$

29. $\dfrac{5x + 2}{(x + 7)(2x - 1)}$

30. $\dfrac{6x^2 + 7}{(3x + 2)(4x - 5)}$

31. $\dfrac{2x^2 + 1}{3x^2 + x}$

32. $\dfrac{5x - 4}{x^2 - 3x}$

Find the domain. SEE EXAMPLE 2. (OBJECTIVE 2)

33. $\dfrac{4x^2 + 3x}{5x - 2}$

34. $\dfrac{12x - 7}{6x + 5}$

35. $\dfrac{2m^2 + 5m}{2m^2 - m - 3}$

36. $\dfrac{5q^2 - 3}{6q^2 - q - 2}$

Write each expression in simplest form. If it is already in simplest form, so indicate. Assume that no denominators are 0. SEE EXAMPLE 3. (OBJECTIVE 3)

37. $\dfrac{4x}{2}$

38. $\dfrac{2x}{4}$

39. $\dfrac{-25y^2}{5y}$

40. $\dfrac{-6x}{18}$

41. $\dfrac{6x^2 y}{6xy^2}$

42. $\dfrac{x^2 y^3}{x^2 y^4}$

43. $\dfrac{2x^2}{3y}$

44. $\dfrac{7y^2}{5x^2}$

Write each expression in simplest form. If it is already in simplest form, so indicate. Assume that no denominators are 0. SEE EXAMPLE 4. (OBJECTIVE 3)

45. $\dfrac{x^2 + 7x}{2x + 14}$

46. $\dfrac{a^2 - 10a}{4a - 40}$

47. $\dfrac{3x + 15}{x^2 - 25}$

48. $\dfrac{x^2 + 3x}{2x + 6}$

49. $\dfrac{3x + 6}{2x + 1}$

50. $\dfrac{x^2 + 6x}{5x^2 + 6}$

51. $\dfrac{10x - 5}{18x - 9}$

52. $\dfrac{2x^2 - 2x}{5x - 5}$

Write each expression in simplest form. If it is already in simplest form, so indicate. Assume that no denominators are 0. SEE EXAMPLE 5. (OBJECTIVE 3)

53. $\dfrac{x + 3}{3(x + 3)}$

54. $\dfrac{2(x + 7)}{x + 7}$

55. $\dfrac{5x + 35}{x + 7}$

56. $\dfrac{x + x}{2}$

57. $\dfrac{5x^2 + 10x}{x - 2}$

58. $\dfrac{8y^2 - 4y}{y - 4}$

59. $\dfrac{3y^2 + 12y}{4 + y}$

60. $\dfrac{6 + x}{5x^2 + 30x}$

Write each expression in simplest form. If it is already in simplest form, so indicate. Assume that no denominators are 0. SEE EXAMPLE 6. (OBJECTIVE 3)

61. $\dfrac{x^2 + 3x + 2}{x^2 + x - 2}$

62. $\dfrac{x^2 + x - 6}{x^2 - x - 2}$

63. $\dfrac{x^2 - 8x + 15}{x^2 - x - 6}$

64. $\dfrac{x^2 - 6x - 7}{x^2 + 8x + 7}$

65. $\dfrac{2x^2 - 8x}{x^2 - 6x + 8}$

66. $\dfrac{3y^2 - 15y}{y^2 - 3y - 10}$

67. $\dfrac{2a^3 - 16}{2a^2 + 4a + 8}$

68. $\dfrac{3y^3 + 81}{y^2 - 3y + 9}$

69. $\dfrac{x^2 - 4x + 4}{x^2 - 4}$

70. $\dfrac{x^2 - 81}{x^2 - 18x + 81}$

71. $\dfrac{30x^2 - 14x - 8}{3x^2 + 4x + 1}$

72. $\dfrac{x^3 - x}{5x^3 - 5x}$

Write each expression in simplest form. If it is already in simplest form, so indicate. Assume that no denominators are 0. SEE EXAMPLES 7–8. (OBJECTIVE 3)

73. $\dfrac{4(x + 3) + 4}{3(x + 2) + 6}$

74. $\dfrac{x^2 - 3(2x - 3)}{x^2 - 9}$

75. $\dfrac{x^2 + 5x + 4}{2(x + 3) - (x + 2)}$

76. $\dfrac{x^2 - 9}{(2x + 3) - (x + 6)}$

Write each expression in simplest form. If it is already in simplest form, so indicate. Assume that no denominators are 0. SEE EXAMPLE 9. (OBJECTIVE 3)

77. $\dfrac{x^3 + 1}{ax + a + x + 1}$

78. $\dfrac{x^3 - 8}{ax + x - 2a - 2}$

79. $\dfrac{ab + b + 2a + 2}{ab + a + b + 1}$

80. $\dfrac{xy + 2y + 3x + 6}{x^2 + 5x + 6}$

Write each expression in simplest form. If it is already in simplest form, so indicate. Assume that no denominators are 0. SEE EXAMPLE 10. (OBJECTIVE 4)

81. $\dfrac{x - y}{y - x}$

82. $\dfrac{d - c}{c - d}$

83. $\dfrac{6x - 3y}{3y - 6x}$

84. $\dfrac{3c - 4d}{4c - 3d}$

ADDITIONAL PRACTICE Write each expression in simplest form. If it is already in simplest form, so indicate. Assume that no denominators are 0.

85. $\dfrac{45}{9a}$

86. $\dfrac{48}{16y}$

87. $\dfrac{15x^2y}{5xy^2}$

88. $\dfrac{12xz}{4xz^2}$

89. $\dfrac{x^2 + 3x + 2}{x^3 + x^2}$

90. $\dfrac{x^2 - 8x + 16}{x^2 - 16}$

91. $\dfrac{3x + 3y}{x^2 + xy}$

92. $\dfrac{xy + 2x^2}{2xy + y^2}$

93. $\dfrac{3y + xy}{3x + xy}$

94. $\dfrac{6x^2 - 13x + 6}{3x^2 + x - 2}$

95. $\dfrac{xz - 2x}{yz - 2y}$

96. $\dfrac{x^2 - 2x - 15}{x^2 + 2x - 15}$

97. $\dfrac{15x - 3x^2}{25y - 5xy}$

98. $\dfrac{x^3 + 1}{x^2 - x + 1}$

99. $\dfrac{4 + 2(x - 5)}{3x - 5(x - 2)}$

100. $\dfrac{x^3 - 1}{x^2 + x + 1}$

101. $\dfrac{x^2 + 4x - 77}{x^2 - 4x - 21}$

102. $\dfrac{x^2 - 10x + 25}{25 - x^2}$

103. $\dfrac{xy + 3y + 3x + 9}{x^2 - 9}$

104. $\dfrac{ab + b^2 + 2a + 2b}{a^2 + 2a + ab + 2b}$

105. $\dfrac{2x^2 - 8}{x^2 - 3x + 2}$

106. $\dfrac{3x^2 - 27}{x^2 + 3x - 18}$

107. $\dfrac{a + b - c}{c - a - b}$

108. $\dfrac{x - y - z}{z + y - x}$

109. $\dfrac{6a - 6b + 6c}{9a - 9b + 9c}$

110. $\dfrac{3a - 3b - 6}{2a - 2b - 4}$

WRITING ABOUT MATH

111. Explain why $\dfrac{x - 7}{7 - x} = -1$.

112. Exercise 99 has two possible answers: $\dfrac{x - 3}{5 - x}$ and $-\dfrac{x - 3}{x - 5}$ Why is either answer correct?

SOMETHING TO THINK ABOUT

113. Find two different-looking but correct answers for the following problem.

Simplify: $\dfrac{y^2 + 5(2y + 5)}{25 - y^2}$

Section 6.2

Multiplying and Dividing Rational Expressions

Objectives

1. Multiply two rational expressions and write the result in simplest form.
2. Multiply a rational expression by a polynomial and write the result in simplest form.
3. Divide two rational expressions and write the result in simplest form.
4. Divide a rational expression by a polynomial and write the result in simplest form.
5. Perform combined operations on three or more rational expressions.

Getting Ready

Multiply or divide the fractions and simplify.

1. $\dfrac{3}{7} \cdot \dfrac{14}{9}$ 2. $\dfrac{21}{15} \cdot \dfrac{10}{3}$ 3. $\dfrac{19}{38} \cdot 6$ 4. $42 \cdot \dfrac{3}{21}$

5. $\dfrac{4}{9} \div \dfrac{8}{45}$ 6. $\dfrac{11}{7} \div \dfrac{22}{14}$ 7. $\dfrac{75}{12} \div \dfrac{50}{6}$ 8. $\dfrac{13}{5} \div \dfrac{26}{20}$

Just like arithmetic fractions, rational expressions can be multiplied, divided, added, and subtracted. In this section, we will show how to multiply and divide rational expressions.

1   Multiply two rational expressions and write the result in simplest form.

Recall that to multiply fractions, we multiply their numerators and multiply their denominators. For example, to find the product of $\frac{4}{7}$ and $\frac{3}{5}$, we proceed as follows.

$$\frac{4}{7} \cdot \frac{3}{5} = \frac{4 \cdot 3}{7 \cdot 5} \qquad \text{Multiply the numerators and multiply the denominators.}$$

$$= \frac{12}{35} \qquad \text{Simplify.}$$

This suggests the rule for multiplying rational expressions.

MULTIPLYING RATIONAL EXPRESSIONS

If a, b, c, and d are polynomials, then

$$\frac{a}{b} \cdot \frac{c}{d} = \frac{ac}{bd} \qquad \text{provided no denominators are 0.}$$

EXAMPLE 1   Multiply. Assume that no denominators are 0.

a. $\dfrac{1}{3} \cdot \dfrac{2}{5}$ **b.** $\dfrac{7}{9} \cdot \dfrac{-5}{3x}$ **c.** $\dfrac{x^2}{2} \cdot \dfrac{3}{y^2}$ **d.** $\dfrac{q+1}{q} \cdot \dfrac{q-1}{q-2}$

Solution   We will multiply the numerators, multiply the denominators, and then simplify, if possible.

a. $\dfrac{1}{3} \cdot \dfrac{2}{5} = \dfrac{1 \cdot 2}{3 \cdot 5}$

$= \dfrac{2}{15}$

b. $\dfrac{7}{9} \cdot \dfrac{-5}{3x} = \dfrac{7(-5)}{9 \cdot 3x}$

$= \dfrac{-35}{27x}$

c. $\dfrac{x^2}{2} \cdot \dfrac{3}{y^2} = \dfrac{x^2 \cdot 3}{2 \cdot y^2}$

$= \dfrac{3x^2}{2y^2}$

d. $\dfrac{q+1}{q} \cdot \dfrac{q-1}{q-2} = \dfrac{(q+1)(q-1)}{q(q-2)}$

SELF CHECK 1   Multiply. Assume that no denominators are 0.

a. $\dfrac{4}{5} \cdot \dfrac{7}{9}$

b. $\dfrac{-5}{3} \cdot \dfrac{2x}{3}$

c. $\dfrac{x}{y^2} \cdot \dfrac{5x}{4}$

d. $\dfrac{y+1}{9y} \cdot \dfrac{y+2}{y-1}$

EXAMPLE 2   Multiply $\dfrac{35x^2y}{7y^2z} \cdot \dfrac{z}{5xy}$. Assume that no denominators are 0.

Solution   We will multiply the numerators, multiply the denominators, and then simplify, if possible.

$\dfrac{35x^2y}{7y^2z} \cdot \dfrac{z}{5xy} = \dfrac{35x^2yz}{35y^3zx}$ Multiply the numerators and multiply the denominators.

$= \dfrac{x}{y^2}$ Simplify using properties of exponents.

SELF CHECK 2   Multiply $\dfrac{a^2b^2}{2a} \cdot \dfrac{9a^3}{3b^3}$. Assume that no denominators are 0.

EXAMPLE 3   Multiply $\dfrac{x^2-x}{2x+4} \cdot \dfrac{x+2}{x}$. Assume that no denominators are 0.

Solution   We will multiply the numerators, multiply the denominators, and then simplify.

$\dfrac{x^2-x}{2x+4} \cdot \dfrac{x+2}{x} = \dfrac{(x^2-x)(x+2)}{(2x+4)(x)}$ Multiply the numerators and multiply the denominators.

$= \dfrac{x(x-1)(x+2)}{2(x+2)x}$ Factor.

$= \dfrac{\overset{1}{x}(x-1)\overset{1}{\cancel{(x+2)}}}{2\underset{1}{\cancel{(x+2)}}\underset{1}{x}}$ Divide out common factors.

$= \dfrac{x-1}{2}$

SELF CHECK 3   Multiply $\dfrac{x^2+x}{3x+6} \cdot \dfrac{x+2}{x+1}$. Assume that no denominators are 0.

Recall from arithmetic that $\dfrac{2}{3} \cdot \dfrac{9}{8}$ could be simplified before multiplying:

$\dfrac{\overset{1}{2}}{\underset{1}{3}} \cdot \dfrac{\overset{3}{9}}{\underset{4}{8}} = \dfrac{3}{4}$

We could approach Example 3 in the same way. Factor each expression that is not in completely factored form and divide out common factors.

$$\frac{x^2 - x}{2x + 4} \cdot \frac{x + 2}{x} = \frac{x(x - 1)}{2(x + 2)} \cdot \frac{x + 2}{x}$$

$$= \frac{\overset{1}{\cancel{x}}(x - 1)}{2\cancel{(x + 2)}} \cdot \frac{\cancel{x + 2}^{\,1}}{\cancel{x}}$$

$$= \frac{x - 1}{2}$$

We will work Example 4 using this method.

EXAMPLE 4 Multiply $\dfrac{x^2 - 3x}{x^2 - x - 6}$ and $\dfrac{x^2 + x - 2}{x^2 - x}$. Assume that no denominators are 0.

Solution We will factor the numerators and denominators, simplify, and then multiply.

$$\frac{x^2 - 3x}{x^2 - x - 6} \cdot \frac{x^2 + x - 2}{x^2 - x}$$

$$= \frac{x(x - 3)}{(x - 3)(x + 2)} \cdot \frac{(x + 2)(x - 1)}{x(x - 1)} \qquad \text{Factor where possible.}$$

$$= \frac{\cancel{x}\cancel{(x - 3)}}{\cancel{(x - 3)}\cancel{(x + 2)}} \cdot \frac{\cancel{(x + 2)}\cancel{(x - 1)}}{\cancel{x}\cancel{(x - 1)}} \qquad \text{Divide out common factors.}$$

$$= 1$$

SELF CHECK 4 Multiply $\dfrac{a^2 + a}{a^2 - 4} \cdot \dfrac{a^2 - a - 2}{a^2 + 2a + 1}$. Assume that no denominators are 0.

2 Multiply a rational expression by a polynomial and write the result in simplest form.

Since any number divided by 1 remains unchanged, we can write any polynomial as a rational expression by writing it with a denominator of 1.

EXAMPLE 5 Multiply $\dfrac{x^2 + x}{x^2 + 8x + 7} \cdot (x + 7)$. Assume that the denominator is not 0.

Solution We will write $x + 7$ as $\frac{x + 7}{1}$, multiply the numerators, multiply the denominators, and then simplify.

$$\frac{x^2 + x}{x^2 + 8x + 7} \cdot (x + 7) = \frac{x^2 + x}{x^2 + 8x + 7} \cdot \frac{x + 7}{1} \qquad \begin{array}{l}\text{Write } x + 7 \text{ as a fraction with a} \\ \text{denominator of 1.}\end{array}$$

$$= \frac{x(x + 1)(x + 7)}{(x + 1)(x + 7)1} \qquad \begin{array}{l}\text{Multiply the fractions and factor} \\ \text{where possible.}\end{array}$$

$$= \frac{x\cancel{(x + 1)}\cancel{(x + 7)}}{1\cancel{(x + 1)}\cancel{(x + 7)}} \qquad \text{Divide out all common factors.}$$

$$= x$$

SELF CHECK 5 Multiply $(a - 7) \cdot \dfrac{a^2 - a}{a^2 - 8a + 7}$. Assume that the denominator is not 0.

3 Divide two rational expressions and write the result in simplest form.

Recall that division by a nonzero number is equivalent to multiplying by the reciprocal of that number. Thus, to divide two fractions, we can invert the *divisor* (the fraction following the $\div$ sign) and multiply. For example, to divide $\frac{4}{7}$ by $\frac{3}{5}$, we proceed as follows:

$$\frac{4}{7} \div \frac{3}{5} = \frac{4}{7} \cdot \frac{5}{3} \qquad \text{Invert } \tfrac{3}{5} \text{ and write as multiplication.}$$

$$= \frac{20}{21} \qquad \text{Multiply the numerators and multiply the denominators.}$$

This suggests the rule for dividing rational expressions.

DIVIDING RATIONAL EXPRESSIONS

If a, b, c, and d are polynomials, then

$$\frac{a}{b} \div \frac{c}{d} = \frac{a}{b} \cdot \frac{d}{c} = \frac{ad}{bc} \qquad \text{provided no denominators are equal to 0.}$$

EXAMPLE 6 Divide, assuming that no denominators are 0.

a. $\dfrac{7}{13} \div \dfrac{21}{26}$ **b.** $\dfrac{-9x}{35y} \div \dfrac{15x^2}{14}$

Solution We will change each division to a multiplication and then multiply the resulting rational expressions.

a. $\dfrac{7}{13} \div \dfrac{21}{26} = \dfrac{7}{13} \cdot \dfrac{26}{21}$ Invert the divisor and multiply.

$$= \frac{\overset{1}{7}}{\underset{1}{13}} \cdot \frac{\overset{2}{26}}{\underset{3}{21}} \qquad \text{Divide out common factors.}$$

$$= \frac{2}{3}$$

b. $\dfrac{-9x}{35y} \div \dfrac{15x^2}{14} = \dfrac{-9x}{35y} \cdot \dfrac{14}{15x^2}$ Invert the divisor and multiply.

$$= \frac{\overset{-3}{-9x}}{\underset{5}{35y}} \cdot \frac{\overset{2}{14}}{\underset{5x}{15x^2}} \qquad \text{Divide out common factors.}$$

$$= -\frac{6}{25xy} \qquad \text{Multiply the remaining factors.}$$

SELF CHECK 6 Divide. Assume that no denominators are 0.

a. $\dfrac{15}{14} \div \dfrac{25}{7}$ **b.** $\dfrac{-8a}{3b} \div \dfrac{16a^2}{9b^2}$

EXAMPLE 7   Divide $\dfrac{x^2 + x}{3x - 15} \div \dfrac{x^2 + 2x + 1}{6x - 30}$. Assume that no denominators are 0.

Solution   We invert the divisor and write the division as multiplication and then multiply the resulting rational expressions.

$$\dfrac{x^2 + x}{3x - 15} \div \dfrac{x^2 + 2x + 1}{6x - 30}$$

$$= \dfrac{x^2 + x}{3x - 15} \cdot \dfrac{6x - 30}{x^2 + 2x + 1} \qquad \textcolor{red}{\text{Invert the divisor and multiply.}}$$

$$= \dfrac{x(x + 1)}{3(x - 5)} \cdot \dfrac{6(x - 5)}{(x + 1)(x + 1)} \qquad \textcolor{red}{\text{Factor where possible.}}$$

$$= \dfrac{x\cancel{(x + 1)}^{\,1}}{\cancel{3}(x\cancel{- 5})} \cdot \dfrac{{}^{2}\cancel{6}(x\cancel{- 5})^{\,1}}{\cancel{(x + 1)}(x + 1)} \qquad \textcolor{red}{\text{Divide out all common factors.}}$$

$$= \dfrac{2x}{x + 1} \qquad \textcolor{red}{\text{Multiply the fractions.}}$$

🌿 **SELF CHECK 7**   Divide $\dfrac{a^2 - 1}{a^2 + 4a + 3} \div \dfrac{a - 1}{a^2 + 2a - 3}$. Assume that no denominators are 0.

---

4 ## Divide a rational expression by a polynomial and write the result in simplest form.

To divide a rational expression by a polynomial, we write the polynomial as a rational expression with a denominator of 1 and then divide the expressions.

EXAMPLE 8   Divide $\dfrac{2x^2 - 3x - 2}{2x + 1} \div (4 - x^2)$. Assume that no denominators are 0.

Solution   We will write $4 - x^2$ as $\dfrac{4 - x^2}{1}$, invert the divisor, and then multiply the resulting rational expressions.

$$\dfrac{2x^2 - 3x - 2}{2x + 1} \div (4 - x^2)$$

$$= \dfrac{2x^2 - 3x - 2}{2x + 1} \div \dfrac{4 - x^2}{1} \qquad \textcolor{red}{\begin{array}{l}\text{Write } 4 - x^2 \text{ as a fraction with a}\\ \text{denominator of 1.}\end{array}}$$

$$= \dfrac{2x^2 - 3x - 2}{2x + 1} \cdot \dfrac{1}{4 - x^2} \qquad \textcolor{red}{\text{Invert the divisor and multiply.}}$$

$$= \dfrac{(2x + 1)(x - 2)}{2x + 1} \cdot \dfrac{1}{(2 + x)(2 - x)} \qquad \textcolor{red}{\text{Factor where possible.}}$$

$$= \dfrac{\cancel{(2x + 1)}\,\overset{-1}{\cancel{(x - 2)}}}{\underset{1}{\cancel{2x + 1}}} \cdot \dfrac{1}{(2 + x)\underset{1}{\cancel{(2 - x)}}} \qquad \textcolor{red}{\text{Divide out common factors: } \tfrac{x - 2}{2 - x} = -1}$$

$$= \dfrac{-1}{2 + x} \qquad \textcolor{red}{\text{Multiply.}}$$

$$\text{OR} \;\; -\dfrac{1}{2 + x} \qquad \textcolor{red}{\dfrac{-a}{b} = -\dfrac{a}{b}}$$

🌿 **SELF CHECK 8**   Divide $(b - a) \div \dfrac{a^2 - b^2}{a^2 + ab}$. Assume that no denominators are 0.

5 Perform combined operations on three or more rational expressions.

Unless parentheses indicate otherwise, we will perform multiplications and divisions in order from left to right.

EXAMPLE 9 Simplify $\dfrac{x^2 - x - 6}{x - 2} \div \dfrac{x^2 - 4x}{x^2 - x - 2} \cdot \dfrac{x - 4}{x^2 + x}$. Assume that no denominators are 0.

Solution Since there are no parentheses to indicate otherwise, we perform the division first.

$$\dfrac{x^2 - x - 6}{x - 2} \div \dfrac{x^2 - 4x}{x^2 - x - 2} \cdot \dfrac{x - 4}{x^2 + x}$$

$$= \dfrac{x^2 - x - 6}{x - 2} \cdot \dfrac{x^2 - x - 2}{x^2 - 4x} \cdot \dfrac{x - 4}{x^2 + x} \qquad \text{Invert the divisor and multiply.}$$

$$= \dfrac{(x + 2)(x - 3)(x + 1)(x - 2)(x - 4)}{(x - 2)x(x - 4)x(x + 1)} \qquad \text{Factor where possible and multiply.}$$

$$= \dfrac{(x + 2)(x - 3)\,\cancel{(x + 1)}^{\,1}\cancel{(x - 2)}^{\,1}\,\cancel{(x - 4)}^{\,1}}{\cancel{(x - 2)}_{\,1}\,x\,\cancel{(x - 4)}_{\,1}\,x\,\cancel{(x + 1)}_{\,1}} \qquad \text{Divide out all common factors.}$$

$$= \dfrac{(x + 2)(x - 3)}{x^2}$$

🌿 **SELF CHECK 9** Simplify $\dfrac{a^2 + ab}{ab - b^2} \cdot \dfrac{a^2 - b^2}{a^2 + ab} \div \dfrac{a + b}{b}$. Assume that no denominators are 0.

EXAMPLE 10 Simplify $\dfrac{x^2 + 6x + 9}{x^2 - 2x}\left(\dfrac{x^2 - 4}{x^2 + 3x} \div \dfrac{x + 2}{x}\right)$. Assume that no denominators are 0.

Solution We perform the division within the parentheses first.

$$\dfrac{x^2 + 6x + 9}{x^2 - 2x}\left(\dfrac{x^2 - 4}{x^2 + 3x} \div \dfrac{x + 2}{x}\right)$$

$$= \dfrac{x^2 + 6x + 9}{x^2 - 2x}\left(\dfrac{x^2 - 4}{x^2 + 3x} \cdot \dfrac{x}{x + 2}\right) \qquad \text{Invert the divisor and multiply.}$$

$$= \dfrac{(x + 3)(x + 3)}{x(x - 2)}\left[\dfrac{(x - 2)(x + 2)}{x(x + 3)} \cdot \dfrac{x}{x + 2}\right] \qquad \text{Factor where possible.}$$

$$= \dfrac{(x + 3)(x + 3)}{x(x - 2)}\left[\dfrac{(x - 2)\cancel{(x + 2)}^{\,1}}{\cancel{x}_{\,1}\cancel{(x + 3)}} \cdot \dfrac{\cancel{x}^{\,1}}{\cancel{x + 2}_{\,1}}\right] \qquad \substack{\text{Divide out all common factors} \\ \text{within the brackets.}}$$

$$= \dfrac{(x + 3)(x + 3)}{x(x - 2)} \cdot \dfrac{x - 2}{x + 3} \qquad \text{Remove brackets.}$$

$$= \dfrac{\cancel{(x + 3)}^{\,1}(x + 3)}{x\,\cancel{(x - 2)}_{\,1}} \cdot \dfrac{\cancel{x - 2}^{\,1}}{\cancel{x + 3}_{\,1}} \qquad \text{Divide out all common factors.}$$

$$= \dfrac{x + 3}{x}$$

🌿 **SELF CHECK 10** Simplify $\dfrac{x^2 - 2x}{x^2 + 6x + 9} \div \left(\dfrac{x^2 - 4}{x^2 + 3x} \cdot \dfrac{x}{x + 2}\right)$. Assume that no denominators are 0.

NOW TRY THIS

Simplify. Assume no division by 0.

1. $(x^2 - 4x - 12) \cdot \dfrac{(x+6)^2}{x^2 - 36}$

2. $\dfrac{x^2 - 9}{x - 2} \div \dfrac{9 - x^2}{3x - 6}$

3. $\dfrac{\dfrac{1}{2}}{\dfrac{3}{4}}$

4. $\dfrac{\dfrac{3}{5} - \dfrac{2}{3}}{\dfrac{7}{3} + \dfrac{2}{5}}$

6.2 Exercises

WARM-UPS *Fill in the boxes. Assume no denominator is 0.*

1. $\dfrac{5}{2} \cdot \dfrac{3}{\Box} = \dfrac{3}{2}$

2. $\dfrac{7}{3} \cdot \dfrac{5}{\Box} = \dfrac{35}{18}$

3. $\dfrac{x}{2} \cdot \dfrac{3}{\Box} = \dfrac{3}{2}$

4. $\dfrac{2}{\Box}{5} \cdot \dfrac{7}{x+1} = \dfrac{14}{5}$

5. $\dfrac{9}{\Box} \div \dfrac{3}{2} = \dfrac{3}{2}$

6. $\dfrac{3}{7} \div \dfrac{}{\Box} = 1$

7. $\dfrac{\Box}{2} \div \dfrac{x}{3} = \dfrac{3x}{2}$

8. $\dfrac{1}{\Box} \div \dfrac{x}{1(x+1)} = \dfrac{1}{x}$

REVIEW *Simplify each expression. Write all answers without using negative exponents. Assume that no denominators are 0.*

9. $2x^3 y^2 (-3x^2 y^4 z)$

10. $\dfrac{8x^4 y^5}{-2x^3 y^2}$

11. $(5y)^{-3}$

12. $(a^{-2} a)^{-3}$

13. $\dfrac{x^{3m}}{x^{4m}}$

14. $(3x^2 y^3)^0$

Perform the operations and simplify.

15. $-4(y^3 - 4y^2 + 3y - 2) + 6(-2y^2 + 4) - 4(-2y^3 - y)$

16. $y - 5 \overline{)5y^3 - 3y^2 + 4y - 1}$ $(y \neq 5)$

VOCABULARY AND CONCEPTS *Fill in the blanks.*

17. To multiply fractions, we multiply their _____ and multiply their _____.

18. Unless parentheses indicate otherwise, do multiplications and divisions in order from ___ to ____.

19. $\dfrac{a}{b} \cdot \dfrac{c}{d} =$ ___

20. To write a polynomial in fractional form, we insert a denominator of _.

21. $\dfrac{a}{b} \div \dfrac{c}{d} = \dfrac{a}{b} \cdot$ ___

22. To divide two fractions, invert the _____ and _____.

GUIDED PRACTICE *Perform the multiplication. Assume that no denominators are 0. SEE EXAMPLE 1. (OBJECTIVE 1)*

23. $\dfrac{5}{7} \cdot \dfrac{9}{13}$

24. $\dfrac{3}{5} \cdot \dfrac{11}{7}$

25. $\dfrac{5x}{y} \cdot \dfrac{4x}{3y^2}$

26. $\dfrac{-8x^2}{z^4} \cdot \dfrac{2x}{5}$

27. $\dfrac{z+7}{7} \cdot \dfrac{z+2}{z}$

28. $\dfrac{a-3}{a} \cdot \dfrac{a+3}{5}$

29. $\dfrac{-3a}{a+2} \cdot \dfrac{a-1}{5}$

30. $\dfrac{-b}{b-1} \cdot \dfrac{b+2}{b-2}$

Perform the multiplication. Assume that no denominator is 0. Simplify the answers. SEE EXAMPLE 2. (OBJECTIVE 1)

31. $\dfrac{2y}{z} \cdot \dfrac{z}{3}$

32. $\dfrac{5x}{y} \cdot \dfrac{4}{x}$

33. $\dfrac{5y}{7} \cdot \dfrac{7x}{5z}$

34. $\dfrac{4x}{3y} \cdot \dfrac{3y}{7x}$

35. $\dfrac{x}{2x^2} \cdot \dfrac{-28xy}{7}$

36. $\dfrac{-2xy}{x^2} \cdot \dfrac{3xy}{2}$

37. $\dfrac{ab^2}{a^2 b} \cdot \dfrac{b^2 c^2}{abc} \cdot \dfrac{abc^2}{a^3 c^2}$

38. $\dfrac{x^3 y}{z} \cdot \dfrac{xz^3}{x^2 y^2} \cdot \dfrac{yz}{xyz}$

Perform the multiplication. Assume that no denominators are 0. Simplify the answers, if possible. SEE EXAMPLE 3. (OBJECTIVE 1)

39. $\dfrac{x-2}{2} \cdot \dfrac{2x}{x-2}$

40. $\dfrac{y+7}{y} \cdot \dfrac{3y}{y+7}$

41. $\dfrac{5y - 5}{y - 1} \cdot \dfrac{y}{10y^2}$

42. $\dfrac{x - 7}{4x + 8} \cdot \dfrac{x + 2}{x^2 - 49}$

43. $\dfrac{3y - 12}{y + 8} \cdot \dfrac{y^2 + 8y}{y - 4}$

44. $\dfrac{y^2 + 3y}{9} \cdot \dfrac{3x}{y + 3}$

45. $\dfrac{5z - 10}{z + 2} \cdot \dfrac{3}{3z - 6}$

46. $\dfrac{4x - 8}{4x - 4} \cdot \dfrac{x^2 - x}{x}$

Perform the multiplication. Assume that no denominators are 0. Simplify the answers, if possible. SEE EXAMPLE 4. (OBJECTIVE 1)

47. $\dfrac{z^2 + 4z - 5}{5z - 5} \cdot \dfrac{5z}{z + 5}$

48. $\dfrac{x^2 + x - 6}{5x} \cdot \dfrac{5x - 10}{x + 3}$

49. $\dfrac{(x + 1)^2}{x + 1} \cdot \dfrac{x + 2}{x + 1}$

50. $\dfrac{(y - 3)^2}{y - 3} \cdot \dfrac{y - 3}{y - 3}$

51. $\dfrac{m^2 - 2m - 3}{2m + 4} \cdot \dfrac{m^2 - 4}{m^2 + 3m + 2}$

52. $\dfrac{p^2 - p - 6}{3p - 9} \cdot \dfrac{p^2 - 9}{p^2 + 6p + 9}$

53. $\dfrac{abc^2}{a + 1} \cdot \dfrac{c}{a^2b^2} \cdot \dfrac{a^2 + a}{ac}$

54. $\dfrac{x^3yz^2}{4x + 8} \cdot \dfrac{x^2 - 4}{2x^2y^2z^2} \cdot \dfrac{8yz}{x - 2}$

Perform the multiplication. Assume that no denominators are 0. Simplify the answers when possible. SEE EXAMPLE 5. (OBJECTIVE 2)

55. $\dfrac{x - 5}{2x - 8} \cdot (x - 4)$

56. $\dfrac{x^2 + x}{x^2 + 4x + 3} \cdot (x + 3)$

57. $(5x - 10) \cdot \dfrac{x^2 + 2x}{x^2 - 4}$

58. $(6x - 8) \cdot \dfrac{x - 2}{9x - 12}$

Perform each division. Assume that no denominators are 0. Simplify answers when possible. SEE EXAMPLE 6. (OBJECTIVE 3)

59. $\dfrac{2}{3} \div \dfrac{1}{2}$

60. $\dfrac{3}{4} \div \dfrac{1}{3}$

61. $\dfrac{21}{14} \div \dfrac{5}{2}$

62. $\dfrac{14}{3} \div \dfrac{10}{3}$

63. $\dfrac{3x}{2} \div \dfrac{x}{2}$

64. $\dfrac{y}{6} \div \dfrac{2}{3y}$

65. $\dfrac{x^2y}{3xy} \div \dfrac{xy^2}{6y}$

66. $\dfrac{2xz}{z} \div \dfrac{4x^2}{z^2}$

Perform each division. Assume that no denominators are 0. Simplify answers when possible. SEE EXAMPLE 7. (OBJECTIVE 3)

67. $\dfrac{x + 2}{3x} \div \dfrac{x + 2}{2}$

68. $\dfrac{z - 3}{3z} \div \dfrac{z + 3}{z}$

69. $\dfrac{x^2 - 4}{3x + 6} \div \dfrac{x - 2}{x + 2}$

70. $\dfrac{x^2 - 16}{3x + 12} \div \dfrac{x - 4}{x + 4}$

71. $\dfrac{y(y + 2)}{y^2(y - 3)} \div \dfrac{y^2(y + 2)}{(y - 3)^2}$

72. $\dfrac{(z - 3)^2}{4z^2} \div \dfrac{z - 3}{8z}$

73. $\dfrac{x^2 - x - 6}{2x^2 + 9x + 10} \div \dfrac{x^2 - 25}{2x^2 + 15x + 25}$

74. $\dfrac{x^2 - 2x - 35}{3x^2 + 27x} \div \dfrac{x^2 + 7x + 10}{6x^2 + 12x}$

75. $\dfrac{5x^2 + 13x - 6}{x + 3} \div \dfrac{5x^2 - 17x + 6}{x - 2}$

76. $\dfrac{2x^2 + 8x - 42}{x - 3} \div \dfrac{2x^2 + 14x}{x^2 + 5x}$

77. $\dfrac{x^2 + 7xy + 12y^2}{x^2 + 2xy - 8y^2} \cdot \dfrac{x^2 - xy - 2y^2}{x^2 + 4xy + 3y^2}$

78. $\dfrac{m^2 + 9mn + 20n^2}{m^2 - 25n^2} \cdot \dfrac{m^2 - 9mn + 20n^2}{m^2 - 16n^2}$

Perform the operations. Assume that no denominators are 0. Simplify answers when possible. SEE EXAMPLE 8. (OBJECTIVE 4)

79. $\dfrac{3x + 9}{x + 1} \div (x + 3)$

80. $\dfrac{2x - 5}{6x + 1} \div (16x - 40)$

81. $(3x + 9) \div \dfrac{x^2 - 9}{6x}$

82. $(x - 2) \div \dfrac{6x - 12}{x + 2}$

Perform the operations. Assume that no denominators are 0. Simplify answers when possible. SEE EXAMPLE 9. (OBJECTIVE 5)

83. $\dfrac{x}{3} \cdot \dfrac{9}{4} \div \dfrac{x^2}{6}$

84. $\dfrac{y^3}{3y} \cdot \dfrac{3y^2}{4} \div \dfrac{15}{20}$

85. $\dfrac{y^2}{2} \div \dfrac{4}{y} \cdot \dfrac{y^2}{8}$

86. $\dfrac{x^2}{18} \div \dfrac{x^3}{6} \div \dfrac{12}{x^2}$

87. $\dfrac{x^2 - 4}{2x + 6} \div \dfrac{x + 2}{4} \cdot \dfrac{x + 3}{x - 2}$

88. $\dfrac{2}{3x - 3} \div \dfrac{2x + 2}{x - 1} \cdot \dfrac{5}{x + 1}$

89. $\dfrac{x^2 + x - 6}{x^2 - 4} \cdot \dfrac{x^2 + 2x}{x - 2} \div \dfrac{x^2 + 3x}{x + 2}$

90. $\dfrac{x^2 - x - 6}{x^2 + 6x - 7} \cdot \dfrac{x^2 + x - 2}{x^2 + 2x} \div \dfrac{x^2 + 7x}{x^2 - 3x}$

Perform the operations. Assume that no denominators are 0. Simplify answers when possible. SEE EXAMPLE 10. (OBJECTIVE 5)

91. $\dfrac{x^2 - 1}{x^2 - 9}\left(\dfrac{x + 3}{x + 2} \div \dfrac{5}{x + 2}\right)$

92. $\dfrac{x^2 - 5x}{x + 1}\left(\dfrac{x + 1}{x^2 + 3x} \div \dfrac{x - 5}{x - 3}\right)$

93. $\dfrac{x - x^2}{x^2 - 4}\left(\dfrac{2x + 4}{x + 2} \div \dfrac{5}{x + 2}\right)$

94. $\dfrac{2}{3x - 3} \div \left(\dfrac{2x + 2}{x - 1} \cdot \dfrac{5}{x + 1}\right)$

ADDITIONAL PRACTICE *Perform the indicated operation(s). Assume that no denominators are 0. Simplify answers when possible.*

95. $\dfrac{8z}{2x} \cdot \dfrac{16x}{3x}$

96. $\dfrac{2x^2y}{3xy} \cdot \dfrac{3xy^2}{2}$

97. $\dfrac{2x^2z}{z} \cdot \dfrac{5x}{z}$

98. $\dfrac{10r^2st^3}{6rs^2} \cdot \dfrac{3r^3t}{2rst} \cdot \dfrac{2s^3t^4}{5s^2t^3}$

99. $\dfrac{3a^3b}{25cd^3} \cdot \dfrac{-5cd^2}{6ab} \cdot \dfrac{10abc^2}{2bc^2d}$

100. $\dfrac{3y}{8} \div \dfrac{2y}{4y}$

101. $\dfrac{3x}{y} \div \dfrac{2x}{4}$

102. $\dfrac{4x}{3x} \div \dfrac{2y}{9y}$

103. $\dfrac{14}{7y} \div \dfrac{10}{5z}$

104. $\dfrac{y - 9}{y + 9} \cdot \dfrac{y}{9}$

105. $\dfrac{(x + 7)^2}{x + 7} \div \dfrac{(x - 3)^2}{x + 7}$

106. $\dfrac{x^2 - 1}{3x - 3} \div \dfrac{x + 1}{3}$

107. $\dfrac{x^2 - 16}{x - 4} \div \dfrac{3x + 12}{x}$

108. $\dfrac{3x^2 + 5x + 2}{x^2 - 9} \cdot \dfrac{x - 3}{x^2 - 4} \cdot \dfrac{x^2 + 5x + 6}{6x + 4}$

109. $\dfrac{a^2 - ab + b^2}{a^3 + b^3} \cdot \dfrac{ac + ad + bc + bd}{c^2 - d^2}$

110. $\dfrac{ab + 4a + 2b + 8}{b^2 + 4b + 16} \div \dfrac{b^2 - 16}{b^3 - 64}$

111. $\dfrac{xw - xz + wy - yz}{x^2 + 2xy + y^2} \cdot \dfrac{x^3 - y^3}{z^2 - w^2}$

112. $\dfrac{s^3 - r^3}{r^2 + rs + s^2} \div \dfrac{pr - ps - qr + qs}{q^2 - p^2}$

113. $\dfrac{p^3 - p^2q + pq^2}{mp - mq + np - nq} \div \dfrac{q^3 + p^3}{q^2 - p^2}$

114. $\dfrac{x^2 - y^2}{x^4 - x^3} \div \dfrac{x - y}{x^2} \div \dfrac{x^2 + 2xy + y^2}{x + y}$

WRITING ABOUT MATH

115. Explain how to multiply two fractions and how to simplify the result.

116. Explain why any mathematical expression can be written as a fraction.

117. To divide fractions, you must first know how to multiply fractions. Explain.

118. Explain how to do the division $\dfrac{a}{b} \div \dfrac{c}{d} \div \dfrac{e}{f}$.

SOMETHING TO THINK ABOUT

119. Let x equal a number of your choosing. Without simplifying first, use a calculator to evaluate

$$\dfrac{x^2 + x - 6}{x^2 + 3x} \cdot \dfrac{x^2}{x - 2}$$

Try again, with a different value of x. If you were to simplify the expression, what do you think you would get?

120. Simplify the expression in Exercise 119 to determine whether your answer was correct.

Section 6.3

Adding and Subtracting Rational Expressions

Objectives

1 Add two rational expressions with like denominators and write the answer in simplest form.

2 Subtract two rational expressions with like denominators and write the answer in simplest form.

3 Find the least common denominator (LCD) of two or more polynomials and use it to write equivalent rational expressions.

4 Add two rational expressions with unlike denominators and write the answer in simplest form.

5 Subtract two rational expressions with unlike denominators and write the answer in simplest form.

Vocabulary

least common denominator
(LCD)

Getting Ready

Add or subtract the fractions and simplify.

1. $\dfrac{1}{5} + \dfrac{3}{5}$ **2.** $\dfrac{3}{7} + \dfrac{4}{7}$ **3.** $\dfrac{3}{8} + \dfrac{4}{8}$ **4.** $\dfrac{18}{19} + \dfrac{20}{19}$

5. $\dfrac{5}{9} - \dfrac{4}{9}$ **6.** $\dfrac{7}{12} - \dfrac{1}{12}$ **7.** $\dfrac{7}{13} - \dfrac{9}{13}$ **8.** $\dfrac{20}{10} - \dfrac{7}{10}$

In this section, we will discuss how to add and subtract rational expressions.

1   **Add two rational expressions with like denominators and write the answer in simplest form.**

To add rational expressions with a common denominator, we follow the same process we use to add arithmetic fractions; add their numerators and keep the common denominator. For example,

$$\frac{2x}{7} + \frac{3x}{7} = \frac{2x + 3x}{7} \qquad \text{Add the numerators and keep the common denominator.}$$

$$= \frac{5x}{7} \qquad 2x + 3x = 5x$$

In general, we have the following result.

ADDING RATIONAL EXPRESSIONS WITH LIKE DENOMINATORS

If a, b, and d represent polynomials, then

$$\frac{a}{d} + \frac{b}{d} = \frac{a + b}{d} \qquad \text{provided the denominator is not equal to 0.}$$

EXAMPLE 1   Perform each addition. Assume that no denominators are 0.

Solution   In each part, we will add the numerators and keep the common denominator.

a. $\dfrac{xy}{8z} + \dfrac{3xy}{8z} = \dfrac{xy + 3xy}{8z}$ Add the numerators and keep the common denominator.

$\qquad = \dfrac{4xy}{8z}$ Combine like terms.

$\qquad = \dfrac{xy}{2z}$ $\dfrac{4xy}{8z} = \dfrac{4 \cdot xy}{4 \cdot 2z} = \dfrac{xy}{2z}$, because $\dfrac{4}{4} = 1$

b. $\dfrac{3x + y}{5x} + \dfrac{x + y}{5x} = \dfrac{3x + y + x + y}{5x}$ Add the numerators and keep the common denominator.

$\qquad = \dfrac{4x + 2y}{5x}$ Combine like terms.

SELF CHECK 1   Perform each addition. Assume no denominators are 0.

a. $\dfrac{x}{7} + \dfrac{y}{7}$ **b.** $\dfrac{3x}{7y} + \dfrac{4x}{7y}$

COMMENT   After adding two fractions, simplify the result if possible.

EXAMPLE 2   Add: $\dfrac{3x + 21}{5x + 10} + \dfrac{8x + 1}{5x + 10}$ $(x \neq -2)$

Solution   Since the rational expressions have the same denominator, we add their numerators and keep the common denominator.

$$\dfrac{3x + 21}{5x + 10} + \dfrac{8x + 1}{5x + 10} = \dfrac{3x + 21 + 8x + 1}{5x + 10} \qquad \text{Add the fractions.}$$

$$= \dfrac{11x + 22}{5x + 10} \qquad \text{Combine like terms.}$$

$$= \dfrac{11\overset{1}{\cancel{(x + 2)}}}{5\underset{1}{\cancel{(x + 2)}}} \qquad \text{Factor and divide out the common}$$
$$\qquad\qquad\qquad \text{factor of } (x + 2).$$

$$= \dfrac{11}{5}$$

 SELF CHECK 2   Add: $\dfrac{x + 4}{6x - 12} + \dfrac{x - 8}{6x - 12}$ $(x \neq 2)$

2   **Subtract two rational expressions with like denominators and write the answer in simplest form.**

To subtract rational expressions with a common denominator, we subtract their numerators and keep the common denominator.

SUBTRACTING RATIONAL EXPRESSIONS WITH LIKE DENOMINATORS
If a, b, and d represent polynomials, then

$$\dfrac{a}{d} - \dfrac{b}{d} = \dfrac{a - b}{d} \qquad \text{provided the denominator is not equal to 0.}$$

EXAMPLE 3   Subtract, assuming no divisions by zero.

a. $\dfrac{5x}{3} - \dfrac{2x}{3}$ **b.** $\dfrac{5x + 1}{x - 3} - \dfrac{4x - 2}{x - 3}$

Solution   In each part, the rational expressions have the same denominator. To subtract them, we subtract their numerators and keep the common denominator.

a. $\dfrac{5x}{3} - \dfrac{2x}{3} = \dfrac{5x - 2x}{3} \qquad \text{Subtract the numerators and keep the common denominator.}$

$$= \dfrac{3x}{3} \qquad \text{Combine like terms.}$$

$$= \dfrac{x}{1} \qquad \text{Divide out the common factor.}$$

$$= x$$

b. $\dfrac{5x+1}{x-3} - \dfrac{4x-2}{x-3} = \dfrac{(5x+1)-(4x-2)}{x-3}$ Subtract the numerators and keep the common denominator.

$\qquad\qquad\qquad\quad = \dfrac{5x+1-4x\,\textbf{+}\,2}{x-3}$ Use the distributive property to remove parentheses.

$\qquad\qquad\qquad\quad = \dfrac{x+3}{x-3}$ Combine like terms.

 SELF CHECK 3   Subtract: **a.** $\dfrac{9y}{4} - \dfrac{5y}{4}$ **b.** $\dfrac{2y+1}{y+5} - \dfrac{y-4}{y+5}$ $(y \neq -5)$

To add and/or subtract three or more rational expressions, we follow the order of operations.

EXAMPLE 4   Simplify: $\dfrac{3x+1}{x-7} - \dfrac{5x+2}{x-7} + \dfrac{2x+1}{x-7}$ $(x \neq 7)$

Solution   This example involves both addition and subtraction of rational expressions. Unless parentheses indicate otherwise, we do additions and subtractions from left to right.

$\dfrac{3x+1}{x-7} - \dfrac{5x+2}{x-7} + \dfrac{2x+1}{x-7}$

$\qquad = \dfrac{(3x+1)-(5x+2)+(2x+1)}{x-7}$ Combine the numerators and keep the common denominator.

$\qquad = \dfrac{3x+1-5x-2+2x+1}{x-7}$ Use the distributive property to remove parentheses.

$\qquad = \dfrac{0}{x-7}$ Combine like terms.

$\qquad = 0$ Simplify.

 SELF CHECK 4   Simplify: $\dfrac{2a-3}{a-5} + \dfrac{3a+2}{a-5} - \dfrac{24}{a-5}$ $(a \neq 5)$

Example 4 is a reminder that if the numerator of a rational expression is 0 and the denominator is not, the value of the expression is 0.

3   **Find the least common denominator (LCD) of two or more polynomials and use it to write equivalent rational expressions.**

Since the denominators of the fractions in the addition $\frac{4}{7} + \frac{3}{5}$ are different, we cannot add the fractions in their present form.

$\qquad\qquad$ four-sevenths $\qquad + \qquad$ three-fifths

$\qquad\qquad\qquad$ ↑—— Different denominators ——↑

To add these fractions, we need to find a common denominator. The smallest common denominator (called the **least** or **lowest common denominator**) is the easiest one to use.

LEAST COMMON DENOMINATOR   The least common denominator (LCD) for a set of fractions is the smallest number that each denominator will divide exactly.

We now review the method of writing two fractions using the LCD. In the addition $\frac{4}{7} + \frac{3}{5}$, the denominators are 7 and 5. The smallest number that 7 and 5 will divide exactly is 35. This is the LCD. We now build each fraction into a fraction with a denominator of 35.

$$\frac{4}{7} + \frac{3}{5} = \frac{4 \cdot 5}{7 \cdot 5} + \frac{3 \cdot 7}{5 \cdot 7} \qquad \text{Multiply numerator and denominator of } \frac{4}{7} \text{ by 5, and multiply}$$
$$\text{numerator and denominator of } \frac{3}{5} \text{ by 7.}$$

$$= \frac{20}{35} + \frac{21}{35} \qquad \text{Do the multiplications.}$$

Now that the fractions have a common denominator, we can add them.

$$\frac{20}{35} + \frac{21}{35} = \frac{20 + 21}{35} = \frac{41}{35}$$

EXAMPLE 5 Write each rational expression as an equivalent expression with a denominator of $30y$ $(y \neq 0)$.

a. $\frac{1}{2y}$ **b.** $\frac{3y}{5}$ **c.** $\frac{7x}{10y}$

Solution To build each rational expression into an expression with a denominator of $30y$, we multiply the numerator and denominator by what it takes to make the denominator $30y$.

a. $\frac{1}{2y} = \frac{1 \cdot 15}{2y \cdot 15} = \frac{15}{30y}$

b. $\frac{3y}{5} = \frac{3y \cdot 6y}{5 \cdot 6y} = \frac{18y^2}{30y}$

c. $\frac{7x}{10y} = \frac{7x \cdot 3}{10y \cdot 3} = \frac{21x}{30y}$

SELF CHECK 5 Write each rational expression as a rational expression with a denominator of $30ab$ $(a, b \neq 0)$.

a. $\frac{1}{5b}$ **b.** $\frac{8b}{3}$ **c.** $\frac{5a}{6b}$

There is a process that we can use to find the least common denominator of several rational expressions.

FINDING THE LEAST COMMON DENOMINATOR (LCD)

1. List the different denominators that appear in the rational expressions.
2. Completely factor each denominator.
3. Form a product using each different factor obtained in Step 2. Use each different factor the *greatest* number of times it appears in any *one* factorization. The product formed by multiplying these factors is the LCD.

EXAMPLE 6 Find the LCD of $\frac{5a}{24b}$, $\frac{11a}{18b}$, and $\frac{35a}{36b}$ $(b \neq 0)$.

Solution We list and factor each denominator into the product of prime numbers.

$$24b = 2 \cdot 2 \cdot 2 \cdot 3 \cdot b = 2^3 \cdot 3 \cdot b$$
$$18b = 2 \cdot 3 \cdot 3 \cdot b = 2 \cdot 3^2 \cdot b$$
$$36b = 2 \cdot 2 \cdot 3 \cdot 3 \cdot b = 2^2 \cdot 3^2 \cdot b$$

We then form a product with factors of 2, 3, and b. To find the LCD, we use each of these factors the *greatest* number of times it appears in any one factorization. We use 2 three times, because it appears three times as a factor of 24. We use 3 twice, because it occurs twice as a factor of 18 and 36. We use b once because it occurs once in each factor of $24b$, $18b$, and $36b$.

$$\text{LCD} = 2 \cdot 2 \cdot 2 \cdot 3 \cdot 3 \cdot b$$
$$= 8 \cdot 9 \cdot b$$
$$= 72b$$

 **SELF CHECK 6**   Find the LCD of $\dfrac{3y}{28z}$, $\dfrac{7xy}{12z}$, and $\dfrac{5x}{21z}$ $(z \neq 0)$.

4 **Add two rational expressions with unlike denominators and write the answer in simplest form.**

The process for adding and subtracting rational expressions with different denominators is the same as the process for adding and subtracting arithmetic expressions with different numerical denominators.

For example, to add $\frac{4x}{7}$ and $\frac{3x}{5}$, we first find the LCD of 7 and 5, which is 35. We then build the rational expressions so that each one has a denominator of 35. Finally, we add the results.

$$\frac{4x}{7} + \frac{3x}{5} = \frac{4x \cdot \mathbf{5}}{7 \cdot \mathbf{5}} + \frac{3x \cdot \mathbf{7}}{5 \cdot \mathbf{7}}$$ Multiply numerator and denominator of $\frac{4x}{7}$ by 5 and numerator and denominator of $\frac{3x}{5}$ by 7.

$$= \frac{20x}{35} + \frac{21x}{35}$$ Do the multiplications.

$$= \frac{41x}{35}$$ Add the numerators and keep the common denominator.

The following steps summarize how to add rational expressions that have unlike denominators.

ADDING RATIONAL EXPRESSIONS WITH UNLIKE DENOMINATORS

To add rational expressions with unlike denominators:

1. Find the LCD.

2. Write each rational expression as an equivalent expression with a denominator that is the LCD.

3. Add the resulting fractions.

4. Simplify the result, if possible.

EXAMPLE 7   Add: $\dfrac{5a}{24b}$, $\dfrac{11a}{18b}$, and $\dfrac{35a}{36b}$ $(b \neq 0)$

Solution   In Example 6, we saw that the LCD of these rational expressions is $2 \cdot 2 \cdot 2 \cdot 3 \cdot 3 \cdot b = 72b$. To add the rational expressions, we first factor each denominator:

$$\frac{5a}{24b} + \frac{11a}{18b} + \frac{35a}{36b} = \frac{5a}{2 \cdot 2 \cdot 2 \cdot 3 \cdot b} + \frac{11a}{2 \cdot 3 \cdot 3 \cdot b} + \frac{35a}{2 \cdot 2 \cdot 3 \cdot 3 \cdot b}$$

In each resulting expression, we multiply the numerator and the denominator by whatever it takes to build the denominator to the lowest common denominator of $2 \cdot 2 \cdot 2 \cdot 3 \cdot 3 \cdot b$.

$$= \frac{5a \cdot 3}{2 \cdot 2 \cdot 2 \cdot 3 \cdot b \cdot 3} + \frac{11a \cdot 2 \cdot 2}{2 \cdot 3 \cdot 3 \cdot b \cdot 2 \cdot 2} + \frac{35a \cdot 2}{2 \cdot 2 \cdot 3 \cdot 3 \cdot b \cdot 2}$$

$$= \frac{15a + 44a + 70a}{72b} \qquad \text{Do the multiplications.}$$

$$= \frac{129a}{72b} \qquad \text{Add the fractions.}$$

$$= \frac{43a}{24b} \qquad \text{Simplify.}$$

 **SELF CHECK 7**   Add: $\dfrac{3x}{28z}, \dfrac{5x}{21z},$ and $\dfrac{7x}{12z}$ $(z \neq 0)$

EXAMPLE 8   Add: $\dfrac{x+1}{x-2} + \dfrac{x+3}{x-1}$ $(x \neq 1, 2)$

Solution   We first find the LCD.

$$\left. \begin{array}{l} x - 2 \\ x - 1 \end{array} \right\} \quad \text{LCD} = (x-2)(x-1)$$

Then build the rational expressions so that each one has a denominator of $(x-2)(x-1)$.

$$\frac{x+1}{x-2} + \frac{x+3}{x-1} = \frac{(x+1)(\boldsymbol{x-1})}{(x-2)(\boldsymbol{x-1})} + \frac{(\boldsymbol{x-2})(x+3)}{(\boldsymbol{x-2})(x-1)}$$

Multiply the numerator and denominator of $\frac{(x+1)}{(x-2)}$ by $\frac{(x-1)}{(x-1)}$ and $\frac{(x+3)}{(x-1)}$ by $\frac{(x-2)}{(x-2)}$.

$$= \frac{x^2 - 1}{(x-2)(x-1)} + \frac{x^2 + x - 6}{(x-2)(x-1)} \qquad \text{Do the multiplications.}$$

$$= \frac{2x^2 + x - 7}{(x-2)(x-1)} \qquad \text{Add the fractions.}$$

This is in simplified form.

 **SELF CHECK 8**   Add: $\dfrac{x+5}{x+1} + \dfrac{x-4}{x-3}$ $(x \neq -1, 3)$

EXAMPLE 9   Add: $\dfrac{1}{x} + \dfrac{x}{y}$ $(x, y \neq 0)$

Solution   By inspection, the LCD is xy.

COMMENT   Building fractions means to multiply each fraction by 1 in the form necessary to obtain an equivalent fraction with a new denominator of the LCD.

$$\frac{1}{x} + \frac{x}{y} = \frac{1(\boldsymbol{y})}{x(\boldsymbol{y})} + \frac{(\boldsymbol{x})x}{(\boldsymbol{x})y} \qquad \text{Build the fractions.}$$

$$= \frac{y}{xy} + \frac{x^2}{xy} \qquad \text{Do the multiplications.}$$

$$= \frac{y + x^2}{xy} \qquad \text{Add the fractions.}$$

 **SELF CHECK 9**   Add: $\dfrac{a}{b} + \dfrac{3}{a}$ $(a, b \neq 0)$

5 Subtract two rational expressions with unlike denominators and write the answer in simplest form.

To subtract rational expressions with unlike denominators, we first write them as expressions with the same denominator and then subtract the numerators.

EXAMPLE 10   Subtract: $\dfrac{x}{x+1} - \dfrac{3}{x}$ $(x \ne 0, -1)$

Solution   Because x and $x + 1$ represent different values and have no common factors, the least common denominator (LCD) is their product, $(x + 1)x$.

$$\frac{x}{x+1} - \frac{3}{x} = \frac{x(\mathbf{x})}{(x+1)\mathbf{x}} - \frac{3(\mathbf{x+1})}{x(\mathbf{x+1})} \qquad \text{Build the fractions to obtain the common denominator.}$$

$$= \frac{x(x) - 3(x+1)}{(x+1)x} \qquad \text{Subtract the numerators and keep the common denominator.}$$

$$= \frac{x^2 - 3x - 3}{(x+1)x} \qquad \text{Do the multiplication in the numerator.}$$

SELF CHECK 10   Subtract: $\dfrac{a}{a-1} - \dfrac{5}{a}$ $(a \ne 0, 1)$

EXAMPLE 11   Subtract: $\dfrac{a}{a-1} - \dfrac{2}{a^2-1}$ $(a \ne 1, -1)$

Solution   To find the LCD, we factor the denominators, where possible.

$$\left. \begin{array}{l} a - 1 = \mathbf{a - 1} \\ a^2 - 1 = (a+1)(\mathbf{a - 1}) \end{array} \right\} \quad \text{LCD} = (a+1)(a-1)$$

After finding the LCD, we proceed as follows:

$$\frac{a}{a-1} - \frac{2}{a^2-1}$$

$$= \frac{a}{(a-1)} - \frac{2}{(a+1)(a-1)} \qquad \text{Factor the denominator.}$$

$$= \frac{a(\mathbf{a+1})}{(a-1)(\mathbf{a+1})} - \frac{2}{(a+1)(a-1)} \qquad \text{Build the first fraction.}$$

$$= \frac{a(a+1) - 2}{(a-1)(a+1)} \qquad \text{Subtract the numerators and keep the common denominator.}$$

$$= \frac{a^2 + a - 2}{(a-1)(a+1)} \qquad \text{Use the distributive property to remove parentheses.}$$

$$= \frac{(a+2)\overset{1}{\cancel{(a-1)}}}{\underset{1}{\cancel{(a-1)}}(a+1)} \qquad \text{Factor and divide out the common factor of } a - 1.$$

$$= \frac{a+2}{a+1} \qquad \text{Simplify.}$$

SELF CHECK 11   Subtract: $\dfrac{b}{b+1} - \dfrac{3}{b^2-1}$ $(b \ne 1, -1)$

EXAMPLE 12   Subtract: $\dfrac{3}{x-y} - \dfrac{x}{y-x}$ $(x \neq y)$

Solution   We note that the second denominator is the negative of the first, so we can multiply both the numerator and the denominator of the second fraction by -1 to obtain

$$\dfrac{3}{x-y} - \dfrac{x}{y-x} = \dfrac{3}{x-y} - \dfrac{-1x}{-1(y-x)} \qquad \text{Multiply both the numerator and the denominator by } -1.$$

$$= \dfrac{3}{x-y} - \dfrac{-x}{-y+x} \qquad \text{Use the distributive property to remove parentheses.}$$

$$= \dfrac{3}{x-y} - \dfrac{-x}{x-y} \qquad \text{Apply the commutative property of addition.}$$

$$= \dfrac{3-(-x)}{x-y} \qquad \text{Subtract the numerators and keep the common denominator.}$$

$$= \dfrac{3+x}{x-y}$$

SELF CHECK 12   Subtract: $\dfrac{5}{a-b} - \dfrac{2}{b-a}$ $(a \neq b)$

EXAMPLE 13   Perform the operations $\dfrac{3}{x^2 - y^2} + \dfrac{2}{x-y} - \dfrac{1}{x+y}$. Assume that no denominator is 0.

Solution   Find the least common denominator.

$$\left. \begin{array}{l} x^2 - y^2 = (x-y)(x+y) \\ x - y = x - y \\ x + y = x + y \end{array} \right\} \qquad \text{Factor each denominator, where possible.}$$

Since the least common denominator is $(x-y)(x+y)$, we build each fraction into a new fraction with that common denominator.

$$\dfrac{3}{x^2 - y^2} + \dfrac{2}{x-y} - \dfrac{1}{x+y}$$

$$= \dfrac{3}{(x-y)(x+y)} + \dfrac{2}{x-y} - \dfrac{1}{x+y} \qquad \text{Factor.}$$

$$= \dfrac{3}{(x-y)(x+y)} + \dfrac{2(x+y)}{(x-y)(x+y)} - \dfrac{1(x-y)}{(x+y)(x-y)} \qquad \text{Build each fraction.}$$

$$= \dfrac{3 + 2(x+y) - 1(x-y)}{(x-y)(x+y)} \qquad \text{Combine the numerators and keep the common denominator.}$$

$$= \dfrac{3 + 2x + 2y - x + y}{(x-y)(x+y)} \qquad \text{Use the distributive property to remove parentheses.}$$

$$= \dfrac{3 + x + 3y}{(x-y)(x+y)} \qquad \text{Combine like terms.}$$

SELF CHECK 13   Perform the operations: $\dfrac{5}{a^2 - b^2} - \dfrac{3}{a+b} + \dfrac{4}{a-b}$ $(a \neq b, -b)$

NOW TRY THIS

Simplify each expression. Assume no denominators are 0.

1. $\dfrac{5}{x+3} - \dfrac{2}{x-3}$

2. $\left(\dfrac{2}{3x} - 1\right) \div \left(\dfrac{4}{9x} - x\right)$

3. $x^{-1} + x^{-2}$

6.3 Exercises

WARM-UPS *Determine whether the expressions are equal.*

1. $\dfrac{2}{3}, \dfrac{12}{18}$

2. $\dfrac{3}{8}, \dfrac{15}{40}$

3. $\dfrac{6}{11}, \dfrac{24}{42}$

4. $\dfrac{5}{10}, \dfrac{15}{30}$

5. $\dfrac{x}{3}, \dfrac{3x}{9}$

6. $\dfrac{5}{3}, \dfrac{5x}{3y}$ $(y \neq 0)$

7. $\dfrac{5}{3}, \dfrac{5x}{3x}$ $(x \neq 0)$

8. $\dfrac{4y}{20}, \dfrac{y}{5}$

REVIEW *Write each number in prime-factored form.*

9. 81

10. 64

11. 136

12. 242

13. 102

14. 315

15. 144

16. 217

VOCABULARY AND CONCEPTS *Fill in the blanks.*

17. The _____ for a set of rational expressions is the smallest number that each denominator divides exactly.

18. When we multiply the numerator and denominator of a rational expression by some number to get a common denominator, we say that we are _____ the fraction.

19. To add two rational expressions with like denominators, we add their _____ and keep the _____.

20. To subtract two rational expressions with _____ denominators, we need to find a common denominator.

GUIDED PRACTICE *Perform the operations. Simplify answers, if possible. Assume that no denominators are 0.* SEE EXAMPLES 1–2. (OBJECTIVE 1)

21. $\dfrac{1}{8a} + \dfrac{1}{8a}$

22. $\dfrac{3}{4y} + \dfrac{3}{4y}$

23. $\dfrac{2x}{y} + \dfrac{2x}{y}$

24. $\dfrac{8y}{7x} + \dfrac{6y}{7x}$

25. $\dfrac{4y-1}{y-4} + \dfrac{5y+3}{y-4}$

26. $\dfrac{2x+3}{x+4} + \dfrac{3x-1}{x+4}$

27. $\dfrac{3x-5}{x-2} + \dfrac{6x-13}{x-2}$

28. $\dfrac{8x-7}{x+3} + \dfrac{2x+37}{x+3}$

Perform the operations. Simplify answers, if possible. Assume that no denominators are 0. SEE EXAMPLE 3. (OBJECTIVE 2)

29. $\dfrac{35}{72} - \dfrac{44}{72}$

30. $\dfrac{5}{48} - \dfrac{11}{48}$

31. $\dfrac{9y}{3x} - \dfrac{6y}{3x}$

32. $\dfrac{9y}{x} - \dfrac{5y}{x}$

33. $\dfrac{4y-5}{2y} - \dfrac{3}{2y}$

34. $\dfrac{6x+3}{5x} - \dfrac{3x}{5x}$

35. $\dfrac{6x-5}{3xy} - \dfrac{3x-5}{3xy}$

36. $\dfrac{7x+7}{5y} - \dfrac{2x+7}{5y}$

37. $\dfrac{3y-2}{y+3} - \dfrac{2y-5}{y+3}$

38. $\dfrac{5x+8}{x+5} - \dfrac{3x-2}{x+5}$

39. $\dfrac{5y+3}{y-4} - \dfrac{4y-1}{y-4}$

40. $\dfrac{2x-1}{x-3} - \dfrac{x+2}{x-3}$

Perform the operations. Simplify answers, if possible. Assume that no denominators are 0. SEE EXAMPLE 4. (OBJECTIVES 1–2)

41. $\dfrac{11x}{12} + \dfrac{7x}{12} - \dfrac{2x}{12}$

42. $\dfrac{13y}{32} + \dfrac{13y}{32} - \dfrac{10y}{32}$

43. $\dfrac{3x+1}{x-2} + \dfrac{5x+2}{x-2} - \dfrac{2x+1}{x-2}$

44. $\dfrac{2y-3}{y+2} - \dfrac{y-3}{y+2} + \dfrac{3y+6}{y+2}$

45. $\dfrac{4b+5}{b+1} - \dfrac{6b-2}{b+1} + \dfrac{b-7}{b+1}$

46. $\dfrac{7a+1}{a-1} + \dfrac{a+1}{a-1} - \dfrac{10a}{a-1}$

47. $\dfrac{x+1}{x-2} - \dfrac{2(x-3)}{x-2} + \dfrac{3(x+1)}{x-2}$

48. $\dfrac{3xy}{x-y} - \dfrac{x(3y-x)}{x-y} - \dfrac{x(x-y)}{x-y}$

Build each fraction into an equivalent fraction with the indicated denominator. Assume that no denominators are 0. SEE EXAMPLE 5. (OBJECTIVE 3)

49. $\dfrac{21}{8}$; 32

50. $\dfrac{7}{x}$; xy

51. $\dfrac{8}{x}$; $x^2 y$

52. $\dfrac{7}{y}$; xy^2

53. $\dfrac{4x}{x+3}$; $(x+3)^2$

54. $\dfrac{5y}{y-2}$; $(y-2)^2$

55. $\dfrac{2y}{x}$; $x^2 + x$

56. $\dfrac{3x}{y}$; $y^2 - y$

57. $\dfrac{z}{z-1}$; $z^2 - 1$

58. $\dfrac{z}{z+3}$; $z^2 - 9$

59. $\dfrac{2}{x+1}$; $x^2 + 3x + 2$

60. $\dfrac{3}{x-1}$; $x^2 + x - 2$

Several denominators are given. Find the LCD. SEE EXAMPLE 6. (OBJECTIVE 3)

61. $2x, 6x$

62. $10y, 15y$

63. $3x, 6y, 9xy$

64. $2x^2, 6y, 3xy$

65. $x^2 - 4, x + 2$

66. $y^2 - 9, y - 3$

67. $x^2 + 6x, x + 6, x$

68. $xy^2 - xy, xy, y - 1$

Perform the operations. Simplify answers, if possible. Assume that no denominators are 0. SEE EXAMPLE 7. (OBJECTIVE 4)

69. $\dfrac{4x}{3y} + \dfrac{2x}{y}$

70. $\dfrac{5x}{2z} + \dfrac{7x}{6z}$

71. $\dfrac{x+2}{2x} + \dfrac{x-1}{3x}$

72. $\dfrac{x+3}{x^2} + \dfrac{x+5}{2x}$

Perform the operations. Simplify answers, if possible. Assume that no denominators are 0. SEE EXAMPLE 8. (OBJECTIVE 4)

73. $\dfrac{x+1}{x-1} + \dfrac{x-1}{x+1}$

74. $\dfrac{2x}{x+2} + \dfrac{x+1}{x-3}$

75. $\dfrac{x}{5x+2} + \dfrac{x-1}{x+2}$

76. $\dfrac{2x-1}{3x+2} + \dfrac{x+4}{2x+3}$

Perform the operations. Simplify answers, if possible. Assume that no denominators are 0. SEE EXAMPLE 9. (OBJECTIVE 4)

77. $\dfrac{x-2}{x} + \dfrac{y+2}{y}$

78. $\dfrac{a+2}{b} + \dfrac{b-2}{a}$

79. $\dfrac{3y}{x} + \dfrac{x+1}{y-1}$

80. $\dfrac{a}{b} + \dfrac{2b-1}{a+2}$

Perform the operations. Simplify answers, if possible. Assume that no denominators are 0. SEE EXAMPLE 10. (OBJECTIVE 5)

81. $\dfrac{5}{x} - \dfrac{x+2}{x+1}$

82. $\dfrac{x+2}{x+1} - \dfrac{5}{x}$

83. $\dfrac{2x+3}{x+5} - \dfrac{x-1}{x+2}$

84. $\dfrac{x-1}{x+5} - \dfrac{2x+3}{x+2}$

Perform the operations. Simplify answers, if possible. Assume that no denominators are 0. SEE EXAMPLE 11. (OBJECTIVE 5)

85. $\dfrac{x}{x-2} + \dfrac{4+2x}{x^2-4}$

86. $\dfrac{y}{y+3} - \dfrac{2y-6}{y^2-9}$

87. $\dfrac{x+1}{x+2} - \dfrac{x^2+1}{x^2-x-6}$

88. $\dfrac{x+1}{2x+4} - \dfrac{x^2}{2x^2-8}$

Perform the operations. Simplify answers, if possible. Assume that no denominators are 0. SEE EXAMPLE 12. (OBJECTIVE 5)

89. $\dfrac{y+3}{y-1} - \dfrac{y+4}{1-y}$

90. $\dfrac{2x+2}{x-2} - \dfrac{2x}{2-x}$

91. $\dfrac{x+5}{2x-y} - \dfrac{x-1}{y-2x}$

92. $\dfrac{2a-b}{a-2b} - \dfrac{a+3b}{2b-a}$

Perform the operations. Simplify answers, if possible. Assume that no denominators are 0. SEE EXAMPLE 13. (OBJECTIVES 4–5)

93. $\dfrac{2x}{x^2-3x+2} + \dfrac{2x}{x-1} - \dfrac{x}{x-2}$

94. $\dfrac{4a}{a-2} - \dfrac{3a}{a-3} + \dfrac{4a}{a^2-5a+6}$

95. $\dfrac{a}{a-1} - \dfrac{2}{a+2} + \dfrac{3(a-2)}{a^2+a-2}$

96. $\dfrac{2x}{x-1} + \dfrac{3x}{x+1} - \dfrac{x+3}{x^2-1}$

ADDITIONAL PRACTICE *Several denominators are given. Find the LCD.*

97. $x^2 - x - 6, x^2 - 9$

98. $x^2 - 4x - 5, x^2 - 25$

99. $\dfrac{15x}{6y} - \dfrac{7x}{8}$

Perform the operations. Assume that no denominators are 0.

100. $\dfrac{2x}{y} - \dfrac{x}{y}$

101. $\dfrac{1}{2} + \dfrac{2}{3}$

102. $\dfrac{2y}{9} + \dfrac{y}{3}$

103. $\dfrac{2}{3} - \dfrac{5}{6}$

104. $\dfrac{8a}{15} - \dfrac{5a}{12}$

105. $\dfrac{2y}{5x} - \dfrac{y}{2}$

106. $\dfrac{x}{x+1} + \dfrac{x-1}{x}$

107. $\dfrac{x+2}{x} + \dfrac{x-5}{x+2}$

108. $\dfrac{x+5}{xy} - \dfrac{x-1}{x^2y}$

109. $\dfrac{y-7}{y^2} - \dfrac{y+7}{2y}$

110. $\dfrac{x}{3y} + \dfrac{2x}{3y} - \dfrac{x}{3y}$

111. $\dfrac{5y}{8x} + \dfrac{4y}{8x} - \dfrac{y}{8x}$

112. $\dfrac{5r^2}{2r} - \dfrac{r^2}{2r}$

113. $\dfrac{3x}{y+2} - \dfrac{3y}{y+2} + \dfrac{x+y}{y+2}$

114. $\dfrac{3y}{x-5} + \dfrac{x}{x-5} - \dfrac{y-x}{x-5}$

115. $\dfrac{-a}{3a^2-27} + \dfrac{1}{3a+9}$

116. $\dfrac{d}{d^2+6d+5} - \dfrac{d}{d^2+5d+4}$

117. $14 + \dfrac{10}{y^2}$

118. $\dfrac{2}{x} - 3x$

WRITING ABOUT MATH

119. Explain how to add rational expressions with the same denominator.

120. Explain how to subtract rational expressions with the same denominator.

121. Explain how to find a lowest common denominator.

122. Explain how to add two rational expressions with different denominators.

SOMETHING TO THINK ABOUT

123. Find the error:

$$\dfrac{2x+3}{x+5} - \dfrac{x+2}{x+5} = \dfrac{2x+3-x+2}{x+5}$$
$$= \dfrac{x+5}{x+5}$$
$$= 1$$

124. Find the error:

$$\dfrac{5x-4}{y} + \dfrac{x}{y} = \dfrac{5x-4+x}{y+y}$$
$$= \dfrac{6x-4}{2y}$$
$$= \dfrac{3x-2}{y}$$

Show that each formula is true.

125. $\dfrac{a}{b} + \dfrac{c}{d} = \dfrac{ad+bc}{bd}$

126. $\dfrac{a}{b} - \dfrac{c}{d} = \dfrac{ad-bc}{bd}$

Simplifying Complex Fractions

Objectives

1 Simplify a complex fraction.
2 Simplify a fraction containing terms with negative exponents.

Vocabulary

complex fraction

Getting Ready

Use the distributive property to remove parentheses, and simplify.

1. $3\left(1 + \dfrac{1}{3}\right)$ **2.** $10\left(\dfrac{1}{5} - 2\right)$ **3.** $4\left(\dfrac{3}{2} + \dfrac{1}{4}\right)$ **4.** $14\left(\dfrac{3}{7} - 1\right)$

5. $x\left(\dfrac{3}{x} + 3\right)$ **6.** $y\left(\dfrac{2}{y} - 1\right)$ **7.** $4x\left(3 - \dfrac{1}{2x}\right)$ **8.** $6xy\left(\dfrac{1}{2x} + \dfrac{1}{3y}\right)$

In this section, we will consider fractions that contain fractions. These complicated fractions are called *complex fractions*.

1 Simplify a complex fraction.

Fractions such as

$$\frac{\frac{1}{3}}{\frac{4}{}}, \qquad \frac{\frac{5}{3}}{\frac{2}{9}}, \qquad \frac{x + \frac{1}{2}}{3 - x}, \qquad \text{and} \qquad \frac{\frac{x+1}{2}}{x + \frac{1}{x}}$$

that contain fractions in their numerators and/or denominators are called **complex fractions**. Complex fractions should be simplified. For example, we can simplify

$$\frac{\frac{5x}{3}}{\frac{2y}{9}}$$

by doing the division:

$$\frac{\frac{5x}{3}}{\frac{2y}{9}} = \frac{5x}{3} \div \frac{2y}{9} = \frac{5x}{3} \cdot \frac{9}{2y} = \frac{5x \cdot 3 \cdot \overset{1}{\cancel{3}}}{\underset{1}{\cancel{3}} \cdot 2y} = \frac{15x}{2y}$$

There are two methods that we can use to simplify complex fractions.

SIMPLIFYING COMPLEX FRACTIONS

METHOD 1

Write the numerator and the denominator of the complex fraction as single fractions. Then divide the fractions and simplify.

METHOD 2

Multiply the numerator and denominator of the complex fraction by the LCD of the fractions in its numerator and denominator. Then simplify the results, if possible.

Hypatia

AD 370–415

Hypatia is the earliest known woman in the history of mathematics. She was a professor at the University of Alexandria. Because of her scientific beliefs, she was considered to be a heretic. At the age of 45, she was attacked by a mob and murdered for her beliefs.

Using Method 1 to simplify $\dfrac{\dfrac{3x}{5} + 1}{2 - \dfrac{x}{5}}$ (assuming no division by 0), we proceed as follows:

$$\dfrac{\dfrac{3x}{5} + 1}{2 - \dfrac{x}{5}} = \dfrac{\dfrac{3x}{5} + \dfrac{5}{5}}{\dfrac{10}{5} - \dfrac{x}{5}}$$

Write 1 as $\frac{5}{5}$ and 2 as $\frac{10}{5}$.

$$= \dfrac{\dfrac{3x + 5}{5}}{\dfrac{10 - x}{5}}$$

Add the fractions in the numerator and subtract the fractions in the denominator.

$$= \dfrac{3x + 5}{5} \div \dfrac{10 - x}{5}$$

Write the complex fraction as an equivalent division problem.

$$= \dfrac{3x + 5}{5} \cdot \dfrac{5}{10 - x}$$

Invert the divisor and multiply.

$$= \dfrac{(3x + 5)\overset{1}{5}}{\underset{1}{5}(10 - x)}$$

Multiply the fractions.

$$= \dfrac{3x + 5}{10 - x}$$

Divide out the common factor of 5: $\frac{5}{5} = 1$.

To use Method 2, we first determine that the LCD of the fractions in the numerator and denominator is 5. We then multiply both the numerator and denominator by 5.

$$\dfrac{\dfrac{3x}{5} + 1}{2 - \dfrac{x}{5}} = \dfrac{5\left(\dfrac{3x}{5} + 1\right)}{5\left(2 - \dfrac{x}{5}\right)}$$

Multiply both numerator and denominator by 5.

$$= \dfrac{5 \cdot \dfrac{3x}{5} + 5 \cdot 1}{5 \cdot 2 - 5 \cdot \dfrac{x}{5}}$$

Use the distributive property to remove parentheses.

$$= \dfrac{3x + 5}{10 - x}$$

Simplify.

With practice, you will be able to see which method is easier to understand in any given situation.

EXAMPLE 1   Simplify $\dfrac{\frac{x}{3}}{\frac{y}{3}}$. Assume that no denominators are 0.

Solution   We will simplify the complex fraction using both methods.

Method 1

$$\frac{\frac{x}{3}}{\frac{y}{3}} = \frac{x}{3} \div \frac{y}{3}$$

$$= \frac{x}{3} \cdot \frac{3}{y}$$

$$= \frac{3x}{3y}$$

$$= \frac{x}{y}$$

Method 2

$$\frac{\frac{x}{3}}{\frac{y}{3}} = \frac{3\left(\frac{x}{3}\right)}{3\left(\frac{y}{3}\right)}$$

$$= \frac{\frac{x}{1}}{\frac{y}{1}}$$

$$= \frac{x}{y}$$

SELF CHECK 1   Simplify $\dfrac{\frac{a}{4}}{\frac{5}{b}}$. Assume no denominator is 0.

EXAMPLE 2   Simplify $\dfrac{\frac{x}{x+1}}{\frac{y}{x}}$. Assume no denominator is 0.

Solution   We will simplify the complex fraction using both methods.

Method 1

$$\frac{\frac{x}{x+1}}{\frac{y}{x}} = \frac{x}{x+1} \div \frac{y}{x}$$

$$= \frac{x}{x+1} \cdot \frac{x}{y}$$

$$= \frac{x^2}{y(x+1)}$$

Method 2

$$\frac{\frac{x}{x+1}}{\frac{y}{x}} = \frac{x(x+1)\left(\frac{x}{x+1}\right)}{x(x+1)\left(\frac{y}{x}\right)}$$

$$= \frac{\frac{x^2}{1}}{\frac{y(x+1)}{1}}$$

$$= \frac{x^2}{y(x+1)}$$

SELF CHECK 2   Simplify $\dfrac{\frac{x}{y}}{\frac{x}{y+1}}$. Assume no denominator is 0.

EXAMPLE 3   Simplify $\dfrac{1+\dfrac{1}{x}}{1-\dfrac{1}{x}}$. Assume no denominator is 0.

Solution   We will simplify the complex fraction using both methods.

Method 1

$$\frac{1+\dfrac{1}{x}}{1-\dfrac{1}{x}}=\frac{\dfrac{x}{x}+\dfrac{1}{x}}{\dfrac{x}{x}-\dfrac{1}{x}}$$

$$=\frac{\dfrac{x+1}{x}}{\dfrac{x-1}{x}}$$

$$=\frac{x+1}{x}\div\frac{x-1}{x}$$

$$=\frac{x+1}{x}\cdot\frac{x}{x-1}$$

$$=\frac{(x+1)x}{x(x-1)}$$

$$=\frac{x+1}{x-1}$$

Method 2

$$\frac{1+\dfrac{1}{x}}{1-\dfrac{1}{x}}=\frac{x\left(1+\dfrac{1}{x}\right)}{x\left(1-\dfrac{1}{x}\right)}$$

$$=\frac{x+1}{x-1}$$

SELF CHECK 3   Simplify $\dfrac{\dfrac{1}{x}+1}{\dfrac{1}{x}-1}$. Assume no denominator is 0.

EXAMPLE 4   Simplify $\dfrac{1}{1+\dfrac{1}{x+1}}$. Assume no denominator is 0.

Solution   We will simplify this complex fraction by using Method 2 only.

$$\frac{1}{1+\dfrac{1}{x+1}}=\frac{(x+1)\cdot 1}{(x+1)\left(1+\dfrac{1}{x+1}\right)}$$ Multiply the numerator and denominator of the complex fraction by $(x+1)$.

$$=\frac{x+1}{(x+1)1+1}$$ Use the distributive property.

$$=\frac{x+1}{x+2}$$ Simplify.

SELF CHECK 4   Simplify $\dfrac{2}{\dfrac{1}{x+2}-2}$. Assume no denominator is 0.

2 Simplify a fraction containing terms with negative exponents.

Many fractions with terms containing negative exponents are complex fractions as the next example illustrates.

EXAMPLE 5 Simplify $\dfrac{x^{-1} + y^{-2}}{x^{-2} - y^{-1}}$. Assume no denominator is 0.

Solution We will write each expression using positive exponents and then simplify the complex fraction using Method 2:

$$\frac{x^{-1} + y^{-2}}{x^{-2} - y^{-1}} = \frac{\dfrac{1}{x} + \dfrac{1}{y^2}}{\dfrac{1}{x^2} - \dfrac{1}{y}} \qquad \text{Write without negative exponents.}$$

$$= \frac{x^2 y^2 \left(\dfrac{1}{x} + \dfrac{1}{y^2} \right)}{x^2 y^2 \left(\dfrac{1}{x^2} - \dfrac{1}{y} \right)} \qquad \begin{array}{l}\text{Multiply the numerator and denominator of the complex}\\ \text{fraction by } x^2 y^2, \text{ the LCD.}\end{array}$$

$$= \frac{xy^2 + x^2}{y^2 - x^2 y} \qquad \text{Distribute.}$$

$$= \frac{x(y^2 + x)}{y(y - x^2)} \qquad \text{Factor the numerator and denominator.}$$

The result cannot be simplified. Therefore either of the last two steps is a correct answer.

SELF CHECK 5 Simplify $\dfrac{x^{-2} - y^{-1}}{x^{-1} + y^{-2}}$. Assume no denominator is 0.

SELF CHECK ANSWERS
1. $\dfrac{ab}{20}$ 2. $\dfrac{y+1}{y}$ 3. $\dfrac{1+x}{1-x}$ 4. $\dfrac{2(x+2)}{-2x-3}$ 5. $\dfrac{y(y-x^2)}{x(y^2+x)}$

NOW TRY THIS

Simplify each complex fraction. Assume no division by 0.

1. $\dfrac{\dfrac{a}{y^2}}{\dfrac{b}{x^3}}$

2. $\dfrac{\dfrac{x}{x+2} + \dfrac{5}{x}}{\dfrac{1}{3x} + \dfrac{x}{2x+4}}$

6.4 Exercises

WARM-UPS *Simplify each complex fraction.*

1. $\dfrac{\frac{2}{3}}{\frac{1}{2}}$

2. $\dfrac{\frac{2}{3}}{\frac{3}{4}}$

3. $\dfrac{\frac{6}{7}}{\frac{8}{21}}$

4. $\dfrac{\frac{4}{5}}{\frac{32}{15}}$

5. $\dfrac{\frac{7}{8}}{\frac{49}{4}}$

6. $\dfrac{2}{\frac{1}{2}}$

7. $\dfrac{\frac{1}{2}}{2}$

8. $\dfrac{1 + \frac{1}{2}}{\frac{1}{2}}$

9. $\dfrac{\frac{2}{3} + 1}{\frac{1}{3} + 1}$

10. $\dfrac{\frac{5}{4} - 3}{\frac{3}{4} - 3}$

11. $\dfrac{\frac{1}{2} + \frac{3}{4}}{\frac{3}{2} + \frac{1}{4}}$

12. $\dfrac{\frac{2}{3} - \frac{5}{2}}{\frac{2}{3} - \frac{3}{2}}$

REVIEW *Write each expression as an expression involving only one exponent. Assume no variable is zero.*

13. $t^5 t^2 t$

14. $(a^0 a^2)^3$

15. $-2r(r^3)^2$

16. $(b^4)^2 (b^6)^0$

Write each expression without parentheses or negative exponents.

17. $\left(\dfrac{3r}{4r^3}\right)^{-4}$

18. $\left(\dfrac{12y^{-3}}{3y^2}\right)^{-3}$

19. $\left(\dfrac{6r^{-2}}{2r^3}\right)^{-2}$

20. $\left(\dfrac{4x^3}{5x^{-3}}\right)^{-2}$

VOCABULARY AND CONCEPTS *Fill in the blanks.*

21. If a fraction contains a fraction in its numerator and/or denominator, it is called a _____.

22. The denominator of the complex fraction
$$\dfrac{\frac{3}{x} + \frac{x}{y}}{\frac{1}{x} + 2} \text{ is } \underline{\hphantom{aa}}.$$

23. In Method 1, we write the numerator and denominator of a complex fraction as _____ fractions and then _____.

24. In Method 2, we multiply the numerator and denominator of the complex fraction by the _____ of the fractions in its numerator and denominator.

GUIDED PRACTICE *Simplify each complex fraction. Assume no division by 0. SEE EXAMPLE 1. (OBJECTIVE 1)*

25. $\dfrac{\frac{2x}{y}}{\frac{4}{xy}}$

26. $\dfrac{\frac{y}{x}}{\frac{x}{xy}}$

27. $\dfrac{\frac{5t^2}{9x^2}}{\frac{3t}{x^2 t}}$

28. $\dfrac{\frac{4w^2}{5t}}{\frac{w}{15t}}$

Simplify each complex fraction. Assume no division by 0. SEE EXAMPLE 2. (OBJECTIVE 1)

29. $\dfrac{\frac{a}{b}}{\frac{a}{a+1}}$

30. $\dfrac{\frac{x}{y-1}}{\frac{x}{y}}$

31. $\dfrac{\frac{x}{y-1}}{\frac{x}{y+1}}$

32. $\dfrac{\frac{x+y}{x-y}}{\frac{x}{x+y}}$

Simplify each complex fraction. Assume no division by 0. SEE EXAMPLE 3. (OBJECTIVE 1)

33. $\dfrac{\frac{1}{y} + 3}{\frac{3}{y} - 2}$

34. $\dfrac{2 + \frac{1}{x}}{2 - \frac{3}{x}}$

35. $\dfrac{5 + \frac{3}{x}}{3 + \frac{2}{x}}$

36. $\dfrac{\frac{3}{x} - 3}{\frac{9}{x} - 3}$

37. $\dfrac{\frac{2}{a+2} + 1}{\frac{3}{a+2}}$

38. $\dfrac{3 - \frac{2}{m-3}}{\frac{4}{m-3}}$

39. $\dfrac{\frac{3}{x} + \frac{4}{x+1}}{\frac{2}{x+1} - \frac{3}{x}}$

40. $\dfrac{\frac{5}{y-3} - \frac{2}{y}}{\frac{1}{y} + \frac{2}{y-3}}$

41. $\dfrac{\frac{3y}{x} - y}{y - \frac{y}{x}}$

42. $\dfrac{\frac{y}{x} + 3y}{y + \frac{2y}{x}}$

43. $\dfrac{1}{\dfrac{1}{x}+\dfrac{1}{y}}$

44. $\dfrac{1}{\dfrac{b}{a}-\dfrac{a}{b}}$

45. $\dfrac{\dfrac{2}{x}}{\dfrac{2}{y}-\dfrac{4}{x}}$

46. $\dfrac{\dfrac{2y}{3}}{\dfrac{2y}{3}-\dfrac{8}{y}}$

47. $\dfrac{\dfrac{3}{x}+\dfrac{2x}{y}}{\dfrac{4}{x}}$

48. $\dfrac{\dfrac{4}{a}-\dfrac{a}{b}}{\dfrac{b}{a}}$

Simplify each complex fraction. Assume no division by 0. SEE
EXAMPLE 4. *(OBJECTIVE 1)*

49. $\dfrac{\dfrac{1}{x+1}}{1+\dfrac{1}{x+1}}$

50. $\dfrac{\dfrac{1}{x-1}}{1-\dfrac{1}{x-1}}$

51. $\dfrac{\dfrac{x}{x+2}}{\dfrac{x}{x+2}+x}$

52. $\dfrac{\dfrac{2}{x-2}}{\dfrac{2}{x-2}-1}$

53. $\dfrac{\dfrac{2}{x}-\dfrac{3}{x+1}}{\dfrac{2}{x+1}-\dfrac{3}{x}}$

54. $\dfrac{\dfrac{5}{y}+\dfrac{4}{y+1}}{\dfrac{4}{y}-\dfrac{5}{y+1}}$

55. $\dfrac{\dfrac{m}{m+2}-\dfrac{2}{m-1}}{\dfrac{3}{m+2}+\dfrac{m}{m-1}}$

56. $\dfrac{\dfrac{2a}{a-3}+\dfrac{1}{a-2}}{\dfrac{a}{a-2}-\dfrac{3}{a-3}}$

Simplify each complex fraction. Assume no division by 0. SEE
EXAMPLE 5. *(OBJECTIVE 2)*

57. $\dfrac{x^{-2}+1}{x^{-1}+1}$

58. $\dfrac{3x^{-1}+2}{3x^{-1}-1}$

59. $\dfrac{y^{-2}+1}{y^{-2}-1}$

60. $\dfrac{1+x^{-1}}{x^{-1}-1}$

61. $\dfrac{a^{-2}+a}{a+1}$

62. $\dfrac{t-t^{-2}}{1-t^{-1}}$

63. $\dfrac{2x^{-1}+4x^{-2}}{2x^{-2}+x^{-1}}$

64. $\dfrac{x^{-2}-3x^{-3}}{3x^{-2}-9x^{-3}}$

ADDITIONAL PRACTICE   *Simplify each complex fraction.
Assume no division by 0.*

65. $\dfrac{\dfrac{y}{x-1}}{\dfrac{y}{x}}$

66. $\dfrac{\dfrac{a}{b}}{\dfrac{a-1}{b}}$

67. $\dfrac{3+\dfrac{3}{x-1}}{3-\dfrac{3}{x}}$

68. $\dfrac{2-\dfrac{2}{x+1}}{2+\dfrac{2}{x}}$

69. $\dfrac{\dfrac{2}{x+2}}{\dfrac{3}{x-3}+\dfrac{1}{x}}$

70. $\dfrac{\dfrac{1}{x-1}-\dfrac{4}{x}}{\dfrac{3}{x+1}}$

71. $\dfrac{\dfrac{1}{x}+\dfrac{2}{x+1}}{\dfrac{2}{x-1}-\dfrac{1}{x}}$

72. $\dfrac{\dfrac{3}{x+1}-\dfrac{2}{x-1}}{\dfrac{1}{x+2}+\dfrac{2}{x-1}}$

73. $\dfrac{\dfrac{1}{y^2+y}-\dfrac{1}{xy+x}}{\dfrac{1}{xy+x}-\dfrac{1}{y^2+y}}$

74. $\dfrac{\dfrac{2}{b^2-1}-\dfrac{3}{ab-a}}{\dfrac{3}{ab-a}-\dfrac{2}{b^2-1}}$

75. $\dfrac{1-25y^{-2}}{1+10y^{-1}+25y^{-2}}$

76. $\dfrac{1-9x^{-2}}{1-6x^{-1}+9x^{-2}}$

WRITING ABOUT MATH

77. Explain how to use Method 1 to simplify

$$\dfrac{1+\dfrac{1}{x}}{3-\dfrac{1}{x}}$$

78. Explain how to use Method 2 to simplify the expression in
Exercise 77.

SOMETHING TO THINK ABOUT

79. Simplify each complex fraction:

$$\dfrac{1}{1+1},\ \dfrac{1}{1+\dfrac{1}{2}},\ \dfrac{1}{1+\dfrac{1}{1+\dfrac{1}{2}}},\ \dfrac{1}{1+\dfrac{1}{1+\dfrac{1}{1+\dfrac{1}{2}}}}$$

80. In Exercise 79, what is the pattern in the numerators and
denominators of the four answers? What would be the next
answer?

Section 6.5

Solving Equations That Contain Rational Expressions

Objectives

1. Solve an equation that contains one or more rational expressions.
2. Identify extraneous solutions.

Vocabulary

extraneous solution

Getting Ready

Simplify.

1. $3\left(x + \dfrac{1}{3}\right)$

2. $8\left(x - \dfrac{1}{8}\right)$

3. $x\left(\dfrac{3}{x} + 2\right)$

4. $3y\left(\dfrac{1}{3} - \dfrac{2}{y}\right)$

5. $6x\left(\dfrac{5}{2x} + \dfrac{2}{3x}\right)$

6. $9x\left(\dfrac{7}{9} + \dfrac{2}{3x}\right)$

7. $(y - 1)\left(\dfrac{1}{y - 1} + 1\right)$

8. $(x + 2)\left(3 - \dfrac{1}{x + 2}\right)$

We will now use our knowledge of rational expressions to solve equations that contain rational expressions with variables in their denominators. To do so, we will use new equation-solving methods that sometimes lead to false solutions. For this reason, it is important to check all apparent answers.

1 Solve an equation that contains one or more rational expressions.

To solve equations containing rational expressions, we use the same process we did with equations containing fractions. To clear the fractions, we multiply both sides of the equation by the LCD of the rational expressions that appear in the equation. To review this process, we will solve an equation containing only numerical denominators.

$$\frac{x}{3} + 1 = \frac{x}{6}$$

$$6\left(\frac{x}{3} + 1\right) = 6\left(\frac{x}{6}\right) \qquad \text{Multiply both sides of the equation by 6, the LCD, to clear fractions.}$$

We then use the distributive property to remove parentheses, simplify, and solve the resulting equation for x.

$$6 \cdot \frac{x}{3} + 6 \cdot 1 = 6 \cdot \frac{x}{6}$$

$$2x + 6 = x$$

$$x + 6 = 0 \qquad \text{Subtract } x \text{ from both sides.}$$

$$x = -6 \qquad \text{Subtract 6 from both sides.}$$

Check: $\dfrac{x}{3} + 1 = \dfrac{x}{6}$

$$\dfrac{-6}{3} + 1 \overset{?}{=} \dfrac{-6}{6} \qquad \text{Substitute } -6 \text{ for } x.$$

$$-2 + 1 \overset{?}{=} -1 \qquad \text{Simplify.}$$

$$-1 = -1$$

Because -6 satisfies the original equation, it is the solution.

EXAMPLE 1   Solve: $\dfrac{4}{x} + 1 = \dfrac{6}{x}$

Solution   Note that $x = 0$ is a restricted value since it creates division by 0. To clear the equation of rational expressions, we multiply both sides of the equation by the LCD of $\frac{4}{x}$, 1, and $\frac{6}{x}$, which is x.

$$\dfrac{4}{x} + 1 = \dfrac{6}{x}$$

$$x\left(\dfrac{4}{x} + 1\right) = x\left(\dfrac{6}{x}\right) \qquad \text{Multiply both sides by } x, \text{ the LCD.}$$

$$x \cdot \dfrac{4}{x} + x \cdot 1 = x \cdot \dfrac{6}{x} \qquad \text{Use the distributive property.}$$

$$4 + x = 6 \qquad \text{Simplify.}$$

$$x = 2 \qquad \text{Subtract 4 from both sides.}$$

Check: $\dfrac{4}{x} + 1 = \dfrac{6}{x}$

$$\dfrac{4}{2} + 1 \overset{?}{=} \dfrac{6}{2} \qquad \text{Substitute 2 for } x.$$

$$2 + 1 \overset{?}{=} 3 \qquad \text{Simplify.}$$

$$3 = 3$$

Because 2 satisfies the original equation, it is the solution.

SELF CHECK 1   Solve: $\dfrac{6}{x} - 1 = \dfrac{3}{x}$

2   Identify extraneous solutions.

If we multiply both sides of an equation by an expression that involves a variable, as we did in Example 1, we must check the apparent solutions. The next example shows why.

EXAMPLE 2   Solve: $\dfrac{x + 3}{x - 1} = \dfrac{4}{x - 1}$

Solution   Find the restricted value of the denominators. In this example $x \neq 1$. To clear the equation of rational expressions, we multiply both sides by $(x - 1)$, the LCD of the fractions contained in the equation.

$$\dfrac{x + 3}{x - 1} = \dfrac{4}{x - 1}$$

$$(x - 1)\dfrac{x + 3}{x - 1} = (x - 1)\dfrac{4}{x - 1} \qquad \text{Multiply both sides by } (x - 1), \text{ the LCD.}$$

$$x + 3 = 4 \quad \text{Simplify.}$$

$$x = 1 \quad \text{Subtract 3 from both sides.}$$

COMMENT Whenever a restricted (excluded) value is a possible solution, it will be extraneous.

Because both sides were multiplied by an expression containing a variable, we must check to see if the apparent solution is a value that must be excluded. If we replace x with 1 in the original equation, both denominators will become 0. Therefore, 1 is not a solution. Such false solutions are often called **extraneous solutions**. Because 1 does not satisfy the original equation, there is no solution. The solution set of the equation is $\varnothing$.

 **SELF CHECK 2** Solve: $\dfrac{x + 5}{x - 2} = \dfrac{7}{x - 2}$

The next two examples suggest the steps to follow when solving equations that contain rational expressions.

SOLVING EQUATIONS CONTAINING RATIONAL EXPRESSIONS

1. Find any restrictions on the variable. Remember that the denominator of a fraction cannot be 0.

2. Multiply both sides of the equation by the LCD of the rational expressions appearing in the equation to clear the equation of fractions.

3. Solve the resulting equation.

4. If an apparent solution of an equation is a restricted value, that value must be excluded. Check all solutions for extraneous roots.

EXAMPLE 3 Solve: $\dfrac{3x + 1}{x + 1} - 2 = \dfrac{3(x - 3)}{x + 1}$

Solution Since the denominator $x + 1$ cannot be 0, $x \neq -1$. To clear the equation of rational expressions, we multiply both sides by $(x + 1)$, the LCD of the rational expressions contained in the equation. We then can solve the resulting equation.

$$\frac{3x + 1}{x + 1} - 2 = \frac{3(x - 3)}{x + 1}$$

$$(x + 1)\left[\frac{3x + 1}{x + 1} - 2\right] = (x + 1)\left[\frac{3(x - 3)}{x + 1}\right] \quad \text{Multiply both sides by } (x + 1), \text{ the LCD.}$$

$$3x + 1 + (x + 1)(-2) = 3(x - 3) \quad \text{Use the distributive property to remove brackets.}$$

$$3x + 1 - 2x - 2 = 3x - 9 \quad \text{Use the distributive property to remove parentheses.}$$

$$x - 1 = 3x - 9 \quad \text{Combine like terms.}$$

$$-2x = -8 \quad \text{On both sides, subtract } 3x \text{ and add 1.}$$

$$x = 4 \quad \text{Divide both sides by } -2.$$

The apparent solution 4 is not an excluded value. We will check the solution to verify our work.

Check: $\dfrac{3x + 1}{x + 1} - 2 = \dfrac{3(x - 3)}{x + 1}$

$$\frac{3(4) + 1}{4 + 1} - 2 \overset{?}{=} \frac{3(4 - 3)}{4 + 1} \quad \text{Substitute 4 for } x.$$

$$\frac{13}{5} - \frac{10}{5} \overset{?}{=} \frac{3(1)}{5}$$

$$\frac{3}{5} = \frac{3}{5}$$

Because 4 satisfies the original equation, it is the solution.

SELF CHECK 3   Solve: $\dfrac{12}{x+1} - 5 = \dfrac{2}{x+1}$

To solve an equation with rational expressions, we often will have to factor a denominator to determine the least common denominator.

EXAMPLE 4   Solve: $1 = \dfrac{3}{x-2} - \dfrac{12}{x^2 - 4}$

Solution   To find the LCD and any restricted values of x, we must factor the second denominator.

$$1 = \frac{3}{x-2} - \frac{12}{x^2 - 4}$$

$$1 = \frac{3}{x-2} - \frac{12}{(x+2)(x-2)} \qquad \text{Factor } x^2 - 4.$$

Since $x + 2$ and $x - 2$ cannot be 0, $x \neq -2, 2$.

To clear the equation of rational expressions, we multiply both sides by $(x + 2)(x - 2)$, the LCD of the fractions contained in the equation.

$$(x+2)(x-2)(1) = (x+2)(x-2)\left[\frac{3}{x-2} - \frac{12}{(x+2)(x-2)}\right]$$

Multiply both sides by $(x + 2)(x - 2)$.

$$(x+2)(x-2) = (x+2)(x-2)\frac{3}{x-2} + (x+2)(x-2)\left(\frac{-12}{(x+2)(x-2)}\right)$$

Use the distributive property to remove brackets.

$$\begin{array}{ll}
(x+2)(x-2) = (x+2)(3) - 12 & \text{Simplify.} \\
x^2 - 4 = 3x + 6 - 12 & \text{Use the distributive property to remove} \\
& \text{parentheses.} \\
x^2 - 4 = 3x - 6 & \text{Simplify.} \\
x^2 - 3x - 4 + 6 = 0 & \text{Subtract } 3x \text{ and add 6 to both sides.} \\
x^2 - 3x + 2 = 0 & \text{Combine like terms.} \\
(x-2)(x-1) = 0 & \text{Factor the left side.} \\
(x-2) = 0 \quad \text{or} \quad (x-1) = 0 & \text{Apply the zero-factor property.} \\
x = 2 \qquad\qquad x = 1 & \text{Solve each equation.}
\end{array}$$

Because 2 is an excluded value, it is an extraneous solution. Verify that 1 is the solution of the given equation.

SELF CHECK 4   Solve: $\dfrac{x-4}{x-3} + \dfrac{x-2}{x-3} = x - 3$

EXAMPLE 5   Solve: $\dfrac{4}{5} + y = \dfrac{4y - 50}{5y - 25}$

Solution   Since $5y - 25$ cannot be 0, $y \neq 5$. Thus, 5 is a restricted value.

$$\frac{4}{5} + y = \frac{4y - 50}{5y - 25}$$

$$\frac{4}{5} + y = \frac{4y - 50}{5(y - 5)} \qquad \text{Factor } 5y - 25.$$

$$\mathbf{5(y - 5)}\left[\frac{4}{5} + y\right] = \mathbf{5(y - 5)}\left[\frac{4y - 50}{5(y - 5)}\right] \qquad \text{Multiply both sides by } 5(y - 5), \text{ the LCD.}$$

$$4\mathbf{(y - 5)} + 5y\mathbf{(y - 5)} = 4y - 50 \qquad \text{Use the distributive property to remove brackets.}$$

$$4y - 20 + 5y^2 - 25y = 4y - 50 \qquad \text{Use the distributive property to remove parentheses.}$$

$$5y^2 - 25y - 20 = -50 \qquad \text{Subtract } 4y \text{ from both sides and rearrange terms.}$$

$$5y^2 - 25y + 30 = 0 \qquad \text{Add 50 to both sides.}$$

$$y^2 - 5y + 6 = 0 \qquad \text{Divide both sides by 5.}$$

$$(y - 3)(y - 2) = 0 \qquad \text{Factor } y^2 - 5y + 6.$$

$$y - 3 = 0 \quad \text{or} \quad y - 2 = 0 \qquad \text{Apply the zero-factor property.}$$

$$y = 3 \qquad \qquad y = 2$$

Verify that 3 and 2 both satisfy the original equation.

 **SELF CHECK 5** Solve: $\dfrac{x - 6}{3x - 9} - \dfrac{1}{3} = \dfrac{x}{2}$

Many formulas are equations that contain rational expressions. The formula $\dfrac{1}{r} = \dfrac{1}{r_1} + \dfrac{1}{r_2}$ is used in electronics to calculate parallel resistances. To solve the formula for r, we eliminate the denominators by multiplying both sides by the LCD, which is rr_1r_2.

$$\frac{1}{r} = \frac{1}{r_1} + \frac{1}{r_2}$$

$$\mathbf{rr_1r_2}\left(\frac{1}{r}\right) = \mathbf{rr_1r_2}\left(\frac{1}{r_1} + \frac{1}{r_2}\right) \qquad \text{Multiply both sides by } rr_1r_2 \text{ the LCD.}$$

$$r_1r_2 = rr_2 + rr_1 \qquad \text{Use the distributive property to remove parentheses.}$$

$$r_1r_2 = r(r_2 + r_1) \qquad \text{Factor out } r, \text{ the GCF.}$$

$$\frac{r_1r_2}{r_2 + r_1} = r \qquad \text{Divide both sides by } r_2 + r_1.$$

or

$$r = \frac{r_1r_2}{r_2 + r_1}$$

SELF CHECK ANSWERS **1.** 3 **2.** ∅, 2 is extraneous **3.** 1 **4.** 5; 3 is extraneous **5.** 1, 2

NOW TRY THIS

1. Solve: $\dfrac{6}{x} = x - 5$

2. Solve: $\dfrac{x - 2}{(x + 3)^2} - \dfrac{5}{x + 3} + 1 = 0$

3. Explain how to identify extraneous solutions.

6.5   Exercises

WARM-UPS *Solve each equation and check the solution.*

1. $\dfrac{y}{3} + 6 = \dfrac{4y}{3}$

2. $\dfrac{2y}{5} - 8 = \dfrac{4y}{5}$

3. $\dfrac{z - 3}{2} = z + 2$

4. $\dfrac{b + 2}{3} = b - 2$

5. $\dfrac{5(x + 1)}{8} = x + 1$

6. $\dfrac{3(x - 1)}{2} + 2 = x$

Indicate the LCD you will use to clear the fractions. Do not solve. Assume no denominators are zero.

7. $\dfrac{x - 3}{x + 5} = \dfrac{x}{2}$

8. $\dfrac{1}{x - 1} = \dfrac{8}{x}$

9. $\dfrac{y}{y - 1} + 5 = \dfrac{y + 1}{y + 2}$

10. $\dfrac{5x - 8}{3x} + 3x = \dfrac{x}{15}$

11. $\dfrac{5}{x^2 - 9} + \dfrac{3x}{x - 3} = \dfrac{4}{x + 3}$

12. $\dfrac{7y}{3y + 6} - \dfrac{2}{y + 2} = 5$

REVIEW *Factor each expression.*

13. $x^2 + 8x$

14. $x^3 - 27$

15. $2x^2 + x - 3$

16. $6a^2 - 5a - 6$

17. $4x^2 + 10x - 6$

18. $x^4 - 81$

VOCABULARY AND CONCEPTS *Fill in the blanks.*

19. False solutions that result from multiplying both sides of an equation by a variable are called _____ solutions.

20. If you multiply both sides of an equation by an expression that involves a variable, you must _____ the solution.

21. To clear an equation of rational expressions, we multiply both sides by the _____ of the expressions in the equation.

22. To clear the equation $\dfrac{x}{x - 2} - \dfrac{x}{x - 1} = 5$ of fractions, we multiply both sides by _____.

GUIDED PRACTICE *Solve each equation and check the solution. SEE EXAMPLE 1. (OBJECTIVE 1)*

23. $\dfrac{3}{x} + 2 = 3$

24. $\dfrac{2}{x} + 9 = 11$

25. $\dfrac{x}{x + 2} + 3 = \dfrac{2x}{x + 2}$

26. $\dfrac{11}{b} + \dfrac{13}{b} = 12$

27. $\dfrac{2}{y + 1} + 5 = \dfrac{12}{y + 1}$

28. $\dfrac{1}{t - 3} = \dfrac{-2}{t - 3} + 1$

29. $\dfrac{1}{x - 1} + \dfrac{3}{x - 1} = 1$

30. $\dfrac{3}{p + 6} - 2 = \dfrac{7}{p + 6}$

Solve each equation and check the solution. Identify any extraneous values. SEE EXAMPLE 2. (OBJECTIVE 2)

31. $\dfrac{a^2}{a + 2} - \dfrac{4}{a + 2} = a$

32. $\dfrac{z^2}{z + 1} + 2 = \dfrac{1}{z + 1}$

33. $\dfrac{x}{x - 5} - \dfrac{5}{x - 5} = 3$

34. $\dfrac{3}{y - 2} + 1 = \dfrac{3}{y - 2}$

Solve each equation and check the solution. Identify any extraneous values. SEE EXAMPLE 3. (OBJECTIVE 2)

35. $\dfrac{2x + 1}{x + 5} - 1 = \dfrac{3x - 2}{x + 5}$

36. $\dfrac{3x + 1}{x + 3} + \dfrac{x - 3}{x + 3} = 2$

37. $\dfrac{x - 4}{x - 3} + \dfrac{x - 2}{x - 3} = x - 3$

38. $\dfrac{4x}{x - 1} + 1 = \dfrac{x + 3}{x - 1}$

Solve each equation and check the solution. Identify any extraneous values. SEE EXAMPLE 4. (OBJECTIVE 2)

39. $\dfrac{v}{v + 2} + \dfrac{1}{v - 1} = 1$

40. $\dfrac{b + 2}{b + 3} + 1 = \dfrac{-7}{b - 5}$

41. $\dfrac{u}{u - 1} + \dfrac{1}{u} = \dfrac{u^2 + 1}{u^2 - u}$

42. $\dfrac{3}{x - 2} + \dfrac{1}{x} = \dfrac{2(3x + 2)}{x^2 - 2x}$

43. $\dfrac{5}{x} + \dfrac{3}{x + 2} = \dfrac{-6}{x(x + 2)}$

44. $\dfrac{x - 3}{x - 2} - \dfrac{1}{x} = \dfrac{x - 3}{x}$

45. $\dfrac{-5}{s^2 + s - 2} + \dfrac{3}{s + 2} = \dfrac{1}{s - 1}$

46. $\dfrac{n}{n^2 - 9} + \dfrac{n + 8}{n + 3} = \dfrac{n - 8}{n - 3}$

Solve each equation and check the solution. Identify any extraneous values. SEE EXAMPLE 5. (OBJECTIVE 2)

47. $y + \dfrac{3}{4} = \dfrac{3y - 50}{4y - 24}$

48. $y + \dfrac{2}{3} = \dfrac{2y - 12}{3y - 9}$

49. $\dfrac{3}{5x - 20} + \dfrac{4}{5} = \dfrac{3}{5x - 20} - \dfrac{x}{5}$

50. $\dfrac{x}{x - 1} - \dfrac{12}{x^2 - x} = \dfrac{-1}{x - 1}$

51. $\dfrac{7}{q^2 - q - 2} + \dfrac{1}{q + 1} = \dfrac{3}{q - 2}$

52. $\dfrac{x - 3}{4x - 4} + \dfrac{1}{9} = \dfrac{x - 5}{6x - 6}$

53. $\dfrac{3y}{3y - 6} + \dfrac{8}{y^2 - 4} = \dfrac{2y}{2y + 4}$

54. $1 - \dfrac{3}{b} = \dfrac{-8b}{b^2 + 3b}$

ADDITIONAL PRACTICE *Solve each equation.*

55. $\dfrac{c - 4}{4} = \dfrac{c + 4}{8}$

56. $\dfrac{x}{2} + 4 = \dfrac{3x}{2}$

57. $\dfrac{x}{5} - \dfrac{x}{3} = -8$

58. $\dfrac{3a}{2} + \dfrac{a}{3} = -22$

59. $\dfrac{x + 2}{2} - 3x = x + 8$

60. $\dfrac{3x - 1}{6} - \dfrac{x + 3}{2} = \dfrac{3x + 4}{3}$

61. $\dfrac{3r}{2} - \dfrac{3}{r} = \dfrac{3r}{2} + 3$ **62.** $\dfrac{2p}{3} - \dfrac{1}{p} = \dfrac{2p - 1}{3}$

63. $\dfrac{1}{3} + \dfrac{2}{x - 3} = 1$ **64.** $\dfrac{3}{5} + \dfrac{7}{x + 2} = 2$

65. $\dfrac{z - 4}{z - 3} = \dfrac{z + 2}{z + 1}$ **66.** $\dfrac{a + 2}{a + 8} = \dfrac{a - 3}{a - 2}$

67. $\dfrac{x - 2}{x - 3} + \dfrac{x - 1}{x^2 - 8x + 15} = 1$

68. $\dfrac{x + 1}{x + 2} + \dfrac{1}{x^2 + x - 2} = 1$

69. $\dfrac{1}{a} + \dfrac{1}{b} = 1$ for a

70. $\dfrac{1}{a} - \dfrac{1}{b} = 1$ for b

71. $\dfrac{a}{b} + \dfrac{c}{d} = 1$ for b

72. $\dfrac{a}{b} - \dfrac{c}{d} = 1$ for a

73. Solve the formula $\dfrac{1}{r} = \dfrac{1}{r_1} + \dfrac{1}{r_2}$ for r_1.

74. Solve the formula $\dfrac{1}{r} = \dfrac{1}{r_1} + \dfrac{1}{r_2}$ for r_2.

APPLICATIONS

75. Optics The focal length f of a lens is given by the formula

$$\dfrac{1}{f} = \dfrac{1}{d_1} + \dfrac{1}{d_2}$$

where d_1 is the distance from the object to the lens and d_2 is the distance from the lens to the image. Solve the formula for f.

76. Solve the formula in Exercise 75 for d_1.

WRITING ABOUT MATH

77. Explain how you would decide what to do first when you solve an equation that involves fractions.

78. Explain why it is important to check your solutions to an equation that contains fractions with variables in the denominator.

SOMETHING TO THINK ABOUT

79. What numbers are equal to their own reciprocals?

80. Solve: $x^{-2} + x^{-1} = 0$.

Section 6.6

Solving Applications of Equations That Contain Rational Expressions

Objectives

1 Solve an application using a rational equation.

Getting Ready

1. If it takes 5 hours to fill a pool, what part could be filled in 1 hour?

2. $x is invested at 5% annual interest. Write an expression for the interest earned in one year.

3. Write an expression for the amount of an investment that earns $y interest in one year at 5%.

4. Express how long it takes to travel y miles at 52 mph.

In this section, we will consider applications whose solutions depend on solving equations containing rational expressions.

1 Solve an application using a rational equation.

EXAMPLE 1   **NUMBERS**   If the same number is added to both the numerator and denominator of the fraction $\frac{3}{5}$, the result is $\frac{4}{5}$. Find the number.

Analyze the problem   We are asked to find a number. We will let n represent the unknown number.

Form an equation   If we add the number n to both the numerator and denominator of the fraction $\frac{3}{5}$, we will get $\frac{4}{5}$. This gives the equation

$$\frac{3 + n}{5 + n} = \frac{4}{5}$$

Solve the equation   To solve the equation, we proceed as follows:

$$\frac{3 + n}{5 + n} = \frac{4}{5}$$

$$5(5 + n)\,\frac{3 + n}{5 + n} = 5(5 + n)\frac{4}{5} \qquad \text{Multiply both sides by } 5(5 + n), \text{ the LCD.}$$

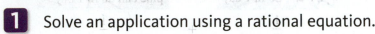

$$5(3 + n) = (5 + n)4 \qquad \text{Use the distributive property.}$$

$$15 + 5n = 20 + 4n \qquad \text{Use the distributive property to remove parentheses.}$$

$$15 + n = 20 \qquad \text{Subtract } 4n \text{ from both sides.}$$

$$n = 5 \qquad \text{Subtract 15 from both sides.}$$

State the conclusion   The number is 5.

Check the result   Add 5 to both the numerator and denominator of $\frac{3}{5}$ and get

$$\frac{3 + 5}{5 + 5} = \frac{8}{10} = \frac{4}{5}$$

The result checks.

SELF CHECK 1   If the same number is subtracted from both the numerator and denominator of the fraction $\frac{4}{5}$, the result is $\frac{5}{6}$. Find the number.

As we did in Example 1, it is important to state the conclusion after solving an application.

EXAMPLE 2   **FILLING AN OIL TANK**   An inlet pipe can fill an oil tank in 7 days, and a second inlet pipe can fill the same tank in 9 days. If both pipes are used, how long will it take to fill the tank?

Analyze the problem   We are asked to find how long it will take to fill the tank, so we let x represent the number of days it will take to fill the tank.

Form an equation   The key is to note what each pipe can do in 1 day. If you add what the first pipe can do in 1 day to what the second pipe can do in 1 day, the sum is what they can do together in 1 day. Since the first pipe can fill the tank in 7 days, it can do $\frac{1}{7}$ of the job in 1 day. Since the second pipe can fill the tank in 9 days, it can do $\frac{1}{9}$ of the job in 1 day. If it takes x days for both pipes to fill the tank, together they can do $\frac{1}{x}$ of the job in 1 day. This gives the equation

What the first inlet pipe can do in 1 day	plus	what the second inlet pipe can do in 1 day	equals	what they can do together in 1 day.
$\frac{1}{7}$	$+$	$\frac{1}{9}$	$=$	$\frac{1}{x}$

Solve the equation   To solve the equation, we proceed as follows:

$$\frac{1}{7} + \frac{1}{9} = \frac{1}{x}$$

$$63x\left(\frac{1}{7} + \frac{1}{9}\right) = 63x\left(\frac{1}{x}\right) \quad \text{Multiply both sides by } 63x, \text{ the LCD.}$$

$$9x + 7x = 63 \qquad \text{Use the distributive property to remove parentheses and simplify.}$$

$$16x = 63 \qquad \text{Combine like terms.}$$

$$x = \frac{63}{16} \qquad \text{Divide both sides by 16.}$$

State the conclusion   It will take $\frac{63}{16}$ or $3\frac{15}{16}$ days for both inlet pipes to fill the tank.

Check the result   In $\frac{63}{16}$ days, the first pipe fills $\frac{1}{7}\left(\frac{63}{16}\right)$ of the tank, and the second pipe fills $\frac{1}{9}\left(\frac{63}{16}\right)$ of the tank. The sum of these efforts, $\frac{9}{16} + \frac{7}{16}$, is equal to one full tank.

 **SELF CHECK 2**   If an inlet pipe can fill the oil tank in 6 days, and a second inlet pipe can fill the same tank in 11 days, how long will it take to fill the tank if both pipes are used?

EXAMPLE 3   **TRACK AND FIELD**   A coach can run 10 miles in the same amount of time that her best student athlete can run 12 miles. If the student can run 1 mph faster than the coach, how fast can the student run?

Analyze the problem   We are asked to find how fast the student can run. Since we know that the student runs 1 mph faster than the coach, we will let r represent the rate of the coach and $r + 1$ represent the rate of the student. In this case, we want to find the rate of the student, which is $(r + 1)$.

Form an equation   This is a uniform motion problem, based on the formula $d = rt$, where d is the distance traveled, r is the rate, and t is the time. If we solve this formula for t, we obtain

$$t = \frac{d}{r}$$

If the coach runs 10 miles at some unknown rate of r mph, it will take $\frac{10}{r}$ hours. If the student runs 12 miles at some unknown rate of $(r + 1)$ mph, it will take $\frac{12}{r + 1}$ hours. We can organize the information of the problem as in Table 6-1.

TABLE 6-1

	d	$=$	r	$\cdot$	t
Student	12		$r + 1$		$\frac{12}{r + 1}$
Coach	10		r		$\frac{10}{r}$

Because the times are given to be equal, we know that $\frac{12}{r + 1} = \frac{10}{r}$. This gives the equation

The time it takes the student to run 12 miles	equals	the time it takes the coach to run 10 miles.
$\dfrac{12}{r + 1}$	$=$	$\dfrac{10}{r}$

Solve the equation　We can solve the equation as follows:

COMMENT　This example could have been set up with r representing the student's rate and $r - 1$ the coach's rate.

$$\frac{12}{r + 1} = \frac{10}{r}$$

$$r(r + 1)\frac{12}{r + 1} = r(r + 1)\frac{10}{r} \quad \text{Multiply both sides by } r(r + 1), \text{ the LCD.}$$

$$12r = 10(r + 1) \quad \text{Simplify.}$$

$$12r = 10r + 10 \quad \text{Use the distributive property to remove parentheses.}$$

$$2r = 10 \quad \text{Subtract } 10r \text{ from both sides.}$$

$$r = 5 \quad \text{Divide both sides by 2.}$$

State the conclusion　The coach can run 5 mph. The student, running 1 mph faster, can run 6 mph.

Check the result　Verify that this result checks.

 **SELF CHECK 3**　The coach runs 8 miles in the same amount of time that her student can run 12 miles. If the student can run 2 mph faster than the coach, how fast can the student run?

EXAMPLE 4　**COMPARING INVESTMENTS**　At one bank, a sum of money invested for one year will earn $96 interest. If invested in bonds, that same money would earn $120, because the interest rate paid by the bonds is 1% greater than that paid by the bank. Find the bank's rate of interest.

Analyze the problem　We are asked to find the bank's rate of interest, so we can let r represent the bank's rate. If the interest on the bonds is 1% greater, then the bonds' interest rate will be $r + 0.01$.

Form an equation　This is an interest problem that is based on the formula $i = pr$, where i is the interest, p is the principal (the amount invested), and r is the annual rate of interest.

If we solve this formula for p, we obtain

$$p = \frac{i}{r}$$

If an investment at a bank earns \$96 interest at some unknown rate r, the principal invested is $\frac{96}{r}$. If an investment in bonds earns \$120 interest at some unknown rate $(r + 0.01)$, the principal invested is $\frac{120}{r + 0.01}$. We can organize the information of the problem as in Table 6-2.

	TABLE 6-2		
	Interest $=$	**Principal** $\cdot$	**Rate**
Bank	96	$\dfrac{96}{r}$	r
Bonds	120	$\dfrac{120}{r + 0.01}$	$r + 0.01$

Because the same principal would be invested in either account, we can set up the following equation:

$$\frac{96}{r} = \frac{120}{r + 0.01}$$

Solve the equation   We can solve the equation as follows:

$$\frac{96}{r} = \frac{120}{r + 0.01}$$

$$\boldsymbol{r(r + 0.01) \cdot \frac{96}{r} = \frac{120}{r + 0.01} \cdot r(r + 0.01)} \quad \text{Multiply both sides by } r(r + 0.01), \text{ the LCD.}$$

$$96(r + 0.01) = 120r \qquad\qquad \text{Use the distributive property.}$$

$$96r + 0.96 = 120r \qquad\qquad \text{Use the distributive property to remove parentheses.}$$

$$0.96 = 24r \qquad\qquad\quad \text{Subtract } 96r \text{ from both sides.}$$

$$0.04 = r \qquad\qquad\quad\; \text{Divide both sides by 24.}$$

State the conclusion   The bank's interest rate is 0.04 or 4%. The bonds pay 5% interest, a rate 1% greater than that paid by the bank.

Check the results   Verify that these rates check.

🌿 SELF CHECK 4   Find the bank's rate of interest if the sum of money will earn \$90 interest. The same money would earn \$120 in bonds, because the interest rate is 1% greater than that paid by the bank.

🌿 SELF CHECK ANSWERS   **1.** The number is −1.   **2.** It will take $\frac{66}{17}$ or $3\frac{15}{17}$ days for both inlet pipes to fill the tank.   **3.** The student can run 6 mph.   **4.** The bank's annual interest rate is 3%. The bonds pay 4% annual interest.

NOW TRY THIS

Chris can clean a house in 3 hours and Cheryl can clean the house in 2 hours. Their son, Tyler, can scatter toys all over the house in 4 hours. If Tyler starts scattering toys at the same time Chris and Cheryl start cleaning, when (if ever) will the house be clean?

6.6 Exercises

WARM-UPS

1. Write the formula that relates the principal p that is invested, the earned interest i, and the rate r for 1 year.
2. Write the formula that relates the distance d traveled at a speed r, for a time t.
3. Write the formula that relates the cost C of purchasing q items that cost $\$d$ each.
4. Write the formula that relates the value v of a mixture of n pounds costing $\$p$ per pound.

REVIEW *Solve each equation.*

5. $x^2 - 5x - 6 = 0$
6. $x^2 - 81 = 0$
7. $(y - 3)(y^2 + 5y + 4) = 0$
8. $(x^2 - 1)(x^2 - 4) = 0$
9. $y^3 - y^2 = 0$
10. $3b^3 - 27b = 0$
11. $4(y - 3) = -y^2$
12. $6t^3 + 35t^2 = 6t$

VOCABULARY AND CONCEPTS

13. List the five steps used in problem solving.
14. Write 3% as a decimal.

APPLICATIONS *Solve and verify your answer.* **SEE EXAMPLE 1.** *(OBJECTIVE 1)*

15. **Numbers**   If the denominator of $\frac{3}{4}$ is increased by a number and the numerator is doubled, the result is 1. Find the number.
16. **Numbers**   If a number is added to the numerator of $\frac{7}{8}$ and the same number is subtracted from the denominator, the result is 2. Find the number.
17. **Numbers**   If a number is added to the numerator of $\frac{3}{4}$ and twice as much is added to the denominator, the result is $\frac{4}{7}$. Find the number.
18. **Numbers**   If a number is added to the numerator of $\frac{4}{9}$ and twice as much is subtracted from the denominator, the result is –9. Find the number.

Solve and verify your answer. **SEE EXAMPLE 2.** *(OBJECTIVE 1)*

19. **Grading papers**   It takes a teacher 60 minutes to grade a set of quizzes and takes her aide twice as long to do the same amount of grading. How long will it take them to grade a set of quizzes if they work together?
20. **Printing schedules**   It takes a printer 12 hours to print the class schedules for all of the students in a college. A faster printer can do the job in 9 hours. How long will it take to do the job if both printers are used?
21. **Filling a pool**   An inlet pipe can fill an empty swimming pool in 5 hours, and another inlet pipe can fill the pool in 4 hours. How long will it take both pipes to fill the pool?

22. **Roofing a house**   A homeowner estimates that it will take 7 days to roof his house. A professional roofer estimates that he could roof the house in 4 days. How long will it take if the homeowner helps the roofer?

Solve and verify your answer. **SEE EXAMPLE 3.** *(OBJECTIVE 1)*

23. **Flying speeds**   On average, a Canada goose can fly 10 mph faster than a Great Blue heron. Find their flying speeds if a goose can fly 180 miles in the same time it takes a heron to fly 120 miles.
24. **Touring**   A tourist can bicycle 28 miles in the same time as he can walk 8 miles. If he can ride 10 mph faster than he can walk, how much time should he allow to walk a 30-mile trail?   (*Hint:* How fast can he walk?)

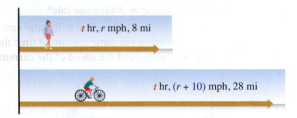

t hr, r mph, 8 mi

t hr, $(r + 10)$ mph, 28 mi

25. **Comparing travel**   A plane can fly 300 miles in the same time as it takes a car to go 120 miles. If the car travels 90 mph slower than the plane, find the speed of the plane.
26. **Wind speed**   A plane can fly 300 miles downwind in the same amount of time as it can travel 210 miles upwind. Find the velocity of the wind if the plane can fly 255 mph in still air.

Solve and verify your answer. **SEE EXAMPLE 4.** *(OBJECTIVE 1)*

27. **Comparing investments**   Two certificates of deposit pay interest at rates that differ by 1%. Money invested for one year in the first CD earns $175 interest. The same principal invested in the other CD earns $200. Find the two rates of interest.
28. **Comparing interest rates**   Two bond funds pay interest at rates that differ by 2%. Money invested for one year in the first fund earns $315 interest. The same amount invested in the other fund earns $385. Find the lower rate of interest.
29. **Comparing interest rates**   Two mutual funds pay interest at rates that differ by 4%. Money invested for one year in the first fund earns $300 interest. The same amount invested in the other fund earns $60. Find the higher rate of interest.
30. **Comparing interest rates**   Two banks pay interest at rates that differ by 1%. Money invested for one year in the first account earns $105 interest. The same amount invested in the other account earns $125. Find the two rates of interest.

ADDITIONAL PRACTICE *Solve and verify your answer.*

31. Filling a pool One inlet pipe can fill an empty pool in 4 hours, and a drain can empty the pool in 8 hours. How long will it take the pipe to fill the pool if the drain is left open?

32. Sewage treatment A sludge pool is filled by two inlet pipes. One pipe can fill the pool in 15 days and the other pipe can fill it in 21 days. However, if no sewage is added, waste removal will empty the pool in 36 days. How long will it take the two inlet pipes to fill an empty pool?

33. Sales A bookstore can purchase several calculators for a total cost of $120. If each calculator cost $1 less, the bookstore could purchase 10 additional calculators at the same total cost. How many calculators can be purchased at the regular price?

34. Furnace repairs A repairman purchased several furnace-blower motors for a total cost of $210. If his cost per motor had been $5 less, he could have purchased 1 additional motor. How many motors did he buy at the regular rate?

35. Boating A boat that can travel 18 mph in still water can travel 22 miles downstream in the same amount of time that it can travel 14 miles upstream. Find the speed of the current in the river.

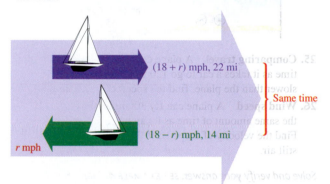

$(18 + r)$ mph, 22 mi

Same time

$(18 - r)$ mph, 14 mi

r mph

36. Conveyor belts The diagram shows how apples are processed for market. Although the second conveyor belt is shorter, an apple spends the same time on each belt because the second belt moves 1 ft/sec slower than the first. Find the speed of each belt.

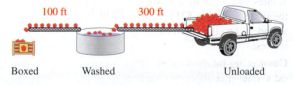

100 ft   300 ft

Boxed   Washed   Unloaded

37. Numbers The sum of a number and its reciprocal is $\frac{13}{6}$. Find the numbers.

38. Numbers The sum of the reciprocals of two consecutive even integers is $\frac{7}{24}$. Find the integers.

39. Road Rallies The Uzbekistan Road Rally from London to Tashkent takes roughly 2 weeks to complete. One leg of the race, from Nurburgring, Germany, to Wroclaw, Poland, covers 512 miles and the leg from Krakow, Poland, to Kiev, Ukraine, covers 528 miles. If the speed from Germany to Poland is 20 mph faster than that from Poland to Ukraine and the difference in times is 4 hours, find the rates of travel for the 2 legs. (*Hint*: Add the 4 hours to the expression for the faster rate.)

40. Pony Express The Pony Express carried mail from St. Joseph, MO, to Sacramento, CA, between 1860 and 1861. Pony Express riders stopped only long enough to change horses and these were saddled and waiting. The distance covered was 1,800 miles. A car travels 52 mph faster than the horse. If it took the Pony Express riders 7.5 times as long to make the trip as by car, find how long it took the Pony Express to make the trip.

41. Sharing costs Some office workers bought a $60 gift for their boss. If there had been five more employees to contribute, everyone's cost would have been $2 less. How many workers contributed to the gift?

42. Sales A dealer bought some radios for a total of $1,200. She gave away 6 radios as gifts, sold each of the rest for $10 more than she paid for each radio, and broke even. How many radios did she buy?

43. River tours A river boat tour begins by going 60 miles upstream against a 5 mph current. Then the boat turns around and returns with the current. What still-water speed should the captain use to complete the tour in 5 hours?

44. Travel time A company president flew 680 miles in a corporate jet but returned in a smaller plane that could fly only half as fast. If the total travel time was 6 hours, find the speeds of the planes.

WRITING ABOUT MATH

45. The key to solving shared work problems is to ask, "How much of the job could be done in 1 unit of time?" Explain.

46. It is difficult to check the solution of a shared work problem. Explain how you could decide if the answer is at least reasonable.

SOMETHING TO THINK ABOUT

47. Create a problem, involving either investment income or shared work, that can be solved by an equation that contains rational expressions.

48. Solve the problem you created in Exercise 47.

Section 6.7 Ratios

Objectives

1. Express a ratio in simplest form.
2. Translate an English sentence to a ratio.
3. Write a ratio as a unit cost.
4. Write a ratio as a rate.

Vocabulary

ratio   equal ratios   unit cost

Getting Ready

Simplify each fraction.

1. $\dfrac{2}{4}$ 2. $\dfrac{8}{12}$ 3. $-\dfrac{20}{25}$ 4. $\dfrac{-45}{81}$

In this section, we will discuss ratios, unit costs, and rates. These ideas are important in many areas of everyday life.

1   Express a ratio in simplest form.

Ratios appear often in real-world situations. For example,

- To prepare fuel for a Lawnboy lawnmower, gasoline must be mixed with oil in the ratio of 50 to 1.
- To make 14-karat jewelry, gold is mixed with other metals in the ratio of 14 to 10.
- At Rock Valley College, the ratio of students to faculty is 16 to 1.

Ratios give us a way to compare numerical quantities.

RATIOS

A **ratio** is a comparison of two numbers by their indicated quotient. In symbols,

if a and b are two numbers, the ratio of a to b is $\dfrac{a}{b}$.

COMMENT The denominator b cannot be 0 in the fraction $\dfrac{a}{b}$, but b can be 0 in the ratio a to b. For example, the ratio of women to men on a women's softball team could be 25 to 0. However, these applications are rare.

Some examples of ratios are

$$\frac{7}{9}, \qquad \frac{21}{27}, \qquad \text{and} \qquad \frac{2{,}290}{1{,}317}$$

- The fraction $\frac{7}{9}$ can be read as "the ratio of 7 to 9."
- The fraction $\frac{21}{27}$ can be read as "the ratio of 21 to 27."
- The fraction $\frac{2{,}290}{1{,}317}$ can be read as "the ratio of 2,290 to 1,317."

Because $\frac{7}{9}$ and $\frac{21}{27}$ represent equal numbers, they are **equal ratios**.

EXAMPLE 1 Express each phrase as a fraction in simplest form.

a. the ratio of 15 to 12   **b.** the ratio of 0.3 to 1.2

Solution **a.** The ratio of 15 to 12 can be written as the fraction $\frac{15}{12}$. After simplifying, the ratio is $\frac{5}{4}$.

b. The ratio of 0.3 to 1.2 can be written as the fraction $\frac{0.3}{1.2}$. We can simplify this fraction as follows:

$$\frac{0.3}{1.2} = \frac{0.3 \cdot \mathbf{10}}{1.2 \cdot \mathbf{10}} \qquad \text{To clear the decimals, multiply both numerator and denominator by 10.}$$

$$= \frac{3}{12} \qquad \text{Multiply.}$$

$$= \frac{1}{4} \qquad \text{Simplify.}$$

 **SELF CHECK 1** Express each phrase as a fraction in simplest form.

a. the ratio of 8 to 12   **b.** the ratio of 3.2 to 16

EXAMPLE 2 Express each phrase as a ratio in simplest form.

a. the ratio of 3 meters to 8 meters   **b.** the ratio of 4 ounces to 1 pound

Solution **a.** The ratio of 3 meters to 8 meters can be written as the fraction $\frac{3 \text{ meters}}{8 \text{ meters}}$, or just $\frac{3}{8}$.

b. The ratio of 4 ounces to 1 pound can be written as the fraction $\frac{4 \text{ ounces}}{1 \text{ pound}}$. Since there are 16 ounces in 1 pound, the ratio is $\frac{4 \text{ ounces}}{16 \text{ ounces}}$, which simplifies to $\frac{1}{4}$.

 **SELF CHECK 2** Express each phrase as a ratio in simplest form.
a. the ratio of 8 ounces to 2 pounds
b. the ratio of 1 foot to 2 yards   (*Hint*: 3 feet = 1 yard.)

2 Translate an English sentence to a ratio.

EXAMPLE 3 **STUDENT/FACULTY RATIOS** At a college, there are 2,772 students and 154 faculty members. Write a fraction in simplified form that expresses the ratio of students per faculty member.

Solution The ratio of students to faculty is 2,772 to 154. We can write this ratio as the fraction $\frac{2{,}772}{154}$ and simplify it.

$$\frac{2{,}772}{154} = \frac{18 \cdot \overset{1}{\cancel{154}}}{1 \cdot \underset{1}{\cancel{154}}}$$

$$= \frac{18}{1} \qquad \frac{154}{154} = 1$$

The ratio of students to faculty is 18 to 1.

SELF CHECK 3   In a college graduating class, 224 students out of 632 went on to graduate school. Write a fraction in simplified form that expresses the ratio of the number of students going on to graduate school to the number in the graduating class.

---

3   Write a ratio as a unit cost.

The **unit cost** of an item is the ratio of its cost to its quantity. For example, the unit cost (the cost per pound) of 5 pounds of gourmet coffee priced at $41.75 is given by

$$\frac{\$41.75}{5 \text{ pounds}} = \$8.35 \text{ per pound} \quad \text{\textcolor{red}{$\$41.75 \div 5 = \$8.35$}}$$

The unit cost is $8.35 per pound.

EXAMPLE 4   **SHOPPING** Pizza sauce comes packaged in a 14-ounce jar, which sells for $1.49, or in a 30-ounce jar, which sells for $3.32. Which is the better buy?

Solution   To find the better buy, we must find each unit cost. The unit cost of the 14-ounce jar is

$$\frac{\$1.49}{14 \text{ ounces}} = \frac{149¢}{14 \text{ ounces}} \quad \text{\textcolor{red}{Change \$1.49 to 149 cents.}}$$

$$= 10.64¢ \text{ per ounce}$$

The unit cost of the 30-ounce jar is

$$\frac{\$3.32}{30 \text{ ounces}} = \frac{332¢}{30 \text{ ounces}} \quad \text{\textcolor{red}{Change \$3.32 to 332 cents.}}$$

$$= 11.07¢ \text{ per ounce}$$

Since the unit cost is less when the pizza sauce is packaged in 14-ounce jars, that is the better buy.

SELF CHECK 4   A fast-food restaurant sells a 12-ounce soft drink for 79¢ and a 16-ounce soft drink for 99¢. Which is the better buy?

---

4   Write a ratio as a rate.

When ratios are used to compare quantities with different units, they often are called rates. For example, if we drive 413 miles in 7 hours, the average rate of speed is the quotient of the miles driven to the length of time of the trip.

$$\text{Average rate of speed} = \frac{413 \text{ miles}}{7 \text{ hours}} = \frac{59 \text{ miles}}{1 \text{ hour}} \quad \text{\textcolor{red}{$\frac{413}{7} = \frac{7 \cdot 59}{7 \cdot 1} = \frac{59}{1}$}}$$

The rate $\frac{59 \text{ miles}}{1 \text{ hour}}$ can be expressed in any of the following forms:

$$59 \frac{\text{miles}}{\text{hour}}, \qquad 59 \text{ miles per hour}, \qquad 59 \text{ miles/hour}, \qquad \text{or} \qquad 59 \text{ mph}$$

EXAMPLE 5 **HOURLY PAY** Find the hourly rate of pay for a student who earns $370 for working 40 hours.

Solution We can write the rate of pay as

$$\text{Rate of pay} = \frac{\$370}{40 \text{ hours}}$$

and simplify by dividing 370 by 40.

$$\text{Rate of pay} = 9.25 \frac{\text{dollars}}{\text{hour}}$$

The rate is $9.25 per hour.

 SELF CHECK 5 Lawanda earns $716 for a 40-hour week managing an office supply store. Find her hourly rate of pay.

EXAMPLE 6 **ENERGY CONSUMPTION** One household used 813.75 kilowatt hours (kWh) of electricity during a 31-day period. Find the rate of energy consumption in kWh per day.

Solution We can write the rate of energy consumption as

$$\text{Rate of energy consumption} = \frac{813.75 \text{ kWh}}{31 \text{ days}}$$

and simplify by dividing 813.75 by 31.

$$\text{Rate of energy consumption} = 26.25 \frac{\text{kWh}}{\text{day}}$$

The rate of consumption is 26.25 kilowatt hours per day.

 SELF CHECK 6 During the month of August the household used 1733 kWh of electricity. Find the rate of energy consumption in kWh per day.

Accent
on technology

▸Computing Gas Mileage

A man drove a total of 775 miles. Along the way, he stopped for gas three times, pumping 10.5, 11.3, and 8.75 gallons of gas. He started with the tank half-full and ended with the tank half-full. To find how many miles he got per gallon, we need to divide the total distance by the total number of gallons of gas consumed.

$$\frac{775}{10.5 + 11.3 + 8.75} \quad \begin{array}{l}\leftarrow \text{ Total distance} \\ \leftarrow \text{ Total number of gallons consumed}\end{array}$$

We can make this calculation by entering these numbers and pressing these keys.

775 ÷ (10.5 + 11.3 + 8.75) =   Using a scientific calculator
775 ÷ (10.5 + 11.3 + 8.75) **ENTER**   Using a TI84 graphing calculator

Either way, the display will read 25.36824877. To the nearest one-hundredth, he got 25.37 mpg.

For instructions regarding the use of a Casio graphing calculator, please refer to the Casio Keystroke Guide in the back of the book.

NOW TRY THIS

1. Express $(x + 2)$ to $(2x + 4)$ as a ratio in simplest form.
2. Express $(x^2 - 25)$ to $(x^2 - 4x - 5)$ as a ratio in simplest form.

6.7   Exercises 🌿

WARM-UPS *State the number of*

1. minutes in an hour.
2. feet in a yard.
3. ounces in a pound.
4. months in a year.
5. inches in a yard.
6. days in the month of July.
7. feet in a mile.
8. days in a non-leap year.

REVIEW *Solve each equation.*

9. $7x + 2 = 44$
10. $\frac{x}{2} - 4 = 38$
11. $3(x + 2) = 24$
12. $\frac{x - 6}{3} = 20$

Factor each expression.

13. $4x + 28$
14. $x^2 - 100$
15. $2x^2 - x - 6$
16. $x^3 + 27$

VOCABULARY AND CONCEPTS *Fill in the blanks.*

17. A ratio is a _____ of two numbers.
18. The _____ of an item is the quotient of its cost to its quantity.
19. The ratios $\frac{3}{5}$ and $\frac{6}{10}$ are _____ ratios.
20. The quotient $\frac{500 \text{ miles}}{15 \text{ hours}}$ is called a ___.
21. Give three examples of ratios that you have encountered this past week.
22. Suppose that a basketball player made 8 free throws out of 12 tries. The ratio of $\frac{8}{12}$ can be simplified as $\frac{2}{3}$. Interpret this result.

GUIDED PRACTICE *Express each phrase as a ratio in simplest form.* SEE EXAMPLE 1. (OBJECTIVE 1)

23. 5 to 7
24. 6 to 11
25. 17 to 34
26. 19 to 38
27. 26 to 39
28. 40 to 50
29. 7 to 24.5
30. 0.65 to 0.15

Express each phrase as a ratio in simplest form. SEE EXAMPLE 2. (OBJECTIVE 1)

31. 8 ounces to 24 ounces
32. 4 inches to 36 inches
33. 12 minutes to 1 hour
34. 8 ounces to 1 pound
35. 3 days to 1 week
36. 4 inches to 2 yards
37. 18 months to 2 years
38. 8 feet to 4 yards

Translate each sentence to a ratio in simplest form. SEE EXAMPLE 3. (OBJECTIVE 2)

39. **Faculty-to-student ratio**   At a college, there are 125 faculty members and 2,000 students. Find the faculty-to-student ratio.
40. **Ratio of men to women**   In a state senate, there are 94 men and 24 women. Find the ratio of men to women.

For Exercises 41–44, refer to the monthly family budget shown in the table.

Item	Amount
Rent	\$750
Food	\$652
Gas and electric	\$188
Phone	\$125
Entertainment	\$110

41. Find the total amount of the budget.
42. Find the ratio of the amount budgeted for rent to the total budget.
43. Find the ratio of the amount budgeted for entertainment to the total budget.
44. Find the ratio of the amount budgeted for phone to the amount budgeted for entertainment.

Find the unit cost. SEE EXAMPLE 4. (OBJECTIVE 3)

45. Unit cost of gasoline A driver pumped 17 gallons of gasoline into his tank at a cost of $53.55. Find the unit cost of gasoline.

46. Unit cost of grass seed A 50-pound bag of grass seed costs $222.50. Find the unit cost of grass seed.

47. Comparative shopping A 6-ounce can of orange juice sells for 89¢, and an 8-ounce can sells for $1.19. Which is the better buy?

48. Comparative shopping A 30-pound bag of fertilizer costs $12.25, and an 80-pound bag costs $30.25. Which is the better buy?

Express each result in simplest form. SEE EXAMPLES 5–6. (OBJECTIVE 3)

49. Rate of pay Ricardo worked for 27 hours to help insulate a hockey arena. For his work, he received $337.50. Find his hourly rate of pay.

50. Real estate taxes The real estate taxes on a summer home assessed at $75,000 were $1,500. Find the tax rate as a percent.

51. Rate of speed A car travels 520 miles in 8 hours. Find its rate of speed in mph.

52. Rate of speed An airplane travels from Chicago to San Francisco, a distance of 1,883 miles, in 3.5 hours. Find the average rate of speed of the plane.

ADDITIONAL PRACTICE *Express each result in simplest form.*

53. Unit cost of cranberry juice A 14-ounce bottle of cranberry juice sells for $1.19. Give the unit cost in cents per ounce.

54. Unit cost of beans A 24-ounce package of green beans sells for $1.29. Give the unit cost in cents per ounce.

55. Comparing speeds A car travels 345 miles in 6 hours, and a truck travels 376 miles in 6.2 hours. Which vehicle travels faster?

56. Comparing reading speeds One seventh-grader read a 54-page book in 40 minutes, and another read an 80-page book in 62 minutes. If the books were equally difficult, which student read faster?

57. Comparing gas mileage One car went 1,235 miles on 51.3 gallons of gasoline, and another went 1,456 miles on 55.78 gallons. Which car had the better mpg rating?

58. Comparing electric rates In one community, a bill for 575 kilowatt hours (kWh) of electricity was $38.81. In a second community, a bill for 831 kWh was $58.10. In which community is electricity cheaper?

59. Emptying a tank An 11,880-gallon tank can be emptied in 27 minutes. Find the rate of flow in gallons per minute.

60. Filling a gas tank It took 2.5 minutes to put 17 gallons of gas in a car. Find the rate of flow in gallons per minute.

For Exercises 61–64 refer to the tax deductions listed in the table. Give each ratio in simplest form.

Item	Amount
Medical	$995
Real estate tax	$1,245
Contributions	$1,680
Mortgage interest	$4,580
Union dues	$225

61. Find the total amount of deductions.

62. Find the ratio of real estate tax deductions to the total deductions.

63. Find the ratio of the contributions to the total deductions.

64. Find the ratio of the mortgage interest deduction to the union dues deduction.

WRITING ABOUT MATH

65. Some people think that the word *ratio* comes from the words *rational number*. Explain why this may be true.

66. In the fraction $\frac{a}{b}$, b cannot be 0. Explain why. In the ratio a to b, b can be 0. Explain why.

SOMETHING TO THINK ABOUT

67. Which ratio is the larger? How can you tell?

$$\frac{17}{19} \quad \text{or} \quad \frac{19}{21}$$

68. Which ratio is the smaller? How can you tell?

$$-\frac{13}{29} \quad \text{or} \quad -\frac{17}{31}$$

Section 6.8

Proportions and Similar Triangles

Objectives

1. Determine whether an equation is a proportion.
2. Solve a proportion.
3. Solve an application using a proportion.
4. Solve an application using the properties of similar triangles.

Vocabulary

proportion
extremes

means
proportional

similar triangles

Getting Ready

Solve each equation.

1. $\dfrac{5}{2} = \dfrac{x}{4}$

2. $\dfrac{7}{9} = \dfrac{y}{3}$

3. $\dfrac{y}{10} = \dfrac{2}{7}$

4. $\dfrac{1}{x} = \dfrac{8}{40}$

5. $\dfrac{w}{14} = \dfrac{7}{21}$

6. $\dfrac{c}{12} = \dfrac{5}{12}$

7. $\dfrac{3}{q} = \dfrac{1}{7}$

8. $\dfrac{16}{3} = \dfrac{8}{z}$

A statement that two ratios are equal is called a *proportion*. In this section, we will discuss proportions and use them to solve problems.

1   Determine whether an equation is a proportion.

Consider Table 6-3, in which we are given the costs of various numbers of gallons of gasoline.

TABLE 6-3	
Number of gallons	**Cost (in $)**
2	7.60
5	19.00
8	30.40
12	45.60
20	76.00

If we find the ratios of the costs to the numbers of gallons purchased, we will see that they are equal. In this example, each ratio represents the cost of 1 gallon of gasoline, which is $3.80 per gallon.

$$\frac{7.60}{2} = 3.80, \qquad \frac{19.00}{5} = 3.80, \qquad \frac{30.40}{8} = 3.80, \qquad \frac{45.60}{12} = 3.80, \qquad \frac{76.00}{20} = 3.80$$

When two ratios such as $\frac{7.60}{2}$ and $\frac{19.00}{5}$ are equal, they form a *proportion*.

PROPORTIONS	A **proportion** is a statement that two ratios are equal.

Some examples of proportions are

$$\frac{1}{2} = \frac{3}{6}, \qquad \frac{7}{3} = \frac{21}{9}, \qquad \frac{8x}{1} = \frac{40x}{5}, \qquad \text{and} \qquad \frac{a}{b} = \frac{c}{d}$$

- The proportion $\frac{1}{2} = \frac{3}{6}$ can be read as "1 is to 2 as 3 is to 6."

- The proportion $\frac{7}{3} = \frac{21}{9}$ can be read as "7 is to 3 as 21 is to 9."

- The proportion $\frac{8x}{1} = \frac{40x}{5}$ can be read as "8x is to 1 as 40x is to 5."

- The proportion $\frac{a}{b} = \frac{c}{d}$ can be read as "a is to b as c is to d."

In the proportion $\frac{1}{2} = \frac{3}{6}$, the numbers 1 and 6 are called the **extremes**, and the numbers 2 and 3 are called the **means**.

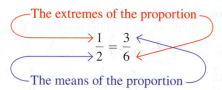

The extremes of the proportion

$$\frac{1}{2} = \frac{3}{6}$$

The means of the proportion

In this proportion, the product of the extremes is equal to the product of the means.

$$1 \cdot 6 = 6 \qquad \text{and} \qquad 2 \cdot 3 = 6$$

This illustrates a fundamental property of proportions.

FUNDAMENTAL PROPERTY OF PROPORTIONS	In any proportion, the product of the extremes is equal to the product of the means.

In the proportion $\frac{a}{b} = \frac{c}{d}$, a and d are the extremes, and b and c are the means. We can show that the product of the extremes (ad) is equal to the product of the means (bc) by multiplying both sides of the proportion by bd to clear the fractions, and observing that $ad = bc$.

$$\frac{a}{b} = \frac{c}{d}$$

$$\boldsymbol{bd} \cdot \frac{a}{b} = \boldsymbol{bd} \cdot \frac{c}{d} \qquad \text{To eliminate the fractions, multiply both sides by } bd, \text{ the LCD.}$$

$$ad = bc \qquad \text{Simplify.}$$

Since $ad = bc$, the product of the extremes equals the product of the means.

To determine whether an equation is a proportion, we can check to see whether the product of the extremes is equal to the product of the means.

EXAMPLE 1 Determine whether each equation is a proportion.

a. $\dfrac{3}{7} = \dfrac{9}{21}$ **b.** $\dfrac{8}{3} = \dfrac{13}{5}$

Solution In each part, we check to see whether the product of the extremes is equal to the product of the means.

a. The product of the extremes is $3 \cdot 21 = 63$. The product of the means is $7 \cdot 9 = 63$. Since the products are equal, the equation is a proportion: $\dfrac{3}{7} = \dfrac{9}{21}$.

b. The product of the extremes is $8 \cdot 5 = 40$. The product of the means is $3 \cdot 13 = 39$. Since the products are not equal, the equation is not a proportion: $\dfrac{8}{3} \neq \dfrac{13}{5}$.

 SELF CHECK 1 Determine whether the equation is a proportion: $\dfrac{6}{13} = \dfrac{24}{53}$

When two pairs of numbers, such as 2 and 3 and 8 and 12, form a proportion, we say that they are **proportional**. To show that 2 and 3, and 8 and 12, are proportional, we check to see whether the equation

$$\frac{2}{3} = \frac{8}{12}$$

is a proportion. To do so, we find the product of the extremes and the product of the means.

$2 \cdot 12 = 24$ The product of the extremes $\qquad 3 \cdot 8 = 24$ The product of the means

Since the products are equal, the equation is a proportion, and the numbers are proportional.

EXAMPLE 2 Determine whether 3 and 7, and 36 and 91, are proportional.

Solution We check to see whether $\dfrac{3}{7} = \dfrac{36}{91}$ is a proportion by finding two products:

$3 \cdot 91 = 273$ This is the product of the extremes.

$7 \cdot 36 = 252$ This is the product of the means.

Since the products are not equal, the numbers are not proportional.

 SELF CHECK 2 Determine whether 6 and 11, and 54 and 99, are proportional.

2 Solve a proportion.

Suppose that we know three of the values in the proportion

$$\frac{x}{5} = \frac{24}{20}$$

To find the unknown value, we multiply the extremes and multiply the means, set them equal, and solve for x:

$\dfrac{x}{5} = \dfrac{24}{20}$

$20x = 5 \cdot 24$ In a proportion, the product of the extremes is equal to the product of the means.

$20x = 120$ Multiply.

$\dfrac{20x}{20} = \dfrac{120}{20}$ Divide both sides by 20.

$x = 6$

The unknown value is 6.

Everyday connections
Golden Ratio

The Golden Ratio is an irrational number whose approximation, 1.618, is found in the measurements of many ancient buildings in Greece. It is believed that this ratio is the most pleasing to the eye, explaining the abundance of monuments with these measurements. More interestingly, this number is found in the comparison of various measurements on the human body (anthropometry), such as the length of the index finger compared to the length

from the fingertip to the big knuckle and the length of the forearm (from the tip of the middle finger to the elbow) compared to the length of the hand (from the tip of the middle finger to the wrist).

1. Measure the length of your index finger and divide that by the length of your finger from the tip to the big knuckle. What is this measurement? Is it close to the Golden Ratio?

2. If a professional basketball player has an index finger the length of which is 7.5 inches, predict the length of his finger from the tip to the large knuckle.

EXAMPLE 3 Solve: $\dfrac{12}{18} = \dfrac{3}{x}$

Solution To solve the proportion, we clear the fractions by multiplying the extremes and multiplying the means. Then, we solve for x.

$$\frac{12}{18} = \frac{3}{x}$$

$12 \cdot x = 18 \cdot 3$ In a proportion, the product of the extremes equals the product of the means.

$12x = 54$ Multiply.

$\dfrac{12x}{12} = \dfrac{54}{12}$ Divide both sides by 12.

$x = \dfrac{9}{2}$

The solution is $\dfrac{9}{2}$.

 SELF CHECK 3 Solve: $\dfrac{15}{x} = \dfrac{25}{40}$

EXAMPLE 4 Solve: $\dfrac{3.5}{7.2} = \dfrac{x}{15.84}$

Solution To solve the proportion, we clear the fractions by multiplying the extremes and multiplying the means. Then, we solve for x.

$$\frac{3.5}{7.2} = \frac{x}{15.84}$$

$3.5(15.84) = 7.2x$ In a proportion, the product of the extremes equals the product of the means.

$55.44 = 7.2x$ Multiply.

$\dfrac{55.44}{7.2} = \dfrac{7.2x}{7.2}$ Divide both sides by 7.2.

$7.7 = x$

The solution is 7.7.

 **SELF CHECK 4** Solve: $\dfrac{6.7}{x} = \dfrac{33.5}{38}$

Accent
on technology

▸ Solving Equations
with a Calculator

To solve the equation in Example 4 with a calculator, we can proceed as follows.

$$\frac{3.5}{7.2} = \frac{x}{15.84}$$

$$\frac{3.5(15.84)}{7.2} = x \qquad \text{Multiply both sides by 15.84.}$$

We can find the value of x by entering these numbers and pressing these keys.

3.5 × 15.84 ÷ 7.2 =   Using a scientific calculator

3.5 × 15.84 ÷ 7.2 **ENTER**   Using a TI84 graphing calculator

Either way, the display will read 7.7. Thus, the solution is 7.7.

For instructions regarding the use of a Casio graphing calculator, please refer to the Casio Keystroke Guide in the back of the book.

EXAMPLE 5   Solve: $\dfrac{2x + 1}{4} = \dfrac{10}{8}$

Solution   To solve the proportion, we clear the fractions by multiplying the extremes and multiplying the means. Then, we solve for x.

$$\frac{2x + 1}{4} = \frac{10}{8}$$

$$8(2x + 1) = 4 \cdot 10 \qquad \text{In a proportion, the product of the extremes equals the product of the means.}$$

$$16x + 8 = 40 \qquad \text{Use the distributive property to remove parentheses.}$$

$$16x + 8 - 8 = 40 - 8 \qquad \text{Subtract 8 from both sides.}$$

$$16x = 32 \qquad \text{Simplify.}$$

$$\frac{16x}{16} = \frac{32}{16} \qquad \text{Divide both sides by 16.}$$

$$x = 2$$

Thus, the solution is 2.

 SELF CHECK 5   Solve: $\dfrac{3x - 1}{2} = \dfrac{12.5}{5}$

3 Solve an application using a proportion.

When solving applications, we often need to set up and solve a proportion.

EXAMPLE 6   If 6 avocados cost $10.38, how much will 16 avocados cost?

Solution   Let c represent the cost of 16 avocados. The ratios of the numbers of avocados to their costs are equal.

6 avocados is to $10.38 as 16 avocados is to c.

6 avocados → $\dfrac{6}{10.38} = \dfrac{16}{c}$ ← 16 avocados
Cost of 6 avocados →   ← Cost of 16 avocados

$$6 \cdot c = 10.38(16) \qquad \text{In a proportion, the product of the extremes is equal to the product of the means.}$$

$$6c = 166.08 \qquad \text{Simplify.}$$

$$\frac{6c}{6} = \frac{166.08}{6} \qquad \text{Divide both sides by 6.}$$

$$c = 27.68$$

Sixteen avocados will cost \$27.68.

 SELF CHECK 6   If 5 tickets to a concert cost \$112.50, how much will 9 tickets cost?

EXAMPLE 7   **MIXING SOLUTIONS**   A solution contains 2 quarts of antifreeze and 5 quarts of water. How many quarts of antifreeze must be mixed with 18 quarts of water to have the same concentration?

Solution   Let q represent the number of quarts of antifreeze to be mixed with the water. The ratios of the quarts of antifreeze to the quarts of water are equal.

2 quarts antifreeze is to 5 quarts water as q quarts antifreeze is to 18 quarts water.

$$\begin{array}{l} \text{2 quarts antifreeze} \rightarrow \\ \text{5 quarts water} \rightarrow \end{array} \quad \frac{2}{5} = \frac{q}{18} \quad \begin{array}{l} \leftarrow q \text{ quarts of antifreeze} \\ \leftarrow 18 \text{ quarts water} \end{array}$$

$$2 \cdot 18 = 5q \qquad \text{In a proportion, the product of the extremes is equal to the product of the means.}$$

$$36 = 5q \qquad \text{Multiply.}$$

$$\frac{36}{5} = \frac{5q}{5} \qquad \text{Divide both sides by 5.}$$

$$\frac{36}{5} = q$$

The mixture should contain $\frac{36}{5}$ or 7.2 quarts of antifreeze.

 SELF CHECK 7   A solution should contain 2 ounces of alcohol for every 7 ounces of water. How much alcohol should be added to 20 ounces of water to get the proper concentration?

EXAMPLE 8   **BAKING**   A recipe for rhubarb cake calls for $1\frac{1}{4}$ cups of sugar for every $2\frac{1}{2}$ cups of flour. How many cups of flour are needed if the baker intends to use 3 cups of sugar?

Solution   Let f represent the number of cups of flour to be mixed with the sugar. The ratios of the cups of sugar to the cups of flour are equal.

$1\frac{1}{4}$ cups sugar is to $2\frac{1}{2}$ cups flour as 3 cups sugar is to f cups flour.

$$\begin{array}{l} 1\frac{1}{4} \text{ cups sugar} \rightarrow \\ 2\frac{1}{2} \text{ cups flour} \rightarrow \end{array} \quad \frac{1\frac{1}{4}}{2\frac{1}{2}} = \frac{3}{f} \quad \begin{array}{l} \leftarrow 3 \text{ cups sugar} \\ \leftarrow f \text{ cups flour} \end{array}$$

$$\frac{\frac{5}{4}}{\frac{5}{2}} = \frac{3}{f} \qquad \text{Express the mixed numbers as fractions.}$$

$$\frac{5}{4}f = \frac{5}{2} \cdot 3 \qquad \text{In a proportion, the product of the extremes is equal to the product of the means.}$$

$$\frac{5}{4}f = \frac{15}{2} \qquad \text{Multiply.}$$

$$\frac{4}{5} \cdot \frac{5}{4}f = \frac{4}{5} \cdot \frac{15}{2} \qquad \text{Multiply both sides by } \tfrac{4}{5}.$$

$$f = 6$$

The baker should use 6 cups of flour.

 SELF CHECK 8   How many cups of sugar will be needed to make several cakes that will require a total of 25 cups of flour?

EXAMPLE 9   **QUALITY CONTROL**   In a manufacturing process, 15 parts out of 90 were found to be defective. How many defective parts will be expected in a run of 120 parts?

Solution   Let d represent the expected number of defective parts. In each run, the ratio of the defective parts to the total number of parts should be the same.

15 defective parts is to 90 as d defective parts is to 120.

$$\begin{array}{l} \text{15 defective parts} \rightarrow \\ \text{90 parts} \rightarrow \end{array} \qquad \frac{15}{90} = \frac{d}{120} \begin{array}{l} \leftarrow d \text{ defective parts} \\ \leftarrow 120 \text{ parts} \end{array}$$

$$15 \cdot 120 = 90d \qquad \text{In a proportion, the product of the extremes is equal to the product of the means.}$$

$$1{,}800 = 90d \qquad \text{Multiply.}$$

$$\frac{1{,}800}{90} = \frac{90d}{90} \qquad \text{Divide both sides by 90.}$$

$$20 = d$$

The expected number of defective parts is 20.

 SELF CHECK 9   How many defective parts will be expected in a run of 3,000 parts?

4   Solve an application using the properties of similar triangles.

If two angles of one triangle have the same measure as two angles of a second triangle, the triangles will have the same shape. Triangles with the same shape are called **similar triangles**. In Figure 6-1, $\triangle ABC \sim \triangle DEF$ (read the symbol $\sim$ as "is similar to").

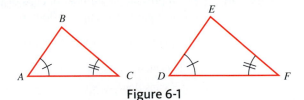

Figure 6-1

PROPERTY OF SIMILAR TRIANGLES   If two triangles are similar, the lengths of all pairs of corresponding sides are in proportion.

In the similar triangles shown in Figure 6-1, the following proportions are true.

$$\frac{AB}{DE} = \frac{BC}{EF}, \qquad \frac{BC}{EF} = \frac{CA}{FD}, \qquad \text{and} \qquad \frac{CA}{FD} = \frac{AB}{DE}$$

EXAMPLE 10   **HEIGHT OF A TREE**   A tree casts a shadow 18 feet long at the same time as a woman 5 feet tall casts a shadow that is 1.5 feet long. Find the height of the tree.

Solution   We let h represent the height of the tree. Figure 6-2 shows the triangles determined by the tree and its shadow and the woman and her shadow.

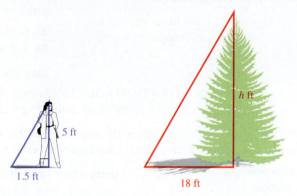

Figure 6-2

Since the triangles have the same shape, they are similar, and the lengths of their corresponding sides are in proportion. We can find h by solving the following proportion.

$$\frac{h}{5} = \frac{18}{1.5} \qquad \frac{\text{Height of the tree}}{\text{Height of the woman}} = \frac{\text{Length of the shadow of the tree}}{\text{Length of the shadow of the woman}}$$

$$1.5h = 5(18) \qquad \text{In a proportion, the product of the extremes is equal to the product of the means.}$$

$$1.5h = 90 \qquad \text{Multiply.}$$

$$h = 60 \qquad \text{Divide both sides by 1.5.}$$

The tree is 60 feet tall.

 SELF CHECK 10   Find the height of the tree if the woman is 5 feet 6 inches tall and her shadow is still 1.5 feet long.

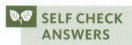 **SELF CHECK ANSWERS**   **1.** no   **2.** yes   **3.** 24   **4.** 7.6   **5.** 2   **6.** $202.50   **7.** $\frac{40}{7}$ oz   **8.** 12.5 cups   **9.** 500   **10.** 66 ft

NOW TRY THIS

To monitor the population of wild animals, the Fish and Wildlife department uses a method called "tag and release." A number of animals are captured, tagged, and returned to the wild. A year later, another group of the animals is captured and the ratio of the tagged animals to the total captured is believed to be approximately the same as the ratio of the original number tagged to the total population.

If 25 wolves were tagged and released and a year later 21 wolves were captured, 3 of which were tagged, predict the total population of the wolves.

6.8   Exercises

WARM-UPS   *Which expressions are proportions?*

1. $\dfrac{3}{5} = \dfrac{6}{10}$

2. $\dfrac{1}{2} = \dfrac{1}{3}$

3. $\dfrac{1}{2} = \dfrac{1}{4}$

4. $\dfrac{1}{x} = \dfrac{2}{2x}$

REVIEW

5. Change $\dfrac{3}{10}$ to a percent.

6. Change $\dfrac{7}{8}$ to a percent.

7. Change $33\dfrac{1}{3}\%$ to a fraction.

8. Change 60% to a fraction.

9. Find 30% of 1,600.

10. Find $\dfrac{1}{2}\%$ of 520.

11. **Shopping**   If Maria bought a dress for 25% off the original price of $98, how much did the dress cost?

12. **Shopping**   Bill purchased a shirt on sale for $17.50. Find the original cost of the shirt if it was marked down 30%.

VOCABULARY AND CONCEPTS   *Fill in the blanks.*

13. A _____ is a statement that two _____ are equal.

14. In the proportion $\dfrac{a}{b} = \dfrac{c}{d}$, a and d are called the _____ of the proportion.

15. In the proportion $\dfrac{a}{b} = \dfrac{c}{d}$, b and c are called the _____ of the proportion.

16. When two pairs of numbers form a proportion, we say that the numbers are _____.

17. If two triangles have the same shape, they are said to be _____.

18. If two triangles are similar, the lengths of their corresponding sides are in _____.

19. The equation $\dfrac{a}{b} = \dfrac{c}{d}$ is a proportion if the product __ is equal to the product __.

20. If $3 \cdot 10 = 17x$, then _____ is a proportion. (Note that answers may differ.)

21. Read $\triangle ABC$ as _____ ABC.

22. The symbol $\sim$ is read as _____.

GUIDED PRACTICE   *Determine whether each statement is a proportion.* SEE EXAMPLE 1. (OBJECTIVE 1)

23. $\dfrac{9}{7} = \dfrac{81}{70}$

24. $\dfrac{6}{5} = \dfrac{18}{15}$

25. $\dfrac{-8}{5} = \dfrac{32}{-20}$

26. $\dfrac{13}{-19} = \dfrac{-65}{95}$

27. $\dfrac{11}{17} = \dfrac{33}{50}$

28. $\dfrac{40}{29} = \dfrac{29}{22}$

29. $\dfrac{10.4}{3.6} = \dfrac{41.6}{14.4}$

30. $\dfrac{13.23}{3.45} = \dfrac{39.96}{11.35}$

Determine whether the given values are proportional. SEE EXAMPLE 2. (OBJECTIVE 1)

31. 6, 10, 15, 25

32. 4, 2, 17, 8.5

33. 3, 7, 4, 8

34. 4.5, 6, 8.5, 10

Solve. SEE EXAMPLE 3. (OBJECTIVE 2)

35. $\dfrac{2}{3} = \dfrac{x}{6}$

36. $\dfrac{3}{6} = \dfrac{x}{8}$

37. $\dfrac{4}{12} = \dfrac{3}{a}$

38. $\dfrac{5}{20} = \dfrac{2}{c}$

Solve. SEE EXAMPLE 4. (OBJECTIVE 2)

39. $\dfrac{3.2}{6.4} = \dfrac{7.5}{x}$

40. $\dfrac{1.5}{2.7} = \dfrac{4.5}{x}$

41. $\dfrac{-5.6}{x} = \dfrac{4.9}{2.8}$

42. $\dfrac{x}{-9.9} = \dfrac{-7.7}{23.1}$

Solve for the variable in each proportion. SEE EXAMPLE 5. (OBJECTIVE 2)

43. $\dfrac{x+1}{5} = \dfrac{3}{15}$

44. $\dfrac{x-1}{7} = \dfrac{2}{21}$

45. $\dfrac{x+2}{14} = \dfrac{-8}{7}$

46. $\dfrac{x+7}{-4} = \dfrac{3}{12}$

ADDITIONAL PRACTICE   *Solve.*

47. $\dfrac{-8}{x} = \dfrac{6}{3}$

48. $\dfrac{4}{x} = \dfrac{2}{8}$

49. $\dfrac{x}{5} = \dfrac{-6}{15}$

50. $\dfrac{x}{9} = \dfrac{-45}{27}$

51. $\dfrac{4-x}{13} = \dfrac{11}{26}$

52. $\dfrac{5-x}{17} = \dfrac{13}{34}$

53. $\dfrac{2x+1}{18} = \dfrac{14}{3}$

54. $\dfrac{2x-1}{18} = \dfrac{9}{54}$

55. $\dfrac{3p-2}{12} = \dfrac{p+1}{3}$

56. $\dfrac{12}{m} = \dfrac{18}{m+2}$

APPLICATIONS   *Set up and solve a proportion.* SEE EXAMPLE 6. (OBJECTIVE 3)

57. **Grocery shopping**   If 3 pints of yogurt cost $1, how much will 51 pints cost?

58. **Shopping for clothes**   If shirts are on sale at two for $25, how much will 5 shirts cost?

Set up and solve a proportion. SEE EXAMPLE 7. (OBJECTIVE 3)

59. **Mixing perfume**   A perfume is to be mixed in the ratio of 3 drops of pure essence to 7 drops of alcohol. How many drops of pure essence should be mixed with 56 drops of alcohol?

60. **Making cologne**   A cologne can be made by mixing 3 drops of pure essence with 7 drops of distilled water. How many drops of water should be used with 42 drops of pure essence?

Set up and solve a proportion. SEE EXAMPLE 8. (OBJECTIVE 3)

61. Making cookies   A recipe for chocolate chip cookies calls for $1\frac{1}{4}$ cups of flour and 1 cup of sugar. The recipe will make $3\frac{1}{2}$ dozen cookies. How many cups of flour will be needed to make 12 dozen cookies?

62. Making brownies   A recipe for brownies calls for 4 eggs and $1\frac{1}{2}$ cups of flour. If the recipe makes 15 brownies, how many cups of flour will be needed to make 130 brownies?

Set up and solve a proportion. SEE EXAMPLE 9. (OBJECTIVE 3)

63. Quality control   In a manufacturing process, 96% of the parts made are to be within specifications. How many defective parts would be expected in a run of 650 pieces?

64. Quality control   Out of a sample of 500 men's shirts, 17 were rejected because of crooked collars. How many crooked collars would you expect to find in a run of 15,000 shirts?

Set up and solve a proportion. SEE EXAMPLE 10. (OBJECTIVE 4)

65. Height of a tree   A tree casts a shadow of 26 feet at the same time as a 6-foot man casts a shadow of 4 feet. The two triangles in the illustration are similar. Find the height of the tree.

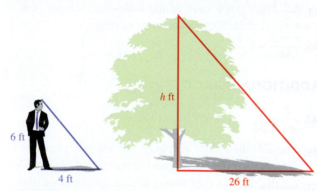

66. Height of a flagpole   A man places a mirror on the ground and sees the reflection of the top of a flagpole, as in the illustration. The two triangles in the illustration are similar. Find the height *h* of the flagpole.

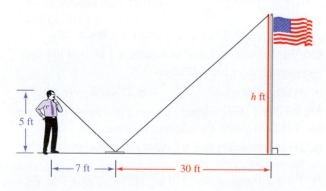

Set up and solve a proportion.

67. Gardening   Garden seed is on sale at 3 packets for 70¢. How much will 39 packets cost?

68. Cooking   A recipe for spaghetti sauce requires four 16-ounce bottles of catsup to make two gallons of sauce. How many bottles of catsup are needed to make 10 gallons of sauce?

69. Gas consumption   If a car can travel 42 miles on 1 gallon of gas, how much gas will it need to travel 315 miles?

70. Gas consumption   If a truck gets 12 mpg, how far can it go on 17 gallons of gas?

71. Computing paychecks   Chen earns $412 for a 40-hour week. If he missed 10 hours of work last week, how much did he get paid?

72. Computing paychecks   Danielle has a part-time job earning $169.50 for a 30-hour week. If she works a full 40-hour week, how much will she earn?

73. Model railroading   An HO-scale model railroad engine is 9 inches long. If HO scale is 87 feet to 1 foot, how long is a real engine?

74. Model railroading   An N-scale model railroad caboose is 3.5 inches long. If N scale is 169 feet to 1 foot, how long is a real caboose to the nearest half inch?

75. Model houses   A model house is built to a scale of 1 inch to 8 inches. If a model house is 36 inches wide, how wide is the real house?

76. Drafting   In a scale drawing, a 280-foot antenna tower is drawn 7 inches high. The building next to it is drawn 2 inches high. How tall is the actual building?

77. Mixing fuel   The instructions on a can of oil intended to be added to lawnmower gasoline read:

Recommended	Gasoline	Oil
50 to 1	6 gal	16 oz

Are these instructions correct?   (*Hint:* There are 128 ounces in 1 gallon.)

78. Mixing fuel   See Exercise 77. How much oil should be mixed with 28 gallons of gas?

79. Width of a river   Use the dimensions in the illustration to find *w*, the width of the river. The two triangles in the illustration are similar.

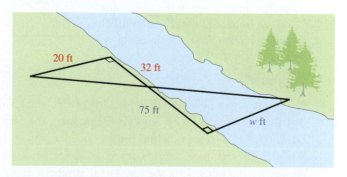

80. Flight path   An airplane ascends 100 feet as it flies a horizontal distance of 1,000 feet. The two triangles in the illustration are similar. How much altitude will it gain as it flies a horizontal distance of 1 mile? (*Hint*: 5,280 feet = 1 mile.)

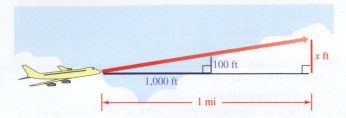

81. Flight path An airplane descends 1,350 feet as it flies a horizontal distance of 2 miles. How much altitude is lost as it flies a horizontal distance of 5 miles?

82. Ski runs A ski course falls 100 feet in every 300 feet of horizontal run. If the total horizontal run is $\frac{1}{2}$ mile, find the height of the hill.

83. Mountain travel A road ascends 750 feet in every 2,500 feet of travel. By how much will the road rise in a trip of 10 miles?

84. Photo enlargements The 3-by-5 photo in the illustration is to be blown up to the larger size. Find x.

WRITING ABOUT MATH

85. Explain the difference between a ratio and a proportion.

86. Explain how to tell whether the equation $\frac{3.2}{3.7} = \frac{5.44}{6.29}$ is a proportion.

SOMETHING TO THINK ABOUT

87. Verify that $\frac{3}{5} = \frac{12}{20} = \frac{3 + 12}{5 + 20}$. Is the following rule always true?

$$\frac{a}{b} = \frac{c}{d} = \frac{a + c}{b + d}$$

88. Verify that since $\frac{3}{5} = \frac{9}{15}$, then $\frac{3 + 5}{5} = \frac{9 + 15}{15}$. Is the following rule always true?

$$\text{If } \frac{a}{b} = \frac{c}{d}, \text{ then } \frac{a + b}{b} = \frac{c + d}{d}$$

Projects

PROJECT 1

If the sides of two similar triangles are in the ratio of 1 to 1, the triangles are said to be congruent. Congruent triangles have the same shape and the same size (area).

a. Draw several triangles with sides of length 1, 1.5, and 2 inches. Are the triangles all congruent? What general rule could you make?

b. Draw several triangles with the dimensions shown in the illustration on the right. Are the triangles all congruent? What general rule could you make?

c. Draw several triangles with the dimensions shown in the illustration on the right. Are the triangles all congruent? What general rule could you make?

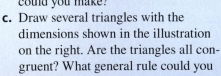

PROJECT 2

If y is equal to a polynomial divided by a polynomial, we call the resulting function a rational function. The simplest of these functions is defined by the equation $f(x) = \frac{1}{x}$. Since the denominator of this fraction cannot be 0, the domain of the function is the set of real numbers, except 0. Since a fraction with a numerator of 1 cannot be 0, the range is also the set of real numbers, except 0.

Construct the graph of this function by making a table of values containing at least eight ordered pairs, plotting the ordered pairs, and joining the points with two curves. Then graph each of the following rational functions and find each one's domain and range.

a. $f(x) = \dfrac{2}{x}$ **b.** $f(x) = -\dfrac{1}{x}$

c. $f(x) = -\dfrac{1}{x + 1}$ **d.** $f(x) = \dfrac{2}{x - 1}$

PROJECT 3

Suppose that the cost of telephone service is $6 per month plus 5¢ per call. If n represents the number of calls made one month, the cost C of phone service that month will be given by $C = 0.05n + 6$. If we divide the total cost C by the number of calls n, we will obtain the average cost per call, which we will denote as $\bar{c}$.

$$(1) \quad \bar{c} = \frac{C}{n} = \frac{0.05n + 6}{n}$$

$\bar{c}$ is the average cost per call, C is the total monthly cost, and n is the number of phone calls made that month.

Use the rational function in Equation 1 to find the average monthly cost per call when

a. 5 calls were made **b.** 25 calls were made.

Assume that a phone company charges $15 a month and 5¢ per phone call and answer the following questions.

a. Write a function that will give the cost C per month for making n phone calls.

b. Write a function that will give the average cost $\bar{c}$ per call during the month.

c. Find the cost if 45 phone calls were made during the month.

d. Find the average cost per call if 45 calls were made that month.

Reach for Success
EXTENSION OF PREPARING FOR A TEST

Now that you have a few strategies for preparing for a test, consider how to be emotionally and physically ready. *Emotional intelligence* is defined as the ability to identify one's emotions and then manage those emotions. A good start to physical fitness is a healthy lifestyle that includes proper nutrition, exercise, and sleep. If you feel good physically you will more likely feel good emotionally.

On a scale of 1 to 10, how would you rate your emotional readiness for college if 10 represents emotionally strong?

1 2 3 4 5 6 7 8 9 10

What strategies can you employ to improve your emotional readiness?

1. _____

2. _____

3. _____

On a scale of 1 to 10, how would you rate your physical health if 10 represents healthy and well?

1 2 3 4 5 6 7 8 9 10

What strategies can you employ to improve your physical health?

1. _____

2. _____

3. _____

Talk with your professor or talk with your advisor for additional strategies for successful test preparation.

6 Review

SECTION 6.1   Simplifying Rational Expressions

Fractions that are the quotient of two integers are *rational numbers*.	Rational numbers: $\dfrac{2}{13}$ $\dfrac{15}{28}$
Fractions that are the quotient of two polynomials are *rational expressions*.	Rational expressions: $\dfrac{x}{x-5}$ $\dfrac{5x-2}{x^2+x-3}$
$\dfrac{a}{0}$ is undefined. To find the values of the variable that make a rational expression undefined, we set the denominator equal to 0 and solve. The domain is the set of all values that can be substituted for the variable.	x cannot be 2 in the rational expression $\dfrac{x+1}{x-2}$ because 2 will cause the denominator to be 0. In interval notation, $(-\infty, 2) \cup (2, \infty)$. $\quad x - 2 = 0 \qquad$ Set the denominator equal to 0. $\quad\quad x = 2 \qquad$ Solve. The domain of $f(x) = \dfrac{x-1}{x-2}$, is $\{x \mid x \in \mathbb{R}, x \neq 2\}$.
If b and c are not 0, then $$\dfrac{a \cdot c}{b \cdot c} = \dfrac{a}{b}$$ $$\dfrac{a}{1} = a$$ $$\dfrac{a-b}{b-a} = -1$$	$\dfrac{22xy^3}{33x^2y} = \dfrac{2 \cdot \overset{1}{\cancel{11}} \cdot \overset{1}{\cancel{x}} \cdot \overset{1}{\cancel{y}} \cdot y \cdot y}{3 \cdot \cancel{11} \cdot x \cdot x \cdot \cancel{y}}$ Factor the numerator and denominator. $\quad = \dfrac{2y^2}{3x}$ Divide out the common factors of 11, x, and y. $\dfrac{2x^3 - 2x^2}{x-1} = \dfrac{2x^2\overset{1}{\cancel{(x-1)}}}{\overset{1}{\cancel{x-1}}}$ Factor the numerator. $\quad = 2x^2$ Divide out the common factor of $x-1$. Because $x-3$ and $3-x$ are opposites (additive inverses), their quotient is -1. $\qquad \dfrac{x-3}{3-x} = -1$

REVIEW EXERCISES

For what values of x is the rational expression undefined?

1. $\dfrac{x+5}{(x+4)(x-2)}$

2. $\dfrac{x-3}{x^2+x-6}$

Write each fraction in lowest terms.

3. $\dfrac{24}{27}$

4. $-\dfrac{12}{18}$

5. $-\dfrac{51}{153}$

6. $\dfrac{105}{45}$

7. $\dfrac{3x^2}{6x^3}$

8. $\dfrac{5xy^2}{2x^2y^2}$

9. $\dfrac{x^2}{x^2+x}$

10. $\dfrac{x+2}{x^2+2x}$

11. $\dfrac{12x^2y}{4x^2y^2}$

12. $\dfrac{8x^2y}{2x(4xy)}$

13. $\dfrac{7a-5}{5-7a}$

14. $\dfrac{x^2-x-56}{x^2-5x-24}$

15. $\dfrac{2x^2-16x}{2x^2-18x+16}$

16. $\dfrac{a^2+2a+ab+2b}{a^2+2ab+b^2}$

SECTION 6.2   Multiplying and Dividing Rational Expressions

DEFINITIONS AND CONCEPTS	EXAMPLES	
$\dfrac{a}{b} \cdot \dfrac{c}{d} = \dfrac{a \cdot c}{b \cdot d} \quad (b, d \neq 0)$	$\dfrac{x+1}{x+4} \cdot \dfrac{x^2-16}{x^2-x-2} = \dfrac{(x+1)(x^2-16)}{(x+4)(x^2-x-2)}$	Multiply the numerators and multiply the denominators.
	$\quad = \dfrac{\cancel{(x+1)}\cancel{(x+4)}(x-4)}{\cancel{(x+4)}\cancel{(x+1)}(x-2)}$	Factor and divide out the common factors.
	$\quad = \dfrac{x-4}{x-2}$	

$$\frac{a}{b} \div \frac{c}{d} = \frac{a}{b} \cdot \frac{d}{c} \quad (b, c, d \neq 0)$$

$$\frac{6x^2}{2x + 8} \div \frac{3x}{x^2 - 2x - 24}$$

$$= \frac{6x^2}{2x + 8} \cdot \frac{x^2 - 2x - 24}{3x} \qquad \text{Invert the denominator and multiply.}$$

$$= \frac{6x^2}{2(x + 4)} \cdot \frac{(x - 6)(x + 4)}{3x} \qquad \text{Factor.}$$

$$= \frac{\overset{3}{\underset{1}{\cancel{6x^2}}}}{2\cancel{(x + 4)}} \cdot \frac{(x - 6)\overset{1}{\cancel{(x + 4)}}}{\cancel{3x}} \qquad \text{Divide out the common factors.}$$

$$= \frac{x(x - 6)}{1}$$

$$= x(x - 6) \qquad \text{Denominators of 1 need not be written.}$$

REVIEW EXERCISES

Perform each multiplication and simplify.

17. $\dfrac{5x^2 y}{3x} \cdot \dfrac{6x}{15y^2}$ **18.** $\dfrac{3x}{x^2 - x} \cdot \dfrac{2x - 2}{x^2}$

19. $\dfrac{x^2 + 3x + 2}{x^2 + 2x} \cdot \dfrac{x}{x + 1}$ **20.** $\dfrac{x^2 + x}{3x - 15} \cdot \dfrac{6x - 30}{x^2 + 2x + 1}$

Perform each division and simplify.

21. $\dfrac{9xy}{14x^2} \div \dfrac{18y^2}{21xy}$ **22.** $\dfrac{x^2 + 5x}{x^2 + 4x - 5} \div \dfrac{x^2}{x - 1}$

23. $\dfrac{x^2 - x - 6}{2x - 1} \div \dfrac{x^2 - 2x - 3}{2x^2 + x - 1}$

24. $\dfrac{x^2 - 3x}{x^2 - x - 6} \div \dfrac{x^2 - x}{x^2 + x - 2}$

25. $\dfrac{x^2 + 4x + 4}{x^2 + x - 6} \left(\dfrac{x - 2}{x - 1} \div \dfrac{x + 2}{x^2 + 2x - 3} \right)$

SECTION 6.3 Adding and Subtracting Rational Expressions

DEFINITIONS AND CONCEPTS	EXAMPLES	
$\dfrac{a}{d} + \dfrac{b}{d} = \dfrac{a + b}{d} \quad (d \neq 0)$	$\dfrac{4x + 3}{2x} + \dfrac{x + 1}{2x} = \dfrac{4x + 3 + x + 1}{2x}$	Add the numerators and keep the common denominator.
	$= \dfrac{5x + 4}{2x}$	Combine like terms.
$\dfrac{a}{d} - \dfrac{b}{d} = \dfrac{a - b}{d} \quad (d \neq 0)$	$\dfrac{4x + 3}{2x} - \dfrac{x + 1}{2x} = \dfrac{4x + 3 - (x + 1)}{2x}$	Subtract the numerators and keep the common denominator.
	$= \dfrac{4x + 3 - x - 1}{2x}$	Use the distributive property to remove the parentheses.
	$= \dfrac{3x + 2}{2x}$	Combine like terms.

The least common denominator (LCD) for a set of fractions is the smallest number that each denominator will divide exactly.

Finding the Least Common Denominator (LCD):

1. List the different denominators that appear in the rational expressions.

2. Completely factor each denominator.

3. Form a product using each different factor obtained in Step 2. Use each different factor the *greatest* number of times it appears in any one factorization. The product formed by multiplying these factors is the LCD.

The LCD of 6 and 9 is 18.
The LCD of 10 and 15 is 30.

Find the LCD of $\dfrac{2x}{3y}, \dfrac{4x}{15y}$, and $\dfrac{5x}{18y}$.

$$3y = 3 \cdot y$$
$$15y = 3 \cdot 5 \cdot y$$
$$18y = 2 \cdot 3 \cdot 3 \cdot y$$

Form a product with factors of 2, 3 and y. We use 2 one time, because it appears only once as a factor of 18. We use 3 two times because it appears twice as a factor of 18. We use 5 one time, because it appears only once as a factor of 15. We use y once because it only occurs once in each factor of $3y$, $15y$, and $18y$.

$$\text{LCD} = 2 \cdot 3 \cdot 3 \cdot 5 \cdot y = 90y$$

To add or subtract rational expressions with unlike denominators, first find the LCD of the expressions. Then express each fraction in equivalent form with this LCD. Finally, add or subtract the expressions. Simplify, if possible.

To perform the subtraction $\frac{x+1}{x^2-9} - \frac{2}{x+3}$, we factor $x^2 - 9$ and find that the LCD is $(x+3)(x-3)$. Then, we proceed as follows:

$$\frac{x+1}{x^2-9} - \frac{2}{x+3}$$

$$= \frac{x+1}{(x+3)(x-3)} - \frac{2(x-3)}{(x+3)(x-3)}$$ Build the second fraction and factor the denominator of the first fraction.

$$= \frac{x+1-2(x-3)}{(x+3)(x-3)}$$ Subtract the numerators and keep the common denominator.

$$= \frac{x+1-2x+6}{(x+3)(x-3)}$$ Distribute.

$$= \frac{-x+7}{(x+3)(x-3)}$$ Combine like terms.

REVIEW EXERCISES

Perform each operation. Simplify all answers.

26. $\dfrac{5x}{x+3} + \dfrac{x-8}{x+3}$

27. $\dfrac{y}{x-y} - \dfrac{x}{x-y}$

28. $\dfrac{x}{x-1} + \dfrac{1}{x}$

29. $\dfrac{1}{7} - \dfrac{1}{x}$

30. $\dfrac{3}{x+1} - \dfrac{2}{x}$

31. $\dfrac{x+2}{2x} - \dfrac{2-x}{x^2}$

32. $\dfrac{x}{x+2} + \dfrac{3}{x} - \dfrac{4}{x^2+2x}$

33. $\dfrac{2}{x-1} - \dfrac{3}{x+1} + \dfrac{x-5}{x^2-1}$

SECTION 6.4   Simplifying Complex Fractions

DEFINITIONS AND CONCEPTS

To simplify a complex fraction, use either of these methods:

Method 1

Write the numerator and denominator of the complex fraction as single fractions, do the division of the fractions, and simplify.

EXAMPLES

Method 1

$$\frac{\dfrac{2x}{x+1} + \dfrac{3}{x}}{\dfrac{1}{x} + \dfrac{2}{x+1}}$$

$$= \frac{\dfrac{2x \cdot x}{x(x+1)} + \dfrac{3(x+1)}{x(x+1)}}{\dfrac{1(x+1)}{x(x+1)} + \dfrac{2x}{x(x+1)}}$$ Build each fraction.

$$= \frac{\dfrac{2x^2}{x(x+1)} + \dfrac{3x+3}{x(x+1)}}{\dfrac{x+1}{x(x+1)} + \dfrac{2x}{x(x+1)}}$$ Simplify wherever possible.

$$= \frac{2x^2+3x+3}{x(x+1)} \div \frac{x+1+2x}{x(x+1)}$$ Add the fractions and write the resulting complex fraction as an equivalent division problem.

$$= \frac{2x^2+3x+3}{x(x+1)} \div \frac{3x+1}{x(x+1)}$$ Simplify the numerators.

$$= \frac{2x^2+3x+3}{x(x+1)} \cdot \frac{x(x+1)}{3x+1}$$ Invert the divisor and multiply.

$$= \frac{2x^2+3x+3}{3x+1}$$ Multiply the fractions and simplify.

Method 2

Multiply both the numerator and the denominator of the complex fraction by the LCD of the fractions that appear in the numerator and the denominator, then simplify.

Method 2

$$\frac{\dfrac{2x}{x+1} + \dfrac{3}{x}}{\dfrac{1}{x} + \dfrac{2}{x+1}}$$

$$= \frac{x(x+1)\left(\dfrac{2x}{x+1} + \dfrac{3}{x}\right)}{x(x+1)\left(\dfrac{1}{x} + \dfrac{2}{x+1}\right)}$$ Multiply the numerator and the denominator by the LCD of $x(x+1)$.

$$= \frac{x(x+1)\left(\dfrac{2x}{x+1}\right) + x(x+1)\left(\dfrac{3}{x}\right)}{x(x+1)\left(\dfrac{1}{x}\right) + x(x+1)\left(\dfrac{2}{x+1}\right)}$$ Distribute.

$$= \frac{x \cdot 2x + 3(x+1)}{x+1+2x}$$ Do the multiplication.

$$= \frac{2x^2 + 3x + 3}{3x+1}$$ Simplify.

REVIEW EXERCISES

Simplify each complex fraction.

34. $\dfrac{\dfrac{9}{4}}{\dfrac{4}{9}}$

35. $\dfrac{\dfrac{3}{2} + 1}{\dfrac{2}{3} + 1}$

36. $\dfrac{\dfrac{1}{x} + 1}{\dfrac{1}{x} - 1}$

37. $\dfrac{2 + \dfrac{7}{x}}{3 - \dfrac{1}{x^2}}$

38. $\dfrac{\dfrac{2}{x-1} + \dfrac{x-1}{x+1}}{\dfrac{1}{x^2-1}}$

39. $\dfrac{\dfrac{a}{b} + c}{\dfrac{b}{a} + c}$

SECTION 6.5   Solving Equations That Contain Rational Expressions

DEFINITIONS AND CONCEPTS

1. Find any restrictions on the variable. Remember that the denominator of a fraction cannot be 0.

2. Multiply both sides of the equation by the LCD of the rational expressions appearing in the equation to clear the equation of fractions.

3. Solve the resulting equation.

4. If an apparent solution of an equation is a restricted value, that value must be excluded. Check all solutions for extraneous roots.

EXAMPLES

Solve: $\dfrac{1}{x+1} + \dfrac{2}{x} = \dfrac{x}{x^2+x}$

$$x(x+1)\left(\dfrac{1}{x+1} + \dfrac{2}{x}\right) = x(x+1)\left[\dfrac{x}{x(x+1)}\right]$$ Multiply both sides by $x(x+1)$, the LCD.

$$x(x+1)\left(\dfrac{1}{x+1}\right) + x(x+1)\left(\dfrac{2}{x}\right) = x(x+1)\left[\dfrac{x}{x(x+1)}\right]$$ Use the distributive property.

$$x + 2(x+1) = x$$ Multiply each term by $x(x+1)$.

$$x + 2x + 2 = x$$ Distribute.

$$3x + 2 = x$$ Combine like terms.

$$2x = -2$$ Subtract x and 2 from both sides.

$$x = -1$$ Divide both sides by 2.

When -1 is substituted for x, the denominators become 0. Since division by 0 is undefined, $x = -1$ is extraneous. Since there is no solution, the solution set is $\varnothing$.

REVIEW EXERCISES

For Exercises 40–45 solve each equation and check all answers.

40. $\dfrac{4}{x} = \dfrac{6}{x-3}$

41. $\dfrac{7}{x+3} = \dfrac{5}{x+1}$

42. $\dfrac{2}{3x} + \dfrac{1}{x} = \dfrac{5}{9}$

43. $\dfrac{2x}{x+4} = \dfrac{3}{x-1}$

44. $\dfrac{2}{x-1} + \dfrac{3}{x+4} = \dfrac{-5}{x^2+3x-4}$

45. $\dfrac{4}{x+2} - \dfrac{3}{x+3} = \dfrac{6}{x^2+5x+6}$

46. Solve for r_1: $\dfrac{1}{r} = \dfrac{1}{r_1} + \dfrac{1}{r_2}$

47. The efficiency E of a Carnot engine is given by the formula

$$E = 1 - \dfrac{T_2}{T_1}$$

Solve the formula for T_1.

48. Nuclear medicine Radioactive tracers are used for diagnostic work in nuclear medicine. The effective half-life H of a radioactive material in a biological organism is given by the formula

$$H = \dfrac{RB}{R+B}$$

where R is the radioactive half-life and B is the biological half-life of the tracer. Solve the formula for R.

SECTION 6.6 Solving Applications of Equations That Contain Rational Expressions

DEFINITIONS AND CONCEPTS	EXAMPLES
1. Analyze the problem. **2.** Form an equation. **3.** Solve the equation. **4.** State the conclusion. **5.** Check the result.	An inlet pipe can fill a pond in 4 days, and a second inlet pipe can fill the same pond in 3 days. If both pipes are used, how long will it take to fill the pond? Let x represent the number of days it takes to fill the pond.

What the first inlet pipe can do in 1 day	plus	what the second inlet pipe can do in 1 day	equals	what they can do together in 1 day.
$\dfrac{1}{4}$	$+$	$\dfrac{1}{3}$	$=$	$\dfrac{1}{x}$

To solve the equation, we proceed as follows:

$$\dfrac{1}{4} + \dfrac{1}{3} = \dfrac{1}{x}$$

$$\mathbf{12x}\left(\dfrac{1}{4} + \dfrac{1}{3}\right) = \mathbf{12x}\left(\dfrac{1}{x}\right) \qquad \text{Multiply both sides by } 12x.$$

$$3x + 4x = 12 \qquad \text{Use the distributive property to remove parentheses and simplify.}$$

$$7x = 12 \qquad \text{Combine like terms.}$$

$$x = \dfrac{12}{7} \qquad \text{Divide both sides by 7.}$$

It will take $\dfrac{12}{7}$ or $1\dfrac{5}{7}$ days for both inlet pipes to fill the pond.

REVIEW EXERCISES

49. Pumping a basement If one pump can empty a flooded basement in 18 hours and a second pump can empty the basement in 20 hours, how long will it take to empty the basement when both pumps are used?

50. Painting houses If a homeowner can paint a house in 12 days and a professional painter can paint it in 8 days, how long will it take if they work together?

51. Jogging A jogger can bicycle 30 miles in the same time as he can jog 10 miles. If he can ride 10 mph faster than he can jog, how fast can he jog?

52. Wind speed A plane can fly 400 miles downwind in the same amount of time as it can travel 320 miles upwind. If the plane can fly at 360 mph in still air, find the velocity of the wind.

SECTION 6.7   Ratios

DEFINITIONS AND CONCEPTS	EXAMPLES
A **ratio** is the comparison of two numbers by their indicated quotient.	Ratios: $\dfrac{1}{2}$ $\dfrac{4}{5}$ $\dfrac{x}{y}$
The **unit cost** of an item is the ratio of its cost to its quantity.	If 3 pounds of peanuts costs $6.75, the unit cost (the cost per pound) is $\dfrac{6.75}{3} =$ $2.25 per pound.
Rates are ratios that are used to compare quantities with different units.	If a student drives 120 miles in 3 hours, her average rate is $\dfrac{120 \text{ miles}}{3 \text{ hours}}$, or 40 mph.

REVIEW EXERCISES

Write each ratio as a fraction in lowest terms.

53. 7 to 21

54. $12x$ to $15x$

55. 2 feet to 1 yard

56. 5 pints to 3 quarts

57. If three pounds of coffee cost $8.79, find the unit cost (the cost per pound).

58. If a factory used 2,275 kWh of electricity in February, what was the rate of energy consumption in kWh per week?

SECTION 6.8   Proportions and Similar Triangles

DEFINITIONS AND CONCEPTS	EXAMPLES
A **proportion** is a statement that two ratios are equal.	Proportions: $\dfrac{6}{7} = \dfrac{12}{14}$ $\dfrac{9}{11} = \dfrac{27}{33}$
In any proportion, the product of the extremes is equal to the product of the means.	Solve: $\dfrac{x-2}{6} = \dfrac{x}{7}$
	$7(x-2) = 6 \cdot x$ In a proportion, the product of the extremes is equal to the product of the means.
	$7x - 14 = 6x$ Do the multiplication.
	$x = 14$ Subtract $6x$ and add 14 to both sides.
The measures of corresponding sides of similar triangles are in proportion.	A tree casts a shadow 14 feet long at the same time as a woman 5.6 feet tall casts a shadow that is 2.8 feet long. Find the height of the tree.
	Let h represent the height of the tree. Since the triangles formed are similar, the lengths of their corresponding sides are in proportion.
	$\dfrac{h}{5.6} = \dfrac{14}{2.8}$ $\dfrac{\text{Height of tree}}{\text{Height of the woman}} = \dfrac{\text{Length of the shadow of the tree}}{\text{Length of the shadow of the woman}}$
	$2.8 \cdot h = 5.6 \cdot 14$ In a proportion, the product of the extremes is equal to the product of the means.
	$2.8h = 78.4$ Do the multiplication.
	$h = 28$ Divide both sides by 2.8.
	The tree is 28 feet tall.

REVIEW EXERCISES

Determine whether the following equations are proportions.

59. $\dfrac{4}{7} = \dfrac{20}{34}$

60. $\dfrac{8}{11} = \dfrac{24}{33}$

Solve each proportion.

61. $\dfrac{3}{x} = \dfrac{6}{9}$

62. $\dfrac{x}{3} = \dfrac{x}{5}$

63. $\dfrac{x-2}{5} = \dfrac{x}{7}$

64. $\dfrac{4x-1}{18} = \dfrac{x}{6}$

65. Height of a pole   A telephone pole casts a shadow 12 feet long at the same time that a man 6 feet tall casts a shadow of 3.6 feet. How tall is the pole?

6 Test

Assume no division by 0.

1. Simplify: $\dfrac{27x^3y^2}{45xy^3}$

2. Simplify: $\dfrac{2x^2 - x - 3}{4x^2 - 9}$

3. Simplify: $\dfrac{3(x + 2) - 3}{2x - 4 - (x - 5)}$

4. Multiply and simplify: $\dfrac{12x^2y}{15xyz} \cdot \dfrac{25y^2z}{16xt}$

5. Multiply and simplify: $\dfrac{x^2 + 3x + 2}{3x + 9} \cdot \dfrac{x + 3}{x^2 - 4}$

6. Divide and simplify: $\dfrac{7ab^2}{24ac} \div \dfrac{21a^2b^3}{40abc^2}$

7. Divide and simplify: $\dfrac{x^2 - x}{3x^2 + 6x} \div \dfrac{3x - 3}{3x^3 + 6x^2}$

8. Simplify: $\dfrac{x^2 + xy}{x - y} \cdot \dfrac{x^2 - y^2}{x^2 - 2x} \div \dfrac{x^2 + 2xy + y^2}{x^2 - 4}$

9. Add: $\dfrac{6x + 5}{x - 2} + \dfrac{3x - 7}{x - 2}$

10. Subtract: $\dfrac{3y + 7}{2y + 3} - \dfrac{3(y - 2)}{2y + 3}$

11. Add: $\dfrac{x + 1}{x} + \dfrac{x - 1}{x + 1}$

12. Subtract: $\dfrac{5x}{x - 2} - 3$

13. Simplify: $\dfrac{\dfrac{9x^3}{xy^2}}{\dfrac{6y^2}{x^3y^4}}$

14. Simplify: $\dfrac{1 + \dfrac{y}{x}}{\dfrac{y}{x} - 1}$

15. Solve for x: $\dfrac{x}{10} - \dfrac{1}{2} = \dfrac{x}{5}$

16. Solve for x: $\dfrac{1}{x + 2} + \dfrac{1}{x - 2} = \dfrac{4}{x^2 - 4}$

17. Solve for x: $\dfrac{7}{x + 4} - \dfrac{1}{2} = \dfrac{3}{x + 4}$

18. Solve for B: $H = \dfrac{RB}{R + B}$

19. Cleaning highways   One highway worker could pick up all the trash on a strip of highway in 7 hours, and his helper could pick up the trash in 9 hours. How long will it take them if they work together?

20. Boating   A boat can motor 28 miles downstream in the same amount of time as it can motor 18 miles upstream. Find the speed of the current if the boat can motor at 23 mph in still water.

21. Flight path   A plane drops 575 feet as it flies a horizontal distance of $\frac{1}{2}$ mile. How much altitude will it lose as it flies a horizontal distance of 7 miles?

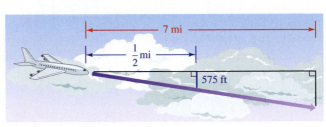

22. Express as a ratio in lowest terms: 8 inches to 3 feet.

23. Is $\dfrac{3xy}{5xy} = \dfrac{3xt}{5xt}$ a proportion? Explain.

24. Solve for y: $\dfrac{y}{y - 1} = \dfrac{y - 2}{y}$

25. A tree casts a shadow that is 30 feet long when a 6-foot-tall man casts a shadow that is 4 feet long. How tall is the tree?

Cumulative Review

Simplify each expression.

1. $x^3 x^6$

2. $(x^2)^5$

3. $\dfrac{x^5}{x^2}$

4. $(6x^4)^0$

5. $(3x^2 - 2x) + (6x^3 - 3x^2 - 1)$

6. $(4x^3 - 2x) - (2x^3 - 2x^2 - 3x + 1)$

7. $3(5x^2 - 4x + 3) + 2(-x^2 + 2x - 4)$

8. $4(3x^2 - 4x - 1) - 2(-2x^2 + 4x - 3)$

Perform each operation. Assume no division by 0.

9. $(2x^4y^3)(-7x^5y)$

10. $-5x^2(7x^3 - 2x^2 - 2)$

11. $(5x + 2)(4x + 1)$

12. $(5x - 4y)(3x + 2y)$

13. $x + 3 \overline{)x^2 + 7x + 12}$

14. $2x - 3 \overline{)2x^3 - x^2 - x - 3}$

Factor each expression.

15. $4xy^2 - 12x^2y^3$

16. $3(a + b) + x(a + b)$

17. $2a + 2b + ab + b^2$

18. $25p^4 - 16q^2$

19. $x^2 - 5x - 14$

20. $x^2 - xy - 6y^2$

21. $6a^2 - 7a - 20$

22. $8m^2 - 10mn - 3n^2$

23. $p^3 - 27q^3$

24. $8r^3 + 64s^3$

Solve each equation.

25. $\dfrac{4}{5}x + 6 = 18$

26. $5 - \dfrac{x + 2}{3} = 7 - x$

27. $6x^2 - x - 2 = 0$

28. $5x^2 = 10x$

29. $x^2 + 6x + 5 = 0$

30. $2y^2 + 5y - 12 = 0$

Solve each inequality and graph the solution set.

31. $4x - 7 > 1$

32. $7x - 9 < 5$

33. $-2 < -x + 3 < 5$

34. $0 \leq \dfrac{4 - x}{3} \leq 2$

Graph each equation.

35. $4x - 3y = 12$

36. $3x + 4y = 4y + 12$

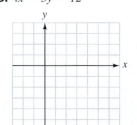

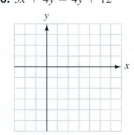

If $f(x) = 2x^2 - 3$, find each value.

37. $f(-1)$

38. $f(3)$

39. $f(-2)$

40. $f(2x)$

Simplify each fraction.

41. $\dfrac{x^2 + 3x + 2}{x^2 - 4}$

42. $\dfrac{x^2 + 2x - 15}{x^2 + 3x - 10}$

Perform the operation(s) and simplify when possible. Assume no division by 0.

43. $\dfrac{x^2 + x - 6}{5x - 5} \cdot \dfrac{5x - 10}{x + 3}$

44. $\dfrac{p^2 - p - 6}{3p - 9} \div \dfrac{p^2 + 6p + 9}{p^2 - 9}$

45. $\dfrac{3x}{x + 2} + \dfrac{5x}{x + 2} - \dfrac{7x - 2}{x + 2}$

46. $\dfrac{x - 1}{x + 1} + \dfrac{x + 1}{x - 1}$

47. $\dfrac{a + 1}{2a + 4} - \dfrac{a^2}{2a^2 - 8}$

48. $\dfrac{\dfrac{1}{x} + \dfrac{1}{y}}{\dfrac{1}{x} - \dfrac{1}{y}}$

Systems of Linear Equations and Inequalities in Two Variables

Careers and Mathematics

MARKET AND RESEARCH ANALYSTS

Market and research analysts gather information about what people think. They help companies understand what types of products people want to buy and at what price. They gather statistical data on competitors and examine prices, sales, and methods of marketing and distribution. They analyze statistical data on past sales to predict future sales. Prospective researchers need at least a bachelor's degree and analysts should study mathematics, statistics, sampling theory, and survey design.

Job Outlook:
Employment of market and research analysts is projected to grow faster than the average for all occupations through the year 2020.

Annual Earnings:
$42,190–$84,070

For More Information:
http://www.bls.gov/oco/ocos013.htm

For a Sample Application:
See Problem 72 in Section 7.4.

REACH FOR SUCCESS

In this chapter

In many fields, applied problems are most easily solved by using several equations, each with several variables. In this chapter, we will consider three methods to solve systems of two linear equations, each with two variables. We then will use these skills to solve many applications.

Reach for Success   Examining Test Results

Now that you have your exam score, what are you going to do with it?

Regardless of whether your score is high or low, it is important that you take the time to find out *why* you missed any of the questions so that you might do better the next time.

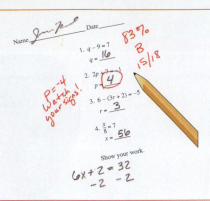

Look at different types of errors that can be made on a test. First, consider the *careless errors*. These are generally sign errors, arithmetic errors, miscopying from one line to another, and misreading the directions.

Go over your test and identify any problems for which you lost points for careless errors.

How many points did you lose? _____

By reviewing your work prior to submitting your test, you may be able to avoid losing points to careless errors.

Now look at *conceptual errors*. These errors reflect a lack of understanding the material. (These will require you to answer some tough questions and to be totally truthful with yourself!)

Review your test and identify any problems for which you lost points because you did not understand the material.

How many points did you lose?_____

Now, think back to the days prior to the test.

Were you in class the day these topics were covered?_____

Did you do *all* the homework assigned for the topics?_____

Did you ask questions in class?_____

Did you go to your professor or the mathematics/tutor center for additional help?_____

Next, look at *test-taking strategies*. These include writing down formulas at the start of the test, using all the allotted time, and previewing questions to work from easier to more difficult.

Did you write down formulas at the start of the test? _____

Did you preview to work questions you understood first? _____

How much of the allotted time did you use? _____

Set up an appointment with your instructor to discuss your exam errors. This is especially critical if your instructor does not allow you to keep the exam.

Describe the result of the consultation with your instructor. _____

_____

To separate anxiety interference from conceptual errors, rework the problems by covering up your test and sliding a blank sheet of paper down as you progress. What was the result of covering the problems and reworking? _____

_____

If you now get them correct, math anxiety might be interfering and you should consult with your instructor.

If you still miss them, it might be a lack of understanding and you should seek tutoring.

A Successful Study Strategy . . .

 Review each test in detail. Consider test-taking as an opportunity to show what you know. This positive attitude can improve performance.

At the end of the chapter you will find an additional exercise to guide you in planning for a successful semester.

438

Section 7.1

Solving Systems of Linear Equations by Graphing

Objectives

1. Determine whether an ordered pair is a solution to a given system of linear equations.
2. Solve a system of linear equations by graphing.
3. Recognize that an inconsistent system has no solution.
4. Express the infinitely many solutions of a dependent system as a general ordered pair.

Vocabulary

system of equations
simultaneous solution

independent equations
consistent system

inconsistent system
dependent equations

Getting Ready

If $y = x^2 - 3$, find the value of y when

1. $x = 0$ **2.** $x = 1$ **3.** $x = -2$ **4.** $x = 3$

The lines graphed in Figure 7-1 approximate the percentages of American households with only a landline phone and those with only a cell phone for the years 2005 to 2010. We can see that over this period, the percentage of those with only a landline decreased while those with only a cell phone increased.

By graphing this information on the same coordinate system, it appears that the percentage of households with a cell phone only was the same as those with only a landline in October 2008—about 19.3% each. In this section, we will work with pairs of linear equations whose graphs often will be intersecting lines.

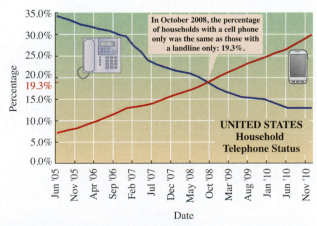

In October 2008, the percentage of households with a cell phone only was the same as those with a landline only: 19.3%.

UNITED STATES
Household
Telephone Status

Source: National Center for Health Statistics

Figure 7-1

1 Determine whether an ordered pair is a solution to a given system of linear equations.

Recall that we have previously graphed equations such as $x + y = 3$ that contain two variables. Because there are infinitely many pairs of numbers whose sum is 3, there are infinitely many pairs (x, y) that will satisfy this equation. Some of these pairs are listed in Table 7-1(a). Likewise, there are infinitely many pairs (x, y) that will satisfy the equation $3x - y = 1$. Some of these pairs are listed in Table 7-1(b).

$x + y = 3$

x	y
0	3
1	**2**
2	1
3	0

(a)

$3x - y = 1$

x	y
0	−1
1	**2**
2	5
3	8

(b)

Table 7-1

Although there are infinitely many pairs that satisfy each of these equations, only the pair $(1, 2)$ satisfies both equations.

The pair of equations

$$\begin{cases} x + y = 3 \\ 3x - y = 1 \end{cases}$$

is called a **system of equations**. Because the ordered pair $(1, 2)$ satisfies both equations, it is called a **simultaneous solution** or just a *solution of the system of equations*. In this chapter, we will discuss three methods for finding the solution of a system of two linear equations. In this section, we consider the graphing method.

2 Solve a system of linear equations by graphing.

To use the method of graphing to solve the system

$$\begin{cases} x + y = 3 \\ 3x - y = 1 \end{cases}$$

we will graph both equations on one set of coordinate axes. Using the intercept method, recall that to find the y-intercept, we let $x = 0$ and solve for y and to find the x-intercept, we let $y = 0$ and solve for x. We will also plot one extra point as a check. See Figure 7-2.

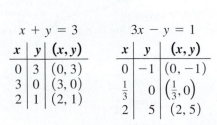

$x + y = 3$

x	y	(x, y)
0	3	$(0, 3)$
3	0	$(3, 0)$
2	1	$(2, 1)$

$3x - y = 1$

x	y	(x, y)
0	−1	$(0, -1)$
$\frac{1}{3}$	0	$\left(\frac{1}{3}, 0\right)$
2	5	$(2, 5)$

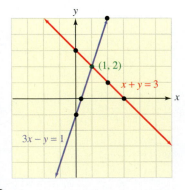

Figure 7-2

To schedule a company's workers, managers must consider several factors to match a worker's ability to the demands of various jobs and to match company resources to the requirements of the job. To design bridges or office buildings, engineers must analyze the effects of thousands of forces to ensure that structures won't collapse. A telephone switching network decides which of thousands of possible routes is the most efficient and then rings the correct telephone in seconds. Each of these tasks requires solving systems of equations—not just two equations in two variables, but hundreds of equations in hundreds of variables. These tasks are common in every business, industry, educational institution, and government in the world. All would be much more difficult without a computer.

One of the earliest computers in use was the Mark I, which resulted from a collaboration between IBM and a Harvard mathematician, Howard Aiken. The Mark I was started in 1939 and finished in 1944. It was 8 feet tall, 2 feet thick, and more than 50 feet long. It contained more than 750,000 parts and performed 3 calculations per second.

Ironically, Aiken could not envision the importance of his invention. He advised the National Bureau of Standards that there was no point in building a better computer, because "there will never be enough work for more than one or two of these machines."

© Courtesy of IBM

Mark I Relay Computer (1944)

Although there are infinitely many pairs (x, y) that satisfy $x + y = 3$ and infinitely many pairs (x, y) that satisfy $3x - y = 1$, only the coordinates of the point where their graphs intersect satisfy both equations. The solution of the system is the ordered pair $(1, 2)$.

To check the solution, we substitute 1 for x and 2 for y in each equation and verify that the pair $(1, 2)$ satisfies each equation.

First equation	*Second equation*
$x + y = 3$	$3x - y = 1$
$1 + 2 \overset{?}{=} 3$	$3(1) - 2 \overset{?}{=} 1$
$3 = 3$	$3 - 2 \overset{?}{=} 1$
	$1 = 1$

When the graphs of two equations in a system are different lines, the equations are called **independent equations**. When a system of equations has a solution, the system is called a **consistent system**.

To solve a system of equations in two variables by graphing, we follow these steps.

THE GRAPHING METHOD

1. Graph each equation on one set of coordinate axes.
2. Find the coordinates of the point where the graphs intersect, if applicable.
3. Check the solution in the equations of the original system, if applicable.

EXAMPLE 1   Solve the system by graphing: $\begin{cases} 2x + 3y = 2 \\ 3x = 2x + 16 \end{cases}$

Solution   Using the intercept method, we graph both equations on one set of coordinate axes, as shown in Figure 7-3. We also plot a third point as a check.

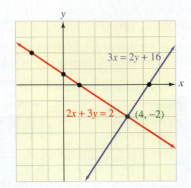

$2x + 3y = 2$		
x	y	(x, y)
0	$\frac{2}{3}$	$\left(0, \frac{2}{3}\right)$
1	0	$(1, 0)$
-2	2	$(-2, 2)$

$3x = 2y + 16$		
x	y	(x, y)
0	-8	$(0, -8)$
$\frac{16}{3}$	0	$\left(\frac{16}{3}, 0\right)$
4	-2	$(4, -2)$

Figure 7-3

Although there are infinitely many pairs (x, y) that satisfy $2x + 3y = 2$ and infinitely many pairs (x, y) that satisfy $3x = 2y + 16$, only the coordinates of the point where the graphs intersect satisfy both equations. The solution is the ordered pair $(4, -2)$.

To check, we substitute 4 for x and -2 for y in each equation and verify that the pair $(4, -2)$ satisfies each equation.

$$2\mathbf{x} + 3\mathbf{y} = 2 \qquad\qquad 3\mathbf{x} = 2\mathbf{y} + 16$$
$$2(\mathbf{4}) + 3(\mathbf{-2}) \stackrel{?}{=} 2 \qquad 3(\mathbf{4}) \stackrel{?}{=} 2(\mathbf{-2}) + 16$$
$$8 - 6 \stackrel{?}{=} 2 \qquad\qquad 12 \stackrel{?}{=} -4 + 16$$
$$2 = 2 \qquad\qquad\qquad 12 = 12$$

The equations in this system are independent equations, and the system is a consistent system of equations.

 **SELF CHECK 1**   Solve the system by graphing: $\begin{cases} 2x = y - 5 \\ x + y = -1 \end{cases}$

COMMENT   Remember that the graph is a tool we use to find the solution; it is not the solution.

When the coefficients of some of the variables are fractions, it may be easier to clear the fractions before proceeding.

EXAMPLE 2   Solve the system by graphing: $\begin{cases} \frac{2}{3}x - \frac{1}{2}y = 1 \\ \frac{1}{10}x + \frac{1}{15}y = 1 \end{cases}$

Solution   We can multiply both sides of the first equation by 6, the LCD, to clear it of fractions.

$$\frac{2}{3}x - \frac{1}{2}y = 1$$

$$\mathbf{6}\left(\frac{2}{3}x - \frac{1}{2}y\right) = \mathbf{6}(1) \qquad \text{Multiply both sides of the equation by 6, the LCD.}$$

(1) $\qquad\qquad 4x - 3y = 6 \qquad\qquad \text{Use the distributive property and simplify.}$

We then multiply both sides of the second equation by 30, the LCD, to clear it of fractions.

$$\frac{1}{10}x + \frac{1}{15}y = 1$$

$$\mathbf{30}\left(\frac{1}{10}\boldsymbol{x} + \frac{1}{15}\boldsymbol{y}\right) = \mathbf{30}\,(1)$$ Multiply both sides of the equation by 30, the LCD.

(2) $3x + 2y = 30$ Use the distributive property and simplify.

Equations 1 and 2 form the following equivalent system of equations, which has the same solution as the original system.

$$\begin{cases} 4x - 3y = 6 \\ 3x + 2y = 30 \end{cases}$$

We can graph each equation of the previous system (see Figure 7-4) and find that their point of intersection has coordinates of $(6, 6)$. The solution of the given system is the ordered pair $(6, 6)$.

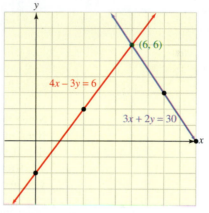

$4x - 3y = 6$				$3x + 2y = 30$		
x	y	(x, y)		x	y	(x, y)
0	-2	$(0, -2)$		10	0	$(10, 0)$
3	2	$(3, 2)$		8	3	$(8, 3)$
6	6	$(6, 6)$		6	6	$(6, 6)$

Figure 7-4

To verify that $(6, 6)$ satisfies each equation of the original system, we substitute 6 for x and 6 for y in each of the original equations and simplify.

$$\frac{2}{3}\boldsymbol{x} - \frac{1}{2}\boldsymbol{y} = 1 \qquad\qquad \frac{1}{10}\boldsymbol{x} + \frac{1}{15}\boldsymbol{y} = 1$$

$$\frac{2}{3}(\mathbf{6}) - \frac{1}{2}(\mathbf{6}) \stackrel{?}{=} 1 \qquad\qquad \frac{1}{10}(\mathbf{6}) + \frac{1}{15}(\mathbf{6}) \stackrel{?}{=} 1$$

$$4 - 3 \stackrel{?}{=} 1 \qquad\qquad \frac{3}{5} + \frac{2}{5} \stackrel{?}{=} 1$$

$$1 = 1 \qquad\qquad\qquad 1 = 1$$

The equations in this system are independent and the system is consistent.

SELF CHECK 2   Solve by graphing: $\begin{cases} -\dfrac{x}{2} = \dfrac{y}{4} \\ \dfrac{1}{4}x - \dfrac{3}{8}y = -2 \end{cases}$

COMMENT Always check your answer in the original equations. If you made an error in simplifying one of the equations, your answer would check in the simplified equations but not in the original.

3 Recognize that an inconsistent system has no solution.

Sometimes a system of equations will have no solution. These systems are called **inconsistent systems**.

EXAMPLE 3   Solve the system by graphing: $\begin{cases} 2x + y = -6 \\ 4x + 2y = 8 \end{cases}$

Solution   We graph both equations on one set of coordinate axes, as in Figure 7-5.

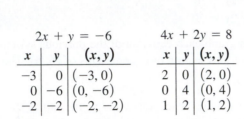

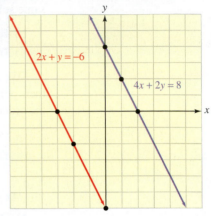

$2x + y = -6$

x	y	(x, y)
-3	0	$(-3, 0)$
0	-6	$(0, -6)$
-2	-2	$(-2, -2)$

$4x + 2y = 8$

x	y	(x, y)
2	0	$(2, 0)$
0	4	$(0, 4)$
1	2	$(1, 2)$

Figure 7-5

Since the graphs are different lines, the equations of the system are independent. The lines in the figure appear to be parallel. To be sure, we can find their slopes. Recall that if two lines have the same slope but different y-intercepts, they will be parallel. To determine the slope of each line, we write each equation in slope-intercept form, $y = mx + b$.

$$2x + y = -6 \qquad\qquad 4x + 2y = 8$$
$$y = -2x - 6 \qquad\qquad 2y = -4x + 8$$
$$y = -2x + 4$$

Because both equations have the same slope (-2) but different y-intercepts $(0, -6)$ and $(0, 4)$, they are parallel. Since parallel lines do not intersect, the system is inconsistent. Its solution set is $\varnothing$.

SELF CHECK 3   Solve the system by graphing: $\begin{cases} 2y = 3x \\ 3x - 2y = 6 \end{cases}$

In Examples 1 and 2, we saw that a system of equations can have a single solution. In Example 3, we saw that a system can have no solution. In Example 4, we will see that a system can have infinitely many solutions.

4 Express the infinitely many solutions of a dependent system as a general ordered pair.

Sometimes a system will have infinitely many solutions. In this case, we say that the equations of the system are **dependent equations**.

EXAMPLE 4 Solve the system by graphing: $\begin{cases} y - 2x = 4 \\ 4x + 8 = 2y \end{cases}$

Solution We graph each equation on one set of axes, as in Figure 7-6.

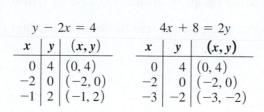

$y - 2x = 4$				$4x + 8 = 2y$		
x	y	(x, y)		x	y	(x, y)
0	4	$(0, 4)$		0	4	$(0, 4)$
-2	0	$(-2, 0)$		-2	0	$(-2, 0)$
-1	2	$(-1, 2)$		-3	-2	$(-3, -2)$

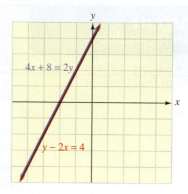

Figure 7-6

The lines in the figure appear to be the same line. To be sure, we can find their slopes and y-intercepts by writing each equation in slope-intercept form. If two lines have the same slope and the same y-intercept, they will be the same line.

$$y - 2x = 4 \qquad\qquad 4x + 8 = 2y$$
$$y = 2x + 4 \qquad\qquad 2x + 4 = y$$

Since the lines in the figure are the same line, they intersect at infinitely many points and there are infinitely many solutions. Every solution to the first equation is a solution to the second equation. To describe these solutions, we can solve either equation for y.

Because $2x + 4$ is equal to y, every solution (x, y) of the system will have the form $(x, 2x + 4)$. This solution can also be written in set-builder notation, $\{(x, y) | y = 2x + 4\}$.

To find some specific solutions, we can substitute 0, -3, and -1 for x in the general ordered pair $(x, 2x + 4)$ to obtain $(0, 4)$, $(-3, -2)$, and $(-1, 2)$. From the graph, we can see that each point lies on the one line that is the graph of both equations.

 **SELF CHECK 4** Solve the system by graphing: $\begin{cases} 6x - 2y = 4 \\ y + 2 = 3x \end{cases}$

Table 7-2 summarizes the possibilities that can occur when two nonvertical linear equations, each with two variables, are graphed.

TABLE 7-2

Possible graph	If the	then
	lines are different and intersect,	the equations are independent and the system is consistent. One solution exists, expressed as (x, y).
	lines are different and parallel,	the equations are independent and the system is inconsistent. No solution exists, expressed as $\varnothing$.
	lines coincide (are the same line),	the equations are dependent and the system is consistent. Infinitely many solutions exist, expressed as $(x, mx + b)$.

COMMENT   Recall that all vertical lines can be written in the form $x = c$, where c is a constant. If two vertical lines coincide, the equations are dependent and the system is consistent. Infinitely many solutions exist, expressed as (c, y).

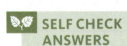

Accent on technology

▶ Solving Systems of Equations

We can use a graphing calculator to solve the system: $\begin{cases} 2x + y = 12 \\ 2x - y = -2 \end{cases}$

However, before we can enter the equations into the calculator, we must solve them for y.

$$2x + y = 12 \qquad\qquad 2x - y = -2$$
$$y = -2x + 12 \qquad\qquad -y = -2x - 2$$
$$y = 2x + 2$$

We can now enter the resulting equations into a TI84 calculator and graph them. If we use standard window settings of $x = [-10, 10]$ and $y = [-10, 10]$, their graphs will look like Figure 7-7(a). We can find the intersection point by using the INTERSECT command found in the CALC (**2nd** **TRACE**) menu. Use the calculator reference card in this textbook for specific keystrokes.

We find that the solution is $(2.5, 7)$. See Figure 7.7(b). Check the solution.

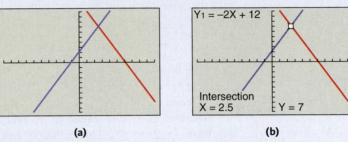

(a)   (b)

Figure 7-7

For instructions regarding the use of a Casio graphing calculator, please refer to the Casio Keystroke Guide in the back of the book.

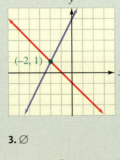
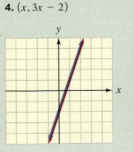
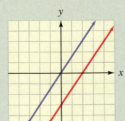
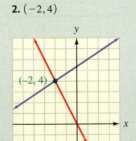

NOW TRY THIS

Solve each system by graphing.

1. $\begin{cases} y = -1 \\ x = 4 \end{cases}$

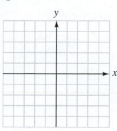

2. $\begin{cases} x = y \\ y = 0 \end{cases}$

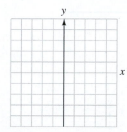

Solve using a graphing calculator.

3. $\begin{cases} y = \dfrac{3}{4}x - 2 \\ 2x + 4y = 24 \end{cases}$

7.1 Exercises

WARM-UPS *Determine whether the ordered pair $(-2, 1)$ is a solution to each equation.*

1. $x + y = -1$

2. $x - 2y = 0$

3. $2x + 5y = 1$

4. $-3x - 8y = 14$

5. $y = -\dfrac{1}{2}x + 3$

6. $y = \dfrac{3}{4}x + \dfrac{5}{2}$

7. $-\dfrac{3}{5}x + \dfrac{2}{3}y = \dfrac{28}{15}$

8. $\dfrac{3}{4}x - \dfrac{1}{5}y = \dfrac{-8}{5}$

REVIEW *Evaluate each expression. Assume that $x = -3$.*

9. $(-x)^4$

10. $-x^4$

11. $5x - 2x^2$

12. $\dfrac{-6 + 5x}{7x}$

VOCABULARY AND CONCEPTS *Fill in the blanks.*

13. The pair of equations $\begin{cases} x - y = -1 \\ 2x - y = 1 \end{cases}$ is called a _____ of equations.

14. Because the ordered pair $(2, 3)$ satisfies both equations in Exercise 13, it is called a _____ of the system.

15. When the graphs of two equations in a system are different lines, the equations are called _____ equations.

16. When a system of equations has a solution, the system is called a _____.

17. If a system of equations is _____, there is no solution and the solution set is __ .

18. When a system has infinitely many solutions, the equations of the system are said to be _____ equations.

GUIDED PRACTICE *Determine whether the ordered pair is a solution of the given system. (OBJECTIVE 1)*

19. $(2, 1)$, $\begin{cases} x + y = 3 \\ x - 2y = 0 \end{cases}$

20. $(1, 4)$, $\begin{cases} 3x + y = 7 \\ 2x - y = -2 \end{cases}$

21. $(-1, 3)$, $\begin{cases} 2x + y = 1 \\ 3x + y = 0 \end{cases}$

22. $(-2, 4)$, $\begin{cases} 2x + 2y = 4 \\ x + 3y = 10 \end{cases}$

23. $(4, 5)$, $\begin{cases} 2x - 3y = -7 \\ 4x - 5y = 25 \end{cases}$

24. $(3, 4)$, $\begin{cases} 4x - 3y = 0 \\ 2x - 3y = 6 \end{cases}$

25. $\left(-2, \frac{1}{2}\right)$, $\begin{cases} 2x + 4y = -2 \\ 3x - 8y = 10 \end{cases}$ **26.** $\left(2, \frac{1}{3}\right)$, $\begin{cases} x - 3y = 1 \\ -2x + 6y = -6 \end{cases}$

Solve each system by graphing. SEE EXAMPLE 1. (OBJECTIVE 2)

27. $\begin{cases} x + y = 2 \\ x - y = 0 \end{cases}$

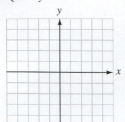

28. $\begin{cases} x + y = 4 \\ x - y = 0 \end{cases}$

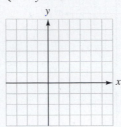

29. $\begin{cases} x + y = 2 \\ x - y = 4 \end{cases}$

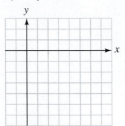

30. $\begin{cases} x + y = 1 \\ x - y = -5 \end{cases}$

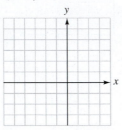

31. $\begin{cases} y = 2x \\ x + y = 0 \end{cases}$

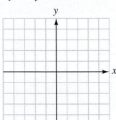

32. $\begin{cases} y = -x \\ x - y = 0 \end{cases}$

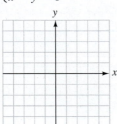

33. $\begin{cases} 3x + 2y = -8 \\ 2x - 3y = -1 \end{cases}$

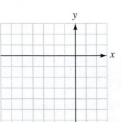

34. $\begin{cases} x + 4y = -2 \\ x + y = -5 \end{cases}$

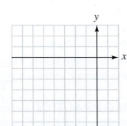

Solve each system by graphing. SEE EXAMPLE 2. (OBJECTIVE 2)

35. $\begin{cases} x + 2y = -4 \\ x - \frac{1}{2}y = 6 \end{cases}$

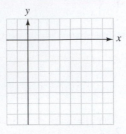

36. $\begin{cases} \frac{2}{3}x - y = -3 \\ 3x + y = 3 \end{cases}$

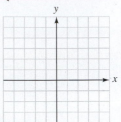

37. $\begin{cases} -\frac{3}{4}x + y = 3 \\ \frac{1}{4}x + y = -1 \end{cases}$

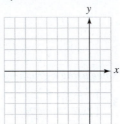

38. $\begin{cases} \frac{1}{3}x + y = 7 \\ \frac{2}{3}x - y = -4 \end{cases}$

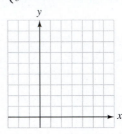

Solve each system by graphing. SEE EXAMPLE 3. (OBJECTIVE 3)

39. $\begin{cases} 3x - 6y = 18 \\ x = 2y + 3 \end{cases}$

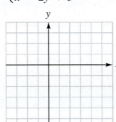

40. $\begin{cases} 5x - 4y = 20 \\ 4y = 5x + 12 \end{cases}$

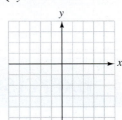

41. $\begin{cases} y = x \\ x - y = 7 \end{cases}$

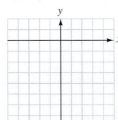

42. $\begin{cases} x = 2y - 8 \\ y = \frac{1}{2}x - 5 \end{cases}$

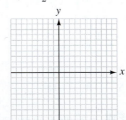

Solve each system by graphing. Give each answer as a general ordered pair. *SEE EXAMPLE 4. (OBJECTIVE 4)*

43. $\begin{cases} 4x - 2y = 8 \\ y = 2x - 4 \end{cases}$

44. $\begin{cases} 2x = 3(2 - y) \\ 3y = 2(3 - x) \end{cases}$

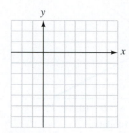
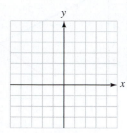

45. $\begin{cases} 6x + 3y = 9 \\ y + 2x = 3 \end{cases}$

46. $\begin{cases} x = y \\ y - x = 0 \end{cases}$

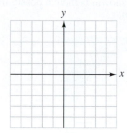
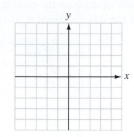

ADDITIONAL PRACTICE *Determine whether the pair is a solution of the system.*

47. $(3, 2)$, $\begin{cases} x + y = 5 \\ x - y = 1 \end{cases}$

48. $(1, 2)$, $\begin{cases} x - y = -1 \\ x + y = 3 \end{cases}$

49. $(4, 1)$, $\begin{cases} x + y = 5 \\ x - y = 2 \end{cases}$

50. $(5, 2)$, $\begin{cases} x - y = 3 \\ x + y = 6 \end{cases}$

51. $(-2, -3)$, $\begin{cases} 4x + 5y = -23 \\ -3x + 2y = 0 \end{cases}$

52. $(-5, 1)$, $\begin{cases} -2x + 7y = 17 \\ 3x - 4y = -19 \end{cases}$

53. $\left(-\frac{2}{5}, \frac{1}{4}\right)$, $\begin{cases} 5x - 4y = -6 \\ 8y = 10x + 12 \end{cases}$

54. $\left(-\frac{1}{3}, \frac{3}{4}\right)$, $\begin{cases} 3x + 4y = 2 \\ 12y = 3(2 - 3x) \end{cases}$

Solve each system by graphing.

55. $\begin{cases} 2x - 3y = -18 \\ 3x + 2y = -1 \end{cases}$

56. $\begin{cases} -x + 3y = -11 \\ 3x - y = 17 \end{cases}$

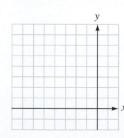
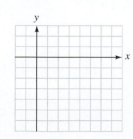

57. $\begin{cases} 4x = 3(4 - y) \\ 2y = 4(3 - x) \end{cases}$

58. $\begin{cases} 8x = 2y - 9 \\ 4y = -x - 16 \end{cases}$

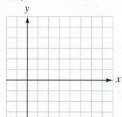
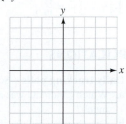

59. $\begin{cases} \frac{1}{2}x + \frac{1}{4}y = 0 \\ \frac{1}{4}x - \frac{3}{8}y = -2 \end{cases}$

60. $\begin{cases} \frac{1}{2}x + \frac{2}{3}y = -5 \\ \frac{3}{2}x - y = 3 \end{cases}$

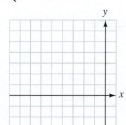
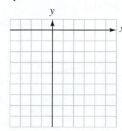

61. $\begin{cases} 2x - 3y = 3 \\ y = \frac{2}{3}x + 3 \end{cases}$

62. $\begin{cases} 3y + x = 1 \\ y = -\frac{1}{3}x + \frac{1}{3} \end{cases}$

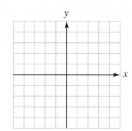
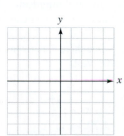

63. $\begin{cases} y + 3x = 2 \\ 6x - 4 = -2y \end{cases}$

64. $\begin{cases} 6x - 2y = 6 \\ x = \frac{1}{3}y - 1 \end{cases}$

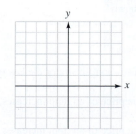
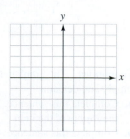

Use a graphing calculator to solve each system.

65. $\begin{cases} 3x - 6y = 4 \\ 2x + y = 1 \end{cases}$

66. $\begin{cases} 4x + 9y = 4 \\ 6x + 3y = -1 \end{cases}$

67. $\begin{cases} \frac{1}{3}x - \frac{1}{2}y = \frac{1}{6} \\ \frac{2}{5}x + \frac{1}{2}y = \frac{13}{10} \end{cases}$

68. $\begin{cases} \frac{3}{4}x + \frac{2}{3}y = -\frac{19}{6} \\ y - x = -\frac{4}{3}x \end{cases}$

APPLICATIONS

69. Daily tracking polls See the illustration.
   a. Which candidate was ahead on October 28 and by how much?
   b. On what day did the challenger pull even with the incumbent?
   c. If the election was held November 4, whom did the poll predict as the winner and by how many percentage points?

Daily Tracking Political Poll

70. Latitude and longitude See the illustration.
   a. Name three American cities that lie on the latitude line of 30° north.
   b. Name three American cities that lie on the longitude line of 90° west.
   c. What city lies on both lines?

71. Economics The graph in the illustration illustrates the law of supply and demand.
   a. Complete this sentence: "As the price of an item increases, the *supply* of the item _____."
   b. Complete this sentence: "As the price of an item increases, the *demand* for the item _____."

c. For what price will the supply equal the demand? How many items will be supplied for this price?

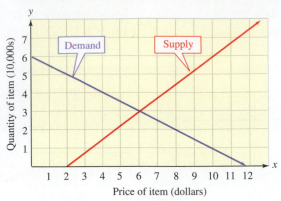

72. Traffic control The equations describing the paths of two airplanes are $y = -\frac{1}{2}x + 3$ and $3y = 2x + 2$. Graph each equation on the radar screen shown. Is there a possibility of a midair collision? If so, where?

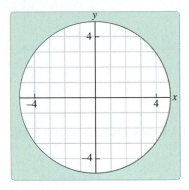

WRITING ABOUT MATH

73. Explain what we mean when we say "inconsistent system."
74. Explain what we mean when we say, "The equations of a system are dependent."

SOMETHING TO THINK ABOUT

75. Use a graphing calculator to solve the system

$$\begin{cases} 11x - 20y = 21 \\ -4x + 7y = 21 \end{cases}$$

What problems did you encounter?

76. Can the equations of an inconsistent system with two equations in two variables be dependent?

Section 7.2

Solving Systems of Linear Equations by Substitution

Objectives

1. Solve a system of linear equations by substitution.
2. Identify an inconsistent system of linear equations.
3. Express the solution of a dependent system of linear equations as a general ordered pair.

Getting Ready

Use the distributive property to remove parentheses.

1. $2(3x + 2)$ **2.** $5(-5 - 2x)$

Substitute $x - 2$ for y and simplify.

3. $2y$ **4.** $3(y - 2)$

The graphing method for solving systems of equations can be cumbersome and, unless you are using a graphing calculator, does not always provide exact solutions. Fortunately, there are other methods that can be quicker to use and provide exact solutions. We now consider one of them, called the *substitution method*.

1 Solve a system of linear equations by substitution.

To solve the system

$$\begin{cases} y = 3x - 2 \\ 2x + y = 8 \end{cases}$$

by the substitution method, we note that $y = 3x - 2$. Because $y = 3x - 2$, we can substitute $3x - 2$ for y in the equation $2x + y = 8$ to obtain

$$2x + \mathbf{y} = 8$$
$$2x + (\mathbf{3x - 2}) = 8$$

The resulting equation has only one variable and can be solved for x.

$$
\begin{aligned}
2x + (3x - 2) &= 8 \\
2x + 3x - 2 &= 8 &&\text{Remove parentheses.} \\
5x - 2 &= 8 &&\text{Combine like terms.} \\
5x &= 10 &&\text{Add 2 to both sides.} \\
x &= 2 &&\text{Divide both sides by 5.}
\end{aligned}
$$

We can find the value of y by substituting 2 for x in either equation of the given system. Because $y = 3x - 2$ is already solved for y, it is easier to substitute in this equation.

$$
\begin{aligned}
y &= 3\mathbf{x} - 2 \\
&= 3(\mathbf{2}) - 2 \\
&= 6 - 2 \\
&= 4
\end{aligned}
$$

The solution of the given system is the ordered pair $(2, 4)$.

Check: $y = 3x - 2$ $2x + y = 8$

$4 \overset{?}{=} 3(2) - 2$ $2(2) + 4 \overset{?}{=} 8$

$4 \overset{?}{=} 6 - 2$ $4 + 4 \overset{?}{=} 8$

$4 = 4$ $8 = 8$

The lines represented by the equations of the given system intersect at the point $(2, 4)$. The equations of this system are independent, and the system is consistent.

To solve a system of linear equations in x and y by the substitution method, we follow these steps.

THE SUBSTITUTION METHOD

1. If necessary, solve one of the equations for x or y, preferably a variable with a coefficient of 1.

2. Substitute the resulting expression for the variable obtained in Step 1 into the other equation, and solve that equation.

3. Find the value of the other variable by substituting the solution found in Step 2 into any equation containing both variables.

4. Check the solution in the equations of the original system.

EXAMPLE 1   Solve the system by substitution: $\begin{cases} 2x + y = -5 \\ 3x + 5y = -4 \end{cases}$

Solution   We first solve one of the equations for one of its variables. Since the term y in the first equation has a coefficient of 1, we solve the first equation for y.

$2x + y = -5$

$y = -5 - 2x$ Subtract $2x$ from both sides.

We then substitute $-5 - 2x$ for y in the second equation and solve for x.

$3x + 5y = -4$

$3x + 5(-5 - 2x) = -4$

$3x - 25 - 10x = -4$ Use the distributive property.

$-7x - 25 = -4$ Combine like terms.

$-7x = 21$ Add 25 to both sides.

$x = -3$ Divide both sides by -7.

We can find the value of y by substituting -3 for x in the original equation $2x + y = -5$.

$2x + y = -5$

$2(-3) + y = -5$

$-6 + y = -5$

$y = 1$

The solution is the ordered pair $(-3, 1)$.

Check: $2x + y = -5$ $3x + 5y = -4$

$2(-3) + 1 \overset{?}{=} -5$ $3(-3) + 5(1) \overset{?}{=} -4$

$-6 + 1 \overset{?}{=} -5$ $-9 + 5 \overset{?}{=} -4$

$-5 = -5$ $-4 = -4$

SELF CHECK 1   Solve the system by substitution: $\begin{cases} 2x - 3y = 13 \\ 3x + y = 3 \end{cases}$

EXAMPLE 2   Solve the system by substitution: $\begin{cases} 2x + 3y = 5 \\ 3x + 2y = 0 \end{cases}$

Solution   Although we can solve either equation for x or y, we will solve the second equation for x:

$$3x + 2y = 0$$

$$3x = -2y \qquad \text{Subtract } 2y \text{ from both sides.}$$

$$x = \frac{-2y}{3} \qquad \text{Divide both sides by 3.}$$

We then substitute $\frac{-2y}{3}$ for x in the other equation and solve for y.

$$2\boldsymbol{x} + 3y = 5 \qquad \text{This is the first equation of the system.}$$

$$2\left(\frac{-2y}{3}\right) + 3y = 5 \qquad \text{Substitute.}$$

$$\frac{-4y}{3} + 3y = 5 \qquad \text{Multiply.}$$

$$\boldsymbol{3}\left(\frac{-4y}{3} + 3y\right) = \boldsymbol{3}(5) \qquad \text{Multiply both sides of the equation by 3 to clear it of fractions.}$$

$$-4y + 9y = 15 \qquad \text{Use the distributive property.}$$

$$5y = 15 \qquad \text{Combine like terms.}$$

$$y = 3 \qquad \text{Divide both sides by 5.}$$

We can find the value of x by substituting 3 for y in the original equation, $3x + 2y = 0$.

$$3x + 2\boldsymbol{y} = 0$$

$$3x + 2(\boldsymbol{3}) = 0 \qquad \text{Substitute.}$$

$$3x + 6 = 0 \qquad \text{Multiply.}$$

$$3x = -6 \qquad \text{Subtract 6 from both sides of the equation.}$$

$$x = -2 \qquad \text{Divide both sides by 3.}$$

Check the solution $(-2, 3)$ in each equation of the original system.

COMMENT   Be certain your answer is in the correct order!

 SELF CHECK 2   Solve the system by substitution: $\begin{cases} 3x - 2y = -19 \\ 2x + 5y = 0 \end{cases}$

EXAMPLE 3   Solve the system by substitution: $\begin{cases} 3(x - y) = 5 \\ x + 3 = -\frac{5}{2}y \end{cases}$

Solution   We begin by solving the second equation for x because it has a coefficient of 1.

$$x + 3 = -\frac{5}{2}y$$

$$x = -\frac{5}{2}y - 3 \qquad \text{Subtract 3 from both sides.}$$

We can now substitute $-\frac{5}{2}y - 3$ for x in the first equation.

$$3(\boldsymbol{x} - y) = 5 \qquad \text{This is the first equation of the system.}$$

$$3\left(-\frac{5}{2}\boldsymbol{y} - \boldsymbol{3} - y\right) = 5 \qquad \text{Substitute.}$$

$$-\frac{15}{2}y - 9 - 3y = 5 \qquad \text{Use the distributive property.}$$

$$2\left(-\frac{15}{2}y - 9 - 3y\right) = (5)\mathbf{2} \qquad \text{Multiply both sides of the equation by 2 to clear the fraction.}$$

$$-15y - 18 - 6y = 10 \qquad \text{Use the distributive property.}$$

$$-21y - 18 = 10 \qquad \text{Combine like terms.}$$

$$-21y = 28 \qquad \text{Add 18 to both sides.}$$

$$y = -\frac{28}{21} \qquad \text{Divide both sides by } -21.$$

$$y = -\frac{4}{3} \qquad \text{Simplify the fraction.}$$

To find the value of x, we substitute $-\frac{4}{3}$ for y in one of the original equations. We'll use the equation $x + 3 = -\frac{5}{2}y$ and simplify.

$$x + 3 = -\frac{5}{2}\mathbf{y} \qquad \text{This is the second equation of the system.}$$

$$x + 3 = -\frac{5}{2}\left(-\frac{\mathbf{4}}{\mathbf{3}}\right) \qquad \text{Substitute.}$$

$$x + 3 = \frac{10}{3} \qquad \text{Multiply.}$$

$$3(x + 3) = 3\left(\frac{10}{3}\right) \qquad \text{Multiply both sides of the equation by 3.}$$

$$3x + 9 = 10 \qquad \text{Use the distributive property.}$$

$$3x = 1 \qquad \text{Subtract 9 from both sides.}$$

$$x = \frac{1}{3} \qquad \text{Divide both sides by 3.}$$

It is important that we check the solution $\left(\frac{1}{3}, -\frac{4}{3}\right)$ in each original equation.

 **SELF CHECK 3**   Solve the system by substitution: $\begin{cases} 2(x + y) = -5 \\ x + 2 = -\frac{3}{5}y \end{cases}$

---

2 Identify an inconsistent system of linear equations.

EXAMPLE 4   Solve the system by substitution: $\begin{cases} x = 4(3 - y) \\ 2x = 4(3 - 2y) \end{cases}$

Solution   Since $x = 4(3 - y)$, we can substitute $4(3 - y)$ for x in the second equation and solve for y.

$$2\mathbf{x} = 4(3 - 2y)$$

$$2[\mathbf{4(3 - y)}] = 4(3 - 2\mathbf{y})$$

$$2(12 - 4y) = 4(3 - 2y) \qquad \text{Simplify within the brackets.}$$

$$24 - 8y = 12 - 8y \qquad \text{Use the distributive property.}$$

$$24 = 12 \qquad \text{Add 8y to both sides of the equation.}$$

This impossible result indicates that the equations in this system are independent, but that the system is inconsistent. If each equation in this system were graphed, these graphs would be parallel lines. Since there are no solutions to this system, the solution set is $\varnothing$.

SELF CHECK 4   Solve the system by substitution: $\begin{cases} 0.1x - 0.4 = 0.1y \\ -2y = 2(2 - x) \end{cases}$

3 Express the solution of a dependent system of linear equations as a general ordered pair.

EXAMPLE 5   Solve the system by substitution: $\begin{cases} 3x = 4(6 - y) \\ 4y + 3x = 24 \end{cases}$

Solution   Because $3x$ is contained in both equations, we can substitute $4(6 - y)$ for $3x$ in the second equation and proceed as follows:

$$4y + \mathbf{3x} = 24$$
$$4y + \mathbf{4(6 - y)} = 24$$
$$4y + 24 - 4y = 24 \qquad \text{Use the distributive property.}$$
$$24 = 24 \qquad \text{Combine like terms.}$$

Although $24 = 24$ is true, we did not find a value for y. This result indicates that the equations of this system are dependent. If both equations were graphed, the same line would result.

Because any ordered pair that satisfies one equation satisfies the other as well, the system has infinitely many solutions. To obtain a general solution, we can solve the second equation of the system for y:

$$4y + 3x = 24$$
$$4y = -3x + 24 \qquad \text{Subtract } 3x \text{ from both sides of the equation.}$$
$$y = \frac{-3x + 24}{4} \qquad \text{Divide both sides of the equation by 4.}$$
$$y = -\frac{3}{4}x + 6 \qquad \text{Write in slope-intercept form.}$$

A general solution (x, y) is $\left(x, -\frac{3}{4}x + 6 \right)$.

SELF CHECK 5   Solve the system by substitution: $\begin{cases} 3y = -3(x + 4) \\ 3x + 3y = -12 \end{cases}$

SELF CHECK ANSWERS   1. $(2, -3)$ 2. $(-5, 2)$ 3. $\left(-\frac{5}{4}, -\frac{5}{4} \right)$ 4. $\varnothing$ 5. $(x, -x - 4)$

NOW TRY THIS

Solve each of the systems.

1. $\begin{cases} \dfrac{1}{3}a + \dfrac{1}{3}b = 2 \\ 5a + 7b = 12 \end{cases}$

2. $\begin{cases} 6x = 5 - 3y \\ y = -2x + 1 \end{cases}$

3. $\begin{cases} 9(y + 1) = \dfrac{27}{4}x \\ 4y + 4 = 3x \end{cases}$

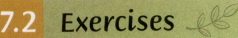

7.2 Exercises

WARM-UPS *Let $y = x - 1$. Write each expression in terms of x only.*

1. $3y$

2. $-5y$

3. $-2x + y$

4. $4x + y$

REVIEW *Let $x = -2$ and $y = 3$ and evaluate each expression.*

5. $y - x^2$

6. $-x^2 + y^3$

7. $\dfrac{3x - 2y}{2x + y}$

8. $-2x^2 y^2$

9. $-x(3y - 4)$

10. $-3x(x - 2y)$

Identify each graphed system as consistent or inconsistent and dependent or independent.

11.
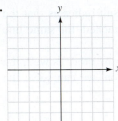

12.
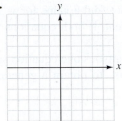

13.
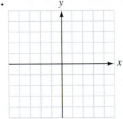

14.
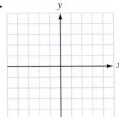

VOCABULARY AND CONCEPTS *Fill in the blanks.*

15. We say the equation $y = 2x + 4$ is solved for _ or that y is expressed in _____ of x.

16. To _____ a solution of a system means to see whether the coordinates of the ordered pair satisfy both original equations.

17. The solution set of $2(x - 6) = 2x - 15$ is _.

18. In mathematics, to _____ means to replace an expression with one that is equivalent to it.

19. A system with dependent equations has _____ solutions. Its solutions can be described by a general ordered pair or in set-builder notation.

20. In the term y, the _____ is understood to be 1.

GUIDED PRACTICE *Use substitution to solve each system.*
SEE EXAMPLE 1. (OBJECTIVE 1)

21. $\begin{cases} y = 2x \\ x + y = 6 \end{cases}$

22. $\begin{cases} y = 3x \\ x + y = 4 \end{cases}$

23. $\begin{cases} y = 3x - 5 \\ 3x + y = 1 \end{cases}$

24. $\begin{cases} y = 2x - 9 \\ x + 3y = 8 \end{cases}$

25. $\begin{cases} y = 2x + 5 \\ x + 2y = -5 \end{cases}$

26. $\begin{cases} y = -2x \\ 3x + 2y = -1 \end{cases}$

27. $\begin{cases} 4x + 5y = 2 \\ 3x - y = 11 \end{cases}$

28. $\begin{cases} 5u + 3v = 5 \\ 4u - v = 4 \end{cases}$

Use substitution to solve each system. SEE EXAMPLE 2. (OBJECTIVE 1)

29. $\begin{cases} 3x + y = 0 \\ 5x + 2y = -1 \end{cases}$

30. $\begin{cases} 3x - y = 7 \\ 2x + 3y = 1 \end{cases}$

31. $\begin{cases} 2x + 3y = 5 \\ 3x + 2y = 5 \end{cases}$

32. $\begin{cases} 3x - 2y = -1 \\ 2x + 3y = -5 \end{cases}$

33. $\begin{cases} 2x + 5y = -2 \\ 4x + 3y = 10 \end{cases}$

34. $\begin{cases} 3x + 4y = -6 \\ 2x - 3y = -4 \end{cases}$

35. $\begin{cases} 2x - 3y = -3 \\ 3x + 5y = -14 \end{cases}$

36. $\begin{cases} 4x - 5y = -12 \\ 5x - 2y = 2 \end{cases}$

Use substitution to solve each system. SEE EXAMPLE 3. (OBJECTIVE 1)

37. $\begin{cases} 3(x - 1) + 3 = 8 + 2y \\ 2(x + 1) = 4 + 3y \end{cases}$

38. $\begin{cases} 4(x - 2) = 19 - 5y \\ 3(x + 1) - 2y = 2y \end{cases}$

39. $\begin{cases} 6a = 5(3 + b + a) - a \\ 3(a - b) + 4b = 5(1 + b) \end{cases}$

40. $\begin{cases} 5(x + 1) + 7 = 7(y + 1) \\ 5(y + 1) = 6(1 + x) + 5 \end{cases}$

Use substitution to solve each system. SEE EXAMPLE 4. (OBJECTIVE 2)

41. $\begin{cases} 3y = -12x + 7 \\ 4x + y = 2 \end{cases}$

42. $\begin{cases} 3a + 6b = -9 \\ a = 4 - 2b \end{cases}$

43. $\begin{cases} a = \frac{3}{2} b + 5 \\ 2a - 3b = 8 \end{cases}$

44. $\begin{cases} 2x - 4y = 10 \\ x = 2y - 6 \end{cases}$

Use substitution to solve each system. SEE EXAMPLE 5. (OBJECTIVE 3)

45. $\begin{cases} 6x = 3y + 15 \\ 5 = 2x - y \end{cases}$

46. $\begin{cases} x = \frac{1}{2} y + \frac{5}{4} \\ 4x - 2y = 5 \end{cases}$

47. $\begin{cases} 3a + 6b = -15 \\ a = -2b - 5 \end{cases}$

48. $\begin{cases} y - 3x = 9 \\ 6x + 18 = 2y \end{cases}$

ADDITIONAL PRACTICE *Use substitution to solve each system.*

49. $\begin{cases} 2x + y = 4 \\ 4x + y = 5 \end{cases}$

50. $\begin{cases} x + 3y = 3 \\ 2x + 3y = 4 \end{cases}$

51. $\begin{cases} 2a = 3b - 13 \\ b = 2a + 7 \end{cases}$

52. $\begin{cases} a = 3b - 1 \\ b = 2a + 2 \end{cases}$

53. $\begin{cases} r + 3s = 9 \\ 3r + 2s = 13 \end{cases}$

54. $\begin{cases} x - 2y = 2 \\ 2x + 3y = 11 \end{cases}$

55. $\begin{cases} 3y + x = 1 \\ y = -\frac{1}{3}x + \frac{1}{3} \end{cases}$

56. $\begin{cases} y + 3x = 2 \\ 6x - 4 = -2y \end{cases}$

57. $\begin{cases} y - x = 3x \\ 2(x + y) = 14 - y \end{cases}$

58. $\begin{cases} y + x = 2x + 2 \\ 2(3x - 2y) = 21 - y \end{cases}$

59. $\begin{cases} 3x + 4y = -7 \\ 2y - x = -1 \end{cases}$

60. $\begin{cases} 4x + 5y = -2 \\ x + 2y = -2 \end{cases}$

61. $\begin{cases} 2x - 3y = 3 \\ y = \frac{2}{3}x + 3 \end{cases}$

62. $\begin{cases} 6x - 2y = 6 \\ x = \frac{1}{3}y - 1 \end{cases}$

63. $\begin{cases} 8x - 3y = 1 \\ -4x + 3y = -5 \end{cases}$

64. $\begin{cases} -8x + 3y = 22 \\ 4x + 3y = -2 \end{cases}$

65. $\begin{cases} 2a + 3b = 2 \\ 8a - 3b = 3 \end{cases}$

66. $\begin{cases} 3a - 2b = 0 \\ 9a + 4b = 5 \end{cases}$

67. $\begin{cases} \frac{1}{2}x + \frac{1}{2}y = -1 \\ \frac{1}{3}x - \frac{1}{2}y = -4 \end{cases}$

68. $\begin{cases} \frac{2}{3}y + \frac{1}{5}z = 1 \\ \frac{1}{3}y - \frac{2}{5}z = 3 \end{cases}$

69. $\begin{cases} 5x = \frac{1}{2}y - 1 \\ \frac{1}{4}y = 10x - 1 \end{cases}$

70. $\begin{cases} \frac{2}{3}x = 1 - 2y \\ 2(5y - x) + 11 = 0 \end{cases}$

71. $\begin{cases} \frac{6x - 1}{3} - \frac{5}{3} = \frac{3y + 1}{2} \\ \frac{1 + 5y}{4} + \frac{x + 3}{4} = \frac{17}{2} \end{cases}$

72. $\begin{cases} \frac{5x - 2}{4} + \frac{1}{2} = \frac{3y + 2}{2} \\ \frac{7y + 3}{3} = \frac{x}{2} + \frac{7}{3} \end{cases}$

APPLICATIONS

73. **Geometry** In the illustration, $x + y = 90°$ and $y = 2x$. Find x and y.

74. **Geometry** In the illustration, $x + y = 180°$ and $y = 5x$. Find x and y.

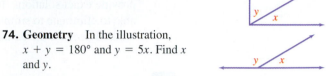

WRITING ABOUT MATH

75. Explain how to use substitution to solve a system of equations.

76. If the equations of a system are written in general form, why is it to your advantage to solve for a variable whose coefficient is 1?

SOMETHING TO THINK ABOUT

77. Could you use substitution to solve the system

$$\begin{cases} y = 2y + 4 \\ x = 3x - 5 \end{cases}$$

How would you solve it?

78. What are the advantages and disadvantages of
   a. the graphing method?
   b. the substitution method?

Section 7.3

Solving Systems of Linear Equations by Elimination (Addition)

Objectives

1. Solve a system of linear equations by elimination (addition).
2. Identify an inconsistent system of linear equations.
3. Express the solution of a dependent system of linear equations as a general ordered pair.

Vocabulary

elimination

Add the left sides and the right sides of the equations in each system.

1. $\begin{cases} 2x + 3y = 4 \\ 3x - 3y = 6 \end{cases}$

2. $\begin{cases} 4x - 2y = 1 \\ -4x + 3y = 5 \end{cases}$

3. $\begin{cases} 6x - 5y = 23 \\ -4x + 5y = 10 \end{cases}$

4. $\begin{cases} -5x + 6y = 18 \\ 5x + 12y = 10 \end{cases}$

We now consider a second algebraic method for solving systems of equations that will provide exact solutions. It is called the **elimination** method because we will select a variable to eliminate to simplify the solving process. It is sometimes referred to as the addition method because the equations will eventually be added together.

1   Solve a system of linear equations by elimination (addition).

In the Getting Ready exercises, each is a system of two linear equations in two variables. When we added the two equations together, the result was one equation with a single variable. The other variable was "eliminated" because the coefficients on one of the variables were opposites (or negatives) of one another.

To solve the system

$$\begin{cases} x + y = 8 \\ x - y = -2 \end{cases}$$

by the elimination (addition) method, we first note that the coefficients of y are 1 and -1, which are negatives (opposites). We then add the left and right sides of the equations to eliminate the variable y.

$x + y = 8$ Equal quantities, $x - y$ and -2, are added to both sides of the
$x - y = -2$ equation $x + y = 8$.

Now, column by column, we add like terms.

Combine like terms.

$x + y = 8$
$\underline{x - y = -2}$
$2x \qquad = 6$ ← Write each result here.

We can then solve the resulting equation for x.

$2x = 6$

$x = 3$ Divide both sides by 2.

To find the value of y, we substitute 3 for x in either equation of the original system and solve it for y.

$x + y = 8$ The first equation of the system

$3 + y = 8$ Substitute.

$y = 5$ Subtract 3 from both sides of the equation.

We check the solution by verifying that the ordered pair $(3, 5)$ satisfies each equation of the original system.

To solve a system of equations in x and y by the elimination (addition) method, we follow these steps.

THE ELIMINATION (ADDITION) METHOD	1. If necessary, write both equations in general form: $Ax + By = C$.

1. If necessary, write both equations in general form: $Ax + By = C$.

2. If necessary, multiply one or both of the equations by nonzero quantities to make the coefficients of x (or the coefficients of y) opposites.

3. Add the equations to eliminate the term involving x (or y).

4. Solve the equation resulting from Step 3.

5. Find the value of the other variable by substituting the solution found in Step 4 into one of the original equations.

6. Check the solution in both equations of the original system.

EXAMPLE 1   Solve the system by elimination: $\begin{cases} 3y = 14 + x \\ x + 22 = 5y \end{cases}$

Solution   We begin by writing the equations in general form:

$$\begin{cases} -x + 3y = 14 \\ x - 5y = -22 \end{cases}$$

When these equations are added, the terms involving x are eliminated and we can solve the resulting equation for y.

$$\begin{aligned} -x + 3y &= 14 \\ \underline{x - 5y} &= \underline{-22} \\ -2y &= -8 \\ y &= 4 \qquad \text{Divide both sides by } -2. \end{aligned}$$

To find the value of x, we substitute 4 for y in either equation of the system. If we substitute 4 for y in the equation $3y = 14 + x$, we have

$$\begin{aligned} 3\mathbf{y} &= 14 + x \\ 3(\mathbf{4}) &= 14 + x \\ 12 &= 14 + x \qquad \text{Simplify.} \\ -2 &= x \qquad \text{Subtract 14 from both sides.} \end{aligned}$$

Since we performed operations on the equations, it is important to verify that $(-2, 4)$ satisfies each original equation.

 **SELF CHECK 1**   Solve the system by elimination: $\begin{cases} 3y = 7 - x \\ 2x - 3y = -22 \end{cases}$

When the coefficients of neither variable in a system are opposites, we need to find an equivalent equation that describes one of the lines. We accomplish this by multiplying both sides of one of the equations by a value that will make the coefficients of one of the variables opposites.

EXAMPLE 2   Solve the system by elimination: $\begin{cases} 3x + y = 7 \\ x + 2y = 4 \end{cases}$

Solution   If we add the equations as they are, neither variable will be eliminated. We must write the equations so that the coefficients of one of the variables are opposites. To eliminate x, we can multiply both sides of the second equation by -3 to obtain

$$\begin{cases} 3x + y = 7 \\ \mathbf{-3}(x + 2y) = \mathbf{-3}(4) \end{cases} \quad \rightarrow \quad \begin{cases} 3x + y = 7 \\ -3x - 6y = -12 \end{cases}$$

The coefficients of the terms $3x$ and $-3x$ are opposites. When the equations are added, x is eliminated.

$$
\begin{array}{rl}
3x + y = & 7 \\
\underline{-3x - 6y = -12} & \\
-5y = & -5 \quad \text{Add the equations.} \\
y = & 1 \quad\;\; \text{Divide both sides by } -5.
\end{array}
$$

To find the value of x, we substitute 1 for y in the equation $3x + y = 7$.

$$
\begin{array}{ll}
3x + \mathbf{y} = 7 & \text{This is the first equation.} \\
3x + (\mathbf{1}) = 7 & \text{Substitute.} \\
3x = 6 & \text{Subtract 1 from both sides of the equation.} \\
x = 2 & \text{Divide both sides by 3.}
\end{array}
$$

Check the solution $(2, 1)$ in the original system of equations.

 **SELF CHECK 2**   Solve the system by elimination: $\begin{cases} 3x + 4y = 25 \\ 2x + y = 10 \end{cases}$

COMMENT   In Example 2, we could have multiplied the first equation by -2 and eliminated y. The result would be the same.

In some instances, it is more efficient to multiply both equations by nonzero quantities to make the coefficients of one of the variables opposites. We find these quantities by considering the least common multiple (LCM) of the coefficients of one of the variables.

EXAMPLE 3   Solve the system by elimination: $\begin{cases} 2a - 5b = 10 \\ 3a - 2b = -7 \end{cases}$

Solution   The equations in the system must be written so that one of the variables will be eliminated when the equations are added.

To eliminate a, we can multiply the first equation by 3 and the second equation by -2 to obtain coefficients that are opposites of the LCM of 2 and 3.

$$
\begin{cases} \mathbf{3}(2a - 5b) = \mathbf{3}(10) \\ \mathbf{-2}(3a - 2b) = \mathbf{-2}(-7) \end{cases} \rightarrow \begin{cases} 6a - 15b = 30 \\ -6a + 4b = 14 \end{cases}
$$

When these equations are added, the terms $6a$ and $-6a$ are eliminated.

$$
\begin{array}{rl}
6a - 15b = 30 & \\
\underline{-6a + 4b = 14} & \\
-11b = 44 & \\
b = -4 & \text{Divide both sides by } -11.
\end{array}
$$

To find the value of a, we substitute -4 for b in the equation $2a - 5b = 10$.

COMMENT   Note that solving Example 3 by the substitution method would involve fractions. In these cases, the elimination method is usually easier.

$$
\begin{array}{ll}
2a - 5\mathbf{b} = 10 & \\
2a - 5(\mathbf{-4}) = 10 & \text{Substitute } -4 \text{ for } b. \\
2a + 20 = 10 & \text{Simplify.} \\
2a = -10 & \text{Subtract 20 from both sides.} \\
a = -5 & \text{Divide both sides by 2.}
\end{array}
$$

Check the solution $(-5, -4)$ in the original system of equations.

 **SELF CHECK 3**   Solve the system by elimination: $\begin{cases} 2a + 3b = 7 \\ 5a + 2b = 1 \end{cases}$

EXAMPLE 4   Solve the system by elimination:
$$\begin{cases} \frac{5}{6}x + \frac{2}{3}y = \frac{7}{6} \\ \frac{10}{7}x - \frac{4}{9}y = \frac{17}{21} \end{cases}$$

Solution   To clear the equations of fractions, we multiply both sides of the first equation by 6 (the LCD of 6 and 3) and both sides of the second equation by 63 (the LCD of 7, 9, and 21). This results in the system

(1) $\begin{cases} 5x + 4y = 7 \\ 90x - 28y = 51 \end{cases}$
(2)

We can solve for x by eliminating the terms involving y. To do so, we multiply Equation 1 by 7 and add the result to Equation 2.

$$\begin{array}{r} 35x + 28y = 49 \\ 90x - 28y = 51 \\ \hline 125x \qquad\;\; = 100 \end{array}$$

$$x = \frac{100}{125} \qquad \text{Divide both sides by 125.}$$

$$x = \frac{4}{5} \qquad \text{Simplify.}$$

To solve for y, we substitute $\frac{4}{5}$ for x in one of the original equations and simplify.

$$\frac{5}{6}\boldsymbol{x} + \frac{2}{3}y = \frac{7}{6} \qquad \text{This is the first equation of the system.}$$

$$\frac{5}{6}\left(\boldsymbol{\frac{4}{5}}\right) + \frac{2}{3}y = \frac{7}{6} \qquad \text{Substitute.}$$

$$\frac{2}{3} + \frac{2}{3}y = \frac{7}{6} \qquad \text{Multiply.}$$

$$\boldsymbol{6}\left(\frac{2}{3} + \frac{2}{3}y\right) = \boldsymbol{6}\left(\frac{7}{6}\right) \qquad \text{Multiply both sides of the equation by 6, the LCD, to clear fractions.}$$

$$4 + 4y = 7 \qquad \text{Use the distributive property.}$$

$$4y = 3 \qquad \text{Subtract 4 from both sides.}$$

$$y = \frac{3}{4} \qquad \text{Divide both sides by 4.}$$

Check the solution of $\left(\frac{4}{5}, \frac{3}{4}\right)$ in the original system of equations.

 SELF CHECK 4   Solve the system by elimination:
$$\begin{cases} \frac{1}{3}x + \frac{1}{6}y = 1 \\ \frac{1}{2}x - \frac{1}{4}y = 0 \end{cases}$$

---

2   **Identify an inconsistent system of linear equations.**

In the next example, we encounter the situation in which both variables are simultaneously eliminated.

EXAMPLE 5   Solve the system by elimination:
$$\begin{cases} x - \frac{2y}{3} = \frac{8}{3} \\ -\frac{3x}{2} + y = -6 \end{cases}$$

Solution   We can multiply both sides of the first equation by 3 and both sides of the second equation by 2 to clear the equations of fractions.

$$\begin{cases} 3\left(x - \dfrac{2y}{3}\right) = 3\left(\dfrac{8}{3}\right) \\ 2\left(-\dfrac{3x}{2} + y\right) = 2(-6) \end{cases} \rightarrow \begin{cases} 3x - 2y = 8 \\ -3x + 2y = -12 \end{cases}$$

We can add the resulting equations to eliminate the terms involving x.

$$\begin{array}{r} 3x - 2y = 8 \\ -3x + 2y = -12 \\ \hline 0 = -4 \end{array}$$

Here, the terms involving both x and y are eliminated, and a false result $(0 = -4)$ is obtained. This shows that the equations of the system are independent, but the system itself is inconsistent. If we graphed these two equations, they would be parallel. This system has no solution; its solution set is $\varnothing$.

 **SELF CHECK 5** Solve the system by elimination: $\begin{cases} x - \dfrac{y}{3} = \dfrac{10}{3} \\ 3x - y = \dfrac{5}{2} \end{cases}$

---

3 Express the solution of a dependent system of linear equations as a general ordered pair.

In the next example, both variables are again simultaneously eliminated but the interpretation of the result is different.

EXAMPLE 6 Solve the system by elimination: $\begin{cases} x - \dfrac{5}{2}y = \dfrac{19}{2} \\ -0.2x + 0.5y = -1.9 \end{cases}$

Solution We can multiply both sides of the first equation by 2 to clear it of fractions and both sides of the second equation by 10 to clear it of decimals.

$$\begin{cases} 2\left(x - \dfrac{5}{2}y\right) = 2\left(\dfrac{19}{2}\right) \\ 10(-0.2x + 0.5y) = 10(-1.9) \end{cases} \rightarrow \begin{cases} 2x - 5y = 19 \\ -2x + 5y = -19 \end{cases}$$

We add the resulting equations to obtain

$$\begin{array}{r} 2x - 5y = 19 \\ -2x + 5y = -19 \\ \hline 0 = 0 \end{array}$$

As in Example 5, both x and y are eliminated. However, this time a true result $(0 = 0)$ is obtained. This indicates that the equations are dependent and the system has infinitely many solutions. If we graphed these two equations, they would be the same line. Any ordered pair that satisfies one equation satisfies the other. To find a general solution, we can solve the equation $-2x + 5y = -19$ for y,

$$-2x + 5y = -19$$

$$5y = 2x - 19 \qquad \text{Add } 2x \text{ to both sides of the equation.}$$

$$y = \dfrac{2x - 19}{5} \qquad \text{Divide both sides by 5.}$$

$$y = \dfrac{2}{5}x - \dfrac{19}{5} \qquad \text{Write in slope-intercept form.}$$

A general solution is given by an ordered pair of the form $\left(x, \frac{2}{5}x - \frac{19}{5}\right)$. Generate several ordered pairs by substituting values for x in the general ordered pair. Verify that these pairs check in both of the original equations.

 SELF CHECK 6 Solve the system by elimination: $\begin{cases} \dfrac{3x + y}{6} = \dfrac{1}{3} \\ -0.3x - 0.1y = -0.2 \end{cases}$

 **SELF CHECK ANSWERS**

1. $(-5, 4)$ **2.** $(3, 4)$ **3.** $(-1, 3)$ **4.** $\left(\frac{3}{2}, 3\right)$ **5.** $\varnothing$ **6.** $(x, -3x + 2)$

NOW TRY THIS

Solve each system by elimination.

1. $\begin{cases} x + y = 8 \\ 0.70x + 0.30y = 3.04 \end{cases}$

2. $\begin{cases} 5x - 4y = 16 \\ 3y + 2x = -12 \end{cases}$

7.3   Exercises

WARM-UPS *Identify the variable whose coefficients are opposites. Do not solve.*

1. $\begin{cases} x + y = 1 \\ x - y = 1 \end{cases}$ **2.** $\begin{cases} 2x + y = 4 \\ x - y = 2 \end{cases}$

3. $\begin{cases} 2x + 3y = 12 \\ -2x + 6y = 9 \end{cases}$ **4.** $\begin{cases} 6x - 3y = 9 \\ 4x + 3y = 12 \end{cases}$

State the value that is needed as a multiplier of the first equation to eliminate the variable x in each system. Do not solve.

5. $\begin{cases} x - 5y = 1 \\ -3x + 4y = 6 \end{cases}$ **6.** $\begin{cases} x + 3y = 4 \\ 4x - 2y = 7 \end{cases}$

7. $\begin{cases} 2x - y = 3 \\ 4x + 5y = 9 \end{cases}$ **8.** $\begin{cases} 3x - y = 1 \\ -12x + y = 7 \end{cases}$

REVIEW *Solve.*

9. $8(3x - 5) - 12 = 4(2x + 3)$

10. $5x - 13 = x - 1$

11. $x - 2 = \dfrac{x + 2}{3}$

12. $\dfrac{3}{2}(y + 4) = \dfrac{20 - y}{2}$

Solve and graph the solution.

13. $7x - 9 \le 5$ **14.** $-3x - 1 > 14$

VOCABULARY AND CONCEPTS *Fill in the blanks.*

15. The numerical _____ of $-3x$ is -3.

16. The _____ of -7 is 7.

17. $Ax + By = C$ is the _____ form of the equation of a line.

18. The process of adding the equations

$$5x - 6y = 10$$
$$-3x + 6y = 24$$

to eliminate the variable y is called the _____ method.

19. To clear the equation $\frac{2}{3}x + 4y = -\frac{4}{5}$ of fractions, we must multiply both sides by __, the LCD.

20. To solve the system

$$\begin{cases} 3x + 12y = 4 \\ 6x - 4y = 8 \end{cases}$$

we would multiply the second equation by _ and add to eliminate the y.

GUIDED PRACTICE *Use elimination to solve each system. SEE EXAMPLE 1. (OBJECTIVE 1)*

21. $\begin{cases} x + y = 5 \\ x - y = -3 \end{cases}$ 22. $\begin{cases} x - y = 1 \\ x + y = 7 \end{cases}$

23. $\begin{cases} -x + y = 4 \\ x + y = 2 \end{cases}$ 24. $\begin{cases} -x + y = -3 \\ x + 2y = 3 \end{cases}$

25. $\begin{cases} 2x + y = -1 \\ -2x + y = 3 \end{cases}$ 26. $\begin{cases} 3x + y = -6 \\ x - y = -2 \end{cases}$

27. $\begin{cases} 2x - 3y = -11 \\ 3x + 3y = 21 \end{cases}$ 28. $\begin{cases} 3x - 2y = 16 \\ -3x + 8y = -10 \end{cases}$

Use elimination to solve each system. SEE EXAMPLE 2. (OBJECTIVE 1)

29. $\begin{cases} x + 3y = 1 \\ x + y = 5 \end{cases}$ 30. $\begin{cases} x + 2y = 0 \\ x - y = -3 \end{cases}$

31. $\begin{cases} 2x + y = 4 \\ 2x + 3y = 0 \end{cases}$ 32. $\begin{cases} 2x + 5y = -13 \\ 2x - 3y = -5 \end{cases}$

33. $\begin{cases} 3x + 29 = 5y \\ 4y - 34 = -3x \end{cases}$ 34. $\begin{cases} 3x - 16 = 5y \\ 33 - 5y = 4x \end{cases}$

35. $\begin{cases} 2x + y = 10 \\ x + 2y = 10 \end{cases}$ 36. $\begin{cases} 3x - y = 9 \\ 5x + 4y = -2 \end{cases}$

Use elimination to solve each system. SEE EXAMPLE 3. (OBJECTIVE 1)

37. $\begin{cases} 3x + 2y = 0 \\ 2x - 3y = -13 \end{cases}$ 38. $\begin{cases} 3x + 4y = -17 \\ 4x - 3y = -6 \end{cases}$

39. $\begin{cases} 4x + 5y = -20 \\ 5x - 4y = -25 \end{cases}$ 40. $\begin{cases} 3x - 5y = 4 \\ 7x + 3y = 68 \end{cases}$

41. $\begin{cases} 6x = -3y \\ 5y = 2x + 12 \end{cases}$ 42. $\begin{cases} 3y = 4x \\ 5x = 4y - 2 \end{cases}$

43. $\begin{cases} 3x - 2y = -1 \\ 2x + 3y = -5 \end{cases}$ 44. $\begin{cases} 2x - 3y = -3 \\ 3x + 5y = -14 \end{cases}$

Use elimination to solve each system. SEE EXAMPLE 4. (OBJECTIVE 1)

45. $\begin{cases} \frac{3}{5}x + \frac{4}{5}y = 1 \\ -\frac{1}{4}x + \frac{3}{8}y = 1 \end{cases}$ 46. $\begin{cases} \frac{1}{2}x - \frac{1}{4}y = 1 \\ \frac{1}{3}x + y = 3 \end{cases}$

47. $\begin{cases} \frac{3}{5}x + y = 1 \\ \frac{4}{5}x - y = -1 \end{cases}$ 48. $\begin{cases} \frac{1}{2}x + \frac{4}{7}y = -1 \\ 5x - \frac{4}{5}y = -10 \end{cases}$

Use elimination to solve each system. SEE EXAMPLE 5. (OBJECTIVE 2)

49. $\begin{cases} 5x = 2(y - 3) \\ 5(x + 2) = 2y \end{cases}$ 50. $\begin{cases} 4(x + 2y) = 15 \\ x + 2y = 4 \end{cases}$

51. $\begin{cases} 4x = 3(4 - y) \\ 3y = 4(2 - x) \end{cases}$ 52. $\begin{cases} 2x - 7y = 10 \\ 2x = 7y - 10 \end{cases}$

Use elimination to solve each system. SEE EXAMPLE 6. (OBJECTIVE 3)

53. $\begin{cases} 10x = 2y + 12 \\ 6 = 5x - y \end{cases}$ 54. $\begin{cases} x - y = 9 \\ \frac{1}{3}x = \frac{1}{3}y + 3 \end{cases}$

55. $\begin{cases} 3(x - 2) = 4y \\ 2(2y + 3) = 3x \end{cases}$ 56. $\begin{cases} 3(x - 2y) = 12 \\ x = 2(y + 2) \end{cases}$

ADDITIONAL PRACTICE *Solve each system using any method.*

57. $\begin{cases} 2x + y = -2 \\ -2x - 3y = -6 \end{cases}$ 58. $\begin{cases} 3x + 4y = 8 \\ 5x - 4y = 24 \end{cases}$

59. $\begin{cases} 3x + y = 5 \\ 3x - 4y = 10 \end{cases}$ 60. $\begin{cases} -x - 5y = 4 \\ 2x - 5y = -8 \end{cases}$

61. $\begin{cases} 2x + 3y = 2 \\ 4x - 9y = -1 \end{cases}$ 62. $\begin{cases} 4x + 5y = 2 \\ 16x - 15y = 1 \end{cases}$

63. $\begin{cases} 4x - 2y = 9 \\ 3x - 8 = 2y \end{cases}$ 64. $\begin{cases} 2(2x + 3y) = 5 \\ 8x = 3(1 + 3y) \end{cases}$

65. $\begin{cases} 5x - 2y = 12 \\ y = \frac{5}{2}x - 6 \end{cases}$ 66. $\begin{cases} 3x + 4y = 10 \\ y = -\frac{3}{4}x - \frac{5}{2} \end{cases}$

67. $\begin{cases} 5(x - 1) = 8 - 3(y + 2) \\ 4(x + 2) - 7 = 3(2 - y) \end{cases}$

68. $\begin{cases} 4(x + 1) = 17 - 3(y - 1) \\ 2(x + 2) + 3(y - 1) = 9 \end{cases}$

69. $\begin{cases} -2(x + 1) = 3(y - 2) \\ 3(y + 2) = 6 - 2(x - 2) \end{cases}$

70. $\begin{cases} 3(x + 3) + 2(y - 4) = 5 \\ 3(x - 1) = -2(y + 2) \end{cases}$

71. $\begin{cases} \frac{1}{2}x - \frac{1}{3}y = -2 \\ x + 2y = 4 \end{cases}$ 72. $\begin{cases} \frac{1}{4}x + \frac{1}{3}y = -\frac{1}{12} \\ \frac{1}{5}x - \frac{1}{2}y = \frac{7}{10} \end{cases}$

73. $\begin{cases} \frac{x - 3}{2} + \frac{y + 5}{3} = \frac{11}{6} \\ \frac{x + 3}{3} - \frac{5}{12} = \frac{y + 3}{4} \end{cases}$ 74. $\begin{cases} \frac{x + 2}{3} = \frac{3 - y}{2} \\ \frac{x + 3}{2} = \frac{2 - y}{3} \end{cases}$

APPLICATIONS

75. **Boating**   Use the information in the table to find x and y.

	Rate	·	Time	=	Distance (mi)
Downstream	$x + y$		2		10
Upstream	$x - y$		5		5

76. Flying Use the information in the table to find x and y.

	Rate	·	Time	=	Distance (mi)
Downwind	$x + y$		3		1,800
Upwind	$x - y$		5		2,400

WRITING ABOUT MATH

77. Why is it usually best to write the equations of a system in general form before using the elimination method to solve it?

78. How would you decide whether to use substitution or elimination to solve a system of equations?

SOMETHING TO THINK ABOUT

79. If possible, find a solution to the system

$$\begin{cases} x + y = 5 \\ x - y = -3 \\ 2x - y = -2 \end{cases}$$

80. If possible, find a solution to the system

$$\begin{cases} x + y = 5 \\ x - y = -3 \\ x - 2y = 0 \end{cases}$$

Section 7.4

Solving Applications of Systems of Linear Equations

Objective

1 Solve an application using a system of linear equations.

Getting Ready

Let x and y represent two numbers. Use an algebraic expression to denote each phrase.

1. The sum of x and y

2. The difference when y is subtracted from x

3. The product of x and y

4. The quotient x divided by y

5. Give the formula for the area of a rectangle.

6. Give the formula for the perimeter of a rectangle.

We have previously set up equations involving one variable to solve applications. In this section, we consider ways to solve word problems by using equations in two variables.

1 ## Solve an application using a system of linear equations.

The following steps are helpful when solving applications involving two unknown quantities.

PROBLEM SOLVING

1. Read the problem and *analyze* the facts. Identify the unknowns by asking yourself "What am I asked to find?"

   a. Select different variables to represent two unknown quantities.

   b. Write a sentence to define each variable.

2. Form two equations involving each of the two variables. This will create a system of two equations in two variables. (This may require reading the problem several times to understand the given facts. What information is given? Is there a formula that applies to this situation? Will a sketch, chart, or diagram help you visualize the facts of the problem?)

3. Solve the system using the most convenient method: graphing, substitution, or elimination.

4. State the conclusion as a sentence.

5. Check the solution in the words of the problem.

EXAMPLE 1   **FARMING**   A farmer raises wheat and soybeans on 215 acres. If he wants to plant 31 more acres in wheat than in soybeans, how many acres of each should he plant?

Analyze the problem   The farmer plants two fields, one in wheat and one in soybeans. We are asked to find how many acres of each he should plant. So, we let w represent the number of acres of wheat and s represent the number of acres of soybeans.

Form two equations   We know that the *number of acres* of wheat planted plus the *number of acres* of soybeans planted will equal a total of 215 *acres*. So we can form the equation

The number of acres planted in wheat	plus	the number of acres planted in soybeans	equals	the total number of acres.
w	$+$	s	$=$	215

Since the farmer wants to plant 31 more acres in wheat than in soybeans, we can form the equation

The number of acres planted in wheat	minus	the number of acres planted in soybeans	equals	31 acres.
w	$-$	s	$=$	31

Solve the system   We can now solve the system

$$(1) \quad \begin{cases} w + s = 215 \\ w - s = 31 \end{cases}$$
$$(2)$$

by the elimination method.

$$\begin{aligned} w + s &= 215 \\ \underline{w - s = 31} \\ 2w &= 246 \\ w &= 123 \quad \text{\color{red}Divide both sides by 2.} \end{aligned}$$

To find s, we substitute 123 for w in Equation 1.

$$w + s = 215$$
$$123 + s = 215 \quad \text{\color{red}Substitute 123 for } w.$$
$$s = 92 \quad \text{\color{red}Subtract 123 from both sides.}$$

State the conclusion   The farmer should plant 123 acres of wheat and 92 acres of soybeans.

Check the result   The total acreage planted is $123 + 92$, or 215 acres. The area planted in wheat is 31 acres greater than that planted in soybeans, because $123 - 92 = 31$. The answers check.

 SELF CHECK 1   A farmer raises wheat and soybeans on 163 acres. If he wants to plant 29 more acres in wheat than in soybeans, how many acres of each should he plant?

EXAMPLE 2   **LAWN CARE**   An installer of underground irrigation systems wants to cut a 20-foot length of plastic tubing into two pieces. The longer piece is to be 2 feet longer than twice the shorter piece. Find the length of each piece.

Analyze the problem   Refer to Figure 7-8, which shows the pipe. We need to find the length of each pipe, so we let s represent the *length* of the shorter piece in feet and l represent the *length* of the longer piece in feet.

Figure 7-8

Form two equations   Since the length of the plastic tube is 20 feet, we can form the equation

The length of the shorter piece	plus	the length of the longer piece	equals	the total length of the pipe.
s	$+$	l	$=$	20

Since the longer piece is 2 feet longer than twice the shorter piece, we can form the equation

The length of the longer piece	equals	2	times	the length of the shorter piece	plus	2 feet.
l	$=$	2	$\cdot$	s	$+$	2

Solve the system   We can use the substitution method to solve the system

$$\begin{cases} s + l = 20 \\ l = 2s + 2 \end{cases}$$

$s + (2s + 2) = 20$ Substitute $2s + 2$ for l in the first equation.

$\qquad 3s + 2 = 20$ Remove parentheses and combine like terms.

$\qquad\quad 3s = 18$ Subtract 2 from both sides.

$\qquad\quad s = 6$ Divide both sides by 3.

State the conclusion   The shorter piece should be 6 feet long. To find the length of the longer piece, we substitute 6 for s in the first equation and solve for l.

$s + l = 20$

$6 + l = 20$ Substitute.

$\quad l = 14$ Subtract 6 from both sides.

The longer piece should be 14 feet long.

Check the result   The sum of 6 and 14 is 20 and 14 is 2 more than twice 6. The answers check.

SELF CHECK 2   If the installer has a 24-foot length of plastic tubing, find the length of each piece if the longer piece is 3 feet longer than twice the shorter piece.

EXAMPLE 3   **GARDENING**   Tom has 150 feet of fencing to enclose a rectangular garden. If the length is to be 5 feet less than 3 times the width, find the area of the garden.

Analyze the problem   To find the area of a rectangle, we need to know its length and width, so we can let l represent the length of the garden in feet and w represent the width in feet. See Figure 7-9.

Figure 7-9

Form two equations   Since the perimeter of the rectangle is 150 feet, and this is two lengths plus two widths, we can form the equation:

2	times	the length of the garden	plus	2	times	the width of the garden	equals	the perimeter of the garden.
2	·	l	+	2	·	w	=	150

Since the length is 5 feet less than 3 times the width, we can form the equation

The length of the garden	equals	3	times	the width of the garden	minus	5 feet.
l	=	3	·	w	−	5

Solve the system   Because one variable is already isolated, we can use the substitution method to solve this system.

$$\begin{cases} 2l + 2w = 150 \\ l = 3w - 5 \end{cases}$$

$2(3w - 5) + 2w = 150$	Substitute $3w - 5$ for l in the first equation.
$6w - 10 + 2w = 150$	Use the distributive property.
$8w - 10 = 150$	Combine like terms.
$8w = 160$	Add 10 to both sides.
$w = 20$	Divide both sides by 8.

The width of the garden is 20 feet. To find the length, we substitute 20 for w in the second equation and simplify.

$$l = 3w - 5$$
$$= 3(20) - 5 \qquad \text{Substitute 20 for } w.$$
$$= 60 - 5$$
$$= 55$$

The length of the garden is 55 feet.

Although we have found the length and the width of the garden, we are asked to find the area. Since the dimensions of the rectangle are 55 feet by 20 feet, and the area of a rectangle is given by the formula

$$A = \textbf{\textit{l}} \cdot \textbf{\textit{w}} \qquad \text{Area = length times width}$$

we have

$$A = \textbf{55} \cdot \textbf{20}$$
$$= 1{,}100$$

State the conclusion   The garden covers an area of 1,100 square feet.

Check the result   Because the dimensions of the garden are 55 feet by 20 feet, the perimeter is

$$P = 2\textbf{\textit{l}} + 2\textbf{\textit{w}}$$
$$= 2(\textbf{55}) + 2(\textbf{20}) \qquad \text{Substitute for } l \text{ and } w.$$
$$= 110 + 40$$
$$= 150$$

It is also true that 55 feet is 5 feet less than 3 times 20 feet. The answers check.

 SELF CHECK 3   Tom has 160 feet of fencing to enclose a rectangular garden. If the length is to be 5 feet more than twice the width, find the area of the garden.

Everyday connections
Paralympic Medals

The International Olympic Committee sets the specifications for all medals awarded in the Olympics and Paralympics. They must be at least 60 millimeters in diameter and at least 3 millimeters thick. Gold medals must be 92.5% pure silver and plated with at least 6 grams of gold. The gold medal for the 2010 Vancouver Winter Paralympics was 100 millimeters in diameter, 6 millimeters thick, and because each medal was unique, each weighed between 500 and 576 grams.

Photo by Adrian Pang, available under a Creative Commons Attribution license.

When the medal was awarded in early 2010, 6 grams of gold and 481 grams of silver were worth $478.75. In late 2011, when the price of silver had doubled and gold increased by 150%, the same medal was worth $850.40.

1. What was the price of one gram of gold and one gram of silver in 2010?

2. What was the price of one gram of gold and one gram of silver in 2011?

Source: http://www.olympic.org/Documents/
Reference_documents_Factsheets/
Winter_Games_Medals_FACTSHEET_EN.pdf

EXAMPLE 4   **INVESTING**   Terri and Juan earned $650 in interest from a one-year investment of $15,000. If Terri invested some of the money at 4% annual interest and Juan invested the rest at 5% annual interest, how much did each invest?

Analyze the problem   We are asked to find how much money Terri and Juan invested. We can let x represent the amount invested by Terri and y represent the amount of money invested by Juan. We are told that Terri invested an unknown part of the $15,000 at 4% interest and Juan invested the rest at 5% interest. Together, these investments earned $650 in interest.

Form two equations   Because the total investment is $15,000, we have

The amount invested by Terri	plus	the amount invested by Juan	equals	the total amount invested.
x	$+$	y	$=$	15,000

Since the interest on x dollars invested at 4% is $0.04x$, the interest on y dollars invested at 5% is $0.05y$, and the combined income is $650, we have

The interest on the 4% investment	plus	the interest on the 5% investment	equals	the total interest earned.
$0.04x$	$+$	$0.05y$	$=$	650

Thus, we have the system

(1)
(2)
$$\begin{cases} x + y = 15{,}000 \\ 0.04x + 0.05y = 650 \end{cases}$$

Solve the system   To solve the system, we use the elimination method.

$$
\begin{array}{rl}
-4x - 4y = -60{,}000 & \text{Multiply both sides of Equation 1 by } -4. \\
\underline{4x + 5y = 65{,}000} & \text{Multiply both sides of Equation 2 by 100 to clear the decimals.} \\
y = 5{,}000 & \text{Add the equations together.}
\end{array}
$$

To find the value of x, we substitute 5,000 for y in Equation 1 and simplify.

$$
\begin{array}{rl}
x + y = 15{,}000 & \\
x + \mathbf{5{,}000} = 15{,}000 & \text{Substitute 5,000 for } y. \\
x = 10{,}000 & \text{Subtract 5,000 from both sides.}
\end{array}
$$

State the conclusion   Terri invested $10,000, and Juan invested $5,000.

Check the result

$$\$10{,}000 + \$5{,}000 = \$15{,}000 \qquad \text{The two investments total \$15,000.}$$
$$0.04(\$10{,}000) = \$400 \qquad \text{Terri earned \$400.}$$
$$0.05(\$5{,}000) = \$250 \qquad \text{Juan earned \$250.}$$

The combined interest is $400 + $250 = $650. The answers check.

SELF CHECK 4   Terri and Juan earned $800 from a one-year investment of $20,000. If Terri invested some of the money at 3% and Juan invested the rest at 5%, how much did each invest? How much interest did each earn?

All of the previous examples could be solved using one equation in one variable as you did in Chapter 2. In Examples 5–7, using two variables to create a system of linear equations will be more efficient.

EXAMPLE 5   **BOATING**   A boat traveled 30 kilometers downstream in 3 hours and made the return trip in 5 hours. Find the speed of the boat in still water.

Analyze the problem   We are asked to find the speed of the boat, so we let s represent the speed of the boat in km/hr in still water. Recall from earlier problems that when traveling upstream or downstream, the current affects that speed. Therefore, we let c represent the speed of the current, in km/hr.

Form two equations   Traveling downstream, the rate of the boat will be the speed of the boat in still water, s, in km/hr plus the speed of the current, c, in km/hr. Thus, the rate of the boat going downstream is $(s + c)$.

Traveling upstream, the rate of the boat will be the speed of the boat in still water, s, minus the speed of the current, c. Thus, the rate of the boat going upstream is $(s - c)$. We can organize the information of the problem as in Table 7-3.

TABLE 7-3

	Distance	=	Rate	·	Time
Downstream	30		$s + c$		3
Upstream	30		$s - c$		5

Because $d = r \cdot t$, the information in the table produces two equations in two variables.

$$\begin{cases} 30 = 3(s + c) \\ 30 = 5(s - c) \end{cases}$$

After distributing and writing the equations in $ax + by = c$ form, we have

$$\begin{cases} 3s + 3c = 30 \\ 5s - 5c = 30 \end{cases}$$

Solve the system   To solve this system by elimination, we multiply the first equation by 5, the second equation by 3, add the equations, and solve for s.

$$\begin{aligned} 15s + 15c &= 150 \\ \underline{15s - 15c = 90} \\ 30s &= 240 \\ s &= 8 \qquad \text{Divide both sides by 30.} \end{aligned}$$

State the conclusion   The speed of the boat in still water is 8 kilometers per hour.

Check the result   If $s = 8$, $3(8) + 3c = 30$

$$24 + 3c = 30$$
$$3c = 6$$
$$c = 2$$

The speed of the current is 2 kilometers per hour. In 3 hours, the boat will travel downstream $3(8 + 2) = 30$ kilometers. In 5 hours, the boat will return upstream $5(8 - 2) = 30$ kilometers. The answers check.

 SELF CHECK 5   A boat traveled 36 kilometers downstream in 3 hours and made the return trip in 4 hours. Find the speed of the boat in still water.

EXAMPLE 6   **MEDICAL TECHNOLOGY**   A laboratory technician has one batch of antiseptic that is 40% alcohol and a second batch that is 60% alcohol. She needs to make 8 liters of solution that is 55% alcohol. How many liters of each batch should she use?

Analyze the problem   We need to know how many liters of each type of alcohol she should use, so we can let x represent the number of liters to be used from batch 1 and let y represent the number of liters to be used from batch 2.

Form two equations   Some 60% alcohol solution must be added to some 40% alcohol solution to make a 55% alcohol solution. We can organize the information of the problem as in Table 7-4 on page 472.

TABLE 7-4					
	Fractional part that is alcohol	·	Number of liters of solution	=	Number of liters of alcohol
Batch 1	0.40		x		0.40x
Batch 2	0.60		y		0.60y
Mixture	0.55		8		0.55(8)

The information in Table 7-4 provides the system.

$$\begin{cases} x + y = 8 \\ 0.40x + 0.60y = 0.55(8) \end{cases}$$

The *number of liters* of batch 1 plus the *number of liters* of batch 2 equals the *total number of liters* in the mixture.

The *amount of alcohol* in batch 1 plus the *amount of alcohol* in batch 2 equals the total *amount of alcohol* in the mixture.

Solve the system   We can use substitution to solve this system. Solve the first equation for x.

$$x = -y + 8$$

$0.40(-y + 8) + 0.60y = 4.4$ Substitute the expression for x in the second equation and $0.55(8) = 4.4$.

$-0.40y + 3.2 + 0.60y = 4.4$ Use the distributive property.

$0.20y + 3.2 = 4.4$ Combine like terms.

$0.20y = 1.2$ Subtract 3.2 from both sides.

$y = 6$ Divide both sides by 0.20.

To find the value of x, we substitute 6 for y in the first equation and simplify:

$x + y = 8$

$x + 6 = 8$ Substitute 6 for y.

$x = 2$ Subtract 6 from both sides.

State the conclusion   The technician should use 2 liters of the 40% alcohol solution and 6 liters of the 60% alcohol solution.

Check the result   Mixing 2 liters of batch 1 with 6 liters of batch 2 produces 8 liters.

The amount of alcohol in each is $0.40(2) + 0.60(6) = 0.55(8)$.

$$0.8 + 3.6 = 4.40$$
$$4.4 = 4.4$$

The answers check.

SELF CHECK 6   What if the technician wanted 10 liters of solution that is 46% alcohol. How many liters of each batch should she use?

EXAMPLE 7   **MANUFACTURING**   The set-up cost of a machine that mills brass plates is $750. After set-up, it costs $0.25 to mill each plate. Management is considering the use of a larger machine that can produce the same plates at a cost of $0.20 per plate. If the set-up cost of the larger machine is $1,200, how many plates would the company have to produce to make the switch worthwhile?

Analyze the problem   We can let p represent the number of brass plates produced. Then, we will let c represent the total cost of milling p plates (set-up cost plus cost per plate).

Form two equations   To determine whether the switch is worthwhile, we need to know if the larger machine can produce the plates cheaper than the old machine and if so, when that occurs. We begin by finding the number of plates that will cost the same to produce on either machine (called the *break-even point*).

If we call the machine currently being used machine 1, and the larger machine 2, we can form the two equations

The cost of making p plates on machine 1	equals	the set-up cost of machine 1	plus	the cost per plate on machine 1	times	the number of plates p to be made.
c_1	$=$	750	$+$	0.25	$\cdot$	p

The cost of making p plates on machine 2	equals	the set-up cost of machine 2	plus	the cost per plate on machine 2	times	the number of plates p to be made.
c_2	$=$	1,200	$+$	0.20	$\cdot$	p

Solve the system   Since the costs at the break-even point are equal $(c_1 = c_2)$, we can use the substitution method to solve the system

$$\begin{cases} c_1 = \mathbf{750 + 0.25p} \\ c_2 = 1{,}200 + 0.20p \end{cases}$$

$750 + 0.25p = 1{,}200 + 0.20p$	Substitute $750 + 0.25p$ for c_2 in the second equation.
$0.25p = 450 + 0.20p$	Subtract 750 from both sides.
$0.05p = 450$	Subtract $0.20p$ from both sides.
$p = 9{,}000$	Divide both sides by 0.05.

State the conclusion   If 9,000 plates are milled, the cost will be the same on either machine. If more than 9,000 plates are milled, the cost will be cheaper on the larger machine, because it mills the plates less expensively than the smaller machine.

Check the solution   Figure 7-10 verifies that the break-even point occurs when 9,000 plates are produced. It also interprets the solution graphically.

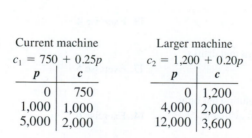

Current machine
$c_1 = 750 + 0.25p$

p	c
0	750
1,000	1,000
5,000	2,000

Larger machine
$c_2 = 1{,}200 + 0.20p$

p	c
0	1,200
4,000	2,000
12,000	3,600

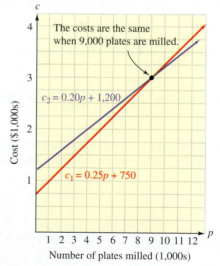

Figure 7-10

SELF CHECK 7   If the set-up cost of the larger machine is $1,500, how many plates would the company have to produce to make the switch worthwhile?

SELF CHECK ANSWERS

1. The farmer should plant 96 acres of wheat and 67 acres of soybeans. **2.** The shorter piece should be 7 feet long. The longer piece should be 17 feet long. **3.** The garden covers an area of 1,375 square feet. **4.** Terri and Juan each invested $10,000. Terri earned $300 in interest while Juan earned $500. **5.** The speed of the boat in still water is 10.5 km/hr. **6.** The technician should use 7 liters of the 40% alcohol solution and 3 liters of the 60% alcohol solution. **7.** If 15,000 plates are milled, the cost will be the same. If more than 15,000 plates are milled, the cost will be cheaper on the larger machine.

NOW TRY THIS

A chemist has 20 ml of a 30% alcohol solution. How much pure alcohol must she add so that the resulting solution contains 50% alcohol?

7.4 Exercises

WARM-UPS *For each of the following, identify the two variables needed to set up a system of equations and write a description of what each variable represents.*

1. Prince's Pizza sold a total of 52 pizzas and calzones.

2. The length of a garden is 8 feet longer than twice its width.

3. Danielle and Kinley invested a total of $15,000.

4. Tony spent $24 on a total purchase of 6 hot dogs and drinks.

5. A nurse needs 25 milliliters of a 15% saline solution but only has 9% saline solution and 20% saline solution available.

6. The cost of 6 shirts and 3 pairs of jeans is $123.

7. The theater department sold a total of 273 adult tickets and student tickets.

8. A candy store owner wants to mix a total of 28 pounds of cashews and Brazil nuts.

Set up an equation for each of the statements in Exercises 1–8 using the variables you identified. State, in words, what the equation means.

9. Exercise 1

10. Exercise 2

11. Exercise 3

12. Exercise 4

13. Exercise 5

14. Exercise 6

15. Exercise 7

16. Exercise 8

REVIEW *Graph.*

17. $x < 4$

18. $x \geq -3$

19. $-1 < x \le 2$ **20.** $-2 \le x \le 0$

Write each product using exponents.

21. $9 \cdot 9 \cdot 9 \cdot 9 \cdot a$ **22.** $5(\pi)(r)(r)$
23. $x \cdot x \cdot x \cdot y \cdot y \cdot y \cdot y$ **24.** $(-5)(-5)$

VOCABULARY AND CONCEPTS *Fill in the blanks.*

25. A _____ is a letter that represents a number.
26. An _____ is a statement indicating that two quantities are equal.
27. $\begin{cases} a + 2b = 9 \\ a = 4b + 3 \end{cases}$ is a _____ of linear equations.
28. A _____ of a system of two linear equations satisfies both equations simultaneously.

GUIDED PRACTICE *Use two equations in two variables to solve each application.* **SEE EXAMPLE 1. (OBJECTIVE 1)**

29. Government   The salaries of the President and Vice President of the United States total $592,600 a year. If the President makes $207,400 more than the Vice President, find each of their salaries.

30. Splitting the lottery   Chayla and Lena pool their resources to buy several lottery tickets. They win $250,000! They agree that Lena should get $50,000 more than Chayla, because she gave most of the money. How much will Chayla get?

31. Figuring inheritances   In his will, a man left his older son $5,000 more than twice as much as he left his younger son. If the estate is worth $742,250, how much did the younger son get?

32. Causes of death   The number of American women that died from cancer is approximately seven times the number that died from diabetes. If the number of deaths from these two causes was 308,000, how many American women died from each cause?

Use two equations in two variables to solve each application. **SEE EXAMPLE 2. (OBJECTIVE 1)**

33. Cutting pipe   A plumber wants to cut the pipe shown in the illustration into two pieces so that one piece is 5 feet longer than the other. How long should each piece be?

25 ft

34. Cutting lumber   A carpenter wants to cut a 20-foot board into two pieces so that one piece is 4 times as long as the other. How long should each piece be?

35. Geometry   The perimeter of the rectangle shown in the illustration is 110 feet. Find its dimensions.

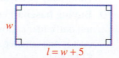
w
$l = w + 5$

36. Geometry   A rectangle is 3 times as long as it is wide, and its perimeter is 80 centimeters. Find its dimensions.

Use two equations in two variables to solve each application. **SEE EXAMPLE 3. (OBJECTIVE 1)**

37. Geometry   The length of a rectangle is 3 feet less than twice its width. If its perimeter is 48 feet, find its area.
38. Geometry   A 50-meter path surrounds a rectangular garden. The width of the garden is two-thirds its length. Find its area.

39. At the movies   At an IMAX theater, the giant rectangular movie screen has a width 26 feet less than its length. If its perimeter is 332 feet, find the area of the screen.
40. Geometry   The length of a residential lot is 20 feet longer than twice the width. If its perimeter is 340 feet, find its area.

Use two equations in two variables to solve each application. **SEE EXAMPLE 4. (OBJECTIVE 1)**

41. Investing money   Bill invested some money at 5% annual interest, and Janette invested some at 7%. If their combined interest was $310 on a total investment of $5,000, how much did Bill invest?
42. Investing money   Peter invested some money at 6% annual interest, and Martha invested some at 12%. If their combined investment was $6,000 and their combined interest was $540, how much money did Martha invest?
43. Buying tickets   Students can buy tickets to a basketball game for $1. The admission for nonstudents is $2. If 350 tickets are sold and the total receipts are $450, how many student tickets are sold?
44. Buying tickets   If receipts for the movie advertised in the illustration were $720 for an audience of 190 people, how many senior citizens attended?

DOLLAR MOVIE THEATER TICKETS
Admissions: $4
Seniors: $3
Showtimes: 7, 9, 11

Use two equations in two variables to solve each application.
SEE EXAMPLE 5. (OBJECTIVE 1)

45. Boating   A boat can travel 24 miles downstream in 2 hours and can make the return trip in 3 hours. Find the speed of the boat in still water.

46. Aviation   With the wind, a plane can fly 3,000 miles in 5 hours. Against the same wind, the trip takes 6 hours. Find the airspeed of the plane (the speed in still air).

47. Aviation   An airplane can fly downwind a distance of 600 miles in 2 hours. However, the return trip against the same wind takes 3 hours. Find the speed of the wind.

48. Finding the speed of a current   It takes a motorboat 4 hours to travel 56 miles down a river, and it takes 3 hours longer to make the return trip. Find the speed of the current.

Use two equations in two variables to solve each application.
SEE EXAMPLE 6. (OBJECTIVE 1)

49. Mixing chemicals   A chemist has one solution that is 40% alcohol and another that is 55% alcohol. How much of each must she use to make 15 liters of a solution that is 50% alcohol?

50. Mixing pharmaceuticals   A nurse has a solution that is 25% alcohol and another that is 50% alcohol. How much of each must he use to make 20 liters of a solution that is 40% alcohol?

51. Mixing nuts   A merchant wants to mix the peanuts with the cashews shown in the illustration to get 48 pounds of mixed nuts to sell at $4 per pound. How many pounds of each should the merchant use?

52. Mixing peanuts and candy   A merchant wants to mix peanuts worth $3 per pound with jelly beans worth $1.50 per pound to make 30 pounds of a mixture worth $2.10 per pound. How many pounds of each should he use?

Use two equations in two variables to solve each application.
SEE EXAMPLE 7. (OBJECTIVE 1)

53. Choosing a furnace   A high-efficiency 90+ furnace costs $2,250 and costs an average of $412 per year to operate in Rockford, IL. An 80+ furnace costs only $1,715 but costs $466 per year to operate. Find the break-even point.

54. Choosing a furnace   See Exercise 53. If you intended to live in a house for seven years, which furnace would you choose?

55. Making tires   A company has two molds to form tires. One mold has a set-up cost of $600 and the other a set-up cost of

$1,100. The cost to make each tire on the first machine is $15, and the cost per tire on the second machine is $13. Find the break-even point.

56. Making tires   See Exercise 55. If you planned a production run of 500 tires, which mold would you use?

ADDITIONAL PRACTICE   *Use two equations in two variables to solve each application.*

57. Buying contact lens cleaner   Two bottles of contact lens cleaner and three bottles of soaking solution cost $29.40, and three bottles of cleaner and two bottles of soaking solution cost $28.60. Find the cost of each.

58. Buying clothes   Two pairs of shoes and four pairs of socks cost $109, and three pairs of shoes and five pairs of socks cost $160. Find the cost of a pair of socks.

59. Raising livestock   A rancher raises five times as many cows as horses. If he has 168 animals, how many cows does he have?

60. Grass seed mixture   A landscaper used 100 pounds of grass seed containing twice as much bluegrass as rye. He added 15 more pounds of bluegrass to the mixture before seeding a lawn. How many pounds of bluegrass did he use?

61. TV programming   The producer of a 30-minute documentary about World War I divided it into two parts. Four times as much program time was devoted to the causes of the war as to the outcome. How long was each part of the documentary?

62. Selling ice cream   At a store, ice cream cones cost $1.90 and sundaes cost $2.65. One day, the receipts for a total of 148 cones and sundaes were $328.45. How many cones were sold?

63. Integers   One integer is three times another, and their sum is 112. Find the integers.

64. Integers   The sum of two integers is 38, and their difference is 12. Find the integers.

65. Integers   Three times one integer plus another integer is 29. If the first integer plus twice the second is 18, find the integers.

66. Integers   Twice one integer plus another integer is 21. If the first integer plus 3 times the second is 33, find the integers.

67. Investing money   An investment of $950 at one rate of interest and $1,200 at a higher rate together generate an annual income of $88.50. If the investment rates differ by 2%, find the lower rate.

68. Motion   A man drives for a while at 45 mph. Realizing that he is running late, he increases his speed to 60 mph and completes his 405-mile trip in 8 hours. How long does he drive at 45 mph?

69. Selling radios   An electronics store put two types of car radios on sale. One model sold for $87, and the other sold for $119. During the sale, the receipts for 25 radios sold were $2,495. How many of the less expensive radios were sold?

70. Buying baseball equipment   One catcher's mitt and ten outfielder's gloves cost $239.50. How much does each cost if one catcher's mitt and five outfielder's gloves cost $134.50?

71. Buying painting supplies Two partial receipts for paint supplies appear in the illustration. How much did each gallon of paint and each brush cost?

Colorf
Paint
Wallpa

8 latex @
gallon
3 brushes @
Total $ 135.00

Colorf
Paint a
Wallpa

6 latex @
gallon
2 brushes @
Total $ 100.00

72. Equilibrium price The number of canoes sold at a marina depends on price. As the price gets higher, fewer canoes will be sold. The equation that relates the price of a canoe to the number sold is called a **demand equation**. Suppose that the demand equation for canoes is

$$p = -\frac{1}{2}q + 1{,}300$$

where p is the price and q is the number sold at that price.

The number of canoes produced also depends on price. As the price gets higher, more canoes will be manufactured. The equation that relates the number of canoes produced to the price is called a **supply equation**. Suppose that the supply equation for canoes is

$$p = \frac{1}{3}q + \frac{1{,}400}{3}$$

where p is the price and q is the number produced at that price. The equilibrium price is the price at which supply equals demand. Find the equilibrium price.

WRITING ABOUT MATH

73. Which problem in the preceding set did you find the hardest? Why?

74. Which problem in the preceding set did you find the easiest? Why?

SOMETHING TO THINK ABOUT

75. In the illustration below, how many nails will balance one nut?

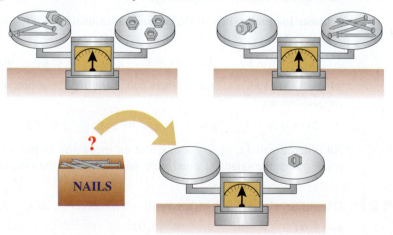

NAILS

Section 7.5

Solving Systems of Linear Inequalities

Objectives

1 Determine whether an ordered pair is a solution to a given linear inequality.

2 Graph a linear inequality in one or two variables.

3 Solve an application involving a linear inequality in two variables.

4 Graph the solution set of a system of linear inequalities in one or two variables.

5 Solve an application using a system of linear inequalities.

linear inequality in two variables	half-plane test point	doubly shaded region

Getting Ready

Graph $y = \frac{1}{3}x + 3$ and determine whether the given point lies on the line, above the line, or below the line.

1. $(0, 0)$ **2.** $(0, 4)$ **3.** $(2, 2)$ **4.** $(6, 5)$

5. $(-3, 2)$ **6.** $(6, 8)$ **7.** $(-6, 0)$ **8.** $(-9, 5)$

In this section, we will discuss how to solve linear inequalities in two variables graphically. Then we will show how to solve systems of inequalities. We conclude this section by using these skills to solve applications.

1 **Determine whether an ordered pair is a solution to a given linear inequality.**

A **linear inequality in two variables** is an inequality that can be written in one of the following forms:

$$Ax + By > C \qquad Ax + By < C \qquad Ax + By \geq C \qquad Ax + By \leq C$$

where A, B, and C are real numbers and A and B are not both 0. Some examples of linear inequalities are

$$2x - y > -3 \qquad y < 3 \qquad x + 47 \geq 6 \qquad x \leq -2$$

An ordered pair (x, y) is a solution of an inequality in two variables if a true statement results when the values of x and y are substituted into the inequality.

EXAMPLE 1   Determine whether each ordered pair is a solution of $y \geq x - 5$.

     a. $(4, 2)$ **b.** $(0, -6)$ **c.** $(5, 0)$

Solution   **a.** To determine whether $(4, 2)$ is a solution, we substitute 4 for x and 2 for y.

$$y \geq x - 5$$
$$2 \geq 4 - 5$$
$$2 \geq -1$$

Since $2 \geq -1$ is a true inequality, $(4, 2)$ is a solution.

b. To determine whether $(0, -6)$ is a solution, we substitute 0 for x and -6 for y.

$$y \geq x - 5$$
$$-6 \geq 0 - 5$$
$$-6 \geq -5$$

Since $-6 \geq -5$ is a false inequality, $(0, -6)$ is not a solution.

c. To determine whether $(5, 0)$ is a solution, we substitute 5 for x and 0 for y.

$$y \geq x - 5$$
$$0 \geq 5 - 5$$
$$0 \geq 0$$

Since $0 \geq 0$ is a true inequality, $(5, 0)$ is a solution.

SELF CHECK 1   Determine whether each ordered pair is a solution of $y < x + 4$.
a. $(2, 8)$ **b.** $(3, -4)$ **c.** $(-4, 0)$

---

2   **Graph a linear inequality in one or two variables.**

The graph of $y = x - 5$ is a line consisting of the points whose coordinates satisfy the equation. The graph of the inequality $y \geq x - 5$ is not a line but rather an area bounded by a line, called a **half-plane**. The half-plane consists of the points whose coordinates satisfy the inequality. The line serves as the *boundary* of the half-plane.

EXAMPLE 2   Graph the inequality: $y \geq x - 5$

Solution   Because equality is included in the original inequality, we begin by graphing the equation $y = x - 5$ with a *solid* line, as in Figure 7-11(a). Because the graph of $y \geq x - 5$ also indicates that y can be greater than $x - 5$, the coordinates of points other than those shown in Figure 7-11(a) satisfy the inequality. To determine which side of the line to shade, we select a **test point** and substitute its value into the inequality. A convenient test point is the origin.

$$y \geq x - 5$$
$$0 \geq 0 - 5 \quad \text{Substitute 0 for } x \text{ and 0 for } y.$$
$$0 \geq -5$$

Because $0 \geq -5$ is true, the coordinates of the origin satisfy the original inequality. In fact, the coordinates of every point on the same side of the line as the origin satisfy the inequality. The graph of $y \geq x - 5$ is the half-plane that is shaded in Figure 7-11(b).

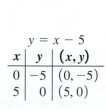

$y = x - 5$

x	y	(x, y)
0	-5	$(0, -5)$
5	0	$(5, 0)$

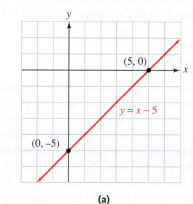

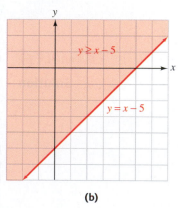

(a)   (b)

Figure 7-11

SELF CHECK 2   Graph: $y \geq -x - 2$

EXAMPLE 3   Graph: $x + 2y < 6$

Solution   We find the boundary by graphing the equation $x + 2y = 6$. Since the symbol $<$ does not include equality, the points on the graph of $x + 2y = 6$ will not be a part of the graph. To indicate this, we draw the boundary line as a *dashed* line. See Figure 7-12.

To determine which half-plane to shade, we substitute the coordinates of a test point that lies on one side of the boundary line into $x + 2y < 6$. The origin is a convenient choice.

$$\boldsymbol{x} + 2\boldsymbol{y} < 6$$
$$\boldsymbol{0} + 2(\boldsymbol{0}) < 6 \qquad \text{Substitute 0 for } x \text{ and 0 for } y.$$
$$0 < 6$$

Since $0 < 6$ is true, we shade the side of the line that includes the origin. The graph is shown in Figure 7-12.

Sophie Germain

1776–1831

Sophie Germain was 13 years old during the French Revolution. Because of dangers caused by the insurrection in Paris, she was kept indoors and spent most of her time reading about mathematics in her father's library. Since interest in mathematics was considered inappropriate for a woman at that time, much of her work was written under the pen name of M. LeBlanc.

$x + 2y = 6$		
x	y	(x, y)
0	3	$(0, 3)$
6	0	$(6, 0)$
4	1	$(4, 1)$

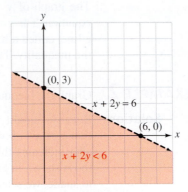

Figure 7-12

SELF CHECK 3   Graph: $-2x + y > -4$

COMMENT   The decision to use a dashed line or solid line is determined by the inequality symbol. If the symbol is $<$ or $>$, the line is dashed. If it is $\leq$ or $\geq$, the line is solid.

Alternatively, the decision to shade above or below the boundary can be determined by the direction of the inequality after it is solved for y. If $y < mx + b$, shade below the boundary and if $y > mx + b$, shade above it.

EXAMPLE 4   Graph: $2x - y < 0$

Solution   To find the boundary line, we graph the equation $2x - y = 0$. Since the symbol $<$ does not include equality, the points on the boundary are not a part of the graph of $2x - y < 0$. To show this, we draw the boundary as a dashed line. See Figure 7-13(a).

To determine which half-plane to shade, we substitute the coordinates of a test point that lies on one side of the boundary into $2x - y < 0$. Because the origin lies on the line, we must choose a different test point. Point $T(0, 2)$, for example, is below the boundary line. See Figure 7-13(a). To see if point $T(2, 0)$ satisfies $2x - y < 0$, we substitute 2 for x and 0 for y in the inequality.

$$2\boldsymbol{x} - \boldsymbol{y} < 0$$
$$2(\boldsymbol{2}) - \boldsymbol{0} < 0$$
$$4 < 0$$

Since $4 < 0$ is false, the coordinates of point T do not satisfy the inequality, and point T is not on the side of the line we want to shade. Instead, we shade the other side of the boundary line. The graph of the solution set of $2x - y < 0$ is shown in Figure 7-13(b).

$2x - y = 0$

x	y	(x, y)
0	0	$(0, 0)$
-1	-2	$(-1, -2)$
3	6	$(3, 6)$

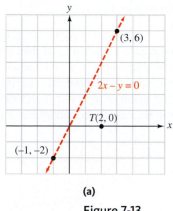

(a)

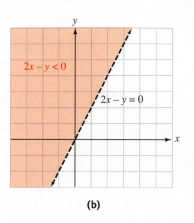

(b)

Figure 7-13

SELF CHECK 4   Graph: $3x - y > 0$

3   Solve an application involving a linear inequality in two variables.

EXAMPLE 5   **EARNING MONEY**   Chen has two part-time jobs, one paying $5 per hour and the other paying $6 per hour. He must earn at least $120 per week to pay his expenses while attending college. Write an inequality and graph it to show the various ways he can schedule his time to achieve his goal.

Solution   If we let x represent the number of hours Chen works on the first job and y the number of hours he works on the second job, we have

The hourly rate on the first job	times	the hours worked on the first job	plus	the hourly rate on the second job	times	the hours worked on the second job	is at least	$120.
$5	·	x	+	$6	·	y	≥	$120

The graph of the inequality $5x + 6y \geq 120$ is shown in Figure 7-14. Since Chen cannot work a negative number of hours, the graph in the figure has no meaning when either x or y is negative. Any point in the shaded region indicates a possible way Chen can schedule his time and earn $120 or more per week. For example, if he works 20 hours on the first job and 10 hours on the second job, he will earn

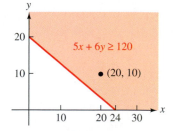

Figure 7-14

$$\$5(20) + \$6(10) = \$100 + \$60$$
$$= \$160$$

SELF CHECK 5   If Chen's expenses are $150 per week and his two jobs pay $9 and $10 per hour, write an inequality and graph it to show how he can schedule his time to achieve his goal.

4   Graph the solution set of a system of linear inequalities in one or two variables.

We have seen that the graph of a linear inequality in two variables is a half-plane. Therefore, we would expect the graph of a system of two linear inequalities to contain two half-planes. For example, to solve the system

$$\begin{cases} x + y \geq 1 \\ x - y \geq 1 \end{cases}$$

we graph each inequality and then superimpose the graphs on one set of coordinate axes.

The graph of $x + y \geq 1$ includes the graph of the equation $x + y = 1$ and all points above it. Because the boundary line is included, we draw it with a solid line. See Figure 7-15(a).

The graph of $x - y \geq 1$ includes the graph of the equation $x - y = 1$ and all points below it. Because the boundary line is included, we draw it with a solid line. See Figure 7-15(b).

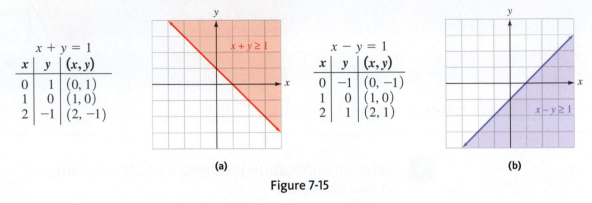

$x + y = 1$		
x	y	(x, y)
0	1	$(0, 1)$
1	0	$(1, 0)$
2	−1	$(2, -1)$

$x - y = 1$		
x	y	(x, y)
0	−1	$(0, -1)$
1	0	$(1, 0)$
2	1	$(2, 1)$

(a)   (b)

Figure 7-15

In Figure 7-16(a), we show the result when the graphs are superimposed on one coordinate system. The area that is shaded twice represents the set of solutions of the given system. Any point in the **doubly shaded region** has coordinates that satisfy both of the inequalities.

Figure 7-16(a) is our "working" graph. Just as we solved compound inequalities in one variable, we shade the various regions of the graph and then interpret the results. The solution is only that part of the graph that is doubly shaded, so we create a new graph, Figure 7-16(b), with only the solution.

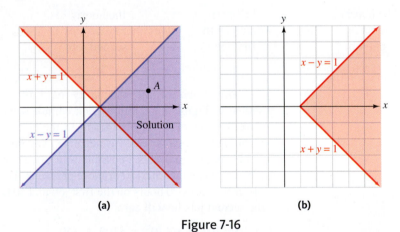

(a)   (b)

Figure 7-16

To see that this is true, we select a test point, such as point A, that lies in the doubly shaded region and show that its coordinates (4, 1) satisfy both inequalities.

$$\begin{array}{ll} \boldsymbol{x + y \geq 1} & \boldsymbol{x - y \geq 1} \\ \boldsymbol{4 + 1} \geq 1 & \boldsymbol{4 - 1} \geq 1 \\ \quad\;\; 5 \geq 1 & \quad\;\; 3 \geq 1 \end{array}$$

Since the coordinates of point A satisfy both inequalities, point A is a solution and therefore all points in the doubly shaded region are solutions.

In general, to solve systems of linear inequalities, we will take the following steps.

> **SOLVING SYSTEMS OF INEQUALITIES**
>
> 1. Graph each inequality in the system on the same coordinate axes using solid or dashed lines as appropriate.
> 2. Find the region where the graphs overlap.
> 3. Select a test point from the region to verify the solution.
> 4. Graph only the solution set, if there is one.

EXAMPLE 6 Graph the solution set: $\begin{cases} 2x + y < 4 \\ -2x + y > 2 \end{cases}$

Solution We graph each inequality on one set of coordinate axes, as in Figure 7-17(a).

- The graph of $2x + y < 4$ includes all points below the line $2x + y = 4$. Since the boundary is not included, we draw it as a dashed line.
- The graph of $-2x + y > 2$ includes all points above the line $-2x + y = 2$. Since the boundary is not included, we draw it as a dashed line.

The area that is shaded twice represents the set of solutions of the given system. We create a new graph that reflects only the solution, Figure 7-17(b).

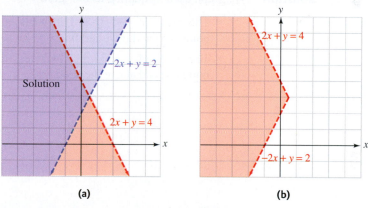

(a)   (b)

Figure 7-17

Select a test point in the doubly shaded region and show that it satisfies both inequalities.

 **SELF CHECK 6** Graph the solution set: $\begin{cases} x + 3y \le 6 \\ -x + 3y < 6 \end{cases}$

EXAMPLE 7 Graph the solution set: $\begin{cases} x \le 2 \\ y > 3 \end{cases}$

Solution We graph each inequality on one set of coordinate axes, as in Figure 7-18(a).

- The graph of $x \le 2$ includes all points on the line $x = 2$ and all points to the left of the line. Since the boundary line is included, we draw it as a solid line.
- The graph $y > 3$ includes all points above the line $y = 3$. Since the boundary is not included, we draw it as a dashed line.

The area that is shaded twice represents the set of solutions of the given system. Pick a point in the doubly shaded region and show that this is true. Graph the solution on a new set of axes as in Figure 7-18(b) on the next page.

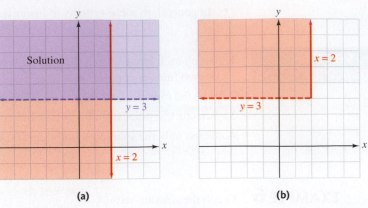

Figure 7-18

 SELF CHECK 7 Graph the solution set: $\begin{cases} y \geq 1 \\ x > 2 \end{cases}$

EXAMPLE 8 Graph the solution set: $\begin{cases} y < 3x - 1 \\ y \geq 3x + 1 \end{cases}$

Solution We graph each inequality, as in Figure 7-19.

- The graph of $y < 3x - 1$ includes all of the points below the dashed line $y = 3x - 1$.
- The graph of $y \geq 3x + 1$ includes all of the points on and above the solid line $y = 3x + 1$.

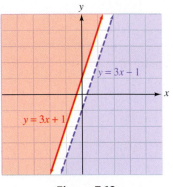

Figure 7-19

Since the graphs of these inequalities do not intersect, the solution set is $\varnothing$.

 SELF CHECK 8 Graph the solution set: $\begin{cases} y \geq -\frac{1}{2} + 1 \\ y \leq -\frac{1}{2}x - 1 \end{cases}$

---

5 Solve an application using a system of linear inequalities.

EXAMPLE 9 **LANDSCAPING** A man budgets from $300 to $600 for trees and shrubs to landscape his yard. After shopping around, he finds that good trees cost $150 and mature shrubs cost $75. What combinations of trees and shrubs can he afford to buy?

Analyze the problem The man wants to spend *at least* $300 but *not more than* $600 for trees and shrubs.

Form two inequalities We can let x represent the number of trees purchased and y the number of shrubs purchased. We can then form the following system of inequalities.

The cost of a tree	times	the number of trees purchased	plus	the cost of a shrub	times	the number of shrubs purchased	should be at least	$300.
$150	·	x	+	$75	·	y	≥	$300

The cost of a tree	times	the number of trees purchased	plus	the cost of a shrub	times	the number of shrubs purchased	should not be more than	$600.
$150	·	x	+	$75	·	y	≤	$600

Solve the system We graph the system

$$\begin{cases} 150x + 75y \geq 300 \\ 150x + 75y \leq 600 \end{cases}$$

as in Figure 7-20. The coordinates of each point shown in the graph give a possible combination of the number of trees (x) and the number of shrubs (y) that can be purchased. These possibilities are

$(0, 4), (0, 5), (0, 6), (0, 7), (0, 8)$

$(1, 2), (1, 3), (1, 4), (1, 5), (1, 6)$

$(2, 0), (2, 1), (2, 2), (2, 3), (2, 4)$

$(3, 0), (3, 1), (3, 2), (4, 0)$

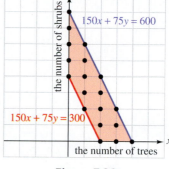

Figure 7-20

Only these points can be used, because the man cannot buy part of a tree or part of a shrub.

SELF CHECK 9 From the graph, what is the maximum number of trees he may purchase? At what geometric point does this occur?

SELF CHECK ANSWERS 1. **a.** no **b.** yes **c.** no 2. 3. 4. 5. 6.

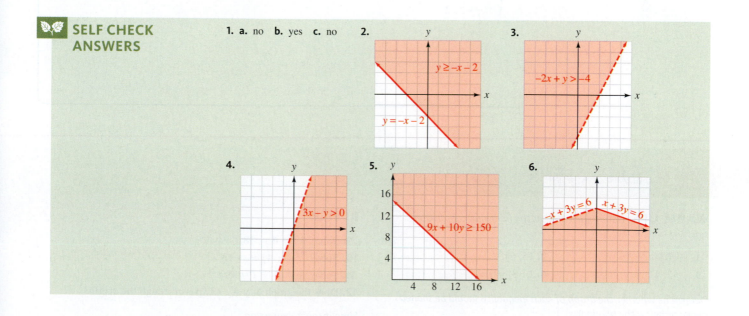

7.

8. ∅

9. Four trees is the maximum number he may purchase and occurs at one of the x-intercepts, $(4, 0)$. He would not be able to purchase any shrubs.

NOW TRY THIS

Solve each system by graphing.

1. $\begin{cases} y \geq x \\ x < -y + 2 \end{cases}$

2. $\begin{cases} x - y > 4 \\ y < x + 5 \end{cases}$

3. $\begin{cases} x - y > 0 \\ \quad x \geq 3 \\ \quad -y > 2 \end{cases}$

7.5 Exercises

WARM-UPS *Determine whether the following coordinates satisfy* $y > 3x + 2$.

1. $(0, 0)$

2. $(5, 5)$

3. $(-2, 4)$

4. $(-3, -6)$

Determine whether the graph of each inequality should be shaded above or below the boundary.

5. $y > x - 2$

6. $y \leq x + 4$

7. $x - y > -1$

8. $x - y < 5$

REVIEW

9. Solve: $4x - 6 = 18$.

10. Solve: $2(x - 4) \leq -12$.

11. Solve: $A = P + Prt$ for t.

12. Does the graph of $y = -3x$ pass through the origin?

Simplify each expression.

13. $7x + 4(x - 6)$

14. $5a - 2(6 - a)$

15. $3(y - x) + 5y + 8x$

16. $3p + 2(q - p) + q$

VOCABULARY AND CONCEPTS *Fill in the blanks.*

17. $2x - y \le 4$ is a linear _____ in x and y.

18. The symbol $\le$ means _____ or _____.

19. In the accompanying graph, the line $2x - y = 4$ is the _____ of the graph $2x - y \le 4$.

20. In the accompanying graph, the line $2x - y = 4$ divides the rectangular coordinate system into two ____ _____.

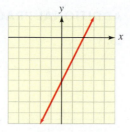

21. $\begin{cases} x + y > 5 \\ x - y < 3 \end{cases}$ is a system of linear _____.

22. The _____ of a system of linear inequalities are all the ordered pairs that make all of the inequalities of the system true at the same time.

23. Any point in the _____ region of the graph of the solution of a system of two linear inequalities has coordinates that satisfy both of the inequalities of the system.

24. To graph a linear inequality such as $2x + y > 4$, first graph the boundary with a dashed line. Then pick a test ____ to determine which half-plane to shade.

25. Determine whether the graph of each linear inequality includes the boundary line.
   a. $x - y > 5$ **b.** $4x + 3y \le 12$

26. If a false statement results when the coordinates of a test point are substituted into a linear inequality, which half-plane should be shaded to represent the solution of the inequality?

GUIDED PRACTICE *Determine whether each ordered pair is a solution of the given inequality. SEE EXAMPLE 1. (OBJECTIVE 1)*

27. Determine whether each ordered pair is a solution of $5x - 3y \ge 0$.
   a. $(1, 1)$ **b.** $(-2, -3)$
   c. $(0, 0)$ **d.** $\left(\frac{1}{5}, \frac{4}{3}\right)$

28. Determine whether each ordered pair is a solution of $x + 3y < -20$.
   a. $(3, -9)$ **b.** $(0, 0)$
   c. $(2, 1)$ **d.** $\left(-\frac{1}{2}, -8\right)$

29. Determine whether each ordered pair is a solution of $x + y > 4$.
   a. $(0, 4)$ **b.** $(1, 5)$
   c. $\left(-1, \frac{1}{2}\right)$ **d.** $\left(-\frac{3}{4}, 7\right)$

30. Determine whether each ordered pair is a solution of $x - 2y < -6$.
   a. $(4, 2)$ **b.** $(0, 2)$
   c. $(-1, 5)$ **d.** $\left(\frac{3}{4}, 6\right)$

Graph each inequality. SEE EXAMPLE 2. (OBJECTIVE 2)

31. $y \le x + 2$

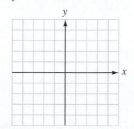

32. $y \le -x + 1$

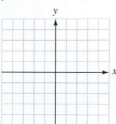

33. $y \le 4x$

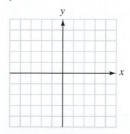

34. $y \ge 3 - x$

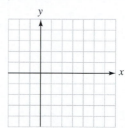

Graph each inequality. SEE EXAMPLE 3. (OBJECTIVE 2)

35. $y > x - 3$

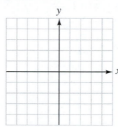

36. $y + 2x < 0$

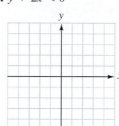

37. $y > 2x - 4$

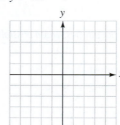

38. $y < 2 - x$

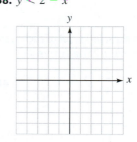

Graph each inequality. SEE EXAMPLE 4. (OBJECTIVE 2)

39. $2x - y \le 4$

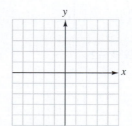

40. $3x - 4y > 12$

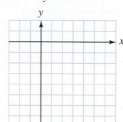

41. $x - 2y \leq 4$

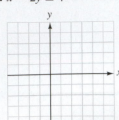

42. $7x - 2y < 21$

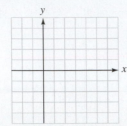

Graph the solution set of each system of inequalities, when possible. SEE EXAMPLE 6. (OBJECTIVE 4)

43. $\begin{cases} x + 2y \leq 3 \\ 2x - y \geq 1 \end{cases}$

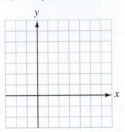

44. $\begin{cases} 2x + y \geq 3 \\ x - 2y \leq -1 \end{cases}$

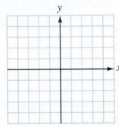

45. $\begin{cases} x + y < -1 \\ x - y > -1 \end{cases}$

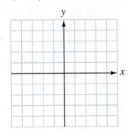

46. $\begin{cases} x + y > 2 \\ x - y < -2 \end{cases}$

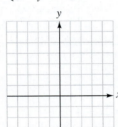

Graph the solution set of each system of inequalities, when possible. SEE EXAMPLE 7. (OBJECTIVE 4)

47. $\begin{cases} x > 2 \\ y \leq 3 \end{cases}$

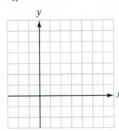

48. $\begin{cases} x \geq -1 \\ y > -2 \end{cases}$

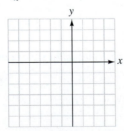

49. $\begin{cases} x \leq 0 \\ y < 0 \end{cases}$

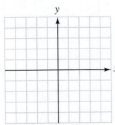

50. $\begin{cases} x < -2 \\ y \geq 3 \end{cases}$

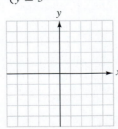

Graph the solution set of each system of inequalities, when possible. If not possible, state $\emptyset$. SEE EXAMPLE 8. (OBJECTIVE 4)

51. $\begin{cases} x + y < 1 \\ x + y > 3 \end{cases}$

52. $\begin{cases} y < 2x - 1 \\ 2x - y < -4 \end{cases}$

53. $\begin{cases} y \leq -\frac{4}{3}x - 2 \\ 4x + 3y > 15 \end{cases}$

54. $\begin{cases} 3x + y < -2 \\ y > 3(1 - x) \end{cases}$

ADDITIONAL PRACTICE *Graph the solution.*

55. $y < 3x$

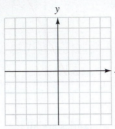

56. $3x + 2y \geq 12$

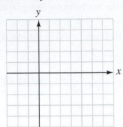

57. $\begin{cases} 2x - y < 4 \\ x + y \geq -1 \end{cases}$

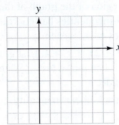

58. $\begin{cases} x - y \geq 5 \\ x + 2y < -4 \end{cases}$

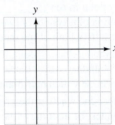

59. $y < 2 - 3x$

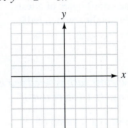

60. $y \geq 5 - 2x$

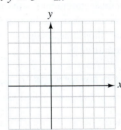

61. $x < 2$

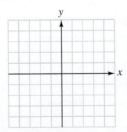

62. $2y - x < 8$

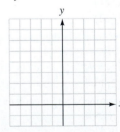

63. $y + 9x \geq 3$

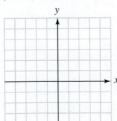

64. $y > -3$

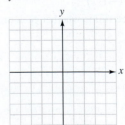

65. $4x + 3y \leq 12$

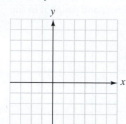

66. $5x + 4y \geq 20$

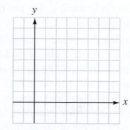

67. $y \leq 1$

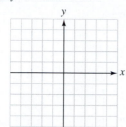

68. $x \geq -4$

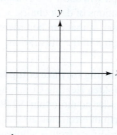

69. $\begin{cases} 3x + 4y > -7 \\ 2x - 3y \geq 1 \end{cases}$

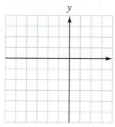

70. $\begin{cases} 3x + y \leq 1 \\ 4x - y > -8 \end{cases}$

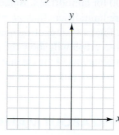

71. $\begin{cases} 2x - 4y > -6 \\ 3x + y \geq 5 \end{cases}$

72. $\begin{cases} 2x - 3y < 0 \\ 2x + 3y \geq 12 \end{cases}$

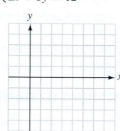

73. $\begin{cases} \frac{x}{2} + \frac{y}{3} \geq 2 \\ \frac{x}{2} - \frac{y}{2} < -1 \end{cases}$

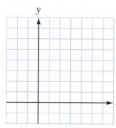

74. $\begin{cases} \frac{x}{3} - \frac{y}{2} < -3 \\ \frac{x}{3} + \frac{y}{2} > -1 \end{cases}$

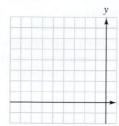

APPLICATIONS *Graph each inequality for nonnegative values of x and y. Then give some ordered pairs that satisfy the inequality.* *SEE EXAMPLE 5. (OBJECTIVE 3)*

75. Production planning It costs a bakery $3 to make a cake and $4 to make a pie. Production costs cannot exceed $120 per day. Find an inequality that shows the possible combinations of cakes, x, and pies, y, that can be made, and graph it in the illustration.

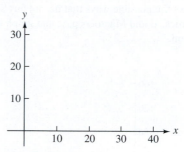

76. Hiring babysitters Tomiko has a choice of two babysitters. Sitter 1 charges $6 per hour, and sitter 2 charges $7 per hour. Tomiko can afford no more than $42 per week for sitters. Find an inequality that shows the possible ways that she can hire sitter 1 (x) and sitter 2 (y), and graph it in the illustration.

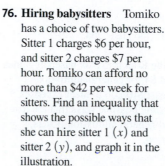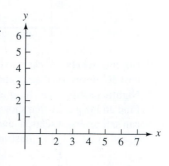

77. Inventory A clothing store advertises that it maintains an inventory of at least $4,400 worth of men's jackets. A leather jacket costs $100, and a nylon jacket costs $88. Find an inequality that shows the possible ways that leather jackets, x, and nylon jackets, y, can be stocked, and graph it in the illustration.

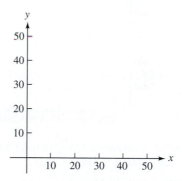

78. Making sporting goods To keep up with demand, a sporting goods manufacturer allocates at least 2,400 units of time per day to make baseballs and footballs. It takes 20 units of time to make a baseball and 30 units of time to make a football. Find an inequality that shows the possible ways to schedule the time to make baseballs, x, and footballs, y, and graph it in the illustration.

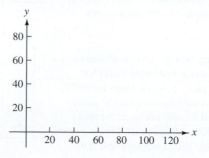

79. Investing Robert has up to $8,000 to invest in two companies. Stock in Robotronics sells for $40 per share, and stock in Macrocorp sells for $50 per share. Find an inequality that shows the possible ways that he can buy shares of Robotronics, x, and Macrocorp, y, and graph it in the illustration.

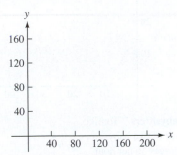

80. Buying tickets Tickets to the Rockford Rox baseball games cost $6 for reserved seats and $4 for general admission. Nightly receipts must be at least $10,200 to meet expenses. Find an inequality that shows the possible ways that the Rox can sell reserved seats, x, and general admission tickets, y, and graph it in the illustration.

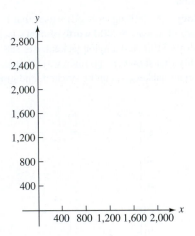

Graph each system of inequalities and give two possible solutions to each problem. SEE EXAMPLE 9. (OBJECTIVE 5)

81. Buying CDs Melodic Music has compact discs on sale for either $10 or $15. A customer wants to spend at least $30 but no more than $60 on CDs. Find a system of inequalities whose graph will show the possible combinations of $10 CDs, x, and $15 CDs, y, that the customer can buy, and graph it in the illustration.

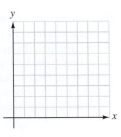

82. Buying boats Dry Boatworks wholesales aluminum boats for $800 and fiberglass boats for $600. Northland Marina wants to order at least $2,400 but no more than $4,800 worth of boats. Find a system of inequalities whose graph will

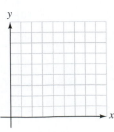

show the possible combinations of aluminum boats, x, and fiberglass boats, y, that can be ordered, and graph it in the illustration.

83. Buying furniture A distributor wholesales desk chairs for $150 and side chairs for $100. Best Furniture wants to order no more than $900 worth of chairs and wants to order more side chairs than desk chairs. Find a system of inequalities whose graph will show the possible combinations of desk chairs, x, and side chairs, y, that can be ordered, and graph it in the illustration.

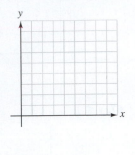

84. Ordering furnace equipment J. Bolden Heating Company wants to order no more than $2,000 worth of electronic air cleaners and humidifiers from a wholesaler that charges $500 for air cleaners and $200 for humidifiers. Bolden wants more humidifiers than air cleaners. Find a system of inequalities whose graph will show the possible combinations of air cleaners, x, and humidifiers, y, that can be ordered, and graph it in the illustration.

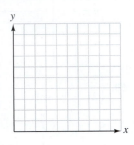

WRITING ABOUT MATH

85. Explain how to find the boundary for the graph of an inequality.

86. Explain how to decide which side of the boundary line to shade.

87. Explain how to use graphing to solve a system of inequalities.

88. Explain when a system of inequalities will have no solutions.

SOMETHING TO THINK ABOUT

89. What are some limitations of the graphing method for solving inequalities?

90. Graph $y = 3x + 1$, $y < 3x + 1$, and $y > 3x + 1$. What do you discover?

91. Can a system of inequalities have
  a. no solutions?
  b. exactly one solution?
  c. infinitely many solutions?

92. Find a system of two inequalities that has a solution of $(2, 0)$ but no solutions of the form (x, y) where $y < 0$.

Projects

PROJECT 1

The graphing method of solving a system of equations is not as accurate as algebraic methods, and some systems are more difficult than others to solve accurately. For example, the two lines in Illustration 1(a) could be drawn carelessly, and the point of intersection would not be far from the correct location. If the lines in Illustration 1(b) were drawn carelessly, the point of intersection could move substantially from its correct location.

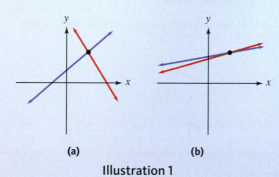

(a)   (b)

Illustration 1

a. Carefully solve each of these systems of equations graphically (by hand, not with a graphing calculator). Indicate your best estimate of the solution of each system.

$$\begin{cases} 2x - 4y = -7 \\ 4x + 2y = 11 \end{cases} \qquad \begin{cases} 5x - 4y = -1 \\ 12x - 10y = -3 \end{cases}$$

b. Solve each system algebraically. How close were your graphical solutions to the actual solutions? Write a paragraph explaining any differences.

c. Create a system of equations with the solutions $x = 3$, $y = 2$ for which an accurate solution could be obtained graphically.

d. Create a system of equations with the solutions $x = 3$, $y = 2$ that is more difficult to solve graphically than the previous system, and write a paragraph explaining why.

PROJECT 2

Find the solutions of the following system of linear inequalities by graphing the inequalities of the given coordinate system.

$$\begin{cases} x + \frac{2}{3}y < \frac{4}{3} \\ y \le \frac{3}{5}x + 2 \end{cases}$$

For each point, A through G on the graph, determine whether its coordinates satisfy the first inequality, the second inequality, neither inequality, or both. Present your results in a table like the one shown below.

Point	Coordinates	1st inequality	2nd inequality
A			

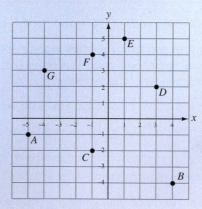

Illustration 2

Reach for Success   REVIEWING YOUR GAME PLAN

In an activity in a previous chapter, you set goals for this course. It is now time to check your progress toward achieving them.

A. Fill in the blanks from your *original goals* for this course.

The grade I am willing to work to achieve is a/an _____.

Considering other course commitments as well as any work and family commitments, state the number of hours *outside of class* you realistically believe you can devote each week to this one course. _____

List at least three things you are willing to add to your game plan that will support your success in this course.

1. _____

2. _____

3. _____

Write down anything else you can do to improve your performance in this class. _____

_____

B. Considering your performance in this course to date, please answer these questions.

What is your approximate grade in this class now? _____ Is this the grade you want to earn? _____ If this is your goal grade, or a higher one, congratulations! It appears you stayed with your game plan and should continue with this plan for the remainder of the semester.

If the grade is lower than what you want, let's look at some factors that may have contributed to this.

List the items on your game plan that you have done to be successful in this course.

_____

Congratulations on following through with these items!

If there are items on your game plan that you have not done, list these and consider why you have not been able to follow through with your plan. Some reasons might be work schedule, unreliable child care, other course demands, content more difficult than anticipated, motivational issues, or personal issues.

_____

C. Now, let's determine if there are changes you can make now to turn the semester around.

Is it necessary to modify the grade you would be working toward? _____ What is that new grade? _____

Of the items listed in Part B above, consider and list any changes you could make now to improve your chance of success in this course before the end of the semester.

_____

_____

D. Is your success in this course a high enough priority to make the changes you listed above? _____

7   Review

SECTION 7.1   Solving Systems of Linear Equations by Graphing

DEFINITIONS AND CONCEPTS	EXAMPLES

DEFINITIONS AND CONCEPTS

An ordered pair is a *solution* of a system of equations if it satisfies both equations.

EXAMPLES

To determine whether $(5, -1)$ is a solution of the following system, we proceed as follows:

$$\begin{cases} x + y = 4 \\ 2x - y = 11 \end{cases}$$

$$x + y = 4$$
$$5 + (-1) \overset{?}{=} 4 \qquad \text{Substitute 5 for } x \text{ and } -1 \text{ for } y.$$
$$4 = 4 \qquad \text{Add.}$$

$$2x - y = 11$$
$$2(5) - (-1) \overset{?}{=} 11 \qquad \text{Substitute 5 for } x \text{ and } -1 \text{ for } y.$$
$$10 + 1 \overset{?}{=} 11 \qquad \text{Multiply 2 and 5, change sign of } -1.$$
$$11 = 11 \qquad \text{Add.}$$

Because the ordered pair $(5, -1)$ satisfies both equations, it is a solution of the system of equations.

The Graphing Method

1. Graph each equation on one set of coordinate axes.
2. Find the coordinates of the point where the graphs intersect, if applicable.
3. Check the solution in the equations of the original system, if applicable.

Remember that the graph is not the solution.

To solve the system $\begin{cases} x + y = 8 \\ 2x - 3y = 6 \end{cases}$ by graphing, we graph both equations on one set of coordinate axes.

$x + y = 8$		
x	y	(x, y)
0	8	$(0, 8)$
8	0	$(8, 0)$

$2x - 3y = 6$		
x	y	(x, y)
0	-2	$(0, -2)$
3	0	$(3, 0)$

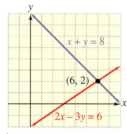

The solution of the system is the ordered pair $(6, 2)$.

If a system of equations has infinitely many solutions, the equations of the system are *dependent equations*.

To solve the system $\begin{cases} x + y = 8 \\ 2x + 2y = 16 \end{cases}$ by graphing, we graph both equations on one set of coordinate axes.

$x + y = 8$		
x	y	(x, y)
0	8	$(0, 8)$
8	0	$(8, 0)$

$2x + 2y = 16$		
x	y	(x, y)
0	8	$(0, 8)$
8	0	$(8, 0)$

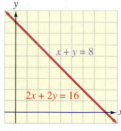

Solve each equation for y.

$$\begin{array}{ll} x + y = 8 & 2x + 2y = 16 \\ y = 8 - x & 2y = 16 - 2x \\ & y = 8 - x \end{array}$$

Since the lines in the illustration are the same line, there are infinitely many solutions, which can be written in the general form $(x, 8 - x)$.

If a system of equations has no solutions, it is an *inconsistent system* and we write the solution set as $\varnothing$.

To solve the system $\begin{cases} x + y = 8 \\ 2x + 2y = 6 \end{cases}$ by graphing, we graph both equations on one set of coordinate axes.

$x + y = 8$

x	y	(x, y)
0	8	$(0, 8)$
8	0	$(8, 0)$

$2x + 2y = 6$

x	y	(x, y)
0	3	$(0, 3)$
3	0	$(3, 0)$

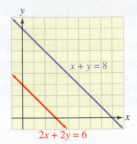

Since the lines in the figure are parallel, there are no solutions and the solution set is $\varnothing$.

REVIEW EXERCISES

Determine whether the ordered pair is a solution of the system.

1. $(2, 3)$, $\begin{cases} 3x + 2y = 12 \\ 2x - y = 1 \end{cases}$

2. $(-3, -1)$, $\begin{cases} 5x - 3y = -12 \\ 2x - 3y = -9 \end{cases}$

3. $\left(14, \frac{1}{2}\right)$, $\begin{cases} 2x + 4y = 30 \\ \frac{x}{4} - y = 3 \end{cases}$

4. $\left(\frac{7}{2}, -\frac{2}{3}\right)$, $\begin{cases} 4x - 6y = 18 \\ \frac{x}{3} + \frac{y}{2} = \frac{5}{6} \end{cases}$

Use the graphing method to solve each system.

5. $\begin{cases} x + y = 7 \\ 2x - y = 5 \end{cases}$

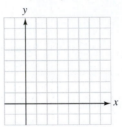

6. $\begin{cases} \frac{x}{3} + \frac{y}{5} = -1 \\ x - 3y = -3 \end{cases}$

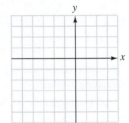

7. $\begin{cases} 3x + 6y = 6 \\ x + 2y = 2 \end{cases}$

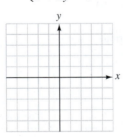

8. $\begin{cases} 6x + 3y = 12 \\ 2x + y = 2 \end{cases}$

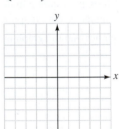

SECTION 7.2   Solving Systems of Linear Equations by Substitution

DEFINITIONS AND CONCEPTS	**EXAMPLES**

The Substitution Method

1. If necessary, solve one of the equations for x or y, preferably a variable with a coefficient of 1.

2. Substitute the resulting expression for the variable obtained in Step 1 into the other equation, and solve that equation.

3. Find the value of the other variable by substituting the solution found in Step 2 into any equation containing both variables.

4. Check the solution in the equations of the original system.

To solve the system $\begin{cases} x + y = 8 \\ 2x - 3y = 6 \end{cases}$ by substitution, we solve one of the equations for one of its variables. If we solve $x + y = 8$ for y, we have

$y = 8 - x$ Subtract x from both sides.

We then substitute $8 - x$ for y in the second equation and solve for x.

$2x - 3y = 6$

$2x - 3(8 - x) = 6$ Substitute $8 - x$ for y.

$2x - 24 + 3x = 6$ Use the distributive property.

$5x - 24 = 6$ Combine like terms.

$5x = 30$ Add 24 to both sides.

$x = 6$ Divide by 5.

We can find y by substituting 6 for x in the equation $y = 8 - x$.

$$y = 8 - \boldsymbol{x}$$
$$y = 8 - \boldsymbol{6} \qquad \text{Substitute 6 for } x.$$
$$y = 2 \qquad \text{Add.}$$

The solution is the ordered pair $(6, 2)$.

REVIEW EXERCISES

Use substitution to solve each system.

9. $\begin{cases} x = 4y + 1 \\ 3x - 2y = -7 \end{cases}$

10. $\begin{cases} 3x - \frac{2y}{5} = 2(x - 2) \\ 2x - 3 = 3 - 2y \end{cases}$

11. $\begin{cases} 8x + 5y = 3 \\ 5x - 8y = 13 \end{cases}$

12. $\begin{cases} 6(x + 2) = y - 1 \\ 5(y - 1) = x + 2 \end{cases}$

SECTION 7.3 Solving Systems of Linear Equations by Elimination (Addition)

DEFINITIONS AND CONCEPTS

The Elimination (Addition) Method

1. If necessary, write both equations in general form: $Ax + By = C$.

2. If necessary, multiply one or both of the equations by nonzero quantities to make the coefficients of x (or the coefficients of y) opposites.

3. Add the equations to eliminate the term involving x (or y).

4. Solve the equation resulting from Step 3.

5. Find the value of the other variable by substituting the solution found in Step 4 into one of the original equations.

6. Check the solution in both equations of the original system.

EXAMPLES

To solve the system $\begin{cases} x + y = 8 \\ 2x - 3y = 6 \end{cases}$ by elimination, we can eliminate x by multiplying the first equation by -2 to get

$$\begin{cases} \boldsymbol{-2}(x + y) = \boldsymbol{-2}(8) \\ 2x - 3y = 6 \end{cases} \quad \rightarrow \quad \begin{cases} -2x - 2y = -16 \\ 2x - 3y = 6 \end{cases}$$

When these equations are added, the terms $-2x$ and $2x$ are eliminated.

$$\begin{array}{r} -2x - 2y = -16 \\ 2x - 3y = 6 \\ \hline -5y = -10 \end{array}$$
$$y = 2 \qquad \text{Divide both sides by } -5.$$

To find x, we substitute 2 for y in the equation $x + y = 8$.

$$x + \boldsymbol{y} = 8$$
$$x + \boldsymbol{2} = 8 \qquad \text{Substitute 2 for } y.$$
$$x = 6 \qquad \text{Subtract 2 from both sides.}$$

The solution is the ordered pair $(6, 2)$.

REVIEW EXERCISES

Use elimination to solve each system.

13. $\begin{cases} 4x + y = 3 \\ 7x - 2y = 39 \end{cases}$

14. $\begin{cases} x - 6y = 7 \\ x + 2y = 3 \end{cases}$

15. $\begin{cases} 5x + y = 2 \\ 3x + 2y = 11 \end{cases}$

16. $\begin{cases} x + y = 3 \\ 3x = 2 - y \end{cases}$

17. $\begin{cases} 11x + 3y = 27 \\ 8x + 4y = 36 \end{cases}$

18. $\begin{cases} 9x + 3y = 5 \\ 3x = 4 - y \end{cases}$

19. $\begin{cases} 9x + 3y = 5 \\ 3x + y = \frac{5}{3} \end{cases}$

20. $\begin{cases} \frac{x}{3} + \frac{y + 2}{2} = 1 \\ \frac{x + 8}{8} + \frac{y - 3}{3} = 0 \end{cases}$

SECTION 7.4 Solving Applications of Systems of Linear Equations

DEFINITIONS AND CONCEPTS

Systems of equations are useful in solving many types of applications.

EXAMPLES

Boating A boat traveled 30 kilometers downstream in 3 hours and traveled 12 kilometers in 3 hours against the current. Find the speed of the boat in still water.

Analyze the problem We can let s represent the speed of the boat in still water in km/hr and let c represent the speed of the current in km/hr.

Form two equations   The rate of speed of the boat while going downstream is $(s + c)$. The rate of the boat while going upstream is $(s - c)$. Because $d = r \cdot t$, the information gives two equations in two variables.

$$\begin{cases} 30 = 3(s + c) \\ 12 = 3(s - c) \end{cases}$$

After removing parentheses and rearranging terms, we have

(1) $\begin{cases} 3s + 3c = 30 \\ 3s - 3c = 12 \end{cases}$
(2)

Solve the system   To solve this system by elimination (addition), we add the equations, and solve for s.

$$\begin{aligned} 3s + 3c &= 30 \\ \underline{3s - 3c} &= \underline{12} \\ 6s &= 42 \\ s &= 7 \qquad \text{\color{red}{Divide both sides by 6.}} \end{aligned}$$

State the conclusion   The speed of the boat in still water is 7 kilometers per hour.

Check the result

If $s = 7$, $3(7) + 3c = 30$
$$21 + 3c = 30$$
$$3c = 9$$
$$c = 3$$

The speed of the current is 3 km/hr. In 3 hours, the boat will travel downstream $3(7 + 3) = 30$ km.
In 3 hours, the boat will travel against the current $3(7 - 3) = 12$ km. The answers check.

REVIEW EXERCISES

21. Integers   One number is 3 times another, and their sum is 84. Find the numbers.

22. Geometry   The length of a rectangle is 3 times its width, and its perimeter is 24 feet. Find its dimensions.

23. Buying grapefruit   A grapefruit costs 15 cents more than an orange. Together, they cost 85 cents. Find the cost of a grapefruit.

24. Utility bills   A man's electric bill for January was $23 less than his gas bill. The two utilities cost him a total of $109. Find the amount of his gas bill.

25. Buying groceries   Two gallons of milk and 3 dozen eggs cost $6.80. Three gallons of milk and 2 dozen eggs cost $7.35. How much does each gallon of milk cost?

26. Investing money   Carlos invested part of $4,500 in a 4% certificate of deposit account and the rest in a 3% passbook account. If the total annual interest from both accounts is $160, how much did he invest at 3%?

27. Boating   It takes a boat 4 hours to travel 56 miles down a river and 3 hours longer to make the return trip. Find the speed of the current.

28. Medical technology   A laboratory technician has one batch of solution that is 10% saline and a second batch that is 60% saline. He would like to make 50 milliliters of solution that is 30% saline. How many liters of each batch should he use?

SECTION 7.5   Solving Systems of Linear Inequalities

DEFINITIONS AND CONCEPTS	EXAMPLES

1. Graph each inequality in the system on the same coordinate axes using solid or dashed lines as appropriate.
2. Find the region where the graphs overlap.
3. Select a test point from the region to verify the solution.
4. Graph only the solution set, if there is one.

If a given inequality is $<$ or $>$, the boundary line is dashed.

If a given inequality is $\leq$ or $\geq$, the boundary line is solid.

Graph the solution set: $\begin{cases} x + y < 4 \\ 2x - y \geq 6 \end{cases}$

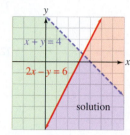 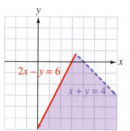

The graph of $x + y < 4$ includes all points below the line $x + y = 4$. Since the boundary is not included, we draw it as a dashed line.

The graph of $2x - y \geq 6$ includes all points below the line $2x - y = 6$. Since the boundary is included, we draw it as a solid line.

The solution is graphed above on the right.

REVIEW EXERCISES

Graph each inequality.

29. $y \geq x + 2$

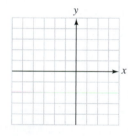

30. $x < 3$

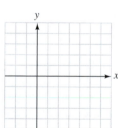

Solve each system of inequalities.

31. $\begin{cases} 5x + 3y < 15 \\ 3x - y > 3 \end{cases}$

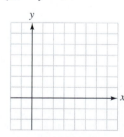

32. $\begin{cases} 5x - 3y \geq 5 \\ 3x + 2y \geq 3 \end{cases}$

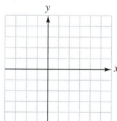

33. $\begin{cases} x \geq 3y \\ y < 3x \end{cases}$

34. $\begin{cases} x \geq 0 \\ x \leq 3 \end{cases}$

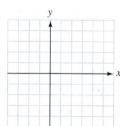

35. Shopping   A mother wants to spend at least $40 but no more than $60 on her child's school uniform. If shirts sell for $10 and pants sell for $20, find a system of inequalities that describe the possible numbers of shirts, x, and pants, y, that she can buy. Graph the system and give two possible solutions.

7   Test

Determine whether the given ordered pair is a solution of the given system.

1. $(-1, 5),$ $\begin{cases} 5x + 3y = 10 \\ 3x - y = -8 \end{cases}$

2. $(-2, -1),$ $\begin{cases} 4x + y = -9 \\ 2x - 3y = -7 \end{cases}$

Solve each system by graphing.

3. $\begin{cases} 3x + y = 7 \\ x - 2y = 0 \end{cases}$

4. $\begin{cases} x + \frac{y}{2} = 1 \\ y = 1 - 3x \end{cases}$

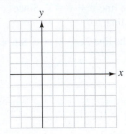   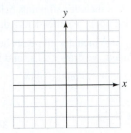

Solve each system by substitution.

5. $\begin{cases} y = 2x + 1 \\ x + y = 10 \end{cases}$

6. $\begin{cases} \frac{x}{6} + \frac{y}{10} = 3 \\ \frac{5x}{16} - \frac{3y}{16} = \frac{15}{8} \end{cases}$

Solve each system by elimination (addition).

7. $\begin{cases} 3x - y = 2 \\ 2x + y = 8 \end{cases}$

8. $\begin{cases} 4x + 3 = -3y \\ \frac{-x}{7} + \frac{4y}{21} = 1 \end{cases}$

Classify each system as consistent or inconsistent and solve.

9. $\begin{cases} 4x + 5(y + 1) = 1 \\ -5y = 2(2x + 3) \end{cases}$

10. $\begin{cases} \frac{1}{4}x + y - 3 = 0 \\ -4y = x - 12 \end{cases}$

Use a system of equations in two variables to solve each application.

11. Numbers   The sum of two numbers is -18. One number is 2 greater than 3 times the other. Find the product of the numbers.

12. Water parks   A father paid $119 for his family of 7 to spend the day at Magic Waters water park. How many adult tickets did he buy?

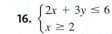

Admission	
Adult ticket	$21
Child ticket	$14

13. Investing   A woman invested some money at 3% annual interest and some at 4% annual interest. The interest on the combined investment of $10,000 was $340 for one year. How much was invested at 4%?

14. Kayaking   A kayaker can paddle 8 miles down a river in 2 hours and make the return trip in 4 hours. Find the speed of the current in the river.

Solve each system of inequalities by graphing.

15. $\begin{cases} x + y < 3 \\ x - y < 1 \end{cases}$

16. $\begin{cases} 2x + 3y \leq 6 \\ x \geq 2 \end{cases}$

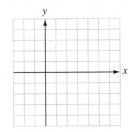   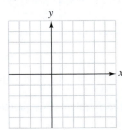

Roots, Radical Expressions, and Radical Equations

8

© iStockphoto.com/David Jones

Careers and Mathematics

CONSTRUCTION MANAGER

Construction managers plan and coordinate a wide variety of construction projects. Large construction projects are too complicated for one person to manage. These projects are divided into many segments, such as site preparation, sewage systems, and road construction. Employers increasingly prefer to hire construction managers with a degree in construction science, construction management, building science, or civil engineering.

Job Outlook:
Employment of construction managers is projected to increase by 17% through 2020. This is about as fast as the average for all occupations.

Annual Earnings:
$50,240–$150,250

For More Information:
http://www.bls.gov/oco/ocos005.htm

For a Sample Application:
See Problem 94 in Section 8.6.

REACH FOR SUCCESS

In this chapter

In this chapter, we reverse the process of squaring or cubing numbers by finding square roots and cube roots of numbers. We will discuss an important theorem, called the Pythagorean theorem, which is used to solve many applications.

We also will discuss how to simplify radical expressions, how to add, subtract, multiply, and divide them, and how to solve equations that contain radicals.

499

Reach for Success   Accessing Academic Resources

When a baseball pitcher starts to struggle with a certain type of pitch, he may fall behind in the ball/strike count. Often, a suggestion from his manager or pitching coach can give him the guidance he needs to correct it.

If you do not feel you are performing as well as you should, or would like, then perhaps it is time to seek help. Using the varied resources available to you can improve your chances of a successful semester.

© Dennis Ku/Shutterstock.com

Obtain a hard copy or view an online version of your college catalog. Where did you find it?

_____

Search for Academic Resources. List the resources that could help with your mathematics class.

_____

_____

Does your college have a mathematics lab or tutoring center? _____

If so, where is it located? _____

What are the hours of operation?
M: _____ T: _____ W: _____
Th: _____ F: _____ S/Su: _____

Write down at least two days and times that you could go to the center to get help for an hour or more.

_____

If not, what other resources are available? _____

In addition to, or in place of, on-campus services, some colleges offer online tutoring.

Does your college offer online tutoring services for mathematics? _____

If so, what is the process to get enrolled? _____

_____

If not, have you tried the software that accompanies your textbook? _____

Colleges usually offer a wide variety of resources to help students succeed.

Are there other services at your college that might support your performance in this class?

If so, what are they? _____

Does your college offer a College Success or Study Skills course? _____

If so, would you consider enrolling? _____ Explain.

_____

A Successful Study Strategy . . .

 Know where to go to get help with your mathematics. In addition to those listed above, remember that your instructor may be your best resource.

At the end of the chapter you will find an additional exercise to guide you in planning for a successful semester.

Section 8.1

Square Roots and the Pythagorean Theorem

Objectives

1. Find the principal square root of a perfect square.
2. Use a calculator to find an approximation of a square root to a specified number of decimal places.
3. Determine whether a number is rational, irrational, or imaginary.
4. Graph a square root function.
5. Apply the Pythagorean theorem to find the length of an unknown side of a right triangle.

Vocabulary

perfect squares	principal square root	hypotenuse
square root	radicand	legs of a triangle
radical sign	integer squares	Pythagorean theorem
radical expression	imaginary numbers	Pythagorean triples

Getting Ready

Find each value.

1. 3^2 **2.** 4^2 **3.** 2^3 **4.** 5^3

5. 3^4 **6.** 4^4 **7.** $(-3)^3$ **8.** $(-2)^5$

In this section, we will reverse the squaring process and find square roots of numbers. We also will discuss an important theorem about right triangles, called the *Pythagorean theorem*.

1 Find the principal square root of a perfect square.

To find the area *A* of the square shown in Figure 8-1, we multiply its length by its width.

$$A = 5 \cdot 5$$
$$= 5^2$$
$$= 25$$

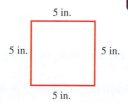

5 in.

5 in.   5 in.

5 in.

Figure 8-1

COMMENT Remember that perimeter is expressed as a linear unit, area is expressed as a square unit, and volume is expressed as a cubic unit.

The area is 25 square inches.

We have seen that the product $5 \cdot 5$ can be denoted by the exponential expression 5^2, where 5 is raised to the second power. Whenever we raise a number to the second power, we are squaring it, or finding its square. This example illustrates that the formula for the area of a square with a side of length *s* is $A = s^2$.

Here are additional **perfect squares** of numbers.

- The square of 3 is 9, because $3^2 = 9$.
- The square of -3 is 9, because $(-3)^2 = 9$.
- The square of $\frac{1}{3}$ is $\frac{1}{9}$, because $\left(\frac{1}{3}\right)^2 = \frac{1}{9}$.

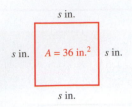

Figure 8-2

- The square of $-\frac{1}{3}$ is $\frac{1}{9}$, because $\left(-\frac{1}{3}\right)^2 = \frac{1}{9}$.
- The square of 0 is 0, because $0^2 = 0$.

Suppose we know that the area of the square shown in Figure 8-2 is 36 square inches. To find the length of each side, we substitute 36 for A in the formula $A = s^2$ and solve for s.

$$A = s^2$$
$$36 = s^2$$

To solve for s, we must find a positive number whose square is 36. Since 6 is such a number, the sides of the square are 6 inches long. The number 6 is called a **square root** of 36, because $6^2 = 36$.

Here are additional examples of square roots.

- 3 is a square root of 9, because $3^2 = 9$.
- -3 is a square root of 9, because $(-3)^2 = 9$.
- $\frac{1}{3}$ is a square root of $\frac{1}{9}$, because $\left(\frac{1}{3}\right)^2 = \frac{1}{9}$.
- $-\frac{1}{3}$ is a square root of $\frac{1}{9}$, because $\left(-\frac{1}{3}\right)^2 = \frac{1}{9}$.
- 0 is a square root of 0, because $0^2 = 0$.

In general, the following is true.

SQUARE ROOTS

The number b is a square root of a if $b^2 = a$.

All positive numbers have two square roots—one positive and one negative. We have seen that the two square roots of 9 are 3 and -3. The two square roots of 144 are 12 and -12, because $12^2 = 144$ and $(-12)^2 = 144$. The number 0 is the only number that has just one square root, which is 0.

The symbol $\sqrt{}$, called a **radical sign**, is used to represent the positive (or **principal**) square root of a number. Any expression involving a radical sign is called a **radical expression**.

PRINCIPAL SQUARE ROOTS

If $a > 0$, the radical expression $\sqrt{a}$ represents the principal (or positive) square root of a.

The principal square root of 0 is 0: $\sqrt{0} = 0$.

The expression under a radical sign is called a **radicand**. For example, in the radical expression $\sqrt{x}$, x is the radicand.

The principal square root of a positive number is always positive. Although 3 and -3 are both square roots of 9, only 3 is the principal square root. The symbol $\sqrt{9}$ represents 3. To represent -3, we place a $-$ sign in front of the radical:

$$\sqrt{9} = 3 \qquad \text{and} \qquad -\sqrt{9} = -3$$

Likewise,

$$\sqrt{144} = 12 \qquad \text{and} \qquad -\sqrt{144} = -12$$

EXAMPLE 1   Find each square root.

 a. $\sqrt{0} = 0$ b. $\sqrt{1} = 1$ c. $-\sqrt{4} = -2$

 d. $-\sqrt{81} = -9$ e. $\sqrt{225} = 15$ f. $\sqrt{169} = 13$

 g. $\sqrt{\dfrac{1}{4}} = \dfrac{1}{2}$ h. $\sqrt{\dfrac{4}{9}} = \dfrac{2}{3}$

 SELF CHECK 1   Find each square root.   a. $\sqrt{121}$ b. $-\sqrt{49}$

 c. $\sqrt{64}$ d. $\sqrt{256}$ e. $\sqrt{\dfrac{1}{25}}$ f. $\sqrt{\dfrac{9}{49}}$

---

2   Use a calculator to find an approximation of a square root to a specified number of decimal places.

Square roots of certain numbers such as 7 cannot be computed. However, we can find an approximation of $\sqrt{7}$ with a calculator. Many times, the directions in exercise sets will specify the number of decimal places to use in the approximation.

Accent on technology

▸ Using a Calculator to Find Square Roots

To find an approximation of the principal square root of 7 with a calculator, we enter 7 and press these keys.

$7\ \sqrt{x}$ Using a scientific calculator

2nd x^2 ($\sqrt{\ }$) 7 **ENTER**   Using a TI84 graphing calculator

Either way, the result is

$\sqrt{7} \approx 2.645751311$ Read $\approx$ as "is approximately equal to."

For instructions regarding the use of a Casio graphing calculator, please refer to the Casio Keystroke Guide in the back of the book.

EXAMPLE 2   The *period of a pendulum* is the time required for the pendulum to swing back and forth to complete one cycle. (See Figure 8-3.) The period t (in seconds) is a function of the pendulum's length l (in feet), which is defined by

$$t = f(l) = 1.11\sqrt{l}$$

Find the period of a pendulum that is 5 feet long.

Solution   We substitute 5 for l in the formula and simplify.

$$t = 1.11\sqrt{l}$$
$$= 1.11\sqrt{5}$$
$$\approx 2.482035455$$

The period is approximately 2.5 seconds for a 5-foot-long pendulum.

Figure 8-3

 SELF CHECK 2   To the nearest tenth, find the period of a pendulum that is 3 feet long.

---

3   Determine whether a number is rational, irrational, or imaginary.

Numbers such as 4, 9, 16, and 49 are called **integer squares**, because each one is the square of an integer. The square root of any integer square is an integer, and therefore a rational number:

$$\sqrt{4} = 2 \qquad \sqrt{9} = 3 \qquad \sqrt{16} = 4 \qquad \text{and} \qquad \sqrt{49} = 7$$

Square roots of positive integers that are not integer squares are *irrational numbers*. For example, $\sqrt{8}$ is an irrational number. Recall that an irrational number is a nonrepeating, nonterminating decimal.

COMMENT Square roots of negative numbers are not real numbers. For example, $\sqrt{-4}$ is nonreal, because the square of no real number is -4. The number $\sqrt{-4}$ is an example from a set of numbers called **imaginary numbers**, which is discussed in the next chapter.

In this chapter, we will assume that *all variable radicands under square root symbols are either positive or* 0. Thus, in this chapter, all square roots will be real numbers.

EXAMPLE 3 Determine whether the following are rational, irrational, or imaginary.

a. $\sqrt{36}$ **b.** $\sqrt{11}$ **c.** $\sqrt{-9}$

Solution **a.** $\sqrt{36}$ is rational because $\sqrt{36} = 6$.

b. $\sqrt{11}$ is irrational because 11 is not an integer square.

c. $\sqrt{-9}$ is imaginary because there is no real number whose square is -9.

SELF CHECK 3 Determine whether the following are rational, irrational, or imaginary.

a. $\sqrt{81}$ **b.** $\sqrt{12}$ **c.** $\sqrt{-6}$

Everyday connections
Launching a Rocket

A rocket launched from Earth's surface must achieve a minimum velocity in order for the rocket to escape Earth's gravitational pull. This escape velocity is given by the formula $v = \sqrt{2gR}$, where g represents the acceleration due to gravity, and R represents Earth's radius.

1. Using $R = 3{,}960$ miles, and $g = \frac{1}{165}$ miles/s^2, calculate the escape velocity in miles per second.

2. Using $R = 6{,}387$ kilometers, and $g = 0.0098$ km/s^2, calculate the escape velocity in kilometers per second.

---

4 Graph a square root function.

Since there is one principal square root for every nonnegative real number x, the equation $f(x) = \sqrt{x}$ determines a square root function.

EXAMPLE 4 Graph: $f(x) = \sqrt{x}$

Solution To graph this function, we make a table of values and plot each pair of points. The graph appears in Figure 8-4(a). A calculator graph appears in Figure 8-4(b).

x	$f(x)$	$(x, f(x))$
0	0	$(0,0)$
1	1	$(1,1)$
4	2	$(4,2)$
9	3	$(9,3)$
16	4	$(16,4)$

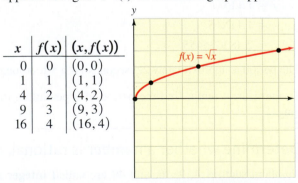

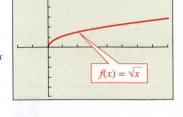

Figure 8-4 (a) **(b)**

SELF CHECK 4 Graph: $f(x) = \sqrt{x} - 1$

The square root function appears in many real-world applications.

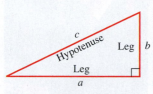

Figure 8-5

5 Apply the Pythagorean theorem to find the length of an unknown side of a right triangle.

A triangle that contains a 90° angle is called a *right triangle*. The longest side of a right triangle is the **hypotenuse**, which is the side opposite the right angle. The remaining two sides are the **legs of the triangle**. In the right triangle shown in Figure 8-5, side c is the hypotenuse, and sides a and b are the legs.

The **Pythagorean theorem** provides a formula relating the lengths of the sides of a right triangle. Natural numbers that satisfy this theorem are called **Pythagorean triples**.

THE PYTHAGOREAN THEOREM	If the length of the hypotenuse of a right triangle is c and the lengths of the two legs are a and b, then $$c^2 = a^2 + b^2$$

SQUARE ROOT PROPERTY OF EQUALITY	Suppose a and b are positive numbers. If $a = b$, then $\sqrt{a} = \sqrt{b}$.

Since the lengths of the sides of a triangle are positive numbers, we can use the square root property of equality (and only consider the positive root) and the Pythagorean theorem to find the length of an unknown side of a right triangle when we know the lengths of the other two sides.

EXAMPLE 5 **BUILDING A HIGH-ROPES ADVENTURE COURSE** The builder of a high-ropes course wants to stabilize the pole shown in Figure 8-6 by attaching a cable from a ground anchor 20 feet from its base to a point 15 feet up the pole. How long will the cable be?

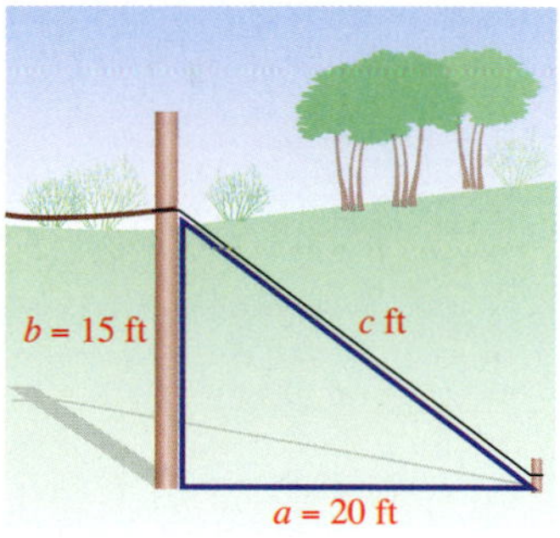

Solution We can use the Pythagorean theorem, with $a = 20$ and $b = 15$.

$$c^2 = \boldsymbol{a}^2 + \boldsymbol{b}^2$$
$$c^2 = \boldsymbol{20}^2 + \boldsymbol{15}^2 \quad \text{Substitute 20 for } a \text{ and 15 for } b.$$
$$c^2 = 400 + 225 \quad \text{Square each term.}$$
$$c^2 = 625 \quad 400 + 225 = 625$$
$$\sqrt{c^2} = \sqrt{625} \quad \text{Take the positive square root of both sides.}$$
$$c = 25 \quad \sqrt{625} = 25 \text{ and } \sqrt{c^2} = c, \text{ because } c \cdot c = c^2$$

b = 15 ft   c ft

a = 20 ft

Figure 8-6

The cable will be 25 feet long.

 SELF CHECK 5 How long must the cable be if the ground anchor is attached 15 feet from its base to a point 8 feet up the pole?

Perspective

Because of the Pythagorean theorem, the ancient Greeks knew that the lengths of the sides of a right triangle could be natural numbers. For example, a right triangle could have sides of lengths 3, 4, and 5, because $3^2 + 4^2 = 5^2$. Similarly, a right triangle could have sides of 5, 12, and 13, because $5^2 + 12^2 = 13^2$. Natural numbers a, b, and c that satisfy the equation $a^2 + b^2 = c^2$ are called Pythagorean triples. The triples 3, 4, 5, and 5, 12, 13 are two of infinitely many possibilities.

In 1637, the French mathematician Pierre de Fermat wrote a note in the margin of a book: There are no natural-number solutions a, b, and c to the equation $a^n + b^n = c^n$ if n is greater than 2. Fermat also mentioned that he had found a marvelous proof that wouldn't fit in the margin. Mathematicians have been trying to prove Fermat's last theorem ever since.

Princeton University mathematician Dr. Andrew Wiles first learned of Fermat's last theorem when he was 10 years old. He was so intrigued by the problem that he decided that he would study mathematics. Dr. Wiles worked on the problem at his Princeton home. "The problem was on my mind all the time," said Dr. Wiles. "When you are really desperate to find an answer, you can't let go." After seven years of concentrated work, Dr. Wiles announced his proof of Fermat's theorem in June of 1993.

Susana Raab/The New York Times
Dr. Andrew Wiles

EXAMPLE 6   **SAVING CABLE**   The builder of a high-ropes course wants to use a 25-foot cable to stabilize the pole shown in Figure 8-7. To be safe, the ground anchor must be more than 18 feet from the base of the pole. Is the cable long enough to use?

Solution   We can use the Pythagorean theorem, with $b = 16$ and $c = 25$.

$$c^2 = a^2 + b^2$$
$$25^2 = a^2 + 16^2 \quad \text{Substitute 25 for } c \text{ and 16 for } b.$$
$$625 = a^2 + 256 \quad \text{Square each term.}$$
$$369 = a^2 \quad \text{Subtract 256 from both sides.}$$
$$\sqrt{369} = \sqrt{a^2} \quad \text{Take the positive square root of both sides.}$$
$$a \approx 19.20937271 \quad \text{Use a calculator to find the approximate value of } \sqrt{369}.$$

$c = 25$ ft
$b = 16$ ft
a ft

Figure 8-7

Since the anchor must be more than 18 feet from the base, the cable is long enough.

 **SELF CHECK 6**   The builder wants to use a 20-foot cable to stabilize the pole but must have the ground anchor more than 17 feet from the base of the pole. It is attached to a point 15 feet up the pole. Is the cable long enough to use?

EXAMPLE 7   **REACH OF A LADDER**   A 26-foot ladder rests against the side of a building. If the base of the ladder is 10 feet from the wall, how far up the side of the building will the ladder reach?

Analyze the problem   We need to find the distance the ladder will reach up the wall. We let d represent this distance in feet.

Form an equation   The wall, the ground, and the ladder form a right triangle, as shown in Figure 8-8. In this triangle, the hypotenuse is 26 feet, and one of the legs is the base-to-wall distance of 10 feet. We can form the equation

The length of the hypotenuse squared	equals	the length of one leg squared	plus	the length of the other leg squared.
26^2	$=$	10^2	$+$	d^2

Solve the equation

$$26^2 = 10^2 + d^2$$
$$676 = 100 + d^2 \qquad \text{Square each term.}$$
$$676 - 100 = d^2 \qquad \text{Subtract 100 from both sides.}$$
$$576 = d^2 \qquad \text{Simplify.}$$
$$\sqrt{576} = \sqrt{d^2} \qquad \text{Take the positive square root of both sides.}$$
$$24 = d \qquad \text{Simplify.}$$

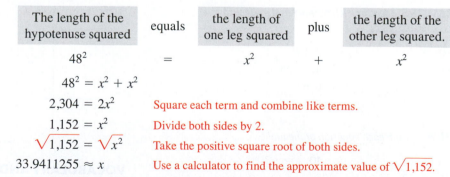

Figure 8-8

State the conclusion   The ladder will reach 24 feet up the side of the building.

 **SELF CHECK 7**   If the ladder is 25 feet and the base is 7 feet from the wall, how far up the side of the building will the ladder reach?

EXAMPLE 8   **MEASURING DISTANCE**   The gable end of the roof shown in Figure 8-9 is an isosceles right triangle with the 90° angle at the peak and a span of 48 feet. Find the distance from the eaves to the peak.

Analyze the problem   We can let x represent the length of each leg in feet, which is the distance from eaves to peak.

Form an equation   The two equal sides of the isosceles right triangle are the two legs of the right triangle, and the span of 48 feet is the length of the hypotenuse.
We can form the equation

Figure 8-9

The length of the hypotenuse squared	equals	the length of one leg squared	plus	the length of the other leg squared.
48^2	$=$	x^2	$+$	x^2

Solve the equation

$$48^2 = x^2 + x^2$$
$$2{,}304 = 2x^2 \qquad \text{Square each term and combine like terms.}$$
$$1{,}152 = x^2 \qquad \text{Divide both sides by 2.}$$
$$\sqrt{1{,}152} = \sqrt{x^2} \qquad \text{Take the positive square root of both sides.}$$
$$33.9411255 \approx x \qquad \text{Use a calculator to find the approximate value of } \sqrt{1{,}152}.$$

State the conclusion   The eaves-to-peak distance of the roof is approximately 34 feet.

 **SELF CHECK 8**   If the span of the gable is 52 feet, find the distance from the eaves to the peak.

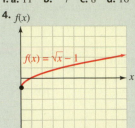
NOW TRY THIS

1. Graph: $f(x) = \sqrt{x - 1}$

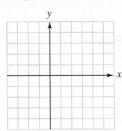

2. If $f(x) = \sqrt{7 - x}$, find
   a. $f(-2)$.   **b.** $f(7)$.

3. Between which two integers will the following be found on a number line?
   a. $\sqrt{5}$ **b.** $-\sqrt{20}$

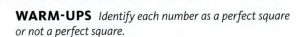

8.1 Exercises

WARM-UPS *Identify each number as a perfect square or not a perfect square.*

1. 100   **2.** 169
3. −8   **4.** −125
5. 20   **6.** 63
7. 64   **8.** 1

REVIEW *Graph each equation or inequality.*

9. $x = 3$ **10.** $y = -3$

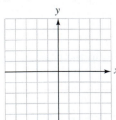

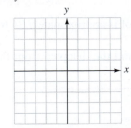

11. $-2x + y = 4$

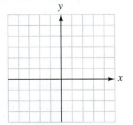

12. $4x - y > 4$

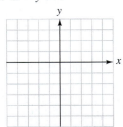

VOCABULARY AND CONCEPTS *Fill in the blanks.*

13. b is a square root of a if _____.
14. The symbol $\sqrt{}$ is called a _____. The symbol $\sqrt{x}$ is called a _____ expression.

15. The principal square root of a positive number is _____.
16. The expression under the radical sign is called the _____.
17. If a triangle has a 90° angle, it is called a ____ triangle.
18. The longest side of a ____ triangle is called the _____ and the other two sides are called legs.
19. The number 25 is an _____ and has ___ square roots. They are _ and ___.
20. $\sqrt{-11}$ is not a ____ number. It is called an _____ number.
21. The formula $A = s^2$ gives the area of a _____.
22. The principal square root of 0 is _.
23. If the length of the _____ of a right triangle is c and the legs are a and b, then $c^2 =$ _____. Numbers such as 3, 4, and 5 that satisfy the Pythagorean theorem are called Pythagorean _____.
24. If a and b are positive numbers and if $a = b$, then _____.

GUIDED PRACTICE *Find each square root. SEE EXAMPLE 1. (OBJECTIVE 1)*

25. $\sqrt{25}$
26. $\sqrt{16}$
27. $\sqrt{81}$
28. $\sqrt{100}$
29. $\sqrt{36}$
30. $\sqrt{64}$
31. $\sqrt{\dfrac{1}{9}}$
32. $\sqrt{\dfrac{1}{121}}$
33. $-\sqrt{4}$
34. $-\sqrt{121}$
35. $-\sqrt{100}$
36. $-\sqrt{196}$
37. $\sqrt{196}$
38. $\sqrt{169}$
39. $\sqrt{\dfrac{9}{256}}$
40. $\sqrt{\dfrac{49}{225}}$

 *Use a calculator to find an approximation of each square root to three decimal places. SEE EXAMPLE 2. (OBJECTIVE 2)*

41. $\sqrt{2}$
42. $\sqrt{3}$
43. $\sqrt{5}$
44. $\sqrt{10}$
45. $\sqrt{6}$
46. $\sqrt{8}$
47. $\sqrt{11}$
48. $\sqrt{17}$
49. $\sqrt{23}$
50. $\sqrt{53}$
51. $\sqrt{95}$
52. $\sqrt{99}$
53. $\sqrt{6,428}$
54. $\sqrt{4,444}$
55. $-\sqrt{9,876}$
56. $-\sqrt{3,619}$

Determine whether each number is rational, irrational, or imaginary. SEE EXAMPLE 3. (OBJECTIVE 3)

57. $\sqrt{25}$
58. $\sqrt{13}$
59. $\sqrt{36}$
60. $\sqrt{-36}$
61. $-\sqrt{8}$
62. $\sqrt{0}$
63. $\sqrt{-100}$
64. $-\sqrt{225}$

Graph each function. Check your work with a graphing calculator. SEE EXAMPLE 4. (OBJECTIVE 4)

65. $f(x) = 1 + \sqrt{x}$

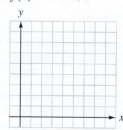

66. $f(x) = -1 + \sqrt{x}$

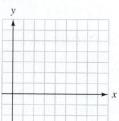

67. $f(x) = -\sqrt{x}$

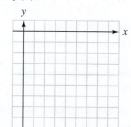

68. $f(x) = 1 - \sqrt{x}$

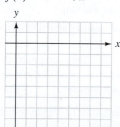

Refer to the right triangle in the illustration. Find the length of the unknown side. (OBJECTIVE 5)

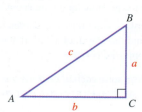

69. $a = 4$ and $b = 3$
70. $a = 6$ and $b = 8$
71. $a = 5$ and $b = 12$
72. $a = 15$ and $c = 17$
73. $a = 21$ and $c = 29$
74. $b = 16$ and $c = 34$
75. $b = 45$ and $c = 53$
76. $a = 14$ and $c = 50$

ADDITIONAL PRACTICE *Find each square root. If it is not exact, give a decimal approximation correct to three decimal places.*

77. $-\sqrt{289}$
78. $\sqrt{900}$
79. $\sqrt{10,000}$
80. $-\sqrt{2,500}$
81. $\sqrt{21.35}$
82. $\sqrt{13.78}$
83. $\sqrt{441}$
84. $-\sqrt{625}$
85. $\sqrt{0.3588}$
86. $\sqrt{0.9999}$
87. $\sqrt{0.9925}$
88. $\sqrt{0.12345}$
89. $-\sqrt{8,100}$
90. $\sqrt{4,900}$
91. $-\sqrt{0.8372}$
92. $-\sqrt{0.4279}$
93. $\sqrt{\dfrac{576}{625}}$
94. $\sqrt{\dfrac{196}{225}}$
95. $-\sqrt{4.6010}$
96. $-\sqrt{33.7212}$

If c is the hypotenuse of a right triangle, and a and b are the legs, find the unknown length.

97. $c = 125$ and $a = 44$

98. $a = 176$ and $b = 57$

 APPLICATIONS *Use a calculator to help solve each. If an answer is not exact, round it to the nearest tenth.* SEE EXAMPLE 5. (OBJECTIVE 5)

99. The formula $D = \sqrt{2h}$ can estimate the distance (in miles) to the horizon from a given height (in feet) above the water. If Kevin's eye-level view from an overlook is 98 feet above the ocean when he sees a ship on the horizon, how far is he from that ship?

100. A tsunami's speed can be estimated by the formula $S = 356\sqrt{d}$ where S is the speed (in kilometers per hour) and d is the average depth of the water (in kilometers). Find the speed of the tsunami if the average water depth is 0.4 kilometer.

Use a calculator to help solve each. If an answer is not exact, round it to the nearest tenth. SEE EXAMPLE 5. (OBJECTIVE 5)

101. Length of guy wires   A 20-foot-tall tower is secured by three guy wires fastened at the top and to anchors 15 feet from the base of the tower. How long is each guy wire?

102. Length of a path   A rectangular garden has sides of 28 and 45 feet. Find the length of a path that extends from one corner to the opposite corner.

Use a calculator to help solve each. If an answer is not exact, round it to the nearest tenth. SEE EXAMPLES 6–8. (OBJECTIVE 5)

103. Adjusting a ladder   A 20-foot ladder reaches a window 16 feet above the ground. How far from the wall is the base of the ladder?

104. Height of a pole   A 34-foot-long wire reaches from the top of a telephone pole to a point on the ground 16 feet from the base of the pole. Find the height of the pole.

105. Football   On first and ten, a coach tells his tight end to go out 6 yards, cut 45° to the right, and run 5 yards. (See the illustration.) The tight end follows instructions, catches a pass, and is tackled immediately. Does he gain the necessary 10 yards for a first down?

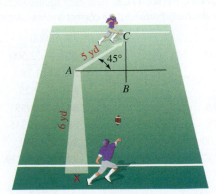

106. Geometry   The legs of a right triangle are equal, and the hypotenuse is $2\sqrt{2}$ units long. Find the length of each leg.

107. Geometry   The sides of a square are 3 feet long. Find the length of each diagonal of the square.

108. Perimeter of a square   The diagonal of a square is 3 feet long. Find its perimeter.

109. Baseball   A baseball diamond is a square, with each side 90 feet long. How far is it from home plate to second base?

110. TV   The size of a TV screen is the diagonal measure of its rectangular screen. To the nearest inch, how large is a screen if it is 37 inches wide and 20 inches high?

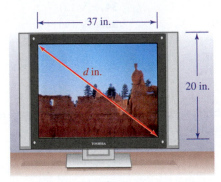

111. Finding locations   A woman drives 4.2 miles east and then 4.0 miles north. How far is she from her starting point?

112. Taking shortcuts   Instead of walking on the sidewalk, students take a diagonal shortcut across the vacant lot shown. How much distance do they save?

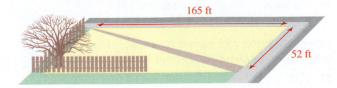

113. Carpentry A square-headed bolt is countersunk into a circular hole drilled in a wooden beam. The corners of the bolt must have $\frac{3}{8}$ inch clearance, as shown in the illustration. Find the diameter of the hole.

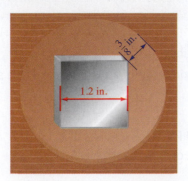

114. Designing tunnels The entrance to a one-way tunnel is a rectangle with a semicircular roof. Its dimensions are given in the illustration. How tall can a 10-foot-wide truck be without getting stuck in the tunnel?

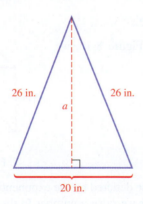

115. Altitude of a triangle Find the altitude of the isosceles triangle shown in the illustration.

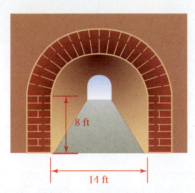

116. Geometry The square in the illustration is inscribed in a circle. The sides of the square are 6 inches long. Find the area of the circle.

117. Carpentry To span the 16-foot by 28-foot room shown in the illustration, a carpenter will use a scissors truss, with the ridge of the vaulted ceiling at the center of the room. The house plans call for the outside walls to be 8 feet high and the ridge of the room to be 12 feet high. How many 4-foot by 8-foot sheets of plaster board will be needed to drywall the entire ceiling?

118. Carpentry How many sheets of plasterboard are needed in Exercise 117 to drywall the entire inside of the building?

WRITING ABOUT MATH

119. Explain why the square root of a negative number cannot be a real number.

120. Explain the Pythagorean theorem.

SOMETHING TO THINK ABOUT

121. To generate Pythagorean triples, pick natural numbers for x and y ($x > y$). Let $a = 2xy$ and $b = x^2 - y^2$, and $c = x^2 + y^2$. Why do you always get a Pythagorean triple?

122. Can you find a Pythagorean triple with $b = 10$? Explain.

Section
8.2

*n*th Roots and Radicands That Contain Variables

Objectives

1. Find the cube root of a perfect cube.
2. Use a calculator to find an approximation of a given radical to a specified decimal place.
3. Graph a cube root function.
4. Find the *n*th root of a perfect *n*th power.
5. Simplify a radical expression that contains variables.

Vocabulary

perfect cubes   *n*th root   even root
cube root   index   odd root
integer cubes

Getting Ready

Find each value.

1. 2^3 2. 4^3 3. $(-5)^3$ 4. $\left(-\dfrac{1}{2}\right)^3$

5. 3^4 6. $\left(-\dfrac{1}{2}\right)^4$ 7. 2^5 8. 2^6

We will now reverse the cubing process and find cube roots of numbers. We also will find *n*th roots of numbers, where *n* is a natural number greater than 3.

1   Find the cube root of a perfect cube.

To find the volume V of the cube shown in Figure 8-10, we multiply its length, width, and height.

$$V = \boldsymbol{l} \cdot \boldsymbol{w} \cdot \boldsymbol{h}$$
$$V = \boldsymbol{5} \cdot \boldsymbol{5} \cdot \boldsymbol{5}$$
$$= 5^3$$
$$= 125$$

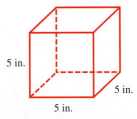

5 in.

5 in.

5 in.

Figure 8-10

The volume is 125 cubic inches.

We have seen that the product $5 \cdot 5 \cdot 5$ can be denoted by the exponential expression 5^3, where 5 is raised to the third power. Whenever we raise a number to the third power, we are cubing it, or finding its *cube*. This example illustrates that the formula for the volume of a cube with each side of length s is $V = s^3$.

Here are some additional **perfect cubes** of numbers.

- The cube of 3 is 27, because $3^3 = 27$.
- The cube of -3 is -27, because $(-3)^3 = -27$.
- The cube of $\frac{1}{3}$ is $\frac{1}{27}$, because $\left(\frac{1}{3}\right)^3 = \frac{1}{27}$.
- The cube of $-\frac{1}{3}$ is $-\frac{1}{27}$, because $\left(-\frac{1}{3}\right)^3 = -\frac{1}{27}$.
- The cube of 0 is 0, because $0^3 = 0$.

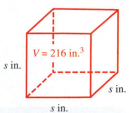

Suppose we know that the volume of the cube shown in Figure 8-11 is 216 cubic inches. To find the length of each side, we substitute 216 for V in the formula $V = s^3$ and solve for s.

$$V = s^3$$
$$216 = s^3$$

To solve for s, we must find a number whose cube is 216. Since 6 is the only number, the sides of the cube are 6 inches long. The number 6 is called a **cube root** of 216, because $6^3 = 216$.

Here are additional examples of cube roots.

- 3 is a cube root of 27, because $3^3 = 27$.
- -3 is a cube root of -27, because $(-3)^3 = -27$.
- $\frac{1}{3}$ is a cube root of $\frac{1}{27}$, because $\left(\frac{1}{3}\right)^3 = \frac{1}{27}$.
- $-\frac{1}{3}$ is a cube root of $-\frac{1}{27}$, because $\left(-\frac{1}{3}\right)^3 = -\frac{1}{27}$.
- 0 is a cube root of 0, because $0^3 = 0$.

In general, the following is true.

Figure 8-11

Pythagoras of Samos

569?–475?BC

Pythagoras is thought to be the world's first pure mathematician. Although he is famous for the theorem that bears his name, he often is called "the father of music," because a society he led discovered some of the fundamentals of musical harmony.

CUBE ROOTS

The number b is a cube root of a if $b^3 = a$.

All real numbers have one real cube root. As the previous examples show, a positive number has a positive cube root, a negative number has a negative cube root, and the cube root of 0 is 0.

CUBE ROOT NOTATION

The cube root of a is denoted by $\sqrt[3]{a}$.

$$\sqrt[3]{a} = b \qquad \text{if} \qquad b^3 = a.$$

EXAMPLE 1   Find each cube root.

   a. $\sqrt[3]{8} = 2$, because $2^3 = 8$.
   b. $\sqrt[3]{343} = 7$, because $7^3 = 343$.
   c. $\sqrt[3]{1{,}728} = 12$, because $12^3 = 1{,}728$.
   d. $\sqrt[3]{-8} = -2$, because $(-2)^3 = -8$.
   e. $\sqrt[3]{-125} = -5$, because $(-5)^3 = -125$.

 SELF CHECK 1   Find each cube root.

   a. $\sqrt[3]{64}$ **b.** $\sqrt[3]{27}$ **c.** $\sqrt[3]{216}$ **d.** $\sqrt[3]{-64}$ **e.** $\sqrt[3]{-216}$

EXAMPLE 2   Find each cube root.

   a. $\sqrt[3]{\dfrac{1}{8}} = \dfrac{1}{2}$, because $\left(\dfrac{1}{2}\right)^3 = \dfrac{1}{2} \cdot \dfrac{1}{2} \cdot \dfrac{1}{2} = \dfrac{1}{8}$.

   b. $\sqrt[3]{-\dfrac{125}{27}} = -\dfrac{5}{3}$, because $\left(-\dfrac{5}{3}\right)^3 = \left(-\dfrac{5}{3}\right)\left(-\dfrac{5}{3}\right)\left(-\dfrac{5}{3}\right) = -\dfrac{125}{27}$.

 SELF CHECK 2   Find each cube root.

   a. $\sqrt[3]{\dfrac{8}{27}}$ **b.** $\sqrt[3]{-\dfrac{8}{125}}$

2   Use a calculator to find an approximation of a given radical to a specified decimal place.

As we discussed with square roots, cube roots of numbers such as 7 are hard to compute by hand. However, we can find a decimal approximation of $\sqrt[3]{7}$ with a calculator.

Accent on technology

▶ Using a Calculator to Find Approximations of Cube Roots

To find an approximation of $\sqrt[3]{7}$ with a calculator, we can press these keys.

7 $\boxed{\sqrt[x]{y}}$ 3 $\boxed{=}$ Using a scientific calculator

$\boxed{\text{MATH}}$ 4 $\left(\sqrt[3]{(\ )}\,7\,\right)$ Using a TI84 graphing calculator

Either way, the result is approximately 1.912931183. To the nearest hundredth, $\sqrt[3]{7} \approx 1.91$.

If your scientific calculator does not have a $\boxed{\sqrt[x]{y}}$ key, you can use the $\boxed{y^x}$ key. We will see later that $\sqrt[3]{7} = 7^{1/3}$. To find the value of $7^{1/3}$, we press these keys.

7 $\boxed{y^x}$ (1 ÷ 3) $\boxed{=}$

For instructions regarding the use of a Casio graphing calculator, please refer to the Casio Keystroke Guide in the back of the book.

Numbers such as 8, 27, 64, 125, -8, and -125 are called **integer cubes**, because each one is the cube of an integer. The cube root of any integer cube is an integer, and therefore a rational number:

$$\sqrt[3]{8} = 2 \quad \sqrt[3]{27} = 3 \quad \sqrt[3]{64} = 4 \quad \sqrt[3]{125} = 5 \quad 3\sqrt{-8} = -2 \quad \text{and} \quad \sqrt[3]{-125} = -5$$

Cube roots of integers that are not integer cubes are irrational numbers. For example, $\sqrt[3]{4}$ and $\sqrt[3]{10}$ are irrational numbers.

3   Graph a cube root function.

Since every real number has one real-number cube root, there is a cube root function $f(x) = \sqrt[3]{x}$.

EXAMPLE 3   Graph: $f(x) = \sqrt[3]{x}$

Solution   To graph this function, we substitute numbers for x, compute $f(x)$, plot the resulting ordered pairs, and connect them with a smooth curve, as shown in Figure 8-12(a). A calculator graph is shown in Figure 8-12(b).

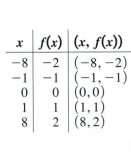

x	$f(x)$	$(x, f(x))$
-8	-2	$(-8, -2)$
-1	-1	$(-1, -1)$
0	0	$(0, 0)$
1	1	$(1, 1)$
8	2	$(8, 2)$

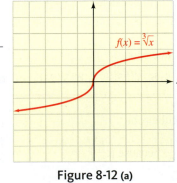

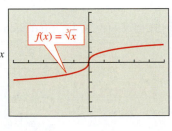

Figure 8-12 (a)   **(b)**

SELF CHECK 3   Graph: $f(x) = \sqrt[3]{x} - 1$

4 Find the *n*th root of a perfect *n*th power.

Just as there are square roots and cube roots, there are also fourth roots, fifth roots, sixth roots, and so on. In general,

*n*TH ROOT OF *a*	The **nth root of a** is denoted by $\sqrt[n]{a}$ and

$$\sqrt[n]{a} = b \quad \text{if} \quad b^n = a$$

The number *n* is called the **index** of the radical.
   If *n* is an even natural number, *a* must be positive or 0.

In the square root symbol $\sqrt{}$, the unwritten index is understood to be 2.

$$\sqrt{a} = \sqrt[2]{a}$$

EXAMPLE 4 Find each root.

   a. $\sqrt[4]{81} = 3$, because $3^4 = 81$.
   b. $\sqrt[5]{32} = 2$, because $2^5 = 32$.
   c. $\sqrt[5]{-32} = -2$, because $(-2)^5 = -32$.
   d. $\sqrt[4]{-81}$ is not a real number, because no real number raised to the fourth power is -81.

 SELF CHECK 4 Find each root.

   a. $\sqrt[4]{16}$ **b.** $\sqrt[5]{243}$ **c.** $\sqrt[5]{-1,024}$

EXAMPLE 5 Find each root.

   a. $\sqrt[4]{\dfrac{1}{81}} = \dfrac{1}{3}$, because $\left(\dfrac{1}{3}\right)^4 = \dfrac{1}{81}$.

   b. $\sqrt[5]{-\dfrac{32}{243}} = -\dfrac{2}{3}$, because $\left(-\dfrac{2}{3}\right)^5 = -\dfrac{32}{243}$.

 SELF CHECK 5 Find each root.   **a.** $\sqrt[4]{\dfrac{1}{16}}$ **b.** $\sqrt[5]{-\dfrac{243}{32}}$

Accent
on technology

▸ Using a Calculator to Find
Approximations of Roots
That Are Not Square
Roots or Cube Roots

To use a TI84 graphing calculator to find an approximation of $\sqrt[6]{512}$, we press these keys:

   6 **MATH** 5 $\left(\sqrt[x]{}\right)$ 512 **ENTER**

The result is approximately 2.828427125.

   In general, to find an approximation for $\sqrt[n]{x}$, we press these keys:

   n **MATH** 5 *x* **ENTER**

For instructions regarding the use of a Casio graphing calculator, please refer to the Casio Keystroke Guide in the back of the book.

---

5 Simplify a radical expression that contains variables.

When *n* is even and $x \geq 0$, we say that the radical $\sqrt[n]{x}$ represents an **even root**. We can find even roots of many quantities that contain variables, provided that these variables represent positive numbers or 0.

EXAMPLE 6   Assume that each variable represents a positive number or 0 and find each root.

    a. $\sqrt{x^2} = x$, because $x^2 = x^2$.

    b. $\sqrt{x^4} = x^2$, because $(x^2)^2 = x^4$.

    c. $\sqrt{x^4 y^2} = x^2 y$, because $(x^2 y)^2 = x^4 y^2$.

    d. $\sqrt[4]{x^{12} y^8} = x^3 y^2$, because $(x^3 y^2)^4 = x^{12} y^8$.

    e. $\sqrt[6]{64 p^{18} q^{12}} = 2p^3 q^2$, because $(2p^3 q^2)^6 = 64 p^{18} q^{12}$.

 SELF CHECK 6   Assume that each variable represents a positive number of 0 and find each root.

    a. $\sqrt{a^4 b^2}$ **b.** $\sqrt{16 m^6 n^8}$ **c.** $\sqrt[4]{16 x^4 y^8}$

When n is odd, we say that the radical $\sqrt[n]{x}$ represents an **odd root**.

EXAMPLE 7   Find each root.

    a. $\sqrt[3]{x^6 y^3} = x^2 y$, because $(x^2 y)^3 = x^6 y^3$.

    b. $\sqrt[5]{32 x^{10} y^5} = 2x^2 y$, because $(2x^2 y)^5 = 32 x^{10} y^5$.

 SELF CHECK 7   Find each root.   **a.** $\sqrt[3]{-27 p^6}$ **b.** $\sqrt[5]{\dfrac{1}{32} m^{10} n^{15}}$

Accent on technology

▸ Radius of a Water Tank

Engineers want to design a spherical tank that will hold 33,500 cubic feet of water. They know that the formula for the radius r of a sphere with volume V is

$$r = \sqrt[3]{\dfrac{3V}{4\pi}} \quad \text{where } \pi \approx 3.141592654.$$

To use a calculator to find the radius r, they will substitute 33,500 for V and enter these numbers and press these keys:

3 ✕ 33,500 ÷ (4 ✕ π) = $\sqrt[x]{y}$ 3 =   *Using a scientific calculator*

MATH 4 (3 ✕ 33,500) ÷ (4 **2ND** ^ (π))) **ENTER**   *Using a TI84 graphing calculator*

Either way, the result is approximately 19.99794636. The engineers should design a tank with a radius of 20 feet.

For instructions regarding the use of a Casio graphing calculator, please refer to the Casio Keystroke Guide in the back of the book.

 SELF CHECK ANSWERS

1. a. 4   **b.** 3   **c.** 6   **d.** -4 **e.** -6 **2. a.** $\frac{2}{3}$ **b.** $-\frac{2}{5}$ **3.**

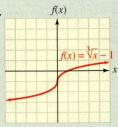

4. a. 2   **b.** 3   **c.** -4 **5. a.** $\frac{1}{2}$ **b.** $-\frac{3}{2}$ **6. a.** $a^2 b$ **b.** $4m^3 n^4$ **c.** $2xy^2$ **7. a.** $-3p^2$ **b.** $\frac{1}{2} m^2 n^3$

> **NOW TRY THIS**
>
> 1. **a.** Simplify: $\sqrt{(x+3)^2}$ **b.** Find the restriction on the variable.
>
> *Hint*: If the index is even, the radicand must be nonnegative.
>
> 2. Simplify: $-\sqrt[3]{-8x^6}$
>
> 3. Simplify: $\sqrt[3]{(-4x^2)^3}$

8.2 Exercises

WARM-UPS *Fill in the blanks.*

1. $2^3 = 8$, so the cube root of 8 is _.
2. $4^3 = 64$, so the cube root of 64 is _.
3. $(-5)^3 = -125$, so the cube root of -125 is ___.
4. $(-3)^3 = -27$, so the cube root of -27 is ___.
5. $3^4 = 81$, so the fourth root of 81 is _.
6. $2^4 = 16$, so the fourth root of 16 is _.
7. $3^5 = 243$, so the fifth root of 243 is _.
8. $2^5 = 32$, so the fifth root of 32 is _.

REVIEW *If $f(x) = 2x^2 - x - 1$, find each value.*

9. $f(-1)$
10. $f(3)$
11. $f(-2)$
12. $f(-t)$

Factor each expression.

13. $x^2 - 49y^2$
14. $x^2 - 3x - 10$
15. $ax + ay + bx + by$
16. $2ax^2 + 2ax - 40a$

VOCABULARY AND CONCEPTS *Fill in the blanks.*

17. If $p^3 = q$, p is called the _____ of q.
18. If $p^4 = q$, p is called a _____ of q.
19. If $p^n = q$, p is called an _____ of q.
20. In the notation $\sqrt[n]{x}$, n is called the _____, and x is called the _____.
21. Numbers such as 8, 27, -64, and -125 are called _____.
22. If the index of a radical is an even number, the root is called an ____ root.
23. If the index of a radical is an odd number, the root is called an ___ root.
24. The formula for the volume of a cube with sides s units long is _____.
25. $\sqrt[n]{a} = b$ if _____.
26. $\sqrt[5]{\dfrac{32x^5}{243}} = \dfrac{2x}{3}$, because _____.

GUIDED PRACTICE *Find each cube root. SEE EXAMPLE 1. (OBJECTIVE 1)*

27. $\sqrt[3]{27}$
28. $\sqrt[3]{64}$

29. $\sqrt[3]{-64}$
30. $\sqrt[3]{-125}$
31. $-\sqrt[3]{-64}$
32. $\sqrt[3]{-27}$
33. $\sqrt[3]{125}$
34. $\sqrt[3]{1{,}000}$
35. $-\sqrt[3]{-8}$
36. $-\sqrt[3]{-216}$
37. $-\sqrt[3]{64}$
38. $-\sqrt[3]{343}$

Find each cube root. SEE EXAMPLE 2. (OBJECTIVE 1)

39. $\sqrt[3]{\dfrac{1}{64}}$
40. $\sqrt[3]{\dfrac{27}{125}}$
41. $\sqrt[3]{-\dfrac{64}{27}}$
42. $\sqrt[3]{-\dfrac{8}{125}}$

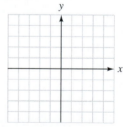 *Use a calculator to find an approximation of each root. Give each answer to the nearest hundredth. (OBJECTIVE 2)*

43. $\sqrt[3]{32{,}100}$
44. $\sqrt[3]{-25{,}713}$
45. $\sqrt[3]{-0.11324}$
46. $\sqrt[3]{0.875}$
47. $\sqrt[4]{125}$
48. $\sqrt[5]{12{,}450}$
49. $\sqrt[5]{-6{,}000}$
50. $\sqrt[6]{0.5}$

Graph each function. SEE EXAMPLE 3. (OBJECTIVE 3)

51. $f(x) = 1 + \sqrt[3]{x}$
52. $f(x) = -\sqrt[3]{x}$

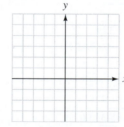

53. $f(x) = \sqrt[3]{x} - 2$
54. $f(x) = \sqrt[3]{x} + 2$

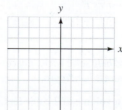
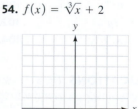

Find each value. If not a real number, so state. SEE EXAMPLE 4.
(OBJECTIVE 4)

55. $\sqrt[4]{16}$
56. $\sqrt[4]{81}$

57. $-\sqrt[5]{32}$
58. $-\sqrt[5]{243}$

59. $\sqrt[4]{-16}$
60. $\sqrt[6]{0}$

61. $\sqrt[5]{-32}$
62. $\sqrt[6]{-64}$

Find each value. SEE EXAMPLE 5. (OBJECTIVE 4)

63. $\sqrt[4]{\dfrac{81}{16}}$
64. $\sqrt[4]{\dfrac{625}{16}}$

65. $\sqrt[5]{-\dfrac{1}{32}}$
66. $\sqrt[5]{-\dfrac{32}{243}}$

Write each expression without a radical sign. Assume all variables represent positive numbers or 0. SEE EXAMPLE 6. (OBJECTIVE 5)

67. $\sqrt{x^2y^2}$
68. $\sqrt{x^4y^6}$

69. $\sqrt{x^4z^4}$
70. $\sqrt{y^6z^8}$

71. $-\sqrt{25x^4z^{12}}$
72. $-\sqrt{16x^2y^4}$

73. $\sqrt[4]{16z^4}$
74. $\sqrt[4]{81t^{12}}$

75. $-\sqrt[4]{x^4y^4z^8}$
76. $-\sqrt[4]{a^{16}b^{12}c^4}$

77. $\sqrt[6]{64x^{12}y^6}$
78. $-\sqrt[6]{x^{18}y^{12}}$

Write each expression without a radical sign. Assume all variables represent positive numbers or 0. SEE EXAMPLE 7. (OBJECTIVE 5)

79. $\sqrt[3]{x^3y^3z^3}$
80. $\sqrt[3]{x^6y^9z^{12}}$

81. $-\sqrt[5]{32x^{15}y^5}$
82. $-\sqrt[5]{100{,}000a^{15}b^{10}}$

ADDITIONAL PRACTICE *Simplify each radical expression. If the answer is not exact, round to the nearest hundredth. All variables represent positive values.*

83. $-\sqrt{81z^4}$
84. $-\sqrt{729x^8y^2}$

85. $\sqrt[5]{32}$
86. $\sqrt[5]{-1}$

87. $\sqrt[5]{\dfrac{1}{32}}$
88. $\sqrt[5]{-\dfrac{1}{32}}$

89. $\sqrt[6]{26.808}$
90. $\sqrt[5]{-39{,}628.27}$

91. $\sqrt[3]{729}$
92. $\sqrt[3]{512}$

93. $\sqrt[3]{27y^3z^6}$
94. $\sqrt[3]{64x^3y^6z^9}$

95. $\sqrt[3]{-1{,}331}$
96. $\sqrt[3]{-1{,}728}$

97. $\sqrt[3]{17{,}313}$
98. $\sqrt[3]{-5{,}313.6}$

99. $\sqrt{36z^{36}}$
100. $\sqrt{64y^{64}}$

101. $\sqrt[3]{-8p^6q^3}$
102. $\sqrt[3]{-r^{12}s^3t^6}$

103. $\sqrt[4]{16a^8b^{12}c^{16}}$
104. $\sqrt[5]{-32x^5y^{15}z^{10}}$

 APPLICATIONS *Use a calculator to help solve each. If necessary, approximate each answer to the nearest hundredth.*

105. Packaging If a cubical box has a volume of 2 cubic feet, how long is each side?

106. Hot air balloons If a hot air balloon approximates the shape of a sphere and has a volume of 15,000 cubic feet, how long is its radius?

© gary718/Shutterstock.com

107. Windmills The power generated by a certain windmill is related to the speed of the wind by the formula

$$S = \sqrt[3]{\dfrac{P}{0.02}}$$

where S is the speed of the wind (in mph) and P is the power (in watts). Find the speed of the wind when the windmill is producing 400 watts of power.

© Carole Castelli/Shutterstock.com

108. Astronomy Johannes Kepler discovered that a planet's mean distance R from the Sun (in astronomical units) is related to its period T (in years) by the formula

$$R = \sqrt[3]{\dfrac{T^2}{k}}$$

Find R when $T = 1.881$ and $k = 1.002$.

109. Geometric sequences The common ratio of a geometric sequence with four terms is given by the formula

$$r = \sqrt[3]{\dfrac{l}{a}}$$

where a is the first term and l is the last term. Find the common ratio of a geometric sequence that has a first term of 3 and a last term of 192.

110. Business The interest rate I after five compoundings is given by the formula

$$\sqrt[5]{\dfrac{FV}{PV}} - 1 = I$$

where FV is the future value and PV is the present value. Find the interest rate I if an investment of \$1,000 grows to \$1,338.23.

112. Explain why a negative number cannot have a real number for its fourth root.

WRITING ABOUT MATH

111. Explain why a negative number can have a real number for its cube root.

SOMETHING TO THINK ABOUT

113. Is $\sqrt{x^2 - 4x + 4} = x - 2$? What are the exceptions?

114. When is $\sqrt{x^2} \neq x$?

Section 8.3

Simplifying Radical Expressions

Objectives

1 Simplify a radical expression using the multiplication property of radicals.
2 Simplify a radical expression using the division property of radicals.
3 Simplify a cube root expression.

Vocabulary

multiplication property of radicals   division property of radicals

Getting Ready

Simplify each radical. Assume that all variables represent positive numbers.

1. $\sqrt{100}$ **2.** $\sqrt{4}$ **3.** $\sqrt{25}$ **4.** $\sqrt{144}$

5. $\sqrt{9x^2}$ **6.** $\sqrt{16x^4}$ **7.** $\sqrt[3]{27x^3y^6}$ **8.** $\sqrt[3]{-8x^6y^9}$

In this section, we will introduce the multiplication and division properties of radicals. These properties can be used to simplify some radical expressions.

1 **Simplify a radical expression using the multiplication property of radicals.**

We introduce the first of two properties of radicals with the following examples:

$$\sqrt{4 \cdot 25} = \sqrt{100} \qquad\qquad \sqrt{4}\sqrt{25} = 2 \cdot 5$$
$$= 10 \qquad\qquad\qquad\qquad\qquad = 10$$

In each case, the answer is 10. Thus, $\sqrt{4 \cdot 25} = \sqrt{4}\sqrt{25}$. Likewise,

$$\sqrt{9 \cdot 16} = \sqrt{144} \qquad\qquad \sqrt{9}\sqrt{16} = 3 \cdot 4$$
$$= 12 \qquad\qquad\qquad\qquad\qquad = 12$$

In each case, the answer is 12. Thus, $\sqrt{9 \cdot 16} = \sqrt{9}\sqrt{16}$. These results suggest the **multiplication property of radicals**.

> **MULTIPLICATION PROPERTY OF RADICALS**
>
> If $a \geq 0$ and $b \geq 0$, then
> $$\sqrt{ab} = \sqrt{a}\sqrt{b}$$

In words, *the square root of the product of two nonnegative numbers is equal to the product of their square roots.*

A square root radical is in simplified form when each of the following statements is true.

> **SIMPLIFIED FORM OF A SQUARE ROOT RADICAL**
>
> 1. Except for 1, the radicand has no perfect-square factors.
> 2. No fraction appears in a radicand.
> 3. No radical appears in the denominator of a fraction.

We can use the multiplication property of radicals to simplify radicals that have perfect-square factors. For example, we can simplify $\sqrt{12}$ as follows:

$$\sqrt{12} = \sqrt{4 \cdot 3} \qquad \text{Factor 12 as } 4 \cdot 3 \text{, because 4 is a perfect square.}$$
$$= \sqrt{4}\sqrt{3} \qquad \text{Use the multiplication property of radicals.}$$
$$= 2\sqrt{3} \qquad \text{Simplify.}$$

To simplify more difficult radicals, we need to know the integers that are perfect squares. For example, 81 is a perfect square, because $9^2 = 81$. The first 20 integer squares are

$$1, 4, 9, 16, 25, 36, 49, 64, 81, 100, 121, 144, 169, 196, 225, 256, 289, 324, 361, 400$$

Expressions with variables also can be perfect squares. For example, $9x^4y^2$ is a perfect square, because

$$9x^4y^2 = (3x^2y)^2$$

EXAMPLE 1 Simplify: $\sqrt{72x^3}$ $(x \geq 0)$

Solution We factor $72x^3$ into two factors, one of which is the greatest perfect square that divides $72x^3$. Since

- 36 is the greatest perfect square that divides 72, and
- x^2 is the greatest perfect square that divides x^3,

the greatest perfect square that divides $72x^3$ is $36x^2$. We can now use the multiplication property of radicals and simplify to get

$$\sqrt{72x^3} = \sqrt{36x^2 \cdot 2x}$$
$$= \sqrt{36x^2}\sqrt{2x} \qquad \text{The square root of a product is equal to the product of the square roots.}$$
$$= 6x\sqrt{2x} \qquad \text{Simplify.}$$

 **SELF CHECK 1** Simplify: $\sqrt{50y^3}$ $(y \geq 0)$

EXAMPLE 2 Simplify: $\sqrt{45x^2y^3}$ $(x \geq 0, y \geq 0)$

Solution We look for the greatest perfect square that divides $45x^2y^3$. Because

- 9 is the greatest perfect square that divides 45,
- x^2 is the greatest perfect square that divides x^2, and
- y^2 is the greatest perfect square that divides y^3,

the factor $9x^2y^2$ is the greatest perfect square that divides $45x^2y^3$.

We now can use the multiplication property of radicals and simplify to get

$$\sqrt{45x^2y^3} = \sqrt{9x^2y^2 \cdot 5y}$$

$$= \sqrt{\mathbf{9x^2y^2}}\sqrt{5y} \qquad \text{The square root of a product is equal to the product of the square roots.}$$

$$= \mathbf{3xy}\sqrt{5y} \qquad \text{Simplify.}$$

🌿 **SELF CHECK 2**   Simplify: $\sqrt{63a^3b^2}$ $(a \geq 0, b \geq 0)$

EXAMPLE 3   Simplify: $3a\sqrt{288a^5b^7}$ $(a \geq 0, b \geq 0)$

Solution   We look for the greatest perfect square that divides $288a^5b^7$. Because

- 144 is the greatest perfect square that divides 288,
- a^4 is the greatest perfect square that divides a^5, and
- b^6 is the greatest perfect square that divides b^7,

the factor $144a^4b^6$ is the greatest perfect square that divides $288a^5b^7$.

We now can use the multiplication property of radicals and simplify to get

$$3a\sqrt{288a^5b^7} = 3a\sqrt{144a^4b^6 \cdot 2ab}$$

$$= 3a\sqrt{\mathbf{144a^4b^6}}\sqrt{2ab} \qquad \text{The square root of a product is equal to the product of the square roots.}$$

$$= 3a\left(\mathbf{12a^2b^3}\sqrt{2ab}\right) \qquad \text{Simplify.}$$

$$= 36a^3b^3\sqrt{2ab} \qquad \text{Multiply.}$$

🌿 **SELF CHECK 3**   Simplify: $5p\sqrt{300p^3q^9}$ $(p \geq 0, q \geq 0)$

---

2   **Simplify a radical expression using the division property of radicals.**

To find the second property of radicals, we consider these examples.

$$\sqrt{\frac{100}{25}} = \sqrt{4} \qquad\qquad \frac{\sqrt{100}}{\sqrt{25}} = \frac{10}{5}$$

$$= \mathbf{2} \qquad\qquad\qquad\qquad = \mathbf{2}$$

Since the answer is 2 in each case, $\sqrt{\frac{100}{25}} = \frac{\sqrt{100}}{\sqrt{25}}$. Likewise,

$$\sqrt{\frac{36}{4}} = \sqrt{9} \qquad\qquad \frac{\sqrt{36}}{\sqrt{4}} = \frac{6}{2}$$

$$= \mathbf{3} \qquad\qquad\qquad\qquad = \mathbf{3}$$

Since the answer is 3 in each case, $\sqrt{\frac{36}{4}} = \frac{\sqrt{36}}{\sqrt{4}}$. These results suggest the **division property of radicals**.

DIVISION PROPERTY OF RADICALS   If $a \geq 0$ and $b > 0$, then

$$\sqrt{\frac{a}{b}} = \frac{\sqrt{a}}{\sqrt{b}}$$

In words, *the square root of the quotient of a nonnegative number and a positive number is the quotient of their square roots.*

We can use the division property of radicals to simplify radicals that have fractions in their radicands. For example,

$$\sqrt{\frac{59}{49}} = \frac{\sqrt{59}}{\sqrt{49}}$$

$$= \frac{\sqrt{59}}{7} \qquad \textcolor{red}{\sqrt{49} = 7}$$

EXAMPLE 4   Simplify: $\sqrt{\dfrac{108}{25}}$

Solution $\sqrt{\dfrac{108}{25}} = \dfrac{\sqrt{108}}{\sqrt{25}}$ The square root of a quotient is equal to the quotient of the square roots.

$\qquad\qquad = \dfrac{\sqrt{36 \cdot 3}}{5}$ Factor 108 using the factorization involving 36, the largest perfect-square factor of 108, and write $\sqrt{25}$ as 5.

$\qquad\qquad = \dfrac{\sqrt{36}\sqrt{3}}{5}$ The square root of a product is equal to the product of the square roots.

$\qquad\qquad = \dfrac{6\sqrt{3}}{5}$

SELF CHECK 4   Simplify: $\sqrt{\dfrac{20}{49}}$

EXAMPLE 5   Simplify: $\sqrt{\dfrac{44x^3}{9xy^2}}$ $(x > 0, y > 0)$

Solution $\sqrt{\dfrac{44x^3}{9xy^2}} = \sqrt{\dfrac{44x^2}{9y^2}}$ Simplify the fraction by dividing out the common factor of x.

$\qquad\qquad = \dfrac{\sqrt{44x^2}}{\sqrt{9y^2}}$ The square root of a quotient is equal to the quotient of the square roots.

$\qquad\qquad = \dfrac{\sqrt{4x^2}\sqrt{11}}{\sqrt{9y^2}}$ The square root of a product is equal to the product of the square roots.

$\qquad\qquad = \dfrac{2x\sqrt{11}}{3y}$ Simplify.

$\qquad\qquad = \dfrac{2x\sqrt{11}}{3y}$

SELF CHECK 5   Simplify: $\sqrt{\dfrac{99b^3}{16a^2b}}$ $(a > 0, b > 0)$

3   Simplify a cube root expression.

The multiplication and division properties of radicals are also true for cube roots and higher. To simplify a cube root, it is helpful to know the following integer cubes:

1, 8, 27, 64, 125, 216, 343, 512, 729, 1,000

Expressions with variables can also be perfect cubes. For example, $27x^6y^3$ is a perfect cube, because

$$27x^6y^3 = (3x^2y)^3$$

EXAMPLE 6   Simplify: **a.** $\sqrt[3]{16x^3y^4}$ **b.** $\sqrt[3]{\dfrac{64n^4}{27m^3}}$ $(m \neq 0)$

Solution   **a.** We look for the greatest perfect cube that divides $16x^3y^4$. Because

- 8 is the greatest perfect cube that divides 16,
- x^3 is the greatest perfect cube that divides x^3, and
- y^3 is the greatest perfect cube that divides y^4,

the greatest perfect cube that divides $16x^3y^4$ is $8x^3y^3$. We now can use the multiplication property of radicals to obtain

$$\sqrt[3]{16x^3y^4} = \sqrt[3]{8x^3y^3 \cdot 2y}$$

$$= \sqrt[3]{8x^3y^3}\,\sqrt[3]{2y} \qquad \text{The cube root of a product is equal to the product of the cube roots.}$$

$$= 2xy\sqrt[3]{2y} \qquad \text{Simplify.}$$

b. $\sqrt[3]{\dfrac{64n^4}{27m^3}} = \dfrac{\sqrt[3]{64n^4}}{\sqrt[3]{27m^3}} \qquad$ The cube root of a quotient is equal to the quotient of the cube roots.

$$= \dfrac{\sqrt[3]{64n^3}\,\sqrt[3]{n}}{3m} \qquad \text{Use the multiplication property of radicals, and write } \sqrt[3]{27m^3} \text{ as } 3m.$$

$$= \dfrac{4n\sqrt[3]{n}}{3m} \qquad \text{Simplify.}$$

SELF CHECK 6   Simplify: **a.** $\sqrt[3]{54a^3b^5}$ **b.** $\sqrt[3]{\dfrac{27q^5}{64p^3}}$ $(p \neq 0)$

COMMENT   Note that $\sqrt{a + b} \neq \sqrt{a} + \sqrt{b}$ and $\sqrt{a - b} \neq \sqrt{a} - \sqrt{b}$. To see that this is true, we consider these correct simplifications:

$$\sqrt{9 + 16} = \sqrt{25} = 5 \qquad \text{and} \qquad \sqrt{25 - 16} = \sqrt{9} = 3$$

Since the radical sign is a grouping symbol, the order of operations requires that we perform the operations under the radicals first.

Remember that it is incorrect to write

$$\sqrt{9 + 16} = \sqrt{9} + \sqrt{16} \qquad\qquad \sqrt{25 - 16} = \sqrt{25} - \sqrt{16}$$
$$= 3 + 4 \qquad\qquad\qquad\qquad = 5 - 4$$
$$= 7 \qquad\qquad\qquad\qquad\qquad = 1$$

SELF CHECK ANSWERS   **1.** $5y\sqrt{2y}$ **2.** $3ab\sqrt{7a}$ **3.** $50p^2q^4\sqrt{3pq}$ **4.** $\dfrac{2\sqrt{5}}{7}$ **5.** $\dfrac{3b\sqrt{11}}{4a}$ **6. a.** $3ab\sqrt[3]{2b^2}$ **b.** $\dfrac{3q\sqrt[3]{q^2}}{4p}$

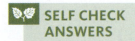

NOW TRY THIS

1. Given a right triangle with hypotenuse c, if $b = 5$ ft and $c = 15$ ft, find the length of a. Give the answer as a radical in simplest form.
2. Simplify (assume the radicand ≥ 0): $\sqrt{x^2 + 10x + 25}$
3. Simplify (assume the radicand ≥ 0): $\sqrt{x^3 + 3x^2 - 9x - 27}$

8.3   Exercises

WARM-UPS *Factor each value so that one factor is a perfect square.*

1. 45

2. 24

3. x^3

4. a^5

Factor each value so that one factor is a perfect cube.

5. 54

6. 24

REVIEW *Simplify each fraction. Assume no division by 0.*

7. $\dfrac{5xy^2z^3}{10x^2y^2z^4}$

8. $\dfrac{35a^3b^2c}{63a^2b^3c^2}$

9. $\dfrac{a^2 - a - 2}{a^2 + a - 6}$

10. $\dfrac{y^2 + 3y - 18}{y^2 - 9}$

VOCABULARY AND CONCEPTS *Fill in the blanks.*

11. Integers such as 4, 9, and 25 are called _____ squares.

12. Integers such as 8, 27, and 125 are called perfect _____.

13. $\sqrt{ab} =$ _____ $(a \geq 0, b \geq 0)$, and this is the _____ property of radicals.

14. $\sqrt{\dfrac{a}{b}} =$ __ $(a \geq 0, b > 0)$, and this is the _____ property of radicals.

Find the error in each solution.

15. $\begin{aligned}\sqrt{13} &= \sqrt{9 + 4} \\ &= \sqrt{9} + \sqrt{4} \\ &= 3 + 2 \\ &= 5\end{aligned}$

16. $\begin{aligned}\sqrt{7} &= \sqrt{16 - 9} \\ &= \sqrt{16} - \sqrt{9} \\ &= 4 - 3 \\ &= 1\end{aligned}$

GUIDED PRACTICE *Simplify each radical. SEE EXAMPLE 1. (OBJECTIVE 1)*

17. $\sqrt{12}$

18. $\sqrt{18}$

19. $\sqrt{45}$

20. $\sqrt{75}$

21. $\sqrt{98}$

22. $\sqrt{200}$

23. $\sqrt{150}$

24. $\sqrt{72}$

25. $\sqrt{27x^2}$

26. $\sqrt{54m^2}$

27. $\sqrt{48b^3}$

28. $\sqrt{128c^3}$

29. $\sqrt{192p^5}$

30. $\sqrt{250q^4}$

31. $\sqrt{88r^7}$

32. $\sqrt{275n^5}$

Simplify each radical. Assume that all variables represent positive numbers. SEE EXAMPLE 2. (OBJECTIVE 1)

33. $\sqrt{9x^2y}$

34. $\sqrt{16xy^2}$

35. $\sqrt{8a^4b}$

36. $\sqrt{20xy^4}$

Simplify each radical. Assume that all variables represent positive numbers. SEE EXAMPLE 3. (OBJECTIVE 1)

37. $4\sqrt{288}$

38. $2\sqrt{800}$

39. $2x\sqrt{245x^2y}$

40. $-4\sqrt{243}$

41. $2\sqrt{245}$

42. $3\sqrt{196}$

43. $3xy\sqrt{196xy^3}$

44. $-4x^5y^3\sqrt{36x^3y^3}$

Simplify each radical. SEE EXAMPLE 4. (OBJECTIVE 2)

45. $\sqrt{\dfrac{25}{9}}$

46. $\sqrt{\dfrac{36}{49}}$

47. $\sqrt{\dfrac{81}{64}}$

48. $\sqrt{\dfrac{121}{144}}$

49. $\sqrt{\dfrac{24}{25}}$

50. $\sqrt{\dfrac{8}{49}}$

51. $\sqrt{\dfrac{20}{49}}$

52. $\sqrt{\dfrac{50}{9}}$

53. $\sqrt{\dfrac{48}{81}}$

54. $\sqrt{\dfrac{27}{64}}$

55. $\sqrt{\dfrac{18}{121}}$

56. $\sqrt{\dfrac{300}{49}}$

Simplify each expression. All variables represent positive numbers. SEE EXAMPLE 5. (OBJECTIVE 2)

57. $\sqrt{\dfrac{50x^3y^2}{a^4b^2}}$

58. $\sqrt{\dfrac{108a^3b^2}{c^2d^4}}$

59. $\sqrt{\dfrac{125m^2n^5}{64n}}$

60. $\sqrt{\dfrac{72p^5q^7}{16pq^3}}$

61. $\sqrt{\dfrac{128m^3n^5}{36mn^7}}$

62. $\sqrt{\dfrac{75p^3q^2}{9p^5q^4}}$

63. $\sqrt{\dfrac{90a^5b^6}{32ab^4}}$

64. $\sqrt{\dfrac{189a^8b^2}{12a^2b^8}}$

Simplify each cube root. Assume no division by 0. SEE EXAMPLE 6. (OBJECTIVE 3)

65. $\sqrt[3]{125a^3}$

66. $\sqrt[3]{27x^3y^3}$

67. $\sqrt[3]{-64x^5}$

68. $\sqrt[3]{-16x^4y^3}$

69. $\sqrt[3]{\dfrac{27m^3}{8n^6}}$

70. $\sqrt[3]{\dfrac{125t^9}{27s^6}}$

71. $\sqrt[3]{\dfrac{16r^4s^5}{1,000t^3}}$

72. $\sqrt[3]{\dfrac{54m^4n^3}{r^3s^6}}$

ADDITIONAL PRACTICE *Simplify each expression. All variables of square root expressions represent positive numbers. Assume no division by 0.*

73. $\sqrt{324}$

74. $\sqrt{405}$

75. $\sqrt{147}$

76. $\sqrt{722}$

77. $-7\sqrt{1,000}$

78. $-3\sqrt{252}$

79. $\sqrt{81x}$

80. $\sqrt{36y}$

81. $-3xyz\sqrt{18x^3y^5}$

82. $15xy^2\sqrt{72x^2y^3}$

83. $\frac{1}{5}x^2y\sqrt{50x^2y^2}$

84. $\frac{1}{5}x^5y\sqrt{75x^3y^2}$

85. $\sqrt{\dfrac{125}{121}}$

86. $\sqrt{\dfrac{250}{49}}$

87. $\sqrt{\dfrac{245}{36}}$

88. $\sqrt{\dfrac{500}{81}}$

89. $\sqrt[3]{54x^3y^4z^6}$

90. $\sqrt[3]{-24x^5y^5z^4}$

91. $\sqrt[3]{-81x^2y^3z^4}$

92. $\sqrt[3]{1,600xy^2z^3}$

93. $\sqrt{\dfrac{180a^3b}{b^3}}$

94. $\sqrt{\dfrac{320x^4y}{xy^3}}$

95. $\sqrt{432}$

96. $\sqrt{720}$

97. $\sqrt{\dfrac{12r^7s^6t}{81r^5s^2t}}$

98. $\sqrt{\dfrac{36m^2n^9}{100mn^3}}$

99. $\frac{3}{4}\sqrt{192a^3b^5}$

100. $-\frac{2}{9}\sqrt{162r^3s^3t}$

101. $-\frac{2}{5}\sqrt{80mn^2}$

102. $\frac{5}{6}\sqrt{180ab^2c}$

103. $\frac{3}{2b}\sqrt{288a^3b^3c}$

104. $\frac{3}{x^2}\sqrt{\dfrac{1}{81}x^5yz^3}$

105. $\sqrt[3]{\dfrac{250a^3b^4}{16b}}$

106. $\sqrt[3]{\dfrac{81p^5q^3}{1,000p^2q^6}}$

WRITING ABOUT MATH

107. Explain the multiplication property of radicals.

108. Explain the division property of radicals.

SOMETHING TO THINK ABOUT

109. Find the errors.

$$\left(\sqrt{a+b}\right)^2 = \left(\sqrt{a}+\sqrt{b}\right)^2$$
$$= \left(\sqrt{a}\right)^2 + \left(\sqrt{b}\right)^2$$
$$= a + b$$

In spite of the errors, is the conclusion correct?

110. Use scientific notation to simplify $\sqrt{0.00000004}$.

Section 8.4

Adding and Subtracting Radical Expressions

Objectives

1. Add or subtract two or more radical expressions containing square roots.
2. Add or subtract two or more radical expressions containing cube roots.

Vocabulary

like radicals

Getting Ready

Combine like terms.

1. $3x + 4x$

2. $5y - 2y$

3. $7xy - 2xy$

4. $7t^2 + 2t^2$

Simplify each radical ($x \geq 0, y \geq 0$).

5. $\sqrt{20}$

6. $\sqrt{45}$

7. $\sqrt{8x^2y}$

8. $\sqrt{18x^2y}$

In this section, we will discuss how to add and subtract radical expressions. To do so, we often will need to simplify them first.

1 Add or subtract two or more radical expressions containing square roots.

When adding monomials, we can combine like terms. For example,

COMMENT The expression $3x + 5y$ cannot be simplified, because $3x$ and $5y$ are not like terms.

$$3x + 5x = (3 + 5)x \quad \text{Use the distributive property.}$$
$$= 8x$$

It is often possible to combine terms that contain *like radicals*.

LIKE RADICALS	Radicals are called **like radicals** when they have the same index and the same radicand.

Since the terms $3\sqrt{2}$ and $5\sqrt{2}$ contain like radicals, they are like terms and can be combined.

$$3\sqrt{2} + 5\sqrt{2} = (3 + 5)\sqrt{2} \quad \text{Use the distributive property.}$$
$$= 8\sqrt{2}$$

Likewise,

$$5x\sqrt{3y} - 2x\sqrt{3y} = (5x - 2x)\sqrt{3y} \quad \text{Use the distributive property.}$$
$$= 3x\sqrt{3y}$$

COMMENT The terms in $3\sqrt{2} + 5\sqrt{7}$ contain unlike radicals and cannot be combined. We cannot combine the terms in the expression $2x\sqrt{5z} + 3y\sqrt{5z}$, because the terms have different variables. However, the expression could be written as $\sqrt{5z}(2x + 3y)$.

Be certain to simplify all radicals prior to deciding whether they can be combined. Radicals such as $3\sqrt{18}$ and $5\sqrt{8}$ can be simplified so that they contain like radicals. They can then be combined.

EXAMPLE 1   Simplify: $3\sqrt{18} + 5\sqrt{8}$

Solution   The radical $\sqrt{18}$ is not in simplest form, because 18 has a perfect-square factor of 9. The radical $\sqrt{8}$ is not in simplest form either, because 8 has a perfect-square factor of 4. To simplify the radicals and add them, we proceed as follows.

$$3\sqrt{18} + 5\sqrt{8} = 3\sqrt{9 \cdot 2} + 5\sqrt{4 \cdot 2} \quad \text{Factor 18 and 8. Look for perfect-square factors.}$$
$$= 3\sqrt{9}\sqrt{2} + 5\sqrt{4}\sqrt{2} \quad \text{The square root of a product is equal to the product of the square roots.}$$
$$= 3(3)\sqrt{2} + 5(2)\sqrt{2} \quad \text{Simplify.}$$
$$= 9\sqrt{2} + 10\sqrt{2} \quad \text{Multiply.}$$
$$= 19\sqrt{2} \quad \text{Combine like terms.}$$

SELF CHECK 1   Simplify: $2\sqrt{50} + \sqrt{32}$

EXAMPLE 2   Simplify: $\sqrt{20} + \sqrt{45} + 3\sqrt{5}$

Solution   We simplify the first two radicals and combine like terms.

$$\sqrt{20} + \sqrt{45} + 3\sqrt{5}$$
$$= \sqrt{4 \cdot 5} + \sqrt{9 \cdot 5} + 3\sqrt{5} \qquad \text{Look for perfect-square factors.}$$
$$= \sqrt{4}\sqrt{5} + \sqrt{9}\sqrt{5} + 3\sqrt{5} \qquad \text{The square root of a product is equal to the product of the square roots.}$$
$$= 2\sqrt{5} + 3\sqrt{5} + 3\sqrt{5} \qquad \text{Simplify.}$$
$$= 8\sqrt{5} \qquad \text{Combine like terms.}$$

SELF CHECK 2   Simplify: $\sqrt{12} + \sqrt{27} + \sqrt{75}$

EXAMPLE 3   Simplify: $\sqrt{8x^2y} + \sqrt{18x^2y}$ $(x \geq 0, y \geq 0)$

Solution   We simplify each radical and combine like terms.

$$\sqrt{8x^2y} + \sqrt{18x^2y}$$
$$= \sqrt{4 \cdot 2x^2y} + \sqrt{9 \cdot 2x^2y} \qquad \text{Factor 8 and 18. Look for perfect-square factors.}$$
$$= \sqrt{4x^2}\sqrt{2y} + \sqrt{9x^2}\sqrt{2y} \qquad \text{The square root of a product is equal to the product of the square roots.}$$
$$= 2x\sqrt{2y} + 3x\sqrt{2y} \qquad \text{Simplify.}$$
$$= 5x\sqrt{2y} \qquad \text{Combine like terms.}$$

SELF CHECK 3   Simplify: $\sqrt{12xy^2} + \sqrt{27xy^2}$ $(x \geq 0, y \geq 0)$

EXAMPLE 4   Simplify: $\sqrt{28x^2y} - 2\sqrt{63y^3}$ $(x \geq 0, y \geq 0)$

Solution   We simplify each radical and combine like terms.

$$\sqrt{28x^2y} - 2\sqrt{63y^3}$$
$$= \sqrt{4 \cdot 7x^2y} - 2\sqrt{9 \cdot 7y^2y} \qquad \text{Factor 28 and 63. Look for perfect-square factors.}$$
$$= \sqrt{4x^2}\sqrt{7} - 2 \cdot \sqrt{9y^2}\sqrt{7y} \qquad \text{The square root of a product is equal to the product of the square roots.}$$
$$= 2x\sqrt{7y} - 2 \cdot 3y\sqrt{7y} \qquad \text{Simplify.}$$
$$= 2x\sqrt{7y} - 6y\sqrt{7y} \qquad \text{Simplify.}$$

COMMENT An alternative way to write the result is in factored form: $2\sqrt{7y}(x - 3y)$.   Since the variables in the terms are different, we cannot simplify further.

SELF CHECK 4   Simplify: $\sqrt{20xy^2} - \sqrt{80x^3}$ $(x \geq 0, y \geq 0)$

EXAMPLE 5   Simplify: $\sqrt{27xy} + \sqrt{20xy}$ $(x \geq 0, y \geq 0)$

Solution   We will simplify each radical and see whether we can combine like terms.

$$\sqrt{27xy} + \sqrt{20xy}$$
$$= \sqrt{9 \cdot 3xy} + \sqrt{4 \cdot 5xy} \qquad \text{Factor 27 and 20. Look for perfect-square factors.}$$
$$= \sqrt{9}\sqrt{3xy} + \sqrt{4}\sqrt{5xy} \qquad \text{The square root of a product is equal to the product of the square roots.}$$
$$= 3\sqrt{3xy} + 2\sqrt{5xy} \qquad \text{Simplify.}$$

Since the terms have unlike radicals, we cannot combine like terms. This expression cannot be simplified further.

 SELF CHECK 5   Simplify: $\sqrt{75ab} + \sqrt{72ab}$ $(a \geq 0, b \geq 0)$

EXAMPLE 6   Simplify: $\sqrt{8x} + \sqrt{3y} - \sqrt{50x} + \sqrt{27y}$ $(x \geq 0, y \geq 0)$

Solution   We simplify the radicals and combine like terms, when possible.

$$\sqrt{8x} + \sqrt{3y} - \sqrt{50x} + \sqrt{27y}$$
$$= \sqrt{4 \cdot 2x} + \sqrt{3y} - \sqrt{25 \cdot 2x} + \sqrt{9 \cdot 3y}$$
$$= \sqrt{4}\sqrt{2x} + \sqrt{3y} - \sqrt{25}\sqrt{2x} + \sqrt{9}\sqrt{3y}$$
$$= 2\sqrt{2x} + \sqrt{3y} - 5\sqrt{2x} + 3\sqrt{3y}$$
$$= -3\sqrt{2x} + 4\sqrt{3y}$$

 SELF CHECK 6   Simplify: $\sqrt{32x} - \sqrt{5y} - \sqrt{200x} + \sqrt{125y}$ $(x \geq 0, y \geq 0)$

2   Add or subtract two or more radical expressions containing cube roots.

It is often possible to combine terms containing like radicals other than square roots.

EXAMPLE 7   Simplify: $\sqrt[3]{81x^4} - x\sqrt[3]{24x}$

Solution   We simplify each radical and combine like terms, when possible.

$$\sqrt[3]{81x^4} - x\sqrt[3]{24x} = \sqrt[3]{27x^3 \cdot 3x} - x\sqrt[3]{8 \cdot 3x}$$
    Factor 81 and 24. Look for perfect-cube factors.

$$= \sqrt[3]{27x^3}\sqrt[3]{3x} - x\sqrt[3]{8}\sqrt[3]{3x}$$
    The cube root of a product is equal to the product of the cube roots.

$$= 3x\sqrt[3]{3x} - 2x\sqrt[3]{3x}$$
    $\sqrt[3]{27x^3} = 3x$ and $\sqrt[3]{8} = 2$.

$$= x\sqrt[3]{3x}$$

 SELF CHECK 7   Simplify: $\sqrt[3]{24a^4} + a\sqrt[3]{81a}$

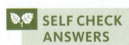 **SELF CHECK ANSWERS**   **1.** $14\sqrt{2}$ **2.** $10\sqrt{3}$ **3.** $5y\sqrt{3x}$ **4.** $2y\sqrt{5x} - 4x\sqrt{5x}$ or $2\sqrt{5x}(y - 2x)$ **5.** $5\sqrt{3ab} + 6\sqrt{2ab}$
6. $-6\sqrt{2x} + 4\sqrt{5y}$ **7.** $5a\sqrt[3]{3a}$

NOW TRY THIS

Simplify. All variables represent positive values.

1. $\sqrt{9x + 9} + \sqrt{x + 1}$

2. $4\sqrt[3]{-16} - 2\sqrt[3]{2}$

3. $\sqrt{49} - 2\sqrt{18}$

4. $\sqrt{(x + 3)^2} + \sqrt{25x^2} - 2\sqrt{x}$

8.4 Exercises

WARM-UPS *Determine whether the terms are like terms.*

1. $5\sqrt{2}, 4\sqrt{2}$
2. $9\sqrt{3}, 2\sqrt{3}$
3. $6\sqrt{2}, 6\sqrt{3}$
4. $2\sqrt{2}, 3\sqrt{3}$
5. $-3\sqrt{2}, 7\sqrt{2}$
6. $-4\sqrt{3}, -9\sqrt{3}$

REVIEW *Solve each proportion.*

7. $\dfrac{a-2}{8} = \dfrac{a+10}{24}$
8. $\dfrac{6}{t+12} = \dfrac{18}{4t}$
9. $\dfrac{-2}{x+14} = \dfrac{6}{x-6}$
10. $\dfrac{y-4}{4} = \dfrac{y+12}{12}$

VOCABULARY AND CONCEPTS *Fill in the blanks.*

11. _____ have the same variables with the same exponents.
12. _____ have the same index and the same radicand.

Determine whether the terms contain like radicals.

13. $2\sqrt{3x}$ and $3\sqrt{2x}$
14. $7\sqrt{3x}$ and $3\sqrt{3x}$
15. $27\sqrt[3]{4a}$ and $-3\sqrt[3]{4a}$
16. $-25\sqrt[4]{2x}$ and $125\sqrt[3]{2x}$
17. Find the error in the following work.

$$7\sqrt{5} - 3\sqrt{2} = 4\sqrt{3}$$

Use a calculator to show that the left side is not equal to the right side.
18. Find the error in the following work.

$$12\sqrt{7} + 20\sqrt{11} = 32\sqrt{18}$$

Use a calculator to show that the left side is not equal to the right side.

GUIDED PRACTICE *Simplify. SEE EXAMPLE 1. (OBJECTIVE 1)*

19. $\sqrt{8} + \sqrt{50}$
20. $\sqrt{20} + \sqrt{45}$
21. $\sqrt{48} + \sqrt{75}$
22. $\sqrt{48} + \sqrt{108}$
23. $\sqrt{45} + \sqrt{80}$
24. $\sqrt{80} + \sqrt{125}$
25. $\sqrt{125} + \sqrt{245}$
26. $\sqrt{72} + \sqrt{200}$

Simplify. SEE EXAMPLE 2. (OBJECTIVE 1)

27. $\sqrt{20} + \sqrt{45} + \sqrt{80}$
28. $\sqrt{48} + \sqrt{27} + \sqrt{75}$
29. $\sqrt{24} + \sqrt{150} + \sqrt{240}$
30. $\sqrt{28} + \sqrt{63} + \sqrt{112}$
31. $\sqrt{12} + \sqrt{18} - \sqrt{27}$
32. $\sqrt{8} - \sqrt{50} + \sqrt{72}$
33. $\sqrt{200} - \sqrt{75} + \sqrt{48}$
34. $\sqrt{20} + \sqrt{80} - \sqrt{125}$

Simplify. All variables represent positive values. SEE EXAMPLE 3. (OBJECTIVE 1)

35. $\sqrt{20x^2} + \sqrt{5x^2}$
36. $\sqrt{3y^2} - \sqrt{12y^2}$
37. $\sqrt{20x^3} + \sqrt{5x^3}$
38. $\sqrt{3y^3} - \sqrt{12y^3}$

39. $\sqrt{18x^2y} - \sqrt{27x^2y}$
40. $\sqrt{81ab} - \sqrt{64ab}$
41. $\sqrt{32x^5} - \sqrt{18x^5}$
42. $\sqrt{27xy^3} - \sqrt{48xy^3}$

Simplify. All variables represent positive values. SEE EXAMPLE 4. (OBJECTIVE 1)

43. $y\sqrt{490y} - 2\sqrt{360y^3}$
44. $3\sqrt{20x} + 2\sqrt{63y}$
45. $\sqrt{20x^3y} + \sqrt{45x^5y^3} - \sqrt{80x^7y^5}$
46. $x\sqrt{48xy^2} - y\sqrt{27x^3} + \sqrt{75x^3y^2}$

Simplify. Assume all variables represent positive values. SEE EXAMPLES 5–6. (OBJECTIVE 1)

47. $\sqrt{48} - \sqrt{8} + \sqrt{27} - \sqrt{32}$
48. $\sqrt{162} + \sqrt{50} - \sqrt{75} - \sqrt{108}$
49. $3\sqrt{54x^2} + 5\sqrt{24x^2}$
50. $\sqrt[3]{24a^5b^4} + \sqrt[3]{81a^5b^4}$

Simplify. SEE EXAMPLE 7. (OBJECTIVE 2)

51. $\sqrt[3]{16} + \sqrt[3]{54}$
52. $\sqrt[3]{24} - \sqrt[3]{81}$
53. $\sqrt[3]{x^{10}} - \sqrt[3]{x^4}$
54. $\sqrt[3]{125x^2} + \sqrt[3]{64x^8}$

ADDITIONAL PRACTICE *Simplify. All variables represent positive values.*

55. $\sqrt{20} + \sqrt{180}$
56. $\sqrt{80} + \sqrt{245}$
57. $\sqrt{160} + \sqrt{360}$
58. $\sqrt{12} + \sqrt{147}$
59. $3\sqrt{45} + 4\sqrt{245}$
60. $2\sqrt{28} + 7\sqrt{63}$
61. $2\sqrt{28} + 2\sqrt{112}$
62. $4\sqrt{63} + 6\sqrt{112}$
63. $\sqrt{49} - \sqrt{12}$
64. $\sqrt{24} - \sqrt{16}$
65. $\sqrt{75} - \sqrt{27}$
66. $\sqrt{50} - \sqrt{32}$
67. $\sqrt{24} - \sqrt{150} - \sqrt{54}$
68. $\sqrt{98} - \sqrt{300} + \sqrt{800}$
69. $\sqrt{200} + \sqrt{300} - \sqrt{75}$
70. $\sqrt{175} + \sqrt{125} - \sqrt{28}$
71. $\sqrt[3]{81} - \sqrt[3]{24}$
72. $\sqrt[3]{32} + \sqrt[3]{108}$
73. $\sqrt[3]{40} + \sqrt[3]{125}$
74. $\sqrt[3]{3,000} - \sqrt[3]{192}$
75. $2\sqrt{80} - 3\sqrt{125}$
76. $3\sqrt{245} - 2\sqrt{180}$
77. $8\sqrt{96} - 5\sqrt{24}$
78. $3\sqrt{216} - 3\sqrt{150}$
79. $\sqrt{72} - \sqrt{32}$
80. $\sqrt{98} - \sqrt{72}$
81. $\sqrt{12} - \sqrt{48}$
82. $\sqrt{27} - \sqrt{147}$
83. $5\sqrt{32} + 3\sqrt{72}$
84. $3\sqrt{72} + 2\sqrt{128}$
85. $3\sqrt{98} + 8\sqrt{128}$
86. $5\sqrt{90} + 7\sqrt{250}$
87. $\sqrt{288} - 3\sqrt{200}$
88. $\sqrt{392} - 2\sqrt{128}$
89. $5\sqrt{250} - 3\sqrt{160}$
90. $4\sqrt{490} - 3\sqrt{360}$

91. $3\sqrt{24x^4y^3} + 2\sqrt{54x^4y^3}$

92. $\sqrt[3]{192x^4y^5} - \sqrt[3]{24x^4y^5}$

93. $\sqrt{108} - \sqrt{75}$

94. $\sqrt{300} - \sqrt{75}$

95. $\sqrt{1,000} - \sqrt{360}$

96. $\sqrt{180} - \sqrt{125}$

97. $\sqrt{147} + \sqrt{216} - \sqrt{108} - \sqrt{27}$

98. $\sqrt{180} - \sqrt{112} + \sqrt{45} - \sqrt{700}$

99. $\sqrt[3]{135x^7y^4} - \sqrt[3]{40x^7y^4}$

100. $\sqrt[3]{56a^4b^5} + \sqrt[3]{7a^4b^5}$

WRITING ABOUT MATH

101. Explain why $\sqrt{3x} + \sqrt{2y}$ cannot be combined.

102. Explain why $\sqrt{4x}$ and $\sqrt[3]{4x}$ cannot be combined.

SOMETHING TO THINK ABOUT

103. Find the error.

$$\sqrt{27} - \sqrt{75} = \sqrt{9}\sqrt{3} - \sqrt{25}\sqrt{3}$$
$$= 3\sqrt{3} - 5\sqrt{3}$$
$$= -2$$

104. Find the error.

$$\sqrt{8} + \sqrt{12} = \sqrt{4}\sqrt{2} + \sqrt{4}\sqrt{3}$$
$$= \sqrt{4}(2) + \sqrt{4}(3)$$
$$= 2\sqrt{4} + 3\sqrt{4}$$
$$= 5\sqrt{4}$$
$$= 5(2)$$
$$= 10$$

Section 8.5

Multiplying and Dividing Radical Expressions

Objectives

1. Multiply two or more radical expressions.
2. Divide two single-term radical expressions.
3. Rationalize an expression with a single-term denominator.
4. Rationalize an expression with a denominator having two terms.

Vocabulary

rationalize the denominator   conjugate

Getting Ready

Perform each operation and simplify, when possible ($x \neq 0, y \neq 0$).

1. x^2x^3 **2.** y^3y^4 **3.** $\dfrac{x^5}{x^2}$ **4.** $\dfrac{y^8}{y^5}$

5. $x(x + 2)$ **6.** $2y^3(3y^2 - 4y)$ **7.** $(x + 2)(x - 3)$ **8.** $(2x + 3y)(3x + 2y)$

We will now discuss how to multiply and divide radical expressions. The process for multiplication is similar to the process of multiplying polynomials.

1   Multiply two or more radical expressions.

Recall that *the product of the square roots of two nonnegative numbers is equal to the square root of the product of those numbers.* For example,

$$\sqrt{3}\sqrt{3} = \sqrt{3 \cdot 3} \qquad \sqrt{3}\sqrt{27} = \sqrt{3 \cdot 27} \qquad \sqrt{x}\sqrt{x^3} = \sqrt{x \cdot x^3}$$
$$= \sqrt{9} \qquad\qquad = \sqrt{81} \qquad\qquad = \sqrt{x^4}$$
$$= 3 \qquad\qquad\quad = 9 \qquad\qquad\quad = x^2$$

Likewise, *the product of the cube roots of two numbers is equal to the cube root of the product of those numbers.* For example,

$$\sqrt[3]{2} \cdot \sqrt[3]{4} = \sqrt[3]{2 \cdot 4} \qquad \sqrt[3]{4} \cdot \sqrt[3]{16} = \sqrt[3]{4 \cdot 16} \qquad \sqrt[3]{3x^2} \cdot \sqrt[3]{9x} = \sqrt[3]{3x^2 \cdot 9x}$$
$$= \sqrt[3]{8} \qquad\qquad = \sqrt[3]{64} \qquad\qquad = \sqrt[3]{27x^3}$$
$$= 2 \qquad\qquad\quad = 4 \qquad\qquad\quad = 3x$$

To multiply radical expressions, we multiply the coefficients and multiply the radicals separately and then simplify the result, when possible.

EXAMPLE 1   Multiply: **a.** $3\sqrt{6}$ by $4\sqrt{3}$ **b.** $-2\sqrt[3]{7x}$ by $6\sqrt[3]{49x^2}$ $(x \geq 0)$

Solution   The commutative and associative properties enable us to multiply the integers and the radicals separately.

a. $3\sqrt{6} \cdot 4\sqrt{3} = 3(4)\sqrt{6}\sqrt{3}$ **b.** $-2\sqrt[3]{7x} \cdot 6\sqrt[3]{49x^2} = -2(6)\sqrt[3]{7x}\sqrt[3]{49x^2}$
$$\qquad\qquad = 12\sqrt{18} \qquad\qquad\qquad\qquad\qquad = -12\sqrt[3]{7 \cdot 49 \cdot x \cdot x^2}$$
$$\qquad\qquad = 12\sqrt{9}\sqrt{2} \qquad\qquad\qquad\qquad\quad = -12\sqrt[3]{\mathbf{343x^3}}$$
$$\qquad\qquad = 12(\mathbf{3})\sqrt{2} \qquad\qquad\qquad\qquad\quad\; = -12(\mathbf{7x})$$
$$\qquad\qquad = 36\sqrt{2} \qquad\qquad\qquad\qquad\qquad\;\; = -84x$$

 SELF CHECK 1   Multiply: **a.** $\left(5\sqrt[3]{2}\right)\left(-2\sqrt[3]{4}\right)$ **b.** $\left(2\sqrt{2x}\right)\left(3\sqrt{3x}\right)$ $(x \geq 0)$

Recall that to multiply a polynomial by a monomial, we use the distributive property. We can use the same process to multiply radical expressions.

EXAMPLE 2   Multiply: **a.** $\sqrt{2x}\left(\sqrt{6x} + \sqrt{8x}\right)$ $(x \geq 0)$ **b.** $\sqrt[3]{3}\left(\sqrt[3]{9} - 2\right)$

Solution   We will use the distributive property to remove parentheses and combine terms whenever possible.

a. $\sqrt{\mathbf{2x}}\left(\sqrt{6x} + \sqrt{8x}\right) = \sqrt{\mathbf{2x}}\sqrt{6x} + \sqrt{\mathbf{2x}}\sqrt{8x}$ Use the distributive property to remove parentheses.

$$\qquad\qquad\qquad\qquad = \sqrt{12x^2} + \sqrt{16x^2}$$ The product of two square roots is equal to the square root of the product.

$$\qquad\qquad\qquad\qquad = \sqrt{4 \cdot 3 \cdot x^2} + \sqrt{16x^2}$$ Factor 12.

$$\qquad\qquad\qquad\qquad = \sqrt{\mathbf{4x^2}}\sqrt{3} + \sqrt{\mathbf{16x^2}}$$ The square root of a product is equal to the product of the square roots.

$$\qquad\qquad\qquad\qquad = \mathbf{2x}\sqrt{3} + \mathbf{4x}$$ Simplify.

b. $\sqrt[3]{\mathbf{3}}\left(\sqrt[3]{9} - 2\right) = \sqrt[3]{\mathbf{3}}\sqrt[3]{9} - \sqrt[3]{\mathbf{3}} \cdot 2$ Use the distributive property to remove parentheses.

$$\qquad\qquad\qquad = \sqrt[3]{\mathbf{27}} - 2\sqrt[3]{3}$$ The product of two cube roots is equal to the cube root of the product.

$$\qquad\qquad\qquad = \mathbf{3} - 2\sqrt[3]{3}$$ $\sqrt[3]{27} = 3$

 SELF CHECK 2   Multiply: **a.** $\sqrt{3}\left(\sqrt{6} - \sqrt{3}\right)$ **b.** $\sqrt[3]{2x}\left(3 - \sqrt[3]{4x^2}\right)$

To multiply two binomials, we multiply each term of one binomial by each term of the other binomial and simplify. We can use this same process to multiply radical expressions, each containing two terms.

EXAMPLE 3   Multiply: $\left(\sqrt{3} + \sqrt{2}\right)\left(\sqrt{3} - \sqrt{2}\right)$

Solution   We can find the product by multiplying each term of the first factor by each term of the second factor and simplifying.

$$\left(\sqrt{3} + \sqrt{2}\right)\left(\sqrt{3} - \sqrt{2}\right)$$

$$= \sqrt{3}\sqrt{3} - \sqrt{3}\sqrt{2} + \sqrt{2}\sqrt{3} - \sqrt{2}\sqrt{2} \qquad \text{Multiply each term of the first factor by each term of the second factor.}$$

$$= \sqrt{9} - \sqrt{6} + \sqrt{6} - \sqrt{4} \qquad \text{Multiply.}$$

$$= 3 - 2 \qquad \text{Combine like terms and simplify.}$$

$$= 1 \qquad \text{Simplify.}$$

🌿 **SELF CHECK 3**   Multiply: $\left(\sqrt{5} + \sqrt{3}\right)\left(\sqrt{5} - \sqrt{3}\right)$

EXAMPLE 4   Multiply: $\left(\sqrt{3x} + 1\right)\left(\sqrt{3x} + 2\right) \quad (x \geq 0)$

Solution   We can find the product by multiplying each term of the first factor by each term of the second factor and simplifying.

$$\left(\sqrt{3x} + 1\right)\left(\sqrt{3x} + 2\right)$$

$$= \sqrt{3x}\sqrt{3x} + 2\sqrt{3x} + \sqrt{3x} + 2 \qquad \text{Multiply each term of the first factor by each term of the second factor.}$$

$$= \sqrt{9x^2} + 2\sqrt{3x} + \sqrt{3x} + 2 \qquad \text{Multiply.}$$

$$= 3x + 3\sqrt{3x} + 2 \qquad \text{Combine like terms and simplify.}$$

🌿 **SELF CHECK 4**   Multiply: $\left(\sqrt{5a} - 2\right)\left(\sqrt{5a} + 3\right) \quad (a \geq 0)$

EXAMPLE 5   Multiply: $\left(\sqrt[3]{4x} - 3\right)\left(\sqrt[3]{2x^2} + 1\right)$

Solution   We can find the product by multiplying each term of the first factor by each term of the second factor and simplifying.

$$\left(\sqrt[3]{4x} - 3\right)\left(\sqrt[3]{2x^2} + 1\right)$$

$$= \sqrt[3]{4x}\sqrt[3]{2x^2} + 1 \cdot \sqrt[3]{4x} - 3\sqrt[3]{2x^2} - 3 \qquad \text{Multiply.}$$

$$= \sqrt[3]{8x^3} + \sqrt[3]{4x} - 3\sqrt[3]{2x^2} - 3 \qquad \text{The product of two cube roots is equal to the cube root of the product.}$$

$$= 2x + \sqrt[3]{4x} - 3\sqrt[3]{2x^2} - 3 \qquad \text{Simplify.}$$

🌿 **SELF CHECK 5**   Multiply: $\left(\sqrt[3]{3x} + 1\right)\left(\sqrt[3]{9x^2} - 2\right)$

2 Divide two single-term radical expressions.

To divide radical expressions, we use the division property of radicals. For example, to divide $\sqrt{108}$ by $\sqrt{36}$, we proceed as follows:

$$\frac{\sqrt{108}}{\sqrt{36}} = \sqrt{\frac{108}{36}} \qquad \text{The quotient of two square roots is the square root of the quotient.}$$

$$= \sqrt{3}$$

EXAMPLE 6 Divide: $\dfrac{\sqrt{22a^2b^7}}{\sqrt{99a^4b^3}}$ $(a > 0, b > 0)$

Solution We can write the quotient of two radicals as the radical of a quotient and then simplify.

$$\frac{\sqrt{22a^2b^7}}{\sqrt{99a^4b^3}} = \sqrt{\frac{22a^2b^7}{99a^4b^3}}$$

$$= \sqrt{\frac{2b^4}{9a^2}} \qquad \text{Simplify the radicand.}$$

$$= \frac{\sqrt{2b^4}}{\sqrt{9a^2}} \qquad \text{The square root of a quotient is equal to the quotient of the square roots.}$$

$$= \frac{\sqrt{b^4}\sqrt{2}}{\sqrt{9a^2}} \qquad \text{The square root of a product is equal to the product of the square roots.}$$

$$= \frac{b^2\sqrt{2}}{3a} \qquad \text{Simplify the radicals.}$$

SELF CHECK 6 Divide: $\dfrac{\sqrt{44x^3y^9}}{\sqrt{99x^5y^5}}$ $(x > 0, y > 0)$

3 Rationalize an expression with a single-term denominator.

The fraction $\dfrac{1}{\sqrt{2}}$ is not written in simplest form because a radical appears in its denominator. We can use a process called **rationalizing the denominator** to simplify such fractions. For example, to eliminate the radical in the denominator of $\dfrac{1}{\sqrt{2}}$, we can multiply the fraction by 1, written in the form $\dfrac{\sqrt{2}}{\sqrt{2}}$. This will result in a denominator that is the rational number 2, because $\left(\sqrt{2}\right)\left(\sqrt{2}\right) = 2$.

COMMENT When rationalizing a denominator, always multiply by a value that will result in the removal of the radical sign in the denominator.

$$\frac{1}{\sqrt{2}} = \frac{1\sqrt{2}}{\sqrt{2}\sqrt{2}} \qquad \text{Multiply the fraction by } 1: \frac{\sqrt{2}}{\sqrt{2}} = 1$$

$$= \frac{\sqrt{2}}{\sqrt{4}} \qquad 1\sqrt{2} = \sqrt{2}, \sqrt{2}\sqrt{2} = \sqrt{4}$$

$$= \frac{\sqrt{2}}{2} \qquad \sqrt{4} = 2$$

EXAMPLE 7　Rationalize each denominator:　**a.** $\dfrac{3}{\sqrt{3}}$　**b.** $\dfrac{2}{\sqrt[3]{3}}$

Solution　**a.** We multiply the numerator and denominator by $\sqrt{3}$ because $\sqrt{3}\cdot\sqrt{3}=3$ and 3 is a rational number. Then we simplify.

$$\frac{3}{\sqrt{3}}=\frac{3\sqrt{3}}{\sqrt{3}\sqrt{3}}\qquad\color{red}\text{Multiply the fraction by 1: }\frac{\sqrt{3}}{\sqrt{3}}=1$$

$$=\frac{3\sqrt{3}}{\sqrt{9}}\qquad\color{red}\sqrt{3}\cdot\sqrt{3}=\sqrt{9}$$

$$=\frac{3\sqrt{3}}{3}\qquad\color{red}\sqrt{9}=3$$

$$=\sqrt{3}\qquad\color{red}\frac{3}{3}=1$$

b. We need to multiply $\sqrt[3]{3}$ by a value that will produce a perfect cube under the radical. Since 27 is a perfect-integer cube, we multiply the numerator and denominator by $\sqrt[3]{9}$ and simplify.

$$\frac{2}{\sqrt[3]{3}}=\frac{2\sqrt[3]{9}}{\sqrt[3]{3}\sqrt[3]{9}}=\frac{2\sqrt[3]{9}}{\sqrt[3]{27}}=\frac{2\sqrt[3]{9}}{3}$$

 SELF CHECK 7　Rationalize each denominator:　**a.** $\dfrac{2}{\sqrt{5}}$　**b.** $\dfrac{5}{\sqrt[3]{5}}$

EXAMPLE 8　Divide:　$\dfrac{5}{\sqrt{20x}}$　$(x>0)$

Solution　To rationalize the denominator, we need to multiply the numerator and denominator by a value that will produce a perfect square under the radical. Such a number is $\sqrt{5x}$, because $5x\cdot20x=100x^2$, which is a perfect square.

$$\frac{5}{\sqrt{20x}}=\frac{5\sqrt{5x}}{\sqrt{20x}\sqrt{5x}}\qquad\color{red}\text{Multiply the fraction by 1: }\frac{\sqrt{5x}}{\sqrt{5x}}=1$$

$$=\frac{5\sqrt{5x}}{\sqrt{100x^2}}\qquad\color{red}\text{The product of two square roots is equal to the square root of the product.}$$

$$=\frac{5\sqrt{5x}}{10x}\qquad\color{red}\text{Simplify.}$$

$$=\frac{\sqrt{5x}}{2x}$$

 SELF CHECK 8　Divide:　$\dfrac{6}{\sqrt{8y}}$　$(y>0)$

EXAMPLE 9　Divide:　$\sqrt{\dfrac{3x^3y^2}{27xy^3}}$　$(x>0,y>0)$

Solution　We will first simplify the rational expression under the radical sign and then simplify.

$$\sqrt{\frac{3x^3y^2}{27xy^3}}=\sqrt{\frac{x^2}{9y}}\qquad\color{red}\text{Simplify the fraction within the radical.}$$

$$=\sqrt{\frac{x^2\cdot y}{9y\cdot y}}\qquad\color{red}\text{Multiply the fraction by 1: }\frac{\sqrt{y}}{\sqrt{y}}=1$$

$$= \frac{\sqrt{x^2 y}}{\sqrt{9y^2}} \qquad \text{The square root of a quotient is the quotient of the square roots.}$$

$$= \frac{x\sqrt{y}}{3y} \qquad \text{Simplify.}$$

SELF CHECK 9   Divide: $\sqrt{\dfrac{3a^5 b}{108ab^2}}$ $(a > 0, b > 0)$

---

4   Rationalize an expression with a denominator having two terms.

Since the denominators of many fractions such as $\frac{2}{\sqrt{3}-1}$ contain radicals, they are not in simplest form. Because the denominator has two terms, multiplying the denominator by $\sqrt{3}$ will not make it a rational number. The key to rationalizing this denominator is to multiply the numerator and denominator by $\sqrt{3}+1$ because the product $(\sqrt{3}+1)(\sqrt{3}-1)$ has no radicals. Radical expressions such as $\sqrt{3}+1$ and $\sqrt{3}-1$ are called **conjugates** of each other.

EXAMPLE 10   Divide: $\dfrac{2}{\sqrt{3}-1}$

Solution   Multiply the numerator and denominator by the conjugate of the denominator.

$$\frac{2}{\sqrt{3}-1} = \frac{2(\sqrt{3}+1)}{(\sqrt{3}-1)(\sqrt{3}+1)} \qquad \begin{array}{l}\text{Multiply numerator and denominator by the}\\ \text{conjugate of the denominator: } \frac{\sqrt{3}+1}{\sqrt{3}+1} = 1\end{array}$$

$$= \frac{2(\sqrt{3}+1)}{3-1} \qquad \text{Multiply the binomials in the denominator.}$$

$$= \frac{2(\sqrt{3}+1)}{2} \qquad \text{Simplify.}$$

$$= \sqrt{3}+1 \qquad \text{Divide out the common factor of 2.}$$

SELF CHECK 10   Divide: $\dfrac{3}{\sqrt{2}+1}$

EXAMPLE 11   Divide: $\dfrac{\sqrt{x}+1}{\sqrt{x}-1}$ $(x \geq 0, x \neq 1)$

Solution   We multiply the numerator and denominator by the conjugate of the denominator, which is $\sqrt{x}+1$.

$$\frac{\sqrt{x}+1}{\sqrt{x}-1} = \frac{(\sqrt{x}+1)(\sqrt{x}+1)}{(\sqrt{x}-1)(\sqrt{x}+1)} \qquad \begin{array}{l}\text{Multiply the fraction by the conjugate}\\ \text{of the denominator: } \frac{\sqrt{x}+1}{\sqrt{x}+1} = 1\\ (x \geq 0).\end{array}$$

$$= \frac{\sqrt{x}\sqrt{x} + \sqrt{x}(1) + 1(\sqrt{x}) + 1}{\sqrt{x}\sqrt{x} + \sqrt{x}(1) - 1(\sqrt{x}) - 1} \qquad \text{Multiply the binomials.}$$

$$= \frac{x + 2\sqrt{x} + 1}{x - 1} \qquad \text{Simplify.}$$

SELF CHECK 11   Divide: $\dfrac{\sqrt{x}-1}{\sqrt{x}+1}$ $(x \geq 0)$

NOW TRY THIS

Simplify.

1. $\dfrac{\sqrt{2} - \sqrt{6}}{\sqrt{6} - \sqrt{2}}$

2. $\left(\sqrt{5} - 6\right)^2$

3. $\left(x - \sqrt{3}\right)^2$

8.5   Exercises

WARM-UPS   *Multiply. All variables represent positive values.*

1. $2\left(x + \sqrt{2}\right)$
2. $3\left(x - \sqrt{2}\right)$
3. $\sqrt{2}(3 - y)$
4. $\sqrt{3}(5 - x)$

What value when multiplied by the numerator and denominator would yield a rational denominator?

5. $\dfrac{1}{\sqrt{5}}$
6. $\dfrac{x}{\sqrt{x}}$

REVIEW   *Factor each polynomial.*

7. $x^2 - 4x - 21$
8. $y^2 + 6y - 27$
9. $6x^2y - 15xy$
10. $x^3 + 8$

VOCABULARY AND CONCEPTS   *Fill in the blanks.*

11. The symbol $\sqrt{}$ is called a _____ sign.
12. The process of changing a radical denominator of a fraction into a rational number is called _____ the denominator.
13. To rationalize the denominator of $\frac{x}{\sqrt{7}}$, we multiply the numerator and denominator by ___.
14. To rationalize the denominator of $\frac{x}{\sqrt{x} + 1}$, we multiply the numerator and denominator by _____, which is the _____ of $\sqrt{x} + 1$.

GUIDED PRACTICE   *Multiply. In square roots, all variables represent positive values. SEE EXAMPLE 1. (OBJECTIVE 1)*

15. $\sqrt{2}\sqrt{8}$
16. $\sqrt{27}\sqrt{3}$
17. $\sqrt[3]{2}\sqrt[3]{4}$
18. $\sqrt[3]{9}\sqrt[3]{3}$
19. $\left(-6\sqrt{5}\right)\left(2\sqrt{5}\right)$
20. $\left(6\sqrt{3}\right)\left(-7\sqrt{3}\right)$
21. $\left(-5\sqrt{6}\right)\left(4\sqrt{3}\right)$
22. $\left(3\sqrt{10}\right)\left(-4\sqrt{2}\right)$

Multiply. In square roots, all variables represent positive values. SEE EXAMPLE 2. (OBJECTIVE 1)

23. $\sqrt{2}\left(\sqrt{2} + 1\right)$
24. $\sqrt{3}\left(\sqrt{3} - 2\right)$
25. $\sqrt{x}\left(\sqrt{x} - 3\right)$
26. $\sqrt{y}\left(\sqrt{y} + 2\right)$
27. $\sqrt{5x}\left(\sqrt{10x} - \sqrt{5x}\right)$
28. $\sqrt{3n}\left(\sqrt{15n} + \sqrt{6n}\right)$
29. $\sqrt[3]{7}\left(\sqrt[3]{49} - 2\right)$
30. $\sqrt[3]{5}\left(\sqrt[3]{25} + 3\right)$

Multiply. SEE EXAMPLE 3. (OBJECTIVE 1)

31. $\left(\sqrt{2} + 1\right)\left(\sqrt{2} - 1\right)$
32. $\left(\sqrt{7} + 5\right)\left(\sqrt{7} - 5\right)$
33. $\left(\sqrt{5} + \sqrt{2}\right)\left(\sqrt{5} - \sqrt{2}\right)$
34. $\left(\sqrt{6} + \sqrt{7}\right)\left(\sqrt{6} - \sqrt{7}\right)$

Multiply. In square roots, all variables represent positive values. SEE EXAMPLE 4. (OBJECTIVE 1)

35. $\left(\sqrt{7} - x\right)\left(\sqrt{7} + x\right)$
36. $\left(\sqrt{2} - \sqrt{x}\right)\left(\sqrt{x} + \sqrt{2}\right)$
37. $\left(\sqrt{6x} + \sqrt{7}\right)\left(\sqrt{6x} - \sqrt{7}\right)$
38. $\left(\sqrt{7y} + 3\right)\left(\sqrt{7y} - 4\right)$

Multiply. SEE EXAMPLE 5. (OBJECTIVE 1)

39. $\left(\sqrt[3]{3x} + 4\right)\left(\sqrt[3]{3x} + 3\right)$
40. $\left(\sqrt[3]{6x} - 1\right)\left(\sqrt[3]{6x} - 4\right)$
41. $\left(\sqrt[3]{4d^2} + 5\right)\left(\sqrt[3]{2d} - 3\right)$
42. $\left(\sqrt[3]{5x} - 1\right)\left(\sqrt[3]{25x^2} + 6\right)$

Simplify. All variables represent positive numbers. SEE EXAMPLE 6. (OBJECTIVE 2)

43. $\dfrac{\sqrt{12x^3}}{\sqrt{27x}}$
44. $\dfrac{\sqrt{32}}{\sqrt{98x^2}}$
45. $\dfrac{\sqrt{18xy^2}}{\sqrt{25x}}$
46. $\dfrac{\sqrt{27y^3}}{\sqrt{75x^2y}}$

47. $\dfrac{\sqrt{196xy^3}}{\sqrt{49x^3y}}$

48. $\dfrac{\sqrt{50xyz^4}}{\sqrt{98xyz^2}}$

49. $\dfrac{\sqrt{3x^2y^3}}{\sqrt{27x}}$

50. $\dfrac{\sqrt{22xy^6}}{\sqrt{99x^3y^2}}$

Rationalize the denominator. SEE EXAMPLE 7. (OBJECTIVE 3)

51. $\dfrac{1}{\sqrt{3}}$

52. $\dfrac{1}{\sqrt{5}}$

53. $\dfrac{9}{\sqrt{27}}$

54. $\dfrac{4}{\sqrt{20}}$

55. $\dfrac{4}{\sqrt[3]{4}}$

56. $\dfrac{7}{\sqrt[3]{10}}$

57. $\dfrac{6}{\sqrt[3]{6}}$

58. $\dfrac{7}{\sqrt[3]{7}}$

Rationalize the denominator. All variables represent positive values. SEE EXAMPLE 8. (OBJECTIVE 3)

59. $\dfrac{10}{\sqrt{x}}$

60. $\dfrac{12}{\sqrt{y}}$

61. $\dfrac{\sqrt{5}}{\sqrt{2x}}$

62. $\dfrac{\sqrt{2}}{\sqrt{3z}}$

63. $\dfrac{\sqrt{2x}}{\sqrt{9y}}$

64. $\dfrac{\sqrt{3xy}}{\sqrt{4x}}$

65. $\dfrac{\sqrt[3]{16x^6}}{\sqrt[3]{54x^3}}$

66. $\dfrac{\sqrt[3]{128a^6b^3}}{\sqrt[3]{16a^3b^6}}$

Rationalize the denominator. Assume no division by 0. SEE EXAMPLE 9. (OBJECTIVE 3)

67. $\sqrt[3]{\dfrac{8}{4x^2y}}$

68. $\sqrt[3]{\dfrac{27}{9xy^2}}$

69. $\sqrt[3]{\dfrac{-5a^2}{25a^4b^2}}$

70. $\sqrt[3]{\dfrac{-16abc}{a^2b^3c^3}}$

Rationalize the denominator. SEE EXAMPLE 10. (OBJECTIVE 4)

71. $\dfrac{3}{\sqrt{3}-1}$

72. $\dfrac{3}{\sqrt{5}-2}$

73. $\dfrac{3}{\sqrt{7}+2}$

74. $\dfrac{5}{\sqrt{8}+3}$

75. $\dfrac{12}{3-\sqrt{3}}$

76. $\dfrac{10}{5-\sqrt{5}}$

77. $\dfrac{\sqrt{2}}{\sqrt{2}+1}$

78. $\dfrac{\sqrt{3}}{\sqrt{3}+1}$

Rationalize the denominator. All variables represent positive values. SEE EXAMPLE 11. (OBJECTIVE 4)

79. $\dfrac{\sqrt{x}+2}{\sqrt{x}-2}$

80. $\dfrac{\sqrt{x}-3}{\sqrt{x}+3}$

81. $\dfrac{\sqrt{3x}-1}{\sqrt{3x}+1}$

82. $\dfrac{\sqrt{2x}-5}{\sqrt{2x}-3}$

83. $\dfrac{\sqrt{3y}+3}{\sqrt{3y}-2}$

84. $\dfrac{\sqrt{5x}-1}{\sqrt{5x}+2}$

ADDITIONAL PRACTICE *Simplify. All variables in square root problems represent positive values. Assume no division by 0.*

85. $\sqrt{x^3}\sqrt{x^3}$

86. $\sqrt{a^7}\sqrt{a^3}$

87. $\left(2\sqrt{5b^3}\right)\left(3\sqrt{2b}\right)$

88. $\left(5\sqrt{3x^2}\right)\left(-4\sqrt{10x^2}\right)$

89. $\sqrt[3]{8x^2}\sqrt[3]{8x}$

90. $\sqrt[3]{4a}\sqrt[3]{250a^2}$

91. $\left(2\sqrt[3]{4p}\right)\left(3\sqrt[3]{16p^2}\right)$

92. $\left(-7\sqrt[3]{5m}\right)\left(2\sqrt[3]{25m^2}\right)$

93. $\sqrt{x}\left(\sqrt{3x}-2\right)$

94. $\sqrt{y}\left(\sqrt{y}+5\right)$

95. $7\sqrt[3]{3x}\left(2+\sqrt[3]{9x^2}\right)$

96. $5\sqrt[3]{2x^2}\left(3-\sqrt[3]{4x}\right)$

97. $\dfrac{2}{\sqrt{7}}$

98. $\dfrac{3}{\sqrt{11}}$

99. $\dfrac{5}{\sqrt{32}}$

100. $\dfrac{3}{\sqrt{18}}$

101. $\sqrt{16}\sqrt{4}$

102. $\sqrt{32}\sqrt{2}$

103. $\dfrac{\sqrt[3]{2x^2}}{\sqrt[3]{2x}}$

104. $\dfrac{\sqrt[3]{3y^4}}{\sqrt[3]{3y}}$

105. $2\sqrt{x}\left(\sqrt{9x}+3\right)$

106. $3\sqrt{z}\left(\sqrt{4z}-\sqrt{z}\right)$

107. $\dfrac{\sqrt{5x}\sqrt{10y^2}}{\sqrt{x^3y}}$

108. $\dfrac{\sqrt{7y}\sqrt{14x}}{\sqrt{8xy}}$

109. $\dfrac{\sqrt{5}}{\sqrt{3}}$

110. $\dfrac{\sqrt{3}}{\sqrt{5}}$

111. $\dfrac{-\sqrt{3}}{\sqrt{3}+1}$

112. $\dfrac{-\sqrt{2}}{\sqrt{2}-1}$

113. $\left(4\sqrt{x}\right)\left(-2\sqrt{x}\right)$

114. $\left(3\sqrt{y}\right)\left(15\sqrt{y}\right)$

115. $\left(-14\sqrt{50x}\right)\left(-5\sqrt{20x}\right)$

116. $\left(12\sqrt{24y}\right)\left(-16\sqrt{2y}\right)$

117. $3\sqrt{x}\left(2+\sqrt{x}\right)$

118. $5\sqrt{y}\left(5-\sqrt{5y}\right)$

119. $\dfrac{\sqrt[3]{5}}{\sqrt[3]{2}}$

120. $\dfrac{\sqrt[3]{2}}{\sqrt[3]{5}}$

121. $\dfrac{5}{\sqrt{3}+\sqrt{2}}$

122. $\dfrac{3}{\sqrt{3}-\sqrt{2}}$

123. $\left(\sqrt{8xy}+1\right)\left(\sqrt{8xy}+1\right)$

124. $\left(\sqrt{5x}+3\sqrt{y}\right)\left(\sqrt{5x}-3\sqrt{y}\right)$

WRITING ABOUT MATH

125. How do you know when a radical has been simplified?

126. Explain how to rationalize a denominator.

SOMETHING TO THINK ABOUT

127. How would you make the numerator of $\dfrac{\sqrt{3}}{2}$ a rational number?

128. Rationalize the numerator of $\dfrac{\sqrt{5}+2}{5}$.

Section 8.6
Solving Equations Containing Radicals; the Distance Formula

Objectives

1. Solve a radical equation containing one square root term.
2. Solve a radical equation containing two square root terms.
3. Solve a radical equation containing a cube root.
4. Find the distance between two points.
5. Solve an application containing a radical equation.

Vocabulary

power rule   squaring property of equality   distance formula

Getting Ready

In each set of numbers, verify that $a^2 + b^2 = c^2$.

1. $a = 3, b = 4, c = 5$ 2. $a = 6, b = 8, c = 10$

3. $a = 5, b = 12, c = 13$ 4. $a = 9, b = 12, c = 15$

In this section, we will discuss how to solve equations that contain radical expressions. We begin by discussing the **squaring property of equality**.

1 Solve a radical equation containing one square root term.

In general, we have a **power rule** to help us solve radical equations.

POWER RULE FOR RADICALS

If $a = b$, then $a^n = b^n$.

This states that if both sides of an equation are raised to the same power, the solutions of the original equation will be part of the solutions of the new equation. However, there could be extraneous solutions among the solutions of the new equation.

Before solving equations containing radicals, we note that if two numbers are equal, their squares are equal.

In particular, the squaring property of equality states: If $a = b$, then $a^2 = b^2$.

If we square both sides of an equation, the resulting equation may or may not have the same solutions as the original one. For example, if we square both sides of the equation

(1) $x = 2$, with the solution 2,

we obtain $x^2 = 2^2$, which simplifies as

(2) $x^2 = 4$, with solutions of 2 and -2. $2^2 = 4$ and $(-2)^2 = 4$

Equations 1 and 2 are not equivalent, because they have different solution sets. The solution -2 of Equation 2 does not satisfy Equation 1. Because squaring both sides of an equation can produce an extraneous solution, *we must always check each possible solution in the original equation.*

To solve an equation containing radical terms, we follow these steps.

SOLVING EQUATIONS CONTAINING RADICALS

1. Whenever possible, isolate a single radical term on one side of the equation.
2. Raise each side of the equation to a power equal to the index.
3. If the equation still contains a radical term, repeat steps 1 and 2.
4. Check the possible solutions in the original equation for extraneous solutions.

EXAMPLE 1   Solve: $\sqrt{x + 2} = 3$

Solution   To solve the equation $\sqrt{x + 2} = 3$, we note that the radical is already isolated on one side. We proceed to Step 2 and square both sides to eliminate the radical. Since this might produce an extraneous solution, we must check each solution.

$$\sqrt{x + 2} = 3$$
$$\left(\sqrt{x + 2}\right)^2 = (3)^2 \quad \text{Square both sides of the equation.}$$
$$x + 2 = 9 \quad \left(\sqrt{x + 2}\right)^2 = x + 2 \text{ and } 3^2 = 9$$
$$x = 7 \quad \text{Subtract 2 from both sides.}$$

We check by substituting 7 for x in the original equation.

$$\sqrt{x + 2} = 3$$
$$\sqrt{7 + 2} \overset{?}{=} 3 \quad \text{Substitute 7 for } x.$$
$$\sqrt{9} \overset{?}{=} 3$$
$$3 = 3$$

The solution checks. Since no solutions are lost in this process, 7 is the only solution of the original equation.

 SELF CHECK 1   Solve: $\sqrt{x - 4} = 9$

EXAMPLE 2   Solve: $\sqrt{x + 1} + 5 = 3$

Solution   We isolate the radical on one side and proceed as follows:

$$\sqrt{x + 1} + 5 = 3$$
$$\sqrt{x + 1} = -2 \quad \text{Subtract 5 from both sides.}$$
$$\left(\sqrt{x + 1}\right)^2 = (-2)^2 \quad \text{Square both sides.}$$
$$x + 1 = 4 \quad \left(\sqrt{x + 1}\right)^2 = x + 1 \text{ and } (-2)^2 = 4$$
$$x = 3 \quad \text{Subtract 1 from both sides.}$$

We check by substituting 3 for x in the original equation.

$$\sqrt{x + 1} + 5 = 3$$
$$\sqrt{3 + 1} + 5 \overset{?}{=} 3 \quad \text{Substitute 3 for } x.$$
$$\sqrt{4} + 5 \overset{?}{=} 3$$
$$2 + 5 \overset{?}{=} 3$$
$$7 \neq 3$$

Since $7 \neq 3$, 3 is an extraneous solution. Since this equation has no solution, its solution set is $\varnothing$.

 SELF CHECK 2 Solve: $\sqrt{x - 2} + 2 = -5$

EXAMPLE 3 The distance d (in feet) that an object will fall in t seconds is given by the formula $t = \sqrt{\frac{d}{16}}$. To find the height of the bridge shown in Figure 8-13, a man drops a stone into the water. If it takes the stone 3 seconds to hit the water, how high is the bridge?

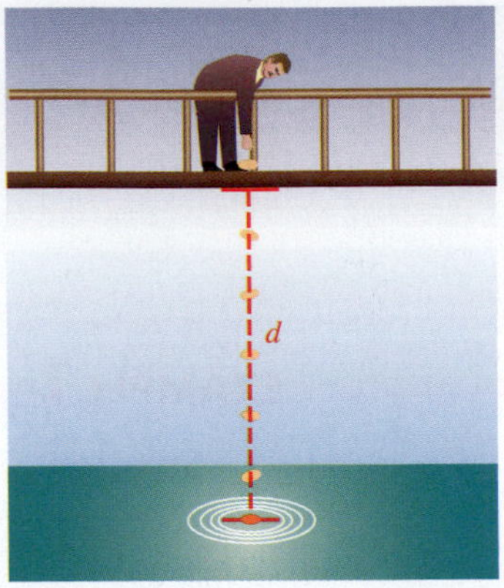

Solution We substitute 3 for t in the formula and solve for d.

$$t = \sqrt{\frac{d}{16}}$$

$$3 = \sqrt{\frac{d}{16}} \qquad \text{Substitute 3 for } t.$$

$$(3)^2 = \left(\sqrt{\frac{d}{16}}\right)^2 \qquad \text{Square both sides.}$$

$$9 = \frac{d}{16} \qquad 3^2 = 9 \text{ and } \left(\sqrt{\frac{d}{16}}\right)^2 = \frac{d}{16}$$

$$144 = d \qquad \text{Multiply both sides by 16.}$$

Figure 8-13

The bridge is 144 feet above the water.

 SELF CHECK 3 If it takes 4 seconds for the stone to hit the water, how high is the bridge?

EXAMPLE 4 Solve: $x = \sqrt{2x + 10} - 1$

Solution We add 1 to both sides to isolate the radical on the right side and then square both sides to eliminate the radical.

$$x = \sqrt{2x + 10} - 1$$

$$x + 1 = \sqrt{2x + 10} \qquad \text{Add 1 to both sides to isolate the radical.}$$

$$(x + 1)^2 = \left(\sqrt{2x + 10}\right)^2 \qquad \text{Square both sides.}$$

$$x^2 + 2x + 1 = 2x + 10 \qquad (x + 1)^2 = x^2 + 2x + 1; \left(\sqrt{2x + 10}\right)^2 = 2x + 10$$

$$x^2 - 9 = 0 \qquad \text{Subtract } 2x \text{ and 10 from both sides.}$$

$$(x - 3)(x + 3) = 0 \qquad \text{Factor.}$$

$$x - 3 = 0 \quad \text{or} \quad x + 3 = 0 \qquad \text{Set each factor equal to 0.}$$

$$x = 3 \quad | \quad x = -3 \qquad \text{Solve each linear equation.}$$

We check each possible solution.

For $x = 3$	*For $x = -3$*
$x = \sqrt{2x + 10} - 1$	$x = \sqrt{2x + 10} - 1$
$3 \overset{?}{=} \sqrt{2(3) + 10} - 1$	$-3 \overset{?}{=} \sqrt{2(-3) + 10} - 1$
$3 \overset{?}{=} \sqrt{16} - 1$	$-3 \overset{?}{=} \sqrt{4} - 1$
$3 \overset{?}{=} 4 - 1$	$-3 \overset{?}{=} 2 - 1$
$3 = 3$	$-3 \neq 1$

Since 3 is the only number that checks, it is the only solution. The false solution -3 is extraneous.

🍃 **SELF CHECK 4**   Solve: $x = \sqrt{3x + 1} + 1$

---

2 Solve a radical equation containing two square root terms.

The next example contains two square root terms.

EXAMPLE 5   Solve: $\sqrt{x + 12} = 3\sqrt{x + 4}$

Solution   We square both sides to eliminate the radicals.

$$\sqrt{x + 12} = 3\sqrt{x + 4}$$
$$\left(\sqrt{x + 12}\right)^2 = \left(3\sqrt{x + 4}\right)^2 \quad \text{Square both sides of the equation.}$$
$$x + 12 = 9(x + 4) \quad \text{Simplify.}$$
$$x + 12 = 9x + 36 \quad \text{Distribute.}$$
$$-8x = 24 \quad \text{Subtract } 9x \text{ and } 12 \text{ from both sides.}$$
$$x = -3 \quad \text{Divide both sides by } -8.$$

We check the solution by substituting -3 for x in the original equation.

$$\sqrt{x + 12} = 3\sqrt{x + 4}$$
$$\sqrt{-3 + 12} \stackrel{?}{=} 3\sqrt{-3 + 4} \quad \text{Substitute } -3 \text{ for } x.$$
$$\sqrt{9} \stackrel{?}{=} 3\sqrt{1}$$
$$3 = 3$$

The value -3 checks and is the only solution.

🍃 **SELF CHECK 5**   Solve: $\sqrt{x - 4} = 2\sqrt{x - 16}$

---

3 Solve a radical equation containing a cube root.

The power rule for radicals can be applied to cube roots. If $a = b$, then $a^3 = b^3$.

To solve an equation involving a cube root, we isolate the radical term and then cube both sides of the equation.

EXAMPLE 6   Solve: $\sqrt[3]{2x + 10} = 2$

Solution   We can cube both sides and proceed as follows:

$$\sqrt[3]{2x + 10} = 2$$
$$\left(\sqrt[3]{2x + 10}\right)^3 = (2)^3 \quad \text{Cube both sides.}$$
$$2x + 10 = 8 \quad \text{Simplify.}$$
$$2x = -2 \quad \text{Subtract 10 from both sides.}$$
$$x = -1 \quad \text{Divide both sides by 2.}$$

Check the result.

🍃 **SELF CHECK 6**   Solve: $\sqrt[3]{3x - 3} = 3$

---

4 Find the distance between two points.

We can use the Pythagorean theorem to derive a formula for finding the distance between two points $P(x_1, y_1)$ and $Q(x_2, y_2)$ on a rectangular coordinate system. The distance d

between points P and Q is the length of the hypotenuse of the triangle in Figure 8-14. The two legs have lengths $x_2 - x_1$ and $y_2 - y_1$.

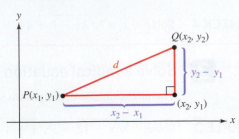

Figure 8-14

By the Pythagorean theorem, we have

$$d^2 = (x_2 - x_1)^2 + (y_2 - y_1)^2$$

We can take the square root of both sides of the equation to obtain the **distance formula**.

$$d = \sqrt{(x_2 - x_1)^2 + (y_2 - y_1)^2}$$

THE DISTANCE FORMULA

The distance d between points $P(x_1, y_1)$ and $Q(x_2, y_2)$ is given by the formula

$$d = \sqrt{(x_2 - x_1)^2 + (y_2 - y_1)^2}$$

EXAMPLE 7   Find the distance between points $P(1, 5)$ and $Q(4, 9)$. (See Figure 8-15.)

Solution   We can use the distance formula by substituting 1 for x_1, 5 for y_1, 4 for x_2, and 9 for y_2.

$$\begin{aligned}
d &= \sqrt{(x_2 - x_1)^2 + (y_2 - y_1)^2} \\
&= \sqrt{(4 - 1)^2 + (9 - 5)^2} \\
&= \sqrt{3^2 + 4^2} \\
&= \sqrt{9 + 16} \\
&= \sqrt{25} \\
&= 5
\end{aligned}$$

The distance between points P and Q is 5 units.

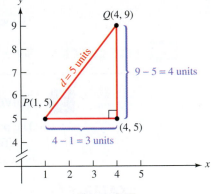

Figure 8-15

🌱 **SELF CHECK 7**   Find the distance between $P(-2, 1)$ and $Q(4, 9)$.

5   Solve an application containing a radical equation.

The distance formula can be used to help solve many applications.

EXAMPLE 8   **BUILDING FREEWAYS**   In a city, streets run north and south, and avenues run east and west. Consecutive streets are 750 feet apart and consecutive avenues are 750 feet apart. The city plans to construct a freeway from the intersection of 21st Street and 4th Avenue to the intersection of 111th Street and 60th Avenue. How long will it be in miles?

Solution   We can represent the roads of the city by the coordinate system shown in Figure 8-16, where each unit on each axis represents 750 feet. We represent the end of the free-way at 21st Street and 4th Avenue by the point $(x_1, y_1) = (21, 4)$. The other end is

$(x_1, y_1) = (111, 60)$. We now can use the distance formula to find the length of the freeway in blocks.

$$d = \sqrt{(x_2 - x_1)^2 + (y_2 - y_1)^2}$$
$$= \sqrt{(111 - 21)^2 + (60 - 4)^2}$$
$$= \sqrt{90^2 + 56^2}$$
$$= \sqrt{8{,}100 + 3{,}136}$$
$$= \sqrt{11{,}236}$$
$$= 106 \qquad\qquad \text{Find the square root.}$$

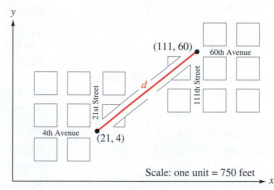

Figure 8-16

The freeway will be 106 blocks in length. Since each block represents 750 feet, the length of the freeway will be $106 \cdot 750 = 79{,}500$ feet. Since there are 5,280 feet in 1 mile, we can divide 79,500 by 5,280 to convert 79,500 feet to 15.056818 miles. The freeway will be about 15 miles long.

 SELF CHECK 8   If the city plans to construct the freeway from the intersection of 15th Street and 20th Avenue to the intersection of 111th Street and 60th Avenue, how long will it be in miles?

 **SELF CHECK ANSWERS**   **1.** 85   **2.** $\varnothing$; 51 is extraneous   **3.** 256 ft   **4.** 5; 0 is extraneous   **5.** 20   **6.** 10   **7.** 10 units   **8.** The freeway will be about 15 miles long.

NOW TRY THIS

1. Explain why you must check your answers when using the power rule.

2. Find the distance between the points $(2a, a)$ and $(2a + 1, a - 3)$.

8.6   Exercises

WARM-UPS   *Is the given value a solution to the problem?*

1. $\sqrt{x} = 4$; 16

2. $\sqrt{x - 1} = 2$; 5

3. $\sqrt{10 - x} = -2$; 6

4. $\sqrt{2x + 1} = -1$; -1

5. $\sqrt{2x + 2} - x = -3$; 1

6. $3\sqrt{2 - 3x} = \sqrt{-7x - 2}$; 1

REVIEW   *Solve each system.*

7. $\begin{cases} x + y = 5 \\ x - y = -1 \end{cases}$

8. $\begin{cases} 2x + y = 0 \\ x + 3y = 5 \end{cases}$

9. $\begin{cases} 2x + 3y = 0 \\ 3x - 2y = 13 \end{cases}$

10. $\begin{cases} 3x - 4y = 11 \\ 4x + y = -17 \end{cases}$

VOCABULARY AND CONCEPTS *Fill in the blanks.*

11. A false solution that occurs because you square both sides of an equation is called an _____ solution.

12. To eliminate a radical, isolate the radical term and then raise each side of the equation to a power equal to the _____.

13. If $a = b$, then $a^n = $ __.

14. The distance formula states that

$d = $ _____.

Find the error in each solution.

15. $\sqrt{x - 2} = 3$
$x - 2 = 3$
$x = 5$

16. $2 = \sqrt{x - 9}$
$4 = x - 9$
$-5 = x$
$x = -5$

GUIDED PRACTICE *Solve each equation. Check all solutions.*
SEE EXAMPLE 1. (OBJECTIVE 1)

17. $\sqrt{x} = 6$ **18.** $\sqrt{x} = 5$

19. $\sqrt{x} = 7$ **20.** $\sqrt{x} = 8$

21. $\sqrt{x + 3} = 2$ **22.** $\sqrt{x - 2} = 3$

23. $\sqrt{x + 4} = 6$ **24.** $\sqrt{x + 8} = 12$

Solve each equation. Check all solutions. SEE EXAMPLE 2. (OBJECTIVE 1)

25. $\sqrt{x + 2} + 4 = 9$ **26.** $\sqrt{x - 5} - 3 = 4$

27. $\sqrt{2x + 10} + 3 = 5$ **28.** $\sqrt{3x + 4} + 7 = 12$

29. $\sqrt{5x + 9} + 4 = 7$ **30.** $\sqrt{9x + 25} - 2 = 3$

31. $\sqrt{7 - 5x} + 4 = 3$ **32.** $\sqrt{7 + 6x} - 4 = -3$

Solve each equation. Check all solutions. SEE EXAMPLE 4. (OBJECTIVE 1)

33. $\sqrt{x + 1} = x - 1$ **34.** $\sqrt{x + 4} = x - 2$

35. $\sqrt{x + 5} = x + 5$ **36.** $\sqrt{x + 9} = x + 7$

37. $\sqrt{2x + 2} - x = -3$ **38.** $\sqrt{3x + 3} + 5 = x$

39. $x - 1 = \sqrt{x - 1}$ **40.** $x - 2 = \sqrt{x + 10}$

Solve each equation. Check all solutions. SEE EXAMPLE 5. (OBJECTIVE 2)

41. $\sqrt{3x + 3} = 3\sqrt{x - 1}$

42. $2\sqrt{4x + 5} = 2\sqrt{x + 4}$

43. $2\sqrt{3x + 4} = \sqrt{5x + 9}$

44. $3\sqrt{2 - 3x} = \sqrt{-7x - 2}$

Solve each equation. Check all solutions. SEE EXAMPLE 6. (OBJECTIVE 3)

45. $\sqrt[3]{x - 1} = 4$ **46.** $\sqrt[3]{2x + 5} = 3$

47. $\sqrt[3]{3x - 5} = 4$ **48.** $\sqrt[3]{-3x - 35} = 4$

Find the distance between the points. Give the exact answer in simplest form. SEE EXAMPLE 7. (OBJECTIVE 4)

49. $(3, -4)$, and $(0, 0)$ **50.** $(0, 0)$ and $(-6, 8)$

51. $(2, 4)$ and $(5, 8)$ **52.** $(5, 9)$ and $(8, 13)$

53. $(-1, -6)$ and $(4, 6)$ **54.** $(-5, -2)$ and $(7, 3)$

55. $(7, -4)$ and $(-8, 4)$ **56.** $(10, 4)$ and $(2, -2)$

ADDITIONAL PRACTICE *Solve each equation. Check all solutions.*

57. $\sqrt{x} = -2$ **58.** $\sqrt{x} = -1$

59. $\sqrt{x + 18} = 4$ **60.** $\sqrt{x + 5} = 7$

61. $\sqrt{3x + 7} = 4$ **62.** $\sqrt{6x + 19} = 7$

63. $\sqrt{7x + 8} = 6$ **64.** $\sqrt{11x - 2} = 3$

65. $\sqrt{13x + 14} = 1$ **66.** $\sqrt{8x + 9} = 1$

67. $\sqrt{10 - x} = -3$ **68.** $\sqrt{7 + x} = -4$

69. $\sqrt{6 + 2x} = 4$ **70.** $\sqrt{5 - x} = 10$

71. $\sqrt[3]{\dfrac{1}{2}x - 3} = 2$ **72.** $\sqrt[5]{x - 245} = -3$

73. $x = \sqrt{3 - x} + 3$ **74.** $x = \sqrt{x - 4} + 4$

75. $\sqrt{3x + 6} = 2\sqrt{2x - 11}$

76. $2\sqrt{9x + 16} = \sqrt{3x + 64}$

77. $\sqrt[4]{x + 4} = 1$ **78.** $\sqrt[4]{x + 18} = 2$

Find the distance between the points. Give the exact answer in simplest form.

79. $(-12, 6)$ and $(8, -9)$ **80.** $(3, -16)$ and $(-7, 8)$

81. $(8, -2)$ and $(6, -12)$ **82.** $(-5, 6)$ and $(-25, -10)$

APPLICATIONS *Use a calculator to help solve each. Give any decimal answer rounded to the nearest tenth. SEE EXAMPLE 3. (OBJECTIVE 2)*

83. Falling objects The distance s (in feet) that an object will fall in t seconds is given by the formula

$$t = \frac{\sqrt{s}}{4}$$

How deep is a shaft if a stone dropped down it hits bottom in 4 seconds?

84. Falling objects How deep would the shaft in Exercise 83 be if the stone hit bottom in 3 seconds?

85. Horology The time t (in seconds) required for a pendulum to swing through one cycle is given by the formula $t = 1.11\sqrt{L}$. Find the length L of a pendulum that completes one cycle in $\frac{3}{2}$ seconds.

86. Foucault pendulums A long pendulum in Chicago's Museum of Science and Industry completes one cycle in 8.91 seconds. How long is it? (See the illustration and Exercise 85.)

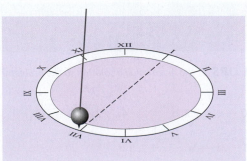

87. Electronics The current I (in amperes), the resistance R (in ohms), and the power P (in watts) are related by the formula

$$I = \sqrt{\frac{P}{R}}$$

Find the power used by an electrical appliance that draws 7 amps when the resistance is 20 ohms.

88. Electronics Find the resistance of a 500-watt space heater that draws 7 amperes. (See Exercise 87.)

89. Road safety The formula $s = k\sqrt{d}$ relates the speed s (in mph) of a car and the distance d (in ft) of the skid when a driver hits the brakes. For wet pavement, $k = 3.24$. To the nearest tenth, how far will a car skid if it is going 56 mph?

90. Road safety To the nearest tenth, how far will the car in Exercise 89 skid if it is going 70 mph?

91. Road safety To the nearest tenth, how far will the car in Exercise 89 skid if it is going 50 mph on dry pavement? On dry pavement, $k = 5.34$.

92. Road safety To the nearest tenth, how far will the car in Exercise 89 skid if it is going 70 mph on dry pavement? On dry pavement, $k = 5.34$.

93. Satellite orbits The orbital speed s of an Earth satellite is related to its distance r from Earth's center by the formula

$$s = \frac{2.029 \times 10^7}{\sqrt{r}}$$

If the satellite's orbital speed is 7×10^3 meters per second, find its altitude a above Earth's surface. (See the illustration.)

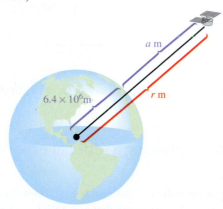

94. Highways A curve banked at 8° will accommodate traffic traveling s miles per hour if the radius of the curve is r feet, according to the equation $s = 1.45\sqrt{r}$. If highway engineers expect 65-mph traffic, what radius should they specify? (See the illustration.)

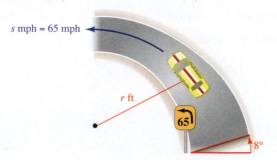

95. Relativity Einstein's theory of relativity predicts that an object moving at speed v will be shortened in the direction of its motion by a factor f given by

$$f = \sqrt{1 - \frac{v^2}{c^2}}$$

where c is the speed of light. Solve this formula for v^2.

96. Relativity Einstein's theory of relativity predicts that a clock moving at speed v will run slower by a factor f given by

$$f = \frac{1}{\sqrt{1 - \frac{v^2}{c^2}}}$$

where c is the speed of light. Solve this equation for v^2.

97. Windmills The power produced by a certain windmill is related to the speed of the wind by the formula

$$s = \sqrt[3]{\frac{P}{0.02}}$$

where P is the power (in watts) and s is the speed of the wind (in mph). How much power will the windmill produce if the wind is blowing at 30 mph?

98. Windmills If the wind is blowing at 20 mph, how much power is the windmill in Exercise 97 producing?

 Use a calculator to help solve each. Give any decimal answer rounded to the nearest tenth. SEE EXAMPLE 8. (OBJECTIVE 5)

99. Road construction If the freeway of Example 8 were to go from 21st Street and 4th Avenue to 120th Street and 70th Avenue, how long would it be?

100. Road construction If the freeway of Example 8 were to go from 10th Street and 3rd Avenue to 100th Street and 60th Avenue, how long would it be?

101. Circles If two endpoints of a diameter of a circle are $(5, -9)$ and $(3, 5)$, what is the radius of the circle?

102. Circles If two endpoints of a diameter of a circle are $(-14, 10)$ and $(-2, -16)$, what is the radius of the circle?

WRITING ABOUT MATH

103. Explain why a check is necessary when solving radical equations.

104. How would you know, without solving it, that the equation $\sqrt{x + 2} = -4$ has no solutions?

SOMETHING TO THINK ABOUT

105. Solve: $\sqrt[4]{3x + 4} = 5$.

106. Solve: $\sqrt{\sqrt{x + 2}} = 3$.

Section 8.7

Rational Exponents

Objectives

1 Simplify a numerical expression containing a rational exponent with a numerator of 1.

2 Simplify a numerical expression containing a rational exponent with a numerator other than 1.

3 Apply the rules of exponents to an expression containing rational exponents.

Vocabulary

rational exponent

Getting Ready

Simplify each expression. Assume no variable is zero.

1. $x^3 x^2$

2. $(x^4)^3$

3. $\dfrac{x^7}{x^2}$

4. $(xy)^0$

5. $(ab)^{-1}$

6. $(a^3 b^2)^4$

7. $\left(\dfrac{a^3}{b^2}\right)^3$

8. $(a^3 a^2 a)^4$

9. $(b^2)^3 (b^3)^2$

So far, we have encountered only integer exponents. In this section, we will consider exponents that are fractions.

1 Simplify a numerical expression containing a rational exponent with a numerator of 1.

We have seen that a positive integer exponent indicates the number of times that a base is to be used as a factor in a product. For example, x^4 means that x is to be used as a factor four times.

4 factors of x

$$x^4 = x \cdot x \cdot x \cdot x$$

Furthermore, we recall the following rules of exponents.

If m and n are natural numbers and $x, y \neq 0$, then

$$x^m x^n = x^{m+n} \qquad (x^m)^n = x^{mn} \qquad (xy)^n = x^n y^n \qquad \left(\frac{x}{y}\right)^n = \frac{x^n}{y^n}$$

$$x^0 = 1 \qquad x^{-1} = \frac{1}{x} \qquad \frac{x^m}{x^n} = x^{m-n}$$

To give meaning to **rational** (fractional) **exponents**, we consider $\sqrt{7}$. Because $\sqrt{7}$ is the positive number whose square is 7, we have

$$\left(\sqrt{7}\right)^2 = 7$$

We now consider the symbol $7^{1/2}$. If fractional exponents are to follow the same rules as integer exponents, the square of $7^{1/2}$ must be 7, because

$$(7^{1/2})^2 = 7^{(1/2)2} \qquad \text{Keep the base and multiply the exponents.}$$
$$= 7^1 \qquad \tfrac{1}{2} \cdot 2 = 1$$
$$= 7$$

Since $(\mathbf{7^{1/2}})^2$ and $\left(\sqrt{\mathbf{7}}\right)^2$ both equal 7, we define $7^{1/2}$ to be $\sqrt{7}$. Similarly, we make these definitions:

$$7^{1/3} = \sqrt[3]{7} \qquad\qquad 7^{1/7} = \sqrt[7]{7}$$

and so on.

RATIONAL EXPONENTS

If n is a positive integer greater than 1 and $\sqrt[n]{x}$ is a real number, then

$$x^{1/n} = \sqrt[n]{x}$$

EXAMPLE 1 Simplify: **a.** $64^{1/2}$ **b.** $64^{1/3}$ **c.** $(-64)^{1/3}$ **d.** $64^{1/6}$

Solution In each case we will change from rational exponent notation to radical notation and simplify.

a. $64^{1/2} = \sqrt{64} = 8$ **b.** $64^{1/3} = \sqrt[3]{64} = 4$
c. $(-64)^{1/3} = \sqrt[3]{-64} = -4$ **d.** $64^{1/6} = \sqrt[6]{64} = 2$

 SELF CHECK 1 Simplify: **a.** $81^{1/2}$ **b.** $125^{1/3}$ **c.** $(-32)^{1/5}$ **d.** $81^{1/4}$

2 Simplify a numerical expression containing a rational exponent with a numerator other than 1.

We can extend the definition of $x^{1/n}$ to cover fractional exponents for which the numerator is not 1. For example, because $4^{3/2}$ can be written as $(4^{1/2})^3$, we have

$$4^{3/2} = (4^{1/2})^3 = \left(\sqrt{4}\right)^3 = 2^3 = 8$$

Because $4^{3/2}$ can also be written as $(4^3)^{1/2}$, we have

$$4^{3/2} = (4^3)^{1/2} = 64^{1/2} = \sqrt{64} = 8$$

In general, $x^{m/n}$ can be written as $(x^{1/n})^m$ or as $(x^m)^{1/n}$. Since $(x^{1/n})^m = \left(\sqrt[n]{x}\right)^m$ and $(x^m)^{1/n} = \sqrt[n]{x^m}$, we make the following definition.

CHANGING FROM RATIONAL EXPONENTS TO RADICALS

If m and n are positive integers, the fraction m/n cannot be simplified, and $x \geq 0$ if n is even, then

$$x^{m/n} = \sqrt[n]{x^m} = \left(\sqrt[n]{x}\right)^m$$

COMMENT Recall that the radicand of an even root cannot be negative if we are working with real numbers. In the rule above, we could state that "x is nonnegative if n is even" to emphasize this fact. There are no such restrictions on odd roots.

EXAMPLE 2 Simplify: **a.** $8^{2/3}$ **b.** $(-27)^{4/3}$

Solution In each case, we will apply the rule $x^{m/n} = \sqrt[n]{x^m} = \left(\sqrt[n]{x}\right)^m$.

a. $8^{2/3} = \left(\sqrt[3]{8}\right)^2$ or $8^{2/3} = \sqrt[3]{8^2}$
$\qquad = 2^2 \qquad\qquad\qquad = \sqrt[3]{64}$
$\qquad = 4 \qquad\qquad\qquad\quad = 4$

b. $(-27)^{4/3} = \left(\sqrt[3]{-27}\right)^4$ or $(-27)^{4/3} = \sqrt[3]{(-27)^4}$
$\qquad\qquad = (-3)^4 \qquad\qquad\qquad\qquad = \sqrt[3]{531{,}441}$
$\qquad\qquad = 81 \qquad\qquad\qquad\qquad\quad = 81$

 SELF CHECK 2 Simplify: **a.** $16^{3/2}$ **b.** $(-8)^{4/3}$

The work in Example 2 suggests that in order to avoid large numbers, it is usually easier to take the root of the base first.

EXAMPLE 3 Simplify: **a.** $125^{4/3}$ **b.** $9^{5/2}$ **c.** $-25^{3/2}$ **d.** $(-27)^{2/3}$

Solution In each case, we will apply the rule $x^{m/n} = \sqrt[n]{x^m} = \left(\sqrt[n]{x}\right)^m$.

a. $125^{4/3} = \left(\sqrt[3]{\mathbf{125}}\right)^4$ **b.** $9^{5/2} = \left(\sqrt{\mathbf{9}}\right)^5$
$\qquad\quad = \mathbf{5}^4 \qquad\qquad\qquad\qquad\quad = \mathbf{3}^5$
$\qquad\quad = 625 \qquad\qquad\qquad\qquad\quad = 243$

c. $-25^{3/2} = -\left(\sqrt{\mathbf{25}}\right)^3$ **d.** $(-27)^{2/3} = \left(\sqrt[3]{\mathbf{-27}}\right)^2$
$\qquad\quad = -\mathbf{5}^3 \qquad\qquad\qquad\qquad\quad = (\mathbf{-3})^2$
$\qquad\quad = -125 \qquad\qquad\qquad\qquad\quad = 9$

 SELF CHECK 3 Simplify: **a.** $100^{3/2}$ **b.** $(-8)^{2/3}$ **c.** $-16^{3/2}$ **d.** $(-125)^{2/3}$

3 Apply the rules of exponents to an expression containing rational exponents.

The familiar rules of exponents are valid for rational exponents. The following example illustrates the use of each rule.

EXAMPLE 4 Use the provided rule to write each expression on the left in a different form.

Problem	*Rule*
a. $4^{2/5}4^{1/5} = 4^{2/5+1/5} = 4^{3/5}$	$x^m x^n = x^{m+n}$
b. $(5^{2/3})^{1/2} = 5^{(2/3)(1/2)} = 5^{1/3}$	$(x^m)^n = x^{mn}$
c. $(3x)^{2/3} = 3^{2/3}x^{2/3}$	$(xy)^m = x^m y^m$
d. $\dfrac{4^{3/5}}{4^{2/5}} = 4^{3/5-2/5} = 4^{1/5}$	$\dfrac{x^m}{x^n} = x^{m-n}$
e. $\left(\dfrac{3}{2}\right)^{2/5} = \dfrac{3^{2/5}}{2^{2/5}}$	$\left(\dfrac{x}{y}\right)^n = \dfrac{x^n}{y^n}$
f. $4^{-2/3} = \dfrac{1}{4^{2/3}}$	$x^{-n} = \dfrac{1}{x^n}$
g. $5^0 = 1$	$x^0 = 1 \quad (x \neq 0)$

SELF CHECK 4   Use the preceding rules to write each expression in simplified form.

    a. $5^{1/3}5^{1/3}$ **b.** $(5^{1/3})^4$ **c.** $(3x)^{1/5}$

    d. $\dfrac{5^{3/7}}{5^{2/7}}$ **e.** $\left(\dfrac{2}{3}\right)^{2/3}$ **f.** $5^{-2/7}$ **g.** 12^0

We often can use the rules of exponents to simplify expressions containing rational exponents.

EXAMPLE 5   Simplify: **a.** $64^{-2/3}$ **b.** $(x^2)^{1/2}$ **c.** $(x^6y^4)^{1/2}$ **d.** $(27x^{12})^{-1/3}$ $(x > 0, y \geq 0)$

Solution   In each case, we apply one or more of the rules listed in Example 4.

a. $64^{-2/3} = \dfrac{1}{64^{2/3}}$ **b.** $(x^2)^{1/2} = x^{2(1/2)}$

$= \dfrac{1}{(64^{1/3})^2}$ $= x^1$

$= \dfrac{1}{4^2}$ $= x$

$= \dfrac{1}{16}$

c. $(x^6y^4)^{1/2} = x^{6(1/2)}y^{4(1/2)}$ **d.** $(27x^{12})^{-1/3} = \dfrac{1}{(27x^{12})^{1/3}}$

$= x^3y^2$ $= \dfrac{1}{27^{1/3}x^{12(1/3)}}$

                                                     $= \dfrac{1}{3x^4}$

 SELF CHECK 5   Simplify. Assume there are no divisions by 0.

    a. $25^{-3/2}$ **b.** $(x^3)^{1/3}$ **c.** $(x^6y^9)^{2/3}$ **d.** $(64x^9)^{-1/3}$

EXAMPLE 6   Write each expression in a form that has only one exponent.

    a. $x^{1/3}x^{1/2}$ **b.** $\dfrac{3x^{2/3}}{6x^{1/5}}$ $(x \neq 0)$ **c.** $\dfrac{2x^{-1/2}}{x^{3/4}}$ $(x > 0)$

Solution   In each case, we will write the rational exponents with a common denominator and then apply one of the rules listed in Example 4.

a. $x^{1/3}x^{1/2} = x^{2/6}x^{3/6}$ Find a common denominator in the fractional exponents.

$= x^{5/6}$ Keep the base and add the exponents.

b. $\dfrac{3x^{2/3}}{6x^{1/5}} = \dfrac{3x^{10/15}}{6x^{3/15}}$ Find a common denominator in the fractional exponents.

$= \dfrac{1}{2}x^{10/15 - 3/15}$ Simplify $\frac{3}{6}$, keep the base, and subtract the exponents.

$= \dfrac{1}{2}x^{7/15}$

c. $\dfrac{2x^{-1/2}}{x^{3/4}} = \dfrac{2x^{-2/4}}{x^{3/4}}$ Find a common denominator in the fractional exponents.

$\quad = 2x^{-2/4-3/4}$ Keep the base and subtract the exponents.

$\quad = 2x^{-5/4}$ Simplify.

$\quad = \dfrac{2}{x^{5/4}}$ $x^{-5/4} = \dfrac{1}{x^{5/4}}$

 SELF CHECK 6   Simplify: **a.** $x^{2/3}x^{1/2}$ **b.** $\dfrac{x^{2/3}}{2x^{1/4}}$ **c.** $\dfrac{3x^{-3/4}}{x^{1/2}}$

 **SELF CHECK ANSWERS**

1. a. 9 **b.** 5 **c.** −2 **d.** 3 **2. a.** 64 **b.** 16 **3. a.** 1,000 **b.** 4 **c.** −64 **d.** 25 **4. a.** $5^{2/3}$
b. $5^{4/3}$ **c.** $3^{1/5}x^{1/5}$ **d.** $5^{1/7}$ **e.** $\dfrac{2^{2/3}}{3^{2/3}}$ **f.** $\dfrac{1}{5^{2/7}}$ **g.** 1 **5. a.** $\dfrac{1}{125}$ **b.** x **c.** x^4y^6 **d.** $\dfrac{1}{4x^3}$ **6. a.** $x^{7/6}$
b. $\dfrac{1}{2}x^{5/12}$ **c.** $\dfrac{3}{x^{5/4}}$

NOW TRY THIS

1. Simplify: $\left(\dfrac{1}{64}\right)^{-1/2}$

2. Explain why $\sqrt[3]{x^6} = x^2$ for all values of x, but $\sqrt{x^6} = x^3$ only if $x \ge 0$.

3. If $3^0 = 1$, $3^1 = 3$, $3^2 = 9$, etc., between which two of the given integers will the following be found?

    a. $3^{1/2}$ **b.** $3^{3/2}$

8.7   Exercises

WARM-UPS *Simplify.*

1. $\left(\sqrt{3}\right)^2$ **2.** $(3^{1/2})^2$
3. $(5^{1/2})^2$ **4.** $\left(\sqrt{5}\right)^2$
5. $\left(\sqrt{16}\right)^3$ **6.** $\left(\sqrt[3]{8}\right)^2$
7. $\left(\sqrt[3]{-27}\right)^2$ **8.** $\left(\sqrt[4]{81}\right)^3$

REVIEW *Factor each expression.*

9. $3z^2 - 15tz + 12t^2$ **10.** $a^4 - b^4$

Solve each equation.

11. $\dfrac{x-5}{7} + \dfrac{2}{5} = \dfrac{7-x}{5}$ **12.** $\dfrac{t}{t+2} - 1 = \dfrac{1}{1-t}$

VOCABULARY AND CONCEPTS *Fill in the blanks. Assume no variable is 0.*

13. A fractional exponent is also called a _____ exponent.
14. In the expression $(2x)^{1/3}$, $2x$ is called the _____, and the exponent is ___.
15. $x^m x^n = $ _____
16. $(x^m)^n = $ _____

17. $\left(\dfrac{x}{y}\right)^n = $ _____   **18.** $x^0 = $ _____

19. $x^{-1} = $ _____   **20.** $\dfrac{x^m}{x^n} = $ _____

GUIDED PRACTICE *Simplify each expression.* SEE EXAMPLE 1. (OBJECTIVE 1)

21. $81^{1/2}$ **22.** $100^{1/2}$
23. $-144^{1/2}$ **24.** $-400^{1/2}$
25. $27^{1/3}$ **26.** $8^{1/3}$
27. $-125^{1/3}$ **28.** $-1{,}000^{1/3}$

Simplify each expression. SEE EXAMPLE 2. (OBJECTIVE 2)

29. $1{,}000^{2/3}$ **30.** $27^{2/3}$
31. $64^{4/3}$ **32.** $8^{4/3}$
33. $(-8)^{2/3}$ **34.** $(-125)^{2/3}$
35. $25^{3/2}$ **36.** $4^{5/2}$

Simplify each expression. SEE EXAMPLE 3. (OBJECTIVE 2)

37. $81^{3/2}$ **38.** $-9^{3/2}$
39. $27^{4/3}$ **40.** $8^{2/3}$

41. $\left(\dfrac{8}{27}\right)^{2/3}$

42. $\left(\dfrac{27}{64}\right)^{2/3}$

43. $-\left(\dfrac{16}{81}\right)^{3/4}$

44. $\left(\dfrac{49}{64}\right)^{3/2}$

Simplify each expression. Write each answer with positive exponents.
SEE EXAMPLE 4. (OBJECTIVE 3)

45. $4^{3/5} \cdot 4^{1/5}$

46. $9^{2/7} 9^{3/7}$

47. $5^{2/3} 5^{4/3}$

48. $2^{7/8} 2^{9/8}$

49. $(7x)^{3/4}$

50. $(5y)^{2/3}$

51. $(5^{2/7})^7$

52. $(3^{3/8})^8$

53. $\dfrac{7^{8/3}}{7^{7/3}}$

54. $\dfrac{11^{9/7}}{11^{2/7}}$

55. $\dfrac{5^{11/3}}{5^{2/3}}$

56. $\dfrac{9^{7/2}}{9^{4/2}}$

57. $4^{-1/2}$

58. $8^{-1/3}$

59. $27^{-2/3}$

60. 36^0

Simplify each expression. Assume that all variables represent positive values. SEE EXAMPLE 5. (OBJECTIVE 3)

61. $(9x^2)^{-1/2}$

62. $(8x^9)^{-1/3}$

63. $(x^{12})^{1/6}$

64. $(x^{18})^{1/9}$

65. $(x^{14}y^{21})^{-1/7}$

66. $(x^8y^{12})^{3/4}$

67. $x^{5/6}x^{7/6}$

68. $x^{2/3}x^{7/3}$

Simplify each expression. Assume that all variables represent positive values. Assume no division by 0. SEE EXAMPLE 6. (OBJECTIVE 3)

69. $\dfrac{10x^{4/5}}{2x^{2/3}}$

70. $\dfrac{5x^{5/3}}{15x^{1/2}}$

71. $\dfrac{x^{1/7}x^{3/7}}{x^{2/7}}$

72. $\dfrac{x^{5/6}x^{5/6}}{x^{7/6}}$

73. $\left(\dfrac{x^{3/5}}{x^{2/5}}\right)^5$

74. $\left(\dfrac{x^{2/9}}{x^{1/9}}\right)^9$

75. $\left(\dfrac{y^{2/7}y^{3/7}}{y^{4/7}}\right)^{49}$

76. $\left(\dfrac{z^{3/5}z^{6/5}}{z^{2/5}}\right)^5$

77. $\left(\dfrac{y^{5/6}y^{7/6}}{y^{1/3}y}\right)^3$

78. $\left(\dfrac{t^{4/9}t^{5/9}}{t^{1/9}t^{2/9}}\right)^9$

79. $x^{2/3}x^{3/4}$

80. $a^{3/5}a^{-1/2}$

81. $(32b^{1/2})^{-2/5}$

82. $(x^{2/5})^{4/7}$

83. $\dfrac{4t^{2/3}}{8t^{-2/5}}$

84. $\dfrac{p^{3/4}}{p^{1/3}}$

85. $\left(\dfrac{1}{4}\right)^{1/2}$

86. $\left(\dfrac{1}{25}\right)^{1/2}$

87. $\left(\dfrac{4}{49}\right)^{1/2}$

88. $\left(\dfrac{9}{64}\right)^{1/2}$

89. $\left(\dfrac{1}{64}\right)^{1/3}$

90. $\left(\dfrac{1}{1,000}\right)^{1/3}$

91. $\left(\dfrac{1}{64}\right)^{1/6}$

92. $\left(\dfrac{32}{243}\right)^{1/5}$

ADDITIONAL PRACTICE *Simplify. Assume that all variables in the radicand of an even root represent positive values. Assume no division by 0. Express each answer with positive exponents only.*

93. $(-8)^{1/3}$

94. $(-125)^{1/3}$

95. $16^{1/4}$

96. $81^{1/4}$

97. $32^{1/5}$

98. $-32^{1/5}$

99. $-243^{1/5}$

100. $\left(-\dfrac{1}{32}\right)^{1/5}$

101. $81^{3/4}$

102. $256^{3/4}$

103. $(-32)^{3/5}$

104. $243^{2/5}$

105. $(-27)^{-4/3}$

106. $(-8)^{-4/3}$

107. $\left(\dfrac{27}{64}\right)^{1/3}$

108. $\left(\dfrac{64}{125}\right)^{1/3}$

109. $16^{-3/2}$

110. $100^{-5/2}$

111. $32^{3/5}$

112. $-243^{3/5}$

113. $y^{4/7}y^{10/7}$

114. $y^{5/11}y^{6/11}$

115. $\dfrac{x^{4/5}x^{1/3}}{x^{2/15}}$

116. $\dfrac{y^{2/3}y^{3/5}}{y^{1/5}}$

117. $\dfrac{a^{2/5}a^{1/5}}{a^{-1/3}}$

118. $\dfrac{q^{3/4}q^{4/5}}{q^{-2/3}}$

119. $(2^{1/2}3^{1/2})^2$

120. $(3^{2/3}5^{1/3})^3$

121. $(4^{3/4}3^{1/4})^4$

122. $(2^{1/5}3^{2/5})^5$

WRITING ABOUT MATH

123. Is $(-4)^{1/2}$ a real number? Explain.

124. Is $(-8)^{1/3}$ a real number? Explain.

SOMETHING TO THINK ABOUT *If $x > y$, which is the larger number in each pair?*

125. $2^x, 2^y$

126. $\left(\dfrac{1}{2}\right)^x, \left(\dfrac{1}{2}\right)^y$

Projects

The Italian mathematician and physicist Galileo Galilei (1564–1642) is best known as the inventor of the telescope and for his discovery of four of the moons of Jupiter, still known as the *Galilean satellites*. Less known is his discovery that a pendulum could be used to keep accurate time. While praying one day in the cathedral, Galileo noticed a suspended candle left swinging after it had been lit. Using his own pulse as a timer, Galileo discovered that the time for one swing remained unchanged as the swings themselves became smaller. By more experimenting, Galileo discovered the relationship between the length of a pendulum and its **period**, the time it takes to complete one swing.

You can discover the relationship, too. You will need a stopwatch, a calculator, a meter stick, and a pendulum. To make the pendulum, try tying a length of string to a rubber band and wrapping the band tightly around a small rock. By tying the free end of the string to a support, you can change the length of the pendulum.

- Start the pendulum swinging and use the stopwatch to determine its period—the time it takes to go from left to right and back to left again. (You might time

10 complete swings and then divide by 10.) Do this for pendulums of at least eight different lengths. Let t be the period (measured in seconds) and let l be the length (measured in centimeters), and record your results in the table of the illustration below.

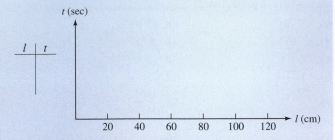

- Plot the points (l, t) on the axes in the figure. Does the graph appear to be a line?
- The pendulum's period t and length l are related by the formula $t = a\sqrt{l}$ for some number a. From your experimental data, find the approximate value of a.
- Use your formula to predict the period of a pendulum 2 meters long.
- Time the period of a 2-meter pendulum. How close was your prediction?

Reach for Success
EXTENSION OF REVIEWING ACADEMIC RESOURCES

Now that you know where to go to get mathematics help, let's move on and look at some other resources.

An advisor can discuss your workload and course load to help you find a good balance. An advisor can also explain the consequences of failing a course or falling below a 2.0 GPA.	Do you have any issues for which an advisor would be good resource? If so, list them. _____ _____ _____
Some colleges assign students to a specific academic advisor while others offer "drop in" service to students.	Were you assigned an academic advisor?_____ If so, what is your advisor's name?_____ What is his/her phone number?_____ Where is his/her office? _____ Have you met with your advisor this semester?_____ If not, where is the advising office located?_____ What is the telephone number?_____ Have you met with any advisor this semester?_____
The following may not apply to all students.	
Financial aid officers can give information on obtaining money for college expenses through scholarships, grants, and loans.	Could you benefit from speaking with a financial aid officer? _____ Write the office phone number here. _____ Where is the office located? _____
The Office of Veteran Affairs is a valuable resource for all students who have served in the armed forces at any time.	Could you benefit from speaking with a Veterans' Affairs officer? _____ Write the office phone number here._____ Where is the office located? _____
Most colleges offer personal counseling on a short-term basis.	Could you benefit from speaking with a personal counselor? _____ Write the phone number of the counseling office. _____ Where is it located?_____

Taking advantage of one or more of these resources may help guide you on the path to success.

8 Review

SECTION 8.1   Square Roots and the Pythagorean Theorem

DEFINITIONS AND CONCEPTS	EXAMPLES
The number b is a **square root** of a if $b^2 = a$.	4 is a square root of 16, because $4^2 = 16$. -4 is a square root of 16, because $(-4)^2 = 16$.
The **principal square root** of a positive number a, denoted by $\sqrt{a}$, is the positive square root of a.	4 is the principal square root of 16 because it is the positive square root of 16.
The principal square root of 0 is 0. The square root of a negative number is not a real number. The expression under a radical sign is called a **radicand**.	0 is a square root of 0, because $0^2 = 0$. $\sqrt{-16}$ is nonreal, because the square of no real number is -16. In the expression $\sqrt{5x}$, $5x$ is the radicand.
The Pythagorean theorem: $a^2 + b^2 = c^2$ provides a formula relating the lengths of the sides of a right triangle where a and b are the lengths of the legs of the triangle and c is the length of the hypotenuse. If $a = b$, then $\sqrt{a} = \sqrt{b}$.	A 13-foot ladder rests against the side of a building. If the base of the ladder is 5 feet from the wall, how far up the side of the building will the ladder reach? We can let d represent the distance in feet that the ladder will reach up the wall. In the right triangle that is formed, the hypotenuse is 13 feet, and the base-to-wall distance is 5 feet. Using the Pythagorean theorem, we can set up and solve the following equation: $\quad\quad 13^2 = 5^2 + d^2$ $\quad\quad 169 = 25 + d^2$ $\quad 13^2 = 169$ and $5^2 = 25$. $169 - 25 = d^2$ $\quad\quad$ Subtract 25 from both sides. $\quad\quad 144 = d^2$ $\quad\quad 169 - 25 = 144$ $\quad\sqrt{144} = \sqrt{d^2}$ $\quad\quad$ Take the positive square root of both sides. $\quad\quad 12 = d$ The ladder will reach 12 feet up the side of the building.
In $\sqrt[n]{x}$, n is called the **index** of the radical.	In $\sqrt{5x}$, 2 is the index. If no other number appears as the index, it is understood to be 2.

REVIEW EXERCISES

Find each value.

1. $\sqrt{81}$ **2.** $\sqrt{36}$

3. $-\sqrt{144}$ **4.** $-\sqrt{121}$

5. $-\sqrt{256}$ **6.** $-\sqrt{64}$

7. $\sqrt{169}$ **8.** $-\sqrt{225}$

Use a calculator to find each root to three decimal places.

9. $\sqrt{21}$ **10.** $-\sqrt{15}$

11. $-\sqrt{57.3}$ **12.** $\sqrt{751.9}$

Graph each function.

13. $f(x) = \sqrt{x}$ **14.** $f(x) = 2 - \sqrt{x}$

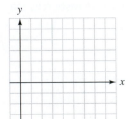

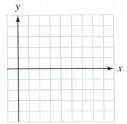

Refer to the right triangle shown in the illustration to find the length of the unknown side.

15. $a = 21$ and $b = 28$

16. $a = 25$ and $c = 65$

17. $a = 1$ and $c = \sqrt{2}$

18. $b = 6$ and $c = 7$

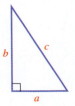

19. Installing windows The window frame shown in the illustration is 32 inches by 60 inches. It is to be shipped with a temporary brace attached diagonally. Find the length of the brace.

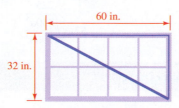

60 in.

32 in.

20. Height of a mast A 53-foot rope runs from the top of the mast shown in the illustration to a point 28 feet from its base. Find the height of the mast.

53 ft

28 ft

SECTION 8.2 *n*th Roots and Radicands That Contain Variables

DEFINITIONS AND CONCEPTS	EXAMPLES
The number b is a **cube root** of a if $b^3 = a$. The cube root of x is denoted $\sqrt[3]{x}$. The number b is an ***n*th root** of a if $b^n = a$.	$\sqrt[3]{216} = 6$, because $6^3 = 216$. $\sqrt[3]{27x^3} = 3x$, because $(3x)^3 = 27x^3$. $\sqrt[6]{64} = 2$, because $2^6 = 64$.

REVIEW EXERCISES

Find each root.

21. $\sqrt[3]{-27}$

22. $-\sqrt[3]{125}$

23. $\sqrt[4]{81}$

24. $\sqrt[5]{32}$

Use a calculator to find each root to three decimal places.

25. $\sqrt[3]{54.3}$

26. $\sqrt[3]{0.003}$

27. $\sqrt[3]{-0.055}$

28. $\sqrt[3]{-63.777}$

Simplify each expression. All variables represent positive values.

29. $\sqrt{x^4y^2}$

30. $\sqrt{81x^6y^8}$

31. $\sqrt[3]{64x^3y^6}$

32. $\sqrt[3]{1,000a^6b^3}$

SECTION 8.3 Simplifying Radical Expressions

DEFINITIONS AND CONCEPTS	EXAMPLES	
$\sqrt{ab} = \sqrt{a}\sqrt{b}$ $(a \geq 0, b \geq 0)$	$\sqrt{72} = \sqrt{36 \cdot 2}$	Factor 72 as $36 \cdot 2$, where 36 is a perfect square.
	$= \sqrt{36}\sqrt{2}$	Use the multiplication property of radicals.
	$= 6\sqrt{2}$	Write $\sqrt{36}$ as 6.
$\sqrt{\dfrac{a}{b}} = \dfrac{\sqrt{a}}{\sqrt{b}}$ $(a \geq 0, b > 0)$	$\sqrt{\dfrac{48}{49}} = \dfrac{\sqrt{48}}{\sqrt{49}}$	The square root of a quotient is equal to the quotient of the square roots.
	$= \dfrac{\sqrt{16 \cdot 3}}{7}$	Factor 48 and write $\sqrt{49}$ as 7.
	$= \dfrac{\sqrt{16}\sqrt{3}}{7}$	The square root of a product is equal to the product of the square roots.
	$= \dfrac{4\sqrt{3}}{7}$	$\sqrt{16} = 4$

REVIEW EXERCISES

Simplify each expression. All variables represent positive values.

33. $\sqrt{48}$

34. $\sqrt{72}$

35. $\sqrt{500}$

36. $\sqrt{112}$

37. $\sqrt{80x^2}$

38. $\sqrt{45y^4}$

39. $-\sqrt{250t^3}$

40. $-\sqrt{700z^5}$

41. $\sqrt{200x^2y}$

42. $\sqrt{75y^2z}$

43. $\sqrt[3]{8x^2y^3}$

44. $\sqrt[3]{250x^4y^3}$

Simplify each expression. All variables represent positive values. Assume no division by 0.

45. $\sqrt{\dfrac{16}{25}}$ **46.** $\sqrt{\dfrac{100}{49}}$ **49.** $\sqrt{\dfrac{60}{49}}$ **50.** $\sqrt{\dfrac{80}{225}}$

47. $\sqrt[3]{\dfrac{81}{125}}$ **48.** $\sqrt[3]{\dfrac{16}{64}}$ **51.** $\sqrt{\dfrac{242x^4}{169x^2}}$ **52.** $\sqrt{\dfrac{450a^6}{196a^2}}$

SECTION 8.4   Adding and Subtracting Radical Expressions

DEFINITIONS AND CONCEPTS	EXAMPLES	
Radical expressions can be added or subtracted if they contain **like radicals**. Radicals are called like radicals when they have the same index and the same radicand.	$2\sqrt{75} - 4\sqrt{147}$	
	$\quad = 2\sqrt{25 \cdot 3} - 4\sqrt{49 \cdot 3}$	Factor 75 and 147. Look for perfect-square factors.
	$\quad = 2\sqrt{\mathbf{25}}\sqrt{3} - 4\sqrt{\mathbf{49}}\sqrt{3}$	The square root of a product is equal to the product of the square roots.
	$\quad = 2(\mathbf{5})\sqrt{3} - 4(\mathbf{7})\sqrt{3}$	Simplify.
	$\quad = 10\sqrt{3} - 28\sqrt{3}$	Multiply.
	$\quad = -18\sqrt{3}$	Combine like terms.

REVIEW EXERCISES

Perform the operations. All variables represent positive values.

53. $\sqrt{2} + \sqrt{8} - \sqrt{18}$

54. $\sqrt{3} + \sqrt{27} - \sqrt{12}$

55. $3\sqrt{5} + 5\sqrt{45}$

56. $5\sqrt{28} - 3\sqrt{63}$

57. $3\sqrt{2x^2y} + 2x\sqrt{2y}$

58. $3y\sqrt{5xy^3} - y^2\sqrt{20xy}$

59. $\sqrt[3]{40} + \sqrt[3]{625}$

60. $\sqrt[3]{250x^3} - \sqrt[3]{54x^3}$

SECTION 8.5   Multiplying and Dividing Radical Expressions

DEFINITIONS AND CONCEPTS	EXAMPLES	
To multiply single-term radical expressions, we multiply the coefficients, multiply the radicals separately, and then simplify the result, if possible.	$\left(5\sqrt{2x}\right)\left(4\sqrt{3x}\right) = 5(4)\sqrt{2x}\sqrt{3x}$	
	$\qquad\qquad\qquad = 20\sqrt{2 \cdot 3 \cdot x \cdot x}$	
	$\qquad\qquad\qquad = 20\sqrt{6x^2}$	
	$\qquad\qquad\qquad = 20(x)\sqrt{6}$	
	$\qquad\qquad\qquad = 20x\sqrt{6}$	
If a radical appears in the denominator of a fraction, **rationalize the denominator** by multiplying the numerator and denominator by some appropriate radical equivalent to 1 to obtain a rational number in the denominator.	$\dfrac{2}{\sqrt[3]{9}} = \dfrac{2\sqrt[3]{3}}{\sqrt[3]{9}\sqrt[3]{3}}$ Multiply by 1: $\dfrac{\sqrt[3]{3}}{\sqrt[3]{3}} = 1$	
	$\quad = \dfrac{2\sqrt[3]{3}}{\sqrt[3]{27}}$	The product of two cube roots is equal to the cube root of the product.
	$\quad = \dfrac{2\sqrt[3]{3}}{3}$ $\sqrt[3]{27} = 3$	
If the denominator of a fraction contains two terms with one or two square roots, multiply numerator and denominator by the **conjugate** of the denominator to rationalize the denominator.	$\dfrac{3}{\sqrt{2} - 1} = \dfrac{3\left(\sqrt{2} + 1\right)}{\left(\sqrt{2} - 1\right)\left(\sqrt{2} + 1\right)}$	Multiply numerator and denominator by the conjugate of the denominator.
	$\quad = \dfrac{3\left(\sqrt{2} + 1\right)}{2 - 1}$	Multiply the terms in the numerator and in the denominator.
	$\quad = \dfrac{3\left(\sqrt{2} + 1\right)}{1}$	Simplify.
	$\quad = 3\left(\sqrt{2} + 1\right)$	

REVIEW EXERCISES

Perform the operations. All variables represent positive values.

61. $\left(6\sqrt{5}\right)\left(-3\sqrt{2}\right)$ **62.** $\left(-5\sqrt{x}\right)\left(-2\sqrt{x}\right)$

63. $\left(3\sqrt{3x}\right)\left(4\sqrt{6x}\right)$ **64.** $\left(-2\sqrt{27y^3}\right)\left(y\sqrt{2y}\right)$

65. $\left(\sqrt[3]{4}\right)\left(2\sqrt[3]{4}\right)$ **66.** $\left(-2\sqrt[3]{32x^2}\right)\left(3\sqrt[3]{2x^2}\right)$

67. $\sqrt{2}\left(\sqrt{8}-\sqrt{18}\right)$ **68.** $\sqrt{6y}\left(\sqrt{2y}+\sqrt{75}\right)$

69. $\left(\sqrt{2}+\sqrt{7}\right)\left(\sqrt{2}-\sqrt{7}\right)$

70. $\left(\sqrt{15}+3x\right)\left(\sqrt{15}+3x\right)$

71. $\left(\sqrt[3]{3}+2\right)\left(\sqrt[3]{3}-1\right)$ **72.** $\left(\sqrt[3]{5}-1\right)\left(\sqrt[3]{5}+1\right)$

Rationalize each denominator.

73. $\dfrac{1}{\sqrt{3}}$ **74.** $\dfrac{3}{\sqrt{18}}$

75. $\dfrac{8}{\sqrt[3]{16}}$ **76.** $\dfrac{10}{\sqrt[3]{32}}$

Rationalize each denominator. All variables represent positive values. Assume no division by 0.

77. $\dfrac{5}{\sqrt{5}+1}$ **78.** $\dfrac{3}{\sqrt{3}-1}$

79. $\dfrac{2\sqrt{5}}{\sqrt{5}+\sqrt{3}}$ **80.** $\dfrac{\sqrt{7x}+\sqrt{x}}{\sqrt{7x}-\sqrt{x}}$

SECTION 8.6   Solving Equations Containing Radicals; the Distance Formula

DEFINITIONS AND CONCEPTS	EXAMPLES
If $a=b$, then $a^n=b^n$. **To solve a radical equation:** 1. Whenever possible, isolate a single radical term on one side of the equation. 2. Raise each side of the equation to a power equal to the index. 3. If the equation still contains a radical term, repeat steps 1 and 2. 4. Check the possible solutions in the original equation for extraneous solutions.	Solve: $\sqrt{x+1}=5$ $\left(\sqrt{x+1}\right)^2=(5)^2$ Square both sides. $x+1=25$ $\left(\sqrt{x+1}\right)^2=x+1$ and $5^2=25$ $x=24$ Subtract 1 from both sides. We check by substituting 24 for x in the original equation. $\sqrt{x+1}=5$ $\sqrt{24+1}\overset{?}{=}5$ $\sqrt{25}\overset{?}{=}5$ $5=5$ The solution checks.
The distance formula: $d=\sqrt{(x_2-x_1)^2+(y_2-y_1)^2}$, where (x_1,y_1) and (x_2,y_2) represent two points.	To find the distance between points $(3,-2)$ and $(-5,4)$, we can substitute the respective values into the distance formula and simplify. $d=\sqrt{(x_2-x_1)^2+(y_2-y_1)^2}$ $\quad=\sqrt{(-5-3)^2+[4-(-2)]^2}$ $\quad=\sqrt{(-8)^2+6^2}$ $\quad=\sqrt{64+36}$ $\quad=\sqrt{100}$ $\quad=10$ The distance is 10 units.

REVIEW EXERCISES

Solve each equation and check all solutions.

81. $\sqrt{x-5}=4$ **82.** $\sqrt{3x+6}=6$

83. $\sqrt{3x+4}=-2\sqrt{x}$ **84.** $\sqrt{2(x+4)}-\sqrt{4x}=0$

85. $\sqrt{x+5}=x-1$ **86.** $\sqrt{2x+9}=x-3$

87. $\sqrt{2x+5}-1=x$ **88.** $\sqrt{4a+13}+2=a$

Find the distance between the points.

89. $(-7,12),(-4,8)$ **90.** $(-15,-3),(-10,-15)$

91. $(2,-4),(-2,-1)$ **92.** $(-10,11),(10,-10)$

SECTION 8.7   Rational Exponents

DEFINITIONS AND CONCEPTS	EXAMPLES
$x^{1/n} = \sqrt[n]{x}$ $(x \geq 0$ if n is even$)$ $x^{m/n} = \sqrt[n]{x^m} = \left(\sqrt[n]{x}\right)^m$ $(x \geq 0$ if n is even$)$	$25^{1/2} = \sqrt{25} = 5$ $(-8)^{1/3} = \sqrt[3]{-8} = -2$ $9^{3/2} = \left(\sqrt{9}\right)^3 = 3^3 = 27$

REVIEW EXERCISES

Simplify each expression. All variables represent positive values. Assume no division by 0. Write answers with positive exponents only.

93. $81^{1/2}$

94. $(-1{,}000)^{1/3}$

95. $16^{3/2}$

96. $\left(\dfrac{4}{9}\right)^{5/2}$

97. $8^{2/3}8^{4/3}$

98. $\dfrac{5^{17/7}}{5^{3/7}}$

99. $\dfrac{x^{4/5}x^{3/5}}{(x^{2/5})^3}$

100. $\left(\dfrac{r^{1/3}r^{2/3}}{r^{4/3}}\right)^3$

101. $7^{8/5}7^{-3/5}$

102. $\dfrac{5^{2/3}}{5^{-1/3}}$

103. $\dfrac{x^{2/5}x^{1/5}}{x^{-2/5}}$

104. $(a^6b^{10})^{-1/2}$

105. $x^{1/3}x^{2/5}$

106. $\dfrac{t^{3/4}}{t^{2/3}}$

107. $\dfrac{x^{-4/5}x^{1/3}}{x^{1/3}}$

108. $\dfrac{r^{1/4}r^{1/3}}{r^{5/6}}$

8 Test

Write each expression without a radical sign.

1. $\sqrt{49}$

2. $-\sqrt{900}$

3. $\sqrt[3]{-27}$

4. $\sqrt{3x}\sqrt{27x}, x \geq 0$

5. Find the length of the hypotenuse of a right triangle with legs of 5 inches and 12 inches.

6. A 26-foot ladder reaches a point on a wall 24 feet above the ground. How far from the wall is the ladder's base?

Simplify each expression. All variables represent positive values. Assume no division by 0.

7. $\sqrt{12x^2}$

8. $\sqrt{54x^3y}$

9. $\sqrt{\dfrac{320}{10}}$

10. $\sqrt{\dfrac{18x^2y^3}{2xy}}$

11. $\sqrt[3]{x^3y^9}$

12. $\sqrt[4]{\dfrac{16x^8}{y^4}}$

Perform each operation and simplify. All variables represent positive values.

13. $\sqrt{12} + \sqrt{27}$

14. $\sqrt{8x^3} - x\sqrt{18x}$

15. $\left(-2\sqrt{8x}\right)\left(3\sqrt{12x}\right)$

16. $\sqrt{3}\left(\sqrt{8} + \sqrt{6}\right)$

17. $\left(\sqrt{2} + \sqrt{3}\right)\left(\sqrt{2} - \sqrt{3}\right)$ **18.** $\left(2\sqrt{x} + 2\right)\left(\sqrt{x} - 3\right)$

Rationalize each denominator. All variables represent positive values. Assume no division by 0.

19. $\dfrac{3}{\sqrt{5}}$

20. $\sqrt{\dfrac{3xy^3}{48x^2}}$

21. $\dfrac{2}{\sqrt{5} - 2}$

22. $\dfrac{\sqrt{3x}}{\sqrt{x} + 2}$

Solve each equation.

23. $\sqrt{x} + 4 = 8$

24. $\sqrt{x - 3} - 2 = 7$

25. $\sqrt{3x + 9} = 2\sqrt{x + 1}$

26. $3\sqrt{x - 3} = \sqrt{2x + 8}$

27. $\sqrt{3x + 1} = x - 1$

28. $\sqrt[3]{x - 2} = 3$

29. Find the distance between the points $(1, 4)$ and $(7, 12)$.

30. Find the distance between points $(-2, -3)$ and $(-5, 1)$.

Simplify each expression and write all answers with positive exponents only. All variables represent positive values. Assume no division by 0.

31. $169^{-1/2}$

32. $64^{2/3}$

33. $(y^{15})^{2/5}$

34. $\left(\dfrac{a^{5/3}a^{4/3}}{(a^{1/3})^2a^{2/3}}\right)^6$

35. $p^{2/3}p^{3/4}$

36. $\dfrac{x^{2/3}x^{-4/5}}{x^{2/15}}$

❧ Cumulative Review ❧

Simplify each expression.

1. $(7x^2 + 3x) + (4x^3 - 5x^2 + 3)$

2. $(6x^3 - 5x) - (x^3 - 4x^2 - 2x + 5)$

3. $3(6x^2 - 3x + 3) + 2(-x^2 + 2x - 5)$

4. $5(3x^2 - 4x - 1) - 2(-2x^2 + 4x + 3)$

Perform each multiplication.

5. $(4x^5y^2)(-2x^3y^4)$

6. $-3x^2(-5x^3 - 3x^2 + 2)$

7. $(4x + 3)(5x + 2)$

8. $(4x - 3y)(3x - 2y)$

Perform each division.

9. $x + 3\overline{)x^2 - x - 12}$ $(x \neq -3)$

10. $2x - 1\overline{)2x^3 + x^2 - x - 3}$ $\left(x \neq \frac{1}{2}\right)$

Factor each expression.

11. $9x^2y^3 - 6xy^2$

12. $7(a + b) + x(a + b)$

13. $3a + 3b + ab + b^2$

14. $49p^4 - 16q^2$

15. $x^2 - 9x - 36$

16. $x^2 - 3xy - 10y^2$

17. $12a^2 + a - 20$

18. $10m^2 - 13mn - 3n^2$

19. $8x^3 - y^3$

20. $2r^3 + 54s^3$

Solve each equation.

21. $a^2 + 3a = -2$

22. $2b^2 - 12 = -5b$

23. $\dfrac{4}{a} = \dfrac{6}{a} - 1$

24. $\dfrac{a + 2}{a + 3} - 1 = \dfrac{-1}{a^2 + 2a - 3}$

Solve each proportion.

25. $\dfrac{4 - a}{13} = \dfrac{11}{26}$

26. $\dfrac{3a - 2}{7} = \dfrac{a}{28}$

Graph each equation.

27. $5x - 4y = 20$ **28.** $y = -x^2$

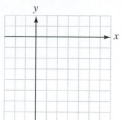 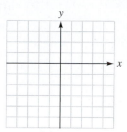

Find the slope of the line with the given properties.

29. passing through $(-2, 4)$ and $(6, 8)$

30. The equation of the line is $2x + 5y = 7$.

Write the equation of the line in slope-intercept form with the following properties.

31. slope of $\frac{2}{3}$, y-intercept of $(0, 5)$

32. passing through $(-2, 4)$ and $(6, 10)$

Are the graphs of the lines parallel, perpendicular, or neither parallel nor perpendicular?

33. $\begin{cases} 3x + 4y = 15 \\ 4x - 3y = 25 \end{cases}$ **34.** $\begin{cases} 3x + 4y = 15 \\ 6x = 15 - 8y \end{cases}$

Graph each inequality.

35. $y \leq -2x + 4$ **36.** $3x + 4y \leq 12$

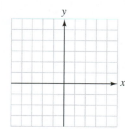 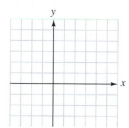

If $f(x) = 2x^2 - 3$, evaluate each expression.

37. $f(0)$ **38.** $f(-3)$

39. $f(-2)$ **40.** $f(2x)$

41. Assume that y varies directly with x. If $y = 4$ when $x = 10$, find y when $x = 30$.

42. Assume that y varies inversely with x. If $y = 8$ when $x = 2$, find the value of y when $x = 8$.

Quadratic Equations, Quadratic Functions, and Complex Numbers

9

©iStockphoto.com/Niko Guido

Careers and Mathematics

RETAIL SALES ASSOCIATES

Retail sales associates assist customers in purchasing merchandise. Because they are dealing with the public, most have shifts in the evenings and work on weekends. Sales associates must be able to describe a product's features and often demonstrate its use.

Some sales involve expensive complex items. For example, sales associates who sell automobiles must explain the features of various models and the types of options available. With experience, some associates move into the finance area of an automobile dealership where they must become familiar with the types of financing available and explain these options to customers.

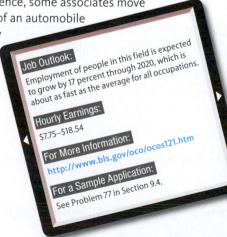

Job Outlook:
Employment of people in this field is expected to grow by 17 percent through 2020, which is about as fast as the average for all occupations.

Hourly Earnings:
$7.75–$18.54

For More Information:
http://www.bls.gov/oco/ocos121.htm

For a Sample Application:
See Problem 77 in Section 9.4.

In this chapter

In Chapter 5, we learned how to solve quadratic equations by a factoring method. In this chapter, we will discuss three other methods used to solve quadratic equations: applying the square root property, completing the square, and the quadratic formula. The chapter concludes with a discussion of how to simplify complex numbers.

Reach for Success   Planning for Your Future

The various courses and services that are needed to attain your academic and career goals should fit together like the pieces of a puzzle. Planning for your future involves more than concentrating on a single piece, such as a single course. It involves all of the pieces necessary to create the picture in your particular puzzle.

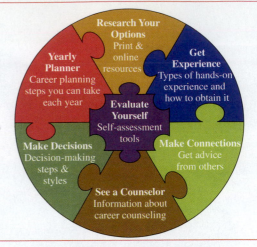

SHORT-TERM EDUCATIONAL PLANNING

This includes ensuring there is sufficient funding to attend college and then selecting the appropriate courses for next semester.

Where can you find your grade point average (GPA)? What factors can affect your GPA?

_____

_____

Taking the entire required mathematics sequence consecutively improves your chance of success by more than 50%.

To be successful, you need to start in the right course and the right format for you. Assessment tests will typically place you in the right course. The format of the course is usually your choice and varies by college.

Some format possibilities include computer-based lecture, mini-sessions, express, weekend, learning communities, and hybrid, blended, or fully online.

Check with your financial aid office to find the requirements for both merit-based and need-based financial aid.

_____

_____

List the importance of a high GPA.

1. _____

2. _____

Familiarize yourself with the mathematics sequence required to meet graduation requirements at your college or transfer school for your major. What courses must you take?

_____

_____

What formats are available at your college for the mathematics courses you need?

_____

_____

_____

Which format is a good fit for you? _____

Explain. _____

A Successful Study Strategy . . .

 Take all of your required courses that have prerequisites without interruption in the semester sequence.

At the end of the chapter you will find an additional exercise to guide you in planning for a successful college experience.

Section 9.1

Solving Quadratic Equations Using the Square Root Property

Objectives

1 Use the zero-factor property to solve a quadratic equation.
2 Use the square root property to solve a quadratic equation.
3 Solve an application involving a quadratic equation.

Getting Ready

Factor each polynomial.

1. $4x^2 - 2x$ **2.** $x^2 - 9$ **3.** $x^2 + x - 6$ **4.** $2x^2 + 3x - 9$

Simplify.

5. $\sqrt{25}$ **6.** $-\sqrt{36}$ **7.** $\sqrt{20}$ **8.** $-\sqrt{50}$

We have previously solved quadratic equations by factoring. In this section, we will review this process. Then, we will discuss another method that uses the *square root property*.

1 ### Use the zero-factor property to solve a quadratic equation.

Recall the definition of a quadratic equation. A quadratic equation in one variable is an equation of the form

$$ax^2 + bx + c = 0 \quad (a \neq 0)$$

where a, b, and c are real numbers.

We have previously solved quadratic equations by factoring and using the zero-factor property. If $ab = 0$, then $a = 0$ or $b = 0$.

The zero-factor property states that when the product of two numbers is 0, at least one of the numbers is 0. For example, the equation $(x - 4)(x + 5) = 0$ indicates that a product is equal to 0. By the zero-factor property, one of the factors must be 0:

$$x - 4 = 0 \quad \text{or} \quad x + 5 = 0$$

We can solve each of these linear equations to get

$$x = 4 \quad \text{or} \quad x = -5$$

The equation $(x - 4)(x + 5) = 0$ has two solutions: 4 and -5.

FACTORING METHOD	To solve a quadratic equation by factoring, we

1. Write the equation in $ax^2 + bx + c = 0$ form (called quadratic or standard form).

2. Factor the left side of the equation.

3. Use the zero-factor property to set each factor equal to 0.

4. Solve each resulting linear equation.

5. Check each solution in the original equation.

We will review the factoring method in Examples 1 and 2.

EXAMPLE 1 Solve: $6x^2 - 3x = 0$

Solution Since the equation is already in quadratic form, we begin by factoring the left side of the equation.

$$6x^2 - 3x = 0$$
$$3x(2x - 1) = 0 \qquad \text{Factor out } 3x, \text{ the GCF.}$$

By the zero-factor property, we have

$$3x = 0 \qquad \text{or} \qquad 2x - 1 = 0$$

We can solve these linear equations to obtain

$$3x = 0 \qquad \text{or} \qquad 2x - 1 = 0$$
$$x = 0 \qquad\qquad\qquad 2x = 1$$
$$x = \frac{1}{2}$$

COMMENT In Example 1, don't make the following error:

$$6x^2 - 3x = 0$$
$$6x^2 = 3x$$
$$2x = 1$$
$$x = \frac{1}{2}$$

In the third step, the division by $3x$ is valid only if $3x \neq 0$. If $x = 0$, then $3x = 0$ and we are dividing both sides of the equation by 0. Because of this invalid step, the solution $x = 0$ is lost.

Check: *For* $x = 0$ *For* $x = \dfrac{1}{2}$

$$6x^2 - 3x = 0 \qquad\qquad\qquad 6x^2 - 3x = 0$$

$$6(0)^2 - 3(0) \overset{?}{=} 0 \qquad\qquad 6\left(\frac{1}{2}\right)^2 - 3\left(\frac{1}{2}\right) \overset{?}{=} 0$$

$$6(0) - 0 \overset{?}{=} 0 \qquad\qquad\qquad 6\left(\frac{1}{4}\right) - \frac{3}{2} \overset{?}{=} 0$$

$$0 - 0 \overset{?}{=} 0 \qquad\qquad\qquad\qquad \frac{3}{2} - \frac{3}{2} \overset{?}{=} 0$$

$$0 = 0 \qquad\qquad\qquad\qquad\qquad\quad 0 = 0$$

The solutions are 0 and $\frac{1}{2}$.

 SELF CHECK 1 Solve: $5x^2 + 10x = 0$

EXAMPLE 2 Solve: $6x^2 - x = 2$

Solution To write the equation in quadratic form, we begin by subtracting 2 from both sides of the equation. Then we can solve the equation by factoring.

$$6x^2 - x = 2$$
$$6x^2 - x - 2 = 0 \qquad\qquad \text{Subtract 2 from both sides of the equation.}$$
$$(3x - 2)(2x + 1) = 0 \qquad\qquad \text{Factor the trinomial.}$$
$$3x - 2 = 0 \qquad \text{or} \qquad 2x + 1 = 0 \qquad \text{Set each factor equal to 0.}$$
$$3x = 2 \qquad\qquad\qquad 2x = -1 \qquad \text{Solve each linear equation.}$$
$$x = \frac{2}{3} \qquad\qquad\qquad x = -\frac{1}{2}$$

Check: *For* $x = \dfrac{2}{3}$

$$6x^2 - x = 2$$

$$6\left(\dfrac{2}{3}\right)^2 - \left(\dfrac{2}{3}\right) \overset{?}{=} 2$$

$$6\left(\dfrac{4}{9}\right) - \dfrac{6}{9} \overset{?}{=} 2$$

$$\dfrac{24}{9} - \dfrac{6}{9} \overset{?}{=} 2$$

$$\dfrac{18}{9} \overset{?}{=} 2$$

$$2 = 2$$

For $x = -\dfrac{1}{2}$

$$6x^2 - x = 2$$

$$6\left(-\dfrac{1}{2}\right)^2 - \left(-\dfrac{1}{2}\right) \overset{?}{=} 2$$

$$6\left(\dfrac{1}{4}\right) + \dfrac{2}{4} \overset{?}{=} 2$$

$$\dfrac{6}{4} + \dfrac{2}{4} \overset{?}{=} 2$$

$$\dfrac{8}{4} \overset{?}{=} 2$$

$$2 = 2$$

The solutions are $\dfrac{2}{3}$ and $-\dfrac{1}{2}$.

SELF CHECK 2 Solve: $2x^2 - 3 = x$

The factoring method doesn't work with many quadratic equations. For example, the trinomial in the equation $x^2 + 5x + 1 = 0$ cannot be factored by using rational coefficients. To solve such equations, we need to develop other methods. The first of these methods uses the square root property.

2 Use the square root property to solve a quadratic equation.

If $x^2 = c$, then x is a number whose square is c. Since $\left(\sqrt{c}\right)^2 = c$ and $\left(-\sqrt{c}\right)^2 = c$, the equation $x^2 = c$ has two solutions.

SQUARE ROOT PROPERTY

The equation $x^2 = c$ has two solutions:

$$x = \sqrt{c} \qquad \text{or} \qquad x = -\sqrt{c}$$

We can write the previous result with double-sign notation. The equation $x = \pm\sqrt{c}$ $\left(\text{read as "}x\text{ equals plus or minus }\sqrt{c}\text{"}\right)$ means that $x = \sqrt{c}$ or $x = -\sqrt{c}$.

EXAMPLE 3 Solve: $x^2 = 16$

Solution The equation $x^2 = 16$ has two solutions:

$$x = \sqrt{16} \qquad \text{or} \qquad x = -\sqrt{16}$$
$$= 4 \qquad\qquad\qquad = -4$$

Using double-sign notation, we have $x = \pm 4$.

Check: *For* $x = 4$

$$x^2 = 16$$

$$4^2 \overset{?}{=} 16$$

$$16 = 16$$

For $x = -4$

$$x^2 = 16$$

$$(-4)^2 \overset{?}{=} 16$$

$$16 = 16$$

The solutions are 4 and -4.

SELF CHECK 3 Solve: $y^2 = 36$

Although Example 3 could be solved by factoring, many times a quadratic equation will not factor. Applying the square root property will enable us to solve.

EXAMPLE 4   Solve: $3x^2 - 9 = 0$

Solution   We can solve the equation using the square root property.

$$3x^2 - 9 = 0$$

$$3x^2 = 9 \qquad \text{Add 9 to both sides of the equation.}$$

$$x^2 = 3 \qquad \text{Divide both sides by 3.}$$

This equation has two solutions:

$$x = \sqrt{3} \qquad \text{or} \qquad x = -\sqrt{3}$$

Check: *For* $x = \sqrt{3}$ $\qquad\qquad$ *For* $x = -\sqrt{3}$

$$3x^2 - 9 = 0 \qquad\qquad\qquad 3x^2 - 9 = 0$$

$$3\left(\sqrt{3}\right)^2 - 9 \overset{?}{=} 0 \qquad\qquad 3\left(-\sqrt{3}\right)^2 - 9 \overset{?}{=} 0$$

$$3(3) - 9 \overset{?}{=} 0 \qquad\qquad\qquad 3(3) - 9 \overset{?}{=} 0$$

$$9 - 9 \overset{?}{=} 0 \qquad\qquad\qquad 9 - 9 \overset{?}{=} 0$$

$$0 = 0 \qquad\qquad\qquad\qquad 0 = 0$$

The solutions can be written as $\pm\sqrt{3}$.

SELF CHECK 4   Solve: $2x^2 - 10 = 0$

As we will see in Example 5, we have a quantity squared set equal to a constant. Thus, we may apply the square root property.

EXAMPLE 5   Solve: $(x + 1)^2 = 9$

Solution   The two solutions are

$$x + 1 = \sqrt{9} \qquad \text{or} \qquad x + 1 = -\sqrt{9}$$

$$x + 1 = 3 \qquad\qquad\qquad x + 1 = -3$$

$$x = 2 \qquad\qquad\qquad\qquad x = -4$$

Check: *For* $x = 2$ $\qquad\qquad$ *For* $x = -4$

$$(x + 1)^2 = 9 \qquad\qquad\qquad (x + 1)^2 = 9$$

$$(2 + 1)^2 \overset{?}{=} 9 \qquad\qquad\qquad (-4 + 1)^2 \overset{?}{=} 9$$

$$3^2 \overset{?}{=} 9 \qquad\qquad\qquad\qquad (-3)^2 \overset{?}{=} 9$$

$$9 = 9 \qquad\qquad\qquad\qquad 9 = 9$$

The solutions are 2 and -4.

SELF CHECK 5   Solve: $(y + 2)^2 = 16$

EXAMPLE 6   Solve: $x^2 - 4x + 4 = 18$

Solution   Since $x^2 - 4x + 4$ factors as $(x - 2)^2$, we can write the equation in the form

$$(x - 2)^2 = 18$$

and solve it as we solved the previous example. The two solutions are

$$x - 2 = \sqrt{18} \quad \text{or} \quad x - 2 = -\sqrt{18}$$
$$x = 2 + \sqrt{18} \qquad\qquad x = 2 - \sqrt{18}$$
$$x = 2 + 3\sqrt{2} \qquad\qquad x = 2 - 3\sqrt{2} \quad {\color{red}\sqrt{18} = \sqrt{9}\sqrt{2} = 3\sqrt{2}}$$

Check: *For $x = 2 + 3\sqrt{2}$*   *For $x = 2 - 3\sqrt{2}$*

$$(\boldsymbol{x} - 2)^2 - 18 = 0 \qquad\qquad (\boldsymbol{x} - 2)^2 - 18 = 0$$
$$\left(\boldsymbol{2 + 3\sqrt{2}} - 2\right)^2 - 18 \overset{?}{=} 0 \qquad \left(\boldsymbol{2 - 3\sqrt{2}} - 2\right)^2 - 18 \overset{?}{=} 0$$
$$\left(3\sqrt{2}\right)^2 - 18 \overset{?}{=} 0 \qquad\qquad \left(-3\sqrt{2}\right)^2 - 18 \overset{?}{=} 0$$
$$18 - 18 \overset{?}{=} 0 \qquad\qquad\qquad 18 - 18 \overset{?}{=} 0$$
$$0 = 0 \qquad\qquad\qquad\qquad 0 = 0$$

The solutions are $2 + 3\sqrt{2}$ and $2 - 3\sqrt{2}$.

 SELF CHECK 6   Solve: $x^2 + 4x + 4 = 12$

EXAMPLE 7   Solve: $3x^2 - 4 = 2(x^2 + 2)$

Solution   We first will solve the equation for x^2 and then solve the result for x.

$$3x^2 - 4 = 2(x^2 + 2)$$
$$3x^2 - 4 = 2x^2 + 4 \qquad\qquad {\color{red}\text{Apply the distributive property.}}$$
$$3x^2 = 2x^2 + 8 \qquad\qquad {\color{red}\text{Add 4 to both sides.}}$$
$$x^2 = 8 \qquad\qquad {\color{red}\text{Subtract } 2x^2 \text{ from both sides.}}$$
$$x = \sqrt{8} \quad \text{or} \quad x = -\sqrt{8}$$
$$= 2\sqrt{2} \qquad\qquad = -2\sqrt{2} \qquad {\color{red}\text{Simplify the radical.}}$$

The solutions are $2\sqrt{2}$ and $-2\sqrt{2}$.

 SELF CHECK 7   Solve: $3x^2 = 2(x^2 + 3) + 6$

---

3   **Solve an application involving a quadratic equation.**

The process of solving many applications involves quadratic equations.

EXAMPLE 8   **INTEGERS**   The product of the first and third of three consecutive odd integers is 77. Find the integers.

Analyze the problem   Some examples of three consecutive odd integers are

     3, 5, and 7   and $-29, -27,$ and -25

These examples illustrate that to obtain the next consecutive odd integer from the previous one, we must add 2.

     If we let x represent the first of three consecutive odd integers, then $x + 2$ represents the second odd integer and $x + 2 + 2$ or $x + 4$ represents the third odd integer.

Form an equation   Since the product of the first integer and the third integer is $x(x + 4)$ and we are given that this product is 77, we have the equation

$$x(x + 4) = 77$$

Solve the equation   We can solve the equation as follows:

$$x(x + 4) = 77$$
$$x^2 + 4x - 77 = 0 \qquad {\color{red}\text{Apply the distributive property and subtract 77 from both sides.}}$$
$$(x + 11)(x - 7) = 0 \qquad {\color{red}\text{Factor.}}$$

$$x + 11 = 0 \qquad \text{or} \qquad x - 7 = 0 \qquad \text{Set each factor equal to 0.}$$
$$x = -11 \qquad\qquad\qquad x = 7$$

State the conclusion Since the first consecutive odd integer x can be either 7 or -11, there are two possible sets of integers. If the first integer is 7, the second is 9, and the third is 11. If the first integer is -11, the second is -9, and the third is -7.

Check the results If the three consecutive odd integers are 7, 9, and 11, the product of the first and third integers is $7(11) = 77$.

If the three consecutive odd integers are -11, -9, and -7, the product of the first and third integers is $-11(-7) = 77$.

The answers check and the integers are 7, 9, and 11 or -11, -9, and -7.

SELF CHECK 8 The product of the first and third of three consecutive even integers is 192. Find the integers.

EXAMPLE 9 **GEOMETRY** The oriental rug shown in Figure 9-1 is 3 feet longer than it is wide. Find the dimensions of the rug if its area is 180 ft^2.

Figure 9-1

Analyze the problem We can let w represent the width of the rug in feet. Then $w + 3$ will represent its length.

Form an equation Since the area of a rectangle is given by the formula $A = lw$ (area = length × width), the area of the rug is $(w + 3)w$, which is equal to 180. This gives the equation

The length of the rug	times	the width of the rug	equals	the area of the rug.
$(w + 3)$	·	w	=	180

Solve the equation We can solve this equation as follows:

$$(w + 3)w = 180$$
$$w^2 + 3w = 180 \qquad \text{Use the distributive property to remove parentheses.}$$
$$w^2 + 3w - 180 = 0 \qquad \text{Subtract 180 from both sides.}$$
$$(w - 12)(w + 15) = 0 \qquad \text{Factor.}$$
$$w - 12 = 0 \qquad \text{or} \qquad w + 15 = 0$$
$$w = 12 \qquad\qquad\qquad w = -15$$

State the conclusion When $w = 12$ feet, the length, $w + 3$, is 15 feet. The dimensions of the rug are 12 feet by 15 feet. We discard the solution $w = -15$, because a rug cannot have a negative width.

Check the result If the dimensions are 12 feet by 15 feet, the area is $12 \cdot 15 = 180$ ft^2. The length, 15 feet, is 3 feet longer than the width of 12 feet.

SELF CHECK 9 Find the dimensions of the rug if its area is 108 ft^2.

EXAMPLE 10 **BALLISTICS** If an object is shot straight up into the air with an initial velocity of 112 feet per second, its height after t seconds is given by the formula

$$h = 112t - 16t^2$$

where h represents the height of the object in feet. After the object has been shot, in how many seconds will it hit the ground?

Solution When the object hits the ground, its height will be 0. Thus, we set h equal to 0 and solve for t.

$$\mathbf{h} = 112t - 16t^2$$
$$\mathbf{0} = 112t - 16t^2$$
$$0 = 16t(7 - t) \qquad \text{Factor out } 16t, \text{ the GCF.}$$
$$16t = 0 \quad \text{or} \quad 7 - t = 0 \qquad \text{Set each factor equal to 0.}$$
$$t = 0 \qquad\qquad\qquad t = 7 \qquad \text{Solve each linear equation.}$$

When $t = 0$, the object's height above the ground is 0 feet, because it has not been released. When $t = 7$, the height is again 0 feet, and the object has hit the ground. The solution is 7 seconds.

 SELF CHECK 10 When will it reach a height of 192 feet?

 **SELF CHECK ANSWERS**

1. 0, −2 **2.** $\frac{3}{2}$, −1 **3.** 6, −6 **4.** $\pm\sqrt{5}$ **5.** 2, −6 **6.** $-2 \pm 2\sqrt{3}$ **7.** $\pm 2\sqrt{3}$ **8.** The integers are 12, 14, and 16 or −16, −14, and −12. **9.** The dimensions of the rug are 9 feet by 12 feet. **10.** The object will reach a height of 192 feet at 3 seconds and again at 4 seconds.

NOW TRY THIS

1. Let $f(x) = (x - 1)^2$. Find all values of x such that $f(x) = 25$.
2. Factor the trinomial square on the left side and then use the square root property to solve
    $$36x^2 - 12x + 1 = 45.$$

9.1 Exercises

WARM-UPS *Factor each expression.*

1. $6x^2 - 9x$

2. $4x^2 - x - 5$

3. $5x^2 + 12x - 9$

4. $25x^2 - 5x$

REVIEW *Write each expression without parentheses.*

5. $(y - 5)^2$

6. $(z + 2)^2$

7. $(x + y)^2$

8. $(a - b)^2$

9. $(2r - s)^2$

10. $(m + 3n)^2$

VOCABULARY AND CONCEPTS *Fill in the blanks.*

11. Any equation that can be written in the form
$ax^2 + bx + c = 0$ $(a \neq 0)$ is called a _____ equation.

12. In the equation $3x^2 - 4x + 5 = 0$, $a = \_$, $b = \_\_$, and $c = \_$.

13. If $ab = 0$, then $a = \_$ or $b = \_$.

14. In the quadratic equation $ax^2 + bx + c = 0$, $a \neq \_$.

15. An equation written in the form $ax^2 + bx + c = 0$ $(a \neq 0)$ is said to be in _____ or _____ form.

16. The square root property states the solutions of $x^2 = c$ $(c \neq 0)$ are ___ and _____.

GUIDED PRACTICE *Solve. (OBJECTIVE 1)*

17. $(x - 2)(x + 3) = 0$ **18.** $(x - 3)(x - 2) = 0$

19. $(x - 4)(x + 1) = 0$ **20.** $(x + 7)(x + 3) = 0$

21. $(2x - 5)(3x + 6) = 0$ **22.** $(5x - 2)(x + 1) = 0$

23. $(x - 1)(x + 2)(x - 3) = 0$
24. $(x + 2)(x + 3)(x - 4) = 0$

Solve. SEE EXAMPLE 1. (OBJECTIVE 1)

25. $x^2 + x = 0$ **26.** $3x^2 + 9x = 0$
27. $9x^2 - 3x = 0$ **28.** $4x^2 + 8x = 0$

Solve. SEE EXAMPLE 2. (OBJECTIVE 1)

29. $x^2 - 5x = -6$ **30.** $x^2 + 10 = -7x$
31. $4x^2 - 5 = -x$ **32.** $2x^2 - x = 6$

Solve. SEE EXAMPLE 3. (OBJECTIVE 2)

33. $x^2 = 9$ **34.** $x^2 = 32$
35. $x^2 = 20$ **36.** $x^2 = 0$

37. $6x^2 = 24$ **38.** $5x^2 - 20 = 0$
39. $3x^2 - 27 = 0$ **40.** $4x^2 - 64 = 0$

Solve. SEE EXAMPLE 4. (OBJECTIVE 2)

41. $3x^2 - 6 = 0$ **42.** $4x^2 - 20 = 0$
43. $5x^2 = 30$ **44.** $2x^2 = 20$

Solve. SEE EXAMPLE 5. (OBJECTIVE 2)

45. $(x + 1)^2 = 25$ **46.** $(x - 1)^2 = 49$
47. $(x + 2)^2 = 81$ **48.** $(x + 3)^2 = 16$
49. $(x + 2)^2 - 8 = 0$ **50.** $(x - 5)^2 - 20 = 0$

51. $(x - 4)^2 - 12 = 0$ **52.** $(x + 3)^2 - 24 = 0$

Solve. SEE EXAMPLE 6. (OBJECTIVE 2)

53. $9x^2 - 12x + 4 = 16$ **54.** $x^2 - 6x + 9 = 9$

55. $x^2 + 4x + 4 = 4$ **56.** $4x^2 - 20x + 25 = 36$

57. $x^2 - 6x + 9 = 15$ **58.** $x^2 + 4x + 4 = 10$

59. $x^2 + 10x + 25 = 8$ **60.** $x^2 - 8x + 16 = 12$

Solve. SEE EXAMPLE 7. (OBJECTIVE 2)

61. $8(x^2 - 6) = 4(x^2 + 13)$
62. $8(x^2 - 1) = 5(x^2 + 10) + 50$
63. $6(x^2 - 1) = 4(x^2 + 3)$ **64.** $5(x^2 - 2) = 2(x^2 + 1)$

65. $5x^2 + 1 = 4x^2 + 6$ **66.** $4x^2 + 1 = 3x^2 + 9$
67. $3(x^2 + 5) = 4(x^2 + 2)$ **68.** $6(x^2 + 7) = 7(x^2 + 1)$

ADDITIONAL PRACTICE *Solve. Variables a, b, c, and y represent positive values.*

69. $6x^2 + 11x + 3 = 0$ **70.** $5x^2 + 13x - 6 = 0$

71. $10x^2 + x - 2 = 0$ **72.** $6x^2 + 37x + 6 = 0$

73. $9x^2 = 81$ **74.** $5x^2 = 125$
75. $x^2 = a$ **76.** $x^2 = 4b$
77. $(x + b)^2 = 16c^2$ **78.** $(x - c)^2 = 25b^2$

79. $(x - a)^2 - 4a^2 = 0$ **80.** $(x + y)^2 - 9y^2 = 0$

Solve. Give the exact answer and a decimal rounded to the nearest tenth.

81. $(x - 2)^2 - 8 = 0$
82. $(x + 2)^2 - 50 = 0$
83. $5(x + 1)^2 = (x + 1)^2 + 32$
84. $6(x - 4)^2 = 4(x - 4)^2 + 36$

85. $4x^2 + 4x + 1 = 20$

86. $9x^2 + 12x + 4 = 12$

APPLICATIONS *Set up an equation to solve. SEE EXAMPLE 8.*
(OBJECTIVE 3)

87. Finding consecutive integers   The product of two consecutive odd integers is 63. Find the integers.

88. Finding consecutive integers   The product of the first and second of three consecutive odd integers is 35. Find the sum of the three integers.

89. Finding the sum of two squares   The sum of the squares of two consecutive even integers is 52. Find the integers.

90. Integers   The sum of an integer and four times its reciprocal is 8.5. Find the integer.

Set up an equation to solve. SEE EXAMPLE 9. (OBJECTIVE 3)

91. Geometry   A rectangular mural is 4 feet longer than it is wide. Find its dimensions if its area is 32 square feet.

92. Geometry   The length of a 220-square-foot rectangular garden is 2 feet more than twice its width. Find its perimeter.

93. Finding the dimensions of a garden   The rectangular garden shown in the illustration is surrounded by a walk of uniform width. Find the dimensions of the garden if its area is 180 square feet.

16 ft

24 ft

94. Building a deck around a pool   The owner of the pool in the illustration wants to surround it with a deck of uniform width. If he can afford 368 square feet of decking, how wide can he make the deck?

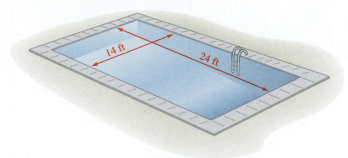

The height of a toy rocket in flight is given by the formula $h = -16t^2 + 144t$, *where t is the time of the flight in seconds and 144 is the initial velocity in feet per second.* SEE EXAMPLE 10. (OBJECTIVE 3)

95. Flight of a rocket   How long will it take for the rocket to hit the ground?

96. Height of a rocket   Find the height of the rocket in 3 seconds.

97. Maximum height of a rocket   If the maximum height of the rocket occurs halfway through its flight, how high will the rocket go? (See Exercise 95.)

98. Height of a rocket   At what time(s) will the rocket be 320 feet high?

A gun is fired straight up with a muzzle velocity of 1,088 feet per second. The height h of the bullet is given by the formula $h = -16t^2 + 1,088t$, *where t is the time in seconds.*

99. Height of a bullet   Find the height of the bullet after 10 seconds.

100. Height of a bullet   When will the bullet hit the ground?

101. Maximum height of a bullet   How high will the bullet go?

102. Height of a rocket   At what time(s) will the bullet be 18,240 feet high?

103. Falling objects   An object will fall s feet in t seconds, where $s = 16t^2$. If a worker 1,454 feet above the ground at the top of the Sears Tower drops a hammer, how long will it take for the hammer to hit the ground?

104. Falling coins   A researcher drops a penny from the observation deck of a skyscraper, 1,377 feet above the ground. How long will it take for the penny to hit the ground? (See Exercise 103.)

105. Finding the height of a triangle   The triangle shown in the illustration has an area of 30 square inches. Find its height.

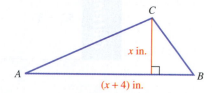

106. Finding the base of a triangle   Find the length of the base of the triangle shown in the above illustration.

WRITING ABOUT MATH

107. Explain how to solve a quadratic equation by the factoring method.

108. Explain how to solve the equation $x^2 = 81$ by the square root property.

SOMETHING TO THINK ABOUT

109. Find the error in the following solution.

$$9x^2 + 3x = 0$$
$$9x^2 = -3x$$
$$3x = -1$$
$$x = -\frac{1}{3}$$

110. What would happen if you solved $x^2 = c$ ($c < 0$) by the square root property? Would the roots be real numbers?

Section
9.2

Solving Quadratic Equations by Completing the Square

Objectives

1 Complete the square of a binomial to create a perfect trinomial square.

2 Solve a quadratic equation by completing the square.

Vocabulary

completing the square

Getting Ready

Find one-half of each number and square the result.

1. 6   **2.** 10   **3.** 2   **4.** 5

5. -8 **6.** -12 **7.** $\dfrac{1}{2}$ **8.** $\dfrac{2}{3}$

When the polynomial in a quadratic equation does not factor easily, we can solve the equation by using a method called **completing the square**. In fact, we can solve any quadratic equation by completing the square.

1 Complete the square of a binomial to create a perfect trinomial square.

The method of completing the square is based on the following special products:

$$x^2 + 2bx + b^2 = (x + b)^2 \qquad \text{and} \qquad x^2 - 2bx + b^2 = (x - b)^2$$

Recall that the trinomials $x^2 + 2bx + b^2$ and $x^2 - 2bx + b^2$ are both perfect trinomial squares, because each one factors as the square of a binomial. In each trinomial, if we take one-half of the coefficient of x and square the result, we get the third term.

$$\left[\frac{1}{2}(\mathbf{2b})\right]^2 = b^2 \qquad \text{and} \qquad \left[\frac{1}{2}(\mathbf{-2b})\right]^2 = b^2$$

To form a perfect trinomial square from the binomial $x^2 + 12x$, we take one-half of the coefficient of x (the 12), square the result, and add it to $x^2 + 12x$.

$$x^2 + 12x + \left[\frac{1}{2}(\mathbf{12})\right]^2 = x^2 + 12x + (\mathbf{6})^2$$

$$= x^2 + 12x + 36$$

This result is a perfect trinomial square, because $x^2 + 12x + 36 = (x + 6)^2$.

EXAMPLE 1   Form perfect trinomial squares using

a. $x^2 + 4x$ **b.** $x^2 - 6x$ **c.** $x^2 - 5x$

Solution   **a.** $x^2 + 4x + \left[\frac{1}{2}(\mathbf{4})\right]^2 = x^2 + 4x + (\mathbf{2})^2$

$$= x^2 + 4x + 4 \qquad \textcolor{red}{\text{This is } (x + 2)^2.}$$

b. $x^2 - 6x + \left[\frac{1}{2}(\mathbf{-6})\right]^2 = x^2 - 6x + (\mathbf{-3})^2$

$$= x^2 - 6x + 9 \qquad \textcolor{red}{\text{This is } (x - 3)^2.}$$

c. $x^2 - 5x + \left[\frac{1}{2}(-5)\right]^2 = x^2 - 5x + \left(-\frac{5}{2}\right)^2$

$$= x^2 - 5x + \frac{25}{4} \qquad \text{This is } \left(x - \frac{5}{2}\right)^2.$$

In each case, note that $\frac{1}{2}$ of the coefficient of x is the second term of the binomial factorization.

SELF CHECK 1   Form perfect trinomial squares using   **a.** $y^2 + 6y$
b. $y^2 - 8y$ **c.** $y^2 + 3y$

2   Solve a quadratic equation by completing the square.

If the quadratic equation $ax^2 + bx + c = 0$ has a leading coefficient of 1 ($a = 1$) and especially if the middle term is even, we can solve by completing the square fairly easily.

EXAMPLE 2   Solve by completing the square: $x^2 - 8x - 20 = 0$

Solution   We can solve the equation by completing the square.

$$x^2 - 8x - 20 = 0$$
$$x^2 - 8x \qquad = 20 \qquad \text{Add 20 to both sides.}$$

We then find one-half of the coefficient of x, square the result, and add it to both sides to make the left side a trinomial square.

$$x^2 - 8x + \left[\frac{1}{2}(-8)\right]^2 = 20 + \left[\frac{1}{2}(-8)\right]^2$$

$$x^2 - 8x + 16 = 20 + 16 \qquad \text{Simplify.}$$

$$(x - 4)^2 = 36 \qquad \text{Factor } x^2 - 8x + 16 \text{ and simplify.}$$

$$x - 4 = \pm\sqrt{36} \qquad \text{Use the square root property to solve for } x - 4.$$

$$x = 4 \pm 6 \qquad \text{Add 4 to both sides, } \pm\sqrt{36} = \pm 6.$$

Because of the $\pm$ sign, there are two solutions.

$$x = 4 + 6 \qquad \text{or} \qquad x = 4 - 6$$
$$= 10 \qquad\qquad\qquad = -2$$

Check each solution. Note that this example could be solved by factoring.

SELF CHECK 2   Solve by completing the square: $x^2 + 4x - 12 = 0$

The previous example illustrates that to solve a quadratic equation by completing the square, we follow these steps.

COMPLETING THE SQUARE

1. If necessary, write the quadratic equation in quadratic form, $ax^2 + bx + c = 0$.

2. If the coefficient of x^2 is not 1, divide both sides of the equation by a, the coefficient of x^2.

3. If necessary, add a number to both sides of the equation to place the constant term on the right side of the equal sign.

4. Complete the square:

   a. Find one-half of the coefficient of x and square it.

   b. Add the square to both sides of the equation.

5. Factor the perfect trinomial square on the left side of the equation and combine any like terms on the right side of the equation.

6. Use the square root property to solve the resulting quadratic equation.

7. Check each solution in the original equation.

EXAMPLE 3   Solve by completing the square: $4x^2 - 3 = -4x$

Solution   We first write the equation in quadratic form

$$4x^2 + 4x - 3 = 0 \qquad \text{Add } 4x \text{ to both sides.}$$

and then divide every term on both sides of the equation by 4 so that the coefficient of x^2 is 1.

$$\frac{4x^2}{4} + \frac{4x}{4} - \frac{3}{4} = \frac{0}{4} \qquad \text{Divide both sides by 4.}$$

$$x^2 + x - \frac{3}{4} = 0 \qquad \text{Simplify.}$$

We then use completing the square to solve the equation.

$$x^2 + x = \frac{3}{4} \qquad \text{Add } \tfrac{3}{4} \text{ to both sides.}$$

$$x^2 + x + \left(\frac{1}{2}\right)^2 = \frac{3}{4} + \left(\frac{1}{2}\right)^2 \qquad \text{Add } \left(\tfrac{1}{2}\right)^2 \text{ to both sides to complete the square.}$$

$$\left(x + \frac{1}{2}\right)^2 = 1 \qquad \text{Factor and simplify.}$$

$$x + \frac{1}{2} = \pm 1 \qquad \text{Solve for } x + \tfrac{1}{2}.$$

$$x = -\frac{1}{2} \pm 1 \qquad \text{Add } -\tfrac{1}{2} \text{ to both sides.}$$

$$x = -\frac{1}{2} + 1 \qquad \text{or} \qquad x = -\frac{1}{2} - 1$$

$$x = \frac{1}{2} \qquad\qquad\qquad x = -\frac{3}{2}$$

Check each solution. Note that this equation also can be solved by factoring.

SELF CHECK 3   Solve by completing the square: $2x^2 - 5x - 3 = 0$

Perspective

Clay tablets that survive from the early period of the Babylonian civilization, 1800 to 1600 BC, show that the Babylonians were accomplished mathematicians. These tablets were compiled for use by the Babylonian merchants. Many of these tablets contain multiplication tables and, for division, lists of reciprocals. For more abstract mathematical purposes, others provide tables of squares, cubes, square roots, and cube roots. Still others contain lists of problems and exercises. Some of these problems are practical, but many are puzzle problems, just for fun. Several problems and their solutions indicate that the Babylonians knew the Pythagorean theorem centuries before the Greeks discovered it.

The Babylonians also could solve certain quadratic equations. For example, one problem from a Babylonian table asks, "What number added to its reciprocal is 5?" Today, we would translate this question into the equation $x + \frac{1}{x} = 5$. Show that this equation is equivalent to the quadratic equation $x^2 - 5x + 1 = 0$.

EXAMPLE 4   Solve by completing the square: $2x^2 - 2 = -4x$

Solution   We add $4x$ to both sides to write the equation in quadratic form and then determine whether the equation can be solved by factoring.

$$2x^2 + 4x - 2 = 0 \quad \text{Add } 4x \text{ to both sides.}$$
$$x^2 + 2x - 1 = 0 \quad \text{Divide both sides by 2.}$$

Since this equation cannot be solved by factoring, we complete the square.

$$x^2 + 2x \qquad = 1 \qquad \text{Add 1 to both sides.}$$
$$x^2 + 2x + (1)^2 = 1 + (1)^2 \qquad \text{Add } 1^2 \text{ to both sides to complete the square.}$$
$$(x + 1)^2 = 2 \qquad \text{Factor and simplify.}$$
$$x + 1 = \pm\sqrt{2} \qquad \text{Solve for } x + 1.$$
$$x = -1 \pm \sqrt{2} \qquad \text{Subtract 1 from both sides.}$$
$$x = -1 + \sqrt{2} \quad \text{or} \quad x = -1 - \sqrt{2}$$

Both solutions check.

 SELF CHECK 4   Solve by completing the square: $3x^2 + 6x = 6$

 **SELF CHECK ANSWERS**   **1. a.** $y^2 + 6y + 9$ **b.** $y^2 - 8y + 16$ **c.** $y^2 + 3y + \frac{9}{4}$ **2.** $-6, 2$ **3.** $3, -\frac{1}{2}$ **4.** $-1 \pm \sqrt{3}$

NOW TRY THIS

Given $x^2 + \frac{b}{a}x = c$, what expression must be added to both sides of the equation to complete the square on the left side?

9.2   Exercises

WARM-UPS *Factor.*

1. $x^2 + 8x + 16$
2. $x^2 + 10x + 25$
3. $x^2 - 12x + 36$
4. $x^2 - 18x + 81$
5. $x^2 + 2x + 1$
6. $x^2 - 4x + 4$

REVIEW *Solve.*

7. $\dfrac{3t(2t + 1)}{2} + 6 = 3t^2$
8. $20r^2 - 11r - 3 = 0$

9. $\dfrac{2}{3x} - \dfrac{5}{9} = -\dfrac{1}{x}$
10. $\sqrt{x + 12} = \sqrt{3x}$

VOCABULARY AND CONCEPTS   *Fill in the blanks.*

11. If the polynomial in the equation $ax^2 + bx + c = 0$ doesn't factor, we can solve the equation by _____ the square.
12. Since $x^2 + 12x + 36 = (x + 6)^2$, we call the trinomial a perfect trinomial _____.
13. $x^2 + 2bx + b^2 =$ _____
14. $x^2 - 2bx + b^2 =$ _____
15. To complete the square on $x^2 + 10x$, we add the _____ of one-half of __, which is 25.
16. To complete the square on $x^2 - 10x$, we add the square of _____ of -10, which is 25.

GUIDED PRACTICE   *Complete the square to form a perfect trinomial square. SEE EXAMPLE 1. (OBJECTIVE 1)*

17. $x^2 + 2x$
18. $x^2 + 16x$

19. $x^2 - 4x$
20. $x^2 - 14x$

21. $y^2 - 18y$
22. $y^2 - 20y$

23. $x^2 + 7x$
24. $x^2 + 21x$

25. $a^2 - 9a$
26. $b^2 - 13b$

27. $b^2 + \dfrac{2}{3}b$
28. $a^2 + \dfrac{8}{5}a$

Solve each equation by completing the square. SEE EXAMPLE 2. (OBJECTIVE 2)

29. $x^2 + 6x + 8 = 0$
30. $x^2 + 8x + 12 = 0$

31. $x^2 - 8x + 12 = 0$
32. $x^2 - 4x + 3 = 0$

33. $x^2 - 5x - 14 = 0$
34. $x^2 - 7x + 10 = 0$

35. $x^2 + 5x - 6 = 0$
36. $x^2 = 9x - 18$

Solve each equation by completing the square. SEE EXAMPLE 3. (OBJECTIVE 2)

37. $2x^2 = 4 - 2x$
38. $3x^2 + 9x + 6 = 0$

39. $3x^2 + 48 = -24x$
40. $3x^2 = 3x + 6$

41. $2x^2 = 3x + 2$
42. $3x^2 = 2 - 5x$
43. $4x^2 = 2 - 7x$
44. $2x^2 = 5x + 3$

Solve each equation by completing the square. SEE EXAMPLE 4. (OBJECTIVE 2)

45. $x^2 + 4x + 1 = 0$
46. $x^2 + 6x + 2 = 0$
47. $x^2 - 2x - 4 = 0$
48. $x^2 - 4x - 2 = 0$

ADDITIONAL PRACTICE   *Complete the square to form a perfect trinomial square.*

49. $c^2 - \dfrac{5}{2}c$
50. $c^2 - \dfrac{11}{3}c$

51. $t^2 - \dfrac{5}{6}t$
52. $z^2 - \dfrac{7}{8}z$

Solve each equation by completing the square.

53. $x^2 - 2x - 15 = 0$
54. $x^2 - 2x - 24 = 0$

55. $x^2 = 4x + 3$
56. $x^2 = 6x - 3$

57. $2x^2 = 2 - 4x$
58. $3x^2 = 12 - 6x$

59. $x(x + 1) = 20$
60. $x(x + 7) = -12$

61. $2x(x + 4) = 10$
62. $3x(x - 2) = 9$

63. $6(x^2 - 3) = 5x$

64. $3(3x^2 - 1) = 5x$

WRITING ABOUT MATH

65. Explain how to complete the square.
66. Explain why the coefficient of x^2 should be 1 before completing the square.

SOMETHING TO THINK ABOUT   *Consider this method of completing the square on x in the binomial $ax^2 + bx$. Multiply the binomial by 4a and then add b^2. Complete the square on x in each binomial.*

67. $2x^2 + 6x$
68. $3x^2 - 4x$

Section 9.3

Solving Quadratic Equations by the Quadratic Formula

Objectives

1. Use the quadratic formula to find all real solutions of a given quadratic equation.
2. Determine whether a quadratic equation has real solutions.
3. Solve an application involving a quadratic equation.

Vocabulary

quadratic formula

Getting Ready

Evaluate $b^2 - 4ac$ when a, b, and c have the following values.

1. $a = 1, b = 2, c = 3$
2. $a = 4, b = 3, c = 1$
3. $a = 1, b = 0, c = -2$
4. $a = 2, b = 4, c = 2$

We can solve any quadratic equation by the method of completing the square, but the work is often tedious. In this section, we will develop a formula, called the *quadratic formula*, that will enable us to solve quadratic equations with much less effort.

1 **Use the quadratic formula to find all real solutions of a given quadratic equation.**

We can solve the standard form of a quadratic equation by completing the square.

$$ax^2 + bx + c = 0 \qquad \text{Standard form of a quadratic equation.}$$

$$\frac{ax^2}{a} + \frac{bx}{a} + \frac{c}{a} = \frac{0}{a} \qquad \text{Divide both sides by } a.$$

$$x^2 + \frac{b}{a}x + \frac{c}{a} = 0 \qquad \text{Simplify.}$$

$$x^2 + \frac{b}{a}x = -\frac{c}{a} \qquad \text{Subtract } \tfrac{c}{a} \text{ from both sides.}$$

We can now complete the square on x by adding $\left(\frac{1}{2} \cdot \frac{b}{a}\right)^2$, or $\frac{b^2}{4a^2}$, to both sides:

$$x^2 + \frac{b}{a}x + \frac{b^2}{4a^2} = \frac{b^2}{4a^2} - \frac{c}{a}$$

After factoring the trinomial on the left side and adding the fractions on the right side, we have

$$\left(x + \frac{b}{2a}\right)\left(x + \frac{b}{2a}\right) = \frac{b^2}{4a^2} - \frac{4ac}{4aa}$$

$$\left(x + \frac{b}{2a}\right)^2 = \frac{b^2 - 4ac}{4a^2}$$

This equation can be solved by the square root method to obtain

$$x + \frac{b}{2a} = \sqrt{\frac{b^2 - 4ac}{4a^2}} \qquad \text{or} \qquad x + \frac{b}{2a} = -\sqrt{\frac{b^2 - 4ac}{4a^2}}$$

$$x + \frac{b}{2a} = \frac{\sqrt{b^2 - 4ac}}{\sqrt{4a^2}} \qquad\qquad x + \frac{b}{2a} = -\frac{\sqrt{b^2 - 4ac}}{\sqrt{4a^2}}$$

$$x + \frac{b}{2a} = \frac{\sqrt{b^2 - 4ac}}{2a} \qquad\qquad x + \frac{b}{2a} = -\frac{\sqrt{b^2 - 4ac}}{2a}$$

$$x = -\frac{b}{2a} + \frac{\sqrt{b^2 - 4ac}}{2a} \qquad\qquad x = -\frac{b}{2a} - \frac{\sqrt{b^2 - 4ac}}{2a}$$

$$x = \frac{-b + \sqrt{b^2 - 4ac}}{2a} \qquad\qquad x = \frac{-b - \sqrt{b^2 - 4ac}}{2a}$$

These solutions usually are written in one expression called the **quadratic formula**.

QUADRATIC FORMULA

The solutions of the quadratic equation $ax^2 + bx + c = 0$ are

$$x = \frac{-b \pm \sqrt{b^2 - 4ac}}{2a} \quad (a \neq 0)$$

COMMENT When you write the quadratic formula, be careful to draw the fraction bar so that it underlines the complete numerator. Do not write

$$x = -b \pm \frac{\sqrt{b^2 - 4ac}}{2a}$$

EXAMPLE 1 Solve using the quadratic formula: $x^2 + 5x + 6 = 0$

Solution In this equation, $a = 1$, $b = 5$, and $c = 6$. We substitute these values into the quadratic formula and simplify.

$$x = \frac{-b \pm \sqrt{b^2 - 4ac}}{2a}$$

$$= \frac{-5 \pm \sqrt{5^2 - 4(1)(6)}}{2(1)} \qquad \text{Substitute 1 for } a\text{, 5 for } b\text{, and 6 for } c.$$

$$= \frac{-5 \pm \sqrt{25 - 24}}{2} \qquad \text{Simplify.}$$

$$= \frac{-5 \pm \sqrt{1}}{2} \qquad \text{Subtract.}$$

$$= \frac{-5 \pm 1}{2}$$

Thus,

$$x = \frac{-5+1}{2} \qquad \text{or} \qquad x = \frac{-5-1}{2}$$

$$= \frac{-4}{2} \qquad\qquad\qquad = \frac{-6}{2}$$

$$= -2 \qquad\qquad\qquad = -3$$

The solutions are -2 and -3. Check both solutions.

Because the solutions were rational, this equation could have been solved by factoring.

SELF CHECK 1   Solve using the quadratic formula: $x^2 - 4x - 12 = 0$

EXAMPLE 2   Solve using the quadratic formula: $2x^2 = 5x + 3$

Solution   We begin by writing the equation in quadratic form.

COMMENT   Be sure to write a quadratic equation in quadratic form before identifying the values of a, b, and c.

$$2x^2 = 5x + 3$$

$$2x^2 - 5x - 3 = 0 \qquad \text{Subtract } 5x \text{ and } 3 \text{ from both sides.}$$

In this equation, $a = 2$, $b = -5$, and $c = -3$. We substitute these values into the quadratic formula and simplify.

$$x = \frac{-b \pm \sqrt{b^2 - 4ac}}{2a}$$

$$= \frac{-(-5) \pm \sqrt{(-5)^2 - 4(2)(-3)}}{2(2)} \qquad \text{Substitute 2 for } a, -5 \text{ for } b, \text{ and } -3 \text{ for } c.$$

$$= \frac{5 \pm \sqrt{25 + 24}}{4} \qquad \text{Simplify.}$$

$$= \frac{5 \pm \sqrt{49}}{4} \qquad \text{Add.}$$

$$= \frac{5 \pm 7}{4}$$

Thus,

$$x = \frac{5+7}{4} \qquad \text{or} \qquad x = \frac{5-7}{4}$$

$$= \frac{12}{4} \qquad\qquad\qquad = \frac{-2}{4}$$

$$= 3 \qquad\qquad\qquad = -\frac{1}{2}$$

The solutions are 3 and $-\frac{1}{2}$. Check both solutions.

Because the solutions were rational, this equation could have been solved by factoring.

SELF CHECK 2   Solve using the quadratic formula: $4x^2 + 4x = 3$

EXAMPLE 3   Solve using the quadratic formula: $3x^2 = 2x + 4$

Solution   We begin by writing the equation in quadratic form.

$$3x^2 = 2x + 4$$

$$3x^2 - 2x - 4 = 0 \qquad \text{Subtract } 2x \text{ and } 4 \text{ from both sides.}$$

In this equation, $a = 3$, $b = -2$, and $c = -4$. We substitute these values into the quadratic formula and simplify.

$$x = \frac{-b \pm \sqrt{b^2 - 4ac}}{2a}$$

$$= \frac{-(-2) \pm \sqrt{(-2)^2 - 4(3)(-4)}}{2(3)} \qquad \text{Substitute 3 for } a, -2 \text{ for } b, \text{ and } -4 \text{ for } c.$$

$$= \frac{2 \pm \sqrt{4 + 48}}{6} \qquad \text{Simplify.}$$

$$= \frac{2 \pm \sqrt{52}}{6} \qquad \text{Add.}$$

$$= \frac{2 \pm 2\sqrt{13}}{6} \qquad \sqrt{52} = \sqrt{4 \cdot 13} = \sqrt{4}\sqrt{13} = 2\sqrt{13}$$

$$= \frac{2(1 \pm \sqrt{13})}{6} \qquad \text{Factor out 2, the GCF.}$$

$$= \frac{1 \pm \sqrt{13}}{3} \qquad \text{Divide out the common factor of 2.}$$

Thus,

$$x = \frac{1}{3} + \frac{\sqrt{13}}{3} \qquad \text{or} \qquad x = \frac{1}{3} - \frac{\sqrt{13}}{3}$$

Both values check. The solutions are $\frac{1}{3} \pm \frac{\sqrt{13}}{3}$.

🌱 **SELF CHECK 3**   Solve using the quadratic formula: $2x^2 - 5x - 5 = 0$

---

2 Determine whether a quadratic equation has real-number solutions.

The next example shows that some quadratic equations have no real-number solutions.

EXAMPLE 4   Determine whether the equation $x^2 + 2x + 5 = 0$ has real-number solutions.

Solution   In this equation, $a = 1$, $b = 2$, and $c = 5$. We substitute these values into the quadratic formula.

$$x = \frac{-b \pm \sqrt{b^2 - 4ac}}{2a}$$

$$= \frac{-2 \pm \sqrt{2^2 - 4(1)(5)}}{2(1)} \qquad \text{Substitute 1 for } a, 2 \text{ for } b, \text{ and } 5 \text{ for } c.$$

$$= \frac{-2 \pm \sqrt{4 - 20}}{2}$$

$$= \frac{-2 \pm \sqrt{-16}}{2}$$

Since $\sqrt{-16}$ is not a real number, there are no real-number solutions.

🌱 **SELF CHECK 4**   Determine whether the equation $2x^2 + 5x + 4 = 0$ has real-number solutions.

3 Solve an application involving a quadratic equation.

EXAMPLE 5 **MANUFACTURING** Samsung Electronics manufactures a 46-inch wide-screen television set. The 46-inch screen is measured along the diagonal. The screens are rectangular in shape and are 17 inches wider than they are high. Find the dimensions of a screen.

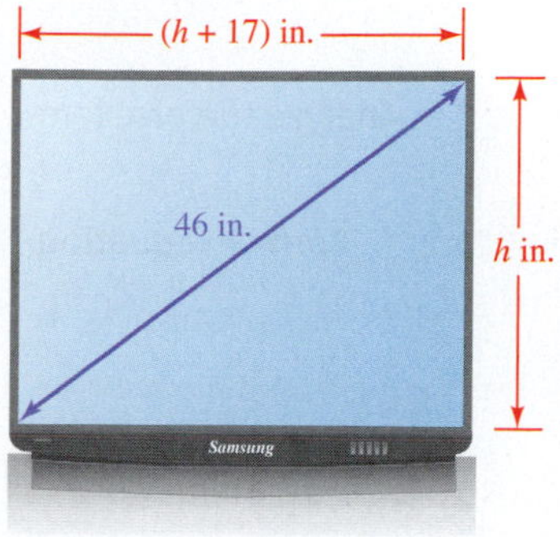

Analyze the problem We can let h represent the height of a screen in inches, as shown in Figure 9-2. Then $h + 17$ will represent the width in inches.

Form an equation Since the sides of the screen and its diagonal form a right triangle, we can use the Pythagorean theorem to form the equation

$$h^2 + (h + 17)^2 = 46^2$$

which we can solve as follows.

Solve the equation

$h^2 + h^2 + 34h + 289 = 2{,}116$ Square each term.

$2h^2 + 34h - 1{,}827 = 0$ Subtract 2,116 from both sides.

We can solve this equation with the quadratic formula.

$$h = \frac{-b \pm \sqrt{b^2 - 4ac}}{2a}$$

$$h = \frac{-34 \pm \sqrt{(34)^2 - 4(2)(-1{,}827)}}{2(2)}$$

$$h = \frac{-34 \pm \sqrt{1{,}156 + 14{,}616}}{4}$$

$$h = \frac{-34 \pm \sqrt{15{,}772}}{4}$$

$$h \approx \frac{-34 \pm 125.5866235}{4}$$

$$h \approx \frac{-34 + 125.5866235}{4} \quad \text{or} \quad h \approx \frac{-34 - 125.5866235}{4}$$

$$\approx \frac{91.58662349}{4} \qquad\qquad\qquad \approx \frac{-159.5866235}{4}$$

$$\approx 22.89665588 \qquad\qquad\qquad \approx -39.89665588$$

State the conclusion The height of each screen will be approximately 22.9 inches, and the width will be approximately 22.9 + 17 or 39.9 inches. We discard the second solution, because the height of a TV screen cannot be negative.

Check the result The width of 39.9 inches is 17 inches wider than the height of 22.9 inches.

$$(22.9)^2 + (39.9)^2 \approx 46^2$$

 **SELF CHECK 5** If the television set is a 40-inch wide-screen set and it is 15 inches wider than it is high, find the dimensions of the screen.

Figure 9-2

(h + 17) in.

46 in.

h in.

Samsung

EXAMPLE 6   **FINANCE**   If P is invested at an annual rate of $r\%$, it will grow to an amount of A in n years according to the formula $A = P(1 + r)^n$. What interest rate is needed for a $5,000 investment to grow to $5,618 after 2 years?

Analyze the problem   Here we are given a formula to find the amount. We will substitute the values into the formula and find r, the interest rate.

Form an equation   We substitute 5,000 for P, 5,618 for A, and 2 for n in the formula.

$$A = P(1 + r)^n$$
$$5{,}618 = 5{,}000(1 + r)^2$$

Solve the equation   We then solve the equation as follows.

$$5{,}618 = 5{,}000(1 + r)^2$$
$$5{,}618 = 5{,}000(1 + 2r + r^2) \qquad \text{Expand } (1 + r)^2.$$
$$5{,}618 = 5{,}000 + 10{,}000r + 5{,}000r^2 \qquad \text{Apply the distributive property.}$$
$$5{,}000r^2 + 10{,}000r - 618 = 0 \qquad \text{Subtract 5,618 from both sides.}$$

We can solve this equation with the quadratic formula.

$$r = \frac{-b \pm \sqrt{b^2 - 4ac}}{2a}$$

$$r = \frac{-10{,}000 \pm \sqrt{10{,}000^2 - 4(5{,}000)(-618)}}{2(5{,}000)}$$

$$r = \frac{-10{,}000 \pm \sqrt{100{,}000{,}000 + 12{,}360{,}000}}{10{,}000}$$

$$r = \frac{-10{,}000 \pm \sqrt{112{,}360{,}000}}{10{,}000}$$

$$r = \frac{-10{,}000 \pm 10{,}600}{10{,}000}$$

$$r = \frac{-10{,}000 + 10{,}600}{10{,}000} \qquad \text{or} \qquad r = \frac{-10{,}000 - 10{,}600}{10{,}000}$$

$$= \frac{600}{10{,}000} \qquad\qquad\qquad = \frac{-20{,}600}{10{,}000}$$

$$= 6\% \qquad\qquad\qquad\qquad = -206\%$$

State the conclusion   The required rate is 6%. The rate of -206% has no meaning in this problem. Check the result.

 SELF CHECK 6   What rate is needed for a $5,000 investment to grow to $5,512.50 after 2 years?

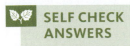 **SELF CHECK ANSWERS**

1. $6, -2$ **2.** $\frac{1}{2}, -\frac{3}{2}$ **3.** $\frac{5}{4} \pm \frac{\sqrt{65}}{4}$ **4.** no   **5.** The height is approximately 19.8 inches and the width approximately 34.8 inches.   **6.** The required rate is 5% annually.

NOW TRY THIS

1. Let $f(x) = 5x^2 - 8x - 3$. Find all values of x such that $f(x) = 0$.

2. Find the x-intercepts of the graph of $f(x) = x^2 - 6x + 2$.

9.3 Exercises

WARM-UPS *Write each in quadratic form. Do not solve.*

1. $5x^2 - x = 1$
2. $x^2 - 9 = -3x$
3. $7x^2 + x = 8$
4. $-5x^2 + 4 = x$
5. $3x^2 = -7x - 4$
6. $4x^2 = 8x - 3$
7. $7x^2 + 21 = -14x$
8. $5y^2 + 25 = -4y$

REVIEW *Solve each formula for the indicated variable.*

9. $A = p + prt$ for r
10. $F = \dfrac{GMm}{d^2}$ for M

Write the equation of the line in slope-intercept form with the given properties.

11. slope of $\frac{3}{5}$ and passing through $(0, 12)$
12. passes through $(6, 8)$ and the origin

Simplify each expression. Assume that all variables represent positive numbers.

13. $\sqrt{80}$
14. $12\sqrt{x^3y^2}$
15. $\dfrac{x}{\sqrt{7x}}$
16. $\dfrac{\sqrt{x} + 2}{\sqrt{x} - 2}$

VOCABULARY AND CONCEPTS *Fill in the blanks.*

17. Write the general quadratic equation. _____
    $(a \neq 0)$
18. Write the quadratic formula for $ax^2 + bx + c = 0$ $(a \neq 0)$.

    _____

19. In the quadratic equation $ax^2 + bx + c = 0$, a cannot equal _.
20. In the quadratic equation $5x^2 - 3 = 0$, $a =$ _, $b =$ _, and $c =$ __.
21. In the quadratic equation $-4x^2 + 8x = -1$, $a =$ __, $b =$ _, and $c =$ _.
22. If a, b, and c are three sides of a right triangle and c is the hypotenuse, then $c^2 =$ _____.

Write each in quadratic form, if necessary, to find the values of a, b, and c. Do not solve the equation.

23. $x^2 + 4x + 3 = 0$
24. $x^2 + x - 4 = 0$
25. $3x^2 - 2x + 7 = 0$
26. $4x^2 + 7x - 3 = 0$
27. $4y^2 = 2y - 1$
28. $2x = 3x^2 + 4$

29. $x(3x - 5) = 2$
30. $y(5y + 10) = 8$
31. $7(x^2 + 3) = -14x$
32. $5(p^2 + 5) = -4p$
33. $(2q + 3)(q - 2) = (q + 1)(q - 1)$
34. $(3m + 2)(m - 1) = (2m + 7)(m - 1)$

GUIDED PRACTICE *Use the quadratic formula to find all real solutions of each equation.* SEE EXAMPLE 1. (OBJECTIVE 1)

35. $x^2 - 5x + 6 = 0$
36. $x^2 + 5x + 4 = 0$
37. $x^2 + 7x + 12 = 0$
38. $x^2 - x - 12 = 0$
39. $2x^2 - x - 1 = 0$
40. $2x^2 + 3x - 2 = 0$
41. $3x^2 + 5x + 2 = 0$
42. $3x^2 - 4x + 1 = 0$

Use the quadratic formula to find all real solutions of each equation. SEE EXAMPLE 2. (OBJECTIVE 1)

43. $5x^2 + x = 4$
44. $3x^2 = -7x - 4$
45. $4x^2 = 8x - 3$
46. $7x^2 - 21 = 14x$
47. $5y^2 - 24 = -7y$
48. $7y^2 + y = 8$
49. $3x^2 + 6 = 11x$
50. $3x^2 + 8x = -5$

Use the quadratic formula to find all real solutions of each equation. SEE EXAMPLE 3. (OBJECTIVE 1)

51. $x^2 + 3x + 1 = 0$
52. $x^2 + 3x - 2 = 0$
53. $x^2 + 5x - 3 = 0$
54. $x^2 + 5x + 3 = 0$
55. $3x^2 - 5x = 1$
56. $4x^2 - x = 2$
57. $2x^2 - 3 = 4x$
58. $3x^2 + 2x = 2$

Use the quadratic formula to determine whether there are real-number solutions to each equation. SEE EXAMPLE 4. (OBJECTIVE 2)

59. $x^2 + 2x + 7 = 0$
60. $2x^2 + 3x = -3$
61. $x^2 + 5 = 2x$
62. $2x^2 - x + 2 = 0$

ADDITIONAL PRACTICE *Solve each equation. Give both the exact answer and a decimal approximation to the nearest tenth.*

63. $4x^2 + 5x - 1 = 0$
64. $4x^2 + 3x - 2 = 0$

65. $5x^2 - 8x - 1 = 0$ **66.** $6x^2 - 8x + 1 = 0$

67. $2x^2 + x = 5$ **68.** $3x^2 - x = 1$

69. $x^2 + 1 = -4x$ **70.** $x^2 + 1 = -8x$

71. $x^2 = 1 - 2x$ **72.** $x^2 = 2 - 2x$

73. $3x^2 = 6x + 2$ **74.** $3x^2 = -8x - 2$

APPLICATIONS *Set up an equation to solve. SEE EXAMPLE 5.*
(OBJECTIVE 3)

75. Finding dimensions   The picture frame in the illustration is 2 inches wider than it is high. Find its dimensions.

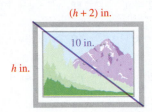

$(h + 2)$ in.

10 in.

h in.

76. Installing sidewalks   A 170-meter-long sidewalk from the mathematics building M to the student center C is shown in the illustration. However, students prefer to walk directly from M to C. How long are the two pieces of the existing sidewalk?

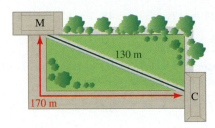

M

130 m

170 m

C

77. Navigation   Two boats left port at the same time, one sailing east and one sailing south. If one boat sailed 10 nautical miles more than the other and they are 50 nautical miles apart, how far did each boat sail?

78. Navigation   One plane heads west from an airport flying at 200 mph. One hour later, a second plane heads north from the same airport, flying at the same speed. When will the planes be 1,000 miles apart?

In Exercises 79–80, use the formula $A = P(1 + r)^2$ to find the amount $A that $P will become when invested at an annual rate of r% for 2 years. SEE EXAMPLE 6. (OBJECTIVE 3)

79. Investing   What interest rate is needed for $5,000 to grow to $5,724.50 in 2 years?

80. Investing   What interest rate is needed for $7,000 to grow to $8,470 in 2 years?

Set up an equation to solve.

81. Manufacturing   An electronics firm has found that its revenue for manufacturing and selling x television sets is given by the formula $R = -\frac{1}{6}x^2 + 450x$. How much revenue will be earned by manufacturing 600 television sets?

82. Wholesale revenue   When a wholesaler sells n portable DVD players, the revenue R is given by the formula $R = 150n - \frac{1}{2}n^2$. How many players would the wholesaler have to sell to receive $11,250?

83. Fabricating metal   A piece of tin, 12 inches on a side, is to have four equal squares cut from its corners, as in the illustration. If the edges are then to be folded up to make a box with a floor area of 64 square inches, find the depth of the box.

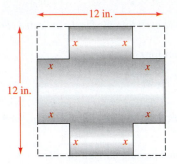

12 in.

12 in.

x x
x x
x x
x x

84. Making gutters   A piece of sheet metal, 18 inches wide, is bent to form the gutter shown in the illustration. If the cross-sectional area is 36 square inches, find the depth of the gutter.

x

x

$18 - 2x$

85. Filling a tank   Two pipes are used to fill a water tank. The first pipe can fill the tank in 4 hours, and the two pipes together can fill the tank in 2 hours less time than the second pipe alone. How long would it take for the second pipe to fill the tank?

86. Filling a pool   A small hose requires 6 more hours to fill a swimming pool than a larger hose. If the two hoses can fill the pool in 4 hours, how long would it take the larger hose alone?

WRITING ABOUT MATH

87. Explain how to use the quadratic formula.

88. Explain the meaning of the $\pm$ symbol.

89. Choose one of the previous applications and list the steps you followed as you worked it.

90. The binomial $b^2 - 4ac$ is called the **discriminant**. From its value, you can predict whether the solutions of a given quadratic equation are real or nonreal numbers. Explain.

SOMETHING TO THINK ABOUT *Use these facts. The two solutions of the equation* $ax^2 + bx + c = 0$ $(a \neq 0)$ *are*

$$x_1 = \frac{-b + \sqrt{b^2 - 4ac}}{2a} \quad \text{and}$$

$$x_2 = \frac{-b - \sqrt{b^2 - 4ac}}{2a}$$

91. Show that $x_1 + x_2 = -\dfrac{b}{a}$.

92. Show that $x_1 x_2 = \dfrac{c}{a}$.

Section 9.4

Graphing Quadratic Functions

Objectives

1. Graph a quadratic function of the form $f(x) = ax^2 + bx + c$ using a table of values and identify the vertex.
2. Find the vertex of a parabola by completing the square.
3. Identify the *x*-intercept, the *y*-intercept, the axis of symmetry, and the vertex of a parabola given a function in the form $f(x) = ax^2 + bx + c$.
4. Solve an application involving a quadratic equation.
5. Solve a quadratic equation using a graphing calculator.

Vocabulary

parabola
minimum point

vertex
maximum point

axis of symmetry

Getting Ready

If $f(x) = 2x^2 - x + 2$, find each value.

1. $f(0)$ **2.** $f(1)$ **3.** $f(-1)$ **4.** $f(-2)$

If $x = -\dfrac{b}{2a}$, find the value of x when a and b have the following values.

5. $a = 2, b = 8$ **6.** $a = 5, b = -20$

The function defined by the equation $f(x) = mx + b$ is a linear function, because its right side is a first-degree polynomial in the variable x. The function defined by $f(x) = ax^2 + bx + c$ $(a \neq 0)$ is a quadratic function, because its right side is a second-degree polynomial in the variable x. In this section, we will discuss many quadratic functions.

1 **Graph a quadratic function of the form $f(x) = ax^2 + bx + c$ using a table of values and identify the vertex.**

A basic quadratic function, first discussed in Section 4.4, is defined by the equation $f(x) = x^2$. Recall that to graph this function, we find several ordered pairs (x, y) that satisfy the equation, plot the pairs, and join the points with a smooth curve. A table of values and the graph appear in Figure 9-3. The graph of a quadratic function is called a **parabola**. The lowest point (or **minimum point**) on the parabola that opens upward is called its **vertex**. The vertex of the parabola shown in Figure 9-3 is the point $V(0, 0)$.

If a parabola opens downward, its highest point (or **maximum point**) is the vertex.

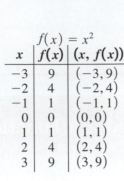

$$f(x) = x^2$$

x	$f(x)$	$(x, f(x))$
-3	9	$(-3, 9)$
-2	4	$(-2, 4)$
-1	1	$(-1, 1)$
0	0	$(0, 0)$
1	1	$(1, 1)$
2	4	$(2, 4)$
3	9	$(3, 9)$

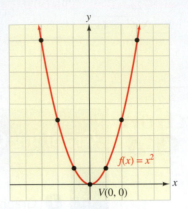

Figure 9-3

EXAMPLE 1   Graph $f(x) = x^2 - 3$. Compare this graph with Figure 9-3.

Solution   To find ordered pairs (x, y) that satisfy the equation, we select several numbers for x and compute the corresponding values of y. Recall that $f(x) = y$. If we let $x = 3$, we have

$y = x^2 - 3$

$y = 3^2 - 3$ Substitute 3 for x.

$y = 6$

The ordered pair $(3, 6)$ and others satisfying the equation appear in the table shown in Figure 9-4. To graph the function, we plot the points and draw a smooth curve passing through them.

The resulting parabola is the graph of $f(x) = x^2 - 3$. The vertex of the parabola is the point $V(0, -3)$.

Note that the graph of $f(x) = x^2 - 3$ looks just like the graph of $f(x) = x^2$, except that it is 3 units lower.

$$f(x) = x^2 - 3$$

x	$f(x)$	$(x, f(x))$
3	6	$(3, 6)$
2	1	$(2, 1)$
1	-2	$(1, -2)$
0	-3	$(0, -3)$
-1	-2	$(-1, -2)$
-2	1	$(-2, 1)$
-3	6	$(-3, 6)$

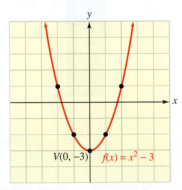

Figure 9-4

SELF CHECK 1   Graph $f(x) = x^2 + 2$. What do you notice?

EXAMPLE 2 Graph $f(x) = x^2 - 4x + 4$ and identify its vertex.

Solution To construct a table like the one shown in Figure 9-5, we select several numbers for x and compute the corresponding values of y. To graph the function, we plot the points and join them with a smooth curve.

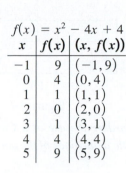

$f(x) = x^2 - 4x + 4$

x	$f(x)$	$(x, f(x))$
-1	9	$(-1, 9)$
0	4	$(0, 4)$
1	1	$(1, 1)$
2	0	$(2, 0)$
3	1	$(3, 1)$
4	4	$(4, 4)$
5	9	$(5, 9)$

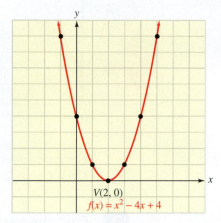

$V(2, 0)$
$f(x) = x^2 - 4x + 4$

Figure 9-5

Since the graph is a parabola that opens upward, the vertex is the minimum point on the graph, the point $V(2, 0)$.

SELF CHECK 2 Graph $f(x) = x^2 - 6x + 9$ and identify its vertex.

EXAMPLE 3 Graph $f(x) = -x^2 + 2x - 1$ and identify its vertex.

Solution We construct the table shown in Figure 9-6, plot the points, and draw the graph.

$f(x) = -x^2 + 2x - 1$

x	$f(x)$	$(x, f(x))$
-2	-9	$(-2, -9)$
-1	-4	$(-1, -4)$
0	-1	$(0, -1)$
1	0	$(1, 0)$
2	-1	$(2, -1)$
3	-4	$(3, -4)$
4	-9	$(4, -9)$

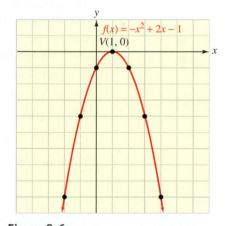

$f(x) = -x^2 + 2x - 1$
$V(1, 0)$

Figure 9-6

Since the parabola opens downward, its maximum point is the vertex, which is the point $V(1, 0)$.

SELF CHECK 3 Graph $f(x) = -x^2 - 4x - 4$ and identify its vertex.

The results of these first three examples suggest the following fact.

GRAPHS OF PARABOLAS The graph of the function $f(x) = ax^2 + bx + c$ $(a \neq 0)$ is a parabola. It opens upward when $a > 0$, and it opens downward when $a < 0$.

Accent
on technology

▸ Graphing Quadratic
Functions

We can use a TI84 graphing calculator to sketch the graphs of Examples 1–3. If we use window values of $x = [-10, 10]$ and $y = [-10, 10]$, enter the quadratic equation, and press the **GRAPH** key, we will obtain the graphs shown in Figure 9-7.

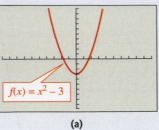

$f(x) = x^2 - 3$

(a)

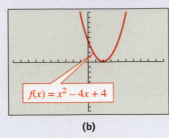

$f(x) = x^2 - 4x + 4$

(b)

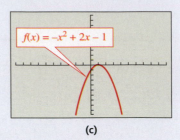

$f(x) = -x^2 + 2x - 1$

(c)

Figure 9-7

For instructions regarding the use of a Casio graphing calculator, please refer to the Casio Keystroke Guide in the back of the book.

2   **Find the vertex of a parabola by completing the square.**

It is easier to graph a parabola when we know the coordinates of its vertex. We can find the coordinates of the vertex of the graph of

$$f(x) = x^2 - 6x + 8$$

if we complete the square in the following way.

$f(x) = x^2 - 6x \mathbf{+ 9 - 9} + 8$ Add 9 to complete the square on $x^2 - 6x$ and then subtract 9.
$f(x) = (x - 3)^2 - 1$ Factor $x^2 - 6x + 9$ and combine like terms.

Since $a > 0$ in the original equation, the graph will be a parabola that opens upward. The vertex will be the minimum point on the parabola, and the y-coordinate of the vertex will be the smallest possible value of y. Because $(x - 3)^2 \geq 0$, the smallest value of y occurs when $(x - 3)^2 = 0$ or when $x = 3$. To find the corresponding value of y, we substitute 3 for x in the equation $f(x) = (x - 3)^2 - 1$ and simplify.

$f(x) = (\mathbf{x} - 3)^2 - 1$
$f(3) = (\mathbf{3} - 3)^2 - 1$ Substitute 3 for x.
$f(3) = 0^2 - 1$
$f(3) = -1$

The vertex of the parabola is the point $V(3, -1)$. The graph appears in Figure 9-8.

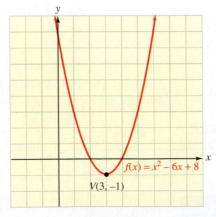

Figure 9-8

A generalization of this discussion leads to the following fact.

GRAPHS OF PARABOLAS WITH VERTEX AT (h, k)

The graph of an equation of the form

$$f(x) = a(x - h)^2 + k$$

is a parabola with its vertex at the point with coordinates (h, k). The parabola opens upward if $a > 0$, and it opens downward if $a < 0$.

Accent on technology

▸ Finding the Vertex of a Parabola

To find the vertex of the parabola determined by the equation $y = x^2 - 6x + 8$, we graph the function as in Figure 9-9(a).

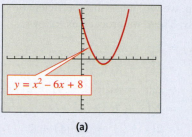

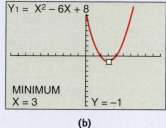

(a)   (b)

Figure 9-9

We can find the exact coordinates of the vertex by using a TI84 graphing calculator using the MINIMUM feature found in the CALC menu. Press

2ND CALC 3 (MINIMUM)

We then can enter an x-value that lies to the left of the vertex such as $x = 0$ for the left bound and press **ENTER**. We then enter an x-value that lies to the right of the vertex such as $x = 5$ for the right bound and press **ENTER**. After pressing **ENTER** one more time, we will see that the coordinates of the vertex are $(3, -1)$ as illustrated in Figure 9-9(b).

For instructions regarding the use of a Casio graphing calculator, please refer to the Casio Keystroke Guide in the back of the book.

EXAMPLE 4   Find the vertex of the parabola determined by $f(x) = -4(x - 3)^2 - 2$. Does the parabola open upward or downward?

Solution   Since $a = -4$ and $-4 < 0$, the parabola opens downward.
   In the equation $y = a(x - h)^2 + k$, the coordinates of the vertex are given by the ordered pair (h, k). In the equation $f(x) = -4(x - \mathbf{3})^2 - \mathbf{2}$, $h = 3$ and $k = -2$. Thus, the vertex is the point $(h, k) = (3, -2)$. In this case, the vertex will be the maximum point on the graph.

 SELF CHECK 4   Confirm the results of Example 4 by using a graphing calculator.

EXAMPLE 5   Find the vertex of the parabola determined by $f(x) = 5(x + 1)^2 + 4$. Does the parabola open upward or downward?

Solution   Since $a = 5$ and $5 > 0$, the parabola opens upward.
   The equation $f(x) = 5(x + 1)^2 + 4$ is equivalent to the equation

$$f(x) = 5[x - (\mathbf{-1})]^2 + \mathbf{4}$$

Since $h = -1$ and $k = 4$, the vertex is the point $(h, k) = (-1, 4)$. In this case, the vertex will be the minimum point on the graph.

SELF CHECK 5   Confirm the results of Example 5 by using a graphing calculator.

EXAMPLE 6   Find the vertex of the parabola determined by $f(x) = 2x^2 + 8x + 2$ and graph the parabola.

Solution   To find the vertex of the parabola, we will write $f(x) = 2x^2 + 8x + 2$ in the form $f(x) = a(x - h)^2 + k$ by completing the square on the right side of the equation. As a first step, we will make the coefficient of x^2 equal to 1 by factoring 2 out of the binomial $2x^2 + 8x$. Then, we proceed as follows:

$$\begin{aligned}
f(x) &= 2x^2 + 8x + 2 \\
&= 2(x^2 + 4x) + 2 && \text{Factor 2 out of } 2x^2 + 8x. \\
&= 2(x^2 + 4x + 4 - 4) + 2 && \text{Complete the square on } x^2 + 4x. \\
&= 2[(x + 2)^2 - 4] + 2 && \text{Factor } x^2 + 4x + 4. \\
&= 2(x + 2)^2 + 2(-4) + 2 && \text{Distribute the multiplication by 2.} \\
&= 2(x + 2)^2 - 6 && \text{Simplify and combine like terms.}
\end{aligned}$$

or

$$f(x) = 2[x - (-2)]^2 + (-6)$$

Since $h = -2$ and $k = -6$, the vertex of the parabola is the point $V(-2, -6)$. Since $a = 2$, the parabola opens upward. In this case, the vertex will be the minimum point on the graph.

We can select numbers on either side of $x = -2$ to construct the table shown in Figure 9-10. To find the y-intercept, we substitute 0 for x in our original equation and solve for y: When $x = 0$, $y = 2$. Thus, the y-intercept is $(0, 2)$. We determine more ordered pairs, plot the points, and draw the parabola.

$y = 2x^2 + 8x + 2$

x	y	(x, y)
0	2	$(0, 2)$
-1	-4	$(-1, -4)$
-2	-6	$(-2, -6)$
-3	-4	$(-3, -4)$
-4	2	$(-4, 2)$

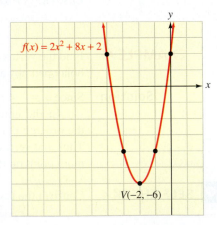

Figure 9-10

SELF CHECK 6   Confirm the results of Example 6 by using a graphing calculator.

3   Identify the *x*-intercept, the *y*-intercept, the axis of symmetry, and the vertex of a parabola given a function in the form $f(x) = ax^2 + bx + c$.

Much can be determined about the graph of $f(x) = ax^2 + bx + c$ from the coefficients a, b, and c. We summarize these results as follows.

GRAPHING THE PARABOLA

$f(x) = ax^2 + bx + c$

1. If $a > 0$, the parabola opens upward and the vertex is the minimum. If $a < 0$, the parabola opens downward and the vertex is the maximum.

2. The coordinates of the vertex are $\left(-\dfrac{b}{2a}, f\left(-\dfrac{b}{2a}\right)\right)$.

3. The **axis of symmetry** is the vertical line $x = -\dfrac{b}{2a}$.

4. The y-intercept is $(0, c)$.

5. The x-intercepts (if any) are determined by the solutions of $ax^2 + bx + c = 0$.

EXAMPLE 7   Graph: $f(x) = x^2 - 2x - 3$

Solution   The equation is in the form $f(x) = ax^2 + bx + c$, with $a = 1$, $b = -2$, and $c = -3$. Since $a > 0$, the parabola opens upward. To find the x-coordinate of the vertex, we substitute the values for a and b into the formula $x = -\dfrac{b}{2a}$.

$$x = -\frac{b}{2a} = -\frac{-2}{2(1)} = 1$$

The x-coordinate of the vertex is $x = 1$. This is also the equation for the axis of symmetry. To find the y-coordinate, we can find $f\left(-\dfrac{b}{2a}\right) = f(1)$ by substituting 1 for x in the equation and solving for y.

$$f(x) = x^2 - 2x - 3$$
$$f(1) = 1^2 - 2 \cdot 1 - 3$$
$$= 1 - 2 - 3$$
$$= -4$$

The vertex of the parabola is the point $(1, -4)$.

To graph the parabola, we identify several other points with coordinates that satisfy the equation. One easy point to find is the y-intercept. It is the value of y when $x = 0$. Thus, the parabola passes through the point $(0, -3)$.

To find the x-intercepts of the graph, we set $f(x)$ equal to 0 and solve the resulting quadratic equation:

$$f(x) = x^2 - 2x - 3$$
$$0 = x^2 - 2x - 3$$
$$0 = (x - 3)(x + 1) \quad \text{Factor.}$$

$x - 3 = 0$ or $x + 1 = 0$ Set each factor equal to 0.

$x = 3$ | $x = -1$

Since the x-intercepts of the graph are $(3, 0)$ and $(-1, 0)$, the graph passes through these points. The graph appears in Figure 9-11.

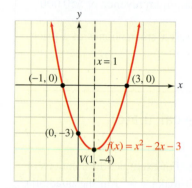

Figure 9-11

COMMENT   If the entire parabola is above or below the x-axis, there will be no x-intercepts.

 SELF CHECK 7   Confirm the results of Example 7 by using a graphing calculator.

4   Solve an application involving a quadratic equation.

EXAMPLE 8   **FINDING MAXIMUM REVENUE**   An electronics firm manufactures a high-quality smartphone. Over the past 10 years, the firm has learned that it can sell x smartphones at a price of $\left(200 - \dfrac{1}{5}x\right)$ dollars. How many smartphones should the firm manufacture and sell to maximize its revenue? Find the maximum revenue.

Solution   The revenue obtained is the product of the number of smartphones that the firm sells (x) and the price of each smartphone $\left(200 - \frac{1}{5}x\right)$. Thus, the revenue R is given by the formula

$$R = x\left(200 - \frac{1}{5}x\right) \qquad \text{or} \qquad R = -\frac{1}{5}x^2 + 200x$$

Since the graph of this function is a parabola that opens downward, the maximum value of R will be the value of R determined by the vertex of the parabola. Because the x-coordinate of the vertex is at $x = \frac{-b}{2a}$, we have

$$x = \frac{-b}{2a} = \frac{-200}{2\left(-\frac{1}{5}\right)} = \frac{-200}{-\frac{2}{5}} = (-200)\left(-\frac{5}{2}\right) = 500$$

If the firm manufactures 500 smartphones, the maximum revenue will be

$$R = -\frac{1}{5}x^2 + 200x$$

$$= -\frac{1}{5}(500)^2 + 200(500)$$

$$= 50{,}000$$

The firm should manufacture 500 smartphones to get a maximum revenue of $50,000.

SELF CHECK 8   Find the maximum revenue if the firm learned it could sell x smartphones at a price of $\left(180 - \frac{1}{4}x\right)$ dollars.

---

5   **Solve a quadratic equation using a graphing calculator.**

Accent on technology

▸ Solving Quadratic Equations

We can use graphing methods to solve quadratic equations. For example, the solutions of the equation $x^2 - x - 3 = 0$ are the values of x that will make $y = 0$ in the quadratic function $f(x) = x^2 - x - 3$. To approximate these values, we graph the quadratic function and identify the x-intercepts.

If we use window values of $x = [-10, 10]$ and $y = [-10, 10]$ and graph the function $f(x) = x^2 - x - 3$, using a TI84 graphing calculator we will obtain the graph shown in Figure 9-12(a).

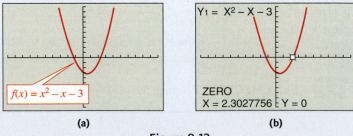

$f(x) = x^2 - x - 3$

$Y_1 = X^2 - X - 3$

ZERO
X = 2.3027756   Y = 0

(a)   (b)

Figure 9-12

We can find the x-intercepts exactly (if they are rational) by using the ZERO command found in the CALC menu. Enter values to the left and right of each x-intercept, similar to the steps you followed to find the minimum/maximum. In this case the x-intercepts are irrational. To find the exact values, we would have to use the quadratic formula.

For instructions regarding the use of a Casio graphing calculator, please refer to the Casio Keystroke Guide in the back of the book.

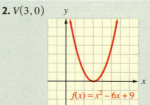

SELF CHECK ANSWERS

1. same shape as the graph of $y = x^2$, but 2 units higher

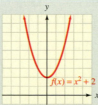

$f(x) = x^2 + 2$

2. $V(3, 0)$

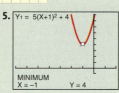

$f(x) = x^2 - 6x + 9$

3. $V(-2, 0)$

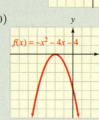

$f(x) = -x^2 - 4x - 4$

4.
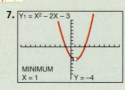

Y₁ = -4(X–3)² – 2
MAXIMUM
X = 3   Y = –2

5.
Y₁ = 5(X+1)² + 4
MINIMUM
X = –1   Y = 4

6.
Y₁ = 2X² + 8X + 2
MINIMUM
X = –2   Y = –6

7.
Y₁ = X² – 2X – 3
MINIMUM
X = 1   Y = –4

8. The maximum revenue will be $32,400 if 360 smartphones are manufactured.

NOW TRY THIS

1. Find the minimum point on the graph of $f(x) = x^2 + 2x - 1$.
2. Explain the process for finding the y-intercept.
3. From the graph, find the solutions of $x^2 - 2x - 3 = 0$.

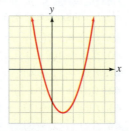

9.4 Exercises

WARM-UPS *Find the y-intercept and any x-intercepts.*

1. $y = 3x + 6$

2. $y = -4x + 12$

3. $2x + y = 4$

4. $3x + 5y = 30$

5. $y = x^2 - 4$

6. $y = x^2 - 25$

7. $y = x^2 + 3x + 2$

8. $y = x^2 - 4x - 5$

REVIEW *Simplify. Assume that all variables represent positive values.*

9. $\sqrt{45} + \sqrt{80}$

10. $3\sqrt{6y}\left(-4\sqrt{3y}\right)$

11. $\left(\sqrt{5} - 1\right)\left(\sqrt{5} + 1\right)$

12. $\dfrac{\sqrt{x} + 2}{\sqrt{x} - 2}$

VOCABULARY AND CONCEPTS *Fill in the blanks.*

13. A function defined by the equation $f(x) = ax^2 + bx + c$ $(a \neq 0)$ is called a _____ function.

14. The lowest (or highest) point on a parabola is called the _____ of the parabola.

15. The point where a parabola intersects the y-axis is called the _____.

16. The graph of $y = ax^2 + bx + c$ $(a \neq 0)$ is a parabola that opens upward when _____. Its vertex is the _____ point on the graph.

17. The graph of $f(x) = ax^2 + bx + c$ $(a \neq 0)$ is a parabola that opens downward when ____. Its vertex is the _____ point on the graph.

18. The vertex of the parabolic graph of $f(x) = a(x - h)^2 + k$ is the point ____.

19. The y-intercept of the graph of $f(x) = ax^2 + bx + c$ is the point ____.

20. The x-coordinate of the vertex of the graph of
$f(x) = ax^2 + bx + c$ is $x =$ ___.

The graph of each equation is a parabola. Does it open upward or downward?

21. $y = 3x^2 - 2x + 4$

22. $y = 2x^2 + x - 5$

23. $y = -x^2 - 3x - 5$

24. $y = -3x^2 + 4x - 1$

25. $y - 4 = -\dfrac{3}{2}x^2 + x$

26. $y + 5 = \dfrac{4}{7}x^2 - x$

 **GUIDED PRACTICE** *Graph each function and compare the graph with the graph of $f(x) = x^2$. Check your work with a graphing calculator.* SEE EXAMPLE 1. (OBJECTIVE 1)

27. $f(x) = x^2 + 1$

28. $f(x) = x^2 - 4$

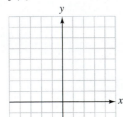   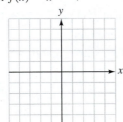

29. $f(x) = -x^2$

30. $f(x) = -(x - 1)^2$

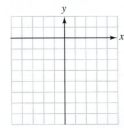   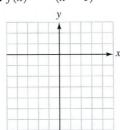

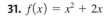 *Graph each function and find the vertex. Check your work with a graphing calculator.* SEE EXAMPLES 2–3. (OBJECTIVE 1)

31. $f(x) = x^2 + 2x$

32. $f(x) = x^2 - 2x$

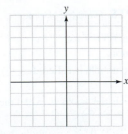   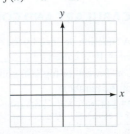

33. $f(x) = -x^2 - 4x$

34. $f(x) = -x^2 + 2x$

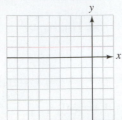   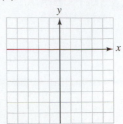

35. $f(x) = x^2 + 4x + 4$

36. $f(x) = x^2 - 6x + 9$

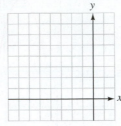   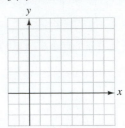

37. $f(x) = x^2 - 4x + 6$

38. $f(x) = x^2 + 2x - 3$

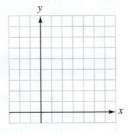   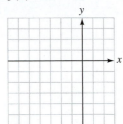

Find the vertex of the graph of each function. Do not sketch the graph. SEE EXAMPLES 4–5. (OBJECTIVE 2)

39. $f(x) = -2(x - 5)^2 + 1$

40. $f(x) = 3(x - 1)^2 + 4$

41. $f(x) = (x - 4)^2$

42. $f(x) = -(x - 2)^2$

43. $f(x) = (x + 2)^2 + 3$

44. $f(x) = (x + 3)^2 - 4$

45. $f(x) = -7x^2 + 4$

46. $f(x) = 5x^2 - 2$

 *Complete the square, if necessary, to determine the vertex of the graph of each function. Then graph the equation. Check your work with a graphing calculator.* SEE EXAMPLE 6. (OBJECTIVE 2)

47. $f(x) = x^2 - 4x + 4$

48. $f(x) = x^2 + 6x + 9$

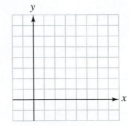   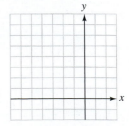

49. $f(x) = -x^2 - 2x - 1$

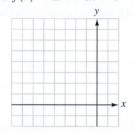

50. $f(x) = -x^2 + 2x - 1$

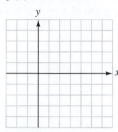

51. $f(x) = 2x^2 + 8x + 6$

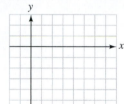

52. $f(x) = 3x^2 - 12x + 9$

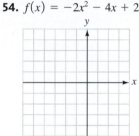

53. $f(x) = -3x^2 + 6x - 2$

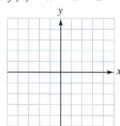

54. $f(x) = -2x^2 - 4x + 2$

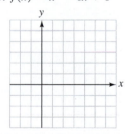

Find the vertex and the x- and y-intercepts of the graph of each function. Then graph each equation and label the axis of symmetry. SEE EXAMPLE 7. (OBJECTIVE 3)

55. $f(x) = x^2 - x - 2$

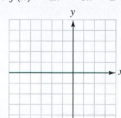

56. $f(x) = x^2 - 6x + 8$

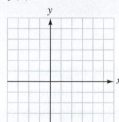

57. $f(x) = 2x^2 + 3x - 2$

58. $f(x) = 3x^2 - 7x + 2$

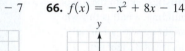
Use a graphing calculator to solve each equation. If an answer is not exact, give the result to the nearest hundredth. (OBJECTIVE 5)

59. $x^2 - 6x + 8 = 0$

60. $6x^2 + 5x - 6 = 0$

61. $x^2 + 5x + 2 = 0$

62. $2x^2 - 5x + 1 = 0$

ADDITIONAL PRACTICE *Graph.*

63. $f(x) = x^2 + 2x - 3$

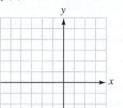

64. $f(x) = x^2 + 6x + 5$

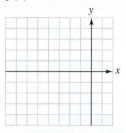

65. $f(x) = -x^2 - 6x - 7$

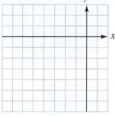

66. $f(x) = -x^2 + 8x - 14$

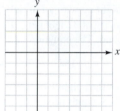

67. $f(x) = -x^2 + 2x + 3$

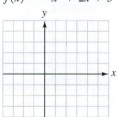

68. $f(x) = -x^2 + 5x - 4$

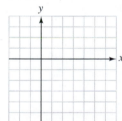

Find the vertex of the graph of each function using any method.

69. $f(x) = -3(x + 6)^2 - 5$

70. $f(x) = 4(x + 3)^2 + 1$

71. $f(x) = x^2 - 8x - 3$

72. $f(x) = x^2 - 8x - 20$

APPLICATIONS *Use a calculator to help solve.* SEE EXAMPLE 8. (OBJECTIVE 4)

73. Selling TVs A company has found that it can sell x TVs at a price of $\$\left(450 - \frac{1}{6}x\right)$. How many TVs must the company sell to maximize its revenue?

74. Finding maximum revenue In Exercise 73, find the maximum revenue.

75. Selling DVD players A wholesaler sells portable DVD players for $150 each. However, he gives discounted prices on purchases of between 500 and 1,000 units according to the formula $\left(150 - \frac{1}{10}n\right)$, where n represents the number of units purchased. How many units would a retailer have to buy for the wholesaler to obtain maximum revenue?

76. **Finding maximum revenue**   In Exercise 75, find the maximum revenue.

77. **Selling refrigerators**   American Appliance has found that it can sell more top-of-the-line refrigerators each year if it lowers the price. Over the years, the sales department has found that the store will sell x refrigerators at a price of $\$\left(5{,}000 - \frac{3}{2}x\right)$. What price should the owner of the store charge for each refrigerator to maximize his revenue?

78. **Selling refrigerators**   See Exercise 77. At that price, how many refrigerators should the store order from the distributor each year?

79. **Selling vases**   A glassworks that makes crystal vases has daily production costs given by the function

$$C(x) = 0.2x^2 - 10x + 650$$

where x is the number of vases made each day. How many vases should be made to minimize the per-day costs? Find the minimum cost.

80. **Ballistics**   From the top of a building, a ball is thrown straight up with an initial velocity of 32 feet per second. The equation

$$h = -16t^2 + 32t + 48$$

gives the height h of the ball t seconds after it is thrown. Find the maximum height reached by the ball.

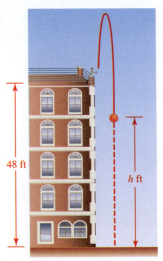

48 ft

h ft

WRITING ABOUT MATH

81. Explain why the y-intercept of the graph of $f(x) = ax^2 + bx + c$ is $(0, c)$.

82. Define the vertex of a parabola and explain how to find its coordinates.

SOMETHING TO THINK ABOUT

83. The graph of $x = y^2$ is a parabola, but the equation does not define a function. Explain.

84. The graph of $x = -y^2$ is a parabola, but the equation does not define a function. Explain.

Section 9.5   Complex Numbers

Objectives

1. Simplify powers of i.
2. Simplify square roots containing negative radicands.
3. Perform operations with complex numbers.
4. Rationalize a denominator, expressing the answer in $a + bi$ form.
5. Solve a quadratic equation that has complex-number solutions.

Vocabulary

imaginary number   complex number   complex conjugates

Getting Ready

Perform the following operations.

1. $(2x + 4) + (3x - 5)$ **2.** $(3x - 4) - (2x + 3)$

3. $(x + 4)(x - 5)$ **4.** $(3x - 1)(2x - 1)$

The solutions of some quadratic equations are not real numbers. For example, in Example 4 of Section 9.3, we solved the equation $x^2 + 2x + 5 = 0$ and obtained the following solutions

$$x = \frac{-2 + \sqrt{-16}}{2} \quad \text{or} \quad x = \frac{-2 - \sqrt{-16}}{2}$$

Each of these solutions involves $\sqrt{-16}$, which is not a real number. Thus, the solutions of this equation are not real numbers. As we will see, the solutions of this equation are from a set called the set of *complex numbers*.

1 Simplify powers of *i*.

For years, mathematicians believed that numbers such as $\sqrt{-1}$, $\sqrt{-9}$, and $\sqrt{-16}$ were nonsense. Even the great English mathematician Sir Isaac Newton (1642–1727) called them impossible. These numbers were called **imaginary numbers** by René Descartes (1596–1650). Today they have important uses such as describing alternating current in electronics.

The imaginary number $\sqrt{-1}$ is usually denoted by the letter *i*. Since

$$i = \sqrt{-1}$$

it follows that

$$i^2 = -1$$

COMMENT Powers of *i*. If *n* is a natural number that has a remainder of *r* when divided by 4, then $i^n = i^r$. When *n* is divisible by 4, the remainder is 0 and $i^0 = 1$.

The powers of *i* produce an interesting pattern:

$$i = \sqrt{-1} = i$$
$$i^2 = \left(\sqrt{-1}\right)^2 = -1$$
$$i^3 = i^2 \cdot i = -1 \cdot i = -i$$
$$i^4 = i^2 \cdot i^2 = (-1)(-1) = 1$$

$$i^5 = i^4 \cdot i = 1 \cdot i = i$$
$$i^6 = i^4 \cdot i^2 = 1(-1) = -1$$
$$i^7 = i^4 \cdot i^3 = 1(-i) = -i$$
$$i^8 = i^4 \cdot i^4 = (1)(1) = 1$$

The pattern continues: $i, -1, -i, 1, \ldots$.

2 Simplify square roots containing negative radicands.

If we assume that multiplication of imaginary numbers is commutative and associative, then

$$(2i)^2 = 2^2 i^2$$
$$= 4(-1)$$
$$= -4$$

Since $(2i)^2 = -4$, it follows that $2i$ is a square root of -4, and we write

$$\sqrt{-4} = 2i$$

This result could have been obtained by the following process:

$$\sqrt{-4} = \sqrt{4(-1)}$$
$$= \sqrt{4}\sqrt{-1}$$
$$= 2i$$

Likewise, we have

$$\sqrt{-25} = \sqrt{25(-1)} = \sqrt{25}\sqrt{-1} = 5i$$
$$\sqrt{-\frac{1}{9}} = \sqrt{\frac{1}{9}(-1)} = \sqrt{\frac{1}{9}}\sqrt{-1} = \frac{1}{3}i$$
$$\sqrt{\frac{-100}{49}} = \sqrt{\frac{100}{49}(-1)} = \frac{\sqrt{100}}{\sqrt{49}}\sqrt{-1} = \frac{10}{7}i$$

In general, we have the following rules.

PROPERTIES OF RADICALS

If at least one of a and b is a nonnegative real number, then

$$\sqrt{ab} = \sqrt{a}\sqrt{b} \qquad \text{and} \qquad \sqrt{\frac{a}{b}} = \frac{\sqrt{a}}{\sqrt{b}} \quad (b \neq 0)$$

3   Perform operations with complex numbers.

Imaginary numbers such as $\sqrt{-1}$, $\sqrt{-3}$, and $\sqrt{-9}$ form a subset of a broader set of numbers called *complex numbers*.

COMPLEX NUMBERS

A **complex number** is any number that can be written in the form $a + bi$ where a and b are real numbers, and $i = \sqrt{-1}$.

The number a is called the *real part*, and the number b is called the *imaginary part* of the complex number $a + bi$.

If $b = 0$, the complex number $a + bi$ is the real number a. If $b \neq 0$ and $a = 0$, the complex number $0 + bi$ (or just bi) is an imaginary number.

Figure 9-13 shows the relationship of the real numbers to the imaginary and complex numbers.

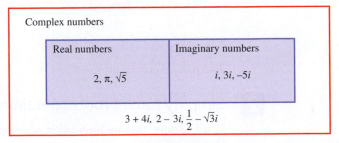

Figure 9-13

EQUALITY OF COMPLEX NUMBERS

The complex numbers $a + bi$ and $c + di$ are equal if and only if

$$a = c \qquad \text{and} \qquad b = d$$

Here are several examples of equal complex numbers.

$$2 + 3i = \sqrt{4} + \frac{6}{2}i, \text{ because } 2 = \sqrt{4} \text{ and } 3 = \frac{6}{2}.$$

$$4 - 5i = \frac{12}{3} - \sqrt{25}\,i, \text{ because } 4 = \frac{12}{3} \text{ and } -5 = -\sqrt{25}.$$

$$x + yi = 4 + 7i \text{ if and only if } x = 4 \text{ and } y = 7.$$

ADDITION AND SUBTRACTION OF COMPLEX NUMBERS

Complex numbers are added and subtracted as if they were binomials:

$$(a + bi) + (c + di) = (a + c) + (b + d)i$$

$$(a + bi) - (c + di) = (a - c) + (b - d)i$$

EXAMPLE 1   Perform each operation: **a.** $(8 + 4i) + (12 + 8i)$ **b.** $(7 - 4i) + (9 + 2i)$
c. $(-6 + i) - (3 - 4i)$ **d.** $(2 - 4i) - (-4 + 3i)$

Solution   We will use the rules for addition and subtraction of complex numbers.

a. $(8 + 4i) + (12 + 8i) = 8 + 4i + 12 + 8i$
$$= 20 + 12i$$

b. $(7 - 4i) + (9 + 2i) = 7 - 4i + 9 + 2i$
$$= 16 - 2i$$

c. $(-6 + i) - (3 - 4i) = -6 + i - 3 + 4i$
$$= -9 + 5i$$

d. $(2 - 4i) - (-4 + 3i) = 2 - 4i + 4 - 3i$
$$= 6 - 7i$$

 SELF CHECK 1   Perform each operation:

a. $(16 - 5i) + (2 + 3i)$ **b.** $(9 - 8i) + (3 + 6i)$
c. $(11 - 3i) - (-2 + 2i)$ **d.** $(-9 - i) - (7 - 4i)$

To multiply a complex number by an imaginary number, we use the distributive property to remove parentheses and then simplify.

EXAMPLE 2   Multiply: $-5i(4 - 8i)$

Solution   We will use the distributive property to remove parentheses.

$$-5i(4 - 8i) = -5i(4) + (-5i)(-8i)$$
$$= -20i + 40i^2$$
$$= -20i + 40(-1) \qquad i^2 = -1$$
$$= -40 - 20i \qquad \text{Write in } a + bi \text{ form.}$$

 SELF CHECK 2   Multiply: $6i(4 - 3i)$

MULTIPLICATION OF COMPLEX NUMBERS	Complex numbers are multiplied as if they were binomials, with $i^2 = -1$:

$$(a + bi)(c + di) = ac + adi + bci + bdi^2$$
$$= (ac - bd) + (ad + bc)i$$

EXAMPLE 3 Perform each multiplication: **a.** $(2 + 3i)(3 - 2i)$ **b.** $(3 + i)(1 + 2i)$
c. $(-4 + 2i)(2 + i)$ **d.** $(-1 - i)(4 - i)$

Solution We will multiply each term of the first factor by each term of the second factor, and simplify.

a. $(2 + 3i)(3 - 2i) = 6 - 4i + 9i - 6i^2$
$$= 6 + 5i + 6$$
$$= 12 + 5i$$

b. $(3 + i)(1 + 2i) = 3 + 6i + i + 2i^2$
$$= 3 + 7i - 2$$
$$= 1 + 7i$$

c. $(-4 + 2i)(2 + i) = -8 - 4i + 4i + 2i^2$
$$= -8 - 2$$
$$= -10$$

d. $(-1 - i)(4 - i) = -4 + i - 4i + i^2$
$$= -4 - 3i - 1$$
$$= -5 - 3i$$

Leonhard Euler
1707–1783
Euler first used the letter *i* to represent $\sqrt{-1}$, the letter *e* for the base of natural logarithms, and the symbol $\sum$ for summation. Euler was one of the most prolific mathematicians of all time, contributing to almost all areas of mathematics. Much of his work was accomplished after he became blind.

 **SELF CHECK 3** Multiply: **a.** $(5 - 6i)(2 + i)$ **b.** $(7 + 6i)(5 + i)$

The next example shows how to write complex numbers in $a + bi$ form. When writing answers, it is acceptable to use $a - bi$ as a substitute for the form $a + (-b)i$.

EXAMPLE 4 Write each number in $a + bi$ form.

a. $7 = 7 + 0i$ **b.** $3i = 0 + 3i$
c. $4 - \sqrt{-16} = 4 - \sqrt{-1(16)}$ **d.** $5 + \sqrt{-11} = 5 + \sqrt{-1(11)}$
$$= 4 - \sqrt{16}\sqrt{-1} \qquad\qquad = 5 + \sqrt{11}\sqrt{-1}$$
$$= 4 - 4i \qquad\qquad = 5 + \sqrt{11}i$$
e. $2i^2 + 4i^3 = 2(-1) + 4(-i) = -2 - 4i$
f. $\sqrt{-4}\sqrt{-25} = 2i \cdot 5i = 10i^2 = 10(-1) = -10 + 0i$

 **SELF CHECK 4** Write each number in $a + bi$ form: **a.** $3 + \sqrt{-25}$ **b.** $3i^3 - 4i^4$
c. $9 - \sqrt{-9}$ **d.** $5 - \sqrt{-10}$ **e.** $5i^3 + 2i^2$
f. $\sqrt{-9}\sqrt{-16}$

4   Rationalize a denominator, expressing the answer in $a + bi$ form.

The fraction $-\dfrac{5}{i}$ is not in simplest form because the denominator contains a radical: $i = \sqrt{-1}$. To simplify the fraction, we must rationalize the denominator just as we did when we simplified rational expressions.

EXAMPLE 5   Rationalize each denominator and write the result in $a + bi$ form.

a. $-\dfrac{5}{i}$ **b.** $\dfrac{3}{2i^3}$

Solution   We can multiply each numerator and denominator by i and simplify.

a. $-\dfrac{5}{i} = -\dfrac{5}{i} \cdot \dfrac{i}{i}$

$\qquad = -\dfrac{5i}{i^2}$

$\qquad = -\dfrac{5i}{-1}$

$\qquad = 5i$

$\qquad = 0 + 5i$

b. $\dfrac{3}{2i^3} = \dfrac{3}{2i^3} \cdot \dfrac{i}{i}$

$\qquad = \dfrac{3i}{2i^4}$

$\qquad = \dfrac{3i}{2(1)}$

$\qquad = \dfrac{3i}{2}$

$\qquad = 0 + \dfrac{3}{2}i$

SELF CHECK 5   Rationalize the denominator and write the result in $a + bi$ form:

a. $\dfrac{7}{i}$ **b.** $\dfrac{4}{5i}$

To rationalize the denominators of fractions such as $\dfrac{1}{3+i}$, $\dfrac{3-i}{2+i}$, and $\dfrac{5+i}{5-i}$, we must multiply the numerator and denominator by the complex conjugate of the denominator.

COMPLEX CONJUGATES   The complex numbers $a + bi$ and $a - bi$ are called **complex conjugates** of each other.

For example,

- $3 + 4i$ and $3 - 4i$ are complex conjugates.
- $5 - 7i$ and $5 + 7i$ are complex conjugates.
- $8 + 17i$ and $8 - 17i$ are complex conjugates.

When you multiply a complex number and its conjugate, the product is a positive real number.

EXAMPLE 6   Find the product of $3 + i$ and its complex conjugate.

Solution   The complex conjugate of $3 + i$ is $3 - i$. We find their product as follows:

$(3 + i)(3 - i) = 9 - 3i + 3i - i^2$

$\qquad\qquad = 9 - i^2$ Combine like terms.

$\qquad\qquad = 9 - (-1)$ $i^2 = -1$

$\qquad\qquad = 10$

SELF CHECK 6   Multiply: $(5 - 4i)(5 + 4i)$

In general, the product of the complex number $a + bi$ and its complex conjugate $a - bi$ is the real number $a^2 + b^2$.

$$(a + bi)(a - bi) = a^2 - abi + abi - b^2 i^2$$
$$= a^2 - b^2(\mathbf{-1})$$
$$= a^2 + b^2$$

EXAMPLE 7   Rationalize the denominator of $\dfrac{1}{3 + i}$ and write the result in $a + bi$ form.

Solution   Since the product of $3 + i$ and its conjugate is a real number, we can rationalize the denominator by multiplying both the numerator and the denominator of the fraction by the complex conjugate of the denominator and simplifying.

$$\frac{1}{3 + i} = \frac{1}{3 + i} \cdot \frac{\mathbf{3 - i}}{\mathbf{3 - i}}$$

$$= \frac{3 - i}{9 - 3i + 3i - i^2} \qquad \text{Use the distributive property.}$$

$$= \frac{3 - i}{9 - (-1)} \qquad \text{Combine like terms; } i^2 = -1.$$

$$= \frac{3 - i}{10} \qquad \text{Subtract.}$$

$$= \frac{3}{10} - \frac{1}{10} i \qquad \text{Write in } a + bi \text{ form.}$$

SELF CHECK 7   Rationalize the denominator of $\dfrac{5}{7 + i}$ and write the result in $a + bi$ form.

EXAMPLE 8   Write $\dfrac{3 - i}{2 + i}$ in $a + bi$ form.

Solution   We rationalize the denominator by multiplying the numerator and denominator by the complex conjugate of the denominator and simplifying.

$$\frac{3 - i}{2 + i} = \frac{3 - i}{2 + i} \cdot \frac{\mathbf{2 - i}}{\mathbf{2 - i}}$$

$$= \frac{6 - 3i - 2i + i^2}{4 - 2i + 2i - i^2} \qquad \text{Use the distributive property.}$$

$$= \frac{5 - 5i}{4 - (-1)} \qquad \text{Combine like terms; } i^2 = -1.$$

$$= \frac{5(1 - i)}{5} \qquad \text{Factor out the GCF, 5, in the numerator.}$$

$$= 1 - i \qquad \text{Divide out the common factor of 5.}$$

SELF CHECK 8   Write $\dfrac{7 + i}{5 - i}$ in $a + bi$ form.

EXAMPLE 9   Divide $(5 + i)$ by $(5 - i)$ and express the quotient in $a + bi$ form.

Solution   The quotient obtained when dividing $(5 + i)$ by $(5 - i)$ is expressed by the fraction $\dfrac{5 + i}{5 - i}$. To express this quotient in $a + bi$ form, we rationalize the denominator by multiplying both the numerator and the denominator by the complex conjugate of the denominator and simplifying.

$$\frac{5 + i}{5 - i} = \frac{5 + i}{5 - i} \cdot \frac{5 + i}{5 + i}$$

$$= \frac{25 + 5i + 5i + i^2}{25 + 5i - 5i - i^2}$$ Use the distributive property.

$$= \frac{25 + 10i - 1}{25 - (-1)}$$ Combine like terms; $i^2 = -1$.

$$= \frac{24 + 10i}{26}$$ Subtract.

$$= \frac{2(12 + 5i)}{26}$$ Factor out the GCF, 2, in the numerator.

$$= \frac{12 + 5i}{13}$$ Divide out the common factor of 2.

$$= \frac{12}{13} + \frac{5}{13}i$$ Write in $a + bi$ form.

 **SELF CHECK 9**   Divide $\frac{7 + i}{7 - i}$ and express the quotient in $a + bi$ form.

COMMENT   Always write complex numbers in $a + bi$ form before performing any operations on them.

Perspective

EULER'S IDENTITY

One of the most celebrated equations in mathematics elegantly combines the numbers $1, 0, \pi, e,$ and i, five of the most fundamental numerical concepts used to describe our world. This equation, known as Euler's Identity in honor of Swiss mathematician Leonhard Euler (1707–1783), usually is written in the form $e^{i\pi} + 1 = 0$ and reveals an unexpected interplay between arithmetic, algebra, trigonometry, geometry, and imaginary numbers. An article posted in the FBI files (http://www.fbi.gov/news/stories/2004/may) reports that in 2003 this formula was used as evidence to help convict an arsonist in a criminal trial.

Perform the given calculations involving complex numbers.

1. $(3 + 5i)(2 - i)$

2. $\dfrac{-4 + 9i}{6 - 7i}$

EXAMPLE 10   Write $\dfrac{4 + \sqrt{-16}}{2 + \sqrt{-4}}$ in $a + bi$ form.

Solution $\dfrac{4 + \sqrt{-16}}{2 + \sqrt{-4}} = \dfrac{4 + 4i}{2 + 2i}$ Change the numerator and the denominator to $a + bi$ form.

$$= \frac{4(1 + i)}{2(1 + i)}$$ Factor out the GCF, 4, in the numerator and the GCF, 2, in the denominator.

$$= 2 + 0i$$ Simplify.

 SELF CHECK 10   Write $\dfrac{3 + \sqrt{-25}}{2 - \sqrt{-9}}$ in $a + bi$ form.

---

5   **Solve a quadratic equation that has complex-number solutions.**

Recall Example 4 of Section 9.3 when we solved $x^2 + 2x + 5 = 0$ and obtained $x = \dfrac{-2 + \sqrt{-16}}{2}$ or $x = \dfrac{-2 - \sqrt{-16}}{2}$. We are now ready to complete the solutions.

$$x = \frac{-2 + 4i}{2} \quad \text{or} \quad x = \frac{-2 - 4i}{2}$$

$$x = -1 + 2i \quad\quad\quad x = -1 - 2i$$

The solutions are the complex numbers $-1 + 2i$ or $-1 - 2i$.

The solutions of many quadratic equations are not real numbers. For example, the solutions of the equation $x^2 + x + 1 = 0$ are not real numbers. We will show that this is true in the next example.

EXAMPLE 11 Solve: $x^2 + x + 1 = 0$

Solution We will solve the equation by using the quadratic formula with $a = 1$, $b = 1$, and $c = 1$.

$$x = \frac{-\boldsymbol{b} \pm \sqrt{\boldsymbol{b}^2 - 4(\boldsymbol{a})(\boldsymbol{c})}}{2(\boldsymbol{a})}$$

$$= \frac{-\boldsymbol{1} \pm \sqrt{\boldsymbol{1}^2 - 4(\boldsymbol{1})(\boldsymbol{1})}}{2(\boldsymbol{1})}$$

$$= \frac{-1 \pm \sqrt{1 - 4}}{2}$$

$$= \frac{-1 \pm \sqrt{-3}}{2}$$

$$x = \frac{-1 + \sqrt{-3}}{2} \quad \text{or} \quad x = \frac{-1 - \sqrt{-3}}{2}$$

Expressed in $a + bi$ form, the solutions are

$$-\frac{1}{2} + \frac{\sqrt{3}}{2}i \quad \text{and} \quad -\frac{1}{2} - \frac{\sqrt{3}}{2}i$$

 SELF CHECK 11 Solve $x^2 + 4x + 29 = 0$ and express the solutions in $a + bi$ form.

 **SELF CHECK ANSWERS**

1. a. $18 - 2i$ **b.** $12 - 2i$ **c.** $13 - 5i$ **d.** $-16 + 3i$ **2.** $18 + 24i$ **3. a.** $16 - 7i$ **b.** $29 + 37i$
4. a. $3 + 5i$ **b.** $-4 - 3i$ **c.** $3 - 3i$ **d.** $5 - \sqrt{10}\,i$ **e.** $-2 - 5i$ **f.** -12 **5. a.** $0 - 7i$ **b.** $0 - \frac{4}{5}i$
6. 41 **7.** $\frac{7}{10} - \frac{1}{10}i$ **8.** $\frac{17}{13} + \frac{6}{13}i$ **9.** $\frac{24}{25} + \frac{7}{25}i$ **10.** $-\frac{9}{13} + \frac{19}{13}i$ **11.** $-2 + 5i, -2 - 5i$

NOW TRY THIS

1. Solve: $x^3 = 64$

2. Multiply and simplify: $\sqrt{-18} \cdot \sqrt{-2}$

9.5 Exercises

WARM-UPS *Find the conjugate for each radical expression.*

1. $2 - \sqrt{3}$

2. $-5 + \sqrt{7}$

3. $\sqrt{2} + \sqrt{5}$

4. $\sqrt{5} - \sqrt{2}$

5. $6 - 2\sqrt{3}$

6. $4 + 5\sqrt{6}$

REVIEW *Factor each expression.*

7. $x^2 - 49$

8. $6x^2 + x - 2$

9. $-2a^2 - 4a + 6$

10. $-12x^2 + 3$

VOCABULARY AND CONCEPTS

11. Numbers that are square roots of negative numbers are called _____ numbers.

12. A _____ number is any number that can be written in the form $a + bi$ where a and b are real numbers and $i = \sqrt{-1}$.

13. $\sqrt{-1} = \_$ and $i^2 = \_\_$.

14. The complex conjugate of $3 - 4i$ is _____.

GUIDED PRACTICE *Simplify each expression. (OBJECTIVE 1)*

15. i^{21}

16. i^{19}

17. i^{27}

18. i^{22}

19. i^{100}

20. i^{42}

21. i^{97}

22. i^{200}

Simplify. (OBJECTIVE 2)

23. $\sqrt{-36}$

24. $\sqrt{-9}$

25. $\sqrt{-\dfrac{1}{81}}$

26. $\sqrt{-\dfrac{49}{16}}$

Perform the operations. Write all answers in $a + bi$ form. **SEE EXAMPLE 1. (OBJECTIVE 3)**

27. $(3 + 4i) + (2 + 3i)$

28. $(5 + 3i) - (6 - 9i)$

29. $(2 + 3i) - (1 - 2i)$

30. $(8 + 3i) + (-7 - 2i)$

Multiply. Write all answers in $a + bi$ form. **SEE EXAMPLE 2. (OBJECTIVE 3)**

31. $4i(3 - i)$

32. $-4i(3 + 4i)$

33. $-2i(3 + 4i)$

34. $8i(-5 + 2i)$

Multiply. Write all answers in $a + bi$ form. **SEE EXAMPLE 3. (OBJECTIVE 3)**

35. $(2 + 3i)(3 - i)$

36. $(3 - 2i)(4 - 3i)$

37. $(5 + 3i)(3 - 2i)$

38. $(7 - 2i)(3 + 4i)$

Write each number in $a + bi$ form. **SEE EXAMPLE 4. (OBJECTIVE 3)**

39. $4 + \sqrt{-49}$

40. $9 - \sqrt{-13}$

41. $3i^2 - 5i^3$

42. $-\dfrac{2}{5}i$

Rationalize the denominator. Write all answers in $a + bi$ form. **SEE EXAMPLE 5. (OBJECTIVE 4)**

43. $\dfrac{1}{i}$

44. $\dfrac{3}{2i}$

45. $\dfrac{3}{7i^3}$

46. $\dfrac{1}{i^3}$

Find the product of each complex number and its conjugate. **SEE EXAMPLE 6.**

47. $9 + i$

48. $5 + 4i$

49. $-8 - 6i$

50. $12 - 2i$

Rationalize the denominator. Write all answers in $a + bi$ form. **SEE EXAMPLE 7. (OBJECTIVE 4)**

51. $\dfrac{1}{6 + i}$

52. $\dfrac{3}{5 + i}$

53. $\dfrac{-2}{2 - i}$

54. $\dfrac{-4i}{2 - 6i}$

Rationalize the denominator. Write all answers in $a + bi$ form. **SEE EXAMPLE 8. (OBJECTIVE 4)**

55. $\dfrac{3 - 2i}{3 + 2i}$

56. $\dfrac{4 + 5i}{4 - 5i}$

57. $\dfrac{5 - 2i}{3 - i}$

58. $\dfrac{5 - i}{2 + 3i}$

Divide and express the quotient in $a + bi$ form. **SEE EXAMPLE 9. (OBJECTIVE 4)**

59. $(8 + 5i) \div (7 + 2i)$

60. $(-7 + 9i) \div (-2 - 8i)$

61. $(4 - i) \div (2 + i)$

62. $(2 - 4i) \div (3 + 2i)$

Simplify. Write all answers in $a + bi$ form. **SEE EXAMPLE 10. (OBJECTIVE 4)**

63. $\dfrac{-12}{7 - \sqrt{-1}}$

64. $\dfrac{4}{3 + \sqrt{-1}}$

65. $\dfrac{3 + \sqrt{-2}}{2 + \sqrt{-5}}$

66. $\dfrac{2 - \sqrt{-5}}{3 + \sqrt{-7}}$

Solve each equation. Write all solutions in bi or $a + bi$ form. **SEE EXAMPLE 11. (OBJECTIVE 5)**

67. $x^2 + 2x + 2 = 0$

68. $x^2 + 3x + 3 = 0$

69. $2x^2 + x + 1 = 0$

70. $3x^2 + 2x + 1 = 0$

ADDITIONAL PRACTICE
Simplify. Write each result in $a + bi$ form.

71. $(8 + 5i) + (7 + 2i)$

72. $(-7 + 9i) - (-2 - 8i)$

73. $(4 - i)(2 + i)$

74. $(2 - 4i)(3 + 2i)$

75. $(7 - 3i) + (-2 + 5i)$

76. $(-8 - 2i) - (-4 + 5i)$

77. $(4 + 3i) - (-4 - i)$

78. $(10 - 3i) + (-12 - 7i)$

79. $\left(-8 - \sqrt{-3}\right) - \left(7 - \sqrt{-27}\right)$

80. $\left(2 + \sqrt{-8}\right) + \left(-3 - \sqrt{-2}\right)$

81. $\left(2 + \sqrt{-3}\right)\left(3 - \sqrt{-2}\right)$

82. $\left(1 + \sqrt{-5}\right)\left(2 - \sqrt{-3}\right)$

83. $\left(8 - \sqrt{-5}\right)\left(-2 - \sqrt{-7}\right)$

84. $\left(-1 + \sqrt{-6}\right)\left(2 - \sqrt{-3}\right)$

85. $\left(2 + \sqrt{-2}\right)\left(3 - \sqrt{-2}\right)$

86. $\left(5 + \sqrt{-3}\right)\left(2 - \sqrt{-3}\right)$

87. $\left(-2 - \sqrt{-16}\right)\left(1 + \sqrt{-4}\right)$

88. $\left(-3 - \sqrt{-81}\right)\left(-2 + \sqrt{-9}\right)$

89. $\dfrac{-3}{5i^5}$

90. $\dfrac{-4}{6i^7}$

91. $\dfrac{3i}{8\sqrt{-9}}$

92. $\dfrac{-7i}{2\sqrt{-16}}$

93. $\dfrac{-6}{\sqrt{-32}}$

94. $\dfrac{5}{\sqrt{-125}}$

Solve each equation. Write the answer in bi or a + bi form.

95. $3x^2 - 4x + 2 = 0$

96. $2x^2 - 3x + 2 = 0$

97. $x^2 + 49 = 0$

98. $x^2 + 16 = 0$

99. $3x^2 = -16$

100. $2x^2 = -25$

WRITING ABOUT MATH *Write a paragraph using your own words.*

101. Explain how to add or subtract two complex numbers.

102. Explain how to multiply or divide two complex numbers.

SOMETHING TO THINK ABOUT

103. Simplify: i^{-1}

104. Simplify: $-i^{-23}$

Projects

PROJECT 1

Solve the equation $x^2 + 5x + 1 = 0$. Are the solutions real numbers? Solve the equation $x^2 + x + 1 = 0$. Are the solutions real numbers? How can you tell without actually solving an equation whether the solutions are real or nonreal? Develop both an algebraic method and a graphing-calculator method.

Test your method on the following quadratic equations.

a. $x^2 - 7x + 2 = 0$ **b.** $x^2 + x - 2 = 0$
c. $x^2 - 3x + 4 = 0$ **d.** $x^2 + 5x + 7 = 0$
e. $3x^2 - 5x + 9 = 0$ **f.** $2x^2 + 5x + 2 = 0$

PROJECT 2

One of the world's greatest scientific geniuses, Sir Isaac Newton (1642–1727), contributed to every major area of science and mathematics known in his time. Because of his early interest in science and mathematics, Newton enrolled at Trinity College, Cambridge, where he studied the mathematics of René Descartes and the astronomy of Galileo but did not show any great talent. In 1665, the plague closed the school, and Newton went home. There, his genius emerged. Within a few years, he had made major discoveries in mathematics, physics, and astronomy.

One of Newton's many contributions to mathematics is known as **Newton's method**, a way of finding better and better estimates of the solutions of certain equations. The process begins with a guess of the solution, and transforms that guess into a better estimate of the solution. When

Newton's method is applied to this better answer, a third solution results, one that is more accurate than any of the previous ones. The method is applied again and again, producing better and better approximations of the solution.

To solve the quadratic equation $x^2 + x - 3 = 0$, for example, Newton's method uses the fraction $\dfrac{x^2 + 3}{2x + 1}$ to generate better solutions. We begin with a guess: 3, for example. We substitute 3 for x in the fraction.

$$\frac{x^2 + 3}{2x + 1} = \frac{3^2 + 3}{2(3) + 1} \approx 1.714286$$

The number 1.714286 is a better estimate of the solution than the original guess, 3. We apply Newton's method again

$$\frac{x^2 + 3}{2x + 1} = \frac{1.714286^2 + 3}{2(1.714286) + 1} \approx 1.341014$$

The number 1.341014 is the best estimate yet of the solution. However, more passes through Newton's method produce even better estimates: 1.303173, and then 1.302776. This final answer is accurate to six decimal places.

- The quadratic equation $x^2 + x - 3 = 0$ has two solutions. Find the other solution by using Newton's method and another initial guess. (Try a negative number.)

- To solve the quadratic equation $ax^2 + bx + c = 0$, Newton's method uses the fraction $\dfrac{ax^2 - c}{2ax + b}$ to generate solutions. Solve $2x^2 - 2x - 1 = 0$ using Newton's method. Find two solutions accurate to six decimal places.

Reach for Success
EXTENSION OF PLANNING FOR YOUR FUTURE

Now that you've considered short-term planning, let's move on and look at your future career goals.

LONG-TERM EDUCATIONAL PLANNING	Your college may have resources to help your choose a major. List a few of these resources.
Take advantage of your college resources to help you identify your area of interest. Ultimately, you will need to declare a major.	_____ _____
Have you chosen a major? _____ If so, consider careers in that field and list at least three. 1. _____ 2. _____ 3. _____	If not, visit your career services area, see an advisor, or talk with a favorite professor. With whom did you consult? _____ What advice did you receive? _____ _____
What are potential employers looking for in your career interest? _____ _____	Circle True or False for the following statement. Most companies want employees with the problem-solving skills developed in mathematics courses. True False
There are certificate programs and associate's, bachelor's, master's, and doctoral degrees.	What type of certificate or degree is required for your field of interest? _____ What additional mathematics courses (if any) are required for your major? _____ _____

Proper planning can help you avoid taking classes that might not be required for your degree, thus saving you time and money.

9 Review

SECTION 9.1 Solving Quadratic Equations Using the Square Root Property

DEFINITIONS AND CONCEPTS	EXAMPLES
Zero-Factor Property: If $ab = 0$, then $a = 0$ or $b = 0$.	$(x - 5)(x + 3) = 0$ $x - 5 = 0$ or $x + 3 = 0$ Set each factor equal to 0. $x = 5$ | $x = -3$ Solve each linear equation.
Square Root Property: The two solutions of $x^2 = c$ are $x = \sqrt{c}$ and $x = -\sqrt{c}$.	$x^2 = 16$ $x = \sqrt{16}$ or $x = -\sqrt{16}$ $x = 4$ | $x = -4$

REVIEW EXERCISES

Solve each quadratic equation by factoring.

1. $3x^2 + 15x = 0$ **2.** $2x^2 - 6x = 0$

3. $x^2 - 9 = 0$ **4.** $y^2 - 81 = 0$

5. $a^2 - 7a + 12 = 0$ **6.** $t^2 - 2t - 15 = 0$

7. $2x - x^2 + 24 = 0$ **8.** $2x^2 + x - 3 = 0$

9. $x^2 - 7x = -12$ **10.** $6p^2 - 5p - 6 = 0$

Use the square root property to solve each quadratic equation.

11. $x^2 = 25$ **12.** $x^2 = 100$

13. $2x^2 = 18$ **14.** $4x^2 = 9$

15. $x^2 = 12$ **16.** $x^2 = 75$

Use the square root property to solve each equation.

17. $(x - 1)^2 = 25$ **18.** $(x + 4)^2 = 49$

19. $2(x + 1)^2 = 18$ **20.** $4(x - 2)^2 = 9$

21. $(x - 8)^2 = 8$ **22.** $(x - 3)^2 = 45$

23. Construction The 45-square-foot base of the preformed concrete panel shown in the illustration is 3 feet longer than twice its height. How long is the base?

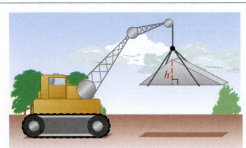

24. Military applications A pilot releases a food container from an altitude of 3,000 feet. The containers's height h above the drop zone t seconds after its release is given by the formula

$$h = 3,600 - 16t^2$$

How long will it be until the container hits the drop zone?

25. Gardening A 27-square-foot rectangular flower bed is 3 feet longer than twice its width. Find its dimensions.

26. Geometry A rectangle is 3 feet longer than it is wide. Its area is numerically equal to its perimeter. Find its dimensions.

SECTION 9.2 Solving Quadratic Equations by Completing the Square

DEFINITIONS AND CONCEPTS	EXAMPLES
To make $x^2 + 2bx$ a perfect trinomial square, add the square of one-half of the coefficient of x: $$x^2 + 2bx + b^2 = (x + b)^2$$	$x^2 + 10x$ $x^2 + 10x + \textbf{?} = (x + \textbf{?})^2$ $x^2 + 10x + \textbf{25} = (x + \textbf{5})^2$ $\left[\frac{1}{2}(10)\right]^2 = 5^2 = 25$
To solve a quadratic equation by completing the square: **1.** If necessary, write the quadratic equation in quadratic form, $ax^2 + bx + c = 0$.	Solve $4x^2 - 8x - 3 = 0$ by completing the square. **1.** $4x^2 - 8x - 3 = 0$ This is quadratic form. **2.** $x^2 - 2x - \dfrac{3}{4} = 0$ Divide both sides by 4.

2. If the coefficient of x^2 is not 1, divide both sides of the equation by a, the coefficient of x^2.

3. If necessary, add a number to both sides of the equation to place the constant term on the right side of the equal sign.

4. Complete the square:

  a. Find one-half of the coefficient of x and square it.

  b. Add the square to both sides of the equation.

5. Factor the perfect trinomial square on the left side of the equation and combine any like terms on the right side of the equation.

6. Use the square root property to solve the resulting quadratic equation.

7. Check each solution in the original equation.

3. $x^2 - 2x = \dfrac{3}{4}$ Add $\dfrac{3}{4}$ to both sides.

4. $x^2 - 2x + (1)^2 = \dfrac{3}{4} + (1)^2$ Add 1 to both sides to complete the square.

5. $(x - 1)^2 = \dfrac{7}{4}$ Factor and simplify.

6. $x - 1 = \pm \dfrac{\sqrt{7}}{2}$ Use the square root property to solve for $(x - 1)$.

$x = 1 \pm \dfrac{\sqrt{7}}{2}$ Add 1 to both sides.

$x = 1 + \dfrac{\sqrt{7}}{2}$ or $x = 1 - \dfrac{\sqrt{7}}{2}$

$x = \dfrac{2 + \sqrt{7}}{2}$ $x = \dfrac{2 - \sqrt{7}}{2}$

7. Both solutions check.

REVIEW EXERCISES

Complete the square to make each expression a perfect trinomial square.

27. $x^2 + 6x$ **28.** $y^2 + 8y$

29. $z^2 - 10z$ **30.** $t^2 - 5t$

31. $a^2 + \dfrac{3}{4}a$ **32.** $b^2 - \dfrac{9}{5}b$

Solve each quadratic equation by completing the square. If an answer is irrational, give the exact answer and then approximate it to the nearest tenth.

33. $x^2 + 5x - 14 = 0$ **34.** $x^2 - 10x + 21 = 0$

35. $x^2 + 4x = 77$ **36.** $x^2 - 2x = 1$

37. $x^2 + 4x - 3 = 0$ **38.** $x^2 - 6x + 4 = 0$

39. $2x^2 + 5x - 3 = 0$ **40.** $2x^2 - 2x - 1 = 0$

SECTION 9.3 Solving Quadratic Equations by the Quadratic Formula

DEFINITIONS AND CONCEPTS

The solution of a quadratic equation $ax^2 + bx + c = 0$ can be found using the quadratic formula:

$$x = \frac{-b \pm \sqrt{b^2 - 4ac}}{2a} \quad (a \neq 0)$$

EXAMPLES

To solve $4x^2 - 8x - 3 = 0$ by the quadratic formula, we note that $a = 4, b = -8, c = -3$, and substitute these values into the quadratic formula.

$$x = \frac{-(-8) \pm \sqrt{(-8)^2 - 4(4)(-3)}}{2(4)}$$

$$x = \frac{8 \pm \sqrt{64 + 48}}{8}$$

$$x = \frac{8 \pm \sqrt{112}}{8}$$

$$x = \frac{8 \pm 4\sqrt{7}}{8}$$

$$x = \frac{2 \pm \sqrt{7}}{2} \qquad \frac{8 \pm 4\sqrt{7}}{8} = \frac{4(2 \pm \sqrt{7})}{4 \cdot 2} = \frac{2 \pm \sqrt{7}}{2}$$

REVIEW EXERCISES

Use the quadratic formula to solve each quadratic equation.

41. $x^2 - 2x - 15 = 0$

42. $x^2 - 4x - 5 = 0$

43. $x^2 - 15x + 26 = 0$

44. $2x^2 + 4x = 0$

45. $x^2 = 288$

46. $2x^2 - 7x + 4 = 0$

47. $6x^2 - 8x = -1$

48. $x^2 + 4x + 1 = 0$

49. $x^2 - 6x + 7 = 0$

50. Geometry The length of the rectangle in the illustration is 14 centimeters greater than the width. Find the perimeter.

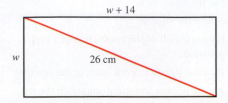

SECTION 9.4 Graphing Quadratic Functions

DEFINITIONS AND CONCEPTS	EXAMPLES
The graph of the equation $f(x) = ax^2 + bx + c$ $(a \ne 0)$ is a parabola. It opens upward when $a > 0$ and downward when $a < 0$. The graph of an equation of the form $f(x) = a(x - h)^2 + k$ is a parabola with vertex at (h, k). It opens upward when $a > 0$ and downward when $a < 0$.	The graph of $f(x) = 2x^2 - 2$ is a parabola that opens upward because $a = 2, 2 > 0$. The graph of $f(x) = -3x^2 + 1$ is a parabola that opens downward because $a = -3, -3 < 0$. The graph of $f(x) = 5(x - 1)^2 + 3$ is a parabola that opens upward with vertex $(1, 3)$ as the minimum point. The graph of $f(x) = -5(x - 1)^2 + 3$ is a parabola that opens downward with vertex $(1, 3)$ as the maximum point.
The x-coordinate of the vertex of the parabola $f(x) = ax^2 + bx + c$ is $x = -\frac{b}{2a}$. $x = -\frac{b}{2a}$ is the equation of the axis of symmetry. To find the y-coordinate of the vertex, substitute the value of $-\frac{b}{2a}$ for x in the equation of the parabola and evaluate.	The x-coordinate of the vertex of $f(x) = 2x^2 - 8x + 1$ is $$x = \frac{-(-8)}{2(2)} = 2$$ The equation $x = 2$ is the axis of symmetry. To find the y-coordinate of the vertex of $f(x) = 2x^2 - 8x + 1$, we substitute 2 for x: $$f(2) = 2(2)^2 - 8(2) + 1 = 8 - 16 + 1 = -7$$ The vertex is $(2, -7)$.

REVIEW EXERCISES

Graph each function.

51. $f(x) = x^2 + 8x + 10$ **52.** $f(x) = -2x^2 - 4x - 6$

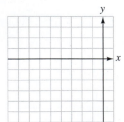   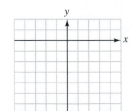

Identify the vertex of the graph of each function. Do not draw the graph.

53. $f(x) = 5(x - 6)^2 + 7$

54. $f(x) = 2(x + 4)^2 - 9$

55. $f(x) = 2x^2 - 4x + 7$

56. $f(x) = -3x^2 + 18x - 11$

SECTION 9.5   Complex Numbers

DEFINITIONS AND CONCEPTS	EXAMPLES
By definition $i = \sqrt{-1}$.	$i^{35} = i^3 = -i$

Powers of i

If n is a natural number that has a remainder of r when divided by 4, then $i^n = i^r$. When n is divisible by 4, the remainder is 0 and $i^0 = 1$.

A **complex number** is any number that can be written in the form $a + bi$, where a and b are real numbers and $i = \sqrt{-1}$.

$3 + 4i \qquad 5 - 0i \qquad 0 + 7i$

The *real part* is a and the *imaginary part* is b.

The real parts of the complex numbers above are 3, 5, and 0.
The imaginary parts of the complex numbers above are 4, 0, and 7.

$$i = \sqrt{-1} \quad i^2 = -1 \quad i^3 = -i \quad i^4 = 1$$

$$\sqrt{-9} = \sqrt{9(-1)} = \sqrt{9}\sqrt{-1} = 3i$$

$$(a + bi) + (c + di) = (a + c) + (b + d)i$$

$$(2 + i) + (7 + 6i) = 2 + i + 7 + 6i$$
$$= 9 + 7i$$

$$(a + bi) - (c + di) = (a - c) + (b - d)i$$

$$(2 + i) - (7 - 6i) = 2 + i - 7 + 6i$$
$$= -5 + 7i$$

$$(a + bi)(c + di) = (ac - bd) + (ad + bc)i$$

$$(2 + i)(7 - 6i) = 14 - 12i + 7i - 6i^2$$
$$= 14 - 5i + 6$$
$$= 20 - 5i$$

The complex numbers $a + bi$ and $a - bi$ are called **complex conjugates** of each other.

The complex conjugate of $-5 + 2i$ is $-5 - 2i$.

To divide by a complex number, multiply the numerator and the denominator by the conjugate of the denominator.

Always write complex numbers in $a + bi$ form before performing any operations on them.

$$\frac{5}{2 + i} = \frac{5}{2 + i} \cdot \frac{2 - i}{2 - i}$$

$$= \frac{5(2 - i)}{4 - 2i + 2i - i^2}$$

$$= \frac{5(2 - i)}{4 - (-1)}$$

$$= \frac{5(2 - i)}{5}$$

$$= 2 - i$$

REVIEW EXERCISES

Solve each equation. Write all solutions in bi or a + bi form.

57. $x^2 + 25 = 0$ **58.** $2x^2 - 3x + 5 = 0$

Simplify each expression.

59. i^{33} **60.** i^{183}

Perform each operation. Give all solutions in a + bi form.

61. $(10 - 7i) + (-4 + 3i)$ **62.** $(12 - 2i) - (5 - i)$

63. $\left(6 - \sqrt{-36}\right) + \left(5 - \sqrt{-9}\right)$

64. $\left(5 + \sqrt{-4}\right) - \left(4 - \sqrt{-81}\right)$

65. $-2i(3 + i)$

66. $7i(-5 + 2i)$

67. $(2 - 5i)(3 + 4i)$

68. $\left(6 - \sqrt{-3}\right)\left(-5 - \sqrt{-5}\right)$

69. $\dfrac{-5}{i^3}$

70. $\dfrac{2}{5 - i}$

71. $\dfrac{3i}{5 - 2i}$

72. $\dfrac{3 - \sqrt{-5}}{2 + \sqrt{-7}}$

9   Test

Solve each equation by factoring.

1. $8x^2 + 2x - 3 = 0$ **2.** $x^2 = 4$

Solve each equation by the square root property.

3. $x^2 = -16$ **4.** $(x - 2)^2 = 3$

Find the number required to complete the square.

5. $x^2 + 14x$ **6.** $x^2 - 9x$

7. Use completing the square to solve $3a^2 + 6a - 12 = 0$.

8. Write the quadratic formula.

Use the quadratic formula to solve each equation.

9. $x^2 + 3x - 10 = 0$

10. $2x^2 - 5x = 12$

11. $2x^2 + 5x + 1 = 2$

12. The base of a triangle with an area of 40 square meters is 2 meters longer than it is high. Find the length of the base.

13. Identify the vertex of the parabola determined by the equation $f(x) = -4(x + 3)^2 + 2$.

14. Graph the function $f(x) = x^2 + 4x + 2$.

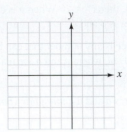

15. Solve: $3x^2 - x + 7 = 0$

16. Simplify: i^{55}

17. Add: $(7 + 5i) + (-2 + 3i)$

18. Subtract: $(-6 - 5i) - (-3 + 2i)$

19. Multiply: $(3 + 5i)(-2 - 3i)$

20. Rationalize the denominator: $\dfrac{2 + i}{3 - 2i}$

GLOSSARY

absolute value The distance between a given number and 0 on a number line, usually denoted with $| \ |$

acute angle An angle with a measure between $0°$ and $90°$

addition property of equality The property that states if a, b, and c are real numbers and $a = b$, then $a + c = b + c$

additive identity The number zero, because for all real numbers a, $a + 0 = 0 + a = a$

additive inverses A pair of numbers, a and b, are additive inverses if $a + b = 0$; also called **negatives** or **opposites**.

algebraic expression A combination of variables, numbers, and arithmetic operations; an algebraic expression does not contain an equality or inequality symbol.

angle A geometric figure consisting of two rays emanating from a common point

area The amount of surface enclosed by a two-dimensional geometric figure; expressed as unit^2

arithmetic expression A combination of numbers and arithmetic operations; an arithmetic expression does not contain an equality or inequality symbol.

ascending order An ordering of the terms of a polynomial such that a given variable's exponents occur in increasing order

ascending powers of a variable An ordering of the terms of a polynomial, such that a given variable's exponents occur in increasing order

associative properties The properties for addition and multiplication that state for real numbers a, b, and c, $(a + b) + c = a + (b + c)$ and $(ab)c = a(bc)$. *Note:* The associative property does not hold for subtraction or division.

axis of symmetry For the parabola $f(x) = ax^2 + bx + c$, a vertical line, $x = -\frac{b}{2a}$

base In the expression x^y, x is the base and y is the exponent. The base, x, will be used as a factor y times.

base angles of an isosceles triangle The two congruent angles of an isosceles triangle opposite the two sides that are of the same length

binomial A polynomial with exactly two terms

boundary line A line that divides the rectangular coordinate plane into two half-planes

Cartesian coordinate system A grid composed of a horizontal axis and a vertical axis that allows us to identify each point in a plane with a unique ordered pair of numbers; also called the **rectangular coordinate system**

Cartesian plane The set of all points in the Cartesian coordinate system

center A fixed point in a plane where all points on the circle are a given distance from this point

circumference The distance around a circle

closure properties The properties that state the sum, difference, product, or quotient of two real numbers is a real number

coefficient The numerical factor of a term; also called **numerical coefficient**

combined variation A relationship among variables x, y, and z of the form $y = \frac{kx}{z}$ $(z \neq 0)$, where k is the constant of proportionality

commutative properties The properties for addition and multiplication that state for real numbers a and b, $a + b = b + a$ and $ab = ba$. *Note:* The commutative property does not hold for subtraction and division.

complementary angles Two angles whose measures sum to $90°$

completing the square The process of forming a perfect-square trinomial from a binomial of the form $x^2 + bx$

complex conjugate The complex conjugate of the number $a + bi$ is $a - bi$

complex fraction A fraction that contains a fraction in its numerator and/or denominator

complex number The sum of a real number and an imaginary number

composite numbers A natural number greater than 1 with factors other than 1 and itself

compound inequality A single statement representing the intersection of two inequalities

conditional equation An equation in one variable that has exactly one solution

conjugate An expression that contains the same two terms as another but with opposite signs between them; for example, $a + \sqrt{b}$ and $a - \sqrt{b}$

consecutive even integers Even integers following in order, can be represented as x and $x + 2$

consecutive integers Integers following in order, can be represented as x and $x + 1$

consecutive odd integers Odd integers following in order, can be represented as x and $x + 2$

consistent system A system of equations with at least one solution

constant A term whose variable factor(s) have an exponent of 0

constant of variation The constant that appears in a variation equation; also called the **constant of proportionality**

contradiction An equation that is false for all values of its variables; its solution set is $\varnothing$.

coordinate The number that corresponds to a given point on a number line

cube root If $a = b^3$, then b is a cube root of a.

cubic function A polynomial function whose equation is of third degree; alternately, a function whose equation is of the form $f(x) = ax^3 + bx^2 + cx + d$

cubic units The units given to a volume measurement

degree A unit of measure for an angle

degree of a monomial The sum of the exponents of all variables in a monomial

degree of a polynomial The largest degree of a polynomial's terms

denominator The expression below the bar of a fraction

dependent equations A system of equations that has infinitely many solutions

dependent variable The variable in an equation of two variables whose value is determined by the independent variable (usually y in an equation involving x and y)

descending powers of a variable An ordering of the terms of a polynomial such that a given variable's exponents occur in decreasing order

diameter A line segment connecting two points on a circle and passing through the center; the length of such a line segment

difference The result of subtracting two expressions

difference of two cubes An expression of the form $a^3 - b^3$

difference of two squares An expression of the form $a^2 - b^2$

direct variation A relationship between two variables x and y of the form $y = kx$, where k is the constant of proportionality

distance formula The distance, d, between two points with coordinates (x_1, y_1) and (x_2, y_2) is $d = \sqrt{(x_2 - x_1)^2 + (y_2 - y_1)^2}$.

distributive property The property that states for real numbers a, b, and c, $a(b + c) = ab + ac$ and $(b + c)a = ab + ac$.

dividend In long division, the expression under the division symbol; in a fraction, the expression in the numerator

division property of radicals The property that states if $a \geq 0$ and $b > 0$, then $\sqrt{\frac{a}{b}} = \frac{\sqrt{a}}{\sqrt{b}}$

divisor In long division, the expression in front of the division symbol; in a fraction, the expression in the denominator

domain of a function The set of all permissible input values of a function

domain of a relation The set of all first elements (components) of a relation

double inequality Two inequalities written together to indicate a set of numbers that lie between two fixed values

doubly shaded region The area of a graph depicting the intersection of two half-planes

elimination (addition) An algebraic method for solving a system of equations where one variable is eliminated when the equations are added

ellipses A symbol consisting of three dots (. . .) meaning "and so forth" in the same manner

empty set A set that has no elements, denoted by the symbol $\varnothing$ or $\{ \}$

equal ratios Two (or more) ratios that represent equal values

equation A statement indicating that two quantities are equal

equivalent equations Two equations that have the same solution set

equivalent fractions Two expressions with different denominators describing the same fraction; for example, $\frac{3}{11}$ and $\frac{6}{22}$ are equivalent fractions.

even integers The set of integers that can be divided exactly by 2 *Note*: 0 is an even integer.

even root If $b = \sqrt[n]{a}$, b is an even root if n is even.

exponent In the expression x^y, x is the base and y is the exponent. The exponent y states the number of times that the base x will be used as a factor.

exponential expression An expression of the form x^y; also called a **power of x**

expression A combination of variables and/or numbers and arithmetic operations. Expressions can be **algebraic**, **arithmetic**, **rational**, **radical**, etc.

extraneous solution A solution to an equation that does not result in a true statement when substituted for the variable in the original equation

extremes In the proportion $\frac{a}{b} = \frac{c}{d}$, the numbers a and d are called the extremes.

factoring The process of finding the individual factors of a product

factoring tree A visual representation of factors of a number, usually used as a tool to express a number in prime-factored form

factors Numbers or polynomials whose product is a given result; for example, the factors of 6 are 1, 2, 3, and 6 and the factors of $x^3 + 6x^2 + 5x$ are x, $(x + 5)$, and $(x + 1)$.

FOIL method An acronym representing the order for multiplying two binomials; First, Outer, Inner, and Last

formula A literal equation in which the variables correspond to defined quantities

function A relation in which to each first component there corresponds exactly one second component, usually denoted by $f(x)$

fundamental theorem of arithmetic The theorem that states there is exactly one prime factorization for any natural number greater than 1

general form of a quadratic equation The equation $ax^2 + bx + c = 0$ $(a \neq 0)$; also known as **quadratic form**

greatest common factor (GCF) The largest expression that is a factor of each of a group of expressions

grouping symbol A symbol (such as a radical sign or a fraction bar) or pair of symbols (such as parentheses or braces) that indicate that these operations should be computed before other operations

half-plane A subset of the coordinate plane consisting of all points on a given side of a boundary line

horizontal line A line parallel to the *x*-axis; a line with the equation $y = b$

hypotenuse The longest side of a right triangle, the side opposite the 90° angle

identity An equation that is true for all values of its variables; its solution set is $\mathbb{R}$.

identity elements The additive identity is 0, because for all real a, $a + 0 = 0 + a = a$. The multiplicative identity is 1, because for all real a, $a \cdot 1 = 1 \cdot a = a$.

imaginary number The square root of a negative number

improper fraction A fraction whose numerator is greater than or equal to its denominator

inconsistent system A system of equations that has no solution; its solution set is $\varnothing$.

independent equations A system of two equations with two variables for which each equation's graph is different

independent variable The variable in an equation of two variables to which we assign input values (usually x in an equation involving x and y)

index If $b = \sqrt[n]{a}$, we say n is the index of the radical.

inequality A mathematical statement indicating that two quantities are not necessarily equal; most commonly a statement involving the symbols $<$, $\leq$, $>$, or $\geq$

inequality symbols A set of six symbols that are used to describe two expressions that are not necessarily equal:
$\approx$ "is approximately equal to"
$\neq$ "is not equal to"
$<$ "is less than"
$>$ "is greater than"
$\leq$ "is less than or equal to"
$\geq$ "is greater than or equal to"

input value A value substituted for the independent variable

integer cube An integer that is the third power of another integer

integer square An integer that is the second power of another integer

integers The set of numbers given by $\{\ldots, -4, -3, -2, -1, 0, 1, 2, 3, 4, \ldots\}$

intersection The set of points common to two or more algebraic statements; the point (or set of points) where two graphs intersect

interval A set of all real numbers between two given real numbers a and b; it must be specified whether or not a and b are included in the interval.

inverse variation A relationship between two variables x and y of the form $y = \frac{k}{x}$, where k is the constant of proportionality

irrational numbers The set of numbers that cannot be put in the form of a fraction with integer numerator and nonzero integer denominator; can be expressed only as nonterminating, nonrepeating decimals

isosceles triangle A triangle that has two sides of the same length

joint variation A relationship among three variables x, y, and z of the form $y = kxz$, where k is the constant of proportionality

least (or lowest) common denominator The smallest expression that is exactly divisible by the denominators of a set of rational expressions

legs of a right triangle One of the two sides adjacent to the right angle

like radicals Radicals that have the same index and radicand

like signs Two numbers that are both positive or both negative

like terms Terms containing the same variables with the same exponents

linear depreciation A model for estimating the decreasing value of aging equipment, based on an initial value and a final value

linear equation in one variable An equation of the form $ax + b = c$, where a, b, and c are real numbers and $a \neq 0$

linear equation in two variables An equation of the form $Ax + By = C$, where A, B, and C are real numbers; A and B cannot both be 0.

linear function A function whose graph is a line; a function whose equation is of the form $f(x) = ax + b$

linear inequality A statement that can be written in one of the following forms: $y > ax + b$, $y \geq ax + b$, $y < ax + b$, or $y \leq ax + b$

literal equation An equation with more than one variable, usually a formula

lowest terms A fraction written in such a way so that no integer greater than 1 divides both its numerator and its denominator; also called **simplest form**

maximum point The highest point of a graph; the vertex of a parabola that opens downward

means In the proportion $\frac{a}{b} = \frac{c}{d}$, the numbers b and c are called the means.

minimum point The lowest point of a graph; the vertex of a parabola that opens upward

minuend The first expression in the difference of two expressions; for example, in the expression $(3x^2 + 2x - 5) - (4x^2 - x + 5)$, the expression $(3x^2 + 2x - 5)$ is the minuend.

mixed number A rational number expressed as an integer plus a proper fraction

monomial A polynomial with exactly one term

multiplication property of equality The property that states if a, b, and c are real numbers and $a = b$, then $ac = bc$.

multiplication property of radicals The property that states if a and b are nonnegative, then $\sqrt{ab} = \sqrt{a}\sqrt{b}$.

multiplicative identity Is 1, because for all real numbers a, $a \cdot 1 = 1 \cdot a = a$

multiplicative inverse Two numbers whose product is 1; also called **reciprocals**

natural numbers The set of numbers given by $\{1, 2, 3, 4, 5, \ldots\}$; also called **positive integers**

negative numbers The set of numbers less than zero

negative reciprocals Two numbers whose product is -1; for example, $\frac{8}{3}$ and $-\frac{3}{8}$ are negative reciprocals.

negatives A pair of numbers, a and b, represented by points that lie on opposite sides of the origin and at equal distances from the origin on a number line; also called **opposites** or **additive inverses**

nonnegative number A number that is positive or 0

nth root If $a = b^n$, then b is an nth root of a and we write $b = \sqrt[n]{a}$.

number line A line that is used to represent real numbers or sets of real numbers

numerator The expression above the bar of a fraction

numerical coefficient The number factor of a term; for example, the numerical coefficient of $8ab$ is 8.

odd integers The set of integers that cannot be divided exactly by 2

odd root If $b = \sqrt[n]{a}$, b is an odd root if n is odd.

opposites A pair of numbers, a and b, represented by points that lie on opposite sides of the origin and at equal distances from the origin on a number line; also called **negatives** or **additive inverses**

ordered pair A pair of real numbers, written as (x, y), that describes a unique point in the Cartesian plane

order of operations The order in which the fundamental operations are performed

origin The point on a number line that represents the number zero; the point on the rectangular coordinate system that represents the point $(0, 0)$

output value The value of the dependent variable, determined by the choice of input value

parabola The graph of a quadratic function

parallel lines Two lines that never intersect; two lines with the same slope or two vertical lines

parallelogram A four-sided polygon with its opposite sides parallel

perfect cubes The exact third power of another integer or polynomial; for example, 8 is a perfect cube (2^3) as is $a^3 + 3a^2b + 3ab^2 + b^3$, which is equal to $(a + b)^3$.

percent The numerator of a fraction whose denominator is 100

perfect squares The exact second power of another integer or polynomial; for example, 9 is a perfect square (3^2) as is $a^2 + 2ab + b^2$, which is equal to $(a + b)^2$.

perfect-square trinomial A trinomial of the form $a^2 + 2ab + b^2$ or of the form $a^2 - 2ab + b^2$

perimeter The distance around a noncircular geometric figure

perpendicular lines Two lines that meet at a 90° angle, two lines whose slopes are negative reciprocals of each other, or a horizontal line and a vertical line

point-slope form The equation of a line in the form $y - y_1 = m(x - x_1)$, where m is the slope and (x_1, y_1) is a point on the line

polynomial An algebraic expression that is a single term or the sum of several terms containing whole-number exponents on the variables

polynomial function A function whose equation is a polynomial; for example, $f(x) = x^2 - 3x + 6$

positive integers The set of numbers given by {1, 2, 3, 4, 5, . . .}, that is, the set of integers that are greater than zero; also called the **natural numbers**

power of x An expression of the form x^y; also called an **exponential expression**

power rule The rule that states if $a = b$, then $a^n = b^n$ where n is a natural number

prime-factored form A natural number written as the product of factors that are prime numbers

prime number A natural number greater than 1 that can be divided exactly only by 1 and itself

prime polynomial A polynomial that does not factor over the rational numbers

principal square root The positive square root of a number

product The result of multiplying two or more expressions

proper fraction A fraction whose numerator is less than its denominator

proportion A statement that two ratios are equal

Pythagorean theorem If the length of the hypotenuse of a right triangle is c and the lengths of the two legs are a and b, then $a^2 + b^2 = c^2$.

Pythagorean triples Natural numbers a, b, and c that satisfy the equation $a^2 + b^2 = c^2$

quadrants The four regions in the rectangular coordinate system formed by the x- and y-axes

quadratic equation An equation that can be written in the form $ax^2 + bx + c = 0$ where $a \neq 0$

quadratic form An equation in the form $ax^2 + bx + c = 0$ where $a \neq 0$; also known as **standard form** or a **general quadratic equation**

quadratic formula A formula that produces the solutions to the general quadratic equation; $x = \dfrac{-b \pm \sqrt{b^2 - 4ac}}{2a}$

quadratic function A polynomial function whose equation in one variable is of second degree; alternately, a function whose equation is of the form $f(x) = ax^2 + bx + c$ $(a \neq 0)$

quotient The result of dividing two expressions

radical expression An expression involving a radical sign

radical sign A symbol used to represent the root of a number

radicand The expression under a radical sign

radius A segment drawn from the center of a circle to a point on the circle; the length of such a line segment

range of a function The set of all output values of a function

range of a relation The set of all second elements of a relation

rate A ratio used to compare two quantities of different units

ratio The comparison of two quantities by their indicated quotient

rational exponent An exponent expressed in the form $\dfrac{b}{c}$ $(c > 0)$; we use $a^{b/c}$ to denote $\sqrt[c]{a^b}$.

rational expression The quotient of two polynomials, where the polynomial in the denominator cannot be equal to 0

rational function A function of the form $f(x) = \frac{p(x)}{q(x)}$ where p and q are polynomial functions of x and $q(x) \neq 0$

rationalizing the denominator The process of simplifying a fraction so that there are no radicals in the denominator

rational numbers The set of fractions that have an integer numerator and a nonzero integer denominator; alternately, any terminating or repeating decimal

real numbers The set that contains the rational numbers and the irrational numbers

reciprocal One number is the reciprocal of another if their product is 1; for example, $\frac{3}{11}$ is the reciprocal of $\frac{11}{3}$; also called **multiplicative inverses**

rectangle A parallelogram with one angle a right angle

rectangular coordinate system A grid that allows us to identify each point in a plane with a unique pair of numbers; also called the **Cartesian coordinate system**

relation A set of ordered pairs

remainder The computation of $x \div y$ will yield an expression of the form $q + \frac{r}{y}$. The quantity r is called the remainder.

repeating decimal A decimal expression of a fraction that eventually falls into an infinitely repeating pattern; for example, $\frac{1}{3} = 0.\overline{3}$.

right angle An angle whose measure is 90°

right triangle A triangle that has an angle whose measure is 90°

rise The vertical change between two points; usually the y value

root A number that makes an equation true when substituted for its variable; if there are several variables, then the set of numbers that make the equation true; also called a **solution**

run The horizontal change between two points; usually the x value

scientific notation The representation of a number as the product of a number between 1 and 10 (1 included), and an integer power of ten, for example, 6.02×10^{23}

set A collection of objects whose members are listed or defined within braces

set-builder notation A method of describing a set that uses a variable (or variables) to represent the elements and a rule to determine the possible values of the variable; for example, we can describe the natural numbers as $\{x \mid x$ is an integer and $x > 0\}$.

similar triangles Two triangles that have the same shape, that is, two triangles whose corresponding angles have the same measure

simplest form The form of a fraction where no expression other than 1 divides both its numerator and its denominator

simplest terms A rational expression that cannot be simplified further

simplify (a fraction) To put a fraction into its simplest form

simultaneous solution Values for the variables that satisfy all of the equations in a system of equations

slope The measure of the slant of a nonvertical line, usually denoted by m

slope-intercept form The equation of the line in the form $y = mx + b$ where m is the slope and b is the y-coordinate of the y-intercept

solution A number that makes an equation true when substituted for its variable; if there are several variables, then the numbers that make the equation true; also called a **root**

solution set The set of all values that makes an equation true when substituted for the variable in the original equation

solution of an inequality The numbers that make a given inequality true

special products Products of two binomials of the form $(a + b)^2$, $(a - b)^2$, or $(a + b)(a - b)$

square root If $a = b^2$, then b is a square root of a.

square units The units given to an area measurement

squaring property of equality The property that states if $a = b$, then $a^2 = b^2$

standard form of a linear equation An equation of the form $Ax + By = C$, where A, B, and C are real numbers and A and B cannot both be 0

standard notation The representation of a given number as an integer part, followed by a decimal; for example, 212.3337012

straight angle An angle whose measure is 180°

subscript notation A method of distinguishing between the x- or y-coordinates of different points; for example, x_1 could be the x-coordinate of one point, and x_2 the x-coordinate of a second point.

subset A set, all of whose elements are included in a different set; for example, the set $\{1, 3, 5\}$ is a subset of the set $\{1, 2, 3, 4, 5\}$.

subtrahend The second expression in the difference of two expressions; for example, in the expression $(3x^2 + 2x - 5) - (4x^2 - x + 5)$, the expression $(4x^2 - x + 5)$ is the subtrahend.

sum The result of adding two expressions

sum of two cubes An expression of the form $a^3 + b^3$

sum of two squares An expression of the form $a^2 + b^2$

supplementary angles Two angles whose measures sum to 180°

system of equations Two or more equations that are solved simultaneously

term An expression that is a number, a variable, or a product of numbers and variables; for example, 37, xyz, and $3x$ are terms.

terminating decimal A decimal expression for a fraction that contains a finite number of decimal places

trinomial A polynomial with exactly three terms

union The set of points that belong to at least one of the given algebraic statements

unit cost The ratio of an item's cost to its quantity; expressed with a denominator of 1

unlike signs Two numbers have unlike signs if one is positive and one is negative

unlike terms Terms that do not have the same variables with the same exponents

variables Letters that are used to represent real numbers

vertex The lowest point of a parabola that opens upward, or the highest point of a parabola that opens downward

vertex angle of an isosceles triangle The angle in an isosceles triangle formed by the two sides that are of the same length

vertical line A line parallel to the y-axis; a line with the equation $x = b$

vertical-line test A method of determining whether the graph of an equation represents a function

volume The amount of space enclosed by a three-dimensional geometric figure

whole numbers The set of numbers given by $\{0, 1, 2, 3, 4, 5, \ldots\}$; that is, the set of integers that are greater than or equal to zero

***x*-axis** The horizontal number line in the rectangular coordinate system

***x*-coordinate** The first number in an ordered pair

***x*-intercept** The point $(a, 0)$ where a graph intersects the x-axis

***y*-axis** The vertical number line in the rectangular coordinate system

***y*-coordinate** The second number in an ordered pair

***y*-intercept** The point $(0, b)$ where a graph intersects the y-axis

zero-factor property The property of real numbers that states if the product of two quantities is 0, then at least one of those quantities must be equal to 0

APPENDIX 1: Permutations, Combinations, and Probability

 **1** Simplify an expression involving factorial notation.

To develop an efficient notation needed for permutations, combinations, and probability, we will indroduce **factorial notation**.

- If n is a natural number, the symbol $n!$ (read as **"n factorial"** or as **"factorial n"**) is defined as

$$n! = n(n - 1)(n - 2)(n - 3) \cdots (3)(2)(1)$$

- Zero factorial is defined as

$$0! = 1$$

EXAMPLE 1 Write each expression without using factorial notation.

a. $2!$ **b.** $5!$ **c.** $-9!$ **d.** $(n - 2)!$ **e.** $4! \cdot 0!$

Solution **a.** $2! = 2 \cdot 1 = 2$

b. $5! = 5 \cdot 4 \cdot 3 \cdot 2 \cdot 1 = 120$

c. $-9! = -9 \cdot 8 \cdot 7 \cdot 6 \cdot 5 \cdot 4 \cdot 3 \cdot 2 \cdot 1 = -362,880$

d. $(n - 2)! = (n - 2)(n - 3)(n - 4) \cdot \cdots \cdot 3 \cdot 2 \cdot 1$

e. $4! \cdot 0! = (4 \cdot 3 \cdot 2 \cdot 1) \cdot 1 = 24$

COMMENT According to the previous definition, part **d** is meaningful only if $n - 2$ is a natural number.

 **SELF CHECK 1** Write each expression without using factorial notation.
a. $6!$ **b.** $x!$

We can find factorials using a calculator. For example, to find $12!$ with a scientific calculator, we enter

12 $x!$ (You may have to use a **2ND** or **SHIFT** key first.) `479001600`

To find $12!$ on a TI-84 graphing calculator, we enter

12 **MATH** ► ► ► 4 **ENTER** `12!`

`479001600`

To discover an important property of factorials, we note that

$5 \cdot 4! = 5 \cdot 4 \cdot 3 \cdot 2 \cdot 1 = 5!$

$7 \cdot 6! = 7 \cdot 6 \cdot 5 \cdot 4 \cdot 3 \cdot 2 \cdot 1 = 7!$

$10 \cdot 9! = 10 \cdot 9 \cdot 8 \cdot 7 \cdot 6 \cdot 5 \cdot 4 \cdot 3 \cdot 2 \cdot 1 = 10!$

These examples suggest the following property.

- If n is a positive integer, then

$$n(n - 1)! = n!$$

A-1

EXAMPLE 2 Simplify each expression. **a.** $\dfrac{6!}{5!}$ **b.** $\dfrac{10!}{8!(10-8)!}$

Solution **a.** If we write 6! as $6 \cdot 5!$, we can simplify the fraction by removing the common factor 5! in the numerator and denominator.

$$\frac{6!}{5!} = \frac{6 \cdot 5!}{5!} = \frac{6 \cdot \cancel{5!}}{\cancel{5!}} = 6$$

b. First, we subtract within the parentheses. Then we write 10! as $10 \cdot 9 \cdot 8!$ and simplify.

$$\frac{10!}{8!(\mathbf{10-8})!} = \frac{10!}{8! \cdot \mathbf{2}!} = \frac{10 \cdot 9 \cdot \cancel{8!}}{\cancel{8!} \cdot 2!} = \frac{5 \cdot \mathbf{2} \cdot 9}{2 \cdot 1} = 45$$

 **SELF CHECK 2** Simplify: **a.** $\dfrac{4!}{3!}$ **b.** $\dfrac{7!}{5!(7-5)!}$

We will now discuss methods of counting the different ways we can accomplish something like arranging books on a shelf or selecting a committee. These kinds of situations are important in statistics, insurance, telecommunications, and other fields.

2 **Use the multiplication principle to determine the number of ways one event can be followed by another.**

Steven goes to the cafeteria for lunch. He has a choice of three different sandwiches (hamburger, hot dog, or ham and cheese) and four different beverages (cola, root beer, orange, or milk). How many different lunches can he choose?

He has three choices of sandwich, and for any one of these choices, he has four choices of drink. The different options are shown in the **tree diagram** in Figure A-1.

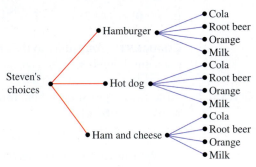

Figure A-1

The tree diagram shows that he has a total of 12 different lunches from which to choose. One of the possibilities is a hamburger with a cola, and another is a hot dog with milk.

A situation that can have several different outcomes—such as choosing a sandwich—is called an **event**. Choosing a sandwich and choosing a beverage can be thought of as two events. The preceding example illustrates the **multiplication principle for events**.

- Let E_1 and E_2 be two events. If E_1 can be done in a_1 ways, and if—after E_1 has occurred—E_2 can be done in a_2 ways, the event "E_1 followed by E_2" can be done in $a_1 \cdot a_2$ ways.

EXAMPLE 3 After dinner, Heidi plans to watch the evening news and then a sitcom on television. If there are choices of four news broadcasts and two sitcoms, in how many ways can she choose to watch television?

Solution Let E_1 be the event "watching the news" and E_2 be the event "watching a sitcom." Because there are four ways to accomplish E_1 and two ways to accomplish E_2, the number of choices that she has is $4 \cdot 2 = 8$.

 SELF CHECK 3 If Jose has 7 shirts and 5 pairs of pants, how many outfits could be created?

The multiplication principle can be extended to any number of events. In Example 4, we use it to complete the number of ways that we can arrange objects in a row.

EXAMPLE 4 In how many ways can we arrange five books on a shelf?

Solution We can fill the first space with any of the 5 books, the second space with any of the remaining 4 books, the third space with any of the remaining 3 books, the fourth space with any of the remaining 2 books, and the fifth space with the remaining 1 (or last) book. By the multiplication principle for events, the number of ways in which the books can be arranged is

$$5 \cdot 4 \cdot 3 \cdot 2 \cdot 1 = 120$$

 SELF CHECK 4 In how many ways can 4 men line up in a row?

EXAMPLE 5 **SENDING SIGNALS** If a sailor has six flags, each of a different color, to hang on a flagpole, how many different signals can the sailor send by using four flags?

Solution The sailor must find the number of arrangements of 4 flags when there are 6 flags from which to choose. The sailor can hang any one of the 6 flags in the top position, any one of the remaining 5 flags in the second position, any one of the remaining 4 flags in the third position, and any one of the remaining 3 flags in the lowest position. By the multiplication principle for events, the total number of signals that can be sent is

$$6 \cdot 5 \cdot 4 \cdot 3 = 360$$

SELF CHECK 5 How many different signals can the sailor send if each signal uses three flags?

3 **Use permutations to find the number of *n* things taken *r* at a time, if order is important.**

When computing the number of possible arrangements of objects such as books on a shelf or flags on a pole, we are finding the number of **permutations** of those objects. In these cases, order is important. A blue flag followed by a yellow flag has a different meaning than a yellow flag followed by a blue flag.

In Example 4, we found that the number of permutations of arranging five books, using all five of them, is 120. In Example 5, we found that the number of permutations of hanging six flags, using four of them, is 360.

The symbol $P(n, r)$, read as "the number of permutations of n things taken r at a time," is often used to express permutation problems. In Example 4, we found that $P(5, 5) = 120$. In Example 5, we found that $P(6, 4) = 360$.

EXAMPLE 6   **SENDING SIGNALS**   If Sarah has 7 flags, each of a different color, to hang on a flagpole, how many different signals can she send by using 3 flags?

Solution   She must find $P(7, 3)$ (the number of permutations of 7 things taken 3 at a time). In the top position Sarah can hang any of the 7 flags, in the middle position any one of the remaining 6 flags, and in the bottom position any one of the remaining 5 flags. According to the multiplication principle for events,

$$P(7, 3) = 7 \cdot 6 \cdot 5 = 210$$

She can send 210 signals using only 3 of the 7 flags.

SELF CHECK 6   How many different signals can Sarah send using 4 flags?

Although it is correct to write $P(7, 3) = 7 \cdot 6 \cdot 5$, there is an advantage in changing the form of this answer to obtain a formula for computing $P(7, 3)$:

$$P(7, 3) = 7 \cdot 6 \cdot 5$$
$$= \frac{7 \cdot 6 \cdot 5 \cdot 4 \cdot 3 \cdot 2 \cdot 1}{4 \cdot 3 \cdot 2 \cdot 1}$$
$$= \frac{7!}{4!}$$
$$= \frac{7!}{(7 - 3)!}$$

The generalization of this idea gives the following formula.

• The number of permutations of n things taken r at a time is given by the formula

$$P(n, r) = \frac{n!}{(n - r)!}$$

EXAMPLE 7   Compute: **a.** $P(8, 2)$ **b.** $P(7, 5)$ **c.** $P(n, n)$ **d.** $P(n, 0)$

Solution   We substitute into the permutation formula $P(n, r) = \frac{n!}{(n - r)!}$.

a. $P(8, 2) = \dfrac{8!}{(8 - 2)!}$
$= \dfrac{8 \cdot 7 \cdot 6!}{6!}$
$= 8 \cdot 7$
$= 56$

b. $P(7, 5) = \dfrac{7!}{(7 - 5)!}$
$= \dfrac{7 \cdot 6 \cdot 5 \cdot 4 \cdot 3 \cdot 2!}{2!}$
$= 7 \cdot 6 \cdot 5 \cdot 4 \cdot 3$
$= 2{,}520$

c. $P(n, n) = \dfrac{n!}{(n - n)!}$
$= \dfrac{n!}{0!}$
$= \dfrac{n!}{1}$
$= n!$

d. $P(n, 0) = \dfrac{n!}{(n - 0)!}$
$= \dfrac{n!}{n!}$
$= 1$

SELF CHECK 7   Compute: **a.** $P(10, 6)$ **b.** $P(10, 0)$

Parts **c** and **d** of Example 7 establish the following formulas.

- The number of permutations of n things taken n at a time and n things taken 0 at a time are given by the formulas

$$P(n, n) = n! \qquad \text{and} \qquad P(n, 0) = 1$$

EXAMPLE 8   **TV PROGRAMMING**   **a.** In how many ways can a television executive arrange the Saturday night lineup of 6 programs if there are 15 programs from which to choose? **b.** If there are only 6 programs from which to choose?

Solution   **a.** To find the number of permutations of 15 programs taken 6 at a time, we will use the formula $P(n, r) = \frac{n!}{(n-r)!}$ with $n = 15$ and $r = 6$.

$$\begin{aligned}
P(15, 6) &= \frac{15!}{(15 - 6)!} \\
&= \frac{15 \cdot 14 \cdot 13 \cdot 12 \cdot 11 \cdot 10 \cdot 9!}{9!} \\
&= 15 \cdot 14 \cdot 13 \cdot 12 \cdot 11 \cdot 10 \\
&= 3{,}603{,}600
\end{aligned}$$

b. To find the number of permutations of 6 programs taken 6 at a time, we use the formula $P(n, n) = n!$ with $n = 6$.

$$P(6, 6) = 6! = 720$$

 SELF CHECK 8   How many ways are there if the executive has 20 programs from which to choose?

4   **Use combinations to find the number of _n_ things taken _r_ at a time.**

Suppose that a student must read 4 books from a reading list of 10 books. The order in which he reads them is not important. For the moment, however, let's assume that order is important and find the number of permutations of 10 things taken 4 at a time:

$$\begin{aligned}
P(10, 4) &= \frac{10!}{(10 - 4)!} \\
&= \frac{10 \cdot 9 \cdot 8 \cdot 7 \cdot 6!}{6!} \\
&= 10 \cdot 9 \cdot 8 \cdot 7 \\
&= 5{,}040
\end{aligned}$$

If order is important, there are 5,040 ways of choosing 4 books when there are 10 books from which to choose. However, because the order in which the student reads the books does not matter, the previous result of 5,040 is too big. Since there are 24 (or 4!) ways of ordering the 4 books that are chosen, the result of 5,040 is exactly 24 (or 4!) times too big. Therefore, the number of choices that the student has is the number of permutations of 10 things 4 at a time, divided by 24:

$$\frac{P(10, 4)}{24} = \frac{5{,}040}{24} = 210$$

The student has 210 ways of choosing 4 books to read from the list of 10 books.

In situations where order is not important, we are interested in **combinations**, not permutations. The symbols $C(n, r)$ and $\binom{n}{r}$ both mean the number of combinations of n things taken r at a time.

If a selection of r books is chosen from a total of n books, the number of possible selections is $C(n, r)$ and there are $r!$ arrangements of the r books in each selection. If we consider the selected books as an ordered grouping, the number of orderings is $P(n, r)$. Therefore, we have

$$(1) \quad r! \cdot C(n, r) = P(n, r)$$

We can divide both sides of Equation 1 by $r!$ to get the formula for finding $C(n, r)$:

$$C(n, r) = \binom{n}{r} = \frac{P(n, r)}{r!} = \frac{n!}{r!(n - r)!}$$

- The number of combinations of n things taken r at a time is given by

$$C(n, r) = \frac{n!}{r!(n - r)!}$$

EXAMPLE 9   Compute: **a.** $C(8, 5)$ **b.** $\binom{7}{2}$ **c.** $C(n, n)$ **d.** $C(n, 0)$

Solution   We will substitute into the combination formula $C(n, r) = \frac{n!}{r!(n - r)!}$.

a. $C(8, 5) = \dfrac{8!}{5!(8 - 5)!}$ **b.** $\dbinom{7}{2} = \dfrac{7!}{2!(7 - 2)!}$

$\qquad\qquad = \dfrac{8 \cdot 7 \cdot 6 \cdot 5!}{5! \cdot 3!}$ $\qquad = \dfrac{7 \cdot 6 \cdot 5!}{2 \cdot 1 \cdot 5!}$

$\qquad\qquad = 8 \cdot 7$ $\qquad = 21$

$\qquad\qquad = 56$

c. $C(n, n) = \dfrac{n!}{n!(n - n)!}$ **d.** $C(n, 0) = \dfrac{n!}{0!(n - 0)!}$

$\qquad\qquad = \dfrac{n!}{n!(0!)}$ $\qquad\qquad = \dfrac{n!}{0! \cdot n!}$

$\qquad\qquad = \dfrac{n!}{n!(1)}$ $\qquad\qquad = \dfrac{1}{0!}$

$\qquad\qquad = 1$ $\qquad\qquad = \dfrac{1}{1}$

$\qquad\qquad\qquad\qquad\qquad\qquad\qquad\quad = 1$

The symbol $C(n, 0)$ indicates that we choose 0 things from the available n things.

SELF CHECK 9   Compute: **a.** $C(9, 6)$ **b.** $C(10, 10)$

Parts **c** and **d** of Example 9 establish the following formulas.

- The number of combinations of n things taken n at a time is 1. The number of combinations of n things taken 0 at a time is 1.

$$C(n, n) = 1 \quad \text{and} \quad C(n, 0) = 1$$

EXAMPLE 10   **PICKING COMMITTEES**   If 15 students want to select a committee of 4 students to plan a party, how many different committees are possible?

Solution   Since the ordering of people on each possible committee is not important, we find the number of combinations of 15 people 4 at a time:

$$C(15, 4) = \frac{15!}{4!(15 - 4)!}$$

$$= \frac{15 \cdot 14 \cdot 13 \cdot 12 \cdot 11!}{4 \cdot 3 \cdot 2 \cdot 1 \cdot 11!}$$

$$= \frac{15 \cdot 14 \cdot 13 \cdot 12}{4 \cdot 3 \cdot 2 \cdot 1}$$

$$= 1{,}365$$

There are 1,365 possible committees.

SELF CHECK 10   In how many ways can 20 students select a committee of 5 students to plan a party?

EXAMPLE 11   **CONGRESS**   A committee in Congress consists of 10 Democrats and 8 Republicans. In how many ways can a subcommittee be chosen if it is to contain 5 Democrats and 4 Republicans?

Solution   There are $C(10, 5)$ ways of choosing the 5 Democrats and $C(8, 4)$ ways of choosing the 4 Republicans. By the multiplication principle for events, there are $C(10, 5) \cdot C(8, 4)$ ways of choosing the subcommittee:

$$C(10, 5) \cdot C(8, 4) = \frac{10!}{5!(10 - 5)!} \cdot \frac{8!}{4!(8 - 4)!}$$

$$= \frac{10 \cdot 9 \cdot 8 \cdot 7 \cdot 6 \cdot 5!}{120 \cdot 5!} \cdot \frac{8 \cdot 7 \cdot 6 \cdot 5 \cdot 4!}{24 \cdot 4!}$$

$$= \frac{10 \cdot 9 \cdot 8 \cdot 7 \cdot 6}{120} \cdot \frac{8 \cdot 7 \cdot 6 \cdot 5}{24}$$

$$= 17{,}640$$

There are 17,640 possible subcommittees.

SELF CHECK 11   In how many ways can a subcommittee be chosen if it is to contain four members from each party?

The **probability** that an event will occur is a measure of the likelihood of that event. A tossed coin, for example, can land in two ways, either heads or tails. Because one of these two equally likely outcomes is heads, we expect that out of several tosses, about half will be heads. We say that the probability of obtaining heads in a single toss of the coin is $\frac{1}{2}$.

If records show that out of 100 days with weather conditions like today's, 30 have received rain, the weather service will report, "There is a $\frac{30}{100}$ or 30% probability of rain today."

5 Find a sample space for an experiment.

Activities such as tossing a coin, rolling a die, drawing a card, and predicting rain are called **experiments**. For any experiment, a list of all possible outcomes is called a **sample space**. For example, the sample space S for the experiment of tossing two coins is the set

$$S = \{(H, H), (H, T), (T, H), (T, T)\}$$

where the ordered pair (H, T) represents the outcome "heads on the first coin and tails on the second coin."

EXAMPLE 12 List the sample space of the experiment "rolling two dice a single time."

Solution We can list ordered pairs and let the first number be the result on the first die and the second number the result on the second die. The sample space S is the following set of ordered pairs:

$$(1, 1)\ (1, 2)\ (1, 3)\ (1, 4)\ (1, 5)\ (1, 6)$$
$$(2, 1)\ (2, 2)\ (2, 3)\ (2, 4)\ (2, 5)\ (2, 6)$$
$$(3, 1)\ (3, 2)\ (3, 3)\ (3, 4)\ (3, 5)\ (3, 6)$$
$$(4, 1)\ (4, 2)\ (4, 3)\ (4, 4)\ (4, 5)\ (4, 6)$$
$$(5, 1)\ (5, 2)\ (5, 3)\ (5, 4)\ (5, 5)\ (5, 6)$$
$$(6, 1)\ (6, 2)\ (6, 3)\ (6, 4)\ (6, 5)\ (6, 6)$$

By counting, we see that the experiment has 36 equally likely possible outcomes.

SELF CHECK 12 How many pairs in the sample space have a sum of 4?

An **event** is a subset of the sample space of an experiment. For example, if E is the event "getting at least one heads" in the experiment of tossing two coins, then

$$E = \{(H, H), (H, T), (T, H)\}$$

Because the outcome of getting at least one heads can occur in 3 out of 4 possible ways, we say that the *probability* of E is $\frac{3}{4}$, and we write

$$P(E) = P(\text{at least one heads}) = \frac{3}{4}$$

- If a sample space of an experiment has n distinct and equally likely outcomes and E is an event that occurs in s of those ways, the *probability of E* is

$$P(E) = \frac{s}{n}$$

Since $0 \leq s \leq n$, it follows that $0 \leq \frac{s}{n} \leq 1$. This implies that all probabilities have a value from 0 to 1. If an event cannot happen, its probability is 0. If an event is certain to happen, its probability is 1.

6 Find the probability of an event.

EXAMPLE 13 Find the probability of the event "rolling a sum of 7 on one roll of two dice."

Solution In the sample space listed in Example 12, the following ordered pairs give a sum of 7:

$$\{(1, 6), (2, 5), (3, 4), (4, 3), (5, 2), (6, 1)\}$$

Since there are 6 ordered pairs whose numbers give a sum of 7 out of a total of 36 equally likely outcomes, we have

$$P(E) = P(\text{rolling a sum of 7}) = \frac{s}{n} = \frac{6}{36} = \frac{1}{6}$$

SELF CHECK 13 Find the probability of rolling a sum of 4.

A standard playing deck of 52 cards has two red suits, hearts and diamonds, and two black suits, clubs and spades. Each suit has 13 cards, including the ace and face cards, king, queen, and jack. We will refer to a standard deck of cards in many examples and exercises.

EXAMPLE 14   Find the probability of drawing an ace on one draw from a standard card deck.

Solution   Since there are 4 aces in the deck, the number of favorable outcomes is $s = 4$. Since there are 52 cards in the deck, the total number of possible outcomes is $n = 52$. The probability of drawing an ace is the ratio of the number of favorable outcomes to the number of possible outcomes.

$$P(\text{an ace}) = \frac{s}{n} = \frac{4}{52} = \frac{1}{13}$$

The probability of drawing an ace is $\frac{1}{13}$.

SELF CHECK 14   Find the probability of drawing a red ace on one draw from a standard card deck.

EXAMPLE 15   Find the probability of drawing 5 cards, all hearts, from a standard card deck.

Solution   The number of ways we can draw 5 hearts from the 13 hearts is $C(13, 5)$, the number of combinations of 13 things taken 5 at a time. The number of ways to draw 5 cards from the deck is $C(52, 5)$, the number of combinations of 52 things taken 5 at a time. The probability of drawing 5 hearts is the ratio of the number of favorable outcomes to the number of possible outcomes.

$$P(5 \text{ hearts}) = \frac{s}{n} = \frac{C(13, 5)}{C(52, 5)}$$

$$P(5 \text{ hearts}) = \frac{\dfrac{13!}{5!8!}}{\dfrac{52!}{5!47!}}$$

$$= \frac{13!}{5!8!} \cdot \frac{5!47!}{52!}$$

$$= \frac{13 \cdot 12 \cdot 11 \cdot 10 \cdot 9 \cdot 8!}{8!} \cdot \frac{47!}{52 \cdot 51 \cdot 50 \cdot 49 \cdot 48 \cdot 47!}$$

$$= \frac{13 \cdot 12 \cdot 11 \cdot 10 \cdot 9}{52 \cdot 51 \cdot 50 \cdot 49 \cdot 48}$$

$$= \frac{33}{66,640}$$

The probability of drawing 5 hearts is $\frac{33}{66,640}$ or $4.951980792 \times 10^{-4}$.

SELF CHECK 15   Find the probability of drawing 6 cards, all spades, from a standard card deck.

SELF CHECK ANSWERS

1. a. 720 **b.** $x(x - 1)(x - 2) \cdot \cdots \cdot 3 \cdot 2 \cdot 1$ **2. a.** 4 **b.** 21 **3.** 35 **4.** 24 **5.** 120 **6.** 840 **7. a.** 151,200 **b.** 1 **8.** 27,907,200 **9. a.** 84 **b.** 1 **10.** 15,504 **11.** 14,700 **12.** 3 **13.** $\frac{1}{12}$ **14.** $\frac{1}{26}$ **15.** $\frac{33}{391,510}$ or $8.428903476 \times 10^{-5}$

A.1 Exercises

Find each value.

1. 1!   **2.** 4!

3. 0!   **4.** 5!

GUIDED PRACTICE *Evaluate each expression.*

5. 3!   **6.** 7!

7. −5!   **8.** −6!

9. 3! + 4!   **10.** 2!(3!)

11. 3!(4!)   **12.** 4! + 4!

13. $\dfrac{9!}{11!}$ **14.** $\dfrac{13!}{10!}$

15. $\dfrac{49!}{47!}$ **16.** $\dfrac{101!}{100!}$

17. $\dfrac{9!}{7! \cdot 0!}$ **18.** $\dfrac{7!}{5! \cdot 0!}$

19. $\dfrac{5!}{3!(5-3)!}$ **20.** $\dfrac{6!}{4!(6-4)!}$

Evaluate using permutations.

21. $P(3,3)$ **22.** $P(4,4)$

23. $P(5,3)$ **24.** $P(3,2)$

25. $P(2,2) \cdot P(3,3)$ **26.** $P(3,2) \cdot P(3,3)$

27. $\dfrac{P(5,3)}{P(4,2)}$ **28.** $\dfrac{P(6,2)}{P(5,4)}$

Evaluate using combinations.

29. $C(5,3)$ **30.** $C(5,4)$

31. $\dbinom{6}{3}$ **32.** $\dbinom{6}{4}$

33. $\dbinom{5}{4}\dbinom{5}{3}$ **34.** $\dbinom{6}{5}\dbinom{6}{4}$

35. $\dfrac{C(38,37)}{C(19,18)}$ **36.** $\dfrac{C(25,23)}{C(40,39)}$

List the sample space of each experiment.

37. Rolling one die followed by tossing one coin

38. Tossing three coins

39. Selecting a letter of the alphabet

40. Picking a one-digit number

An ordinary die is rolled once. Find the probability of each event.

41. Rolling a 2

42. Rolling a number greater than 4

43. Rolling a number larger than 1 but less than 6

44. Rolling an odd number

Balls numbered from 1 to 42 are placed in a container and stirred. If one is drawn at random, find the probability of each result.

45. The number is less than 20.

46. The number is less than 50.

47. The number is a prime number.

48. The number is less than 10 or greater than 40.

Refer to the following spinner. If the spinner is spun, find the probability of each event. Assume that the spinner never stops on a line.

49. The spinner stops on red.

50. The spinner stops on green.

51. The spinner stops on brown.

52. The spinner stops on yellow.

Find the probability of each event.

53. Drawing a diamond on one draw from a standard card deck

54. Drawing a face card from a standard card deck

55. Drawing a red face card from a standard card deck

56. Drawing a face card from a standard deck followed by a 10 after replacing the first card

57. Drawing an ace from a standard deck followed by a 10 after replacing the first card

58. Drawing 6 diamonds from a standard card deck without replacing the cards after each draw

59. Drawing 5 aces from a standard card deck without replacing the cards after each draw

60. Drawing 5 clubs from the black cards in a standard card deck

ADDITIONAL PRACTICE *Evaluate each expression.*

61. 8(7!)   **62.** 4!(5)

63. $\dfrac{7!}{5!(7-5)!}$ **64.** $\dfrac{8!}{6!(8-6)!}$

65. $\dfrac{5!(8-5)!}{4! \cdot 7!}$ **66.** $\dfrac{6! \cdot 7!}{(8-3)!(7-4)!}$

Use a calculator to find each factorial.

67. 11!   **68.** 13!

69. 20!   **70.** 55!

Find the probability of each event.

71. Rolling a sum of 4 on one roll of two dice

72. Drawing a red egg from a basket containing 5 red eggs and 7 blue eggs

73. Drawing a yellow egg from a basket containing 5 red eggs and 7 yellow eggs

Assume that the probability that a backup generator will fail a test is $\frac{1}{2}$ and that the college in question has 4 backup generators. In Exercises 75–79, find each probability.

74. Construct a sample space for the test.

75. All generators will fail the test.

76. Exactly 1 generator will fail.

77. Exactly 2 generators will fail.

78. Exactly 3 generators will fail.

79. No generator will fail.

80. Find the sum of the probabilities in Exercises 75–79.

81. Arranging an evening   Kyoro plans to go to dinner and see a movie. In how many ways can she arrange her evening if she has a choice of five movies and seven restaurants?

82. Travel choices   Paula has five ways to travel from New York to Chicago, three ways to travel from Chicago to Denver, and four ways to travel from Denver to Los Angeles. How many choices are available to Paula if she travels from New York to Los Angeles?

83. Arranging books   In how many ways can seven books be placed on a shelf?

84. Lining up   In how many ways can the people shown be placed in a line?

85. Making license plates   How many six-digit license plates can be manufactured? Note that there are ten choices— 0, 1, 2, 3, 4, 5, 6, 7, 8, 9—for each digit.

86. Making license plates   How many six-digit license plates can be manufactured if no digit can be repeated?

87. Making license plates   How many six-digit license plates can be manufactured if no license can begin with 0 and if no digit can be repeated?

88. Making license plates   How many license plates can be manufactured with two letters followed by four digits?

89. Phone numbers   How many seven-digit phone numbers are available in area code 815 if no phone number can begin with 0 or 1?

90. Phone numbers   How many ten-digit phone numbers are available if area codes 000 and 911 cannot be used and if no local number can begin with 0 or 1?

91. Arranging books   In how many ways can four novels and five biographies be arranged on a shelf if the novels are placed on the left?

92. Making a ballot   In how many ways can six candidates for mayor and four candidates for the county board be arranged on a ballot if all of the candidates for mayor must be placed on top?

93. Combination locks   How many permutations does a combination lock have if each combination has three numbers, no two numbers of any combination are equal, and the lock has 25 numbers?

94. Combination locks   How many permutations does a combination lock have if each combination has three numbers, no two numbers of any combination are equal, and the lock has 50 numbers?

95. Arranging appointments   The receptionist at a dental office has only three appointment times available before Tuesday, and ten patients have toothaches. In how many ways can the receptionist fill those appointments?

96. Computers   In many computers, a word consists of 32 *bits*—a string of thirty-two 1's and 0's. How many different words are possible?

97. Palindromes   A palindrome is any word, such as *madam* or *radar*, that reads the same backward and forward. How many five-digit numerical palindromes (such as 13531) are there?   (*Hint*: A leading 0 would be dropped.)

98. Call letters   The call letters of U.S. commercial radio stations have 3 or 4 letters, and the first is always a W or a K. How many radio stations could this system support?

99. Planning a picnic   A class of 14 students wants to pick a committee of 3 students to plan a picnic. How many committees are possible?

100. Choosing books   Jeff must read 3 books from a reading list of 15 books. How many choices does he have?

101. Forming committees   The number of three-person committees that can be formed from a group of people is ten. How many people are in the group?

102. Forming committees   The number of three-person committees that can be formed from a group of people is 20. How many people are in the group?

103. Winning a lottery   In one state lottery, anyone who picks the correct 6 numbers (in any order) wins. With the numbers 0 through 99 available, how many choices are possible?

104. Taking a test   The instructions on a test read: "Answer any ten of the following fifteen questions. Then choose one of the remaining questions for homework and turn in its solution tomorrow." In how many ways can the questions be chosen?

105. Forming a committee In how many ways can we select a committee of two men and two women from a group containing three men and four women?

106. Forming a committee In how many ways can we select a committee of three men and two women from a group containing five men and three women?

107. Choosing clothes In how many ways can we select 2 shirts and 3 neckties from a group of 12 shirts and 10 neckties?

108. Choosing clothes In how many ways can we select five dresses and two coats from a wardrobe containing nine dresses and three coats?

109. Quality control In a batch of 10 tires, 2 are known to be defective. If 4 tires are chosen at random, find the probability that all 4 tires are good.

110. Medicine Out of a group of 9 patients treated with a new drug, 4 suffered a relapse. If 3 patients are selected at random from the group of 9, find the probability that none of the 3 patients suffered a relapse.

A survey of 282 people is taken to determine the opinions of doctors, teachers, and lawyers on a proposed piece of legislation, with the results shown in the table. A person is chosen at random from those surveyed. Refer to the table to find each probability.

111. The person favors the legislation.

112. A doctor opposes the legislation.

113. A person who opposes the legislation is a lawyer.

	Number that favor	Number that oppose	Number with no opinion	Total
Doctors	70	32	17	119
Teachers	83	24	10	117
Lawyers	23	15	8	46
Total	176	71	35	282

APPENDIX 2: Measurement Conversions

Every day we are confronted with the need to quantify things. We drive a number of miles (distance) to work or school. We fill our cars with a number of gallons (volume) of fuel. We purchase a quantity of deli lunch meats (weight).

In this appendix we will look at the two systems of measurement most commonly used in the United States: the American system and the metric system. We will examine the units of measure in each system for

Length (distance)   Volume (capacity)   Weight (mass)

A.2.1 Measurement in the American System

The American system continues to find wide usage in the United States, particularly in consumer applications such as retail and grocery stores. But even those areas are slowly converting to the metric system. As an example, milk is still sold by the gallon, but many carbonated beverages are now sold by the liter.

1 Identify the units of length in the American system and convert from one unit of length to another.

The standard units of length in the American system are the **inch**, **foot**, **yard**, and **mile**. They are related to one another in the following ways:

1 foot (ft) = 12 inches (in.)

1 yard (yd) = 3 feet = 36 inches

1 mile (mi) = 5,280 feet = 1,760 yards = 63,360 inches

To convert a measurement from one unit of length to another, we use a **unit conversion factor**, a special fraction that is equal to 1 (the unit). It is written with the desired conversion units in the numerator and the initial units in the denominator:

$$\frac{\text{the numerator is written in units that you wish to find}}{\text{the denominator is written in units that you wish to remove}}$$

For example, because there are three feet in one yard, the unit conversion for yards to feet is written as:

$$\frac{3 \text{ feet}}{1 \text{ yard}}$$

Read as "3 feet per yard," it is algebraically equal to 1. Therefore, to convert units of a 100-yard football field to feet, we multiply the initial quantity (100 yards) by the yards-to-feet conversion factor.

$$100 \text{ yards} = 100 \text{ yards} \cdot \frac{3 \text{ feet}}{1 \text{ yard}}$$

$$= 100 \cdot 3$$

$$= 300 \text{ feet}$$

To convert the units of measure among inches, feet, yards and miles, we use the following unit conversion factors.

To convert from	Unit conversion factor	To convert from	Unit conversion factor
Feet to inches	12 in. / 1 ft	Inches to feet	1 ft / 12 in.
Yards to feet	3 ft / 1 yd	Feet to yards	1 yd / 3 ft
Yards to inches	36 in. / 1 yd	Inches to yards	1 yd / 36 in.
Miles to feet	5,280 ft / 1 mi	Feet to miles	1 mi / 5,280 ft

EXAMPLE 1   Determine the height (length) of a 20-foot flagpole in yards.

Solution

$$20 \text{ feet} = 20 \text{ feet} \cdot \frac{1 \text{ yard}}{3 \text{ feet}}$$

$$= \frac{20}{3}$$

$$= 6\frac{2}{3} \text{ yards}$$

SELF CHECK 1   Determine the height of a 25-foot tower in yards.

EXAMPLE 2   Express the length of a 1.5-mile horse race track in feet.

Solution

$$1.5 \text{ miles} = 1.5 \text{ miles} \cdot \frac{5,280 \text{ feet}}{1 \text{ mile}}$$

$$= 1.5 \cdot 5,280 \text{ feet}$$

$$= 7,920 \text{ feet}$$

SELF CHECK 2   Express the length of a 2-mile horse race track in feet.

EXAMPLE 3   What is the height in miles, to the nearest tenth of a mile, of an airplane flying at 30,000 feet?

Solution

$$30,000 \text{ feet} = 30,000 \text{ feet} \cdot \frac{1 \text{ mile}}{5,280 \text{ feet}}$$

$$= \frac{30,000 \text{ miles}}{5,280}$$

$$= 5.7 \text{ miles}$$

SELF CHECK 3   What is the height in miles, to the nearest tenth of a mile, of an airplane flying at 35,000 feet?

2 Identify the units of volume in the American system and convert from one unit of volume to another.

The standard units of volume (sometimes called capacity) in the American system are the fluid **ounce**, **cup**, **pint**, **quart**, and **gallon**. They are related to one another in the following ways:

> 1 cup (c) = 8 fluid ounces (fl oz)
>
> 1 pint (pt) = 2 cups
>
> 1 quart (qt) = 2 pints
>
> 1 gallon (gal) = 4 quarts

In the same manner as shown in converting units of length, we use a unit conversion factor to convert units of volume. To convert the units of measure among ounces, cups, pints, quarts, and gallons, we use the following unit conversion factors.

To convert from	Unit conversion factor	To convert from	Unit conversion factor
Cups to fluid ounces	8 fl oz / 1 c	Fluid ounces to cups	1 c / 8 fl oz
Pints to cups	2 c / 1 pt / 16 fl oz	Cups to pints	1 pt / 2 c / 16 fl oz
Quarts to pints	2 pt / 1 qt / 32 fl oz	Pints to quarts	1 qt / 2 pt / 32 fl oz
Gallons to quarts	4 qt / 1 gal / 128 fl oz	Quarts to gallons	1 gal / 4 qt / 128 fl oz

EXAMPLE 4   How many quarts are in a 20-gallon automobile gasoline tank?

Solution
$$20 \text{ gallons} = 20 \text{ gallons} \cdot \frac{4 \text{ quarts}}{1 \text{ gallon}}$$
$$= 20 \cdot 4 \text{ quarts}$$
$$= 80 \text{ quarts}$$

 SELF CHECK 4   How many quarts are in a 30-gallon automobile gasoline tank?

EXAMPLE 5   How many fluid ounces are in 6 pints of ice cream?

Solution
$$6 \text{ pints} = 6 \text{ pints} \cdot \frac{2 \text{ cups}}{1 \text{ pint}} \cdot \frac{8 \text{ fluid ounces}}{1 \text{ cup}}$$
$$= 6 \cdot 2 \cdot 8$$
$$= 96 \text{ fluid ounces}$$

 SELF CHECK 5   How many fluid ounces are in 4 pints of ice cream?

EXAMPLE 6   How many cups of milk are in a half-gallon container?

Solution
$$\frac{1}{2} \text{ gallon} = \frac{1}{2} \text{ gallon} \cdot \frac{4 \text{ quarts}}{1 \text{ gallon}} \cdot \frac{2 \text{ pints}}{1 \text{ quart}} \cdot \frac{2 \text{ cups}}{1 \text{ pint}}$$
$$= \frac{1}{2} \cdot 4 \cdot 2 \cdot 2$$
$$= 8 \text{ cups}$$

 SELF CHECK 6   How many cups of milk are in a $\frac{3}{4}$-gallon container?

3 Identify the units of weight in the American system and convert from one unit of weight to another.

The standard units of weight in the American system are the **ounce**, **pound**, and **ton**. They are related to one another in the following ways:

1 pound (lb) = 16 ounces (oz)

1 ton (T) = 2,000 pounds

In the same manner as shown in converting units of length and units of volume, we use a unit conversion factor to convert units of weight. To convert the units of measure among ounces, pounds, and tons, we use the following unit conversion factors.

To convert from	Unit conversion factor		To convert from	Unit conversion factor
Pounds to ounces	16 oz / 1 lb		Ounces to pounds	1 lb / 16 oz
Tons to pounds	2,000 lb / 1 T		Pounds to tons	1 T / 2,000 lb

EXAMPLE 7   How many pounds are in a 256-ounce bowling ball?

Solution

$$256 \text{ ounces} = 256 \text{ ounces} \cdot \frac{1 \text{ pound}}{16 \text{ ounces}}$$

$$= \frac{256 \text{ pounds}}{16}$$

$$= 16 \text{ pounds}$$

 SELF CHECK 7   How many pounds are in a 192-ounce bowling ball?

EXAMPLE 8   How many pounds of material are in a 3-ton automobile?

Solution

$$3 \text{ tons} = 3 \text{ tons} \cdot \frac{2,000 \text{ pounds}}{1 \text{ ton}}$$

$$= 3 \cdot 2,000 \text{ pounds}$$

$$= 6,000 \text{ pounds}$$

 SELF CHECK 8   How many pounds of material are in a 2-ton automobile?

EXAMPLE 9   How much does a 25-pound Atlantic salmon weigh in ounces?

Solution

$$25 \text{ pounds} = 25 \text{ pounds} \cdot \frac{16 \text{ ounces}}{1 \text{ pound}}$$

$$= 25 \cdot 16 \text{ ounces}$$

$$= 400 \text{ ounces}$$

 SELF CHECK 9   How much does a 20-pound Atlantic salmon weigh in ounces?

A.2.2 Measurement in the Metric System

The metric system is widely used internationally. In the United States, it is primarily used in scientific, medical, and engineering applications, but its use is slowly expanding. The metric system, like the decimal numeration system, is based on the number 10. This makes the conversions from one metric unit to another much easier than in the American system.

Identify the units of length in the metric system and convert from one unit of length to another.

The primary unit of length in the metric system is the **meter**. It is slightly longer than one American yard, or about 39 inches. The other units of length are obtained by multiplying or dividing the meter by powers of 10. The names of the units are given by adding a prefix to the basic unit, the meter.

1 kilometer (km) = 1,000 meters

1 hectometer (hm) = 100 meters

1 decameter (dam) = 10 meters

1 meter (m) = primary metric unit of length

1 decimeter (dm) = 0.1 meter

1 centimeter (cm) = 0.01 meter

1 millimeter (mm) = 0.001 meter

For conversions in the metric system, we could use a unit conversion factor in the same manner as in the American system. Instead, we use a much simpler method of shifting the decimal point of a measure left or right to change to larger or smaller units. The following chart illustrates how to convert among the metric units of length.

	mm	cm	dm	m	dam	hm	km
1 meter equals	1,000.0	100.0	10.0	1.0	0.1	0.01	0.001
Example	5,250.0			5.25			
	Initial unit	3 places to the *right* >-------------------->		Larger desired unit	If desired unit is 3 places to the *right*, then move the initial decimal point 3 places to the *left*.		
	Smaller desired unit	3 places to the *left* <--------------------<		Initial unit	If desired unit is 3 places to the *left*, then move the initial decimal point 3 places to the *right*.		

The conversion requires counting the number of places—and noting the direction—between the desired unit and the initial unit. For example, if the initial measurement is expressed in millimeters and the desired unit is meters, we note the number of places and the direction (in the chart) between millimeters and meters. Meter units are 3 places *to the right* of millimeter units, so the decimal point must be shifted 3 places *to the left*. In the chart, 5,250.0 millimeters is converted to meters by moving its decimal point 3 places *to the left*, resulting in 5.25 meters.

COMMENT   To convert a measure in any unit to a *larger* unit, move the decimal point of the initial measure to the *left*. To convert a measure in any unit to a *smaller* unit, move the decimal point of the initial measure to the *right*.

EXAMPLE 10   How high, in centimeters, is a 3-meter ceiling in a room?

Solution   There are 2 places between centimeters and meters. To convert from a larger unit to a smaller unit, we move the decimal point 2 places to the right.

3 meters = 300 centimeters

 **SELF CHECK 10**   How high, in centimeters, is a 4-meter ceiling in a room?

EXAMPLE 11   How long is a 5-kilometer race in meters?

Solution   There are 3 places between kilometers and meters. To convert from a larger unit to a smaller unit, we move the decimal point 3 places to the right.

$$5 \text{ kilometers} = 5{,}000 \text{ meters}$$

 SELF CHECK 11   How long is a 10-kilometer race in meters?

EXAMPLE 12   How many 6-centimeter-long nails can be cut from an 18-meter length of nail wire?

Solution   There are 2 places between centimeters and meters. To convert from a larger unit to a smaller unit, we move the decimal point 2 places to the right.

$$18 \text{ meters} = 1{,}800 \text{ centimeters}$$

There will be $1{,}800/6 = 300$ nails cut from each nail wire.

 **SELF CHECK 12**   How many 5-centimeter-long nails can be cut from an 18-meter length of nail wire?

2   **Identify the units of volume in the metric system and convert from one unit of volume to another.**

The primary unit of volume (capacity) in the metric system is the **liter**. It is defined as the volume of a cube whose sides are each 10 centimeters long. (A special unit of volume, used in medical fields, is the **cubic centimeter**. It is defined as the volume of a cube whose sides are each 1 centimeter long and it is equivalent to 1 millimeter.) In the same manner as metric units of length, we obtain the other units of volume by multiplying or dividing the liter by powers of 10. The names of the units are given by adding a prefix to the basic unit, the liter.

1 kiloliter (kl) = 1,000 liters

1 hectoliter (hl) = 100 liters

1 decaliter (dal) = 10 liters

1 liter (l) = primary metric unit of volume

1 deciliter (dl) = 0.1 liter

1 centiliter (cl) = 0.01 liter

1 milliliter (ml) = 0.001 liter = 1 cm^3 (cubic centimeter)

Conversions of metric volume are performed in the same manner as conversions of metric length, by moving the decimal point left or right. The following chart illustrates how to convert among metric units of volume.

	ml	cl	dl	l	dal	hl	kl
1 liter equals	1,000.0	100.0	10.0	1.0	0.1	0.01	0.001
Example		800.0		8.0			
		Initial unit	2 places to the *right*	Larger desired unit	If desired unit is 2 places to the *right*, then move the initial decimal point 2 places to the *left*.		
			>---------->				
		Smaller desired unit	2 places to the *left*	Initial unit	If desired unit is 2 places to the *left*, then move the initial decimal point 2 places to the *right*.		
			<----------<				

For example, to convert from centiliters to liters, we note that liter units are 2 places *to the right* of centiliter units, so we move the decimal point of the initial measure 2 places *to the left*. In the chart, 800.0 centiliters is converted to liters by moving its decimal point 2 places to the left, resulting in 8 liters.

COMMENT   Remember that to convert a measure in any unit to a *larger* unit, move the decimal point of the initial measure to the *left*. To convert a measure in any unit to a *smaller* unit, move the decimal point of the initial measure to the *right*.

EXAMPLE 13   How many centiliters are in a 2-liter bottle of juice?

Solution   There are 2 places between centiliters and liters. To convert from a larger unit to a smaller unit, we move the decimal point 2 places to the right.

$$2 \text{ liters} = 200 \text{ centiliters}$$

 SELF CHECK 13   How many centiliters are in a 1.5-liter bottle of juice?

EXAMPLE 14   How many kiloliters are in a 4,000-centiliter gas tank?

Solution   There are 5 places between kiloliters and centiliters. To convert from a smaller unit to a larger unit, we move the decimal point 5 places to the left.

$$4,000 \text{ centiliters} = 0.04 \text{ kiloliter}$$

 SELF CHECK 14   How many kiloliters are in a 3,000-centiliter gas tank?

EXAMPLE 15   How many milliliters (or cubic centimeters) are in $\frac{1}{2}$ liter of saline solution?

Solution   There are 3 places between milliliters and liters. To convert from a larger unit to a smaller unit, we move the decimal point 3 places to the right.

$$0.5 \text{ liter} = 500 \text{ milliliters}$$

 SELF CHECK 15   How many milliliters (or cubic centimeters) are in $\frac{3}{4}$ liter of saline solution?

3   Identify the units of mass (or weight) in the metric system and convert from one unit of mass to another.

The primary unit of mass in the metric system is the **gram**. It is defined as the weight of water in a cube whose sides are each 1 centimeter long. The names of the units are given by adding a prefix to the basic unit, the gram.

1 kilogram (kg) = 1,000 grams

1 hectogram (hg) = 100 grams

1 decagram (dag) = 10 grams

1 gram (g) = primary metric unit of mass

1 decigram (dg) = 0.1 gram

1 centigram (cg) = 0.01 gram

1 milligram (mg) = 0.001 gram

In the same manner as the metric units of length and volume, we obtain the other units of mass by multiplying or dividing the gram by powers of 10. We convert metric mass in the same manner as metric length and volume, by moving the decimal point left or right. The following chart illustrates how to convert among the metric units of mass.

	mg	cg	dg	g	dag	hg	kg
1 gram equals	1,000.0	100.0	10.0	1.0	0.1	0.01	0.001
Example				4,600.0			4.60
	If desired unit is 3 places to the *right*, then move the initial decimal point 3 places to the *left*.			Initial unit	3 places to the *right* >-------------------->		Larger desired unit
	If desired unit is 3 places to the *left*, then move the initial decimal point 3 places to the *right*.			Smaller desired unit	3 places to the *left* <--------------------<		Initial unit

COMMENT   Remember that to convert a measure in any unit to a *larger* unit, move the decimal point of the initial measure to the *left*. To convert a measure in any unit to a *smaller* unit, move the decimal point of the initial measure to the *right*.

EXAMPLE 16   Four raisins weigh 5 grams. How many milligrams do they weigh?

Solution   There are 3 places between milligrams and grams. To convert from a larger unit to a smaller unit, we move the decimal point 3 places to the right.

   1 gram = 1,000 milligrams

Thus, four raisins weigh 5 grams or 5,000 milligrams.

 SELF CHECK 16   Four raisins weigh 6 grams. How many milligrams do they weigh?

EXAMPLE 17   How many grams in a 6-kilogram bowling ball?

Solution   There are 3 places between kilograms and grams. To convert from a larger unit to a smaller unit, we move the decimal point 3 places to the right.

   6 kilograms = 6,000 grams

 SELF CHECK 17   How many grams in a 4-kilogram bowling ball?

EXAMPLE 18   How many hectograms are in 27 centigrams?

Solution   There are 4 places between kilograms and grams. To convert from a smaller unit to a larger unit, we move the decimal point 4 places to the left.

   27 centigrams = 0.0027 hectogram

 SELF CHECK 18   How many hectograms are in 30 centigrams?

A.2.3 Conversions Between Metric and American Systems

Consider the following questions:

- Is a 4-mile race longer than a 5-kilometer run?
- Is a 4-millimeter closed-end wrench larger than a $\frac{1}{2}$-inch socket wrench?
- Is a 5.2-liter engine smaller than a 352-cubic inch Hemi?

Because we may need to understand what a given quantity in one system actually means in the other system, we will look at the methods used for such conversions.

COMMENT   The labels on many consumer goods often will show both American and metric measures for the package content. A soft conversion lists the American system first, 18 oz (510 g) for example, whereas a hard conversion lists the metric system first, such as 500 ml (16.9 fl oz).

1 Convert units of length from one system to another.

The following table illustrates common conversions of length between the American and metric systems. These can be used to create unit conversion factors.

Conversions of Length	
1 in. $\approx$ 2.54 cm	1 cm $\approx$ 0.39 in.
1 ft $\approx$ 0.30 m	1 m $\approx$ 3.28 ft
1 yd $\approx$ 0.91 m	1 m $\approx$ 1.09 yd
1 mi $\approx$ 1.61 km	1 km $\approx$ 0.62 mi

For any conversion, the unit conversion factor will be written as

$$\frac{\text{the numerator is written in units that you wish to find}}{\text{the denominator is written in units that you wish to remove}}$$

For example, to convert a measure of 87 feet to meters, the process would be

$$87 \text{ feet} \approx 87 \text{ feet} \cdot \frac{1 \text{ meter}}{3.28 \text{ feet}}$$

$$\approx \frac{87 \text{ meters}}{3.28}$$

$$\approx 26.524 \text{ meters}$$

EXAMPLE 19   A bluefin tuna is about 6.5 feet long. How long is that in meters?

Solution
$$6.5 \text{ feet} \approx 6.5 \text{ feet} \cdot \frac{1 \text{ meter}}{3.28 \text{ feet}}$$

$$\approx \frac{6.5 \text{ meters}}{3.28}$$

$$\approx 2 \text{ meters}$$

 **SELF CHECK 19**   A bluefin tuna is about 6 feet long. How long is that in meters?

Converting lengths is important when working with miles per hour and kilometers per hour. Fortunately, for practical purposes, most speedometers have the equivalents posted for us.

2 Convert units of volume (or capacity) from one system to another.

The following table illustrates common conversions of volume between the American and metric systems. These can be used to create unit conversion factors.

Conversions of Volume (Capacity)	
1 fl oz ≈ 29.57 ml	1 l ≈ 33.81 fl oz
1 pt ≈ 0.47 l	1 l ≈ 2.11 pt
1 qt ≈ 0.95 l	1 l ≈ 1.06 qt
1 gal ≈ 3.79 l	1 l ≈ 0.264 gal

We construct a unit conversion factor for volumes in the same manner as for length. Remember that for any conversion, the unit conversion factor will be written as

$$\frac{\text{the numerator is written in units that you wish to find}}{\text{the denominator is written in units that you wish to remove}}$$

For example, to convert a measure of 7 gallons to liters, the process would be

$$7 \text{ gallons} \approx 7 \text{ gallons} \cdot \frac{1 \text{ liter}}{0.264 \text{ gallon}}$$

$$\approx \frac{7 \text{ liters}}{0.264}$$

$$\approx 26.515 \text{ liters}$$

EXAMPLE 20   A 2-liter soft drink is equivalent to how many gallons?

Solution
$$2 \text{ liters} \approx 2 \text{ liters} \cdot \frac{1 \text{ gallon}}{3.79 \text{ liters}}$$

$$\approx \frac{2 \text{ gallons}}{3.79}$$

$$\approx \frac{1}{2} \text{ gallon}$$

 SELF CHECK 20   A 3-liter bottle is equivalent to how many gallons?

3   ## Convert units of weight (or mass) from one system to another.

The following table illustrates common conversions of weight between the American and metric systems. These can be used to create unit conversion factors.

Conversions of Weight (Mass)	
1 oz ≈ 28.35 g	1 g ≈ 0.035 oz
1 lb ≈ 0.5 kg	1 kg ≈ 2.20 lb

We construct a unit conversion factor for weight in the same manner as for length and volume. For example, to convert a measure of 4 pounds to kilograms, the process would be

$$4 \text{ pounds} \approx 4 \text{ pounds} \cdot \frac{1 \text{ kilogram}}{2.20 \text{ pounds}}$$

$$\approx \frac{4 \text{ kilograms}}{2.20}$$

$$\approx 1.818 \text{ kilograms}$$

EXAMPLE 21   A bluefin tuna weighs about 250 kilograms. How much is that in pounds?

Solution $250 \text{ kilograms} \approx 250 \text{ kilograms} \cdot \dfrac{2.20 \text{ pounds}}{\text{kilograms}}$

$\approx 250 \cdot 2.20 \text{ pounds}$

$\approx 550 \text{ pounds}$

SELF CHECK 21   A bluefin tuna weighs about 440 pounds. How much is that in kilograms?

SELF CHECK ANSWERS

1. $8\frac{1}{3}$ yd   **2.** 10,560 ft   **3.** 6.6 mi   **4.** 120 qt   **5.** 64 fl oz   **6.** 12 c   **7.** 12 lb   **8.** 4,000 lb
9. 320 oz   **10.** 400 cm   **11.** 10,000 m   **12.** 360 nails   **13.** 150 cl   **14.** 0.03 kl   **15.** 750 cm³
16. 6,000 mg   **17.** 4,000 g   **18.** 0.003 hg   **19.** ≈1.8 m   **20.** ≈0.8 gallon   **21.** ≈200 kg

---

A.2 Exercises

Make the appropriate conversions.

1. Find, in feet, the height of a 27.3-yard tree.

2. Find the length of an Olympic-size swimming pool (50 yards) in feet.

3. Find the height of Mt. McKinley to the nearest tenth of a mile if it is 20,320 feet high.

4. The Atlanta Motor Speedway Nascar Track is a 1.54-mile oval. A major 500-mile race is generally run there each fall. How may laps does the race take and how many feet are covered in each lap?

5. In the Space Shuttle approximately 1,214,400 feet of wiring is necessary to maintain control of, and communication with, the spacecraft. How many miles of wire is this?

6. Grass sod is sold in 1 foot × 1 yard rectangles. If a front yard is 85 feet long, how many rectangles will it take to make 1 row lengthwise?

7. FAA rules require pilots flying in unpressurized aircraft to use supplemental oxygen for flights above 12,500 feet. What is that altitude to the nearest tenth of a mile?

8. How many gallons of oil are there in a 120-quart container?

9. A gasoline-powered lawn mower holds $\frac{1}{2}$ gallon of gas. How many cups is that?

10. A bottle of sports drink contains 32 ounces and comes in a case of 4 bottles. How many gallons are there in a case?

11. A case of soft drinks contains 24 12-ounce cans. How many 32-ounce bottles would that fill?

12. A pint of premium ice cream costs about $3.50. If the price per gallon were consistent with the price per pint, how much would $\frac{1}{2}$ gallon cost?

13. A woman bought 5 gallons of shelled pecans. If a pint freezer bag truly holds a pint, how many freezer bags will she need to store the 5 gallons of pecans?

14. A recipe for custard calls for 3 pints of milk. How many cups is that?

15. The Endeavor, the last Space Shuttle flown, weighed 4.5 million pounds at liftoff. Two minutes into the flight, it had already lost half that weight. What was its weight at the 2-minute mark in tons?

16. How many 4-ounce packages can be obtained from a 17.2-pound chunk of tuna?

17. To be granted agricultural status for tax purposes in Texas, land must support at least 1,000 pounds per acre of hoofed livestock. The XIT Ranch in Texas was once the largest in the United States, with 3,500,000 acres. To maintain its agricultural tax status, how many tons of cattle would have to be raised on the ranch? If an average grown cow weighs approximately 1,650 pounds, how many head of cattle would be needed?

18. The European Grand Prix is a race of 308.883 kilometers with speeds up to 360 kilometers per hour. If the race is 57 laps, how long is each lap in meters?

19. A heavy-duty trash bag is 2 millimeters thick. How thick is this in meters?

20. A 2-liter bottle of a diet cola contains how many milliliters?

21. A quarter-bottle of champagne holds 18.75 centiliters while the Nebuchadnezzar bottle holds 15 liters. How many centiliters does the Nebuchadenezzar hold?

22. In medicine, 1 cubic centimeter is equivalent to 1 milliliter. If a dose calls for 150 cubic centimeters, how many liters is that?

23. One Troy ounce of gold weighs 31.1034768 grams. If Chris bought 5 Troy ounces of gold, how many kilograms would he have?

24. A baby kangaroo weighs just 2 grams at birth but an adult can weigh up to 90 kilograms. How many grams can an adult weigh?

25. Approximately how many feet are there in one kilometer?

26. If gasoline is 69.4¢ per liter, about how much is gas per gallon?

27. A recipe calls for 340.2 grams of butter to make 12 croissants. How many ounces of butter is in each croissant?

1. How many prime numbers are there between 20 and 30?
   a. 1   b. 2   c. 3   d. 4   e. none of the above

2. If $x = 3, y = -2$, and $z = -1$, find the value of $\dfrac{x + z}{y}$.
   a. 1   b. -1   c. -2   d. 2   e. none of the above

3. If $x = -2, y = -3$, and $z = -4$, find the value of $\dfrac{|x - z|}{|y|}$.
   a. -2   b. 2   c. $\dfrac{2}{3}$ d. $-\dfrac{2}{3}$ e. none of the above

4. The distributive property is written symbolically as
   a. $a + b = b + a$ b. $ab = ba$ c. $a(b + c) = ab + ac$
   d. $(a + b) + c = a + (b + c)$ e. none of the above

5. Solve for x: $7x - 4 = 24$
   a. $x = 4$ b. $x = 5$ c. $x = -4$ d. $x = -5$
   e. none of the above

6. Solve for z: $6z - (9 - 3z) = -3(z + 2)$
   a. $\dfrac{4}{3}$ b. $\dfrac{3}{4}$ c. $-\dfrac{4}{3}$ d. $-\dfrac{1}{4}$
   e. none of the above

7. Solve for r: $\dfrac{r}{5} - \dfrac{r - 3}{10} = 0$
   a. -3   b. -2   c. 3   d. 1   e. none of the above

8. Solve for x: $\dfrac{ax}{b} + c = 4$
   a. $\dfrac{b(4 - x)}{c}$ b. $\dfrac{4b - 4c}{b}$ c. $\dfrac{b(4 - c)}{a}$
   d. none of the above

9. A man bought 25 pencils, some at 10 cents and some at 15 cents. The 25 pencils cost \$3. How many 10-cent pencils did he buy?
   a. 5   b. 10   c. 15   d. 20   e. none of the above

10. Solve the inequality: $-3(x - 2) + 3 \geq 6$
    a. $x \geq -1$ b. $x \geq 1$ c. $x \leq -1$ d. $x \leq 1$
    e. none of the above

11. Simplify: $x^2 x^3 x^7$
    a. $12x$ b. x^{12} c. x^{42} d. x^{35}
    e. none of the above

12. Simplify: $\dfrac{(x^2)^7}{x^3 x^4}$
    a. 0   b. 1   c. x^7 d. x^2 e. none of the above

13. Simplify: $\dfrac{x^{-2} y^3}{x y^{-1}}$
    a. $x^3 y^4$ b. $\dfrac{x^3}{y^4}$ c. xy d. $\dfrac{y^4}{x^3}$
    e. none of the above

14. Write 73,000,000 in scientific notation.
    a. 7.3×10^7 b. 7.3×10^{-7} c. 73×10^6
    d. 0.73×10^9 e. none of the above

15. If $f(x) = 2x^2 + 3x - 4$, find $f(-2)$.
    a. 2   b. -2   c. -18   d. 10
    e. none of the above

16. Simplify: $2(y + 3) - 3(y - 2)$
    a. $-y$ b. $5y$ c. $5y + 12$ d. $-y + 12$
    e. none of the above

17. Multiply: $-3x^2 y^2 (2xy^3)$
    a. $6x^3 y^5$ b. $5x^2 y^5$ c. $-6x^3 y^5$ d. $-6xy^5$
    e. none of the above

18. Multiply: $2a^3 b^2 (3a^2 b - 2ab^2)$
    a. $6a^6 b^3 - 4a^4 b^4$ b. $6a^5 b^3 - 4a^4 b^4$
    c. $6a^5 b^3 - 4a^4 b^2$ d. $6a^5 b^4 - 4a^4 b^4$
    e. none of the above

19. Multiply: $(x + 7)(2x - 3)$
    a. $2x^2 - 11x - 21$ b. $2x^2 + 11x + 21$
    c. $2x^2 + 11x - 21$ d. $-2x^2 + 11x - 21$
    e. none of the above

20. Divide: $(x^2 + 5x - 14) \div (x + 7)$
    a. $x + 2$ b. $x - 2$ c. $x + 1$ d. $x - 1$
    e. none of the above

21. Factor completely: $r^2 h + r^2 a$
    a. $r(rh + ra)$ b. $r^2 h(1 + a)$ c. $r^2 a(h + 1)$
    d. $r^2(h + a)$ e. none of the above

22. Factor completely: $m^2 n^4 - 49$
    a. $(mn^2 + 7)(mn^2 + 7)$ b. $(mn^2 - 7)(mn^2 - 7)$
    c. $(mn + 7)(mn - 7)$ d. $(mn^2 + 7)(mn^2 - 7)$
    e. none of the above

23. One of the factors of $x^2 - 5x + 6$ is
    a. $x + 3$ b. $x + 2$ c. $x - 6$ d. $x - 2$
    e. none of the above

24. One of the factors of $36x^2 + 12x + 1$ is
    a. $6x - 1$ b. $6x + 1$ c. $x + 6$ d. $x - 6$
    e. none of the above

25. One of the factors of $2x^2 + 7xy + 6y^2$ is
    a. $2x + y$ b. $x - 3y$ c. $2x - y$ d. $x + 6y$
    e. none of the above

26. One of the factors of $8x^3 - 27$ is
    a. $2x + 3$ b. $4x^2 - 6x + 9$ c. $4x^2 + 12x + 9$
    d. $4x^2 + 6x + 9$ e. none of the above

27. One of the factors of $2x^2 + 2xy - 3x - 3y$ is
    a. $x - y$ b. $2x + 3$ c. $x - 3$ d. $2x - 3$
    e. none of the above

28. Solve for x: $x^2 + x - 6 = 0$
    a. $x = -2, x = 3$ b. $x = 2, x = -3$
    c. $x = 2, x = 3$ d. $x = -2, x = -3$
    e. none of the above

29. Solve for x: $6x^2 - 7x - 3 = 0$

a. $x = \dfrac{3}{2}, x = \dfrac{1}{3}$ **b.** $x = -\dfrac{3}{2}, x = \dfrac{1}{3}$

c. $x = \dfrac{3}{2}, x = -\dfrac{1}{3}$ **d.** $x = -\dfrac{3}{2}, x = -\dfrac{1}{3}$

e. none of the above

30. Simplify: $\dfrac{x^2 - 16}{x^2 - 8x + 16}$

a. $\dfrac{x + 4}{x - 4}$ **b.** 1   **c.** $\dfrac{x - 4}{x + 4}$ **d.** $-\dfrac{1}{8x}$

e. none of the above

31. Multiply: $\dfrac{x^2 + 11x - 12}{x - 5} \cdot \dfrac{x^2 - 5x}{x - 1}$

a. $-x(x + 12)$ **b.** $x(x + 12)$ **c.** $\dfrac{x^2 + 11x - 12}{(x - 5)(x - 1)}$

d. 1   **e.** none of the above

32. Divide: $\dfrac{t^2 + 7t}{t^2 + 5t} \div \dfrac{t^2 + 4t - 21}{t - 3}$

a. $\dfrac{1}{t + 5}$ **b.** $t + 5$ **c.** 1   **d.** $\dfrac{(t + 7)(t + 7)}{t + 5}$

e. none of the above

33. Simplify: $\dfrac{3x}{2} - \dfrac{x}{4}$

a. $-x$ **b.** $\dfrac{5}{4}$ **c.** $2x$ **d.** $\dfrac{5x}{4}$

e. none of the above

34. Simplify: $\dfrac{a + 3}{2a - 6} - \dfrac{2a + 3}{a^2 - 3a} + \dfrac{3}{4}$

a. $\dfrac{5a + 4}{4a}$ **b.** $\dfrac{4a + 4}{5}$ **c.** 0   **d.** $\dfrac{9 - a}{5a - a^2 - 2}$

e. none of the above

35. Simplify: $\dfrac{x + \dfrac{1}{y}}{\dfrac{1}{x} + y}$

a. $\dfrac{x^2 y + x}{y + xy^2}$ **b.** 1   **c.** $\dfrac{xy + x}{y + xy}$ **d.** $\dfrac{x}{y}$

e. none of the above

36. Solve for x: $\dfrac{1}{x} + \dfrac{1}{2x} = \dfrac{1}{4}$

a. 4   **b.** 5   **c.** 6   **d.** 7   **e.** none of the above

37. Solve for s: $\dfrac{2}{s + 1} + \dfrac{1 - s}{s} = \dfrac{1}{s^2 + s}$

a. 1   **b.** 2   **c.** 3   **d.** 4   **e.** none of the above

38. Find the x-intercept of the graph of $3x - 4y = 12$.

a. $(3, 0)$ **b.** $(-3, 0)$ **c.** $(4, 0)$ **d.** $(-4, 0)$

e. none of the above

39. The graph of $y = -3x + 12$ does not pass through

a. quadrant I   **b.** quadrant II   **c.** quadrant III

d. quadrant IV   **e.** any of the above

40. Find the slope of the line passing through $P(-2, 4)$ and $Q(8, -6)$.

a. -1 **b.** 1   **c.** $-\dfrac{1}{3}$ **d.** 3   **e.** none of the above

41. The equation of the line passing through $P(-2, 4)$ and $Q(8, -6)$ is

a. $y = -x + 2$ **b.** $y = -x + 6$ **c.** $y = -x - 6$

d. $y = -x - 2$ **e.** none of the above

42. If $f(x) = x^2 - 2x$, find $f(a + 2)$.

a. $a^2 - 2a$ **b.** $a^2 + 2a$ **c.** $a + 2$ **d.** $a + 2a^2$

e. none of the above

43. Solve the system for x.

$\begin{cases} 2x - 5y = 5 \\ 3x - 2y = -16 \end{cases}$

a. $x = 8$ **b.** $x = -8$ **c.** $x = 4$ **d.** $x = -4$

e. none of the above

44. Solve the system for y.

$\begin{cases} 8x - y = 29 \\ 2x + y = 11 \end{cases}$

a. $y = -4$ **b.** $y = 4$ **c.** $y = 3$ **d.** $y = -3$

e. none of the above

45. Simplify: $\sqrt{12}$

a. $2\sqrt{3}$ **b.** $4\sqrt{3}$ **c.** $6\sqrt{2}$ **d.** $4\sqrt{2}$

e. none of the above

46. Simplify: $\sqrt{\dfrac{3}{4}}$

a. $\dfrac{\sqrt{3}}{4}$ **b.** $\dfrac{3}{2}$ **c.** $\dfrac{\sqrt{3}}{2}$ **d.** $\dfrac{9}{16}$

e. none of the above

47. Simplify: $\sqrt{75x^3}$

a. $5\sqrt{x^3}$ **b.** $5x\sqrt{x}$ **c.** $x\sqrt{75x}$ **d.** $25x\sqrt{3x}$

e. none of the above

48. Simplify: $3\sqrt{5} - \sqrt{20}$

a. $3\sqrt{-15}$ **b.** $2\sqrt{5}$ **c.** $-\sqrt{5}$ **d.** $\sqrt{5}$

e. none of the above

49. Rationalize the denominator: $\dfrac{11}{\sqrt{11}}$

a. $\dfrac{1}{11}$ **b.** $\sqrt{11}$ **c.** $\dfrac{\sqrt{11}}{11}$ **d.** 1

e. none of the above

50. Rationalize the denominator: $\dfrac{7}{3 - \sqrt{2}}$

a. $3 - \sqrt{2}$ **b.** $7(3 - \sqrt{2})$ **c.** $3 + \sqrt{2}$

d. $7(3 + \sqrt{2})$ **e.** none of the above

51. Solve for x: $\sqrt{\dfrac{3x - 1}{5}} = 2$

a. 7   **b.** 4   **c.** -7 **d.** -4

e. none of the above

52. Solve for n: $3\sqrt{n} - 1 = 1$

a. $\dfrac{2}{3}$ **b.** $\sqrt{\dfrac{2}{3}}$ **c.** $\dfrac{4}{9}$ **d.** $n\sqrt{n + 1}$

e. none of the above

53. Simplify: $(a^6 b^4)^{1/2}$

a. $(ab)^5$ **b.** $a^3 b^2$ **c.** $\dfrac{1}{a^3 b^2}$ **d.** $\dfrac{1}{a^6 b^4}$

e. none of the above

54. Simplify: $\left(\dfrac{8}{125}\right)^{2/3}$

   a. $\dfrac{4}{25}$ **b.** $\dfrac{25}{4}$ **c.** $\dfrac{2}{5}$ **d.** $\dfrac{5}{2}$

   e. none of the above

55. What number must be added to $x^2 + 12x$ to make it a perfect trinomial square?

   a. 6   **b.** 12   **c.** 144   **d.** 36

   e. none of the above

56. Which is the correct quadratic formula?

   a. $x = \dfrac{b \pm \sqrt{b^2 - 4ac}}{2a}$

   b. $x = \dfrac{-b \pm \sqrt{b^2 - 4ac}}{2a}$

   c. $x = \dfrac{-b \pm \sqrt{b^2 + 4ac}}{2a}$

   d. $x = \dfrac{-b \pm \sqrt{b^2 - 4ac}}{2b}$

   e. none of the above

57. One solution of the equation $x^2 - 2x - 2 = 0$ is

   a. $2\sqrt{3}$ **b.** $2 + 2\sqrt{3}$ **c.** $1 - \sqrt{3}$

   d. $2 - 2\sqrt{3}$ **e.** none of the above

58. The vertex of the graph of the equation $y = x^2 - 2x + 1$ is

   a. $(1, 0)$ **b.** $(0, 1)$ **c.** $(-1, 0)$ **d.** $(0, -1)$

   e. none of the above

59. The graph of $y = -x^2 + 2x - 1$

   a. does not intersect the y-axis

   b. does not intersect the x-axis

   c. passes through the origin

   d. passes through $(2, 2)$

   e. none of the above

60. The formula that expresses the sentence "x varies directly with the square of y and inversely with t" is

   a. $x = ky^2 + t$ **b.** $x = \dfrac{ky^2}{t}$ **c.** $x = \dfrac{kt}{y^2}$

   d. $x = ky^2t$ **e.** none of the above

61. Add: $(4 + 5i) + (2 - 4i)$

   a. $6 + 9i$ **b.** $6 - i$ **c.** $6 + i$

   d. $6 - 9i$ **e.** none of the above

62. Subtract: $(3 - i) - (2 + 3i)$

   a. $1 - 4i$ **b.** $1 + 2i$ **c.** $5 - 2i$

   d. $1 - 2i$ **e.** none of the above

63. Multiply: $\left(3 + \sqrt{-4}\right)\left(2 - \sqrt{-9}\right)$

   a. $0 - 5i$ **b.** $0 + 5i$ **c.** $12 + 5i$

   d. $12 - 5i$ **e.** none of the above

64. Rationalize the denominator: $\dfrac{3i}{2 - i}$

   a. $1 + 2i$ **b.** $-\dfrac{3}{5} + \dfrac{6}{5}i$ **c.** $\dfrac{3}{5} + \dfrac{6}{5}i$

   d. $1 - 2i$ **e.** none of the above

n	n^2	$\sqrt{n}$	n^3	$\sqrt[3]{n}$	n	n^2	$\sqrt{n}$	n^3	$\sqrt[3]{n}$
1	1	1.000	1	1.000	51	2,601	7.141	132,651	3.708
2	4	1.414	8	1.260	52	2,704	7.211	140,608	3.733
3	9	1.732	27	1.442	53	2,809	7.280	148,877	3.756
4	16	2.000	64	1.587	54	2,916	7.348	157,464	3.780
5	25	2.236	125	1.710	55	3,025	7.416	166,375	3.803
6	36	2.449	216	1.817	56	3,136	7.483	175,616	3.826
7	49	2.646	343	1.913	57	3,249	7.550	185,193	3.849
8	64	2.828	512	2.000	58	3,364	7.616	195,112	3.871
9	81	3.000	729	2.080	59	3,481	7.681	205,379	3.893
10	100	3.162	1,000	2.154	60	3,600	7.746	216,000	3.915
11	121	3.317	1,331	2.224	61	3,721	7.810	226,981	3.936
12	144	3.464	1,728	2.289	62	3,844	7.874	238,328	3.958
13	169	3.606	2,197	2.351	63	3,969	7.937	250,047	3.979
14	196	3.742	2,744	2.410	64	4,096	8.000	262,144	4.000
15	225	3.873	3,375	2.466	65	4,225	8.062	274,625	4.021
16	256	4.000	4,096	2.520	66	4,356	8.124	287,496	4.041
17	289	4.123	4,913	2.571	67	4,489	8.185	300,763	4.062
18	324	4.243	5,832	2.621	68	4,624	8.246	314,432	4.082
19	361	4.359	6,859	2.668	69	4,761	8.307	328,509	4.102
20	400	4.472	8,000	2.714	70	4,900	8.367	343,000	4.121
21	441	4.583	9,261	2.759	71	5,041	8.426	357,911	4.141
22	484	4.690	10,648	2.802	72	5,184	8.485	373,248	4.160
23	529	4.796	12,167	2.844	73	5,329	8.544	389,017	4.179
24	576	4.899	13,824	2.884	74	5,476	8.602	405,224	4.198
25	625	5.000	15,625	2.924	75	5,625	8.660	421,875	4.217
26	676	5.099	17,576	2.962	76	5,776	8.718	438,976	4.236
27	729	5.196	19,683	3.000	77	5,929	8.775	456,533	4.254
28	784	5.292	21,952	3.037	78	6,084	8.832	474,552	4.273
29	841	5.385	24,389	3.072	79	6,241	8.888	493,039	4.291
30	900	5.477	27,000	3.107	80	6,400	8.944	512,000	4.309
31	961	5.568	29,791	3.141	81	6,561	9.000	531,441	4.327
32	1,024	5.657	32,768	3.175	82	6,724	9.055	551,368	4.344
33	1,089	5.745	35,937	3.208	83	6,889	9.110	571,787	4.362
34	1,156	5.831	39,304	3.240	84	7,056	9.165	592,704	4.380
35	1,225	5.916	42,875	3.271	85	7,225	9.220	614,125	4.397
36	1,296	6.000	46,656	3.302	86	7,396	9.274	636,056	4.414
37	1,369	6.083	50,653	3.332	87	7,569	9.327	658,503	4.431
38	1,444	6.164	54,872	3.362	88	7,744	9.381	681,472	4.448
39	1,521	6.245	59,319	3.391	89	7,921	9.434	704,969	4.465
40	1,600	6.325	64,000	3.420	90	8,100	9.487	729,000	4.481
41	1,681	6.403	68,921	3.448	91	8,281	9.539	753,571	4.498
42	1,764	6.481	74,088	3.476	92	8,464	9.592	778,688	4.514
43	1,849	6.557	79,507	3.503	93	8,649	9.644	804,357	4.531
44	1,936	6.633	85,184	3.530	94	8,836	9.695	830,584	4.547
45	2,025	6.708	91,125	3.557	95	9,025	9.747	857,375	4.563
46	2,116	6.782	97,336	3.583	96	9,216	9.798	884,736	4.579
47	2,209	6.856	103,823	3.609	97	9,409	9.849	912,673	4.595
48	2,304	6.928	110,592	3.634	98	9,604	9.899	941,192	4.610
49	2,401	7.000	117,649	3.659	99	9,801	9.950	970,299	4.626
50	2,500	7.071	125,000	3.684	100	10,000	10.000	1,000,000	4.642

APPENDIX 5: Answers to Selected Exercises

Getting Ready (page 3)

1. 1, 2, 3, etc.　**2.** $\frac{1}{2}, \frac{2}{3}$, etc.　**3.** $-3, -21$, etc.

Exercises 1.1 (page 11)

11. -7　**13.** set　**15.** whole　**17.** integers　**19.** subset
21. rational　**23.** real　**25.** natural, prime　**27.** odd
29. $<$　**31.** variables　**33.** 7　**35.** parenthesis,
open　**37.** distance, 6　**39.** 1, 2, 6, 9　**41.** 1, 2, 6, 9
43. $-3, -1, 0, 1, 2, 6, 9$　**45.** $-3, -\frac{1}{2}, -1, 0, 1, 2, \frac{5}{3}, \sqrt{7}$,
3.25, 6, 9　**47.** $-3, -1, 1, 9$　**49.** 6, 9　**51.** $<$
53. $>$　**55.** $>$　**57.** $>$
59. ; 4, 4　**61.** ; 11, 11
63. ; $-2, -2$　**65.** ; 8, 8
67.
69.　**71.**
73.
75. 36　**77.** 0　**79.** -23　**81.** 8　**83.** 9; natural,
odd, composite, and whole number　**85.** 0; even integer, whole
number　**87.** 24; natural, even, composite, and whole number
89. 3; natural, odd, prime, and whole number　**91.** $<$
93. $=$　**95.** $<$　**97.** $>$　**99.** $=$　**101.** $9 > 4$
103. $8 \le 8$　**105.** $3 + 4 = 7$　**107.** $\sqrt{2} \approx 1.41$
109. $7 \ge 3$　**111.** $0 < 6$　**113.** $8 < 3 + 8$
115. $10 - 4 > 6 - 2$　**117.** $3 \cdot 4 > 2 \cdot 3$　**119.** $\frac{24}{6} > \frac{12}{4}$
121.
123.
125.　**127.** 2

Getting Ready (page 13)

1. 250　**2.** 148　**3.** 16,606　**4.** 105

Exercises 1.2 (page 25)

1. 3　**3.** 6　**5.** $\frac{3}{8}$　**7.** $\frac{9}{16}$　**9.** $\frac{11}{9}$　**11.** $\frac{1}{6}$
13. 5.72　**15.** 0.5　**17.** 5.17　**19.** true　**21.** false
23. false　**25.** true　**27.** $=$　**29.** $=$　**31.** numerator
33. undefined　**35.** prime　**37.** improper　**39.** 1
41. multiply　**43.** numerators, denominator　**45.** least
common denominator, equivalent　**47.** terminating, 2
49. divisor, dividend, quotient　**51.** $2 \cdot 2 \cdot 2 \cdot 3$
53. $2 \cdot 2 \cdot 2 \cdot 2 \cdot 3$　**55.** $\frac{1}{2}$　**57.** $\frac{3}{4}$　**59.** $\frac{3}{2}$　**61.** $\frac{9}{8}$

63. $\frac{2}{15}$　**65.** $\frac{8}{5}$　**67.** 10　**69.** $\frac{20}{3}$　**71.** $\frac{4}{15}$　**73.** $\frac{5}{8}$
75. 24　**77.** $\frac{9}{10}$　**79.** $\frac{6}{5}$　**81.** $\frac{2}{17}$　**83.** $\frac{4}{21}$　**85.** $\frac{22}{35}$
87. $5\frac{1}{5}$　**89.** $1\frac{2}{3}$　**91.** $1\frac{1}{4}$　**93.** $\frac{5}{9}$　**95.** 0.6, terminating
97. $0.40\overline{9}$, repeating　**99.** 359.24　**101.** 44.785
103. 112.32　**105.** 4.55　**107.** 496.26; 496.258
109. 6,025.40; 6,025.398　**111.** $\frac{3}{2}$　**113.** $\frac{1}{4}$　**115.** $\frac{1}{4}$
117. $\frac{14}{5}$　**119.** $\frac{19}{15}$　**121.** $\frac{17}{12}$　**123.** $\frac{9}{4}$　**125.** $\frac{29}{3}$
127. 498.26　**129.** 3,337.52　**131.** 10.02　**133.** 55.21
135. $31\frac{1}{6}$ acres　**137.** 65 yd　**139.** $68.45 million
141. $12,240　**143.** 13,475　**145.** $20,944,000
147. $2,201.95　**149.** $1,170　**151.** the high-capacity boards
153. 205,200 lb　**155.** the high-efficiency furnace

Getting Ready (page 29)

1. 4　**2.** 9　**3.** 27　**4.** 8　**5.** $\frac{1}{4}$　**6.** $\frac{1}{27}$　**7.** $\frac{8}{125}$
8. $\frac{27}{1,000}$

Exercises 1.3 (page 36)

1. 32　**3.** 64　**5.** $\frac{8}{27}$　**7.** y　**9.** $4x$
11.
13. prime number　**15.** exponent　**17.** grouping
19. perimeter, circumference　**21.** $P = 4s$; units
23. $P = 2l + 2w$; units　**25.** $P = a + b + c$; units
27. $P = a + b + c + d$; units　**29.** $C = \pi D$ or $C = 2\pi r$;
units　**31.** $V = lwh$; cubic units　**33.** $V = \frac{1}{3}Bh$; cubic units
35. $V = \frac{4}{3}\pi r^3$; cubic units　**37.** $6 \cdot 6$; 36
39. $\left(-\frac{1}{5}\right)\left(-\frac{1}{5}\right)\left(-\frac{1}{5}\right)\left(-\frac{1}{5}\right); \frac{1}{625}$　**41.** $x \cdot x \cdot x$
43. $8 \cdot z \cdot z \cdot z \cdot z$　**45.** $4x \cdot 4x \cdot 4x$　**47.** $3 \cdot 6y \cdot 6y$
49. 36　**51.** 10,000　**53.** 80　**55.** 216　**57.** 11
59. 3　**61.** 13　**63.** 8　**65.** 17　**67.** 8　**69.** $\frac{1}{144}$
71. 11　**73.** 1　**75.** 1　**77.** 20 in.　**79.** 15 m
81. 36 m^2　**83.** 55 ft^2　**85.** approx. 88 m
87. approx. 1,386 ft^2　**89.** 6 cm^3　**91.** approx. 905 m^3
93. approx. 1,056 cm^3　**95.** 36　**97.** 18　**99.** 36
101. 2　**103.** 16　**105.** 21　**107.** 11　**109.** 8
111. $\frac{8}{9}$　**113.** 493.039　**115.** 640.09　**117.** $(3 \cdot 8) + (5 \cdot 3)$
119. $(3 \cdot 8 + 5) \cdot 3$　**121.** 40,764.51 ft^3　**123.** $121\frac{3}{5}$ m
125. 480 ft^3　**127.** 8　**131.** bigger

Getting Ready (page 39)

1. 17.52　**2.** 2.94　**3.** 2　**4.** 1　**5.** 96　**6.** 382

Exercises 1.4 (page 45)

1. 5　**3.** 3　**5.** 4　**7.** 2　**9.** 20　**11.** 24
13. arrows　**15.** unlike　**17.** subtract, greater

19. add, opposite   **21.** 14   **23.** -9 **25.** $\frac{12}{35}$
27. 78   **29.** 4   **31.** 0.5   **33.** $\frac{5}{12}$ **35.** -34.58
37. 7   **39.** -1 **41.** 0   **43.** -19 **45.** 1   **47.** 2.2
49. 4   **51.** 12   **53.** 5   **55.** $\frac{1}{2}$ **57.** 4   **59.** -7
61. 11   **63.** 12   **65.** 1   **67.** 1   **69.** -1.52
71. -7.08 **73.** 3   **75.** 2.45   **77.** -8 **79.** -7
81. -8 **83.** 3   **85.** 1.3   **87.** $-8\frac{3}{4}$ **89.** -4.2
91. -6 **93.** $-\frac{29}{30}$ **95.** \$235   **97.** $+9$ **99.** $-4°$
101. 2,000 yr   **103.** 1,325 m   **105.** 4,000 ft   **107.** 5°
109. 12,187   **111.** 700   **113.** \$422.66   **115.** \$83,425.57

Getting Ready (page 47)

1. 56   **2.** 54   **3.** 72   **4.** 63   **5.** 9   **6.** 6
7. 8   **8.** 8

Exercises 1.5 (page 52)

1. 3   **3.** 24   **5.** 2   **7.** 9   **9.** 9   **11.** 1,125 lb
13. -45 **15.** positive   **17.** positive   **19.** positive
21. a **23.** 0   **25.** 36   **27.** 56   **29.** -90 **31.** 448
33. -24 **35.** 25   **37.** -64 **39.** 90   **41.** 5
43. -3 **45.** -8 **47.** -10 **49.** -24 **51.** 4
53. -4 **55.** 2   **57.** -4 **59.** -20 **61.** -9
63. -16 **65.** 1   **67.** -4 **69.** 2   **71.** undefined
73. 9   **75.** 88   **77.** 3   **79.** -8 **81.** -96
83. -420 **85.** 49   **87.** -9 **89.** 5   **91.** 9
93. 5   **95.** undefined   **97.** 20   **99.** $-\frac{11}{12}$ **101.** $-\frac{1}{6}$
103. $-\frac{7}{36}$ **105.** $-\frac{25}{144}$ **107.** \$5,100   **109.** $\frac{-18}{-3} = +6$
111. a. $-$ \$2,400   **b.** $-$ \$969   **c.** $-$ \$1,044   **d.** $-$ \$4,413
113. 2-point loss per day   **115.** yes

Getting Ready (page 54)

1. sum   **2.** product   **3.** quotient   **4.** difference
5. quotient   **6.** difference   **7.** product   **8.** sum

Exercises 1.6 (page 59)

1. sum   **3.** product   **5.** quotient   **7.** difference
9. 532   **11.** $\frac{1}{2}$ **13.** sum   **15.** multiplication
17. algebraic   **19.** term, coefficient   **21.** $x + y$
23. $x - 3$ **25.** $(2x)y$ **27.** $3xy$ **29.** $\frac{y}{x}$ **31.** $\frac{3z}{4x}$
33. 3   **35.** 120   **37.** -5 **39.** 60   **41.** undefined
43. 5   **45.** 1; -7 **47.** 3; -1 **49.** 4; -3 **51.** 3; 9
53. 4; 5   **55.** $z + \frac{x}{y}$ **57.** $z - xy$ **59.** $\frac{xy}{x+z}$
61. $\frac{x-4}{3y}$ **63.** the sum of y and 4   **65.** the product of x, y,
and the sum of x and y **67.** the quotient obtained when the
sum of x and 2 is divided by z **69.** the quotient obtained
when y is divided by z **71.** the product of 2, x, and y
73. the quotient obtained when 5 is divided by the sum of x and y
75. $x + z$; 10   **77.** $y - z$; 2   **79.** $yz - 3$; 5   **81.** $\frac{xy}{z}$; 16
83. 19 and x **85.** x **87.** 3, x, y, and z **89.** 17, x, and z
91. 5, 1, and 8   **93.** x and y **95.** 75   **97.** x and y
99. $c + 6$ **101. a.** $h - 20$ **b.** $c + 20$ **103.** \$35,000n
105. $500 - x$ **107.** \$$(3d + 5)$ **109.** 49,995,000

Getting Ready (page 62)

1. 17   **2.** 17   **3.** 38.6   **4.** 38.6   **5.** 56   **6.** 56
7. 0   **8.** 1   **9.** 777   **10.** 777

Exercises 1.7 (page 67)

7. $x + y^2 \geq z$ **9.** $\geq$ **11.** positive   **13.** real   **15.** a
17. $(b + c)$ **19.** ac **21.** a **23.** element, multiplication
25. $a, \frac{1}{a}$; multiplicative   **27.** 10   **29.** -24 **31.** 144
33. 3   **35.** Both are 12.   **37.** Both are 29.   **39.** Both
are 60.   **41.** Both are 175.   **43.** $3x + 15$ **45.** $5z - 20$
47. $-6x - 2y$ **49.** $x^2 + 3x$ **51.** $-ax - bx$
53. $-4x^2 - 4x - 8$ **55.** $-5, \frac{1}{5}$ **57.** $-\frac{1}{3}, 3$ **59.** 0, none
61. $\frac{2}{3}, -\frac{3}{2}$ **63.** $0.2, -5$ **65.** $-\frac{5}{4}, \frac{4}{5}$ **67.** $8x + 16$
69. y^3x **71.** $(y + x)z$ **73.** $x(yz)$ **75.** Both are 0.
77. Both are -6.   **79.** $-6a - 24$ **81.** $-3x^2 + 3xa$
83. comm. prop. of add.   **85.** comm. prop. of mult.
87. distrib. prop.   **89.** comm. prop. of add.   **91.** identity
for mult.   **93.** add. inverse   **95.** identity for add.

Chapter Review (page 71)

1. 1, 2, 3, 4, 5   **2.** 2, 3, 5   **3.** 1, 3, 5   **4.** 4   **5.** $-6, 0, 5$
6. $-6, -\frac{2}{3}, 0, 2.6, 5$ **7.** 5   **8.** all of them   **9.** $-6, 0$
10. 5   **11.** $\sqrt{2}, \pi$ **12.** $-6, -\frac{2}{3}$ **13.** $<$
14. $<$ **15.** $>$ **16.** $=$ **17.** 9   **18.** -8
19. (number line marked 14 15 16 17 18 19 20)
20. (number line marked 19 20 21 22 23 24 25)
21. (number line, -3 to 2)   **22.** (number line, -4 to 3)
23. 5   **24.** 25   **25.** $\frac{5}{3}$ **26.** $\frac{8}{3}$ **27.** $\frac{1}{3}$ **28.** $\frac{1}{3}$
29. 1   **30.** $\frac{5}{2}$ **31.** $\frac{4}{3}$ **32.** $\frac{1}{3}$ **33.** $\frac{9}{20}$ **34.** $\frac{73}{63}$
35. $\frac{11}{21}$ **36.** $\frac{2}{15}$ **37.** $8\frac{11}{12}$ **38.** $2\frac{11}{12}$ **39.** 80.19
40. 20.99   **41.** 6.48   **42.** 3.7   **43.** 4.70   **44.** 26.36
45. 3.57   **46.** 3.75   **47.** $66\frac{3}{4}$ acres   **48.** 6.85 hr
49. 85   **50.** 40.2 ft   **51.** 81   **52.** $\frac{4}{9}$ **53.** 0.25
54. 33   **55.** 25   **56.** 49   **57.** $22\frac{3}{4}$ sq ft
58. 15,133.6 ft^3 **59.** 34   **60.** 14   **61.** 8   **62.** 7
63. 98   **64.** 38   **65.** 3   **66.** 15   **67.** 58
68. 4   **69.** 3   **70.** 3   **71.** 22   **72.** 1
73. 24   **74.** -33 **75.** -6.5 **76.** $\frac{1}{3}$ **77.** -12
78. 16   **79.** 1.2   **80.** -3.54 **81.** 19   **82.** 7
83. -5 **84.** -7 **85.** $-\frac{3}{10}$ **86.** 1   **87.** 1
88. $-\frac{1}{7}$ **89.** 40   **90.** 60   **91.** $\frac{1}{4}$ **92.** 1.3875
93. -35 **94.** -105 **95.** $-\frac{2}{3}$ **96.** -17.22
97. 3   **98.** 7   **99.** $\frac{7}{2}$ **100.** 6
101. -6 **102.** 18   **103.** 26   **104.** 7   **105.** 6
106. $\frac{3}{2}$ **107.** xz **108.** $x + 2y$ **109.** $2(x + y)$
110. $x - yz$ **111.** the product of 5, x, and z
112. 5 decreased by the product of y and z **113.** 4 less than
the product of x and y **114.** the sum of x, y, and z, divided
by twice their product   **115.** 1   **116.** -2 **117.** 8
118. 4   **119.** 4   **120.** 6   **121.** 3   **122.** 6
123. 4   **124.** undefined   **125.** -7 **126.** 39

127. undefined   **128.** -2 **129.** 3   **130.** 7
131. 1   **132.** 9   **133.** closure prop. of add.   **134.** comm. prop. of mult.   **135.** assoc. prop. of add.   **136.** distrib. prop.
137. comm. prop. of add.   **138.** assoc. prop. of mult.
139. comm. prop. of add.   **140.** identity for mult.
141. add. inverse   **142.** identity for add.

Chapter 1 Test (page 77)

1. 31, 37, 41, 43, 47   **2.** 2
3.

4.

     5. -17 **6.** 0   **7.** $=$ **8.** $<$
9. $>$ **10.** $=$ **11.** $\frac{13}{20}$ **12.** $\frac{4}{5}$ **13.** $\frac{3}{14}$ **14.** $\frac{9}{2}$
15. -1 **16.** $-\frac{1}{13}$ **17.** 33.3   **18.** 401.63 ft^2 **19.** 64 cm^2
20. 1,539 in.^{3} **21.** -2 **22.** -14 **23.** -4 **24.** 12
25. 5   **26.** -23 **27.** $\frac{xy}{x+y}$ **28.** $5y - (x + y)$
29. $24x + 14y$ **30.** $\$(12a + 8b)$ **31.** -5 **32.** 4
33. $3x + 6$ **34.** $-pr + pt$ **35.** 0   **36.** 5
37. comm. prop. of mult.   **38.** distrib. prop.
39. comm. prop. of add.   **40.** mult. inverse prop.

Getting Ready (page 81)

1. -3 **2.** 7   **3.** x **4.** 1   **5.** $\frac{1}{x}$ **6.** 1   **7.** 4
8. 4   **9.** 3

Exercises 2.1 (page 91)

1. addition   **3.** subtraction   **5.** division
7. multiplication   **9.** $\frac{22}{15}$ **11.** $\frac{25}{27}$ **13.** 33   **15.** -317
17. equation, expression   **19.** equivalent   **21.** equal
23. equal   **25.** regular price   **27.** 100   **29.** equation
31. expression   **33.** equation   **35.** expression   **37.** yes
39. no   **41.** yes   **43.** yes   **45.** yes   **47.** yes
49. 15   **51.** -14 **53.** 13   **55.** $\frac{1}{2}$ **57.** 3   **59.** -21
61. -4 **63.** $-\frac{7}{15}$ **65.** 18   **67.** 15   **69.** $\frac{1}{3}$
71. $-\frac{3}{2}$ **73.** 4   **75.** -11 **77.** -9 **79.** 27
81. $\frac{1}{8}$ **83.** 98   **85.** 85   **87.** 4,912   **89.** $\$9,345$
91. $\$90$ **93.** 80   **95.** 19   **97.** 520   **99.** 320
101. 150   **103.** 20%   **105.** 3.3   **107.** -60 **109.** -28
111. -5.58 **113.** $\frac{5}{2}$ **115.** 36   **117.** -33 **119.** $-\frac{1}{3}$
121. $-\frac{1}{12}$ **123.** $-\frac{1}{5}$ **125.** 5   **127.** -3 **129.** $-\frac{1}{2}$
131. 6%   **133.** 200%   **135.** $\$270$ **137.** 6%
139. $\$260$ **141.** 234   **143.** 55%   **145.** 2,760
147. 370   **149.** $\$17,750$ **151.** $\$2.22$ **153.** $\$4.95$
155. $\$145,149$ **159.** about 3.16

Getting Ready (page 94)

1. 22   **2.** 36   **3.** 5   **4.** $\frac{13}{2}$ **5.** -1 **6.** -1
7. $\frac{7}{9}$ **8.** $-\frac{19}{3}$

Exercises 2.2 (page 100)

1. add 9   **3.** add 3   **5.** multiply by 3   **7.** 3   **9.** 50 cm
11. 80.325 in.^{2} **13.** cost   **15.** percent
17. percent of increase   **19.** 1   **21.** -2 **23.** -1

25. 3   **27.** 3   **29.** -6 **31.** -3 **33.** -54
35. -16 **37.** 27   **39.** -33 **41.** 28   **43.** 3
45. 10   **47.** 4   **49.** 10   **51.** -4 **53.** 10   **55.** -4
57. 2   **59.** -5 **61.** $\frac{3}{2}$ **63.** 20   **65.** $\frac{15}{4}$ **67.** $\frac{16}{5}$
69. 11   **71.** -7 **73.** $-\frac{2}{3}$ **75.** 0   **77.** 6   **79.** $\frac{4}{3}$
81. $\$325$ **83.** 6 days   **85.** $\$50$ **87.** 6% to 15%
89. 5   **91.** 3   **93.** 30 min   **95.** $\$7,400$
97. no chance; he needs 112   **101.** $\frac{7x + 4}{22} = \frac{1}{2}$

Getting Ready (page 102)

1. $3x + 4x$ **2.** $7x + 2x$ **3.** $8w - 3w$ **4.** $10y - 4y$
5. $7x$ **6.** $9x$ **7.** $5w$ **8.** $6y$

Exercises 2.3 (page 107)

1. expression   **3.** equation   **5.** equation   **7.** identity
9. contradiction   **11.** 0   **13.** 2   **15.** $\frac{13}{56}$ **17.** $\frac{48}{35}$
19. variables, like, unlike, coefficient   **21.** identity, contradiction
23. $20x$ **25.** $3x^2$ **27.** unlike terms   **29.** $2x + 12$
31. $7z - 15$ **33.** $12x + 121$ **35.** $3y - 15$ **37.** $7x + 2y$
39. -41 **41.** 3   **43.** -2 **45.** 3   **47.** 1   **49.** 2
51. 6   **53.** 12   **55.** -9 **57.** -20 **59.** 9   **61.** 16
63. 5   **65.** 4   **67.** identity, $\mathbb{R}$ **69.** contradiction, $\varnothing$
71. contradiction, $\varnothing$ **73.** identity, $\mathbb{R}$
75. expression, $x - 2y$ **77.** equation, 8
79. expression, $5x + 7$ **81.** equation, -20
83. equation, $\mathbb{R}$ **85.** equation, $\frac{1}{2}$ **87.** equation, 2
89. expression, $5x + 24$ **91.** equation, -3
93. expression, $0.7m + 22.16$ **95.** equation, 1
97. equation, 0   **99.** 0.9   **105.** 0

Getting Ready (page 109)

1. 3   **2.** -5 **3.** r **4.** $-a$ **5.** 7   **6.** 12
7. d **8.** s

Exercises 2.4 (page 113)

1. $a = -\frac{c}{b}$ **3.** $b = ac$ **5.** $3x - 4y$ **7.** $-a - 8$
9. literal   **11.** isolate   **13.** subtract   **15.** $I = \frac{E}{R}$
17. $w = \frac{V}{lh}$ **19.** $y = x - 12$ **21.** $h = \frac{3V}{B}$ **23.** $h = \frac{3V}{\pi r^2}$
25. $x = 2y - 2$ **27.** $B = \frac{2}{5}A - 3$ or $B = \frac{2A - 15}{5}$
29. $q = \frac{2p - hr}{h}$ or $q = \frac{2p}{h} - r$ **31.** $r = \frac{G}{2b} + 1$ or $r = \frac{G + 2b}{2b}$
33. $t = \frac{d}{r}; t = 7$ **35.** $b = P - a - c; b = 16$
37. $y = \frac{5 - 3x}{2}$ **39.** $d = \frac{C}{\pi}$ **41.** $w = \frac{P - 2l}{2}$
43. $t = \frac{A - P}{Pr}$ **45.** $w = \frac{2gK}{v^2}$ **47.** $g = \frac{wv^2}{2K}$
49. $M = \frac{Fd^2}{Gm}$ **51.** $p = \frac{i}{rt}; p = 750$ **53.** $h = \frac{2K}{a + b}; h = 8$
55. $h = \frac{3V}{\pi r^2}; h = 3$ in.   **57.** $I = \frac{E}{R}; I = 4$ amp
59. $R = \frac{P}{I^2}; R = 13.78$ ohms   **61.** $m = \frac{Fd^2}{GM}$
63. $D = \frac{L - 3.25r - 3.25R}{2}; D = 6$ ft   **65.** $C \approx 0.1304T$, about 13% of taxable income   **69.** 90,000,000,000 joules

Getting Ready (page 116)

1. $(x + x + 2)$ ft or $(2x + 2)$ ft   **2.** $(x + 3x)$ ft or $4x$ ft
3. $P = 2l + 2w$

Exercises 2.5 (page 122)

1. $12 - x$ **3.** $0.07(18,000 - x)$ **5.** 200 cm^3
7. $8x + 2$ **9.** $-\frac{3}{2}$ **11.** $1,488$ **13.** $2l + 2w$
15. vertex **17.** degrees **19.** straight **21.** supplementary
23. 4 ft and 8 ft **25.** 5 ft, 10 ft, 15 ft **27.** 6 ft, 8 ft,
10 ft **29.** 14,000 hardcovers, 196,000 paperbacks
31. $10°$ **33.** $47°$ **35.** $159°$ **37.** $87°$ **39.** $44°$
41. $130°$ **43.** 19 cm by 26 cm **45.** 12 in. by 18 in.
47. 19 ft **49.** $20°$ **51.** $6,000 **53.** $4,500 at 9% and
$19,500 at 14% **55.** $3,750 in each account **57.** $5,000
59. 6% and 7%

Getting Ready (page 124)

1. 60 mi **2.** 385 mi **3.** 5.6 gal **4.** 9 lb

Exercises 2.6 (page 130)

1. $60h$ mi **3.** 5 oz **5.** -9 **7.** 2 **9.** 2
11. 6 **13.** $d = rt$ **15.** $v = pn$ **17.** 3 hr **19.** 3.5 days
21. 6.5 hr **23.** $\frac{2}{7}$ hr or approximately 17 min
25. 10 hr **27.** 7.5 hr **29.** 65 mph and 45 mph
31. 4 hr **33.** 2.5 liters **35.** 30 gal **37.** 50 gal
39. 7.5 oz **41.** 40 lb lemon drops and 60 lb jelly
beans **43.** $1.20 **45.** 45 lb **47.** $5.60

Getting Ready (page 132)

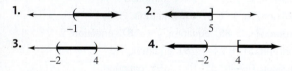

Exercises 2.7 (page 137)

1. same **3.** reverses **5.** same **7.** $5x^2 - 2y^2$
9. $-x + 14$ **11.** is less than; is greater than **13.** double
inequality **15.** inequality **17.** $x > 3$;
19. $x \le -1$; **21.** $x \le 4$;
23. $x > -3$; **25.** $x > -1$;
27. $x \ge 3$; **29.** $x \ge -10$;
31. $x > -3$; **33.** $x < -4$;
35. $x < -2$; **37.** $x > 3$;
39. $x \ge 10$;
41. $7 < x < 10$;
43. $-3 < x \le 5$;
45. $-10 \le x \le 0$;
47. $-4 < x < 1$;

49. $x > -1$; **51.** $x \ge 2$;
53. $x \le 20$; **55.** $x > -7$;
57. $x \ge 4$;
59. $-5 < x < -3$;
61. $-6 \le x \le 10$;
63. $2 \le x < 3$;
65. $-1 \le x < 2$;
67. $-6 < x < 14$;
69. $s \ge 98$ **71.** $s \ge 17$ cm **73.** $r \ge 27$ mpg
75. 0.1 mi $\le x \le 2.5$ mi **77.** 3.3 mi $< x < 4.1$ mi
79. $73.4° < F < 78.8°$ **81.** 37.052 in. $< C < 38.308$ in.
83. 68.18 kg $< w < 86.36$ kg **85.** 5 ft $< w < 9$ ft

Chapter Review (page 141)

1. expression **2.** equation **3.** no **4.** yes **5.** -1
6. -13 **7.** 4 **8.** -15 **9.** -1 **10.** 0 **11.** $-\frac{1}{3}$
12. $-\frac{1}{2}$ **13.** $105.40 **14.** $97.70 **15.** 5 **16.** -2
17. $-\frac{1}{2}$ **18.** $\frac{3}{2}$ **19.** 18 **20.** -35 **21.** $-\frac{1}{2}$ **22.** 6
23. 245 **24.** 1,300 **25.** 37% **26.** 12.5% **27.** 3
28. 2 **29.** 1 **30.** 1 **31.** 1 **32.** -2 **33.** 2
34. 7 **35.** -2 **36.** -1 **37.** 5 **38.** 3 **39.** -6
40. -4 **41.** 8 **42.** 30 **43.** 15 **44.** 4 **45.** $\frac{21}{2}$
46. 44 **47.** $320 **48.** 6.5% **49.** 96.4%
50. 52.8% **51.** $14x$ **52.** $19a$ **53.** $5b$ **54.** $-2x$
55. $-2y$ **56.** not like terms **57.** $9x$ **58.** $6 - 7x$
59. 2 **60.** -8 **61.** 7 **62.** 13 **63.** -3
64. -41 **65.** 9 **66.** -7 **67.** 7 **68.** 4
69. 14 **70.** 39 **71.** identity, $\mathbb{R}$ **72.** contradiction, $\varnothing$
73. identity, $\mathbb{R}$ **74.** contradiction, $\varnothing$ **75.** $R = \frac{E}{I}$
76. $t = \frac{i}{pr}$ **77.** $R = \frac{P}{I^2}$ **78.** $B = \frac{3V}{h}$
79. $c = p - a - b$ **80.** $m = \frac{y - b}{x}$ **81.** $h = \frac{V}{\pi r^2}$
82. $B = \frac{2}{3}A - 4$ or $B = \frac{2A - 12}{3}$ **83.** $G = \frac{Fd^2}{Mm}$
84. $m = \frac{RT}{PV}$ **85.** 5 ft from one end **86.** $15°$ **87.** $45°$
88. $78°$ **89.** $105°$ **90.** 13 in. **91.** $16,000 at 7%,
$11,000 at 9% **92.** 30 min **93.** 2 hr **94.** 9 hr
95. 24 liters **96.** 1 liter **97.** 10 lb of each
98. **99.**
100. **101.**
102. **103.**
104. **105.**
106. 6 ft $< l \le 20$ ft

Chapter 2 Test (page 146)

1. solution   **2.** not a solution   **3.** -36 **4.** 47
5. -12 **6.** -7 **7.** -2 **8.** 1   **9.** 7   **10.** -3
11. $6x - 15$ **12.** $8x - 10$ **13.** -18 **14.** $-36x + 13$
15. $\mathbb{R}$ **16.** $\varnothing$ **17.** -2 **18.** 0   **19.** $t = \frac{d}{r}$
20. $h = \frac{S - 2\pi r^2}{2\pi r}$ **21.** $h = \frac{A}{2\pi r}$ **22.** $y = \frac{-x + 5}{2}$
23. 75°   **24.** 101°   **25.** \$6,000 at 6%, \$4,000 at 5%
26. $\frac{3}{5}$ hr or 36 min   **27.** $7\frac{1}{2}$ liters   **28.** 40 lb
29. 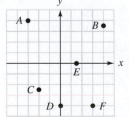   **30.**
31.   **32.**

Cumulative Review (page 147)

1. integer, rational, real   **2.** rational, real
3.
4.   **5.** 0   **6.** $\frac{7}{10}$ **7.** $8\frac{1}{10}$
8. 35.65   **9.** 45   **10.** -2 **11.** 16   **12.** 0
13. 9.9   **14.** 5,275   **15.** 5   **16.** 37, y
17. $-5x + 7y$ **18.** $x - 5$ **19.** $-2x^2 y^3$ **20.** $-9x + 3$
21. -9 **22.** 41   **23.** $\frac{7}{4}$ **24.** $\mathbb{R}$ **25.** $h = \frac{2A}{b + B}$
26. $x = \frac{y - b}{m}$ **27.** \$22,814.56   **28.** \$900   **29.** \$22,690
30. 125 lb   **31.** no   **32.** 7.3 ft and 10.7 ft   **33.** 540 kWh
34. 185 ft   **35.** -9 **36.** 1   **37.** 280   **38.** -564
39.   **40.**

Getting Ready (page 151)

1.
2.   **3.**
4.

Exercises 3.1 (page 158)

1. III   **3.** IV   **5.** 15   **7.** 8   **9.** 7   **11.** -49
13. ordered pair   **15.** origin   **17.** rectangular, Cartesian
19. coordinates   **21.** no   **23.** origin, left, up   **25.** II
27.

29. 2, -2, 4, 3, 0   **31.** 10 min before the workout, her heart rate was 60 beats per min.   **33.** 150 beats per min
35. approximately 5 min and 50 min after starting
37. 10 beats per min faster after cooldown   **39.** \$2
41. \$7

43.

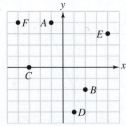

$A(-1, 4), B(2, -2), C(-3, 0),$
$D(1, -4), E(4, 3), F(-4, 4)$
45. 45¢; 85¢   **47.** 40¢
49. Carbondale $\left(5\frac{1}{2}, J\right)$, Champaign $\left(6\frac{1}{2}, D\right)$, Chicago $(8, B)$, Peoria $\left(5\frac{1}{2}, C\right)$, Rockford $\left(5\frac{1}{2}, A\right)$
51. a. 60°; 4 ft   **b.** 30°; 4 ft
53.

Distance (mi) / Gasoline (gal)

a. 30 mi   **b.** 8 gal
c. 32.5 mi

55.

Value (\$1,000s) / Age of car (years)

a. A 3-yr-old car is worth \$7,000.   **b.** \$1,000   **c.** 6 yr

Getting Ready (page 161)

1. 1   **2.** 5   **3.** -3 **4.** 2

Exercises 3.2 (page 171)

1. 2   **3.** 5   **5.** vertical (except $x = 0$)   **7.** -18
9. an expression   **11.** 1.25   **13.** 0.1   **15.** linear, two
17. independent, dependent   **19.** linear
21. y-intercept   **23.** yes   **25.** no
27. $-3, -2, -5, -7; (0, -3), (1, -2), (-2, -5), (-4, -7)$
29. $0, -2, -6, 4; (0, 0), (1, -2), (3, -6), (-2, 4)$
31.   **33.**

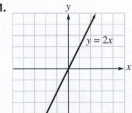

$y = 2x$

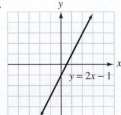

$y = 2x - 1$

35.
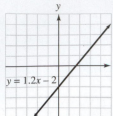
$y = 1.2x - 2$

37.

$y = 2.5x - 5$

39.
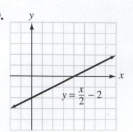
$y = \dfrac{x}{2} - 2$

41.
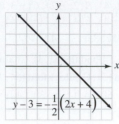
$y - 3 = -\dfrac{1}{2}(2x + 4)$

43.
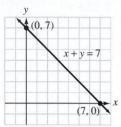
$(0, 7)$
$x + y = 7$
$(7, 0)$

45.
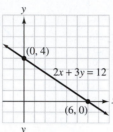
$(0, 4)$
$2x + 3y = 12$
$(6, 0)$

47.
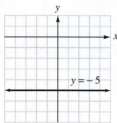
$y = -5$

49.
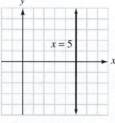
$x = 5$

51.
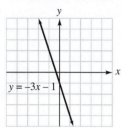
$y = -3x - 1$

53.
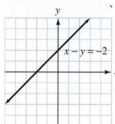
$x - y = -2$

55.
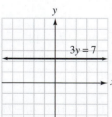
$3y = 7$

57.
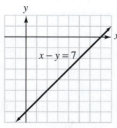
$x - y = 7$

59.
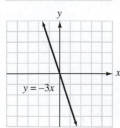
$y = -3x$

61.
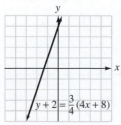
$y + 2 = \dfrac{3}{4}(4x + 8)$

63.
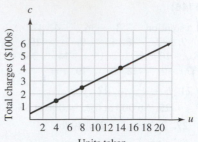
Total charges (\$100s) / Units taken

a. $c = 50 + 25u$ **b.** $150, 250, 400; (4, 150), (8, 250), (14, 400)$
c. The service fee is \$50.   **d.** \$850

65.
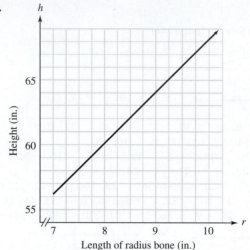
Height (in.) / Length of radius bone (in.)

a. $56.2, 62.1, 64.0; (7, 56.2), (8.5, 62.1), (9, 64.0)$ **b.** taller the
woman is.   **c.** 68 in.   **73.** $(6, 6)$
75. $\left(-\dfrac{1}{2}, \dfrac{5}{2}\right)$ **77.** $(7, 6)$ **79.** $\left(\dfrac{2x - 1}{2}, -\dfrac{1}{2}\right)$

Getting Ready (page 175)

1. 1   **2.** -5 **3.** 0   **4.** undefined

Exercises 3.3 (page 185)

1. $\dfrac{2}{5}$ **3.** undefined   **5.** $6a - 12$ **7.** $-2z - 4w$
9. $a - 5b$ **11.** y, x **13.** rise, run, rise, run
15. hypotenuse   **17.** perpendicular
19. increasing, decreasing   **21.** $\dfrac{2}{3}$ **23.** $\dfrac{4}{3}$ **25.** $-\dfrac{7}{8}$
27. 3   **29.** 3   **31.** 1   **33.** $-\dfrac{1}{3}$ **35.** $\dfrac{13}{5}$
37. $-\dfrac{3}{2}$ **39.** $\dfrac{2}{5}$ **41.** $\dfrac{1}{2}$ **43.** 5   **45.** 0
47. undefined   **49.** 0   **51.** undefined
53. 0   **55.** undefined   **57.** perpendicular
59. neither   **61.** parallel   **63.** perpendicular
65. $-\dfrac{3}{5}$ **67.** -1 **69.** negative   **71.** positive
73. undefined   **75.** not the same line
77. not the same line   **79.** same line   **81.** parallel
83. perpendicular   **85.** neither   **87.** $y = 0, m = 0$
89. $\dfrac{1}{220}$ **91.** $\dfrac{1}{5}$ **93.** 3.5 students per yr
95. \$642.86 per year   **99.** 4

Getting Ready (page 188)

1. $y = -2x + 12$ 2. $y = 3x - 7$ 3. $y = -2x + \frac{9}{2}$
4. $y = \frac{5}{4}x - 3$

Exercises 3.4 (page 192)

1. $2, (3, 5)$ 3. $\frac{7}{8}, (-4, -5)$ 5. 6   7. -1
9. 7,950   11. $y - y_1 = m(x - x_1)$ 13. $(1, 2), 2, 3$
15. $y - 0 = 3(x - 0)$ 17. $y - (-2) = -7[x - (-1)]$
19. $y - 3 = 2[x - (-5)]$ 21. $y - 5 = -\frac{6}{7}(x - 6)$
23. $y = -5x + 7$ 25. $y = 5x + 7$ 27. $y = -3x + 6$
29. $y = \frac{1}{3}x - 4$ 31. $y = \frac{2}{3}x + \frac{11}{3}$ 33. $y = -\frac{1}{2}x - 1$
35. 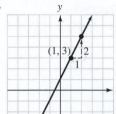   37.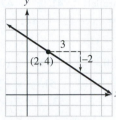
39. $y + 8 = 0.5(x + 1)$ 41. $y - 2 = -4(x + 3)$
43. $y = x$ 45. $y = \frac{7}{3}x - 3$
47. 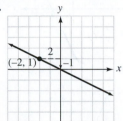   49.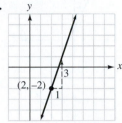
51. $y = -\frac{2}{3}x + 1$ 53. $y = 0.5x + 3$ 55. $y = \frac{3}{2}x + \frac{1}{2}$
57. $206.25   59. $890   61. $17.50   63. $137,200

Getting Ready (page 195)

1. $y = -\frac{1}{2}x + \frac{3}{2}$ 2. $y = \frac{6}{5}x - \frac{7}{5}$

Exercises 3.5 (page 201)

1. $y = -\frac{1}{2}x + 3$ 3. $b = -4$ 5. 2   7. $\frac{7}{2}$
9. $y = mx + b, -3, (0, 7)$ 11. reciprocals   13. $7, (0, -5)$
15. $-\frac{2}{5}, (0, 6)$ 17. $\frac{3}{2}, (0, -4)$ 19. $-\frac{1}{3}, (0, -\frac{5}{6})$
21. $y = 12x$ 23. $y = -5x - 4$ 25. $y = -7x + 54$
27. $y = -5$ 29. $y = -3x - 2$ 31. $y = -\frac{1}{2}x + 6$
33. $1, (0, -1)$ 35. $\frac{2}{3}, (0, 2)$

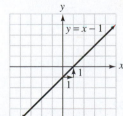   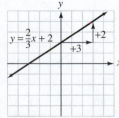

37. $-\frac{2}{3}, (0, 6)$

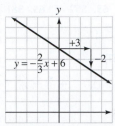

39. parallel   41. perpendicular   43. parallel
45. perpendicular   47. $y = 4x$ 49. $y = 4x - 3$
51. $y = -\frac{1}{4}x$ 53. $y = -\frac{1}{4}x + \frac{11}{2}$ 55. $\frac{7}{2}, (0, 2)$
57. $y = \frac{2}{3}x + 6$ 59. $y = -\frac{4}{3}x + 6$ 61. $y = \frac{4}{3}x - 5$
63. $y = \frac{4}{5}x - \frac{26}{5}$ 65. $y = \frac{3}{4}x - \frac{23}{4}$ 67. $x = -2$
69. $x = 5$ 71. perpendicular   73. perpendicular
75. perpendicular   77. parallel
79. $y = -3,200x + 24,300$ 81. $y = 37,500x + 450,000$
83. $y = -\frac{710}{3}x + 1,900$ 85. $-$120 per year
87. $25   89. about $64,331
91. $372,000; -2,325, y = -2,325x + 465,000$
95. $y = -\frac{A}{B}x + \frac{C}{B}$ 99. $a < 0, b > 0$

Getting Ready (page 204)

1. -1 2. 3   3. -3 4. -5 5. 2   6. -4
7. 5   8. 8

Exercises 3.6 (page 209)

1. 1   3. -1 5. 6   7. -1 9. relation
11. input, function   13. range   15. independent
17. cannot   19. $\{-3, 1, 3\}, \{-1, 2, 7\}$
21. $\{-3, 0, 2, 4\}, \{-8, 0, 5, 7\}$ 23. yes   25. no
27. $-9, 0, 3, -1$ 29. $3, -3, -5, 3$ 31. $22, 7, 2, -\frac{4}{5}$
33. $3, 9, 11, 3$ 35. $3, \frac{3}{2}, 1, 3$
37. 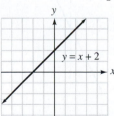   39.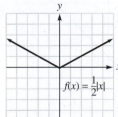

domain: $\mathbb{R}$; range: $\mathbb{R}$ domain: $\mathbb{R}$; range: $\{y \mid y$ is a real number and $y \geq 0\}$

41. yes   43. no   45. $2, 5, 10$ 47. $0, -9, 26$
49. $0, 9, 4$ 51. $1, 16, 21$ 53. $1.5, 9, 1.5$
55. 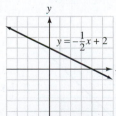   57.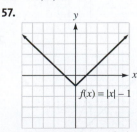

domain: $\mathbb{R}$; range: $\mathbb{R}$ domain: $\mathbb{R}$; range: $\{y \mid y$ is a real number and $y \geq -1\}$

59. $14   **61.** $22   **63.** $57   **65.** $81   **69.** yes
71. yes

Getting Ready (page 211)

1. 4   **2.** 16   **3.** $k = \frac{A}{bh}$ **4.** $k = \frac{PV}{T}$

Exercises 3.7 (page 216)

1. inverse   **3.** combined   **5.** direct   **7.** joint
9. $x > 7$ 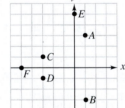   **11.** $x \le 2$
13. direct   **15.** constant   **17.** joint   **19.** $d = kn$
21. $l = \frac{k}{w}$ **23.** $A = kr^2$ **25.** $d = kst$ **27.** $I = \frac{kV}{R}$
29. 33   **31.** 42   **33.** 3   **35.** $\frac{80}{3}$ **37.** 4   **39.** 36
41. 5   **43.** $\frac{28}{15}$ **45.** 180   **47.** 24   **49.** 8
51. 360   **53.** $d = 5.4$ **55.** $p = \frac{405}{4}$ **57.** 576 ft
59. $2,450   **61.** $2\frac{1}{2}$ hr   **63.** $53\frac{1}{3}$ m^3 **65.** $270
67. $3\frac{9}{11}$ amp

Chapter Review (page 220)

1–6.

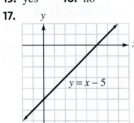

7. $(3, 1)$ **8.** $(-4, 5)$ **9.** $(-3, -4)$ **10.** $(2, -3)$
11. $(0, 0)$ **12.** $(0, 4)$ **13.** $(-5, 0)$ **14.** $(0, -3)$
15. yes   **16.** no

17.   **18.**

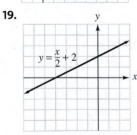

19.   **20.**

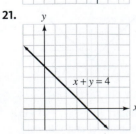

21.   **22.**

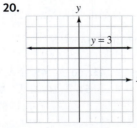

23.   **24.**

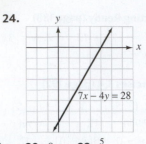

25. $-\frac{3}{2}$ **26.** $-\frac{5}{4}$ **27.** -7 **28.** 0   **29.** $\frac{5}{2}$
30. undefined   **31.** positive   **32.** 0   **33.** undefined
34. negative   **35.** perpendicular   **36.** parallel
37. neither   **38.** neither   **39.** perpendicular
40. parallel   **41.** $\frac{1}{3}$ **42.** $20,500 per year
43. $y = 5x + 13$ **44.** $y = -\frac{1}{3}x + 1$ **45.** $y = \frac{1}{9}x + 1$
46. $y = -\frac{3}{5}x + \frac{2}{5}$ **47.** $-\frac{1}{3}, (0, 6)$
48. $\frac{1}{2}, \left(0, -\frac{3}{2}\right)$ **49.** $-\frac{2}{5}, \left(0, \frac{1}{5}\right)$
50. $-\frac{1}{3}, \left(0, \frac{1}{3}\right)$ **51.** $y = -3x + 2$
52. $y = -7$ **53.** $y = 7x$ **54.** $y = \frac{1}{2}x - \frac{3}{2}$

55.   **56.**

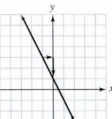

57.   **58.**

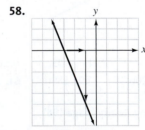

59. neither   **60.** perpendicular   **61.** parallel
62. perpendicular   **63.** $y = 7x - 9$ **64.** $y = -\frac{3}{2}x + \frac{1}{2}$
65. $y = -\frac{5}{2}x$ **66.** $y = -3x - 4$ **67.** $y = -2x + 1$
68. $y = \frac{1}{2}x + 3$ **69.** $1,200
70. domain: $\{-3, 0\}$; range: $\{-2, -1, 5\}$
71. function   **72.** 5   **73.** 15   **74.** 6   **75.** -7
76.

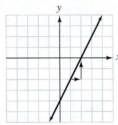

domain: $\mathbb{R}$;
range: $\{y \mid y$ is a real number and $y \ge -3\}$

77. no   **78.** yes
79. 288   **80.** 600   **81.** 81   **82.** $\frac{5}{7}$

Chapter 3 Test (page 225)

1.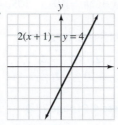

2.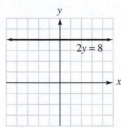

3.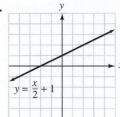

4.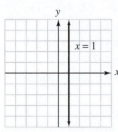

5. $\frac{5}{6}$ **6.** -1 **7.** $\frac{3}{4}$ **8.** $(0, 21)$ **9.** 0

10. undefined **11.** equal **12.** -1

13. perpendicular **14.** parallel **15.** $\frac{1}{4}$

16. $16,666.67 **17.** 2 **18.** $-\frac{1}{2}$ **19.** $y - 5 = 7(x + 2)$

20. $y = \frac{3}{4}x - 5$ **21.** $x = -7$ **22.** $y = -3x + 4$

23. 4 **24.** -11 **25.**

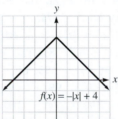

domain: $\mathbb{R}$; range: $\{y \mid y$ is a real number and $y = \ \leq 4\}$

26. no **27.** yes **28.** 1 **29.** 400 **30.** 144

Getting Ready (page 229)

1. 8 **2.** 9 **3.** 6 **4.** 6 **5.** 12 **6.** 32
7. 18 **8.** 3

Exercises 4.1 (page 235)

1. 64 **3.** -343 **5.** 36 **7.** 48

9.

11. the product of 3 and the sum of x and y

13. $|2x| + 3$ **15.** base, $-5, 3$ **17.** $(4y)(4y)(4y)$

19. $y \cdot y \cdot y \cdot y \cdot y$ **21.** $x^n y^n$ **23.** a^{bc}

27. base x, exponent 4 **29.** base 7, exponent 2

31. base $2y$, exponent 3 **33.** base x, exponent 4

35. base x, exponent 1 **37.** base x, exponent 3

39. -25 **41.** 13 **43.** -84 **45.** -55 **47.** $5 \cdot 5 \cdot 5$

49. $-5 \cdot x \cdot x \cdot x \cdot x \cdot x$ **51.** $-2 \cdot y \cdot y \cdot y \cdot y$

53. $(3t)(3t)(3t)(3t)(3t)$ **55.** 2^3 **57.** x^4 **59.** $(2x)^3$

61. $-4t^4$ **63.** x^7 **65.** x^{10} **67.** a^{12} **69.** y^9

71. $12x^7$ **73.** $-10x^5$ **75.** 3^8 **77.** y^{15} **79.** x^{25}

81. a^{27} **83.** x^{31} **85.** r^{36} **87.** $x^3 y^3$ **89.** $r^6 s^4$

91. $16a^2 b^4$ **93.** $-8r^6 s^9 t^3$ **95.** $\frac{a^3}{b^3}$ **97.** $\frac{16x^4}{81y^8}$

99. x^2 **101.** y^4 **103.** $3a$ **105.** ab^4 **107.** t^3

109. $6x^9$ **111.** $-8a^{15}$ **113.** $243z^{30}$ **115.** s^{33}

117. $\frac{-32a^5}{b^5}$ **119.** $\frac{b^6}{27a^3}$ **121.** $\frac{x^{12}y^{16}}{2}$ **123.** $\frac{y^3}{8}$ **125.** $-\frac{8r^3}{27}$

127. $\frac{10r^{13}s^3}{3}$ **129.** 2 ft **131.** $16,000 **133.** $45,947.93

Getting Ready (page 237)

1. $\frac{1}{3}$ **2.** $\frac{1}{y}$ **3.** 1 **4.** $\frac{1}{xy}$

Exercises 4.2 (page 240)

1. x **3.** $(5x)$ **5.** $\frac{1}{a}$ **7.** $\frac{a}{b}$ **9.** 2 **11.** $\frac{6}{5}$

13. $s = \frac{f(P - L)}{i}$ or $s = \frac{fP - fL}{i}$ **15.** $1, \frac{1}{x^n}$ **17.** $\frac{1}{8^2}$

19. 1 **21.** 5 **23.** 1 **25.** 1 **27.** 1 **29.** -2

31. $\frac{1}{625}$ **33.** $\frac{1}{a^5}$ **35.** $\frac{1}{16y^4}$ **37.** $-\frac{1}{125p^3}$ **39.** $\frac{1}{y^{12}}$

41. $\frac{-5}{x^4}$ **43.** $\frac{1}{y}$ **45.** $\frac{1}{a^5}$ **47.** x^5 **49.** $\frac{5b^4}{a}$

51. $\frac{1}{a^{24}}$ **53.** $\frac{1}{b^{32}}$ **55.** x^{3m} **57.** $\frac{1}{x^{3n}}$ **59.** y^{2m+2}

61. x^{3n+12} **63.** u^{5m} **65.** y^{4n-8} **67.** 8 **69.** 1

71. 8 **73.** 512 **75.** $\frac{1}{a^9}$ **77.** y^m **79.** $\frac{1}{64t^3}$

81. $\frac{1}{a^3 b^6}$ **83.** $\frac{1}{x^4 y^2}$ **85.** b^2 **87.** $\frac{1}{m^9 n^{12}}$ **89.** $\frac{1}{x^2}$

91. $\frac{1}{a^2 b^4}$ **93.** $-\frac{y^{10}}{32x^{15}}$ **95.** $a^8 b^{12}$ **97.** $\frac{1}{b^{14}}$

99. $\frac{1}{9a^2 b^2}$ **101.** $\frac{c^{15}}{216a^9 b^3}$ **103.** $-\frac{27}{r^9}$ **105.** $\frac{16u^4 v^8}{81}$

107. $\frac{1}{x^{3n}}$ **109.** $6,678.04 **111.** $95,060.40

Getting Ready (page 242)

1. 100 **2.** $1,000$ **3.** 10 **4.** $\frac{1}{100}$ **5.** 500
6. $8,000$ **7.** 30 **8.** $\frac{7}{100}$

Exercises 4.3 (page 247)

1. 3.9 **3.** 8.37 **5.** 1.052 **7.** 3

9. comm. prop. of multiplication **11.** 7

13. scientific notation **15.** 4.5×10^5 **17.** 1.7×10^6

19. 5.9×10^{-3} **21.** 2.75×10^{-6} **23.** 4.25×10^3

25. 3.7×10^{-5} **27.** 230 **29.** $812,000$

31. 0.00115 **33.** 0.000976 **35.** $714,000$

37. $30,000$ **39.** $200,000$ **41.** 0.000075

43. 5.1×10^{-6} **45.** 8.63×10^8 **47.** 4×10^{-22}

49. $370,000,000$ **51.** 0.000032 **53.** 3.72×10^2

55. 4.72×10^3 **57.** 3.72×10^{-1} **59.** 2.57×10^{13} mi

61. $114,000,000$ mi **63.** 6.22×10^{-3} mi

65. 1.9008×10^{11} ft **67.** 1.64512×10^{12};
$1,645,120,000,000$ **69.** 3.3×10^{-1} km/sec

71. X-rays, visible light, infrared

73. 1.5×10^{-4}; 2.5×10^{13}

Getting Ready (page 249)

1. $2x^2y^3$ **2.** $3xy^3$ **3.** $2x^2 + 3y^2$ **4.** $x^3 + y^3$
5. $6x^3y^3$ **6.** $5x^2y^2z^4$ **7.** $5x^2y^2$ **8.** $x^3y^3z^3$

Exercises 4.4 (page 257)

1. $5a^2 + 2b^3$ **3.** $4a^3b^2$ **5.** $x^3 + 3x^2 + x$ **7.** $a^3 + b^3$
9. 8 **11.** 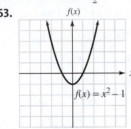 **13.** x^{32} **15.** y^9
17. algebraic **19.** monomial; binomial; trinomial
21. sum **23.** polynomial **25.** cubic **27.** descending; 5
29. function **31.** yes **33.** no **35.** binomial
37. trinomial **39.** monomial **41.** binomial **43.** 7th
45. 3rd **47.** 8th **49.** 6th **51.** -4 **53.** -5
55. $1, -2, -3, -2, 1$ **57.** $-6, 1, 2, 3, 10$
59. -14 **61.** $-\frac{3}{2}$
63. **65.**

D: $\mathbb{R}$, R: $[-1, \infty)$ D: $\mathbb{R}$, R: $\mathbb{R}$

67. 7 **69.** -8 **71.** trinomial **73.** none of these
75. binomial **85.** 12th **87.** 0th **89.** 18 **91.** 11
93. 2.25 **95.** 64 ft **97.** $28,362 **99.** 63 ft

Getting Ready (page 259)

1. $5x$ **2.** $2y$ **3.** $25x$ **4.** $5z$ **5.** $12r$
6. not possible **7.** 0 **8.** not possible

Exercises 4.5 (page 263)

1. $9x$ **3.** $12a$ **5.** unlike **7.** like **9.** -8
11. -3 **13.**
15. monomial **17.** coefficients, variables
19. like terms **21.** $7y$ **23.** unlike terms **25.** $13x^3$
27. $8x^3y^2$ **29.** $10x^4y^2$ **31.** unlike terms **33.** $9y$
35. $25x^2$ **37.** $-4t^6$ **39.** $29x^2y^4$ **41.** $-21a$
43. $16u^3$ **45.** $7x^5y^2$ **47.** $-20ab^3$ **49.** $7x + 4$
51. $8y^2 + 3y - 3$ **53.** $5x^2 + x + 11$
55. $-7x^3 - 7x^2 - x - 1$ **57.** $2a + 7$ **59.** $5a^2 - 2a - 2$
61. $5x^2 + 6x - 8$ **63.** $-x^3 + 6x^2 + x + 14$
65. $-11x - 9y$ **67.** $-2x^2 - 1$ **69.** $6x - 2$
71. $5x^2 - 25x - 20$ **73.** $14rst$ **75.** $-6a^2bc$
77. $-28x^3y^6$ **79.** $216x^5y^{10}$ **81.** $3x - 3y$
83. $-3z^2 + z - 1$ **85.** $-13x^3z + 5x^2z^2 - 14z^3$
87. $13x^3 - 72x^2z + 48xz^2$ **89.** $xy^2 + 13y^2$
91. $6x^2 - 2x - 1$ **93.** $-z^3 - 2z^2 + 5z - 17$
95. $114,000 **97.** $y = 1,900x + 225,000$

99. a. $y = -1,100x + 6,600$ **b.** $y = -1,700x + 9,200$
103. $6x + 3h - 10$ **105.** 49

Getting Ready (page 266)

1. $6x$ **2.** $3x^4$ **3.** $5x^3$ **4.** $8x^5$ **5.** $3x + 15$
6. $-2x - 10$ **7.** $4y - 12$ **8.** $-2y^2 + 6$

Exercises 4.6 (page 273)

1. $15x^3$ **3.** $-8xy$ **5.** $-5x + 20$ **7.** $8x^2 - 18$
9. comm. prop. of add. **11.** comm. prop. of mult.
13. 0 **15.** monomial **17.** special products **19.** $6x^2$
21. $15x$ **23.** $12x^5$ **25.** $-10t^7$ **27.** $6x^5y^5$
29. $-24b^6$ **31.** $3x + 12$ **33.** $-4t - 28$ **35.** $3x^2 - 6x$
37. $-6x^4 + 2x^3$ **39.** $3x^2y + 3xy^2$
41. $-12x^4 - 18x^3 - 30x^2$ **43.** $2x^7 - x^2$
45. $-6r^3t^2 + 2r^2t^3$ **47.** $a^2 + 9a + 20$
49. $3x^2 + 10x - 8$ **51.** $8a^2 - 16a + 6$
53. $6x^2 - 7x - 5$ **55.** $6s^2 + 7st - 3t^2$
57. $u^2 + 2tu + uv + 2tv$ **59.** $2x^2 - 6x - 8$
61. $3a^3 - 3ab^2$ **63.** $3x^2 + xy - 2y^2$
65. $-3x^2 - 11x - 3$ **67.** $x^2 + 10x + 25$
69. $x^2 - 8x + 16$ **71.** $16t^2 + 24t + 9$
73. $x^2 - 4xy + 4y^2$ **75.** $r^2 - 16$ **77.** $16x^2 - 25$
79. $2x^3 + 11x^2 + 10x - 3$ **81.** $4t^3 + 11t^2 + 18t + 9$
83. $4x^2 + 11x + 6$ **85.** $12x^2 + 14xy - 10y^2$ **87.** -3
89. -8 **91.** -1 **93.** 0 **95.** $-3x^4y^7z^8$
97. $2x^2 + 3x - 9$ **99.** $t^2 - 6t + 9$
101. $-4r^2 - 20rs - 21s^2$ **103.** $9x^2 - 12x + 4$
105. $3x^3 + 8x^2y - 6xy^2 + y^3$ **107.** $x^2y + 3xy^2 + 2x^2$
109. -1 **111.** 0 **113.** $2x^2 + xy - y^2$
115. $4x^2 - 5x - 11$ **117.** $-\frac{1}{2}$ **119.** 4 m **121.** 90 ft

Getting Ready (page 275)

1. $2xy^2$ **2.** y **3.** $\frac{3xy}{2}$ **4.** $\frac{x}{y}$ **5.** xy **6.** 3

Exercises 4.7 (page 279)

1. $\frac{1}{7}$ **3.** $-\frac{8}{9}$ **5.** $\frac{1}{6}$ **7.** -1 **9.** binomial
11. none of these **13.** 2 **15.** polynomial **17.** 1
19. $\frac{a}{b}$ **21.** $\frac{x}{z}$ **23.** $\frac{r^2}{s}$ **25.** $\frac{2x^2}{y}$ **27.** $-\frac{3u^3}{v^2}$
29. $\frac{2}{y} + \frac{3}{x}$ **31.** $\frac{x}{3} + \frac{2}{y}$ **33.** $\frac{1}{5y} - \frac{2}{5x}$ **35.** $\frac{1}{y^2} + \frac{2y}{x^2}$
37. $\frac{1}{y} - \frac{1}{2x} + \frac{2z}{xy}$ **39.** $3x^2y - 2x - \frac{1}{y}$ **41.** $5x - 6y + 1$
43. $3a - 2b$ **45.** $\frac{10x^2}{y} - 5x$ **47.** $-\frac{4x}{3} + \frac{3x^2}{2}$
49. $xy - 1$ **51.** 2 **53.** yes **55.** yes **57.** $\frac{3}{4}$
59. $\frac{42}{19}$ **61.** $\frac{4r}{y^2}$ **63.** $-\frac{13}{3rs}$ **65.** $\frac{x^4}{y^6}$ **67.** a^8b^8
69. $\frac{2}{b} + \frac{1}{3a}$ **71.** $\frac{2}{3y} - \frac{1}{3x}$ **73.** $4x - \frac{3y}{2}$
75. $2x^2 + 4x - 1$ **77.** $\frac{xy^2}{3}$ **79.** a^8 **81.** $\frac{x}{y} - \frac{11}{6} + \frac{y}{2x}$

Getting Ready (page 281)

1. 13   **2.** 21   **3.** 19   **4.** 13

Exercises 4.8 (page 285)

1. 24   **3.** $14\frac{5}{19}$ **5.** $x \neq -2$ **7.** $x \neq \frac{7}{2}$

9. 21, 22, 24, 25, 26, 27, 28   **11.** 5   **13.** -5

15. $18x^2 - 2x - 1$ **17.** divisor, dividend   **19.** remainder

21. $2x^3 + 8x^2 + 3x - 5$ **23.** $5x^4 + 6x^3 - 4x^2 + 7x$

25. $0x^3$ and $0x$ **27.** $x + 4$ **29.** $x + 2$ **31.** $x - 3$

33. $a - 4$ **35.** $2a - 1$ **37.** $b + 3$ **39.** $x + 1 + \frac{-1}{2x + 3}$

41. $2x + 2 + \frac{-3}{2x + 1}$ **43.** $a + b$ **45.** $2x - y$

47. $x - 3y$ **49.** $4a + b$ **51.** $2x + 1$ **53.** $x - 7$

55. $x + 4$ **57.** $2x - 3$ **59.** $x^2 + 2x + 4$

61. $x^2 + 2x - 1$ **63.** $x^2 + 2x + 1$

65. $5x^2 - x + 4 + \frac{16}{3x - 4}$ **67.** $2x^2 + 2x + 1$

69. $x^2 + 2x - 1 + \frac{-1}{2x + 3}$ **71.** $2x^2 + 8x + 14 + \frac{31}{x - 2}$

73. $3x + 2y$

Chapter Review (page 289)

1. $(-3x)(-3x)(-3x)(-3x)$ **2.** $\left(\frac{1}{2}pq\right)\left(\frac{1}{2}pq\right)\left(\frac{1}{2}pq\right)$

3. 125   **4.** 243   **5.** 36   **6.** -36 **7.** 13   **8.** 25

9. x^{10} **10.** x^9 **11.** y^{21} **12.** x^{42} **13.** a^3b^3

14. $81x^4$ **15.** b^{12} **16.** $-y^2z^5$ **17.** $256s^3$ **18.** $-3y^6$

19. $8x^{12}y^6$ **20.** $25x^6y^2$ **21.** x^4 **22.** $\frac{x^2}{y^2}$ **23.** $\frac{2y^2}{x^2}$

24. $5yz^4$ **25.** 1   **26.** 1   **27.** 9   **28.** $9x^4$ **29.** $\frac{1}{x^3}$

30. x **31.** y **32.** x^{10} **33.** $\frac{1}{x^2}$ **34.** $\frac{a^6}{b^3}$ **35.** $\frac{1}{x^5}$

36. $\frac{1}{9z^2}$ **37.** 7.28×10^2 **38.** 6.23×10^3

39. 2.75×10^{-2} **40.** 9.42×10^{-3} **41.** 7.73×10^0

42. 7.53×10^5 **43.** 1.8×10^{-4} **44.** 6×10^4

45. 38,700   **46.** 0.0000798   **47.** 2.68   **48.** 57.6

49. 7.39   **50.** 0.000437   **51.** 0.03   **52.** 160

53. 8th, monomial   **54.** 2nd, binomial   **55.** 5th, trinomial

56. 5th, binomial   **57.** 23   **58.** -10 **59.** -4 **60.** 4

61. 402   **62.** 0   **63.** 82   **64.** 0.3405   **65.** -4

66. 12   **67.** 0   **68.** $-\frac{15}{4}$

69.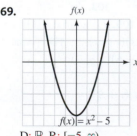
f(x)

$f(x) = x^2 - 5$

D: $\mathbb{R}$, R: $[-5, \infty)$

70.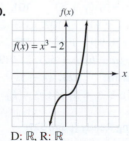
f(x)

$f(x) = x^3 - 2$

D: $\mathbb{R}$, R: $\mathbb{R}$

71. $3x$ **72.** not possible   **73.** $4x^2y^2$ **74.** x^2yz

75. $8x^2 - 6x$ **76.** $4a^2 + 4a - 6$ **77.** $5x^2 + 19x + 3$

78. $6x^3 + 8x^2 + 3x - 72$ **79.** $12x^5y^6$ **80.** x^7yz^5

81. $8x + 12$ **82.** $6x + 12$ **83.** $3x^4 - 5x^2$

84. $2y^4 + 10y^3$ **85.** $-x^2y^3 + x^3y^2$ **86.** $-3x^2y^2 + 3x^2y$

87. $x^2 + 9x + 20$ **88.** $2x^2 - x - 1$ **89.** $6a^2 - 6$

90. $6a^2 - 6$ **91.** $2a^2 - ab - b^2$ **92.** $6x^2 + xy - y^2$

93. $x^2 + 12x + 36$ **94.** $x^2 - 25$ **95.** $y^2 - 49$

96. $x^2 + 8x + 16$ **97.** $x^2 - 6x + 9$ **98.** $y^2 - 4y + 4$

99. $9y^2 + 12y + 4$ **100.** $y^4 - 1$

101. $3x^3 + 7x^2 + 5x + 1$ **102.** $8a^3 - 27$ **103.** 1

104. -1 **105.** 7   **106.** 5   **107.** 1   **108.** 0

109. $\frac{3}{2y} + \frac{3}{x}$ **110.** $3xy - 1$ **111.** $-3a - 4b + 5c$

112. $-\frac{x}{y} - \frac{y}{x}$ **113.** $x + 1 + \frac{3}{x + 2}$ **114.** $x - 5$

115. $3x + 1$ **116.** $x + 5 + \frac{3}{3x - 1}$

117. $3x^2 + 2x + 1 + \frac{2}{2x - 1}$ **118.** $3x^2 - x - 4$

Chapter 4 Test (page 294)

1. $2x^3y^4$ **2.** 134   **3.** y^9 **4.** $6b^7$ **5.** $32x^{21}$

6. $8r^{18}$ **7.** -7 **8.** $\frac{5}{y^3}$ **9.** y^3 **10.** $\frac{64a^3}{b^3}$

11. 5.4×10^5 **12.** 2.5×10^{-3}

13. 7,400   **14.** 0.00067   **15.** binomial

16. 10th degree   **17.** 0

18.

f(x)

$f(x) = x^2 + 2$

D: $(-\infty, \infty)$, R: $[2, \infty)$

19. $-7x + 2y$ **20.** $-3x + 6$

21. $5x^3 + 2x^2 + 2x - 5$

22. $-x^2 - 5x + 4$

23. $-4x^5y$ **24.** $-20x^9y$

25. $6x^2 - 7x - 20$

26. $2x^3 - 7x^2 + 14x - 12$

27. $\frac{y}{2x}$ **28.** $\frac{a}{4b} - \frac{b}{2a}$

29. $x - 2$ **30.** $\frac{1}{2}$

Cumulative Review (page 294)

1. 14   **2.** 71   **3.** $-\frac{11}{10}$ **4.** 7   **5.** 15   **6.** 4

7. -10 **8.** -6 **9.** (number line with open circle at 2, arrow right)

10. (number line with open circle at 2, shaded left)   **11.** (number line, open circles at -2 and 5, shaded outside)

12. (number line, brackets at -2 and 4, shaded between)   **13.** $r = \frac{A - p}{pt}$ **14.** $h = \frac{2A}{b}$

15.

y

$3x - 4y = 12$

16.

y

$y - 2 = \frac{1}{2}(x - 4)$

17. 18   **18.** -7 **19.** -12 **20.** -1 **21.** y^{14}

22. $\frac{x}{y}$ **23.** $\frac{a^7}{b^6}$ **24.** x^2y^2 **25.** $x^2 + 4x - 14$

26. $12x^2 - 7x - 10$ **27.** $x^3 - 8$ **28.** $2x + 1$

29. 4.8×10^{18} m   **30.** 4 units   **31.** 879.6 square units

32. $512

Getting Ready (page 299)

1. $5x + 15$ **2.** $7y - 56$ **3.** $3x^2 - 2x$

4. $5y^2 + 9y$ **5.** $3x + 3y + ax + ay$

6. $xy + x + 5y + 5$ **7.** $5x + 5 - yx - y$

8. $x^2 + 2x - yx - 2y$

Exercises 5.1 (page 305)

1. 3   **3.** 4   **5.** 2   **7.** 6   **9.** 7   **11.** 11
13. prime-factored   **15.** largest   **17.** grouping
19. $2^2 \cdot 3$ **21.** $2^3 \cdot 5$ **23.** $3^2 \cdot 5^2$ **25.** $2^5 \cdot 3^2$
27. $3x$ **29.** $5x$ **31.** $2b$ **33.** $6xy$ **35.** $3a$
37. 4   **39.** 4   **41.** $4, x$ **43.** $3x, x$ **45.** r^2
47. $3(x + 2)$ **49.** $4(x - 2)$ **51.** $3x(2x - 3)$
53. $2b^2(2b - 5)$ **55.** $t^2(t + 2)$ **57.** $5xy^3(2x + 3y)$
59. $a^2b^3z^2(az - 1)$ **61.** $8xy^2z^3(3xyz + 1)$
63. $3(x + y - 2z)$ **65.** $a(b + c - d)$
67. $r(s - t + u)$ **69.** $a^2b^2x^2(1 + a - abx)$
71. $-(x + 2)$ **73.** $-(a + b)$ **75.** $-(2x - 5y)$
77. $-(3ab + 5ac - 9bc)$ **79.** $-3xy(x + 2y)$
81. $-2ab^2c(2ac - 7a + 5c)$ **83.** $(a + b)$ **85.** $(m - n)$
87. $(r - 2s)$ **89.** $(x + 3 - y)$ **91.** $(y + 1)(x - 5)$
93. $(x^2 - 2)(3x - y + 1)$ **95.** $(x + y)(x + y + b)$
97. $(x - 3)(x - 2)$ **99.** $5(2a - 1)(a - 2b)$
101. $3(c - 3d)(x + 2y)$ **103.** $(x + y)(2 + a)$
105. $(p - q)(9 + m)$ **107.** $(3x^2 + 4)(3x + 1)$
109. $(2a^2 - 1)(4a - 1)$ **111.** $(a + b)(x - 1)$
113. $(a - b)(x - y)$ **115.** $r^2(r^2 + 1)$
117. $6uvw^2(2w - 3v)$ **119.** $-7ab(2a^5b^5 - 7ab^2 + 3)$
121. $(a + b + c)(3x - 2y)$ **123.** $7xy(r + 2s - t)(2x - 3)$
125. $(v - 3w)(3t + u)$ **127.** $-2c(b + c)(2a - 1)$
129. $x^2(a + b)(x + 2y)$ **131.** $(y - 3)(y^2 - 5)$
133. $(r - s)(2 + b)$ **135.** $r(r + s)(a - b)$

Getting Ready (page 307)

1. $a^2 - b^2$ **2.** $4r^2 - s^2$ **3.** $9x^2 - 4y^2$ **4.** $16x^4 - 9$

Exercises 5.2 (page 310)

1. $(x - 3)$ **3.** $(z + 2)$ **5.** $(5 - t)$
7. $(9 + y)$ **9.** $(2m - 3n)$ **11.** $(5y^3 + 8x^2)$
13. $p = w\left(k - h - \dfrac{v^2}{2g}\right)$ **15.** difference of two squares
17. prime   **19.** $(x + 6)(x - 6)$ **21.** $(y + 7)(y - 7)$
23. $(2y + 7)(2y - 7)$ **25.** $(5x^2 + 9)(5x^2 - 9)$
27. $(3x + y)(3x - y)$ **29.** $(5t + 6u)(5t - 6u)$
31. $(10a + 7b)(10a - 7b)$ **33.** $(x^2 + 3y)(x^2 - 3y)$
35. $8(x + 2y)(x - 2y)$ **37.** $2(a + 2y)(a - 2y)$
39. $4(5x + 2y)(5x - 2y)$ **41.** $x(x + y)(x - y)$
43. $x(2a + 3b)(2a - 3b)$ **45.** $3m(m + n)(m - n)$
47. $(a^2 + 4)(a + 2)(a - 2)$ **49.** $(a^2 + b^2)(a + b)(a - b)$
51. $2(x^2 + y^2)(x + y)(x - y)$
53. $b(a^2 + b^2)(a + b)(a - b)$
55. $2y(x^2 + 16y^2)(x + 4y)(x - 4y)$
57. $(a + 3)^2(a - 3)$ **59.** $(7y + 15z^2)(7y - 15z^2)$
61. $x^2(2x + y)(2x - y)$ **63.** $(x^2 + 9)(x + 3)(x - 3)$
65. $(4y^4 + 9z^2)(2y^2 + 3z)(2y^2 - 3z)$ **67.** prime
69. $(x^4y^4 + 1)(x^2y^2 + 1)(xy + 1)(xy - 1)$
71. $2ab(a + 11b)(a - 11b)$
73. $3a^2(a^4 + b^2)(a^2 + b)(a^2 - b)$

75. $3a^4(a^2 + 9b^4)(a + 3b^2)(a - 3b^2)$
77. $a^2b^3(b^2 + 25)(b + 5)(b - 5)$ **79.** $3ay(a^4 + 2y^4)$
81. prime   **83.** $2xy(x^8 + y^8)$
85. $(7y + 7 + x)(7y + 7 - x)$ **87.** $(y + 4)(y - 4)(y - 3)$
89. $2c^2d^2(5c + 2d)(5c - 2d)$

Getting Ready (page 312)

1. $x^2 + 12x + 36$ **2.** $y^2 - 14y + 49$ **3.** $a^2 - 6a + 9$
4. $x^2 + 9x + 20$ **5.** $r^2 - 7r + 10$ **6.** $m^2 - 4m - 21$
7. $a^2 + ab - 12b^2$ **8.** $u^2 - 8uv + 15v^2$
9. $x^2 - 2xy - 24y^2$

Exercises 5.3 (page 319)

1. $1, 4$ **3.** $-2, 3$ **5.** $-1, -4$ **7.** $-2, 9$
9. ⟵———(———⟶   **11.** ⟵———————|
            8   -3
13. ⟵———————)⟶   **15.** ⟵———(———|
            17   -2 4
17. $(x + y)^2$ **19.** 3   **21.** $-, 3$ **23.** $6, 1$ **25.** $4, 2$
27. $y, 2y$ **29.** $(x + 4)(x + 1)$ **31.** $(z + 2)(z + 4)$
33. $(x + 3)(x + 5)$ **35.** $(x + 2)(x + 10)$
37. $(t - 7)(t - 2)$ **39.** $(x - 2)(x - 6)$
41. $(x - 4)(x - 5)$ **43.** $(r - 1)(r - 7)$
45. $(q + 9)(q - 1)$ **47.** $(s + 13)(s - 2)$
49. $(c + 5)(c - 1)$ **51.** $(y + 6)(y - 2)$
53. $(b - 6)(b + 1)$ **55.** $(a - 13)(a + 3)$
57. $(m - 5)(m + 2)$ **59.** $(x - 8)(x + 5)$
61. $(m + 7n)(m - 2n)$ **63.** $(a - 6b)(a + 2b)$
65. $(a + 9b)(a + b)$ **67.** $(m - 10n)(m - n)$
69. $-(x + 5)(x + 2)$ **71.** $-(y + 5)(y - 3)$
73. $-(t + 8)(t - 4)$ **75.** $-(r - 10)(r - 4)$
77. prime   **79.** prime   **81.** $2(x + 3)(x + 7)$
83. $3y(y - 6)(y - 1)$ **85.** $3(z - 4t)(z - t)$
87. $-4x(x + 3y)(x - 2y)$ **89.** $(x + 2 + y)(x + 2 - y)$
91. $(b - 3 + c)(b - 3 - c)$ **93.** $(x + 5)(x + 1)$
95. $(t - 7)(t - 2)$ **97.** $(a + 8)(a - 2)$
99. $(y - 6)(y + 5)$ **101.** $(x + 3)^2$ **103.** $(y - 4)^2$
105. $(u - 9)^2$ **107.** $(x + 2y)^2$ **109.** $(x - 4)(x - 1)$
111. $(y + 9)(y + 1)$ **113.** $-(r - 2s)(r + s)$ **115.** prime
117. $-(a + b)(a + 5b)$ **119.** $4y(x + 6)(x - 3)$
121. prime   **123.** $(r - 5s)^2$
125. $(a + 3 + b)(a + 3 - b)$ **127.** $(t + 9)^2$

Getting Ready (page 321)

1. $6x^2 + 7x + 2$ **2.** $6y^2 - 19y + 10$
3. $8t^2 + 6t - 9$ **4.** $4r^2 + 4r - 15$
5. $6m^2 - 13m + 6$ **6.** $16a^2 + 16a + 3$

Exercises 5.4 (page 328)

1. 2   **3.** $2, 1$ **5.** 4   **7.** $n = \dfrac{l - f + d}{d}$
9. the same as   **11.** ac method   **13.** $+, -$ **15.** $3, 1$

17. y, y **19.** $(3a + 1)(a + 2)$ **21.** $(3a + 1)(a + 3)$
23. $(5t + 3)(t + 2)$ **25.** $(4x + 1)(4x + 3)$
27. $(5y - 3)(y - 4)$ **29.** $(2y - 1)(y - 3)$
31. $(8m - 3)(2m - 1)$ **33.** $(3x - 2)(2x - 1)$
35. $(3a + 2)(a - 2)$ **37.** $(4y + 1)(3y - 1)$
39. $(3y - 2)(4y + 1)$ **41.** $(5y + 1)(2y - 1)$
43. $(4q - 1)(2q + 3)$ **45.** $(2x + 5)(5x - 2)$
47. $(5x + 2)(6x - 7)$ **49.** $(2x - 5)(4x + 3)$
51. $(2x + y)(x + y)$ **53.** $(3x - y)(x - y)$
55. $(5p + 2q)(2p - 3q)$ **57.** $(2p + q)(3p - 2q)$
59. $(2a - 5)(4a - 3)$ **61.** $(4x - 5y)(3x - 2y)$
63. $(2x + 1)(3x + 2)$ **65.** $2(3x + 2)(x - 5)$
67. $(3x - 2)^2$ **69.** $(5x + 3)^2$ **71.** $(3a + 4)^2$
73. $(4x - 5)^2$ **75.** $(4x + y + 3)(4x - y - 3)$
77. $(3 + a + 2b)(3 - a - 2b)$ **79.** $-(4y - 3)(3y - 4)$
81. prime   **83.** $2(2x - 1)(x + 3)$
85. $y(y + 12)(y + 1)$ **87.** $3y(3y - 2)(y + 1)$
89. $2s^3(3s + 2)(s - 5)$ **91.** $(4x + 3y)(3x - y)$
93. prime   **95.** $(5x + 2y)^2$ **97.** $-2x^2y^3(8x + y)(x - 2y)$
99. $2(4a + b)(3a + b)$

Getting Ready (page 330)

1. $x^3 - 27$ **2.** $x^3 + 8$ **3.** $y^3 + 64$ **4.** $r^3 - 125$
5. $a^3 - b^3$ **6.** $a^3 + b^3$

Exercises 5.5 (page 333)

1. 2^3 **3.** $(-3)^3$ **5.** $(y^4)^3$ **7.** $(-y^3)^3$ **9.** 9
11. $16y^4$ **13.** $4x^2$ **15.** x^{10} **17.** 0.0000000000001 cm
19. sum of two cubes   **21.** $(x^2 - xy + y^2)$
23. $(a + 2)(a^2 - 2a + 4)$ **25.** $(5x + 2)(25x^2 - 10x + 4)$
27. $(y + 1)(y^2 - y + 1)$ **29.** $(5 + a)(25 - 5a + a^2)$
31. $(m + n)(m^2 - mn + n^2)$ **33.** $(x + y)(x^2 - xy + y^2)$
35. $(2u + w)(4u^2 - 2uw + w^2)$ **37.** $(x - y)(x^2 + xy + y^2)$
39. $(x - 2)(x^2 + 2x + 4)$ **41.** $(s - t)(s^2 + st + t^2)$
43. $(5p - q)(25p^2 + 5pq + q^2)$
45. $(3a - b)(9a^2 + 3ab + b^2)$
47. $2(x + 3)(x^2 - 3x + 9)$ **49.** $-(x - 6)(x^2 + 6x + 36)$
51. $8x(2m - n)(4m^2 + 2mn + n^2)$
53. $xy(x + 6y)(x^2 - 6xy + 36y^2)$
55. $(x + 1)(x^2 - x + 1)(x - 1)(x^2 + x + 1)$
57. $(x^2 + y)(x^4 - x^2y + y^2)(x^2 - y)(x^4 + x^2y + y^2)$
59. $(y + 2)(y^2 - 2y + 4)$ **61.** $(3x + 5)(9x^2 - 15x + 25)$
63. $(4 - z)(16 + 4z + z^2)$
65. $(4x + 3y)(16x^2 - 12xy + 9y^2)$
67. $(x + y)(x^2 - xy + y^2)(3 - z)$
69. $(3y - z)(9y^2 + 3yz + z^2)(x + 5)$
71. $(a + b)(a^2 - ab + b^2)(x - y)$
73. $(y + 1)(y - 1)(y - 3)(y^2 + 3y + 9)$

Getting Ready (page 334)

1. $3ax^2 + 3a^2x$ **2.** $x^2 - 9y^2$ **3.** $x^3 - 8$ **4.** $2x^2 - 8$

5. $x^2 - 3x - 10$ **6.** $6x^2 - 13x + 6$ **7.** $6x^2 - 14x + 4$
8. $ax^2 + bx^2 - ay^2 - by^2$

Exercises 5.6 (page 336)

1. common factor   **3.** sum of two cubes   **5.** none, prime
7. difference of two squares   **9.** $\mathbb{R}$ **11.** -9
13. factors   **15.** binomials   **17.** $3(2x + 1)$
19. $(x + 9)(x + 1)$ **21.** $(4t + 3)(2t - 3)$ **23.** $(t - 1)^2$
25. $2(x + 5)(x - 5)$ **27.** prime
29. $-2x^2(x - 4)(x^2 + 4x + 16)$ **31.** $2t^2(3t - 5)(t + 4)$
33. prime   **35.** $a(6a - 1)(a + 6)$ **37.** $x^2(4 - 5x)^2$
39. $-3x(2x + 7)^2$
41. $8(x - 1)(x^2 + x + 1)(x + 1)(x^2 - x + 1)$
43. $-5x^2(x^3 - x - 5)$ **45.** prime
47. $2a(b + 6)(b - 2)$ **49.** $-4p^2q^3(2pq^4 + 1)$
51. $(2a - b + 3)(2a - b - 3)$
53. $(2a - b)(4a^2 + 2ab + b^2)$
55. $(y^2 - 2)(x + 1)(x - 1)$ **57.** $(a + b + y)(a + b - y)$
59. $(x - 3)(a + b)(a - b)$
61. $(2p^2 - 3q^2)(4p^4 + 6p^2q^2 + 9q^4)$
63. $(5p - 4y)(25p^2 + 20py + 16y^2)$
65. $-x^2y^2z(16x^2 - 24x^3yz^3 + 15yz^6)$
67. $(9p^2 + 4q^2)(3p + 2q)(3p - 2q)$
69. $2(3x + 5y^2)(9x^2 - 15xy^2 + 25y^4)$
71. $(x + y)(x - y)(x + y)(x^2 - xy + y^2)$
73. $2(a + b)(a - b)(c + 2d)$

Getting Ready (page 338)

1. 1   **2.** 13   **3.** 3   **4.** 2

Exercises 5.7 (page 343)

1. $7, 8$ **3.** $2, -3$ **5.** $4, -1$ **7.** $\frac{5}{2}, -2$ **9.** u^9
11. $\frac{a}{b}$ **13.** quadratic   **15.** second   **17.** $0, -7$
19. $1, 1$ **21.** $0, 3$ **23.** $0, -\frac{7}{5}$ **25.** $0, 7$ **27.** $0, -\frac{8}{3}$
29. $5, -5$ **31.** $\frac{2}{3}, -\frac{2}{3}$ **33.** $12, 1$ **35.** $5, -3$
37. $6, -3$ **39.** $5, -4$ **41.** $\frac{1}{2}, -\frac{2}{3}$ **43.** $\frac{1}{2}, 2$
45. $7, -7$ **47.** $\frac{9}{2}, -\frac{9}{2}$ **49.** $-\frac{3}{2}, \frac{2}{3}$ **51.** $\frac{1}{8}, 1$
53. $-4, 5, 7$ **55.** $1, -2, -3$ **57.** $0, -1, -2$
59. $0, 9, -3$ **61.** $0, -3, -\frac{1}{3}$ **63.** $0, -3, -3$
65. $0, 4$ **67.** $0, -\frac{5}{9}$ **69.** $3, -3, -5$ **71.** $-3, 7$
73. $-4, 2$ **75.** $9, -9, -2$ **77.** $\frac{2}{3}, -\frac{1}{5}$ **79.** $8, 1$
81. $-3, -5$ **83.** $-\frac{1}{3}, 3$ **85.** $-\frac{3}{2}, 1$ **87.** $-\frac{1}{7}, -\frac{3}{2}$
89. $2, -5, 4$ **91.** $9, -9$ **93.** $\frac{1}{2}, -\frac{1}{2}$ **95.** $0, \frac{1}{5}, -\frac{3}{2}$

Getting Ready (page 344)

1. s^2 sq in.   **2.** $(2w + 4)$ cm   **3.** $x(x + 1)$
4. $w(w + 3)$ sq in.

Exercises 5.8 (page 348)

1. $A = lw$ **3.** $A = s^2$ **5.** $P = 2l + 2w$ **7.** -10
9. 605 ft^2 **11.** analyze   **13.** 4, 8 or $-4, -8$ **15.** 9 or 1

17. 9 sec   **19.** 4 sec and 10 sec   **21.** 2 sec   **23.** 21 ft
25. 4 m by 9 m   **27.** 48 ft   **29.** $b = 4$ in., $h = 18$ in.
31. 18 sq units   **33.** 1 m   **35.** 3 cm   **37.** 4 cm by 7 cm
39. 845π m^2 **41.** 6, 16 ft

Chapter Review (page 353)

1. $2^3 \cdot 3$ **2.** $3^2 \cdot 5$ **3.** $2^5 \cdot 3$ **4.** $2 \cdot 3 \cdot 17$
5. $3 \cdot 29$ **6.** $3^2 \cdot 11$ **7.** $2 \cdot 5^2 \cdot 41$ **8.** 2^{12}
9. $4(x + 3y)$ **10.** $5a(x^2 + 3)$ **11.** $7x(x + 2)$
12. $3x(3x - 1)$ **13.** $2x(x^2 + 2x - 4)$
14. $-a(x + y - z)$ **15.** $a(x + y - 1)$
16. $xyz(x + y)$ **17.** $(a - b)(x + y)$
18. $(x + y)(x + y + 1)$ **19.** $2x(x + 2)(x + 3)$
20. $5x(a + b)(a + b - 2)$ **21.** $(p + 3q)(3 + a)$
22. $(r - 2s)(a + 7)$ **23.** $(x + a)(x + b)$
24. $(y + 2)(x - 2)$ **25.** $(x + y)(a + b)$
26. $(x - 4)(x^2 + 3)$ **27.** $(x + 5)(x - 5)$
28. $(xy + 4)(xy - 4)$ **29.** $(x + 2 + y)(x + 2 - y)$
30. $(z + x + y)(z - x - y)$ **31.** $2y(x^2 + 9y^2)$
32. $(x + y + z)(x + y - z)$ **33.** $(x + 2)(x + 5)$
34. $(x - 3)(x - 5)$ **35.** $(x + 6)(x - 4)$
36. $(x - 6)(x + 2)$ **37.** $(2x + 1)(x - 3)$
38. $(3x + 1)(x - 5)$ **39.** $(5x + 2)(3x - 1)$
40. $3(2x - 1)(x + 1)$ **41.** $x(x + 3)(6x - 1)$
42. $x(4x + 3)(x - 2)$ **43.** $-2x(x + 2)(2x - 3)$
44. $-4a(a - 3b)(a + 2b)$ **45.** $(c - 5)(c^2 + 5c + 25)$
46. $(d + 2)(d^2 - 2d + 4)$ **47.** $2(x + 3)(x^2 - 3x + 9)$
48. $2ab(b - 1)(b^2 + b + 1)$ **49.** $y(3x - y)(x - 2)$
50. $5(x + 2)(x - 3)$ **51.** $a(a + b)(2x + a)$ **52.** prime
53. $(x + 3)(x - 3 + a)$ **54.** $10(x - 2y)(x^2 + 2xy + 4y^2)$
55. $0, -5$ **56.** $0, 3$ **57.** $0, \frac{2}{3}$ **58.** $0, -5$
59. $7, -7$ **60.** $5, -5$ **61.** $4, 5$ **62.** $-5, 2$
63. $-4, 6$ **64.** $2, 8$ **65.** $3, -\frac{1}{2}$ **66.** $1, -\frac{3}{2}$
67. $\frac{3}{4}, -\frac{3}{4}$ **68.** $\frac{2}{3}, -\frac{2}{3}$ **69.** $0, 3, 4$ **70.** $0, -2, -3$
71. $0, \frac{1}{2}, -3$ **72.** $0, -\frac{2}{3}, 1$ **73.** 5 and 7   **74.** $\frac{1}{3}$
75. 6 ft by 8 ft   **76.** 3 ft by 9 ft   **77.** 3 ft by 6 ft
78. 9 ft

Chapter 5 Test (page 358)

1. $2^3 \cdot 3 \cdot 5$ **2.** $2^2 \cdot 3^3$ **3.** $5a(12b^2c^3 + 6a^2b^2c - 5)$
4. $3x(a + b)(x - 2y)$ **5.** $(x + y)(a + b)$
6. $(x + 8)(x - 8)$ **7.** $2(a + 4b)(a - 4b)$
8. $(4x^2 + 9y^2)(2x + 3y)(2x - 3y)$ **9.** $(x + 6)(x - 1)$
10. $(x - 11)(x + 2)$ **11.** $-1(x + 9y)(x + y)$
12. $6(x - 4y)(x - y)$ **13.** $(3x + 1)(x + 4)$
14. $(2a - 3)(a + 4)$ **15.** prime   **16.** $(4x - 3)(3x - 4)$
17. $6(2a - 3b)(a + 2b)$ **18.** $(x - 2y)(x^2 + 2xy + 4y^2)$
19. $8(3 + a)(9 - 3a + a^2)$
20. $z^3(x^3 - yz)(x^6 + x^3yz + y^2z^2)$ **21.** $0, -10$
22. $-1, -\frac{3}{2}$ **23.** $2, -2$ **24.** $3, -6$ **25.** $\frac{9}{5}, -\frac{1}{2}$
26. $-\frac{9}{10}, 1$ **27.** $\frac{1}{5}, -\frac{9}{2}$ **28.** $-\frac{1}{10}, 9$ **29.** 12 sec
30. 10 m

Getting Ready (page 361)

1. $\frac{3}{4}$ **2.** 2   **3.** $\frac{5}{11}$ **4.** $\frac{1}{2}$

Exercises 6.1 (page 368)

1. $\frac{2}{3}$ **3.** $\frac{3}{4}$ **5.** $\frac{3}{7}$ **7.** $-\frac{1}{3}$
9. $(a + b) + c = a + (b + c)$ **11.** 0   **13.** $\frac{7}{5}$
15. numerator   **17.** 0   **19.** negatives   **21.** $\frac{a}{b}$
23. factor, common   **25.** -4 **27.** $2, -1$
29. $\frac{1}{2}, -7$ **31.** $0, -\frac{1}{3}$ **33.** $\left\{ x \mid x \in \mathbb{R}, x \neq \frac{2}{5} \right\}$
35. $\left\{ m \mid m \in \mathbb{R}, m \neq \frac{3}{2}, -1 \right\}$ **37.** $2x$ **39.** $-5y$ **41.** $\frac{x}{y}$
43. in simplest form   **45.** $\frac{x}{2}$ **47.** $\frac{3}{x - 5}$
49. in simplest form   **51.** $\frac{5}{9}$ **53.** $\frac{1}{3}$ **55.** 5
57. in simplest form   **59.** $3y$ **61.** $\frac{x + 1}{x - 1}$ **63.** $\frac{x - 5}{x + 2}$
65. $\frac{2x}{x - 2}$ **67.** $a - 2$ **69.** $\frac{x - 2}{x + 2}$ **71.** $\frac{2(5x - 4)}{x + 1}$
73. $\frac{4}{3}$ **75.** $x + 1$ **77.** $\frac{x^2 - x + 1}{a + 1}$ **79.** $\frac{b + 2}{b + 1}$
81. -1 **83.** -1 **85.** $\frac{5}{a}$ **87.** $\frac{3x}{y}$ **89.** $\frac{x + 2}{x^2}$
91. $\frac{3}{x}$ **93.** in simplest form   **95.** $\frac{x}{y}$ **97.** $\frac{3x}{5y}$
99. $\frac{x - 3}{5 - x}$ or $-\frac{x - 3}{x - 5}$ **101.** $\frac{x + 11}{x + 3}$ **103.** $\frac{y + 3}{x - 3}$
105. $\frac{2(x + 2)}{x - 1}$ **107.** -1 **109.** $\frac{2}{3}$

Getting Ready (page 370)

1. $\frac{2}{3}$ **2.** $\frac{14}{3}$ **3.** 3   **4.** 6   **5.** $\frac{5}{2}$ **6.** 1
7. $\frac{3}{4}$ **8.** 2

Exercises 6.2 (page 376)

1. 5   **3.** x **5.** 4   **7.** x^2 **9.** $-6x^5y^6z$
11. $\frac{1}{125y^3}$ **13.** $\frac{1}{x^m}$ **15.** $4y^3 + 4y^2 - 8y + 32$
17. numerators, denominators   **19.** $\frac{ac}{bd}$ **21.** $\frac{d}{c}$
23. $\frac{45}{91}$ **25.** $\frac{20x^2}{3y^3}$ **27.** $\frac{(z + 7)(z + 2)}{7z}$ **29.** $\frac{-3a(a - 1)}{5(a + 2)}$
31. $\frac{2y}{3}$ **33.** $\frac{yx}{z}$ **35.** $-2y$ **37.** $\frac{b^3c}{a^4}$ **39.** x
41. $\frac{1}{2y}$ **43.** $3y$ **45.** $\frac{5}{z + 2}$ **47.** z **49.** $x + 2$
51. $\frac{(m - 2)(m - 3)}{2(m + 2)}$ **53.** $\frac{c^2}{ab}$ **55.** $\frac{x - 5}{2}$ **57.** $5x$ **59.** $\frac{4}{3}$
61. $\frac{3}{5}$ **63.** 3   **65.** $\frac{2}{y}$ **67.** $\frac{2}{3x}$ **69.** $\frac{x + 2}{3}$ **71.** $\frac{y - 3}{y^3}$
73. $\frac{x - 3}{x - 5}$ **75.** $\frac{x - 2}{x - 3}$ **77.** 1   **79.** $\frac{3}{x + 1}$ **81.** $\frac{18x}{x - 3}$
83. $\frac{9}{2x}$ **85.** $\frac{y^5}{64}$ **87.** 2   **89.** $\frac{x + 2}{x - 2}$ **91.** $\frac{(x + 1)(x - 1)}{5(x - 3)}$
93. $\frac{2x(1 - x)}{5(x - 2)}$ **95.** $\frac{64z}{3x}$ **97.** $\frac{10x^3}{z}$ **99.** $-\frac{a^3}{2d^2}$
101. $\frac{6}{y}$ **103.** $\frac{z}{y}$ **105.** $\frac{(x + 7)^2}{(x - 3)^2}$ **107.** $\frac{x}{3}$ **109.** $\frac{1}{c - d}$
111. $\frac{-(x - y)(x^2 + xy + y^2)}{(x + y)(w + z)}$ **113.** $-\frac{p}{m + n}$

Getting Ready (page 379)

1. $\frac{4}{5}$ **2.** 1   **3.** $\frac{7}{8}$ **4.** 2   **5.** $\frac{1}{9}$ **6.** $\frac{1}{2}$
7. $-\frac{2}{13}$ **8.** $\frac{13}{10}$

Exercises 6.3 (page 387)

1. equal   **3.** not equal   **5.** equal   **7.** equal

9. 3^4 **11.** $2^3 \cdot 17$ **13.** $2 \cdot 3 \cdot 17$ **15.** $2^4 \cdot 3^2$

17. LCD   **19.** numerators, common denominator   **21.** $\frac{1}{4a}$

23. $\frac{4x}{y}$ **25.** $\frac{9y+2}{y-4}$ **27.** 9   **29.** $-\frac{1}{8}$ **31.** $\frac{y}{x}$

33. $\frac{2(y-2)}{y}$ **35.** $\frac{1}{y}$ **37.** 1   **39.** $\frac{y+4}{y-4}$ **41.** $\frac{4x}{3}$

43. $\frac{6x+2}{x-2}$ **45.** $-\frac{b}{b+1}$ **47.** $\frac{2(x+5)}{x-2}$ **49.** $\frac{84}{32}$

51. $\frac{8xy}{x^2 y}$ **53.** $\frac{4x(x+3)}{(x+3)^2}$ **55.** $\frac{2y(x+1)}{x^2+x}$ **57.** $\frac{z(z+1)}{z^2-1}$

59. $\frac{2(x+2)}{x^2+3x+2}$ **61.** $6x$ **63.** $18xy$ **65.** $(x-2)(x+2)$

67. $x(x+6)$ **69.** $\frac{10x}{3y}$ **71.** $\frac{5x+4}{6x}$ **73.** $\frac{2x^2+2}{(x-1)(x+1)}$

75. $\frac{6x^2-x-2}{(5x+2)(x+2)}$ **77.** $\frac{2xy+2x-2y}{xy}$

79. $\frac{3y^2-3y+x^2+x}{x(y-1)}$ **81.** $\frac{-x^2+3x+5}{x(x+1)}$

83. $\frac{x^2+3x+11}{(x+5)(x+2)}$ **85.** $\frac{x+2}{x-2}$ **87.** $-\frac{2}{x-3}$

89. $\frac{2y+7}{y-1}$ **91.** $\frac{2x+4}{2x-y}$ **93.** $\frac{x}{x-2}$ **95.** $\frac{a+4}{a+2}$

97. $(x-3)(x+2)(x+3)$ **99.** $24y$

101. $\frac{7}{6}$ **103.** $-\frac{1}{6}$ **105.** $\frac{4y-5xy}{10x}$ **107.** $\frac{2x^2-x+4}{x(x+2)}$

109. $-\frac{y^2+5y+14}{2y^2}$ **111.** $\frac{y}{x}$ **113.** $\frac{4x-2y}{y+2}$

115. $\frac{-1}{(a+3)(a-3)}$ **117.** $\frac{14y^2+10}{y^2}$

Getting Ready (page 390)

1. 4   **2.** -18 **3.** 7   **4.** -8 **5.** $3+3x$

6. $2-y$ **7.** $12x-2$ **8.** $3y+2x$

Exercises 6.4 (page 395)

1. $\frac{4}{3}$ **3.** $\frac{9}{4}$ **5.** $\frac{1}{14}$ **7.** $\frac{1}{4}$ **9.** $\frac{5}{4}$ **11.** $\frac{5}{7}$

13. t^8 **15.** $-2r^7$ **17.** $\frac{256r^8}{81}$ **19.** $\frac{r^{10}}{9}$

21. complex fraction   **23.** single, divide   **25.** $\frac{x^2}{2}$

27. $\frac{5t^2}{27}$ **29.** $\frac{a+1}{b}$ **31.** $\frac{y+1}{y-1}$ **33.** $\frac{1+3y}{3-2y}$

35. $\frac{5x+3}{3x+2}$ **37.** $\frac{a+4}{3}$ **39.** $\frac{7x+3}{-x-3}$ **41.** $\frac{3-x}{x-1}$

43. $\frac{xy}{y+x}$ **45.** $\frac{y}{x-2y}$ **47.** $\frac{3y+2x^2}{4y}$ **49.** $\frac{1}{x+2}$

51. $\frac{1}{x+3}$ **53.** $\frac{x-2}{x+3}$ **55.** $\frac{m^2-3m-4}{m^2+5m-3}$ **57.** $\frac{1+x^2}{x+x^2}$

59. $\frac{1+y^2}{1-y^2}$ **61.** $\frac{a^2-a+1}{a^2}$ **63.** 2   **65.** $\frac{x}{x-1}$

67. $\frac{x^2}{(x-1)^2}$ **69.** $\frac{2x(x-3)}{(4x-3)(x+2)}$ **71.** $\frac{(3x+1)(x-1)}{(x+1)^2}$

73. -1 **75.** $\frac{y-5}{y+5}$ **79.** $\frac{1}{2}, \frac{2}{3}, \frac{3}{5}, \frac{5}{8}$

Getting Ready (page 397)

1. $3x+1$ **2.** $8x-1$ **3.** $3+2x$ **4.** $y-6$

5. 19   **6.** $7x+6$ **7.** y **8.** $3x+5$

Exercises 6.5 (page 402)

1. 6   **3.** -7 **5.** -1 **7.** $2(x+5)$

9. $(y-1)(y+2)$ **11.** $(x-3)(x+3)$ **13.** $x(x+8)$

15. $(2x+3)(x-1)$ **17.** $2(x+3)(2x-1)$

19. extraneous   **21.** LCD   **23.** 3

25. -3 **27.** 1   **29.** 5   **31.** $\varnothing$; -2 is extraneous

33. $\varnothing$; 5 is extraneous   **35.** -1 **37.** 5; 3 is extraneous

39. 4   **41.** 2   **43.** $\varnothing$; -2 is extraneous   **45.** 5

47. 2, 4   **49.** -4; 4 is extraneous   **51.** 1

53. $\varnothing$; -2 is extraneous   **55.** 12   **57.** 60   **59.** -2

61. -1 **63.** 6   **65.** 1   **67.** $\varnothing$; 3 is extraneous

69. $a = \frac{b}{b-1}$ **71.** $b = \frac{ad}{d-c}$ **73.** $r_1 = \frac{rr_2}{r_2-r}$

75. $f = \frac{d_1 d_2}{d_1+d_2}$ **79.** $1, -1$

Getting Ready (page 404)

1. $\frac{1}{5}$ **2.** $\$(0.05x)$ **3.** $\$\left(\frac{y}{0.05}\right)$ **4.** $\frac{y}{52}$ hr

Exercises 6.6 (page 408)

1. $i = pr$ **3.** $C = qd$ **5.** $-1, 6$ **7.** $-1, -4, 3$

9. $0, 0, 1$ **11.** $2, -6$ **15.** 2   **17.** 5   **19.** 40 min

21. $2\frac{2}{9}$ hr   **23.** heron, 20 mph; goose, 30 mph   **25.** 150 mph

27. 7% and 8%   **29.** 5%   **31.** 8 hr   **33.** 30

35. 4 mph   **37.** $\frac{2}{3}$ and $\frac{3}{2}$ **39.** 44 mph and 64 mph

41. 10   **43.** 25 mph

Getting Ready (page 410)

1. $\frac{1}{2}$ **2.** $\frac{2}{3}$ **3.** $-\frac{4}{5}$ **4.** $-\frac{5}{9}$

Exercises 6.7 (page 414)

1. 60   **3.** 16   **5.** 36   **7.** 5,280   **9.** 6   **11.** 6

13. $4(x+7)$ **15.** $(2x+3)(x-2)$ **17.** quotient

19. equal   **23.** $\frac{5}{7}$ **25.** $\frac{1}{2}$ **27.** $\frac{2}{3}$ **29.** $\frac{2}{7}$ **31.** $\frac{1}{3}$

33. $\frac{1}{5}$ **35.** $\frac{3}{7}$ **37.** $\frac{3}{4}$ **39.** $\frac{1}{16}$ **41.** $\$1,825$ **43.** $\frac{22}{365}$

45. $\$3.15$/gal   **47.** the 6-oz can   **49.** $\$12.50$/hr

51. 65 mph   **53.** 8.5¢/oz   **55.** the truck

57. the second car   **59.** 440 gal/min   **61.** $\$8,725$

63. $\frac{336}{1,745}$

Getting Ready (page 416)

1. 10   **2.** $\frac{7}{3}$ **3.** $\frac{20}{7}$ **4.** 5   **5.** $\frac{14}{3}$ **6.** 5

7. 21   **8.** $\frac{3}{2}$

Exercises 6.8 (page 424)

1. proportion   **3.** not a proportion   **5.** 30%

7. $\frac{1}{3}$ **9.** 480   **11.** $\$73.50$ **13.** proportion, ratios

15. means   **17.** similar   **19.** ad, bc **21.** triangle

23. no   **25.** yes   **27.** no   **29.** yes   **31.** yes

33. no   **35.** 4   **37.** 9   **39.** 15   **41.** -3.2 **43.** 0

45. -18 **47.** -4 **49.** -2 **51.** $-\frac{3}{2}$ **53.** $\frac{83}{2}$

55. -6 **57.** $\$17$ **59.** 24   **61.** about $4\frac{1}{4}$

63. 26   **65.** 39 ft   **67.** $\$9.10$ **69.** $7\frac{1}{2}$ gal

71. $\$309$ **73.** 65 ft 3 in.   **75.** 288 in. or 24 ft

77. not exactly, but close   **79.** $46\frac{7}{8}$ ft

81. 3,375 ft   **83.** 15,840 ft

Chapter Review (page 428)

1. $-4, 2$ **2.** $2, -3$ **3.** $\frac{8}{9}$ **4.** $-\frac{2}{3}$ **5.** $-\frac{1}{3}$

6. $\frac{7}{3}$ **7.** $\frac{1}{2x}$ **8.** $\frac{5}{2x}$ **9.** $\frac{x}{x+1}$ **10.** $\frac{1}{x}$ **11.** $\frac{3}{y}$

12. 1 **13.** -1 **14.** $\frac{x+7}{x+3}$ **15.** $\frac{x}{x-1}$ **16.** $\frac{a+2}{a+b}$

17. $\frac{2x^2}{3y}$ **18.** $\frac{6}{x^2}$ **19.** 1 **20.** $\frac{2x}{x+1}$ **21.** $\frac{3}{4}$ **22.** $\frac{1}{x}$

23. $x+2$ **24.** 1 **25.** $x+2$ **26.** $\frac{2(3x-4)}{x+3}$

27. -1 **28.** $\frac{x^2+x-1}{x(x-1)}$ **29.** $\frac{x-7}{7x}$ **30.** $\frac{x-2}{x(x+1)}$

31. $\frac{x^2+4x-4}{2x^2}$ **32.** $\frac{x+1}{x}$ **33.** 0 **34.** $\frac{81}{16}$

35. $\frac{3}{2}$ **36.** $\frac{1+x}{1-x}$ **37.** $\frac{x(2x+7)}{3x^2-1}$ **38.** x^2+3

39. $\frac{a(a+bc)}{b(b+ac)}$ **40.** -6 **41.** 4 **42.** 3 **43.** $4, -\frac{3}{2}$

44. -2 **45.** 0 **46.** $r_1 = \frac{rr_2}{r_2-r}$

47. $T_1 = \frac{T_2}{1-E}$ or $T_1 = \frac{-T_2}{E-1}$ **48.** $R = \frac{HB}{B-H}$ or $R = \frac{-HB}{H-B}$

49. $9\frac{9}{19}$ hr **50.** $4\frac{4}{5}$ days **51.** 5 mph **52.** 40 mph

53. $\frac{1}{3}$ **54.** $\frac{4}{5}$ **55.** $\frac{2}{3}$ **56.** $\frac{5}{6}$ **57.** \$2.93/lb

58. 568.75 kWh per week **59.** no **60.** yes

61. $\frac{9}{2}$ **62.** 0 **63.** 7 **64.** 1 **65.** 20 ft

Chapter 6 Test (page 434)

1. $\frac{3x^2}{5y}$ **2.** $\frac{x+1}{2x+3}$ **3.** 3 **4.** $\frac{5y^2}{4t}$ **5.** $\frac{x+1}{3(x-2)}$

6. $\frac{5c}{9a}$ **7.** $\frac{x^2}{3}$ **8.** $x+2$ **9.** $\frac{9x-2}{x-2}$ **10.** $\frac{13}{2y+3}$

11. $\frac{2x^2+x+1}{x(x+1)}$ **12.** $\frac{2x+6}{x-2}$ **13.** $\frac{3x^5}{2}$ **14.** $\frac{x+y}{y-x}$

15. -5 **16.** $\varnothing$ **17.** 4 **18.** $B = \frac{RH}{R-H}$ **19.** $3\frac{15}{16}$ hr

20. 5 mph **21.** 8,050 ft **22.** $\frac{2}{9}$ **23.** yes **24.** $\frac{2}{3}$

25. 45 ft

Cumulative Review (page 434)

1. x^9 **2.** x^{10} **3.** x^3 **4.** 1 **5.** $6x^3 - 2x - 1$

6. $2x^3 + 2x^2 + x - 1$ **7.** $13x^2 - 8x + 1$

8. $16x^2 - 24x + 2$ **9.** $-14x^9y^4$

10. $-35x^5 + 10x^4 + 10x^2$ **11.** $20x^2 + 13x + 2$

12. $15x^2 - 2xy - 8y^2$ **13.** $x+4$ **14.** $x^2 + x + 1$

15. $4xy^2(1 - 3xy)$ **16.** $(a+b)(3+x)$

17. $(a+b)(2+b)$ **18.** $(5p^2 + 4q)(5p^2 - 4q)$

19. $(x-7)(x+2)$ **20.** $(x-3y)(x+2y)$

21. $(3a+4)(2a-5)$ **22.** $(4m+n)(2m-3n)$

23. $(p-3q)(p^2 + 3pq + 9q^2)$

24. $8(r+2s)(r^2 - 2rs + 4s^2)$ **25.** 15 **26.** 4

27. $\frac{2}{3}, -\frac{1}{2}$ **28.** $0, 2$ **29.** $-1, -5$ **30.** $\frac{3}{2}, -4$

31. 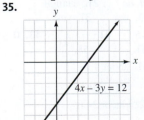 **32.**

33. **34.**

35. 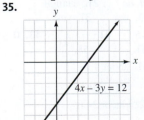 **36.**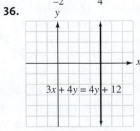

37. -1 **38.** 15 **39.** 5 **40.** $8x^2 - 3$ **41.** $\frac{x+1}{x-2}$

42. $\frac{x-3}{x-2}$ **43.** $\frac{(x-2)^2}{x-1}$ **44.** $\frac{(p+2)(p-3)}{3(p+3)}$ **45.** 1

46. $\frac{2(x^2+1)}{(x+1)(x-1)}$ **47.** $\frac{-1}{2(a-2)}$ **48.** $\frac{y+x}{y-x}$

Getting Ready (page 439)

1. -3 **2.** -2 **3.** 1 **4.** 6

Exercises 7.1 (page 447)

1. yes **3.** yes **5.** no **7.** yes **9.** 81 **11.** -33

13. system **15.** independent **17.** inconsistent; $\varnothing$

19. yes **21.** yes **23.** no **25.** no

27. $(1, 1)$ **29.** $(3, -1)$

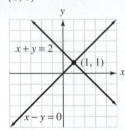 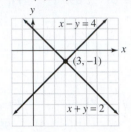

31. $(0, 0)$ **33.** $(-2, -1)$

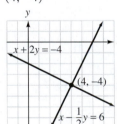 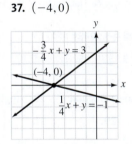

35. $(4, -4)$ **37.** $(-4, 0)$

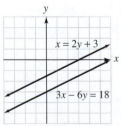 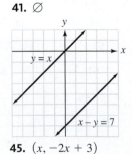

39. $\varnothing$ **41.** $\varnothing$

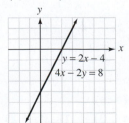 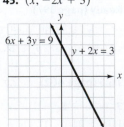

43. $(x, 2x - 4)$ **45.** $(x, -2x + 3)$

47. yes   **49.** no   **51.** yes   **53.** no

55. $(-3, 4)$ **57.** $(3, 0)$

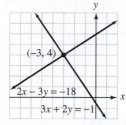

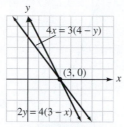

59. $(-2, 4)$ **61.** $\varnothing$

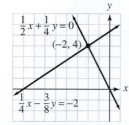

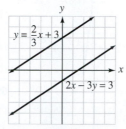

63. $(x, -3x + 2)$ **65.** $\left(\frac{2}{3}, -\frac{1}{3}\right)$ **67.** $(2, 1)$

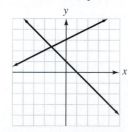

69. a. the incumbent; 7%
b. November 2
c. the challenger; 3 points

71. a. increases
b. decreases
c. $6; 30,000

Getting Ready (page 451)

1. $6x + 4$ **2.** $-25 - 10x$ **3.** $2x - 4$ **4.** $3x - 12$

Exercises 7.2 (page 456)

1. $3x - 3$ **3.** $-x - 1$ **5.** -1 **7.** 12 **9.** 10
11. consistent, independent   **13.** consistent, dependent

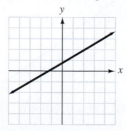

15. y, terms   **17.** $\varnothing$ **19.** infinitely many   **21.** $(2, 4)$
23. $(1, -2)$ **25.** $(-3, -1)$ **27.** $(3, -2)$
29. $(1, -3)$ **31.** $(1, 1)$ **33.** $(4, -2)$ **35.** $(-3, -1)$
37. $(4, 2)$ **39.** $(-5, -5)$ **41.** $\varnothing$ **43.** $\varnothing$
45. $(x, 2x - 5)$ **47.** $\left(a, -\frac{1}{2}a - \frac{5}{2}\right)$ **49.** $\left(\frac{1}{2}, 3\right)$
51. $(-2, 3)$ **53.** $(3, 2)$ **55.** $\left(x, -\frac{1}{3}x + \frac{1}{3}\right)$
57. $(1, 4)$ **59.** $(-1, -1)$ **61.** $\varnothing$ **63.** $(-1, -3)$
65. $\left(\frac{1}{2}, \frac{1}{3}\right)$ **67.** $(-6, 4)$ **69.** $\left(\frac{1}{5}, 4\right)$ **71.** $(5, 5)$
73. $30°, 60°$

Getting Ready (page 458)

1. $5x = 10$ **2.** $y = 6$ **3.** $2x = 33$ **4.** $18y = 28$

Exercises 7.3 (page 463)

1. y **3.** x **5.** 3 **7.** -2 **9.** 4 **11.** 4
13.   **15.** coefficient   **17.** general
19. 15 **21.** $(1, 4)$ **23.** $(-1, 3)$ **25.** $(-1, 1)$
27. $(2, 5)$ **29.** $(7, -2)$ **31.** $(3, -2)$ **33.** $(2, 7)$
35. $\left(\frac{10}{3}, \frac{10}{3}\right)$ **37.** $(-2, 3)$ **39.** $(-5, 0)$ **41.** $(-1, 2)$
43. $(-1, -1)$ **45.** $(-1, 2)$ **47.** $(0, 1)$ **49.** $\varnothing$
51. $\varnothing$ **53.** $(x, 5x - 6)$ **55.** $\left(x, \frac{3}{4}x - \frac{3}{2}\right)$ **57.** $(-3, 4)$
59. $(2, -1)$ **61.** $\left(\frac{1}{2}, \frac{1}{3}\right)$ **63.** $\left(1, -\frac{5}{2}\right)$ **65.** $\left(x, \frac{5}{2}x - 6\right)$
67. $(2, -1)$ **69.** $\left(x, -\frac{2}{3}x + \frac{4}{3}\right)$ **71.** $(-2, 3)$
73. $(2, 2)$ **75.** $x = 3, y = 2$ **79.** $(1, 4)$

Getting Ready (page 465)

1. $x + y$ **2.** $x - y$ **3.** xy **4.** $\frac{x}{y}$ **5.** $A = lw$
6. $P = 2l + 2w$

Exercises 7.4 (page 474)

1. Let p represent the number of pizzas sold. Let c represent the number of calzones sold.   **3.** Let d represent the amount of money Danielle invested. Let k represent the amount of money Kinley invested.   **5.** Let n represent the number of milliliters of the 9% saline solution. Let t represent the number of milliliters of the 20% saline solution.   **7.** Let a represent the cost of one adult ticket. Let s represent the cost of one student ticket.
9. $p + c = 52$; The number of pizzas sold plus the number of calzones sold equals the total number of items sold (52).
11. $d + k = 15,000$; The amount of money Danielle invested plus the amount of money Kinley invested equals the amount of money invested ($15,000).   **13.** $n + t = 25$; The number of milliliters of 9% saline solution plus the number of milliliters of 20% saline solution equals the total number of milliliters needed (25).
15. $a + s = 273$; The number of adult tickets sold plus the number of student tickets sold equals the total number of tickets sold (273).
17. [number line: arrow at 4]   **19.** [number line: -1 to 2]   **21.** 9^4a
23. x^3y^4 **25.** variable   **27.** system   **29.** President: $400,000; Vice President: $192,600   **31.** $245,750
33. 10 ft, 15 ft   **35.** 25 ft by 30 ft   **37.** 135 ft^2
39. 6,720 ft^2 **41.** $2,000   **43.** 250   **45.** 10 mph
47. 50 mph   **49.** 5 L of 40% solution, 10 L of 55% solution
51. 32 lb peanuts, 16 lb cashews   **53.** 9.9 yr   **55.** 250 tires   **57.** cleaner: $5.40; soaking solution: $6.20   **59.** 140
61. causes: 24 min; outcome: 6 min   **63.** 28, 84   **65.** 8, 5
67. 3%   **69.** 15   **71.** paint: $15, brush: $5

Getting Ready (page 478)

1. below   **2.** above   **3.** below   **4.** on
5. on   **6.** above   **7.** below   **8.** above

Exercises 7.5 (page 486)

1. no **3.** yes **5.** above **7.** below **9.** 6

11. $t = \dfrac{A - P}{Pr}$ **13.** $11x - 24$ **15.** $5x + 8y$

17. inequality **19.** boundary **21.** inequalities

23. doubly shaded **25. a.** no **b.** yes **27. a.** yes **b.** no

c. yes **d.** no **29. a.** no **b.** yes **c.** no **d.** yes

31.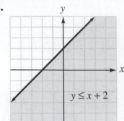
$y \le x + 2$

33.
$y \le 4x$

35.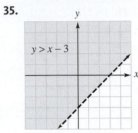
$y > x - 3$

37.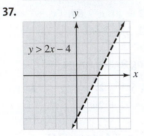
$y > 2x - 4$

39.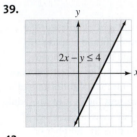
$2x - y \le 4$

41.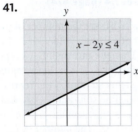
$x - 2y \le 4$

43.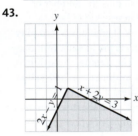
$x + 2y \le 3$ $2x - y = 1$

45.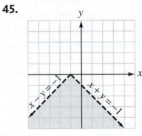
$x - y = -1$ $x + y = -1$

47.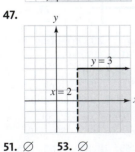
$y = 3$ $x = 2$

49.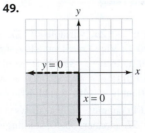
$y = 0$ $x = 0$

51. ∅ **53.** ∅

55.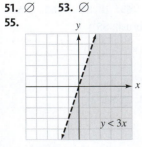
$y < 3x$

57.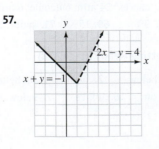
$2x - y = 4$ $x + y = -1$

59.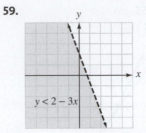
$y < 2 - 3x$

61.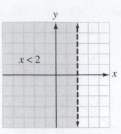
$x < 2$

63.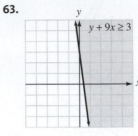
$y + 9x \ge 3$

65.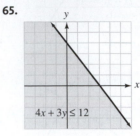
$4x + 3y \le 12$

67.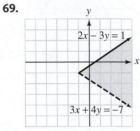
$y \le 1$

69.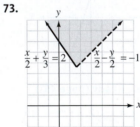
$2x - 3y = 1$ $3x + 4y = -7$

71.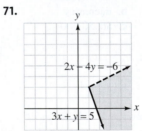
$2x - 4y = -6$ $3x + y = 5$

73.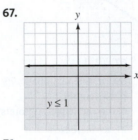
$\dfrac{x}{2} + \dfrac{y}{3} = 2$ $\dfrac{x}{2} - \dfrac{y}{2} = -1$

75. $(10, 10), (20, 10), (10, 20)$

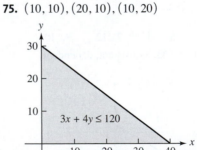
$3x + 4y \le 120$

77. $(50, 50), (30, 40), (40, 40)$

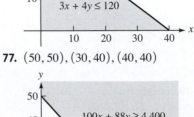
$100x + 88y \ge 4,400$

79. $(80, 40), (80, 80), (120, 40)$

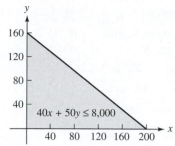

81.   **83.**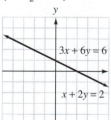

1 \$10 CD and 2 \$15 CDs; 2 desk chairs and 4 side chairs;
4 \$10 CDs and 1 \$15 CD 1 desk chair and 5 side chairs

Chapter Review (page 494)

1. yes **2.** no **3.** yes **4.** yes
5. $(4, 3)$ **6.** $(-3, 0)$

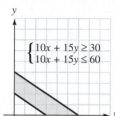 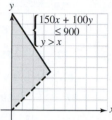

7. $\left(x, -\frac{1}{2}x + 1\right)$ **8.** $\varnothing$

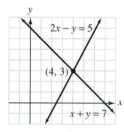 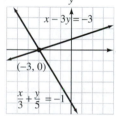

9. $(-3, -1)$ **10.** $(-2, 5)$ **11.** $(1, -1)$ **12.** $(-2, 1)$
13. $(3, -9)$ **14.** $\left(4, -\frac{1}{2}\right)$ **15.** $(-1, 7)$ **16.** $\left(-\frac{1}{2}, \frac{7}{2}\right)$
17. $(0, 9)$ **18.** $\varnothing$ **19.** $\left(x, -3x + \frac{5}{3}\right)$ **20.** $(0, 0)$
21. $21, 63$ **22.** 3 ft by 9 ft **23.** 50¢ **24.** \$66
25. \$1.69 **26.** \$2,000 **27.** 3 mph **28.** 30 milliliters
of 10% saline, 20 milliliters of 60% saline
29. 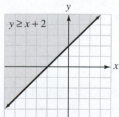 **30.**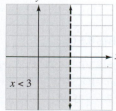

31. 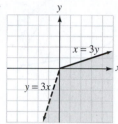 **32.**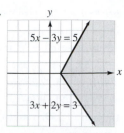

33. 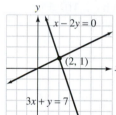 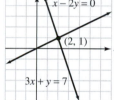 **34.**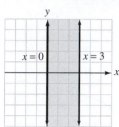

35. 3 shirts and 1 pair of pants; 1 shirt and 2 pairs of pants

Chapter 7 Test (page 498)

1. yes **2.** no
3. $(2, 1)$ **4.** $(-1, 4)$

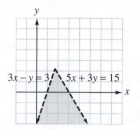

5. $(3, 7)$ **6.** $(12, 10)$ **7.** $(2, 4)$ **8.** $(-3, 3)$
9. inconsistent, $\varnothing$ **10.** consistent, $\left(x, -\frac{1}{4}x + 3\right)$
11. 65 **12.** 3 adult **13.** \$4,000 **14.** 1 mph
15. **16.**

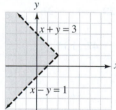 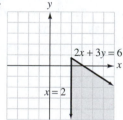

Getting Ready (page 501)

1. 9 **2.** 16 **3.** 8 **4.** 125 **5.** 81 **6.** 256
7. -27 **8.** -32

Exercises 8.1 (page 508)

1. perfect square **3.** not a perfect square
5. not a pefect square **7.** perfect square
9. **11.**

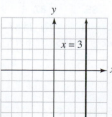 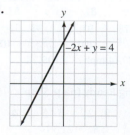

13. $b^2 = a$ **15.** positive   **17.** right
19. integer square, two, 5, -5 **21.** square
23. hypotenuse, $a^2 + b^2$, triples   **25.** 5   **27.** 9   **29.** 6
31. $\frac{1}{3}$ **33.** -2 **35.** -10 **37.** 14   **39.** $\frac{3}{16}$
41. 1.414   **43.** 2.236   **45.** 2.449   **47.** 3.317
49. 4.796   **51.** 9.747   **53.** 80.175   **55.** -99.378
57. rational   **59.** rational   **61.** irrational
63. imaginary
65.

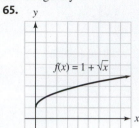

67.

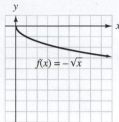

69. 5   **71.** 13   **73.** 20   **75.** 28   **77.** -17
79. 100   **81.** 4.621   **83.** 21   **85.** 0.599   **87.** 0.996
89. -90 **91.** -0.915 **93.** $\frac{24}{25}$ **95.** -2.145
97. 117   **99.** 14 miles   **101.** 25 ft   **103.** 12 ft
105. no   **107.** 4.2 ft   **109.** 127.3 ft   **111.** 5.8 mi
113. 2.4 in.   **115.** 24 in.   **117.** 15 sheets

Getting Ready (page 512)

1. 8   **2.** 64   **3.** -125 **4.** $-\frac{1}{8}$ **5.** 81
6. $\frac{1}{16}$ **7.** 32   **8.** 64

Exercises 8.2 (page 517)

1. 2   **3.** -5 **5.** 3   **7.** 3   **9.** 2   **11.** 9
13. $(x + 7y)(x - 7y)$ **15.** $(a + b)(x + y)$
17. cube root   **19.** nth root   **21.** integer cubes
23. odd   **25.** $b^n = a$ **27.** 3   **29.** -4 **31.** 4
33. 5   **35.** 2   **37.** -4 **39.** $\frac{1}{4}$
41. $-\frac{4}{3}$ **43.** 31.78   **45.** -0.48 **47.** 3.34
49. -5.70
51.

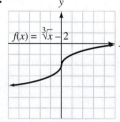

53.

55. 2   **57.** -2 **59.** not a real number   **61.** -2
63. $\frac{3}{2}$ **65.** $-\frac{1}{2}$ **67.** xy **69.** x^2z^2
71. $-5x^2z^6$ **73.** $2z$ **75.** $-xyz^2$ **77.** $2x^2y$
79. xyz **81.** $-2x^3y$ **83.** $-9z^2$ **85.** 2   **87.** $\frac{1}{2}$
89. 1.73   **91.** 9   **93.** $3yz^2$ **95.** -11 **97.** 25.87
99. $6z^{18}$ **101.** $-2p^2q$ **103.** $2a^2b^3c^4$ **105.** 1.26 ft
107. 27.14 mph   **109.** 4

Getting Ready (page 519)

1. 10   **2.** 2   **3.** 5   **4.** 12   **5.** $3x$ **6.** $4x^2$
7. $3xy^2$ **8.** $-2x^2y^3$

Exercises 8.3 (page 524)

1. $9 \cdot 5$ **3.** $x^2 \cdot x$ **5.** $27 \cdot 2$ **7.** $\frac{1}{2xz}$ **9.** $\frac{a+1}{a+3}$
11. perfect   **13.** $\sqrt{a}\sqrt{b}$, multiplication   **17.** $2\sqrt{3}$
19. $3\sqrt{5}$ **21.** $7\sqrt{2}$ **23.** $5\sqrt{6}$ **25.** $3x\sqrt{3}$
27. $4b\sqrt{3b}$ **29.** $8p^2\sqrt{3p}$ **31.** $2r^3\sqrt{22r}$
33. $3x\sqrt{y}$ **35.** $2a^2\sqrt{2b}$ **37.** $48\sqrt{2}$ **39.** $14x^2\sqrt{5y}$
41. $14\sqrt{5}$ **43.** $42xy^2\sqrt{xy}$ **45.** $\frac{5}{3}$ **47.** $\frac{9}{8}$
49. $\frac{2\sqrt{6}}{5}$ **51.** $\frac{2\sqrt{5}}{7}$ **53.** $\frac{4\sqrt{3}}{9}$ **55.** $\frac{3\sqrt{2}}{11}$
57. $\frac{5xy\sqrt{2x}}{a^2b}$ **59.** $\frac{5mn^2\sqrt{5}}{8}$ **61.** $\frac{4m\sqrt{2}}{3n}$
63. $\frac{3a^2b\sqrt{5}}{4}$ **65.** $5a$ **67.** $-4x\sqrt[3]{x^2}$ **69.** $\frac{3m}{2n^2}$
71. $\frac{rs\sqrt[3]{2rs^2}}{5t}$ **73.** 18   **75.** $7\sqrt{3}$ **77.** $-70\sqrt{10}$
79. $9\sqrt{x}$ **81.** $-9x^2y^3z\sqrt{2xy}$ **83.** $x^3y^2\sqrt{2}$
85. $\frac{5\sqrt{5}}{11}$ **87.** $\frac{7\sqrt{5}}{6}$ **89.** $3xyz^2\sqrt[3]{2y}$
91. $-3yz\sqrt[3]{3x^2z}$ **93.** $\frac{6a\sqrt{5a}}{b}$ **95.** $12\sqrt{3}$
97. $\frac{2rs^2\sqrt{3}}{9}$ **99.** $6ab^2\sqrt{3ab}$ **101.** $-\frac{8n\sqrt{5m}}{5}$
103. $18a\sqrt{2abc}$ **105.** $\frac{5ab}{2}$

Getting Ready (page 525)

1. $7x$ **2.** $3y$ **3.** $5xy$ **4.** $9t^2$ **5.** $2\sqrt{5}$
6. $3\sqrt{5}$ **7.** $2x\sqrt{2y}$ **8.** $3x\sqrt{2y}$

Exercises 8.4 (page 529)

1. yes   **3.** no   **5.** yes   **7.** 8   **9.** -9
11. like terms   **13.** no   **15.** yes   **19.** $7\sqrt{2}$
21. $9\sqrt{3}$ **23.** $7\sqrt{5}$ **25.** $12\sqrt{5}$ **27.** $9\sqrt{5}$
29. $7\sqrt{6} + 4\sqrt{15}$ **31.** $3\sqrt{2} - \sqrt{3}$
33. $10\sqrt{2} - \sqrt{3}$ **35.** $3x\sqrt{5}$ **37.** $3x\sqrt{5x}$
39. $3x\sqrt{2y} - 3x\sqrt{3y}$ **41.** $x^2\sqrt{2x}$ **43.** $-5y\sqrt{10y}$
45. $2x\sqrt{5xy} + 3x^2y\sqrt{5xy} - 4x^3y^2\sqrt{5xy}$
47. $7\sqrt{3} - 6\sqrt{2}$ **49.** $19x\sqrt{6}$ **51.** $5\sqrt[3]{2}$
53. $x^3\sqrt[3]{x} - x\sqrt[3]{x}$ **55.** $8\sqrt{5}$ **57.** $10\sqrt{10}$
59. $37\sqrt{5}$ **61.** $12\sqrt{7}$ **63.** $7 - 2\sqrt{3}$ **65.** $2\sqrt{3}$
67. $-6\sqrt{6}$ **69.** $10\sqrt{2} + 5\sqrt{3}$ **71.** $\sqrt[3]{3}$
73. $2\sqrt[3]{5} + 5$ **75.** $-7\sqrt{5}$ **77.** $22\sqrt{6}$ **79.** $2\sqrt{2}$
81. $-2\sqrt{3}$ **83.** $38\sqrt{2}$ **85.** $85\sqrt{2}$
87. $-18\sqrt{2}$ **89.** $13\sqrt{10}$ **91.** $12x^2y\sqrt{6y}$
93. $\sqrt{3}$ **95.** $4\sqrt{10}$ **97.** $6\sqrt{6} - 2\sqrt{3}$
99. $x^2y\sqrt[3]{5xy}$

Getting Ready (page 530)

1. x^5 **2.** y^7 **3.** x^3 **4.** y^3 **5.** $x^2 + 2x$
6. $6y^5 - 8y^4$ **7.** $x^2 - x - 6$ **8.** $6x^2 + 13xy + 6y^2$

Exercises 8.5 (page 536)

1. $2x + 2\sqrt{2}$ **3.** $3\sqrt{2} - \sqrt{2}y$ **5.** $\sqrt{5}$
7. $(x - 7)(x + 3)$ **9.** $3xy(2x - 5)$ **11.** radical
13. $\sqrt{7}$ **15.** 4   **17.** 2   **19.** -60 **21.** $-60\sqrt{2}$

23. $2 + \sqrt{2}$ **25.** $x - 3\sqrt{x}$ **27.** $5x\sqrt{2} - 5x$

29. $7 - 2\sqrt[3]{7}$ **31.** 1 **33.** 3 **35.** $7 - x^2$

37. $6x - 7$ **39.** $\sqrt[3]{9x^2} + 7\sqrt[3]{3x} + 12$

41. $2d - 3\sqrt[3]{4d^2} + 5\sqrt[3]{2d} - 15$ **43.** $\frac{2x}{3}$ **45.** $\frac{3y\sqrt{2}}{5}$

47. $\frac{2y}{x}$ **49.** $\frac{y\sqrt{xy}}{3}$ **51.** $\frac{\sqrt{3}}{3}$ **53.** $\sqrt{3}$

55. $2\sqrt[3]{2}$ **57.** $\sqrt[3]{36}$ **59.** $\frac{10\sqrt{x}}{x}$ **61.** $\frac{\sqrt{10x}}{2x}$

63. $\frac{\sqrt{2xy}}{3y}$ **65.** $\frac{2x}{3}$ **67.** $\frac{\sqrt[3]{2xy^2}}{xy}$ **69.** $-\frac{\sqrt[3]{25ab}}{5ab}$

71. $\frac{3(\sqrt{3} + 1)}{2}$ **73.** $\sqrt{7} - 2$ **75.** $6 + 2\sqrt{3}$

77. $2 - \sqrt{2}$ **79.** $\frac{x + 4\sqrt{x} + 4}{x - 4}$ **81.** $\frac{3x - 2\sqrt{3x} + 1}{3x - 1}$

83. $\frac{3y + 5\sqrt{3y} + 6}{3y - 4}$ **85.** x^3 **87.** $6b^2\sqrt{10}$ **89.** $4x$

91. $24p$ **93.** $x\sqrt{3} - 2\sqrt{x}$ **95.** $14\sqrt[3]{3x} + 21x$

97. $\frac{2\sqrt{7}}{7}$ **99.** $\frac{5\sqrt{2}}{8}$ **101.** 8 **103.** $\sqrt[3]{x}$

105. $6x + 6\sqrt{x}$ **107.** $\frac{5\sqrt{2y}}{x}$ **109.** $\frac{\sqrt{15}}{3}$

111. $\frac{\sqrt{3} - 3}{2}$ **113.** $-8x$ **115.** $700x\sqrt{10}$

117. $6\sqrt{x} + 3x$ **119.** $\frac{\sqrt[3]{20}}{2}$ **121.** $5\sqrt{3} - 5\sqrt{2}$

123. $8xy + 4\sqrt{2xy} + 1$

Exercises 8.6 (page 543)

1. yes   **3.** no   **5.** no   **7.** $(2, 3)$ **9.** $(3, -2)$

11. extraneous   **13.** b^n **17.** 36 **19.** 49 **21.** 1

23. 32 **25.** 23 **27.** -3 **29.** 0

31. $\varnothing$; $\frac{6}{5}$ is extraneous   **33.** 3; 0 is extraneous

35. $-4, -5$ **37.** 7; 1 is extraneous   **39.** $2, 1$

41. 2 **43.** -1 **45.** 65 **47.** 23 **49.** 5

51. 5 **53.** 13 **55.** 17 **57.** $\varnothing$; 4 is extraneous

59. -2 **61.** 3 **63.** 4 **65.** -1

67. $\varnothing$; 1 is extraneous   **69.** 5 **71.** 22

73. 3; 2 is extraneous   **75.** 10 **77.** -3 **79.** 25

81. $2\sqrt{26}$ **83.** 256 ft   **85.** about 1.8 ft   **87.** 980 W

89. 298.7 ft   **91.** 87.7 ft   **93.** 2×10^6 m

95. $v^2 = c^2 - f^2c^2$ **97.** 540 W   **99.** 16.9 mi

101. 7.1 units   **105.** 207

Getting Ready (page 546)

1. x^5 **2.** x^{12} **3.** x^5 **4.** 1 **5.** $\frac{1}{ab}$

6. $a^{12}b^8$ **7.** $\frac{a^9}{b^6}$ **8.** a^{24} **9.** b^{12}

Exercises 8.7 (page 550)

1. 3 **3.** 5 **5.** 64 **7.** 9 **9.** $3(z - 4t)(z - t)$

11. 5 **13.** rational   **15.** x^{m+n} **17.** $\frac{x^n}{y^n}$

19. $\frac{1}{x}$ **21.** 9 **23.** -12 **25.** 3 **27.** -5

29. 100 **31.** 256 **33.** 4 **35.** 125 **37.** 729

39. 81 **41.** $\frac{4}{9}$ **43.** $-\frac{8}{27}$ **45.** $4^{4/5}$ **47.** 25

49. $7^{3/4}x^{3/4}$ **51.** 25 **53.** $7^{1/3}$ **55.** 125 **57.** $\frac{1}{2}$

59. $\frac{1}{9}$ **61.** $\frac{1}{3x}$ **63.** x^2 **65.** $\frac{1}{x^2y^3}$ **67.** x^2

69. $5x^{2/15}$ **71.** $x^{2/7}$ **73.** x **75.** y^7 **77.** y^2

79. $x^{17/12}$ **81.** $\frac{1}{4b^{1/5}}$ **83.** $\frac{1}{2}t^{16/15}$ **85.** $\frac{1}{2}$ **87.** $\frac{2}{7}$

89. $\frac{1}{4}$ **91.** $\frac{1}{2}$ **93.** -2 **95.** 2 **97.** 2 **99.** -3

101. 27 **103.** -8 **105.** $\frac{1}{81}$ **107.** $\frac{3}{4}$ **109.** $\frac{1}{64}$

111. 8 **113.** y^2 **115.** x **117.** $a^{14/15}$ **119.** 6

121. 192 **125.** 2^x

Chapter Review (page 554)

1. 9 **2.** 6 **3.** -12 **4.** -11 **5.** -16 **6.** -8

7. 13 **8.** -15 **9.** 4.583 **10.** -3.873 **11.** -7.570

12. 27.421

13.

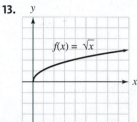

$f(x) = \sqrt{x}$

14.

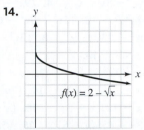

$f(x) = 2 - \sqrt{x}$

15. 35 **16.** 60 **17.** 1 **18.** $\sqrt{13}$ **19.** 68 in.

20. 45 ft   **21.** -3 **22.** -5 **23.** 3 **24.** 2

25. 3.787 **26.** 0.144 **27.** -0.380 **28.** -3.995

29. x^2y **30.** $9x^3y^4$ **31.** $4xy^2$ **32.** $10a^2b$

33. $4\sqrt{3}$ **34.** $6\sqrt{2}$ **35.** $10\sqrt{5}$ **36.** $4\sqrt{7}$

37. $4x\sqrt{5}$ **38.** $3y^2\sqrt{5}$ **39.** $-5t\sqrt{10t}$

40. $-10z^2\sqrt{7z}$ **41.** $10x\sqrt{2y}$ **42.** $5y\sqrt{3z}$

43. $2y\sqrt[3]{x^2}$ **44.** $5xy\sqrt[3]{2x}$ **45.** $\frac{4}{5}$ **46.** $\frac{10}{7}$

47. $\frac{3\sqrt{3}}{5}$ **48.** $\frac{\sqrt[3]{2}}{2}$ **49.** $\frac{2\sqrt{15}}{7}$ **50.** $\frac{4\sqrt{5}}{15}$

51. $\frac{11x\sqrt{2}}{13}$ **52.** $\frac{15a^2\sqrt{2}}{14}$ **53.** 0 **54.** $2\sqrt{3}$

55. $18\sqrt{5}$ **56.** $\sqrt{7}$ **57.** $5x\sqrt{2y}$ **58.** $y^2\sqrt{5xy}$

59. $7\sqrt[3]{5}$ **60.** $2x\sqrt[3]{2}$ **61.** $-18\sqrt{10}$ **62.** $10x$

63. $36x\sqrt{2}$ **64.** $-6y^3\sqrt{6}$ **65.** $4\sqrt[3]{2}$

66. $-24x\sqrt[3]{x}$ **67.** -2 **68.** $2y\sqrt{3} + 15\sqrt{2y}$

69. -5 **70.** $15 + 6x\sqrt{15} + 9x^2$

71. $\sqrt[3]{9} + \sqrt[3]{3} - 2$ **72.** $\sqrt[3]{25} - 1$ **73.** $\frac{\sqrt{3}}{3}$

74. $\frac{\sqrt{2}}{2}$ **75.** $2\sqrt[3]{4}$ **76.** $\frac{5\sqrt[3]{2}}{2}$ **77.** $\frac{5\sqrt{5} - 5}{4}$

78. $\frac{3\sqrt{3} + 3}{2}$ **79.** $5 - \sqrt{15}$ **80.** $\frac{4 + \sqrt{7}}{3}$

81. 21 **82.** 10 **83.** $\varnothing$; 4 is extraneous   **84.** 4

85. 4; -1 is extraneous   **86.** 8; 0 is extraneous

87. 2; -2 is extraneous   **88.** 9; -1 is extraneous

89. 5 units   **90.** 13 units   **91.** 5 units   **92.** 29 units

93. 9 **94.** -10 **95.** 64 **96.** $\frac{32}{243}$ **97.** 64

98. 25 **99.** $x^{1/5}$ **100.** $\frac{1}{r}$ **101.** 7 **102.** 5

103. x **104.** $\frac{1}{a^3b^5}$ **105.** $x^{11/15}$ **106.** $t^{1/12}$

107. $\frac{1}{x^{4/5}}$ **108.** $\frac{1}{r^{1/4}}$

Chapter 8 Test (page 558)

1. 7 **2.** -30 **3.** -3 **4.** $9x$ **5.** 13 in.

6. 10 ft   **7.** $2x\sqrt{3}$ **8.** $3x\sqrt{6xy}$ **9.** $4\sqrt{2}$

10. $3y\sqrt{x}$ **11.** xy^3 **12.** $\frac{2x^2}{y}$ **13.** $5\sqrt{3}$

14. $-x\sqrt{2x}$ **15.** $-24x\sqrt{6}$ **16.** $2\sqrt{6} + 3\sqrt{2}$

17. -1 **18.** $2x - 4\sqrt{x} - 6$ **19.** $\frac{3\sqrt{5}}{5}$

20. $\frac{y\sqrt{xy}}{4x}$ **21.** $2\sqrt{5} + 4$ **22.** $\frac{x\sqrt{3} - 2\sqrt{3x}}{x - 4}$

23. 16   **24.** 84   **25.** 5   **26.** 5

27. 5; 0 is extraneous   **28.** 29   **29.** 10 units

30. 5 units   **31.** $\frac{1}{13}$ **32.** 16   **33.** y^6

34. a^{10} **35.** $p^{17/12}$ **36.** $\frac{1}{x^{4/15}}$

Cumulative Review (page 559)

1. $4x^3 + 2x^2 + 3x + 3$ **2.** $5x^3 + 4x^2 - 3x - 5$

3. $16x^2 - 5x - 1$ **4.** $19x^2 - 28x - 11$

5. $-8x^8y^6$ **6.** $15x^5 + 9x^4 - 6x^2$

7. $20x^2 + 23x + 6$ **8.** $12x^2 - 17xy + 6y^2$

9. $x - 4$ **10.** $x^2 + x + \frac{-3}{2x - 1}$

11. $3xy^2(3xy - 2)$ **12.** $(a + b)(7 + x)$

13. $(a + b)(3 + b)$ **14.** $(7p^2 + 4q)(7p^2 - 4q)$

15. $(x - 12)(x + 3)$ **16.** $(x - 5y)(x + 2y)$

17. $(3a + 4)(4a - 5)$ **18.** $(5m + n)(2m - 3n)$

19. $(2x - y)(4x^2 + 2xy + y^2)$

20. $2(r + 3s)(r^2 - 3rs + 9s^2)$ **21.** $-1, -2$

22. $\frac{3}{2}, -4$ **23.** 2   **24.** 2   **25.** $-\frac{3}{2}$ **26.** $\frac{8}{11}$

27.

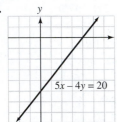

28.

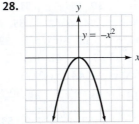

29. $\frac{1}{2}$ **30.** $-\frac{2}{5}$ **31.** $y = \frac{2}{3}x + 5$

32. $y = \frac{3}{4}x + \frac{11}{2}$ **33.** perpendicular   **34.** parallel

35.

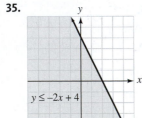

36.

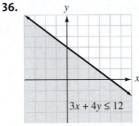

37. -3 **38.** 15   **39.** 5   **40.** $8x^2 - 3$

41. 12   **42.** 2

Getting Ready (page 563)

1. $2x(2x - 1)$ **2.** $(x + 3)(x - 3)$ **3.** $(x + 3)(x - 2)$

4. $(2x - 3)(x + 3)$ **5.** 5   **6.** -6 **7.** $2\sqrt{5}$

8. $-5\sqrt{2}$

Exercises 9.1 (page 569)

1. $3x(2x - 3)$ **3.** $(5x - 3)(x + 3)$ **5.** $y^2 - 10y + 25$

7. $x^2 + 2xy + y^2$ **9.** $4r^2 - 4rs + s^2$ **11.** quadratic

13. $0, 0$ **15.** quadratic, standard   **17.** $2, -3$ **19.** $4, -1$

21. $\frac{5}{2}, -2$ **23.** $1, -2, 3$ **25.** $0, -1$ **27.** $0, \frac{1}{3}$ **29.** $2, 3$

31. $-\frac{5}{4}, 1$ **33.** ± 3 **35.** $\pm 2\sqrt{5}$ **37.** ± 2 **39.** ± 3

41. $\pm\sqrt{2}$ **43.** $\pm\sqrt{6}$ **45.** $-6, 4$ **47.** $7, -11$

49. $-2 \pm 2\sqrt{2}$ **51.** $4 \pm 2\sqrt{3}$ **53.** $2, -\frac{2}{3}$ **55.** $0, -4$

57. $3 \pm \sqrt{15}$ **59.** $-5 \pm 2\sqrt{2}$ **61.** ± 5 **63.** ± 3

65. $\pm\sqrt{5}$ **67.** $\pm\sqrt{7}$ **69.** $-\frac{1}{3}, -\frac{3}{2}$ **71.** $\frac{2}{5}, -\frac{1}{2}$

73. ± 3 **75.** $\pm\sqrt{a}$ **77.** $-b \pm 4c$ **79.** $3a, -a$

81. $2 \pm 2\sqrt{2}$; 4.8 and -0.8 **83.** $-1 \pm 2\sqrt{2}$; 1.8 and -3.8

85. $\frac{-1 \pm 2\sqrt{5}}{2}$; 1.7 and -2.7 **87.** $7, 9$ and $-9, -7$

89. $-6, -4$ and $6, 4$ **91.** 4 ft by 8 ft   **93.** 10 ft by 18 ft

95. 9 sec   **97.** 324 ft   **99.** 9,280 ft   **101.** 18,496 ft

103. about 9.5 sec   **105.** 6 in.

Getting Ready (page 572)

1. 9   **2.** 25   **3.** 1   **4.** $\frac{25}{4}$ **5.** 16   **6.** 36

7. $\frac{1}{16}$ **8.** $\frac{1}{9}$

Exercises 9.2 (page 576)

1. $(x + 4)^2$ **3.** $(x - 6)^2$ **5.** $(x + 1)^2$ **7.** -4 **9.** 3

11. completing   **13.** $(x + b)^2$ **15.** square, 10

17. $x^2 + 2x + 1$ **19.** $x^2 - 4x + 4$ **21.** $y^2 - 18y + 81$

23. $x^2 + 7x + \frac{49}{4}$ **25.** $a^2 - 9a + \frac{81}{4}$ **27.** $b^2 + \frac{2}{3}b + \frac{1}{9}$

29. $-2, -4$ **31.** $2, 6$ **33.** $7, -2$ **35.** $1, -6$

37. $1, -2$ **39.** $-4, -4$ **41.** $2, -\frac{1}{2}$ **43.** $-2, \frac{1}{4}$

45. $-2 \pm \sqrt{3}$ **47.** $1 \pm \sqrt{5}$ **49.** $c^2 - \frac{5}{2}c + \frac{25}{16}$

51. $t^2 - \frac{5}{6}t + \frac{25}{144}$ **53.** $5, -3$ **55.** $2 \pm \sqrt{7}$

57. $-1 \pm \sqrt{2}$ **59.** $4, -5$ **61.** $1, -5$ **63.** $\frac{5}{12} \pm \frac{\sqrt{457}}{12}$

Getting Ready (page 577)

1. -8 **2.** -7 **3.** 8   **4.** 0

Exercises 9.3 (page 583)

1. $5x^2 - x - 1 = 0$ **3.** $7x^2 + x - 8 = 0$

5. $3x^2 + 7x + 4 = 0$ **7.** $7x^2 + 14x + 21 = 0$

9. $r = \frac{A - p}{pt}$ **11.** $y = \frac{3}{5}x + 12$ **13.** $4\sqrt{5}$ **15.** $\frac{\sqrt{7x}}{7}$

17. $ax^2 + bx + c = 0$ **19.** 0   **21.** $-4, 8, 1$

23. $a = 1, b = 4, c = 3$ **25.** $a = 3, b = -2, c = 7$

27. $a = 4, b = -2, c = 1$ **29.** $a = 3, b = -5, c = -2$

31. $a = 7, b = 14, c = 21$ **33.** $a = 1, b = -1, c = -5$

35. $2, 3$ **37.** $-3, -4$ **39.** $1, -\frac{1}{2}$ **41.** $-1, -\frac{2}{3}$

43. $\frac{4}{5}, -1$ **45.** $\frac{1}{2}, \frac{3}{2}$ **47.** $\frac{8}{5}, -3$ **49.** $\frac{2}{3}, 3$

51. $\frac{-3 \pm \sqrt{5}}{2}$ **53.** $\frac{-5 \pm \sqrt{37}}{2}$ **55.** $\frac{5 \pm \sqrt{37}}{6}$

57. $\frac{2 \pm \sqrt{10}}{2}$ **59.** no real solutions   **61.** no real solutions

63. $\frac{-5 \pm \sqrt{41}}{8}$; 0.2, -1.4 **65.** $\frac{4 \pm \sqrt{21}}{5}$; 1.7, -0.1

67. $\frac{-1 \pm \sqrt{41}}{4}$; 1.4, -1.9 **69.** $-2 \pm \sqrt{3}$; $-0.3, -3.7$

71. $-1 \pm \sqrt{2}$; 0.4, -2.4 **73.** $\frac{3 \pm \sqrt{15}}{3}$; 2.3, -0.3

75. 6 in. by 8 in.   **77.** 30 and 40 nautical miles

79. 7%   **81.** $210,000   **83.** 2 in.   **85.** 4 hr

Getting Ready (page 585)

1. 2   **2.** 3   **3.** 5   **4.** 12   **5.** -2 **6.** 2

Exercises 9.4 (page 593)

1. $(0, 6); (-2, 0)$ **3.** $(0, 4); (2, 0)$ **5.** $(0, -4); (2, 0), (-2, 0)$
7. $(0, 2); (-1, 0), (-2, 0)$ **9.** $7\sqrt{5}$ **11.** 4   **13.** quadratic
15. y-intercept   **17.** $a < 0$, maximum   **19.** $(0, c)$
21. upward   **23.** downward   **25.** downward

27.

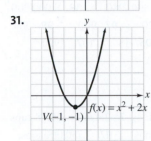

29.

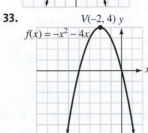

31.

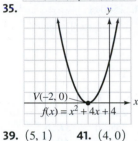

33.

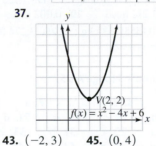

35.

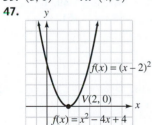

37.

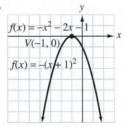

39. $(5, 1)$ **41.** $(4, 0)$ **43.** $(-2, 3)$ **45.** $(0, 4)$

47.

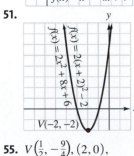

49.

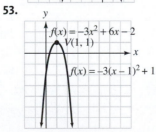

51.

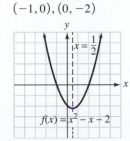

53.

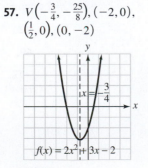

55. $V\left(\frac{1}{2}, -\frac{9}{4}\right), (2, 0),$
$(-1, 0), (0, -2)$

57. $V\left(-\frac{3}{4}, -\frac{25}{8}\right), (-2, 0),$
$\left(\frac{1}{2}, 0\right), (0, -2)$

59. $2, 4$ **61.** $-0.44, -4.56$
63.

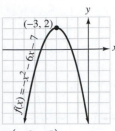

65.

67.

69. $V(-6, -5)$
71. $V(4, -19)$
73. $1,350$
75. 750
77. $\$2,499.50$
79. 25 vases, $525

Getting Ready (page 597)

1. $5x - 1$ **2.** $x - 7$ **3.** $x^2 - x - 20$
4. $6x^2 - 5x + 1$

Exercises 9.5 (page 604)

1. $2 + \sqrt{3}$ **3.** $\sqrt{2} - \sqrt{5}$ **5.** $6 + 2\sqrt{3}$
7. $(x + 7)(x - 7)$ **9.** $-2(a + 3)(a - 1)$ **11.** imaginary
13. $i, -1$ **15.** i **17.** $-i$ **19.** 1 **21.** i **23.** $6i$
25. $\frac{1}{9}i$ **27.** $5 + 7i$ **29.** $1 + 5i$ **31.** $4 + 12i$
33. $8 - 6i$ **35.** $9 + 7i$ **37.** $21 - i$ **39.** $4 + 7i$
41. $-3 + 5i$ **43.** $0 - i$ **45.** $0 + \frac{3}{7}i$ **47.** 82
49. 100 **51.** $\frac{6}{37} - \frac{1}{37}i$ **53.** $-\frac{4}{5} - \frac{2}{5}i$
55. $\frac{5}{13} - \frac{12}{13}i$ **57.** $\frac{17}{10} - \frac{1}{10}i$ **59.** $\frac{66}{53} + \frac{19}{53}i$ **61.** $\frac{7}{5} - \frac{6}{5}i$
63. $-\frac{42}{25} - \frac{6}{25}i$ **65.** $\frac{6 + \sqrt{10}}{9} + \frac{2\sqrt{2} - 3\sqrt{5}}{9}i$ **67.** $-1 \pm i$
69. $-\frac{1}{4} \pm \frac{\sqrt{7}}{4}i$ **71.** $15 + 7i$ **73.** $9 + 2i$ **75.** $5 + 2i$
77. $8 + 4i$ **79.** $-15 + 2\sqrt{3}i$
81. $6 + \sqrt{6} + \left(3\sqrt{3} - 2\sqrt{2}\right)i$
83. $-16 - \sqrt{35} + \left(2\sqrt{5} - 8\sqrt{7}\right)i$ **85.** $8 + \sqrt{2}i$
87. $6 - 8i$ **89.** $0 + \frac{3}{5}i$ **91.** $\frac{1}{8} + 0i$ **93.** $0 + \frac{3\sqrt{2}}{4}i$
95. $\frac{2}{3} \pm \frac{\sqrt{2}}{3}i$ **97.** $\pm 7i$ **99.** $\pm\frac{4\sqrt{3}}{3}i$

Chapter Review (page 608)

1. $0, -5$ **2.** $0, 3$ **3.** $3, -3$ **4.** $9, -9$ **5.** $3, 4$
6. $5, -3$ **7.** $6, -4$ **8.** $1, -\frac{3}{2}$ **9.** $3, 4$ **10.** $\frac{3}{2}, -\frac{2}{3}$
11. ± 5 **12.** ± 10 **13.** ± 3 **14.** $\pm\frac{3}{2}$ **15.** $\pm 2\sqrt{3}$
16. $\pm 5\sqrt{3}$ **17.** $-4, 6$ **18.** $3, -11$ **19.** $2, -4$
20. $\frac{7}{2}, \frac{1}{2}$ **21.** $8 \pm 2\sqrt{2}$ **22.** $3 \pm 3\sqrt{5}$ **23.** 15 ft
24. 15 sec   **25.** 3 ft by 9 ft   **26.** 3 ft by 6 ft
27. $x^2 + 6x + 9$ **28.** $y^2 + 8y + 16$ **29.** $z^2 - 10z + 25$
30. $t^2 - 5t + \frac{25}{4}$ **31.** $a^2 + \frac{3}{4}a + \frac{9}{64}$ **32.** $b^2 - \frac{9}{5}b + \frac{81}{100}$
33. $2, -7$ **34.** $3, 7$ **35.** $7, -11$
36. $1 \pm \sqrt{2}; 2.4$ and -0.4 **37.** $-2 \pm \sqrt{7}; 0.6$ and -4.6

38. $3 \pm \sqrt{5}$; 5.2 and 0.8   **39.** $\frac{1}{2}$, -3

40. $\frac{1 \pm \sqrt{3}}{2}$; 1.4 and -0.4 **41.** 5, -3 **42.** 5, -1

43. 13, 2   **44.** 0, -2 **45.** $\pm 12\sqrt{2}$ **46.** $\frac{7 \pm \sqrt{17}}{4}$

47. $\frac{4 \pm \sqrt{10}}{6}$ **48.** $-2 \pm \sqrt{3}$ **49.** $3 \pm \sqrt{2}$ **50.** 68 cm

51. 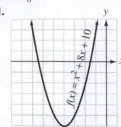   **52.**

$f(x) = x^2 + 8x + 10$

$f(x) = -2x^2 - 4x - 6$

53. $V(6, 7)$ **54.** $V(-4, -9)$ **55.** $V(1, 5)$

56. $V(3, 16)$ **57.** $\pm 5i$ **58.** $\frac{3}{4} \pm \frac{\sqrt{31}}{4}i$ **59.** i

60. $-i$ **61.** $6 - 4i$ **62.** $7 - i$ **63.** $11 - 9i$

64. $1 + 11i$ **65.** $2 - 6i$ **66.** $-14 - 35i$ **67.** $26 - 7i$

68. $-30 - \sqrt{15} + (5\sqrt{3} - 6\sqrt{5})i$ **69.** $-5i$

70. $\frac{5}{13} + \frac{1}{13}i$ **71.** $-\frac{6}{29} + \frac{15}{29}i$ **72.** $\frac{6 - \sqrt{35}}{11} + \frac{-3\sqrt{7} - 2\sqrt{5}}{11}i$

Chapter 9 Test (page 612)

1. $\frac{1}{2}$, $-\frac{3}{4}$ **2.** ± 2 **3.** $4i$, $-4i$ **4.** $2 \pm \sqrt{3}$ **5.** 49

6. $\frac{81}{4}$ **7.** $-1 \pm \sqrt{5}$ **8.** $x = \frac{-b \pm \sqrt{b^2 - 4ac}}{2a}$ **9.** 2, -5

10. $-\frac{3}{2}$, 4   **11.** $\frac{-5 \pm \sqrt{33}}{4}$ **12.** 10 m   **13.** $V(-3, 2)$

14.

$f(x) = x^2 + 4x + 2$

15. $\frac{1}{6} \pm \frac{\sqrt{83}}{6}i$

16. $-i$ **17.** $5 + 8i$

18. $-3 - 7i$ **19.** $9 - 19i$

20. $\frac{4}{13} + \frac{7}{13}i$

Appendix 1 (page A-1)

1. 1   **3.** 1   **5.** 6   **7.** -120 **9.** 30   **11.** 144

13. $\frac{1}{110}$ **15.** 2,352   **17.** 72   **19.** 10   **21.** 6

23. 60   **25.** 12   **27.** 5   **29.** 10   **31.** 20   **33.** 50

35. 2   **37.** $\{(1, H), (2, H), (3, H), (4, H), (5, H), (6, H), (1, T), (2, T), (3, T), (4, T), (5, T), (6, T)\}$

39. $\{a, b, c, d, e, f, g, h, i, j, k, l, m, n, o, p, q, r, s, t, u, v, w, x, y, z\}$

41. $\frac{1}{6}$ **43.** $\frac{2}{3}$ **45.** $\frac{19}{42}$ **47.** $\frac{13}{42}$ **49.** $\frac{3}{8}$ **51.** 0

53. $\frac{1}{4}$ **55.** $\frac{3}{26}$ **57.** $\frac{1}{169}$ **59.** 0   **61.** 40,320   **63.** 21

65. $\frac{1}{168}$ **67.** 39,916,800   **69.** $2.432902008 \times 10^{18}$

71. $\frac{1}{12}$ **73.** $\frac{7}{12}$ **75.** $\frac{1}{16}$ **77.** $\frac{3}{8}$ **79.** $\frac{1}{16}$ **81.** 35

83. 5,040   **85.** 1,000,000   **87.** 136,080   **89.** 8,000,000

91. 2,880   **93.** 13,800   **95.** 720   **97.** 900   **99.** 364

101. 5   **103.** 1,192,052,400   **105.** 18   **107.** 7,920

109. $\frac{1}{3}$ **111.** $\frac{88}{141}$ **113.** $\frac{15}{71}$

Appendix 2 (page A-13)

1. 81.9 ft   **3.** 3.8 mi   **5.** 230 mi   **7.** 2.4 mi   **9.** 8 cups

11. 9   **13.** 40 bags   **15.** 1,125 tons   **17.** 1,750,000 tons; 2,121,212 head   **19.** 0.002 m   **21.** 1500 cl

23. 0.155517384 kg   **25.** 3,274 ft   **27.** 1 oz (1 stick of butter is 4 oz)

Appendix 3 (page A-25)

1. b   **3.** c   **5.** a   **7.** a   **9.** c   **11.** b   **13.** d

15. b   **17.** c   **19.** c   **21.** d   **23.** d   **25.** e

27. d   **29.** c   **31.** b   **33.** d   **35.** d   **37.** b

39. c   **41.** a   **43.** e   **45.** a   **47.** e   **49.** b

51. a   **53.** b   **55.** d   **57.** c   **59.** e   **61.** c

63. d

INDEX

Important Concepts Guide *(continued)*

1.7 PROPERTIES OF REAL NUMBERS

If a, b, and c are real numbers, then

Closure properties:

$a + b$ is a real number.

$a - b$ is a real number.

ab is a real number.

$\dfrac{a}{b}$ is a real number $(b \neq 0)$.

Commutative properties for addition and multiplication:

$a + b = b + a$

$ab = ba$

Associative properties for addition and multiplication:

$(a + b) + c = a + (b + c)$

$(ab)c = a(bc)$

Distributive property:

$a(b + c) = ab + ac$

Identity elements:

0 is the additive identity.

1 is the multiplicative identity.

Additive and multiplicative inverses:

a and $-a$ are additive inverses.

a and $\dfrac{1}{a}$ $(a \neq 0)$ are multiplicative inverses.

2.1 SOLVING BASIC LINEAR EQUATIONS IN ONE VARIABLE

Let a, b, and c be real numbers.

If $a = b$, then $a + c = b + c$.

If $a = b$, then $a - c = b - c$.

If $a = b$, then $ca = cb$ $(c \neq 0)$.

If $a = b$, then $\dfrac{a}{c} = \dfrac{b}{c}$ $(c \neq 0)$.

If r is the rate, b is the base, and a is the amount, then

$rb = a$

2.2 SOLVING MORE LINEAR EQUATIONS IN ONE VARIABLE

Retail price = cost + markup

Markup = percent of markup $\cdot$ cost

2.3 SIMPLIFYING EXPRESSIONS TO SOLVE LINEAR EQUATIONS IN ONE VARIABLE

To combine like terms, add their numerical coefficients and keep the same variables with the same exponents.

2.7 SOLVING LINEAR INEQUALITIES IN ONE VARIABLE

Let a, b, and c be real numbers.

If $a < b$, then $a + c < b + c$.

If $a < b$, then $a - c < b - c$.

If $a < b$, and $c > 0$, then $ac < bc$.

If $a < b$, and $c < 0$, then $ac > bc$.

If $a < b$, and $c > 0$, then $\dfrac{a}{c} < \dfrac{b}{c}$.

If $a < b$, and $c < 0$, then $\dfrac{a}{c} > \dfrac{b}{c}$.

3.2 GRAPHS OF LINEAR EQUATIONS IN TWO VARIABLES

General form of the equation of a line:

$Ax + By = C$, where A and B are not both 0

Equation of a vertical line: $x = a$

Equation of a horizontal line: $y = b$

3.3 SLOPE OF A LINE

If $P(x_1, y_1)$ and $Q(x_2, y_2)$ are two points on a line, the **slope** of the line is

$$m = \frac{y_2 - y_1}{x_2 - x_1} \quad (x_2 \neq x_1)$$

A horizontal line has a slope of 0.

A vertical line has an undefined slope.

Two distinct lines with the same slope are parallel.

Two nonvertical lines with slopes that are negative reciprocals are perpendicular.

The product of the slopes of two nonvertical perpendicular lines is -1.

3.4 POINT-SLOPE FORM

Point-slope form of the equation of a line:

$y - y_1 = m(x - x_1)$

3.5 SLOPE-INTERCEPT FORM

Slope-intercept form of the equation of a line:

$y = mx + b$

(continues)

Important Concepts Guide (continued)

3.6 FUNCTIONS

$y = f(a)$ is the value of $y = f(x)$ when $x = a$.
Domain: The set of all first elements of an ordered pair
Range: The set of all second elements of an ordered pair
Function: Any set of ordered pairs in which each first element determines exactly one second element

3.7 VARIATION

Direct variation: $y = kx$

Inverse variation: $y = \dfrac{k}{x}$

Joint variation: $y = kxz$

Combined variation: $y = \dfrac{kx}{z}$

4.1 NATURAL-NUMBER EXPONENTS

Properties of exponents: If x is a natural number, then

$$\overbrace{x^n = x \cdot x \cdot x \cdots\cdots x}^{n \text{ factors of } x}$$

If m and n are natural numbers and there are no divisions by 0, then

$$x^m x^n = x^{m+n} \qquad (x^m)^n = x^{mn}$$

$$(xy)^n = x^n y^n \qquad \left(\frac{x}{y}\right)^n = \frac{x^n}{y^n}$$

$$\frac{x^m}{x^n} = x^{m-n} \quad \text{provided } m > n$$

4.2 ZERO AND NEGATIVE-INTEGER EXPONENTS

If x is any nonzero real number and n is a natural number, then

$$x^0 = 1, \qquad x^{-n} = \frac{1}{x^n}, \qquad \frac{x^m}{x^n} = x^{m-n}$$

4.3 SCIENTIFIC NOTATION

A number is written in scientific notation if it is written as the product of a number between 1 (including 1) and 10 and an integer power of 10.

4.6 MULTIPLYING POLYNOMIALS

$$(x + y)^2 = x^2 + 2xy + y^2$$
$$(x - y)^2 = x^2 - 2xy + y^2$$
$$(x + y)(x - y) = x^2 - y^2$$

4.7 DIVIDING POLYNOMIALS BY MONOMIALS

$$\frac{a + b}{d} = \frac{a}{d} + \frac{b}{d} \qquad (d \neq 0)$$

5.1–5.6 FACTORING POLYNOMIALS

$$ax + bx = x(a + b)$$
$$(a + b)x + (a + b)y = (a + b)(x + y)$$
$$ax + ay + cx + cy = a(x + y) + c(x + y)$$
$$= (x + y)(a + c)$$
$$x^2 - y^2 = (x + y)(x - y)$$
$$x^2 + 2xy + y^2 = (x + y)(x + y)$$
$$x^2 - 2xy + y^2 = (x - y)(x - y)$$
$$x^3 + y^3 = (x + y)(x^2 - xy + y^2)$$
$$x^3 - y^3 = (x - y)(x^2 + xy + y^2)$$

Zero-factor property: Let a and b be real numbers.

If $ab = 0$, then $a = 0$ or $b = 0$.

6.1–6.3 PROPERTIES OF RATIONAL EXPRESSIONS

If there are no divisions by 0, then

$$\frac{ax}{bx} = \frac{a}{b} \qquad\qquad \frac{a}{b} \cdot \frac{c}{d} = \frac{ac}{bd}$$

$$\frac{a}{b} \div \frac{c}{d} = \frac{ad}{bc} \qquad \frac{a}{d} + \frac{b}{d} = \frac{a + b}{d}$$

$$\frac{a}{d} - \frac{b}{d} = \frac{a - b}{d} \qquad -\frac{a}{b} = \frac{-a}{b} = \frac{a}{-b}$$

6.8 PROPORTIONS AND SIMILAR TRIANGLES

If $\dfrac{a}{b} = \dfrac{c}{d}$, then $ad = bc$.

7.1–7.3 SOLVING SYSTEMS OF LINEAR EQUATIONS

See the table on page 445 of the text.

8.1 SQUARE ROOTS AND THE PYTHAGOREAN THEOREM

b is a square root of a if $b^2 = a$.
$\sqrt{0} = 0$

Pythagorean theorem: If the lengths of two legs of a right triangle are a and b, and the length of the hypotenuse is c, then

$$c^2 = a^2 + b^2$$

Square root property of equality: If a and b are positive numbers,

If $a = b$, then $\sqrt{a} = \sqrt{b}$.

(continues)